Entering an Unseen World

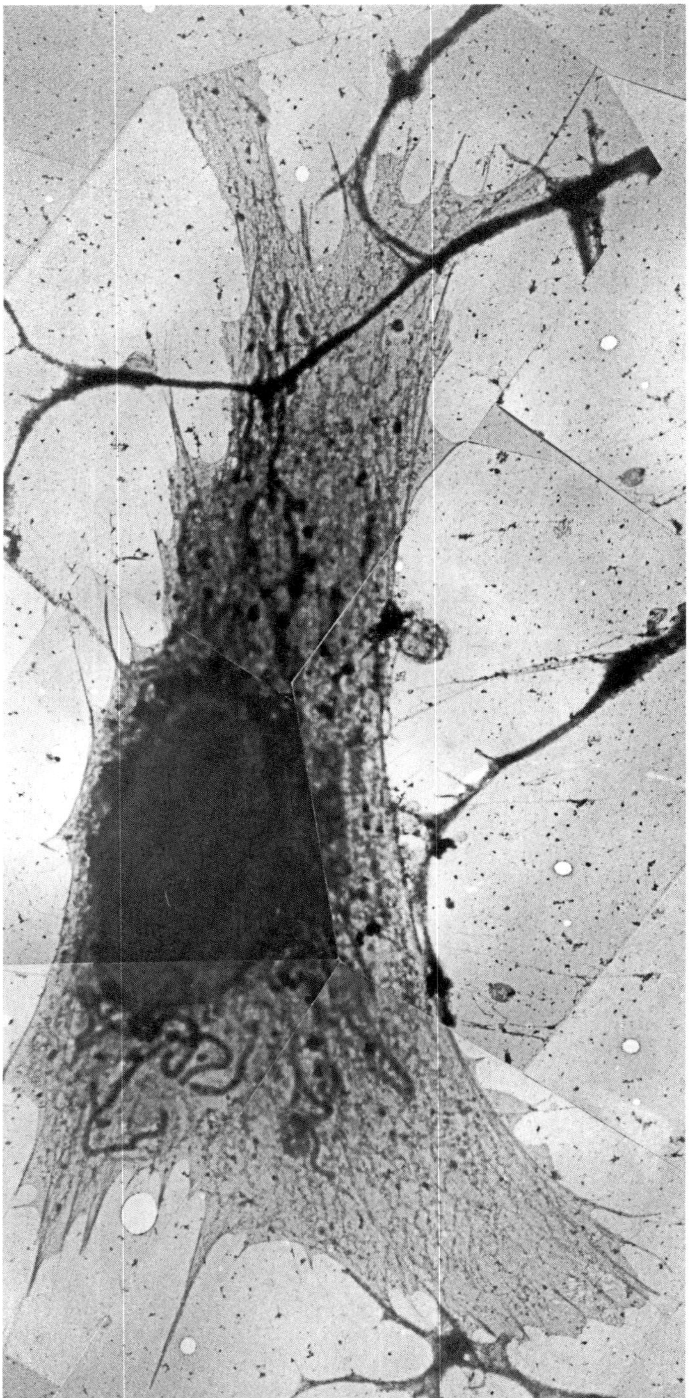

This first electron micrograph of an intact cell was published in 1945 by Keith Porter, Albert Claude, and Ernest Fullam in *The Journal of Experimental Medicine.* It was taken 6 July 1944 with an RCA electron microscope, type B. The image shown here was reprinted in 1994 by Lee D. Peachey from the original montage of five micrographs.

Entering an Unseen World

A Founding Laboratory and Origins of Modern Cell Biology 1910–1974

Carol L. Moberg
Editor and Contributor

THE ROCKEFELLER UNIVERSITY PRESS
NEW YORK

The Rockefeller University Press, New York

Library of Congress Cataloging-in-Publication Data

Entering an unseen world : a founding laboratory and origins of modern cell biology 1910-1974 / Carol L. Moberg, editor and contributor.

pages cm

Includes bibliographical references and index.

ISBN 978-0-87470-063-3 (hardcover : alk. paper) -- ISBN 978-0-87470-105-0 (ebook) (print)

1. Cytology--History. 2. Medicine--Research--United States--History--20th century. 3. Rockefeller Institute for Medical Research--History. 4. Rockefeller University--History. I. Moberg, Carol L.

QH577.E58 2012

571.6--dc23

2012039533

There is no end to our research . . . its pursuits are boundless and without form; its food is wonder, the chase, ambiguity . . . its inventions excite, pursue, and produce one another.

Michel de Montaigne, 1580

Contents

Preface

In 1974, the Nobel Prize in Physiology or Medicine was awarded to Albert Claude, Christian de Duve, and George Palade for their discoveries that linked structures inside cells to their functions. The Nobel citation proclaimed that these discoveries were responsible for the founding of a science called modern cell biology.

The story told here for the first time is much richer than only the workings of fate that singled out these three pioneering scientists. Its human details reveal how their celebrated discoveries fit into a larger, more significant context of what it takes to create a science. To gain new knowledge of the cell was a lengthy endeavor involving innumerable experiments and dozens of fellow scientists, many gradual steps, unusual collaborations, and interdisciplinary skills.

This book presents a singular account of this science from its roots in cancer research in 1910, to its origins in the 1940s, to tentative beginnings in the 1950s, and to its flowering by the mid-1970s. Neither a history of all modern cell biology nor that of the institution from which it emerged, The Rockefeller Institute for Medical Research, these pages approach history from a different angle. This is the story of the laboratory where these founding events coalesced—the pioneering scientists, the experimental procedures they created, the new instruments they adapted and adopted, the discoveries they made, and the standards they established to form a new science. Modern cell biology emerged from a wealth of previously unconnected activities that a few individuals put together.

The history of old cell biology spans two and a half centuries, from the invention of the light microscope to the beginning of the twentieth century. In the mid-seventeenth century, Anton van Leeuwenhoek in the Netherlands handcrafted microscopes that magnified objects more than 200 times and used them to observe and describe single-cell organisms found in animal tissues and pond water; he called them "animalcules." About the same time in England, Robert Hooke published many drawings in *Micrographia* based on his microscopic observations of various biological objects. In particular, his description of cork noted "these pores, or cells, were not very deep, but consisted of a great many little boxes, separated out of one continued long pore." Since these boxes reminded Hooke of a monk's small simple room, or *cella*, he used the term "cell" for the first time in biology.

Another 150 years passed before microscopes with better resolving power became available. By 1830, opticians were able to correct the chromatic and spherical lens aberrations of the earlier microscopes that distracted the eye and blurred images. The results achieved by producing achromatic lenses with higher magnifying power, as well as the introduction of new staining methods for biological samples, were so striking that microscopists could demonstrate that all living organisms are made up of cells.

Only after this technological improvement could ideas be proposed about the nature of cells. Consequently, during the 1830s, botanist Matthias Schleiden and his student zoologist Theodor Schwann advanced the theory that cells are the basic units of life. This theory was not merely created by these two microscopists but was the product of numerous inquiries by other scientists into the nature of organic structures. Schwann also thought that new cells originated from a fluid inside cells, but this notion was soon dashed by Robert Remak, who worked with animal tissues, and Barthélemy Dumortier, who worked with plants, when they showed cells are formed from division by preexisting cells. This view was championed in 1858 by pathologist Rudolf Virchow, whose well-known aphorism *omnis cellula e cellula* (every cell originates from another like it) led to studying how organisms develop and grow from one cell to many. Importantly, for the starting point of this book, Virchow introduced the idea that the cell is also the basic element of pathological processes in which diseased cells arise directly from changes in normal, healthy cells in the organ where they first appear.

The old cell biology effectively ended at the close of the nineteenth century when light microscopes reached the limits of their resolution. Even so, scientists had gotten tantalizing glimpses of what might be inside a cell and offered enticing, tentative, and often contradictory descriptions of the contents that we know now as the nucleus, mitochondria, and the Golgi complex. However, most advances during this century came from studying cell reproduction. In 1896, Edmund B. Wilson, an American cytologist, published a monograph, *The Cell in Development and Inheritance*, describing these known components but focusing on factors at work during the growth of a single-celled egg to its segregation into tissues and organs of a multicelled organism. Wilson also considered the term cell to be biologically inaccurate because by this time cytologists knew that cells were much more than barren, walled chambers. He believed that every biological problem must ultimately be sought within the cell. Yet, this world was still largely inaccessible.

At the turn of the twentieth century, a large gap opened between the world of structures and the world of molecules. Anatomists, histologists, pathologists, and biochemists often disputed the realities of the existence of components in a cell. Many viewed the cell as a bag of enzymes filled with formless protoplasm and devoid of structure. Forty years later, this gap was bridged by the scientists profiled in this book who were trained in cytology, microscopic anatomy, biochemistry, and biophysics. As their work slowly gathered strength and direction, they developed techniques allowing them to enter a previously unseen world inside the cell, a process that led to a successful correlation of structure and function of the newly discovered subcellular components. Mapping the cell's interior parts made it possible to understand how a cell is structured and to recognize the remarkable uniformity of its organization in all living beings.

The birth and flowering of a new science is a peculiar and wonderful happening. This revolution in cell biology has already received some specialized interpretations that include selective events and individuals described more fully in this book. Two philosophers of science have examined theoretical aspects. Hans-Jörg Rheinberger in *Toward a History of Epistemic Things: Synthesizing Proteins in the Test Tube* (1997) presents an in-depth case study of protein synthesis in the Harvard laboratory of Paul Zamecnik to show how the research process creates and develops knowledge. William Bechtel in *Discovering Cell Mechanisms* (2006) analyzes experimental details to characterize how modern cell biology became a discipline through establishing mechanistic explanations of cell functions. Historians of science have focused on the instruments introduced into biology. Boelie Elzen in *Scientists and Rotors: The Development of Biochemical Ultracentrifuges* (1988) and Nicolas Rasmussen in *Picture Control: The Electron Microscope and the Transformation of Biology in America, 1940–1960* (1997) address how the creation of knowledge depends on technology, how a new device is taken into experimental systems, and how scientific facts (and artifacts) are settled through standardized methods of inquiry. Social historian of science Hannah Landecker in *Culturing Life: How Cells Became Technologies* (2007) explores the biological practices, ethical implications, and transformed meanings of life that have been brought about by manipulating cells outside the body in the laboratory. Even a scientific pioneer of modern cell biology who has written a chapter in this book, Christian de Duve, has published *A Guided Tour of the Living Cell* (1984), and in *Singularities* (2005) he has given a guided philosophical tour of biochemistry and phylogeny from the basic molecular building blocks to the origin of humans.

This book differs substantially from these types of histories. Instead, in keeping with a tradition of narrative histories, this book is first of all a story, one that chronicles the particulars of persons, events, and changes within time and place, including many first-person accounts about work done beginning in the 1940s.

The order of presentation in this book is chronological, spanning six decades, from 1910 to 1974. The first chapter introduces The Rockefeller Institute laboratory of Peyton Rous and James B. Murphy and their controversy over a cancer problem, in particular, the nature of the chicken sarcoma agent (now known as the Rous Sarcoma Virus or RSV). The next three chapters introduce the scientists as they joined Murphy's laboratory, the techniques they established to try to solve this problem, and how cancer biology was transformed into cell biology: first, Albert Claude and the centrifuge; followed by Keith Porter and the electron microscope, tissue culture, and the microtome; and then by George Hogeboom, Walter Schneider, George Palade, and Philip Siekevitz, who combined chemical analyses of cell structures with the other techniques. The fifth and sixth chapters describe the emergence and tentative beginnings of modern cell biology during the 1950s as it moved decisively away from controversial studies of cancer cells and began an independent exploration of normal cells. The seventh chapter describes the process from 1950 to 1965 of combining all these techniques to discover a new organelle, the endoplasmic reticulum, establish its role in secreting proteins, and formally inaugurate the process of combining a structure with its function.

The remainder of the book contains personal accounts by scientists who either joined the Porter/Palade/Siekevitz laboratory or had independent laboratories at Rockefeller between 1945 and 1974. Importantly, these scientist-authors, except four, were trained in diverse fields at other institutions worldwide, including Canada, South America, the Middle East, Asia, and Europe as well as the United States. With appointments ranging from a few months to a few years, they came to gain experience working directly with the pioneering biologists and their techniques. Their accounts describe, in their own styles, their background and reasons for coming, their inventions and accomplishments as well as dead ends and surprises, and their collaborators. While they cite their original scientific articles, for the most part they eschew experimental details in favor of ideas and impressions of research that never appear in peer-reviewed publications. These scientists represent a few hundred others who came to Rockefeller from 1923 to 1974 and returned to positions of leadership in medical schools and universities throughout the world. Their stories in this book recount early experiences rather than entire careers.

These chapters also take a more thematic approach and reflect personal research interests: some scientists explored cells of the nerve system or immune system, some examined viruses or protozoa, some studied plant cells, and some studied such cell components as membranes, cilia, mitochondria, ribosomes, lysosomes, or nuclei. This core group of scientists secured the foundation on which modern cell biology became readily accepted and quickly extended far beyond Rockefeller. During these years of great technical, institutional, and personal ferment, these scientists also participated in many non-laboratory activities to widen the horizons of cell biology. They were founding members of two new scientific societies and a premier journal, and, when The Rockefeller Institute became a university, they trained the first PhDs in cell biology.

While the overall focus of the book is scientific origins, there are several secondary themes, including the role of cancer biology, interactions and styles of interdisciplinary scientists, new instruments and techniques, the nurturing institutional environment, as well as historical antecedents and biographical details about people, places, and events. For example, in the second chapter, a curious interaction is depicted between Claude, who used an ordinary commercial centrifuge for his classic experiments, and Edward Pickels, who was in a nearby laboratory developing the ultracentrifuge; the outcome of their eventual exchange was vital to the new science. In other chapters, scientific, biographical, and cultural backgrounds are provided when introducing the centrifuge, electron microscope, and microtomes.

These chapters also offer fresh, personal attitudes about how science works. From one perspective, it is readily apparent that an interdisciplinary group of scientists pursuing research with new technologies at several levels does not produce a linear accretion of discoveries; instead, scientists take erratic paths that are strewn with surprises as they encounter reversals, setbacks, adversaries, and conflicts. That advances were made regardless affirms Louis Pasteur's observation that chance favors the prepared mind.

From another perspective, the work of science demands a wealth of human, technical, and environmental resources as well as many years to produce pioneers who achieve pathbreaking research. Unlike textbooks, reviews, and science histories that present and interpret scientific data or hypotheses about the cell or one of its components, this book presents a complex human story of individuals who did the work, how facts and ideas were wrought and hard-won, how new tools fostered experiments, and how results were gained. In short, the book exposes the precarious nature and human dimension of how science is conducted. A reader can follow in time the formulation of problems, recapture the keen sense of exploring, and take part vicariously in the discoveries themselves.

"For a scientist," George Palade once said, "it is a unique experience to live through a period in which his field of endeavor comes to bloom—to be witness to those rare moments when the dawn of understanding finally descends upon what appeared to be confusion only a while ago—to listen to the sound of darkness crumbling."

This story of modern cell biology hopefully conveys some of the poetry and excitement of science.

Acknowledgments

From the beginning, this book has been encouraged and supported by cell biologists who wanted a written account of how their own science took shape. The idea grew out of a conversation with Zanvil A. Cohn in 1992 about documenting the sciences that originated with Rockefeller scientists, including immunochemistry, addictive diseases, environmental biomedicine, peptide chemistry, and his own field of cell biology. The idea to start with the science of cell biology was immediately endorsed by Ralph M. Steinman, Philip Siekevitz, and Samuel Dales, all of whom were mentors, consultants, contributors, and special friends throughout the project.

When I visited the ailing Keith Porter in 1994, his jolly greeting on opening his apartment door set the book's enthusiastic spirit: "I've been waiting for someone to tell this story." Although no longer able to write, he instantly offered access to his voluminous personal papers, then stashed in uncataloged boxes at the universities of Colorado, Maryland, and Pennsylvania. These documents made a significant difference in telling the human side of this science.

A few months later, in 1995, Günter Blobel and David Sabatini organized a three-day symposium called "Journey into the Cell." Two dozen pioneers of cell biology gathered to celebrate the fiftieth anniversary of the first electron micrograph of an intact cell, published in 1945 in *The Journal of Experimental Medicine* by Porter, Albert Claude, and Ernest Fullam. As the aging scientists reminisced about their work in this emerging field, it was apparent that this oral history was about to be lost. After the symposium, we arranged with nearly all the participants to fully develop and write detailed, annotated, and personal accounts of their research. This book now offers a durable record of what happened when they first entered the previously unseen world inside cells and discovered their secrets.

Tracing the scientific threads that culminated in that first electron micrograph, however, demanded other efforts. An initial survey of available materials revealed they were widely scattered. The Rockefeller University Archives had few primary resources, in part because it was established in the mid-1970s after most scientists featured in this book were pursuing careers in other institutions. Among its holdings related to cell biology are the valuable *Annual Scientific Reports to the Board of Scientific Directors*. These forty-two unpublished volumes, from 1901 to 1954, contain detailed and insightful reports written by the laboratory heads for the scientific director about the ongoing research of each scientist on his staff. Collectively, they describe every laboratory's scientific history. I thank University archivists Renee Mastrocco, Lee Hiltzik, and Margaret Hogan for aiding other research in the personal papers of Herbert Gasser as well as the records of the Institute's Instrument Shop and The Rockefeller Foundation's International Health Division. The papers belonging to the Merrill W. Chase Historical Instrument Collection helped document the development of centrifuges and microtomes.

A research grant from the American Philosophical Society enabled me to study primary resources in several collections. These sources included nearly 400 linear feet of Keith Porter's correspondence, notebooks, manuscripts, and electron micrographs that documented his major and minor discoveries as well as his initiatives in the founding of the Tissue Culture

Association, *The Journal of Cell Biology*, and the American Society for Cell Biology. For extensive help in navigating Porter's papers (then unprocessed), I thank archivists Marty Covey and Bruce Montgomery at the Norlin Library at the University of Colorado in Boulder; John Beck at the Albin O. Kuhn Library at the University of Maryland, Baltimore County; and Lee Peachey, a former student of Porter, whose laboratory then housed additional Porter papers in the biology department at the University of Pennsylvania.

I am most grateful to Philippa Claude, daughter of Albert Claude, who is herself a cell biologist. She graciously provided hospitality and many insights into her father's life and laboratory career. She made available about a hundred linear feet of his correspondence and manuscripts in Madison, Wisconsin, that describe his techniques for centrifuges and microtomes, his views of cancer causation, and details of James B. Murphy's evolving laboratory of cancer pathology.

I thank archivists Scott DeHaven and Beth Carroll-Horrocks at the American Philosophical Society for their skillful and efficient assistance. The Society holds the papers of James B. Murphy and Peyton Rous that detail the early years of research on viruses, the uneven progress of cancer research, and the institutional and laboratory environment in which cell biology emerged. It also holds the papers of Thomas F. Anderson, RCA's skilled electron microscopist, that shed light on the selective use of this instrument and the difficulties of adapting it to biology in the early 1940s.

All these collections turned up many significant primary documents, so what began as a monograph became a comprehensive history. The Richard Lounsbery Foundation headed by Frederick Seitz, former president of The Rockefeller University, gave grants to develop and explain the background and experiments of the science from 1911 to 1950. In 2002, an R01 National Institutes of Health grant from the National Library of Medicine provided additional support for writing, editing, and integrating the manuscript.

In addition, Mary Bonneville, a former student and colleague of Porter, loaned videotapes and transcriptions of oral histories that she recorded between 1984 and 1988, called "Pioneers in Modern Biology." They include seventeen hours of conversations with Porter and twenty-two hours with Palade along with their colleagues Marilyn Farquhar, Günter Blobel, David Sabatini, Philip Siekevitz, James Jamieson, Nicolae Simionescu, Lee Peachey, and Peter Satir. Other helpful oral interviews were those of Ernest Fullam taken by Diane Fullam and of George Palade taken by Robert Silver. Despite the inconsistencies, conflated events, and lack of documentation for these interviews, some fascinating glimpses into places, people, and influences have added color and personality in many sections of the book.

I also wish to thank these individuals for offering perspectives on many aspects of the history of cell biology: David Allis, Olga Amsterdamska, Blair Bowers, Sabine Brauckmann, Merrill W. Chase, Hannah Landecker, David Luck, Bruce McEwen, Richard McIntosh, David Mauzerall, Ira Mellman, James Slater Murphy, Robert Olby, Van Potter, Michel Rabinovitch, Michael Rout, Frederick Seitz, Walter Schneider, Abigail Tierney, and Ton van Helvoort.

Former Rockefeller University president Torsten Wiesel and Günter Blobel read an early version of the introductory chapters and offered enthusiastic support for the project. I thank Samuel Dales and Judith Schwartz, who read the entire manuscript, made detailed comments, and improved it. Any interpretations, errors, and omissions in this book are mine.

The Rockefeller University Press has an exceptional staff for publishing scientific works. Mike Rossner, the executive editor and a cell biologist himself, provided the enthusiasm and commitment to publish this history in print and electronic formats. It was a great pleasure working with him and expert ever cheerful editors Camille Clowery, Erinn Grady, Rob O'Donnell, and Suzanne Runyan.

This book was long in the making, and the historical research for it took even longer. During this time, some elements that appear in this volume were included in articles published elsewhere. Portions of Chapters 1–5 contain revised passages from "James B. Murphy, the Rous sarcoma agent, and origins of modern cell biology" in *Creating a Tradition of Biomedical Research*, edited by Darwin H. Stapleton (2004, The Rockefeller University Press, New York, 259–269). Chapters 3 and 5 contain revised passages from their original publication in "The electron microscope enters the realm of the intact cell" in *The Journal of Experimental Medicine* (1995, 181:829–837) and in "Keith Porter and the founding of the Tissue Culture Association" in *In Vitro Cellular and Developmental Biology, Animal* (1996, 32:663–669). Chapter 10C includes revised passages first published with Philip Siekevitz in "Founding the American Society for Cell Biology" in the *ASCB Newsletter* (2000, 23:2–7).

While working on this book, I have been a member of Ralph Steinman's Laboratory of Cellular Physiology and Immunology, where I benefitted greatly from daily exchanges with him and my colleagues, especially Svetlana Mojsov, Maggi Witmer-Pack, and Judy Adams. Most of all, I am deeply grateful to Ralph for providing a stimulating environment with ample space and freedom to do this book.

Lastly, I thank my husband Lawrence Moberg, who helped to clarify my thoughts and always boosted my morale.

Carol Larsen Moberg
1 March 2012

Timeline for Featured Scientists at Rockefeller 1910–1974

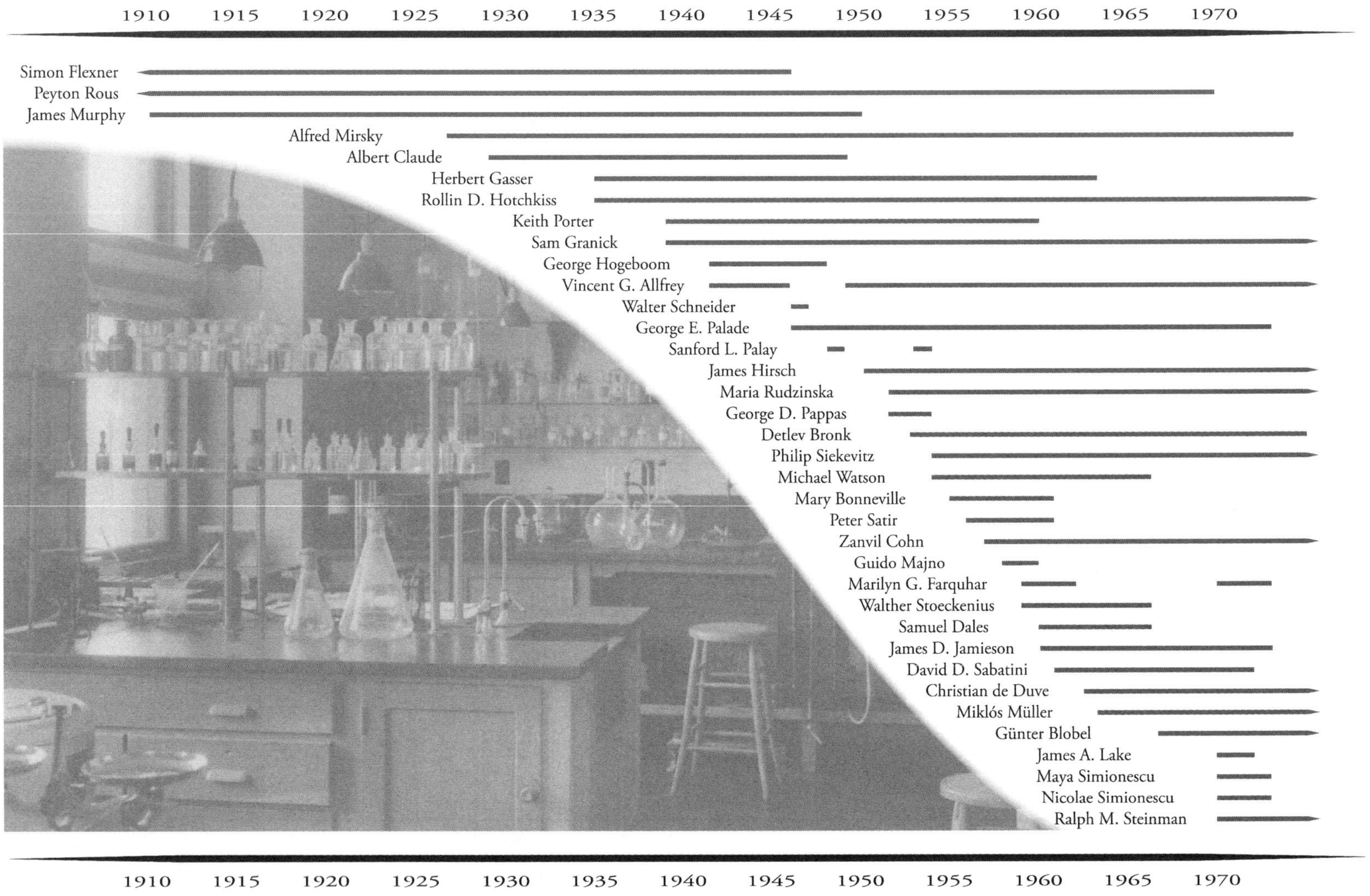

Contributors

Affiliations are at time of writing

2011

Carol L. Moberg
The Rockefeller University
New York, NY
Rockefeller 1965–present

1950

Vincent G. Allfrey
The Rockefeller University
New York, NY
Rockefeller 1941–46, 49–91

1967

Günter Blobel
The Rockefeller University
New York, NY
Rockefeller 1967–present

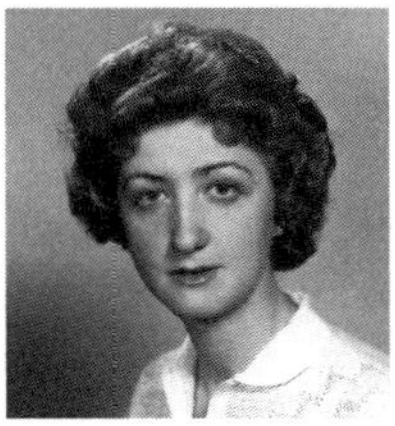

1959

Mary Bonneville
University of Colorado
Boulder, CO
Rockefeller 1955–61

1960

Samuel Dales
The Rockefeller University
New York, NY
Rockefeller 1960–66,
96–present

1961

Christian de Duve
The Rockefeller University
New York, NY
Rockefeller 1962–present

1959

Marilyn G. Farquhar
University of California
San Diego, CA
Rockefeller 1958–62, 70–73

1924

Walther F. Goebel
The Rockefeller University
New York, NY
Rockefeller 1924–93

1935

Rollin D. Hotchkiss
The Rockefeller University
New York, NY
Rockefeller 1935–2004

1960

James D. Jamieson
Yale School of Medicine
New Haven, CT
Rockefeller 1960–73

1970

James A. Lake
University of California
Los Angeles, CA
Rockefeller 1970–72

1975

Guido Majno
University of Massachusetts
Medical School
Worcester, MA
Rockefeller 1958–60

1964

Miklós Müller
The Rockefeller University
New York, NY
Rockefeller 1964–present

1948

George E. Palade
University of California
San Diego, CA
Rockefeller 1946–73

1948

Sanford L. Palay
Harvard Medical School
Boston, MA
Rockefeller 1948, 53

1952

George D. Pappas
University of Illinois
Chicago, IL
Rockefeller 1952–54

Contributors

1961

David D. Sabatini
New York University School of Medicine
New York, NY
Rockefeller 1961–72

1959

Peter Satir
Albert Einstein College of Medicine
Bronx, NY
Rockefeller 1956–61

1954

Philip Siekevitz
The Rockefeller University
New York, NY
Rockefeller 1954–2009

1970

Maya Simionescu
Institute of Cellular Biology and Pathology
Bucharest, Romania
Rockefeller 1970–73

1970

Ralph M. Steinman
The Rockefeller University
New York, NY
Rockefeller 1970–2011

1959

Walther Stoeckenius
University of California
San Francisco, CA
Rockefeller 1959–66

1933

William Trager
The Rockefeller University
New York, NY
Rockefeller 1933–2005

These photographs were taken by the Rockefeller Illustration Service and are courtesy of The Rockefeller Archive Center, except for the following: Majno's photo is courtesy of the University of Massachusetts Medical School Archives, Lamar Soutter Library, University of Massachusetts Medical School; and Moberg's photo is courtesy of Lawrence R. Moberg.

Milestones in Modern Cell Biology at Rockefeller

1910

Peyton Rous discovers a virus in chickens that causes tumors (later known as the Rous Sarcoma Virus [RSV]). 1911

Rous and James Murphy disband research on RSV tumor after disagreeing on its cause: a virus or chemical? 1913

1920

Murphy begins studies to purify chicken tumor agent to disprove claims that William Gye cultivated an organism from the tumor. 1925

Albert Claude joins Murphy to study chemical differences between normal and tumor cells. 1929

1930

Murphy and Claude propose a transmissable mutagen theory of cancer causation. 1932

Claude uses a high speed centrifuge (17,000 rpm) to separate cells into fractions and isolates "microsomes" from both normal and cancer cells. 1937

1940

Claude shows microsomes contain most of the ribonucleoprotein in cells. 1941

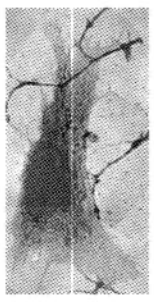

Keith Porter, Claude, and Ernest Fullam publish the first electron micrograph of an intact cell in *The Journal of Experimental Medicine*. 1945

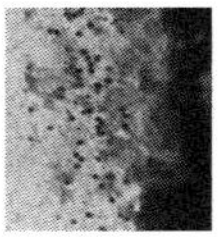

Claude, Porter, and Edward Pickels publish electron micrographs of RSV-infected cells, and controversy ends work on this virus. 1947

1950

George Palade introduces a new fixative for whole tissues containing osmium tetroxide buffered with veronal acetate, initially known as "Palade's Pickle," which produces outstanding resolution of cells in electron micrographs. 1952

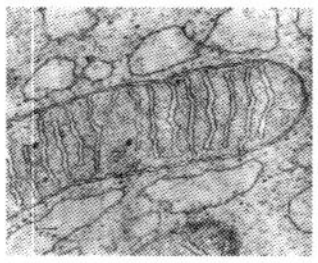

Using cell fractionation, biochemistry, and electron microscopy (EM), Palade and Porter report mitochondria are distinct cell organelles with double membranes, one separating them from the cytoplasm and the other with internal ridges or *cristae mitochondriales* that contain the cell's oxidative enzymes. 1953

Porter with Josef Blum designs and builds an improved microtome to obtain ultrathin tissue sections for EM, which soon becomes the commercial Sorvall MT-1 microtome. 1953

EM of ribosomes by Porter and Palade. 1953

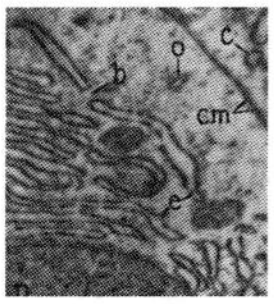

Porter and Palade establish the endoplasmic reticulum (ER) as a cell organelle having a cell-wide system of interconnected tubules and vesicles. 1953–55

First issue of *The Journal of Biophysical and Biochemical Cytology* published by The Rockefeller University Press (renamed *The Journal of Cell Biology* in 1962). 1955

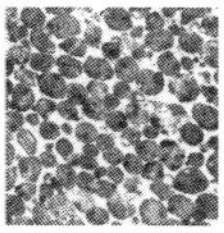

Christian de Duve characterizes the lysosome, an organelle containing the cell's digestive enzymes. 1955–63

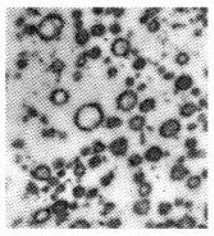

Palade and Philip Siekevitz establish the origin of the microsome in the ER and as the site of protein synthesis. 1956–58

Walther Stoeckenius discovers the purple cell membrane and develops the model of lipid bilayer of cell membranes as the basic permeability barrier. 1959–66

Marilyn Farquhar and Palade, in studies of cell permeability of the glomerulus, discover tight junctions and junctional complexes and their role in sealing intercellular spaces. 1959–62, 70–73

1960

The American Society for Cell Biology is founded, spearheaded by Porter and Palade. 1960

James Hirsch and Zanvil Cohn reveal the process of endocytosis used by leukocytes and macrophages to capture and destroy pathogens. 1960–65

Lucien Caro introduces autoradiography in EM to trace movement of proteins through the secretory pathway. 1962

Palade and James Jamieson elucidate steps in the secretory pathway of proteins, from the ER to their exit from the cell. 1963–68

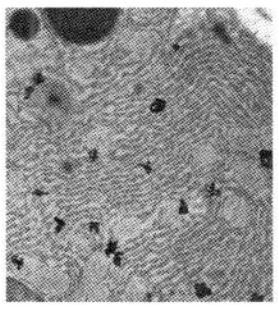

David Luck and Edward Reich discover mitochondria contain their own DNA that is distinct from DNA in the nucleus. 1964

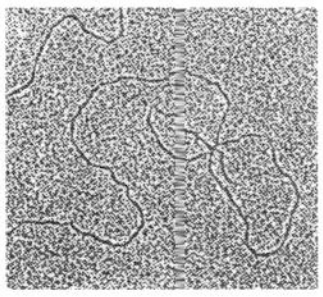

Vincent Allfrey shows how histone proteins coiled around DNA control gene activation. 1964

de Duve discovers the peroxisome, an organelle containing enzymes that catalyze the production and breakdown of hydrogen peroxide. 1965

1970

Günter Blobel and David Sabatini postulate the signal hypothesis of protein secretion for transport to and across the ER membrane. 1971

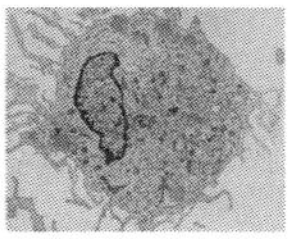

Ralph Steinman and Cohn identify a novel white blood cell that they name the dendritic cell. 1973

Miklós Müller discovers the hydrogenosome, an organelle that releases hydrogen during energy generation under anaerobic conditions. 1973

Nobel Prize in Physiology or Medicine awarded to Claude, de Duve, and Palade. 1974

Part 1

Roots of a New Science
1910–1949

Chapter 1

Early Cancer Research

Carol L. Moberg

> The cell is an *organic system,* and one in which we must recognize the existence of some kind of ordered structure or *organization.*
>
> Edmund B. Wilson, 1928

The scientific origins of modern cell biology can be traced to a search for the cause of a malignant tumor. The story begins in 1910 with a brief collaboration of two pathologists, Peyton Rous and James Bumgardner Murphy, who studied breast tumors in chickens. Only after young postdoctoral biologists, Albert Claude, Keith Porter, and George Palade, brought fresh approaches to the cancer problem did they reveal what had not been seen before—a neatly organized world inside the cell. With their discoveries in the mid-1940s came the pivotal realization that new knowledge about normal cells was needed before any attempt could be made to decipher their cancerous forms. Thus, a new science was born.

What will become apparent in the following chapters is how many years it took for this science to take hold, what new methods and instruments were needed, and how much fundamental knowledge was generated. These years were permeated with controversy, multiple beginnings, and endless frustrations as the initial objectives of pursuit remained beyond reach. Several themes that surface from the collaboration between Rous and Murphy, recounted in this chapter, go far beyond the practical aim of finding the cause of cancer. They account for the initiation and continuing motivation behind experiments, often providing clues about why certain experiments were done and, more peculiarly, why others were not even conceived of. Emerging from this lengthy, consuming quest for a cancer-causing agent are insights into the fabric of basic research itself.

Scientific Milieu

The fertile environment for the birth of a science emerged from a convergence of two experimental ventures at the turn of the twentieth century: a fledgling institute and an open agenda for cancer research.

The Rockefeller Institute for Medical Research in New York City, founded in 1901, was the first American institute devoted to scientific medicine. Difficult as it is to imagine a century later, the Institute was unprecedented in having several laboratories staffed by scientists with interdisciplinary interests and expertise, supported entirely from private funds and free to pursue their research interests. They addressed perplexing medical dilemmas and created novel ways to attack disease while attracting and nurturing a continual flow of young postdoctoral fellows who brought fresh ideas and special talents to tackle these problems.

Until the 1890s, those who wanted to study the most advanced approaches to medicine or to train for a career in research had to go to Europe. Bacteriology, pathology, physiology, micro-anatomy, and many other scientific approaches to disease were not taught in American medical schools, and only a few exceptional physicians had the luxury of establishing a personal laboratory in their office or home. In Europe, however, medical research was advancing at a number of universities and private institutes, especially those created for scientists Louis Pasteur in France and Robert Koch and Rudolph Virchow in Germany.

Among the catalysts for a change in America were two individuals. William Welch, one of the original scientific directors of The Rockefeller Institute, played a great role in bringing achievements of German medical science to America. From 1876 to 1878, Welch learned methods of pathology in the Breslau laboratory of Julius Cohnheim (1839–1884), who had been a student of Virchow. Back in America, Welch accepted the first chair of pathology at Johns Hopkins University, but while his laboratory was under construction he spent another year of study in Germany (1884–1885), where he learned the science of bacteriology from, among others, Robert Koch in Berlin. In a seventieth birthday tribute to Virchow in 1891, Welch wrote that the time was ripe in America to "call into existence pathological institutes, akin to the one Virchow conducted in Berlin," one characterized not by dogma but by the method of the natural and physical sciences. This method, Welch noted, was to "observe; experiment; seek the aid of allied sciences, chemistry, physics, general biology; collect by systematic and purposeful investigation . . . a body of facts, and from them deduce general pathological principles and laws." He also noted that whereas Virchow's institute was funded by the government, in America "we must look in general to private beneficence to endow hospitals and medical schools with such pathological institutes."[1–3]

Six years later, another catalyst for establishing America's first privately endowed research institute was neither a physician nor a scientist but an ordained Baptist minister, Frederick T. Gates, who was an adviser to John D. Rockefeller in matters of philanthropy. In 1897, Gates had read William Osler's textbook, *The Principles and Practice of Medicine*, which made him understand why "the best medical practice did not, and did not pretend to, cure more than four or five diseases." He quickly realized that medicine could hardly hope to become a science until there were facilities for intensive research where "qualified men could give themselves to uninterrupted study and investigation, on ample salary, entirely independent of practice." He proposed to Rockefeller that he should establish a new institution for medical research that would investigate the nature and cause of disease as well as methods for prevention and treatment.[4]

Unlike European institutes built around strong personalities in universities, government, or industry, The Rockefeller Institute was organized around a few investigators in diverse sciences who were given autonomy and flexibility in their choice and methods for research. Unlike their obligations in medical schools, the Institute scientists were free from formal teaching, clinical duties, examinations, and administrative responsibilities. Its laboratories and a research hospital went beyond pathology and bacteriology to include chemistry, biophysics, general physiology, and experimental surgery. Such a founding philosophy that allowed a self-directed approach to science would inevitably prove to combine frequent failure with spectacular rewards. The genius in conceiving the Institute was to foster a broad outlook that welcomed research in any field that gave promise of a return in medical science.

From its beginning, the Institute was a sequestered, unpretentious, and self-sufficient community. It was located on Manhattan's Upper East Side, first in a temporary house and then on a former farm purchased by Rockefeller. At the time, it was far from New York's noisy,

The Rockefeller Institute for Medical Research, 1911. Courtesy of The Rockefeller Archive Center.

bustling centers of commercial activity. An imposing gate in the iron fence surrounding its fifteen acres opened onto a drive that climbed a hill to its first beige brick laboratory building, now called Founder's Hall, situated on a rocky cliff forty feet above the East River.

When this laboratory was formally opened on 11 May 1906, Welch paid tribute to two scientific achievements on which the Institute was founded: the discovery by Virchow of the principles and factors of cellular pathology and the establishment by Pasteur and Koch of the germ theory of infectious diseases. Yet, Welch continued, new fields needed to be opened and explored, noting that "of the cause of cancer and other malignant tumors nothing is known." Anticipating the progress in scientific knowledge and the practical results to come from scientific discovery in medicine, he advised that while "the army contains a multitude of recruits drawn from the most various fields," their final victories might not always result in immediate attacks on disease. Rather, the advance of science would be "made by many and devious and gradual steps, leading often, it might appear, in quite different directions, and mounted more frequently than not to secure a wider prospect, but without any thought of the final goal."[5]

This concept of a deliberate yet rewarding pace in the advance of science at The Rockefeller Institute is nowhere better exemplified than in the story of cancer research leading to modern cell biology.

The Experimental Science of Cancer

From the outset, cancer research at the new institute was controversial and was thereby orphaned as secondary among laboratories that were more intent on studying acute infectious diseases. Research on degenerative, chronic, and malignant diseases such as cancer was deemed unproductive and not yet ripe for solutions. According to Rockefeller historian George Corner, the Institute "never put cancer research in the forefront of its program . . . cancer was too complete a mystery for profitable attack." Reasons for this statement are understandable since, historically, cancer was considered a disease of older people who had escaped the ravages of infectious disease, malnutrition, or accidents.[6]

In the first third of the nineteenth century, cancer was generally believed to be many spontaneous diseases, starting at one spot and spreading from there. It was thought that tumors arose from mucus and plasma and accumulated into large masses. Some forms were believed to be hereditary or to have occupational or environmental causes; for example, cancers were found more frequently in people working with soot, arsenic, tar, or dyes than in the general population. Then in 1838, Johannes Müller (1801–1858), the preceptor of both Theodor Schwann and Virchow, made a fruitful application of the microscope to distinguish clinically benign from highly malignant tumors and found both types had the same cellular structures.[7,8] By midcentury, the idea that tumors arose directly from normal cells was prescribed by Virchow in his dictum that *omnis cellula e cellula* (all cells arise from other cells). By the end of the century, pathologists had established a basis for all diagnostic endeavors in neoplastic processes, including metastases.[9] Nonetheless, according to Osler's medical textbook that Gates had read, cancer was described simply as a growth of cells for which the prognosis was hopeless, and its only treatments were symptomatic.[10]

Other scientists refused to reduce cancer to a disease of misbehaving cells and considered it a contagious disease spread by invisible agents. Fueled by the breakthrough discoveries of Pasteur and Koch in the 1880s and 1890s, specialized cancer hospitals were built and cancer patients were quarantined. The assumption was that this would cause the cancer plague to wither away on its own. However, this notion of cancer as an infectious disease lost ground in the first decades of the new century when it became clear that cancer did not spread from person to person like flu or smallpox. The time for experimental cancer research had come.

Simon Flexner, the Institute's first director, had trained in pathology and bacteriology under Welch at Johns Hopkins and then trained in physiological chemistry in Europe. As a result, he strongly supported a mission of applying physical chemistry to biological processes. One of the first scientists he recruited was German-born physiologist Jacques Loeb, who arrived in 1910. Calling medical science a "contradiction in terms," Loeb advocated the study of greatly simplified biological systems by methods of chemistry and physics. In a letter to Flexner in 1909, Loeb advocated creating a new department of "experimental biology of the cell." Loeb believed this science would form the basis not only of physiology but also of general pathology and therapeutics, and he called Rockefeller "the only place in America where such a new departure could be made for the cause of medicine."[11] He further advocated that experiments should be reduced to analyses of chemical substances that could be separated from biological materials. Loeb's philosophy, examined in his book *The Mechanistic Conception of Life* (1912), put an indelible stamp on the Institute's approach to research.[12]

Flexner did not immediately set up a separate division in the Institute for malignant disease. In matters of cancer, as taught by Welch, cells were classified as normal, tumor, neoplastic, and malignant; cancer lay within the growth mechanisms of normal cells, a view clearly derived from Virchow. Despite Welch's professorial warning, "Whatever you do, don't commit yourself to the cancer problem," Flexner was undaunted by its challenges.[13]

In his own laboratory, Flexner and James W. Jobling very early discovered an epithelial tumor of rat and were able to transplant it to other rats; at the time, this was a considerable technical achievement for the experimental study of tumors, although it shed no light on the cause of cancer.[14] The Flexner-Jobling tumor served as a unique test material in cancer research and was used in laboratories as late as the 1960s. When Jobling left the Institute in 1909, Flexner sought someone to expand the work on this discovery and perpetuate this carcinoma because he was abandoning the project to study poliomyelitis, then an epidemic

Simon Flexner's Laboratory of Pathology and Bacteriology, 1911. Courtesy of The Rockefeller Archive Center.

crippling many American children. The position was offered to Peyton Rous, a young physician who had also studied pathology with Welch and in Germany. Rous had just published three articles on lymphocytes in *The Journal of Experimental Medicine* that had impressed Flexner, the journal's editor.

Introducing Chicken Tumor I (the Rous Sarcoma Virus)

The method of using transplantable tumors was very limiting because they could grow successfully only in blood relatives of the species from which they came. While trying to understand this limitation, Rous made a control experiment to compare the transplantability of normal versus tumor tissues in mice. His critical analysis of the results overturned a long-held fallacy and showed that both tissues are rejected when transplanted in foreign hosts. Rous had discovered that grafts of alien tissues are unrelated to whether or not they are neoplastic; in other words, transplantation was a tissue problem and not a cancer problem.[15]

Rous barely finished this work when a chance event turned his cancer research onto a new path. A poultry breeder brought Rous a purebred Plymouth Rock chicken bearing a large breast tumor. Its growth, a spindle cell sarcoma, could be transplanted into healthy blood relatives of this hen, causing the same local sarcoma, followed by remote metastases and death. Rous now had a novel animal model in which to study cancer. His paper describing this first transplantable solid tumor in birds appeared in September 1910, a month after James Bumgardner Murphy joined the laboratory.[16]

Murphy was also a student of William Welch at Johns Hopkins, where his interest in research was kindled. Initially he wanted to follow his father's successful career in psychiatry and had published two papers before finishing his medical training. However, his father's death from cancer in 1907 turned his interests toward a career in pathology. After an

James B. Murphy, 1911. Courtesy of James Slater Murphy.

Peyton Rous, 1911. Courtesy of The Rockefeller Archive Center.

internship at the Pathological Institute on Wards Island, New York, he decided to pursue cancer problems. He applied for a fellowship with Rous and quickly became absorbed in a startling new dilemma over the cause of cancer.[17]

Since propagation and transmission of tumor tissue was technically difficult, one of Murphy's initial tasks was to find other ways to keep tumors alive in the laboratory. Within a year, he made two substantial contributions to this problem that eliminated the need to transplant tumors from one hen to another.

Implanting sarcoma on the developing embryo. In 1911, Murphy discovered one technique that used the chicken egg as an incubator for tumor tissue. He implanted the sarcoma on the chorioallantoic membrane of chicken embryos, where it grew rapidly and produced a sarcoma having all the characteristics of the one in adult hens. The chick embryo provided a way to keep the tumor growing as well as supplying fresh sarcomatous tissue for research. Using the egg was simpler and easier than handling recalcitrant adult hens. It was possible to inoculate the eggs with either living tumor tissue or with a freeze-dried or lyophilized material (see next section) of the tumor.[18]

This method was amenable to laboratory operations and was easily reproducible. In contrast to complex experiments on intact animals, the egg was cheap and readily obtainable. Also, in contrast to tissue culture methods just being introduced by Alexis Carrel at the Institute, the egg membrane was relatively easy to keep sterile. Murphy extended this technique to show the embryo could support tumor growths from other species of birds and even mammals.[19] The embryo thus became a more favorable host for studying the sarcoma than the adult chicken. After research on viruses became an established field in the 1930s, this chick embryo culture technique became a universal assay for the diagnosis and vaccine production for many viral diseases, such as measles, mumps, and yellow fever, and lasted well into the 1950s, when it was superseded by new tissue culture methods.

Freeze-dried extracts of sarcoma tissue. At the same time, Murphy created a second technique to obtain a cell-free extract containing the active tumor agent. After he observed that freezing and thawing live tumor tissue did not kill the active agent, he tried drying the tissue in a vacuum over sulfuric acid and then further drying it in a freezing box. Murphy's freeze-drying method killed the cells but not the active agent, and he also showed the desiccated extract could be injected either into chickens or onto his new model, chicken embryonic membranes, and the sarcoma grew successfully in both.[20–22] This technique preceded by almost a quarter century the method described in 1935 by Earl Flosdorf and Stuart Mudd, who are often credited with its discovery. More importantly, almost all of Murphy's subsequent research with the chicken sarcoma was performed using lyophilized material. When American, European, and Japanese scientists requested samples of the sarcoma to test in their laboratories, Murphy and Rous mailed them small vials of the lyophilized tumor agent.[23]

The Nature of Chicken Tumor I: Viral or Chemical?

Murphy's lyophilization method also provided a second way to transmit cell-free tumor material. A filtration technique had been reported independently in 1892 by Dimitri Ivanovski in Russia and in 1899 by Martinus Beijerinck in Holland as a means of extracting the active agent, now known to be a virus, that causes tobacco mosaic disease.[24,25] This technique involved grinding up malignant tumor tissue and passing it through an earthenware filter that holds back not only the cells but also bacteria while the ultramicrobes, or viruses, pass through. Ivanovski's observation that an infectious, living agent responsible for disease could pass through a bacterial filter is generally considered the vigorous initiation of the "filterable-virus era." Viruses eventually became distinguished from other pathogenic microbes by three negative properties: they are invisible under ordinary microscopes, filterable through bacterial filters, and unable to multiply except in susceptible cells.[26–28]

Since no effort to implicate bacteria in cancer had been found, it seemed possible that the discovery of viruses at the end of the nineteenth century would inspire research about whether these agents might cause cancer. One success came in 1908, when Danish scientists Vilhelm Ellerman and Oluf Bang succeeded in transmitting cell-free filtrates of a chicken leukemia from one chicken to another.[29] Their work was regarded as unimportant, however, because at that time leukemia was not regarded as a malignancy, and chickens were not considered suitable models for mammalian cancer.

Rous suspected an ultramicrobe might be harbored by the chicken sarcoma, and he immediately tried the filtration method with great success. He prepared an extract of the tumor, passed it through a filter, and injected the filtered material into healthy chickens, which then developed tumors similar to those in the original hen. This technique was the basis of his classic 1911 paper, "A sarcoma of the fowl transmissible by an agent separable from the tumor cells."[20]

Rous also used Murphy's technique and injected a desiccated extract of the tumor; it confirmed the filtration results. In this single-author paper, however, for reasons discussed in the following paragraphs, Rous relegated Murphy's lyophilization method to a four-line footnote. It was not until three years later that Rous and Murphy published full reports on three methods of preparing cell-free extracts—based on filtration, lyophilization, and glycerination. All were equally successful in inducing malignant sarcomas in fowl that resembled the original sarcoma as faithfully as transplanting tumor tissue. Furthermore, they found that cell-free transmission could be transferred from fowl to fowl without losing the activity of the original tumor or the extract.[21]

Rous cautiously reported his 1911 discovery in the rather vague context of a filterable virus. An ostensible reason for emphasizing the filtration method—and nearly omitting Murphy's desiccate method—was "to regard the self-perpetuating agent active in this sarcoma of the fowl as a minute parasitic organism." The filtration method seemed to suggest the tumor agent was still alive in the material that passed through the filter. To emphasize this point, Rous did not give equal emphasis to lyophilization or desiccated material. He was likely unsure whether the freeze-dried material was entirely free of cellular material because it took longer to produce a tumor, and its slow growth may have somehow changed the agent.

Rous, however, did not rule out the possibility that an agent of another sort might be at work: "It is conceivable that a chemical stimulant, elaborated by the neoplastic cells, might cause the tumor in another host and bring about in consequence a further production of the same stimulant. For the moment we have not adopted either hypothesis."[20]

With these two statements, Rous exposed a dividing opinion over cancer causation that would prevail for the next half century. Was the sarcoma agent a virus, or was it a chemical? This controversy took place not only among scientists worldwide, but more remarkably, it separated Rous and Murphy.

The immediate scientific reception to Rous's discovery, similar to the one given Ellerman and Bang, was considerable disbelief. Transmitting a tumor without cells was a wholly unexpected finding, and this suggested that Rous had demonstrated a cause for a cancerous tumor for the first time. In an interview with T. Mitchell Prudden, one of Rockefeller's scientific directors, Rous said that Welch himself examined the chicken growth. While Welch admitted it was a tumor, like the others, Rous claimed he "got around it" as a case peculiar to chickens. In recounting the prevailing negative attitude, Rous said, the first reaction of many pathologists "was to rule it out of court. This was an instinctive reaction; so much did [the] findings on the growth's transmission interfere with customary thought. As [E.F.] Bashford of the Imperial Cancer Research Fund put it, 'Why, my dear fellow, it can't be a tumor. You've found the cause!' "[23] Although ironic for the time, Bashford was correct in retrospect.

A viral theory of cancer challenged all assumptions, going back to Hippocrates, that cancer was not infectious but rather a spontaneous, uncontrolled growth of cells. Rous was convinced that if an agent could pass through bacterial filters, then this was evidence of a virus multiplying during tumor growth. More to the point, he wrote, "a study of the effects of chemical and physical agents on the agent producing the chicken sarcoma gives one every reason to suppose that this agent lives and that it belongs in the animal kingdom."[30]

Rous appeared scientifically conservative, even noncommittal, in publications where he referred to the causative factor simply as the "agent." Yet, in his private 1911 annual report to the Board of Scientific Directors, where he first described this discovery, he dared to call the agent a "virus." According to Rous's interpretation, the agent showed several characteristics of a virus: it increased in the growing tumor, it produced a specific kind of tumor and only in fowls, and a preliminary tissue injury greatly hastened the disease. The agent was found to elicit specific inactivating antibodies that did not retard tumor growth or multiplication of the virus, an early finding that led to the understanding that virus diseases are essentially cell infections.[31] To Rous, the growths were specific responses of the host to a living microorganism.

Murphy, in contrast, opposed a viral cause of cancer. His interpretation of the tumor agent, reflecting Welch's teaching, was that the perverse phenomenon involved an internal mechanism controlling cell growth. He maintained that the evidence he and Rous produced indicated they were dealing with endogenous chemical substances (in the cell) rather than with exogenous living virus. As he later noted, there were features of the agent that resembled a chemical. He saw no evidence that the host's cells became malignant by contact with the inoculated agent, only that the host supplied blood vessels and supporting framework. He observed a strong relationship between chronic irritation and cancer development where injury rather than an inoculated agent seemed more important; the tumor grew at the exact point where the needle was thrust into the tissue. There was no evidence of a contact infection with the disease, only that they were transmitting the disease artificially. High doses of ultraviolet light or x-rays that normally killed tumor cells and pathogenic microbes did not kill the agent. And, no ultramicroscopic living organism or any similar agent associated with mammals had yet been found.[32]

To Murphy and most scientists, these findings seemed to indicate a disease peculiar to chickens. Even though he and Rous identified other *chicken* tumors from which they obtained cell-free extracts that could transmit tumors, each one caused a different kind of tumor.[21,33] He summarized their differences in a 1925 symposium on "The Cancer Problem," held in Connecticut by the American Society of Zoologists. Murphy argued that if such tumors were so highly specific and if viruses were responsible for all tumors, it would be necessary to postulate thousands of different types to account for all the different varieties of tumors known to exist. If the agent were living, then it should cause similar tumors: "Even admitting a specific chemical factor as a necessary accompaniment to the infecting organism, it would strain the imagination to fit this conception to the facts as we know them."[32]

Murphy recalled his tumor studies with Rous were "interpreted as indicating the presence of a so-called filterable or ultramicroscopic living organism, or indicating an agent more closely related to the cell, something in the nature of an abnormal growth enzyme. Exhaustive study of these tumors over several years failed to establish either of these points of view." The phrases "by one of us" and "by the other" appear in his manuscript, but they were deleted in the published version.[34] Since his views were shared by many other scientists, there was no need for attribution. Furthermore, given the still meager knowledge about viruses at the time, research on the tumor agent kept both hypotheses alive for decades.

Keeping Cancer Research Alive

In June 1913, after three years of working together, Murphy and Rous stopped reporting on the chicken tumor work to the Board of Scientific Directors. Their collaboration had reached an impasse. According to Rous's report in 1914, he spent a little more time looking at mammalian tumors without success; a year later he changed fields and began to work with pernicious anemia and developed methods of blood transfusion that were implemented on World War I battlefields. A relinquishing note by Rous summarized the prevailing frustration over the chicken tumor work: "Despite many facts which indicate that the agents causing chicken sarcomata are living entities some authors have preferred to consider them as chemical substances thus reconciling their existence with established views of cancer etiology. For this reason a further study of the nature of the agent seems advisable."[35]

After joint papers published in 1914, the two men separated, though cordially enough. Both stayed in the same nurturing small community of researchers and kept in touch with each other's work. Their basic disagreement, however, was a positive force in propelling research on cancer.

Rous did not return to cancer work for two decades, not until a mammalian tumor discovered by Rockefeller colleague Richard Shope rekindled his interest in tumor causation. Rous's many contributions to the discipline of viral oncology have been well chronicled elsewhere and are not recounted here.[36–40] He abandoned work on the chicken tumor forever, although he never lost sight or interest in work on "his virus" by others, especially Murphy and his colleagues. Predictably, Rous enters the previously untold story in this book on several occasions throughout the following decades with a decided provocation of keeping a viral theory of cancer alive. The story we follow here is Murphy's equally strong belief in the chemical nature of cancer that framed a twenty-five–year search for the "nature" of the chicken tumor agent. Although the majority of Murphy's personal laboratory research during this same period focused on clinical, immunological, genetic, and environmental approaches to malignancy, it was the search for the cause of the sarcoma that led to a different discipline, one devoted to the biology of cells.*

An unusual coincidence in 1913 revived cancer research at Rockefeller. Henry Rutherford, a Vermont resident, left a $200,000 bequest to the Institute, a sum equal to the Institute's founding bequest from John D. Rockefeller Sr. in 1901. Rutherford specified it was expressly for investigating "the causes and nature of cancer and methods of its prevention and treatment."[41] Consequently, this meant a new project had to be launched.

Although Murphy considered leaving the Institute when his collaboration with Rous ended, the combination of this bequest and collegial encouragement induced him to stay and begin a new course of research. Florence Sabin, his Hopkins anatomy professor and a sympathetic friend,[42] and colleague Jacques Loeb were interested in stimulating his shift to study cellular resistance to tissue grafts. Loeb also encouraged Flexner to set up Murphy in an independent laboratory.[43]

Before the Rutherford funds became available in 1915, Murphy expanded his research based on a curious observation that he and Rous had made in 1912. In animals that resisted the transplanted cancer, they found abundant lymphocytes surrounding the foreign graft, a reaction that appeared early and lasted until the transplant was destroyed. Their published version referred to this reaction as a "secondary phenomena."[44] Murphy again disagreed with Rous and began testing this observation on a broader scale. He was able to either create resistance to grafts by implanting lymphoid tissue at the same time or abolish resistance by eliminating lymphocytes with x-rays.[45,46] Loeb referred to these as "remarkable experiments" in *The Organism as a Whole* for showing that lymphocytes could "rid the host of the graft."[47]

*Despite work by Rous and Murphy, the identification and characterization of the agent of this chicken tumor would not come for another half century. Today, the Rous Sarcoma Virus, as it is known, is classed as an RNA retrovirus that is used as a laboratory model for studying how viruses act by invading genes of normal cells. In 1914, however, they could not have realized how a virus accidentally picks up a cancer-inducing gene, an oncogene, from a normal chicken cell where it not only infects new cells in the recipient chick but uses its host genes to prompt them to divide incessantly.

In June 1915, Murphy and his assistant John J. Morton presented this research before the National Academy of Sciences and concluded, "the lymphocyte is a necessary factor in anti-cancer immunity."[48] Although based on a few rodent studies, their report caused sensational press coverage about a "breakthrough" in cancer research, with headlines such as "John D. Doctors Believe Cancer Prevention Found."[49] Murphy's finding so impressed Flexner that he told the scientific directors the same month that the Rutherford funds should be used for "the extension of Dr. Murphy's work to human patients."[50] A budget and proposal were immediately set up for a "Study of Cancer in the Hospital of the Rockefeller Institute."[41] Thus, Murphy organized a new laboratory called Biophysics in which his work with lymphocytes would be applied to cancer patients using the newest instruments in cancer treatment, radium and x-radiation. Of historical note, the interest from the Rutherford funds partially supported the work of Murphy from 1915 until his retirement in 1950, when they were assigned to support work by Peyton Rous.[51]

Murphy was a strong believer that a physician doing clinical work should also carry out his own laboratory investigations. Yet, in a 1917 letter to Flexner, he expressed some cautionary views about studying patients with cancer. He wrote that pathological studies had been devoted "almost entirely to the cancer cell and little to changes in the individual which makes this new growth possible." For this reason, he was uneasy because following "a chronic disease clinically is of course much more difficult than is that of an acute one. . . . The problems are difficult and the points of attack are less evident." Nevertheless, he noted that Rockefeller had a special opportunity to establish an experimental foundation for understanding cancer and testing cures. Rather than specializing in mouse cancer, he was encouraged that his lymphocyte research might offer valuable leads in human disease as well as the basis for a new "theory which will explain many of the phenomena and gives us a direct mode of attack which should be of primary importance in the solution of any problem."[52]

Unfortunately, as Murphy sensed, three years of clinical experiments using x-rays in breast cancer patients did not provide enough reliable information about lymphocytes in cancer that was worthy of publication. In his annual reports to Flexner between 1916 and 1922, he concluded that tracking cellular reactions in chronic illness was the unwieldy pursuit he had foreseen. At this point Murphy opened a more productive approach that he would follow for the next thirty years, namely, biochemical analyses of changes in malignant cells. He had suggested this approach to Flexner five years earlier: "Outside of the gross histological changes, there have been no studies on the physiologico-chemical side . . . such a piece of work would require both patients and animals, as well as a trained chemist."[52] To carry out this investigation, he hired chemist Ernest Sturm, who from 1919 to 1950 was Murphy's invaluable aide and coauthor of many publications.

In 1926, Murphy published an impressive monograph summarizing nearly sixty publications from his laboratory on the role of lymphocytes in tumors, cancer, and tuberculosis. His cellular research implicated a strong relationship between stimulating or inhibiting lymphocytes and the development of experimental tumors. He maintained the lymphocyte is an "important link in the chain of processes which determines resistance to cancer." But he could say little more on the subject because his findings did not extend to the role of the lymphocyte in anti-tumor activity.[53]

Historians of science are only partly correct in their reasons why Murphy abandoned this line of research in biophysics. Ilana Löwy speculated that Murphy's work was premature and awaited the formulation of the concept of the immune system and the rediscovery by others of the critical role of the "immunocompetent" lymphocyte in cancer.[54] Arthur Silverstein, in *A History of Immunology*, noted that "the lymphocyte was then a cell of mysterious ancestry and function," so much so that after Murphy's monograph there was very little knowledge added about its function until the 1950s.[55]

What Murphy clearly abandoned, on behalf of the Institute and with Flexner's urging, was a cumbersome, costly staff of physicists who were using radiation technology to prevent and treat cancer.[56] A more immediate and plausible reason for Murphy's seeming abandonment reflects his focused view of cancer causation. He came to believe the animal experiments on defensive forces of the body (the lymphocytes) were being marshaled against implanted cancer cells that created an artificial disease. To him, this disease state shed no light on the events underlying the "natural" origin of cancer, which he continued to believe arose within normal cells. What he retained in 1925 were two chemists and a geneticist who helped him return to more basic studies of experimental medicine to pursue this view.

Reviving Research on the Chicken Tumor Agent

By the time Murphy's monograph appeared in 1926, three notable influences helped to reorient and then fortify his biochemical approach to cancer biology.

The first event coincided with the twentieth anniversary of the Institute's first permanent laboratory. Flexner spent two years reviewing the strength of Rockefeller's research programs and reported to the Board in 1926 that he wanted a minor reorganization to meet the needs of "new knowledge accruing with ever increasing speed from chemistry . . . and physics."[57] Among the changes he proposed was to make the field of biophysics rival that of biochemistry, and to do this he wanted a physicist to introduce new laboratory (not clinical) technologies that would help scientists extend their methods of discovery. Murphy's biological and pathological expertise with radiation instruments, he added, had set a precedent, but it was only the "merest beginning" of this field. A year later, physicist Ralph W.B. Wyckoff joined the Institute and went on to develop several instruments and introduce major new technologies to biomedical sciences, including some applicable to the cancer problem (discussed in the following chapters).

Another event in the mid-1920s relieved the Institute of the burden of testing new clinical technologies and reinforced its mission of pursuing basic research into the cause and nature of disease. "It is not profitable," Flexner wrote, "to attempt solutions of practical problems of disease far in advance of the possession of fundamental knowledge of facts on which this solution rests."[57] Welch's earlier warning to avoid cancer studies and Murphy's subsequent inconclusive experimental evidence indicated that this field was still too premature for practical solutions.

Nonetheless, John D. Rockefeller Jr., the Institute's chairman of the board, was still very interested in ways to prevent and treat cancer. In an anonymous, poignant, and personal story, *The Legacy of Bessie Dashiell*, which discusses the origins of New York City's Memorial Hospital, Rockefeller's lifelong commitment to rid humanity of cancer is traced to his childhood sweetheart Bessie Dashiell, who died of the disease in 1891. While emphasizing that Rockefeller's "own scientists at the Institute were hard at work" on the basic nature of important medical problems, the story relates that he believed not enough was being done to relieve the scourge of cancer: "He could not shake off the almost obsessive memory of human suffering and sudden, useless death. He wanted to get back to people: to treatment, to the relief of suffering. He sought out James B. Murphy at the Institute and asked his advice. Without hesitation, Murphy advised, 'Back Ewing at Memorial.' "[58]

Pathologist James Ewing had been medical director of Memorial Hospital since 1913. He was even more outspoken than Murphy that cancer was not caused by a microorganism. In a clash of medical cultures for treating cancer at Memorial, Ewing's strong advocacy of radiation therapy brought him into acrimonious conflict with William Coley, one of the Hospital's physicians whose therapy with bacterial vaccines occasionally induced tumor regression. Ewing prevailed with his approach to scientific medicine and to testing such new technologies as radiation therapy.[59]

Consequently, John D. Rockefeller Jr. divided his commitments to support research at The Rockefeller Institute into the cause and nature of the disease and research at Memorial Hospital into its prevention and treatment. By 1934, also with his support, Memorial built a new hospital across York Avenue from the Institute, where it was shaped into a hospital and research institution devoted to cancer research. The working relationship between Murphy and Ewing constituted a strong collegial bond between the two institutions, particularly through the two scientists' considerable efforts during the formative years of the American Cancer Society.

The third compelling event caused Murphy to take up a challenge and begin new studies on the chicken tumor agent. In 1925, William E. Gye in England reported a highly controversial set of experiments in which he claimed to have cultivated an organism from a variety of animal and human tumors, including the chicken tumor of Rous and Murphy.[60] He tried to show two factors were needed to produce tumors, one a microbe that could be cultivated in a test tube and the other a specific chemical that determines the species and tissue that will be attacked by the living organism. Gye's theory suggested there were as many specific factors as there are varieties of cancer and that cancers derive their peculiar properties from the very cells in which the tumor grows. In essence, Gye revived both a viral and chemical theory of cancer causation.

Murphy's laboratory, as well as many others throughout the world, subjected Gye's theory and experiments to severe tests but failed to confirm his results. A few months later in the *Journal of the American Medical Association*, Murphy became the first to criticize Gye, concluding that "finding a cultivated living organism in the interpretation of Gye's results is eliminated."[61] Even Rous, who was on a sabbatical in Cambridge, England, expressed skepticism of Gye's hypothesis. In January 1927, he wrote Murphy, saying, "The better men over here are skeptical as concerns Gye's work," and a month later he called Gye's hypothesis "confused enough to take years for any certainty."[62,63] Murphy reported to Rous that many scientists had confirmed his negative findings against Gye and told him, "It seems to me that in this chaos it is better to hold off and perhaps try to get at the problem from another angle."[64]

Murphy had already outlined his new attack on cancer in 1925 before a symposium on "The Cancer Problem" (mentioned previously). He elaborated on earlier clues in cancer that had been misleading and contradictory, set out his disagreements with Rous about the cause of cancer, and further criticized Gye's experiments. His conclusion held to his original belief that "when the cause of cancer is discovered it will be found to be some agent or force most intimately associated with the mechanism of growth of the cell and will be a discovery of equal importance for those studying normal cell phenomena and to the cancer investigator."[32] A year later, Murphy informed Harvard bacteriologist Hans Zinsser that his new line of attack "seems to indicate without very much doubt that we are not dealing with a living agent in this group of tumors. It falls more in the enzyme group, and I think we can eventually definitely place it there."[65]

Chapter 2

Taking Cells Apart: Centrifuges and Cell Fractionation

Carol L. Moberg

> A present-day scientific session on cancer is like an assemblage of Aladdins met to discuss the best methods to evoke genii by the rubbing of lamps, what sorts of rubbing and what lamps are most rewarding in this respect, and how the genii themselves look and behave—with little said on the intrinsic reason for these apparitions, for the sound reason that little can as yet be said. The old scholastic disputations about the cause of cancer are no longer heard. Facts hold men's thought.
>
> Peyton Rous, 1937

During the four years following the 1925 "Cancer Problem" symposium, James Murphy tested and then discarded several techniques for looking into the nature of the chicken tumor agent. Nevertheless, these techniques helped to advance the understanding of normal cell physiology as the basis for understanding abnormal cells.

As early as 1911, Alexis Carrel and Montrose Burrows at the Institute reported they could grow tumors in culture medium, including the Rous chicken sarcoma.[1] The technical difficulties of keeping tissues alive at this time were substantial because cultures were easily contaminated and growth mechanisms of cells and tissues were poorly understood. Peyton Rous tried to duplicate Carrel's techniques. While he succeeded in using trypsin to separate the cells from tissues, in his hands the cells lived only a few days.[2] As Rous told T. Mitchell Prudden, his failures during the early years of trying to cultivate what he believed to be the causative organism "nearly broke me down."[3] So it is not surprising that Rous's main objections to William Gye's controversial work in 1925 were based on tissue culture techniques. Rous cautioned Hans Zinsser at Harvard, who was also subjecting Gye's work to scrutiny, "I worked intensively for a long time at cultivating the thing, but to begin with ruled out such methods as he has employed, those involving the addition of living and proliferating tissue, for the reason that such tissue can readily be 'infected' when brought in contact in vitro with the tumor-producing agent. How does Gye know that the transfer from tube to tube of his cultures is not merely reinfection after reinfection of living cells? The tumor-producing agent gets into the fluid surrounding such cells, as I also found."[4]

In 1926, Carrel reported another success in cultivating the Rous sarcoma agent in culture using embryo pulp.[5] This time Murphy tried to repeat Carrel's tissue culture experiment, but even in his hands the cells lasted only for three days.[6] After this, Murphy abandoned tissue culture methods for twenty years until Keith Porter developed new methods for growing cells for electron microscopy (see Chapter 3).

In 1924, German biochemist Otto Warburg proposed that cancer results from a defect in cell respiration.[7] He postulated that when oxygenated blood is poor, as in the vicinity of a tumor, cells replace the respiration of oxygen in normal cells by fermenting the sugar glucose, a process called glycolysis, and thus become cancerous. Normal body cells, on the other hand, meet their energy needs by respiration of oxygen. Among the tumors on which Warburg tested this thesis was the Rous sarcoma. He obtained a sample from Rous after a lecture visit to the Institute in late 1924 and found the sarcoma could live without oxygen.[8,9] However, while Murphy and James Hawkins were able to confirm Warburg's original results, when they extended the tests to many other tissues, they found several normal tissues also showed "abnormal" respiration.[10] They concluded that glycolytic activity of a tissue is more a function of its growth rate and not a characteristic of malignancy. Murphy therefore dismissed this unusual theory, saying, "Till normal cells are definitely transformed into malignant cells by depriving them of oxygen, this theory cannot be considered seriously."[11]

The laboratory also performed other types of experiments aimed at characterizing properties of the tumor agent. A working hypothesis for these new experiments, as Murphy told James Ewing, was that "the agent was more closely related to the cell and something in the nature of an abnormal growth enzyme."[12] He used the term "enzyme" in an indefinite sense to indicate unknown chemicals produced by cells. The experiments he performed to test this enzyme hypothesis were in keeping with his cellular views of cancer as well as further discrediting any implication of a viral cause.

Some tests were related to cell physiology. With Francisco Duran-Reynals, Murphy showed that the tumor agent was bound in vitro by muscle tissue from susceptible fowls but not by tissues of non-susceptible animals such as rabbit or pigeon or by organs such as brain or liver of susceptible fowl. This affinity of tumor agent and tissue had no parallel among known viruses and bacteria and was generally interpreted as evidence of an enzymatic reaction.[13] Murphy concluded that "the tumor agent behaves like many cell products, in that it is fixed by the specific substance on which it acts."[14] In one comparison, they tried to inactivate the known vaccinia virus and the tumor agent with chloroform, but only the tumor agent remained active.[15] This finding must have disappointed Duran-Reynals, who also maintained a strong belief in a viral cause of cancer. He had to conclude in 1929 that there was a "fundamental difference between the behavior of the chicken tumor agent and the filterable agents of the virus group."[13]

Other tests failed to find other filterable viruses in two tumors that were also transplantable. Louise Pearce and Murphy studied a mammalian tumor, a rabbit neoplasm.[16] Karl Landsteiner and Murphy created a chicken sarcoma by injecting tar in adult hens.[17] In neither case could they extract an agent from a cell-free filtrate or a desiccate of the tumors.

Still other tests focused on the agent's chemical nature. In collaboration with Frederick Gates Jr., Murphy applied ultraviolet light to various tissues and found the tumor agent needed a hundred times more energy to be inactivated than did bacteria, known viruses, and bacteriophages.[18] With Ernest Sturm and Oscar Helmer, Murphy compared the agent with known viruses and enzymes. They tried electrodialysis and reaction to solvents and found both methods produced precipitates that retained tumor-producing power. The vigorous chemical handling in these experiments would have normally inactivated or killed most living organisms. By this time, the fraction containing the agent had become so purified, or at least concentrated, by repeated precipitations that Murphy concluded "it is inconceivable that it should carry living organisms or virus through all these processes."[19]

While trying to purify the tumor agent chemically, Murphy and Sturm tried a control experiment and performed the same separation and purification techniques on normal cells as they used on tumor cells. These tests followed Murphy's reasoning that cancer arises from normal cells, as he noted in his annual report to director Simon Flexner in 1929: "If the tumor producing agents, as they occur in chickens, are in the nature of hormones or enzyme like substances which normally control tissue growth and differentiation, there might be a possibility of isolating such a substance from active, normal tissues."[20] These unpublished experiments involved separating normal cell precipitates from testicles of roosters, which they assumed had no contact with tumor-bearing hens, injecting them into normal hens, and discovering they produced sarcomatous tumors in fifteen of sixteen hens.[20] Murphy was not surprised by these results. In fact, he felt confident this was sufficient evidence for an enzyme- or ferment-like nature of the tumor agent from normal cells.

Murphy presented these preliminary experiments before the International Conference on Cancer in London on 17 July 1928. It was the first time he argued publicly *against* a viral and *for* an enzyme cause of cancer.[19] His inference that a natural ferment existing in healthy tissues can cause cancer brought him immediately to public attention. The next day, British and American newspapers produced sensational headlines, such as "World-Wide Medical Revolution Looms if Murphy Theory that Cancer is Result of Chemical Ferment, Not Virus, Is Proved." *The New York Times* hailed a "New Cancer Theory Revealed in London" that "startles scientists."[21]

Rous's delayed reaction to this report was politely cautious. As he wrote Murphy: "The report which you have kindly sent me is painstakingly written and gives the established facts, yet inevitably it differs as a presentation from that I would have made. . . . My own notions about the meaning [of the nature of the tumor agent], as you know, do not fit in with yours. But this does not matter as the end will tell the true story."[22]

Further, Rous and Simon Flexner became concerned as editors of *The Journal of Experimental Medicine*. When Murphy submitted the first of eventually fourteen papers on the nature of the tumor agent to the *Journal*, Rous felt the work seemed too premature and advised Flexner more experiments should be done "to validate the present work."[23] But he left the judgment up to Flexner, who hesitated to interfere, although he wondered, "the preliminary statement was made in 1928! When will the definitive statement be made?"[24]

On repeating experiments with testicle extract from healthy animals, there were cycles of failure and success over the ensuing year. The next steps were to test whether the testicle extract enhanced infections or tumors, purify the tumor agent further, and then resume research on normal cells as the source of the tumor agent. Since 1925, most research in Murphy's laboratory used chemicals to study malignant transformation of cells. Returning to work on the sarcoma agent was more time consuming and complicated than planned. Murphy needed an assistant who could devote himself to its mysterious properties. This person was Albert Claude.

Purifying the Chicken Tumor Agent: Enter Albert Claude in 1929

Claude's early views of cancer causation are unknown. However, his motivation, research experience, and interests in cancer had prepared him for somewhat different experiments than he found when he got to New York City.

Claude was born in the Belgian village of Longlier in 1899. He watched his mother die of breast cancer when he was seven years old. Though he dreamed of becoming a physician, he never attended high school. As a teenager during World War I, he was involved in underground activities for the British Intelligence Service and earned several military distinctions. Then a great piece of luck intervened. A dispensation was granted to war veterans from the Allied armies to enter medical school without a diploma or examination. Claude thus enrolled and obtained his MD from the University of Liege in 1928.[25–27]

By his final year of medical school, Claude had acquired a strong resolve to make a career in cancer research. He took on an independent project in which he learned to transplant the S-37 mouse sarcoma into the brains and skin of rats to observe the tissue reactions. In one of his five published papers, he repeated and disagreed with the work of Murphy and Sturm at Rockefeller: they reported lymphocytes only surround cerebral tumor tissue, while Claude observed the cells were invading it.[28] This project won him a government fellowship to work at the Cancer Institute at the University of Berlin in 1929. Claude was promptly expelled after he showed that the bacterial theory of cancer genesis, propounded by the Institute's director Ferdinand Blumenthal, was the result of faulty inoculation procedures. He then found refuge at the new Kaiser Wilhelm Institute (now the Max Planck Institute) headed by Otto Warburg. He worked with Danish scientist Albert Fischer, a pioneer of tissue culture who spent 1921–1922 with Alexis Carrel at Rockefeller. Fischer was just publishing the third edition of a German monograph on these techniques.

Even before taking up research in Berlin, Claude had applied to work at Rockefeller. On 1 December 1928, Jules Duesberg, rector of the University of Liege, wrote to Simon Flexner, saying, "a doctor of medicine is applying for a CRB fellowship."[29] The Commission for Relief in Belgium was organized at the beginning of World War I by Herbert Hoover (who was also its president). In his application, Claude appeared well informed about the Rous sarcoma and clear in his ambitions to study its cancerous properties. He characterized the complex nature of the tumor "at a crossroad which connects the classic entity of cancer (by its properties of growth, virulence, metastases, histological characteristics), bacteriology (by its contagiousness; its filterable principle), and physiochemistry (cancer arising from arsenic and Indole in vivo by Carrel; arsenic acid in vitro by Fischer)."[30] Claude specified that he wanted to study with Rous "on the tumor which carries his name." But he added that if Rous declined he "would be delighted to work under the direction of other *savants*" and named Carrel, Murphy, and Pearce.

On 18 April 1929, Flexner wrote to the CRB that "Dr. Murphy is willing to take Dr. Claude as a voluntary assistant."[31] Before his arrival in September, Claude informed Murphy that "having decided to devote myself to the study of cancerology, I thought it would be good for me to make myself familiar with the most important methods acquired up to the present time by this branch of pathology."[32] During his fellowship in Germany, Claude acquainted himself with tumor histology and pathology, tissue culture of the S-37 sarcoma, metabolism of cancerous tissues, and how microbes probably had little to do with causing cancer. Ironically, while these methods were ones that Murphy had already tested and eliminated, the fellowship training enhanced Claude's familiarity with ongoing research in Murphy's laboratory. Claude worked as a volunteer for two years before being promoted to a salaried position of assistant on the scientific staff; he remained in Murphy's laboratory until 1949, when he returned to Belgium. During these twenty years, Claude's views on the nature of the Rous sarcoma agent were tested dramatically.

Albert Claude (left), 1930. Courtesy of Philippa Claude.

Initial experiments on the tumor agent. Claude immersed himself in two chemical aspects of the tumor research that Murphy had identified as critical: extending work on the testicular extract and purifying the tumor agent. Throughout the next eight years, a great deal of controversy and confusion existed about the nature of sarcoma agent, particularly against a background of knowledge about viruses that was emerging from laboratories at Rockefeller and elsewhere.

By the time Claude got involved with the testicular extract, Duran-Reynals had found the extract enhanced the infectivity of many viruses, bacteria, and mammalian tumors but had no effect on the chicken tumor.[33] Nevertheless, his observation led to a fundamental discovery of a "spreading factor," also known as Reynals factor, a substance that promotes intracellular diffusion and damage of pathogens through all tissues and cells. Duran-Reynals and Claude characterized this factor as a mucolytic enzyme, which is known today as hyaluronidase.[34,35]

Claude's attempts to purify the agent further involved treating tumor filtrates with aluminum hydroxide, a technique used to purify enzymes by removing protein. He found after centrifuging the material that the supernatant, that is, the uppermost clear liquid layer above the sedimented material, was far more active than the full-strength tumor extract. This experiment indicated that the low activity of extracts was not due to the absence of the agent but rather to the presence of what they assumed was an inhibiting agent.[36,37] Other laboratory members pursued the nature of the inhibiting substance for several years based on Murphy's assumption that cells normally contain chemicals to balance and control the tumor growth

James Murphy's laboratory on the second floor in the New North Building (now Theobald Smith Hall), 1930. Courtesy of James Slater Murphy.

agents; a specific inhibitor was never identified, and the search was abandoned by 1939. Claude's isolation techniques, however, allowed them to separate the tumorigenicity of the agent from its antigenicity, once again confirming to Murphy the inanimate nature of the agent.[38,39]

Transmissible mutagen theory of cancer causation. In 1932, Murphy and Claude published the first review of the chicken tumor agent, entitled "Transmissible tumors of the fowl."[40] An ostensible reason was to assemble some 270 papers synthesizing work on the agent done since 1910 by them and other laboratories. More explicitly, they wanted to propose a new hypothesis of the nature of the tumor agent, which was conveyed in the title word "transmissible." Murphy modified his earlier vague enzyme theory with the view that malignant cells produce a substance that acts on normal ones and causes them to assume their distinct malignant characteristics. With this new hypothesis, they implied that cells transformed by the agent must produce more of the agent and that they in turn induce permanent, inherited changes. To account for the agent's mutagenic effect, it was designated a "transmissible mutagen."[41]

The phenomenon of transmissible mutagens had been reported in England in 1928. Fred Griffith had used a filterable extract to transform the avirulent non-capsulated form of pneumococcus into its capsulated, pathogenic form.[42] At the time of the 1932 review, Institute colleagues Oswald Avery and James L. Alloway were attempting to identify the chemical nature of this transforming substance,[43] a search ending in 1944, when Avery reported one of the major discoveries of the twentieth century, that DNA is the material responsible for heredity.[44] Murphy and Claude concluded their review by suggesting that the transmitting agent of the chicken tumor might belong with the mutagens rather than with the viruses. The tenor of discussion in Murphy's annual reports to Flexner during this period bolstered his belief that a chemical would be found in the cell, comparable to the pneumococcal transforming substance, that was responsible for malignant transformation.

In 1933, Flexner congratulated Murphy on the "excellent review" of the tumor agent. He tried to be encouraging, stating that as a result of forceful and consistent research "your view of the 'enzyme like' nature of the specific agent seems to gain more and more support."[45] In fact, Flexner was not convinced there was a real difference between the chicken tumor agent and a virus. Even though he approved Murphy's attack on the agent "in so fundamental and individual a way," Flexner's personal knowledge of research on polio and other viruses at the Institute prompted him to get Murphy to try other ways of looking at the tumor agent. As a corrective, he stated, "I know of things now, of which I am not free to write, that make me very uneasy about your chemical organization."[45] He suggested they should "work up much larger amounts of crude material" as John Northrop did before crystallizing enzymes and should try growing the agent in the fluid medium that Thomas Rivers was using for several viruses. Primarily, he wanted Murphy to pursue the chemical aspects with "a really high trained chemist of the right sort." More critically, Flexner told Murphy, "I cannot imagine Claude ever having solved any of the enzyme or vitamin problems or the soluble specific substance problem; and what I fear is that he will get so far and then some really first-class chemist will turn the real trick (as Northrop has) to your and our great disappointment."[46]

Flexner actually wanted Claude to return to Belgium, just as Murphy's first Belgian postdoctoral fellow Joseph Maisin had done in 1925 after three years of studying x-ray effects on tumors. This time, Murphy countered strongly that the tumor agent had been isolated and purified to a degree not even approached for an animal virus. "All of the recent work on the purification of the agent has been done by Claude," he told Flexner, "and he has shown considerable ingenuity in developing and modifying methods so that they are suitable for this particular material. There is every indication that this line of investigation has not reached its limit. . . . If Claude is eliminated this work will have to be abandoned."[47] With this exchange, the chicken tumor work was rescued, and the research of Albert Claude was propelled forward.

Two "things" that Flexner knew but could not say to Murphy in 1933 concerned other work on viruses at the Institute. Wendell Stanley was in the process of crystallizing the tobacco mosaic virus, a major feat that he achieved first using chemical methods and then later using a centrifuge.[48–51] This success meant filterable viruses could be isolated by methods designed for proteins and that viruses have properties of chemical molecules. Also in 1933, Richard Shope reported a filterable virus that causes a mammalian tumor (in rabbits), and this discovery stimulated Rous to renew his own research on a viral cause of cancer.[52] Both discoveries had the potential to change Murphy's experimental approach and theory of cancer causation, yet he held steadfast to his theory of cellular mutagens.

The Shope virus presented a curious phenomenon because while it gave rise to papillomas in rabbits inoculated with strains of the virus, it could not be recovered from either the neoplasm or metastatic tissues. This became apparent both to Rous, who studied the vital activities of tumor growths themselves, and to Murphy, who made limited, although unpublished, comparisons of it with the chicken tumor.[53] This phenomenon, and the fact there were still no known human cancer viruses, led Murphy and Rous to reassess their views of cancer causation, and they reached strikingly similar conclusions. In 1936, Murphy regarded the Shope virus as similar to coal tar or other carcinogenic agents he was using to initiate cancers in the laboratory. "In both conditions," he said, "the agents start a proliferative change in the tissues which may go into malignancy, but once the process has been started, the initiating agents have no part in maintaining the malignant state."[54] Rous, in his Harvey Lecture, admitted, "the virus may do no more than start the papillomas." While he still argued that viruses were the nearest causes of tumors still known, he also compared them with the chemical carcinogens. "They bring cells to the brink of neoplastic change . . . like the catapult that springs an airplane from the deck of a ship. It provides the plane with the necessary initial velocity and that is all."[55] Two years later he expanded this concept, reasoning that "if human cancer viruses exist, they must be carried everywhere that man goes, perhaps in or upon his body, as are bacteria; yet plainly they cannot produce tumors unless a multitude of conditions are favorable—a happening so infrequent as to remove them from the infectious category."[56] Rous thus acknowledged a difference between the exciting agent of a cancer and factors that promote malignant growth.

There is a notable convergence of Murphy's concepts of "transmissible mutagen" and "initiating agents" with Rous's view of the virus or chemical carcinogen as a catapult. Both views presage the resolution of the Rous Sarcoma Virus (RSV) mystery when the oncogene theory of cancer was introduced in the 1970s. Based on studies with RSV, an oncogene was defined as a mutated gene from a normal cell that was picked up by the virus during infection, and this event led to malignancy. Oncogenes were found not only in tumors induced by virus but also in cells originating from chemically induced or spontaneously arising tumors.[57] The oncogene from RSV was the first one found and was named SRC (short for sarc or sarcoma). However, despite a similarity between oncogene mechanisms, transmissible mutagens, and initiating agents, these views of Murphy and Rous in the 1930s lay dormant so long that they exerted no influence on the oncogene work.[41,58]

Claude Adopts the Centrifuge

Murphy resisted Flexner's advice for two years while Claude continued to purify the chicken tumor agent using various chemical fractionation methods, combined with precipitation techniques and electrodialysis. His results were not satisfactory because they gave varying

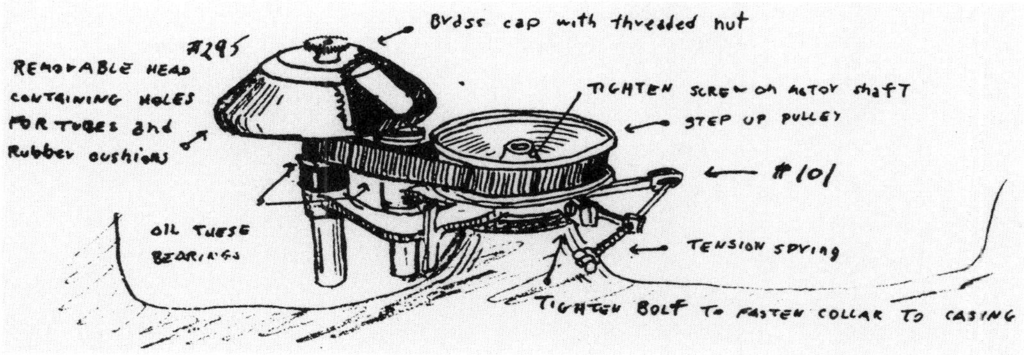

Albert Claude's drawing of his tabletop centrifuge. Courtesy of Philippa Claude.

proportions of protein, phosphorous, and phospholipids. By 1935, he concluded that "the evidence points to a protein and phospholipid as the principal constituents of the active residue."[59–61] Again, the laboratory attributed the problem to inhibitory substance that rendered the tumor activity variable.[38,39] The chemical analyses had become confusing and the concentrated active material too limited for accurate testing.

Also in 1935, three laboratories in England reported spinning the Rous sarcoma filtrates in a high speed centrifuge: William J. Elford and Christopher Andrewes at 11,000 rpm, John C.G. Ledingham and William Gye at 15,000 rpm, and James McIntosh at the remarkable speed of 40,000 rpm.[62–64] They used the centrifuge to learn whether it could sediment what they regarded was a small virus, compared with known large viruses, and to estimate its size.* McIntosh also reported that under the microscope the sedimented pellet contained "numerous round bodies with the appearance of elementary bodies," terms used at the time to designate viruses. Moreover, he also found these bodies in filtrates of normal muscle. Murphy was greatly interested by the latter observation and wrote McIntosh concerning their nature, adding, "Claude has some new information about these [bodies] which I hope he will be ready to publish soon. We do not think they are associated with the agent."[65] McIntosh replied on the troubling nature of the granules and advised Murphy that "the agent may only be attached to the particles, but on the other hand, the behaviour in this spinning approaches that of a virus such as a vaccinia."[66]

That same year, 1935, Claude turned from chemical to physical methods, whereupon a centrifuge became the first critical instrument in the nascent cell biology. Initially, he used the centrifuge to secure a higher concentration of the agent. Claude's method involved an ordinary commercial laboratory instrument, a tabletop electrically driven International Equipment Corporation centrifuge, model B size 1, having a 51° angle head rotor. He operated it in a cold room to reduce the frictional heat that might cause the material to become denatured or inactivated. The centrifuge had a normal speed of nearly 4,000 rpm, although he could greatly increase it to 17,000 rpm by connecting a small rotor to the main shaft of the

*During the span of this story, three common abbreviations for derived units of measure were used by scientists: the symbol "μ" denoted micron or 10^{-6} meters, and the symbol "mμ" denoted millimicron or 10^{-9} meters. In the mid-twentieth century, "μ" was generally replaced by "μm," also 10^{-6}, for micrometer, and "mμ," also 10^{-9}, was replaced by "nm" for nanometer. An ångström, 10^{-10}, is equal to 0.1 nanometers.

centrifuge with a pulley. He tried glass tubes but switched to thinner cellulose tubes in order to increase the centrifuge's capacity up to 84 cubic centimeters of liquid. He began by spinning the Berkefeld tissue filtrate for only five minutes before decanting the extraneous material.[38]

Unlike the English scientists, he tested several spinning episodes for various lengths of time and used an infectivity assay to check the tumorigenic activity of each fraction, including the several layers of the supernatant above the sedimented pellet and the pellet itself. He obtained a 2,800-fold increase in infectivity of the agent in the final pellet over that in the starting material, which he ascribed to the removal of the inhibitory fraction, mentioned earlier. The sediment was opalescent due to light scattering, and under a dark-field microscope he saw small particles, many in aggregates. However, even after four hours of centrifugation, about ten percent of the infectivity still remained in the supernatant, which he attributed to a dissociation of the sediment into smaller-sized particles.[67] This stop-and-start process with careful measurements at each step marked the beginning of the method known as differential centrifugation.

Also unlike the English workers, in addition to using the centrifuge as a preparative instrument, Claude ingeniously went on to use it as an analytical instrument to calculate the weight of the particles he obtained. When he compared the parameters of force and time needed to sediment the tumor agent with another substance, hemocyanin, whose size had been previously calculated by Theodor Svedberg using his ultracentrifugation methods (discussed in the next section), Claude determined the tumor agent had a molecular weight of 140×10^6 and a particle size of 70 nanometers. For the first time, a more robust physical knowledge of the agent had been obtained.[38]

The amount of purification and concentration secured by Claude by centrifugation at high speed was vastly superior to that obtained by chemical means, but there was not enough material to further purify the tumor agent. However, the technique encouraged Murphy and Claude to concentrate on recovering granules from normal cells and to compare them with granules from tumor cells.

Claude Discovers Subcellular Particles Using a Centrifuge

In 1937 Claude began using the centrifuge to sediment normal chick embryos. The same year, Murphy commented to Robert Chambers at New York University that the "cancer field needs cell physiologists more than any other group . . . the full understanding of the process can only come with a fuller knowledge of the normal cell."[68] Murphy had wanted to begin this experiment a decade earlier but had deferred it while Claude attempted to purify the tumor agent. As expected, the sedimented fraction of normal tissue was found very similar in its physical and chemical characteristics to that of the chick tumor tissue. The new data fit remarkably with their transmissible mutagen hypothesis that normal tissues contain a factor that can be "mutagenized" into a tumor factor.[69]

Perhaps overconfident in this expected similarity between normal and tumor fractions, Claude did not go on to test whether the material he obtained from normal tissues produced tumors. In 1939, he reported to Murphy that "the simple principles of Robert Koch, which were adequate to establish the necessary relation of causality between a microorganism and a typical disease are of no value in the case of the filter passing agents."[70] He thought the agents were too small and impure, so he proposed instead introducing "rigorous quantitative" rather than qualitative methods. The infectivity of material from normal tissues was tested, however,

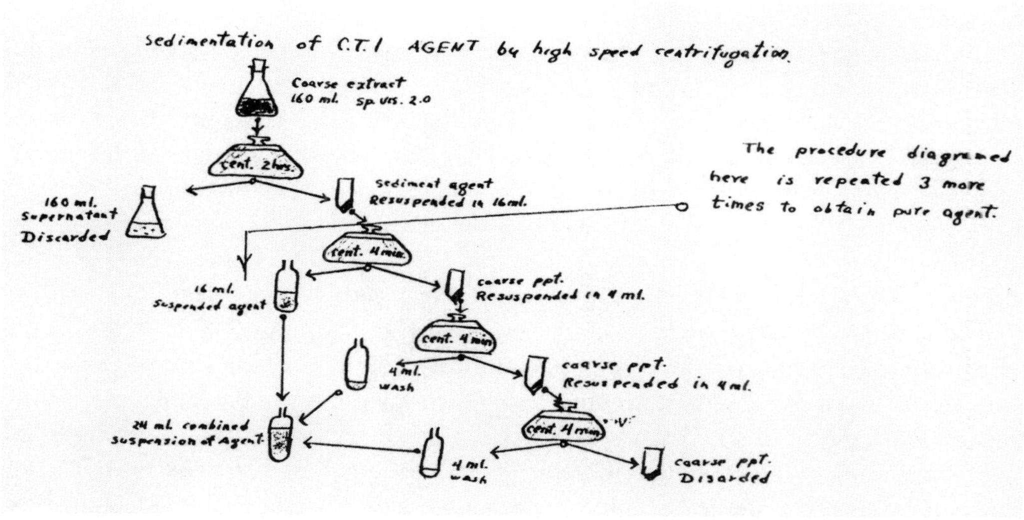

Albert Claude's drawing of his cell fractionation scheme of chicken tumor 1. Courtesy of Philippa Claude.

by Kurt Stern in the laboratory of Duran-Reynals, who had moved to Yale. After leaving Murphy's laboratory in 1938, Duran-Reynals did much to advance the theory and research on cancer viruses, including RSV, while the topic still remained in general disfavor. Stern found that whereas the material from normal tissues failed to produce tumors when inoculated into the breast muscle of healthy chickens, it nonetheless had a growth-stimulating effect in vitro.[71]

Claude's first major finding was not one related to cancer but instead was a revolutionary new way to look at cells themselves. He quickly adapted his techniques with the tumor agent when he realized that centrifugation was a first step in making chemical studies of individual fractions. A centrifuge could achieve a systematic separation of cell contents because the individual elements were different in size and specific gravity. By 1941, Claude had established a basic process with the small International centrifuge in which he was obtaining four separate although somewhat crude fractions.[72] The first fraction to be spun down contained tissue debris and the nuclei, and the last one was the supernatant fluid that remained after the final spinning. The middle two fractions were granular and differed in particle size.

The immediate question was determining whether these granules actually represented cellular constituents. Making such associations was neither an obvious nor an easy task, as Claude learned through an initial misidentification: he proposed the large granule fraction was what he called "secretory granules" and the small particle fraction, from both normal and tumor tissue, was mitochondria or mitochondrial fragments.[73,74] He based this misidentification of mitochondria in tumor and normal cells on their similar size in the granular sediment and their similar composition of phospholipid and nucleoprotein. He also offered the rather bold idea that "the fact that at least one of these fractions produces tumor producing activity should lead to interesting developments in the study of the chemical nature of mitochondria and their possible role in the evolution of the malignant cell."[75]

Claude further postulated that under certain conditions the small particles he assumed were mitochondria would differentiate into the larger secretory granules. During a formal presentation to the Institute's weekly staff meeting in 1940, he argued, "The important fact research has to face is that, once the malignant change has taken place, the agent which was instrumental in inducing the tumor may be removed, without effect on the future course of the malignant process . . . the cancer problem is within the cell and it is probably in the study of the cell, in the analysis of its physico-chemical peculiarities that we shall come to uncover the real cause of cancer." Indulging in another bold speculation before his scientific colleagues, he concluded, "Would it not be possible then that what we know as malignancy would consist in some aberration in the functioning of the mitochondria? Genes have been made responsible for the malignant state. Why not give a chance to mitochondria?"[76]

Claude also presented this hypothesis before the American Association for Cancer Research in 1941. Its impact reached William L. Laurence, science reporter of *The New York Times*, where a lengthy article on Claude's research was headed "Hints Cell Itself is a Cancer Cause: Rockefeller Scientist Reports Constituent Plays Big Role as Villain in the Drama."[77] Laurence reported that these studies pointed "an accusing finger at a fundamental constituent in all living cells as the guilty agent playing a major role, if not the chief one, in the causation of cancer." At this point, despite Claude's speculations, there is clear indication that he was becoming interested in the nature and function of normal subcellular organelles while veering away from pure cancer research. In his personal annual report to Murphy in 1942, he suggested it was the "orderly multiplication and growth of the 'small particles' which would constitute the living cytoplasm" and that malignancy was the result of "excessive multiplication or lack of differentiation of the small units."[78]

The Mysterious Mitochondria

In 1940, Claude assumed the granules in his smaller purified fraction of cells were the mitochondria that were capable of giving rise to cancer. It took another three years before he accepted evidence from new experiments and criticism by others that mitochondria were in the large fraction. He had not realized his small fraction contained previously unknown particles that were in fact ultramicroscopic; that is, beyond the visual powers of the light microscope.

Historically, Claude cannot be blamed for this misidentification of the small particle fraction as mitochondria. His novel experiments to obtain separate fractions of cells were based on his own inventions, and he had little experience with anatomy and biochemistry of cells. Furthermore, not that much was known about mitochondria. Even though mitochondria, as well as nuclei and the Golgi apparatus, had been recognized as independent elements in the cytoplasm of cells since the late nineteenth century, both their shape and function remained subjects of controversy. Due to limited techniques of biochemistry, tissue culture, and light microscopy, very little new information had been gained about mitochondria since 1918 when Edmund V. Cowdry produced his classic monograph "The mitochondrial constituents of protoplasm."[79] As Claude acknowledged, this old document was his major source of information.

An overview of the mitochondrial lore available to Claude suggests how he arrived at some of his erroneous assumptions.[80] Before 1890, many Germany cytologists and anatomists, among them Albert von Kölliker and Walther Flemming, had observed small granules in the cell's clear ground substance, then called hyaloplasm, that surrounded the nucleus.[81,82] The same year, Richard Altmann described granular structures that he found in all cells and called them "bioblasts," observing they resembled bacteria and functioned as "elementary organisms" to carry out vital functions.[83] In 1899, Carl Benda introduced the name mitochondrion, from the Greek *mitos* for thread and *chondros* for granule, a term that captures its characteristic plasticity as it changes from an elongated rod to a sphere when viewed in cultures of living cells.[84] During these early years, each investigator was coining his own name for these bodies, and among the nearly hundred other names listed by Cowdry, the term "chondriosome" was still in use when Claude also suggested it as an alternative term for his small fraction.

In 1899, Leonor Michaelis introduced the vital dye Janus Green B that specifically stained mitochondria.[85] This stain dispelled any doubt about whether mitochondria were artifacts, demonstrated they existed and were widely distributed in living cells, and opened a valuable method of study. Michaelis's drawing of mitochondria in pancreas cells under the light microscope was often referred to by George Palade as their "official portrait" until he provided one with an electron microscope in 1952.[86] Curiously, even though Michaelis was interested in oxidation-reduction processes, he did not relate these processes to the mitochondria.

There were other controversies regarding the function of mitochondria. Both Friedrich Meves in 1908 and Claude Regaud in 1909 suggested mitochondria were "bearers of genes" involved in heredity and in cellular differentiation.[87,88] In 1912, Benjamin F. Kingsbury introduced the theory that mitochondria were "the structural expression of the reducing substances concerned in cellular respiration."[89] This theory was consistent with the fact that respiration is a fundamental attribute of living things and that the mitochondria are omnipresent. It also presaged the confluence of Claude's isolation of mitochondria with biochemical associations that they are the principal sites of respiration (see Chapter 4).

Cowdry concluded his 1918 monograph by stating there was a pressing need for chemical knowledge of the mitochondria and even suggested that it might be possible "to partly separate out mitochondria by means of the centrifuge and then collect them for analyses." The credit for having first used a centrifuge to isolate a cell organelle usually goes to Friedrich Miescher, who in 1871 reported separating nuclei from human cells.[90] In 1913, Otto Warburg segregated but did not isolate particles from liver cells, which he called "grana," and showed they were involved in oxidative processes, but he did not go on to link the particles to mitochondria.[91] Biochemist Albert Lehninger, in *The Mitochondrion*, explained that between Warburg and Claude, "very few biochemists concerned themselves with the possible importance of the fact that respiratory enzymes were found to be associated with particulate matter of cells and tissues." To these biochemists, Lehninger commented, the particles were a nuisance because they stood in the way of purifying the respiratory enzymes.[92]

Cowdry managed to deepen the mitochondrial mystery even more when Simon Flexner brought him to Rockefeller in 1921. Flexner wanted to determine whether a laboratory devoted to histology or microscopic anatomy was a desirable addition.[93] During his seven-year tenure, which ended the year before Claude arrived, Cowdry overturned to his satisfaction the hypothesis that mitochondria were bacteria living within cells of higher organisms by showing a clear difference between the two when lying side by side inside the same cell.[94] He then introduced a surface film theory to account for the uniform diameter and variable length of the mitochondria, suggesting that their function might be to select substances from the surrounding cytoplasm, condense them on their surface or interior, and transform them into useful cell products.[95] His proposals, like many others, were based on morphological observations without direct chemical evidence.*

*Altmann would have been pleased and Cowdry dismayed to know that a theory of the symbiotic origin of mitochondria (through the initial engulfing of a free living bacterium) was revived and championed by Lynn Margulis in 1970, a theory generally accepted today by many scientists.

With great advances in enzyme chemistry during the 1930s, a newer approach to mitochondria may be said to begin with their isolation from liver by anatomists Robert R. Bensley and Normand L. Hoerr at the University of Chicago. In 1934, they isolated mitochondria by a method of partial maceration followed by centrifugation. Bensley had been studying the "Altmann granules" as early as 1910 and had shown that mitochondria were just as abundant in tumors as in normal tissues.[96,97] Lehninger noted that although Bensley failed to extract intact mitochondria, "for lack of appropriate suspending media and centrifugation procedures, he was fully aware of the potential value of having relatively large amounts of isolated mitochondria available for direct biochemical analysis."[92]

Cowdry provided a crucial link for Claude between the old and new cell biologies. As Claude began serious work on mitochondria, his studies struck a discordant note. In a 1940 article in *Science*, he estimated the size of the particles as 0.05–0.2 micrometers (or 50–200 nanometers) in diameter, based on his own calculations, and noted they conformed to Cowdry's 1918 measurements.[75] Shortly thereafter, Murphy met his former Rockefeller colleague Cowdry at a meeting in Pittsburgh and mentioned Claude's work. Cowdry responded that his own 1918 estimate was too low because granules of this size were "beyond the resolution power of our microscope" and that in 1924 he had revised the mitochondria size to be 0.5–1 micrometers in diameter.[98,99] He suggested that Claude compare his mitochondria with those obtained by the Bensley-Hoerr method.[100]

Cowdry thus initiated a lengthy correspondence between Bensley, his former teacher, and Claude, in which they exchanged reprints and results of their centrifugation experiments. On 24 May 1940, Bensley wrote Claude that he also determined the size of the mitochondria particles to be 0.6–1.5 micrometers. However, the main focus of this letter, as well as their continuing correspondence, concerned the nature of the fraction, which Claude thought was mitochondrial. Bensley told Claude his small particles were more likely "a submicroscopic particulate substance rather than mitochondria."[101] From 1940 to 1943, there was no agreement, despite Claude's visit to Bensley's lab in Chicago during April 1941. That year Claude published another paper, in which he compared his large granules to the Bensley fraction, called "Bensley granules," while referring to his small particles as mitochondria.[72]

In 1942, Normand Hoerr invited Claude to participate in a seventy-fifth birthday symposium honoring Bensley in November.[102] A month before this gathering in Chicago, Bensley wrote Claude even more forcefully about his mistaken identification of the small particle fraction, saying, "The fundamental importance of your discovery of the submicroscopic particles is such that I wish to avoid the discussion of your incorrect interpretation of mitochondria on which you have been wrong from the very first. I still hope that it will be possible for you to correct those mistakes in print before it becomes necessary for me to do so. It would be unfortunate to detract in any respect from the importance of the discovery of the cytoplasmic particles."[103] The result was that Bensley did correct Claude in a *Science* paper that appeared the month of the symposium and was subsequently reprinted in the symposium proceedings. Bensley charged that Claude's fractions "by time fractionation at high speed were mixtures." Furthermore, since he and his colleague Arnold Lazarow had also obtained "a glassy cherry-red pellet, composed of particles so minute that they were quite invisible under the microscope," Bensley insisted that what Claude thought were mitochondria "clearly could not be the case." He called for more refined methods of fractionating submicroscopic particles.[104]

At the birthday symposium "Frontiers of Cytochemistry," Claude became acquainted with many esteemed cytologists and their new techniques. Two of Bensley's colleagues continued the criticisms of Claude that had appeared in *Science*. Hoerr suggested Claude had been confused by the earlier work of Robert Noël, who claimed that secretory granules arise from mitochondria, whereas later research showed this was not the case. He further argued that Claude's method led to the large and small particle fractions contaminating one another.[105] Lazarow insisted that the small particles Claude called mitochondria were "newly discovered cell particulates, submicroscopic in size," because the true mitochondria were microscopically visible. He blamed Claude for complicating the terminology of the large granule fraction isolated from liver cells that he labeled "secretory granules." The large microscopic granules in the liver cell, Lazarow corrected Claude, "correspond to the mitochondria of the cytologist."[106]

Reconsidering Mitochondria and Refining Methods

Claude's presentation at the symposium did not touch on mitochondria. He later admitted to Hoerr that he had "no worthwhile opinion on the subject for the moment."[107] Instead, Claude spoke about his small particle fraction, which he tentatively named microsomes, wanting to demonstrate that they were not artifacts of grinding or centrifugation. Since these particles were too small to be seen in the light microscope, he resorted to indirect means to prove his point. He presented a dazzling experiment using very large cells from the frog *Amphiuma tridactylum*. After spinning its liver tissue for sixty minutes at 18,000 times gravity (rpm not given), he stained the resulting pellet with two different dyes and examined the cells under the microscope. Centrifugation had segregated the various components *within* the intact cells just as the components segregated in bulk in centrifuge tubes, and in the same order. The stunning photographs of a whole centrifuged *Amphiuma* cell showed nuclei and glycogen accumulated at the heavier pole, and just above were the stained mitochondria. Above these large granules were two layers, a stained one having no visible structure where the submicroscopic particles were expected to segregate and above it a fainter layer containing the remaining cytoplasm. By comparing the strong affinity of this stained but invisible particulate substance for basic dyes, both in the intact cell and in vitro, he concluded the two substances were identical.[108] (The full story of identifying the nature and function of the microsomes would take another dozen years, and it is told in Chapter 7.)

By the time Claude corrected his misidentification of the small particle fraction in a 1943 *Science* article, he also incorporated other refinements into the published version of his symposium paper. He was helped in part through correspondence with Hoerr, who was editing the proceedings. Responding to Hoerr's objections, Claude defended the term microsomes, claiming the word as used by Edmund B. Wilson in 1925 had lost its precise meaning and his use of it was to emphasize that the "nature and the role of the small particles are still so uncertain."[107] Even today, the term microsomes refers to a fraction spun down at high speed and whose precise function depends on the cell type from which it was obtained. In contrast, mitochondria have a precise structural and biochemical meaning irrespective of the cell type from which they are obtained.

Hoerr also objected to Claude's use of "secretion granule," but Claude also defended its use in *Amphiuma* liver, noting that "granules may be caused to accumulate around the bile ducts by fasting or made to disappear almost completely by a single meal." However,

Claude conceded that, since he and Bensley, among others, had failed to find nucleic acid in mitochondria, "we have been dealing with two quite different fractions" and that he would investigate this point by repeating the Bensley-Hoerr technique. A month later, Claude acknowledged to Hoerr that his fractions were mixtures of secretory granules and mitochondria, noting that "your fraction and mine were really different since the techniques were not the same and the chemical composition is not comparable," and acquiesced that it was wise to adhere to the Bensley designation "large granules."[109]

Claude acknowledged in his *Science* article that Bensley-Hoerr's mitochondria "seem to correspond to our liver secretory granules." This cautious statement implied that mitochondria were still not isolated in pure or concentrated form. More important, he now defined the small fraction as newly discovered submicroscopic particles, which could be isolated by the lengthy centrifugation and had certain chemical staining properties, a high phospholipid content, and a nucleic acid absorption spectrum. He observed that because he found ribonucleoprotein present in normal fractions, as well as in the Shope papilloma virus and chicken tumor agent, these particles might play a "fundamental role, perhaps of an enzymatic nature, in the process which enables the structure to reproduce itself." His working hypothesis offered a biochemical basis for the view that each element that contributes to the life of the cell has the power to reproduce its kind. This remarkable statement was shown years later to be partially true in the case of mitochondria and also relevant to the function of microsomes.[110]

With the benefits of criticisms, reworked experiments, and a new hypothesis, Claude moved on to refining his centrifugation techniques for isolating organelles that would yield better samples for chemical characterization. Several problems had to be resolved. One concerned why only a small portion of his large granule fraction was stained by the specific mitochondrial stain, Janus Green. He decided to test this question on leukemic tissue that has no known secretory granules in order to obtain a better mitochondrial fraction. This time he used a gentler means of breaking the cells with a mortar and pestle, instead of sand as he had used with tumors, and preserved the nuclei in a slightly alkaline medium. He then altered his centrifugation procedure from using a single high speed to testing various speeds and lengths of time. This was in part a response to Hoerr's advice that the purest mitochondria were obtained by running the original emulsion at a low speed for an hour.[111] Consequently, using a low speed, Claude obtained some unbroken cells, debris, and nuclei; at a slightly higher speed he obtained mitochondria; and at an even higher speed he obtained microsomes. His biochemical analyses measured phosphorous, nitrogen content, and ultraviolet absorption in each of these fractions and determined that ribose nucleic acid occurs "in the cytoplasm only in association" with mitochondria and microsomes.[112]

By 1944, Claude was gaining considerable experience with his cell fractionation method. In addition to isolating mitochondria, he used the centrifuge to isolate other *visible* components of cells, including nuclei, chromatin strands, zymogen granules, and melanin granules, as well as the still *invisible* submicroscopic particles he called microsomes. Significantly, no matter whether he fractionated cells from chicken, mouse, rat, rabbit, or frog or from tumors or embryo tissues, each cell type yielded similar granule fractions. From this point forward, he had a systematic way to study the biological significance and function of these individual cell components. After nearly a decade of working with a centrifuge, in 1946 Claude published a full description of his cell fractionation technique and a study of the properties of the various fractions he obtained. He based his definitive report on using the small tabletop International centrifuge in a cold room.[113,114]

Centrifuges and Claude's Cell Fractionation

Claude's choice of an ordinary laboratory centrifuge is curious. In 1935 when he began using this instrument, ultracentrifuges of much higher performance were being developed at The Rockefeller Institute to study viruses and proteins. Even though the more powerful instruments eventually became a major tool in modern cell biology, Claude used a commercial tabletop model for a decade. While an extensive literature about centrifuges exists, the following story is linked in key ways to Claude's achievements and to the early development of the emerging science.[115–123]

Precursor electric and oil-driven centrifuges. The scientific centrifuge era began in Stockholm during the 1920s with Theodor Svedberg. Previously, simple hand or electrically driven centrifuges were available for laboratory use in separating biological materials. As mentioned earlier, in 1871 Miescher used a primitive machine to break up pus by various methods and then to isolate nuclei from cells.[90] Svedberg first built an electrically driven instrument spinning up to 10,000 rpm. Since it could not attain the necessary speed to counteract the natural tendency of complex molecules, like proteins, to diffuse rapidly in solutions, he began to develop higher speed instruments. In 1927 he produced the first oil-driven centrifuge, modified in 1938, to determine more precisely the molecular weight of complex substances.[124] He further developed it to measure the rate of sedimentation by modifying the rotor and adding optical devices that could visualize and record the movement of boundaries as molecules were sedimented through a liquid. By increasing the speed to 67,000 rpm, he drastically reduced experiments to a few hours. Due to the instrument's high cost (around $6,000 in 1938) and its tremendous intricacy of manufacturing, installation, and operation, its commercial production was not successful.[125]

Svedberg also devised the mathematics through which the sedimentation rate could be related to molecule size. The constant S (for Svedberg) value that was measured was derived from the angular velocity, the distance of the boundary from the axis of rotation, and the movement of the observed sedimentation boundary between two time points. S values could be obtained for small proteins as well as for bacteriophages, and later even for multiprotein complexes of cellular organelles such as ribosomes. Further mathematics were developed to convert the S value to actual size.[115,116]

Svedberg originally defined "ultracentrifuge" as an instrument that quantitatively measures sedimentation in a centrifugal field. He reserved this term for the analytical model and to distinguish it from the more typical preparative centrifuge in laboratory use that separates and concentrates substances of different particle weights without determining weight or size. Today, either type of centrifuge that generates very high speeds, whatever its purpose, is called an ultracentrifuge.

Air-driven centrifuges and Edward G. Pickels. The first high speed air-driven machines were produced in 1925 and 1927 by two Belgian scientists, Emile Henriot and Emile Huguenard.[126,127] Their centrifuge had a small spinning top, with a viewing mirror mounted above the rotor to record speeds using a beam of light. It was both supported and driven by compressed air and with later modifications could spin up to 65,000 rpm. As mentioned earlier, this was the centrifuge that James McIntosh used in England in 1935 to sediment the Rous sarcoma agent.[64]

Interest in the air-driven model caught the attention of American graduate student Edward Pickels and his physics professor Jesse Beams at the University of Virginia. In 1935, they built their own spinning-top centrifuge and placed it in a vacuum to reduce the heat generated by the spinning rotor. Essentially, the Pickels-Beam ultracentrifuge consisted of a small compressed-air turbine from which a large rotor was suspended by a piano wire. The rotor holding the tubes of biological samples hung in a sealed chamber, from which the air was evacuated. The piano wire served both as a support for the rotor and as a vertical driving shaft.[128,129] The Sharples Company, known for its centrifugal cream separators, manufactured this Beams/Pickels instrument for $3,000, but within a few years it experienced a commercial failure similar to that of Svedberg's instrument.[117]

In 1935, Johannes Bauer, a yellow fever researcher of the International Health Division (IHD) of The Rockefeller Foundation, hired Pickels to develop a centrifuge to study viruses the way Svedberg had studied proteins. Although independently funded and organized, the IHD was housed at Rockefeller. It shared the Institute's facilities, and there were many fruitful collaborations with its scientists. Pickels once commented on this special collaboration of applying physics to important biological problems that "it is mainly through the intimate involvement in such problems that one comes to appreciate the need or possibility (which may never occur to a biologist or medical man whose mind is prejudiced by traditional methods) of developing a new approach or improving an old one."[130] Within a year, Pickels developed two versions of the air-driven vacuum centrifuge, a preparative model for concentrating and purifying viruses and an analytical model with a suitable optical system for studying the size, weight, and shape of viruses. Both models underwent extensive design changes while they were actively used in biological research. And both were built in Rockefeller's advanced Instrument Shop.

The preparative Pickels and Bauer "high speed" model was used to sediment the yellow fever virus. It had larger rotors and spun in a vacuum up to speeds of 30,000 rpm.[131] When Svedberg visited the IHD laboratories in 1936 to see this instrument, he told Bauer it was "among the best memories from my visit to America. You and your group certainly have accomplished wonderful things. The centrifuge drive you have developed is really extremely elegant."[132] He was so impressed that he invited Pickels and Bauer to contribute a substantial chapter on the air-driven model in his 1940 monograph on ultracentrifuges.[133]

Naturally, many Institute staff members were immediately interested in having their own ultracentrifuge to measure and characterize proteins and viruses. Biophysicists Ralph Wyckoff and Jonathan Biscoe joined Pickels and Bauer to improve the analytical model. This model incorporated the Svedberg optical system and a larger rotor made of duralumin into an air-driven instrument, which permitted more accurate measurements due to the increased distance between the rotating sample and the axis of rotation.[134–136] These air-driven ultracentrifuges with speeds up to 60,000 rpm were rapidly adopted for laboratory use: Wyckoff and Wendell Stanley in 1937 used this model to purify and to analyze proteins of tobacco mosaic virus strains,[49] and Wyckoff and Joseph Beard in 1939 reported the isolation of the Shope papilloma virus.[137]

Fortunately, there was great cooperation between the IHD's Pickels and Josef Blum, the Institute's instrument maker. An immigrant from Germany in 1926, Blum worked as a tool and die maker and built radios and loudspeakers, repaired airplane motors, and made automobile parts. He came to the Institute in 1935 highly recommended for his "ability for precision work, but also because of his ingenuity and his capacity to develop apparatus."[138] He began as a part-time designer and builder, following expansion of the Instrument Shop by the new scientific director Herbert Gasser to accommodate the extra workload of the IHD. Blum proved so adept that he became head of the shop in 1939 and, according to Wyckoff, built four types of centrifuges, two for quantity work (to handle large volumes of liquid) and two for different analytical measurements. Wyckoff put the cost of building a quantitative centrifuge at slightly under $500 and an analytical instrument at just under $1,000.[139] By 1944, IHD director Bauer reported twenty ultracentrifuges were at work in Institute and IHD laboratories.[140]

A new electrically driven centrifuge. The ultracentrifuges of Pickels proved useful in sedimenting viruses and in characterizing proteins by determining size, shape, homogeneity, association-dissociation phenomena, and conformational changes of various molecules. However, this instrument did not easily separate and purify components, whether molecular, macromolecular, or subcellular organelles. Despite numerous improvements, even switching to electrical drives that increased speeds to 44,000 rpm, the ultracentrifuge was still inconvenient and difficult to operate.[141]

Pickels felt a simpler centrifuge was needed that approached the practicality of the low speed ordinary laboratory model. He was prompted by Claude's first paper on centrifugation in 1937 in which he used the small International Equipment tabletop instrument to estimate the size of the chicken tumor agent. Claude achieved this measurement by making comparisons with the small hemocyanin molecule that formed a sharp sedimentation boundary. Since he found that particles from the larger size tumor agent formed a blurry boundary, he was unsure whether blurring was due to heat, vibration, or long centrifugation times. Claude concluded there was need for a "greater centrifugal force" to achieve better separation. Pickels was amazed that Claude could even obtain a sharp boundary with such a small centrifuge. So he repeated Claude's experiments.[142]

In unpublished experiments during 1940, Pickels examined Claude's purified tumor fraction in an analytical ultracentrifuge, and he too got a rapid blurring of the boundary, indicating the material was not homogeneous.[143] Claude attributed this to the fact that either the tumor agent was composed of particles of various sizes or that their uniform size caused them to aggregate. Thus, a greater centrifugal force did not achieve better separation than the underpowered International. Nonetheless, Pickels and Claude made a rough estimate that the particles were about 70 nanometers, a value that agreed with the one derived by Elford and Andrewes by filtering the sarcoma virus through graded collodion membranes.[62] In contrast, physical chemist Alexandre Rothen examined the purified tumor material in a Tiselius apparatus, a newly improved instrument for analyzing proteins by electrophoresis. He found the sharp boundaries indicated a relatively homogenous material, at least regarding its electrical charge. Claude was not convinced, telling Murphy that the "behavior of a particle in an electrical field is a surface phenomenon and does not necessarily reflect the inner constitution of the granule."[144]

This confusing evidence led Pickels to design and build a simpler tabletop centrifuge in 1942–1943 to examine Claude's results further. This instrument spun in free air at 17,000 rpm, nearly the same speed as Claude's International. Pickels used this instrument to investigate problems of the larger ultracentrifuges. While investigating the problem of stresses, he found the centrifuge could safely accommodate either the light, high tensile, and machinable aluminum or magnesium alloys, could spin down several hundred millimeters of fluid, and could operate at 50,000 rpm for thousands of runs. Since his earlier air-driven ultracentrifuges held tubes only at 45°, he also tested various angle heads in this smaller model and found a slightly greater angle produced a sharper boundary and could reliably determine particle size. To counteract disturbances from warmed air due to thermal convection and from molecules sinking at angles, Pickels introduced a sucrose-density gradient into the centrifuge tubes. While this is perhaps the earliest use of sucrose to improve sedimentation, the material became an important refinement of cell fractionation by George Palade in Murphy's laboratory a few years later.[142,145]

By the time Pickels perfected this new instrument, Claude had abandoned experiments to sediment the chicken tumor agent, so he did not resolve the mystery of its blurred boundary. This work was done, nonetheless, by Kurt Stern at Yale, who reported in 1943 that on further fractionating the "polydisperse" purified Rous agent, he and others obtained several crystalline fractions from the phospholipid-ribonucleoprotein complex, a finding that further indicated the agent was a virus.[71]

Two other versions of small centrifuges were developed by Blum to meet special needs of Institute scientists. An air-driven model had an electromagnetic speed control to maintain a precise speed of 60,000 rpm for several hours and a freezing coil so it could be operated at 0°C in a vacuum. This instrument allowed Rothen to measure more accurately such large molecules as crystalline ribonuclease.[146] Blum also designed a motor-driven laboratory tabletop model that operated at 20,000 rpm and used a Swedish patented angle head.[147] An early version was similar to Claude's International and Pickels's tabletop models. By 1945, Blum added the critical feature of a self-centering direct drive that permitted slightly unbalanced rotors to find their natural center of rotation. This improvement thus freed laboratory workers from the exacting task of balancing tubes of samples.[148]

In 1946, the year Claude reported his method of cell fractionation, he finally asked Blum to build a centrifuge with features to facilitate and replicate his research method. The new version incorporated several components of the small electrically driven Blum model described above. This prototype operated in a vacuum, could attain a centrifugal force up to 50,000 times gravity (rpm not given), and could spin down large quantities of fluid. There also was a refrigeration system, consisting of a cooling coil around the rotor connected to an ordinary refrigerator compressor; this feature avoided the detrimental condensation that had been caused by moving the International centrifuge in and out of a cold room. Claude was so pleased that he highly recommended this refrigerated, evacuated Blum centrifuge to two friends in Belgium, Jean Brachet and Victor Desreux, calling it *plus parfait* for breaking cells apart.[149,150] Only two papers specifying this centrifuge were published before it went into commercial production.[151,152]

From experimental to commercial centrifuges. During its decade of development, The Rockefeller Foundation's IHD and the Institute's Instrument Shop received numerous requests from outside researchers about purchasing an ultracentrifuge. These queries were answered with the statement that the Institute was not a manufacturer of a product and its few experimental instruments were for use by their own investigators.[153] Answering a letter

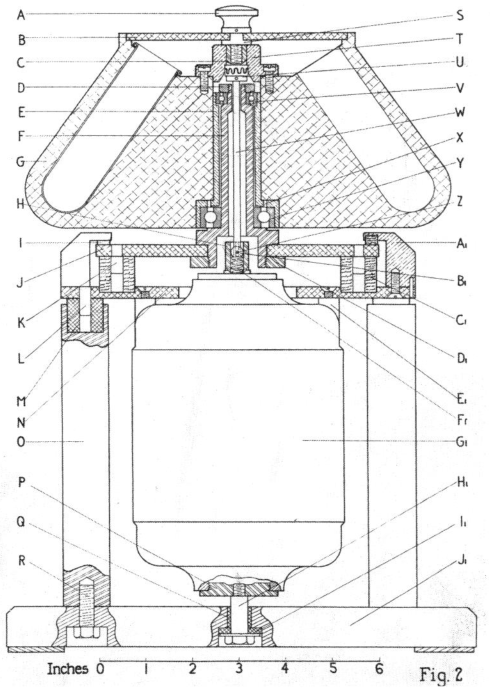

Blueprint of Josef Blum centrifuge. From Blum, 1945. *Rev. Sci. Instrum.* 16:31–34. ©American Institute of Physics.

Josef Blum centrifuge with the angle rotor. From Blum, 1945. *Rev. Sci. Instrum.* 16:31–34. ©American Institute of Physics.

from a California microbiologist about how to construct his own centrifuge, Pickels wrote, "We do not have blueprints for all our equipment, simply because it has been built from rough sketches in our own shop and continually modified during several years of development." Instead these scientists were furnished with copies of review articles, photographic copies of drawings with detailed dimensions, and advice. In particular, Pickels recommended hiring a technically trained person for such an undertaking because the ultracentrifuge was "a naughty, even dangerous child if taken for granted."[154]

The experimental development of these instruments ceased at the end of World War II. The two centrifuges designed and built by Blum quickly went into commercial production. In 1945, Ivan Sorvall, who owned American rights to the Swedish patent on centrifuges using tubes at angles of 0–50°, began producing the tabletop model.[155] The refrigerated, evacuated model used by Claude went into production by Sorvall in late 1947.[150]

In 1945, Pickels acknowledged his air-driven ultracentrifuge was outdated: "If we were starting from scratch we could now build an instrument which would be much more convenient and certainly neater. . . . The somewhat critical design of the mechanical parts is fairly well standardized. But, by and large our ultracentrifuge looks very much like the grandfather it is."[154] Pickels left the IHD in 1946 to build a commercial ultracentrifuge that was more convenient and efficient, forming his own company in California called Specialized Instruments Corporation, or Spinco, which eventually became part of Beckman Instruments. Initially, he produced analytical centrifuges, replacing the air drive with an electric motor that could attain speeds up to 60,000 rpm using rotors of duralumin. Its components were in a refrigerated vacuum chamber, and it included an optical system of the refractive index type with a viewing screen and automatic camera. The new modifications and ease of use for hundreds of hours made the model E a standard analytical centrifuge worldwide.

Spinco's model E could also operate a preparative rotor, but it was not much used. By 1952 Pickels was producing the model L, a preparative machine that was concise, simpler, and easier to operate. The model L was similar to the Blum-Sorvall centrifuge made for Claude: its angle heads were electrically driven with electronic controls, its rotors were spun in an armored, refrigerated vacuum chamber, and its speed could reach 40,000 rpm.[156] Almost immediately, the new Rockefeller laboratory called Cytology, headed by Keith Porter and George Palade, bought the twentieth model L produced by Spinco, and it became their standard centrifuge for studying subcellular particles.

A First Milestone in Modern Cell Biology

The development of cell fractionation involved a great deal of experimentation on Claude's part. During a ten-year process of tinkering and evaluating, Claude refrained from using the centrifuges developed by Pickels, Bauer, Wyckoff, and Blum. His small International in fact achieved a fairly high speed compared with the *ultra*centrifuges being developed in the late 1930s, although he eventually admitted a need for greater centrifugal force.

Initially, Claude did not believe he was sedimenting a virus, and therefore he had no reason to duplicate methods and test the centrifuges being used by IHD and Institute virologists. In order to break fragile cells apart, he needed to overcome distortions caused by non-physiologic solutions, harsh fixatives, and violent centrifugal forces. He kept working with his familiar instrument that he felt would be gentle enough on the cells.

More importantly, the experiment Claude tried with his small centrifuge in 1937 changed the nature of his research as well as the course of this history. His work in separating cells was clearly slowed by using an underpowered centrifuge that initially produced mixed fractions. He was further hampered by imperfect ways to identify physical and chemical characteristics of the particles. His early misidentification of small particles as mitochondria can be attributed partly to Murphy's focus on finding an entity in normal cells responsible for tumor growth and partly to his hypothesis that mitochondria might be involved in cell differentiation and cancer. Keith Porter, his laboratory colleague, admired Claude's persistence during this evolving period of difficult research while taking criticism from others because "when he started tearing cells apart, taking pieces out and examining them, everybody who called himself a decent cytologist or cell biologist was at him . . . [asking] what was the good of doing that, breaking up that gorgeous structure? And he got a lot of bad, bad press from that."[157]

The significance of cell fractionation was to establish a basis for making biochemical analyses of subcellular particles. Claude's special contribution was to determine quantitatively the distribution of substances among each of the fractions. Recognizing as early as 1937 that an analysis of a fraction was only as good as its purity, he kept a strict accounting of the weight, volume, chemical contents, and activity of each fraction as a percent of the total substance or total activity of the initial homogenate. In refining this process, he continued to see that nothing was thrown away during these steps, including the wash waters.

The turning point occurred, according to Christian de Duve, when Claude came to fully appreciate that "tissue fractionation owes its main power to the fact that it renders the different parts of the cell accessible to the whole battery of analytical techniques contained in the biochemists' arsenal."[122] Bensley and Hoerr simply isolated mitochondria from the rest of the cells with a centrifuge; Lazarow and Barron studied uptake of oxygen and succin-oxidase activity only of isolated mitochondria; and others used microtechniques to measure enzyme activities in a few cells at a time. But, as Claude himself observed in 1946, not since Warburg in 1913 had there been a deliberate attempt "to determine, on a quantitative basis, the distribution of enzyme activities among the various cell components which remain in a tissue extract after the unbroken cells and free nuclei have been removed."[113]

For this work, Claude was not only a founder but also leader of the field that can be called the structural-biochemical dissection of the cell. Fractionation, electron microscopy (see Chapter 3), and biochemical assays (see Chapter 4) formed the triad of techniques for studying subcellular particles. Isolating cell structures with centrifuges provided a new beginning in cell biology.

Chapter 3

Entering the Realm of the Cell with an Electron Microscope

Carol L. Moberg

> It was wonderful, believe me, we had never seen anything like it. Men have visited the moon . . . but we were the first . . . to see particles, to see structures that the light microscope had not been able to resolve.
>
> Keith Porter, 1970

Prior to 1935, James Murphy participated in research that had undermined a theory of the spontaneous origin of cancer, enhanced a theory of environmental induction and promotion of cancer from chemical and physical irritants, and contributed to the premise that susceptibility to cancer could be inherited.

In a lecture at Johns Hopkins University that year, he argued that the science of cancer had moved from one solid stepping stone to another over steadily deepening waters without finding a cause or cure. Turning to future prospects, he suggested a new experimental approach based on the fact that "well understood physiological functions are balanced phenomena." In his view, now axiomatic, every cell contains both inhibitors and stimulators of growth, and therefore malignancy is a universal *cell* potentiality activated after a breakdown in internal control mechanisms. He described ongoing studies in his laboratory showing normal tissues yield an unknown inhibiting substance that retards and prevents growth in animal cancers. Yet he was not satisfied with this work because whatever the cause of malignancy, the laboratory research was dealing only with a *body's reaction* to control the overgrowth. He urged instead different studies directed to internal functions of the cell.[1]

This approach was in keeping with The Rockefeller Institute's change of leadership in 1935 from pathologist Simon Flexner to physiologist Herbert Gasser. The new director was moving away from an emphasis on infectious diseases and conferring a new mission to pursue "fundamental life processes at the level of the cell and its constituents."[2] During the following decade, this focus also contributed to Murphy's considerable influence on cancer research both outside and inside the Institute.

Murphy and Trends in Cancer Research

Perhaps unknown and underappreciated today, until the 1930s "disgrace and a certain amount of social opprobrium were connected with cancer. It was concealed by patients, their families, and sympathetic doctors. It could not be mentioned on the radio. In addition there was grave doubt on the part of many—and perhaps most—doctors as to its curability."[3] As a physician and scientist, Murphy made a long-standing commitment to relieve this gloomy climate by advocating the importance of basic research in understanding such a complex disease. He took on tasks that entailed long absences from bench work and kept in regular contact through annual meetings, international congresses, and correspondence with cancer researchers all over the world.

Murphy was active in the American Association for Cancer Research (AACR), founded in 1907, whose moving spirit was his friend James Ewing, director of Memorial Hospital. During his long membership, Murphy served as president (1921–1922) and editor of its research journals while he witnessed cancer treatment evolve from surgical oncology to radiotherapy to chemotherapy.

From 1929 to 1945, Murphy also served on the Board of Directors and Executive Committee of the American Society for the Control of Cancer (ASCC), founded in 1913, which became the American Cancer Society (ACS) in 1944. Its director was Clarence C. Little, also Murphy's personal friend, who established the Jackson Laboratory in Bar Harbor, Maine, and became a leader in genetic research on cancer. The ASCC was organized initially to educate the public about the dangers of cancer and the importance of early diagnosis, a mission that led directly to funding cancer research. In 1945, Murphy was appointed an original member of the Committee on Growth of the National Research Council that advised the ACS on its allocation of grants and fellowships.[4,5]

Both the AACR and ASCC were instrumental in persuading the federal government to fund research on cancer. Franklin D. Roosevelt signed the National Cancer Act on 5 August 1937, which established the National Cancer Institute (NCI) as a division of the Public Health Service. This act also directed the surgeon general to begin a concerted effort to control the disease and seek a cure. Murphy was selected to head a committee that would formulate "the fundamental aspects of the cancer problem and to suggest various lines of work which merit investigation." The committee included well-known scientists C.C. Little, Stanhope Bayne-Jones of Yale, Ross Harrison, chairman of the National Research Council, and John Northrop from The Rockefeller Institute. Twenty years after the founding of the NCI, John R. Heller, a former director, called the committee's 1938 report "a classic in its field" for furnishing basic guidelines and viewpoints for cancer research.[6]

According to Little, "Murphy's intelligence, vision, and confidence made him one of the really great statesmen of that period," a role he continued to perform on the NCI's Advisory Council from 1938 to 1944.[3] Not surprisingly, the 1938 report strongly reflected Murphy's long-standing views and research experiences. The committee evaluated three main lines of previous research—experiments on transplanting tumors, genetic constitution, and carcinogenic agents—and concluded that all evidence showed malignancy is "the result of a fundamental change in cell physiology," regardless of the chemical, biological, or physical agent that stimulates malignancy. Once altered, the cell process becomes automatic, and this new property "becomes a fixed character, and is transmitted to its descendants. It gives evidence of being a somatic mutation."[7]

The committee recommended against further research into the *causal genesis* of cancer, for which they gave a formidable list of known agents, including viruses and hereditary factors, that do not "appear to play any part in continuation of the malignant state." Instead, they recommended several new research objectives focused on the *formal genesis* of cancer in the cell. As presented, the core of the cancer problem lay in understanding the structure, composition, and function of the cell itself. The National Cancer Act thus provided a critical transition from hopelessness to courage as it moved scientific medicine from isolated research programs to a planned attack against the disease.

Subsequent to the report, Murphy publicly promoted the committee's views and new objectives in cancer research. His Barnard Hospital Lecture before the St. Louis Medical Society in November 1941 was published in the *Journal of the American Medical Association*. In particular, he advocated three new experimental approaches: the mechanical separation of various cell components; analysis of toxic actions against enzyme systems of the cell; and a search to identify factors that affect the growth and differentiation of cells. These investigations, he said, would provide "a fuller knowledge of the formal genesis or the mechanisms by which the growth capacity of cancer is maintained."[8]

Although not a member of Murphy's NCI committee, Peyton Rous was coming to agree with its view of cancer. Perhaps not coincidentally, Rous gave the Barnard Hospital Lecture the following November. He described many known carcinogens but maintained that viruses are "nearer causes of cancer" because they maintain a working relationship with the cells they have infected. However, in agreement with the committee report, he also urged a cellular approach: "Workers at the cancer problem have begun to see that they must look beyond the provocative carcinogens for the actuating causes of tumors . . . one must look to some factor accompanying the cells, or to an intrinsic alteration in them, in accounting for their new state."[9]

By 1941, the three approaches that Murphy advocated in his lecture were under way in his own laboratory: physical, chemical, and biological.

A physical approach was extending Albert Claude's mechanical separation of cells with the centrifuge. Claude, who had discovered the alluring physical and chemical similarity between particles from normal and malignant cells, irrespective of the tissue or animal species from which they came, was also able to remove an inhibiting factor from chicken tumor cells using a centrifuge and then to restore it by mixing the two fractions back together.[10] This work was reinforcing Murphy's 1936 hypothesis that cancer arose from a breakdown in the equilibrium between inhibitors and stimulators of growth: "Certain cancers would appear to be due to a decrease or the destruction of inhibiting principles, others, on the contrary, to a hyperproduction of principles promoting growth."[11]

A chemical approach, initiated a few months before Murphy's lecture, involved making enzymatic analyses of the cell components that Claude obtained after spinning the cells in a centrifuge. The first success of this approach came several years later when the mitochondrial fraction was linked to cell energy; this investigation is the subject of the following chapter.

A biological approach dealt with forces that engender, promote, or inhibit cell growth. Murphy believed that Claude's microsome fraction from normal and tumor cells played a role in normal development—the elements known at that time as the induction factor in embryology. This research was assigned in 1939 to a new laboratory member, embryologist Keith Porter.

All the while, Murphy continued to respond to new opportunities, support new methods and ideas, and nurture young cancer scientists. In essence, he kept his laboratory poised to plunge into the inner workings of the cell.

New Techniques with Cultured Cells: Enter Keith Porter

Murphy admitted that bringing together the diverse interests of a cell physiologist, cytologist, and embryologist to take an interest in cancer cells was almost an incongruent task. As he knew from personal experience, growth and differentiation of cells was such a difficult subject to investigate that few, if any, cancer workers qualified. Nonetheless, he was immediately receptive and had "great hopes that some advance can be made by a

Keith Porter, 1939. Courtesy of The Rockefeller Archive Center.

young biologist who . . . has the right training and is thoroughly interested in the problem."[12]

On 15 June 1939, Murphy received an unusual letter from Keith Roberts Porter asking for a position in his department. He had earned his PhD in biology from Harvard University in 1938, as a student of Lee Hoadley, and was then a National Research Council fellow at Princeton University under Gerhard Fankhauser. He offered his expertise to produce hybrid embryos by fertilizing the enucleated eggs of one race of frogs with sperm from another race.

How could a frog embryologist be interested in cancer research? Porter had read an article by Murphy suggesting that malignancy in tumor progression was a developmental problem. In his letter, he wrote, "It has occurred to me that an embryologist should be well fitted to study tumour development . . . and the facts that all differentiation may be hormonally controlled make one wonder what influence, if any, the embryonic organism might have on the undifferentiated cells of tumours. Even though it is contended that the tumour cell is an altered cell, it does not seem impossible for it to be a very primitive (from the standpoint of ontogeny) type of cell which responds to the presence of or lack of certain environmental influences by multiplying and forming a comparatively unorganized mass of cells." He asked Murphy whether this hypothesis was worth investigating and admitted, "Frankly my knowledge of the subject is slim but I think I have the eagerness, determination and sincerity of purpose which makes rapid learning possible."[13]

Five days later, Murphy invited Porter to talk about an opportunity for cancer research in the Institute. And on 30 June, Alphonse R. Dochez, on behalf of the Board of Scientific Directors, offered Porter the position of assistant in the laboratory of Pathology and Bacteriology at a salary of $2,400 a year. He began work on 1 September 1939 and remained at Rockefeller for twenty-two years.

Born in 1912 in Yarmouth, Nova Scotia, where his father built houses, Porter acquired a love of structure while drawing maps and houses in grade school. He became passionately interested in lichens and small mosses and spent hours studying their shapes on the faces of large Canadian boulders. His first interest in biology came from a high school teacher who allowed a group of boys to do laboratory experiments of their own choosing after school. Following graduation from Acadia University in 1934, with a major in biology and chemistry, he attended Harvard University with special encouragement and modest funding from Horace Greeley Perry, his professor of biology at Acadia.[14,15]

Porter attributed his excitement about cells to his Acadia days, when he began to wonder about the exchange between nucleus and cytoplasm: "In those days DNA hadn't been discovered and very little was known about the influence of the nucleus on development or any differentiation. . . . I chose to mix different nuclei with the same cytoplasm or different cytoplasms with the same nucleus and they didn't develop exactly the same which means that the cytoplasm was having a rather profound influence on the early morphogenesis of the embryo."[14]

How was it that a laboratory devoted to cancer was quick to acquire a young investigator interested in frogs? Just before Porter arrived, Murphy and Ernest Sturm were testing chemical induction of tumors and leukemia in chicken embryos, and Francisco Duran-Reynals was performing similar tests on frog embryos.[16] Porter inserted himself in these ongoing experiments by trying to produce cancer in chick embryos with methylcholanthrene as well as with the small particles Claude was isolating with the centrifuge; neither substance led to malignancy. Using frog instead of chick embryos, on the other hand, Murphy felt Porter's in vivo system of analyzing cytoplasm versus nuclear influences would be a base for comparing normal and malignant cells, a system he hoped would show especially the cytoplasm's influence on tumor progression.

Thus, Porter continued his frog experiments by adding sperm from one species to the enucleated eggs of another and then comparing the visible traits of the resulting embryos. This work demanded an especially deft, steady hand, so for this procedure he used a favorite glass needle to remove chromosomes from the egg cells. His research progressed so well that he embarked on a frog-collecting expedition in the summer of 1941 with Rollin D. Hotchkiss, a good friend from Oswald Avery's laboratory. Hotchkiss was hired by Simon Flexner in 1935 upon receiving his PhD in organic chemistry from Yale University. By 1940, Hotchkiss had crystallized and chemically analyzed the two antibiotics discovered by René Dubos that launched the antibiotic era.[17,18] With $175 in expense money from Murphy, Porter and Hotchkiss drove throughout the eastern United States, where they captured 175 specimens of four subspecies of frogs.[19,20]

Looking for Cytoplasmic Markers: Adventures in Field and Stream

by Rollin D. Hotchkiss

When Keith Porter came to The Rockefeller Institute from Princeton in 1939, we soon became friends, and his interest in experimental biology added stimulus to my own movement from chemistry toward biochemistry. In a year or two, as he enjoyed some autonomy at his post in the Murphy laboratory, he began some new steps toward a goal that had come to occupy his thoughts. With expert micro-manipulation, he was able to remove the maternal nucleus from a newly fertilized frog egg before the incoming sperm nucleus fused with it. The remaining *paternal* nucleus diploidizes and stimulates the *maternal* cytoplasm to develop into a haplo-diploid embryo, offering a chance to learn which heritable traits are conveyed through the egg cytoplasm and cortex as distinct from the new resident nucleus.

My knowledge of heredity, limited to three Woods Hole lectures and a self-imposed systematic study of an elementary genetics textbook I had bought, was enough to excite me to participate. The cytoplasmic traits Keith hoped to find were what he believed were racial adaptations (within the species *Rana pipiens*) to the climatic gradients of temperature and vegetation along a North-South range in the eastern U.S. Frog biology was not a particular interest of mine, but I had a used car and a used view camera, and I volunteered to help him collect and photograph frogs around the East. I would like to record some of our adventures—more logistical than scientific—that ensued from our field trips.

Our shake-down runs were for two or three days in late spring 1941, going north from New York City to the Catskills and on well into the Adirondacks. We learned what gear was useful for wading in streams and swamps. We slept out on folding cots in forest spaces, that is, after we became accustomed to the manifold night noises in the wild. I remember the crashing progress of an enormous deer (estimated by the tumult to be at least thirty feet high) who came near us, snorted, and then veered away in the night without ever becoming visible to us. Another inordinately loud rasping noise right beside our cots turned out to be from nothing more than a field mouse noisily cleaning up a can of tinned salmon from which we had made a late meal at nightfall. For the rest of the summer, after our camp-style meals, we washed up our dishes before turning in for sleep.

The strategy of stalking a sitting frog with a butterfly net with its four-foot handle had its little triumphs. A frog's bulbous eyes staring placidly at the world could monitor a wide area, but the sitting creature could only jump away within a certain acute angle. We had to maneuver the net slowly enough so that the frog's short-term memory would forget that the net was moving. Then one fell pounce could enter and encompass that escape angle and might even achieve a bravura capture, intercepting him/her in midair, if not still at the perch. A wild rush with the net did not work, and if approaching too fast, the suspicious motion would precipitate a two-step process—a quick turn to a new angle, then a leap into one of a dozen avenues and escape passages among stalks, reeds, and stones.

We chilled the captured specimens in ice water some few minutes, so I would have a minute or so to focus on them in my ground glass and photograph them on a piece of white cardboard before they hopped off, to be recaptured (now sluggish and in an arena of our choice) by Keith.

Frogs collected in the North rode back to The Rockefeller Institute comfortably in my old Pontiac. But, facing a month down in the South, we had to take along materials and tools for crating the frogs in wet sphagnum moss and expressing them back to accomplices at the laboratory. So for the South we acquired large Thermos jars for the ice cubes purchased daily in towns to pacify the country frogs. Insects and weather obliged us to use sleeping bags and small pup tents with improvised netting for camping on the longer trip. Outfitted thusly, we traveled from New York down to the edge of Florida in a route that included some coastal sites and a detour inland into the Smoky Mountains and the Carolina country.

Finding outdoor sites for camping was not difficult, and weather was mostly good in July and August when we went south. Knowing that among other specialists looking for frogs in Southern swamps might be the poisonous cottonmouth moccasin snake, we had invested in two broad spectrum snakebite kits for safety. At the first campsite in the Delmarva tri-state peninsula, we opened by lantern light for the first time instructions for snakebite treatment that still stand out in my memory. After applying anti-venom, a next step was called for: "Cuts of about one half to three quarter inches are made with a razor blade at the base of the swelling as it progresses, to allow free bleeding. In one such case, for example, 110 cuts were made; it was considered a successful treatment." After some (admittedly weak) laughter, we closed the kits, and with the bravado of youth, managed to get to sleep. The kits were never reopened; though we did see a few snakes in the swamps, they were either draped on tree branches or swimming away from us.

Our camping days usually began with a tasty breakfast of porridge, cooked over a gasoline camp stove by the early rising Nova Scotian, who generously allowed me to consume my portion thinking of it as New England "oatmeal." In gratitude, I could slice bread for toast and the lunch sandwiches and wash our dishes. Shopping was simplified to the modest few groceries and the frog-fixative ice cubes. Most evenings we would eat in some small town with little range of choice in either restaurant or food. Occasionally the food could be superb—and if there was a witty waitress, that made any meal a little more fun. Every few days we would rent a motel cabin and refresh ourselves and our laundry, perhaps avoiding a bit of bad weather.

In the Northeast, we often found *R. pipiens* in meadowland, bordering the streams and ponds, but in the Southeast they were always in the water or on the closely adjacent banks. Well aware of the notorious nighttime serenading of the *R. catesbiana* (bullfrogs), we made several forays into nighttime swamps and ponds. It is some kind of test of one's poise, balancing between imagination and reality, to wade hip deep into a dark Southern cypress swamp at night, snap off the flashlight, and wait for the denizens of the swamp to declare themselves! Almost always we were soon surprised to hear a loud, rasping call from the shrubbery within a few feet of us. Shivering a bit as we snapped on our lights, the shrieks would cease, and we could only with difficulty locate the pretentious source of all the clamor, the tiny tree frog *Hyla ocularis*, no bigger than a child's thumb. We learned to enjoy these chilling episodes but called them off since the *pipiens* seemed to be sleeping through it all—or perhaps carousing elsewhere.

Keith was the perfect companion for such a trip; observant about all natural life and its traces when out of doors, and during the hours when night collecting in dark swamps seemed not worth the trouble and we were indoors, he was a witty idea-man and scholar, ready to partake in any discussion of books and literature. I learned much, and perhaps he did, from our frank talks about our diverse career interests.

Porter's chief conceptions about geographical differences within the frog species had centered upon two or three characters: the size and frequency of skin pigment spotting, e.g., the contour ("pointedness") of the snout. I did not pretend to be a judge of such refinements, but I did sometimes pretend a loyal opposition to his theses, offering over time a number of alternative hypotheses involving such things as illegitimate (for all I know, even impossible) miscegenation between the *Rana pipiens* and such neighbors as (much spotted) *Rana palustris* or the (unspotted) *Rana clamitans*. Keith good-naturedly passed off my chaffing with scornful comments on the occluded vision of would-be microbial biochemists. Actually, while in the field we had too many immediate problems to be concerned with the sophisticated biology, and certainly not the cytogenetics that was its ultimate purpose. Because of that and the incomplete state of the whole project, this narrative will leave the larger questions untouched.

If racial-climatic differences were observed, only one of them was conclusively demonstrable: the "cryophobic" response of Southern frogs to our photo-prep process. Southern frogs—certified *Rana pipiens* or not—lost all self-respect when put into our ice bath. The chilling that rendered a Northern frog just a little deliberate plunged a Southern frog into profound anesthesial disarray: we could cross their legs, drape them over their backs, or extend them out behind, with no offense. In fact, the young photographer had to earn his title of assistant naturalist by assembling most of them into his notion of a frog-like pose.

My sense of fancy and justice is pleased with the idea that a biologist who faced these early problems of making a living frog sit still for the collector's down-swooping net, or the photographer's shutter, would later solve a far advanced aspect of the same problem. Porter, with Claude, made some of their famous electron microscopy possible because they devised a fixative adequate to hold the ever evasive and elastic thin layers of the biological cells themselves quiet before the cutting blades, which made their anatomy visible for all of us.

Our travels included a visit in Virginia to the brother of Dr. Murphy, a retired gentleman and gracious host who had many stories of fads and fancies (he referred to them as conceits) of the old South. Some of them concerned the difficult accommodations made by his family to the Northern armies during the War Between the States. Curtain material in their home was commandeered, for example, to supply bandages for wounded soldiers.

Some of the frogs we shipped to New York arrived intact, and Keith worked with them for a period. I did not participate at all in this research. His annual laboratory reports to the Institute describe continuing concern with frog embryo traits. They seemed to suggest that some recognizable material property survived in the enucleated cytoplasm to modulate the incoming paternal nucleus in a heritable or semi-heritable way. This work came to an end in 1942, when Keith left New York to overcome a tubercular infection in a sanatorium. His cheerfulness, satiric wit, and modest self-irony stayed with him throughout, I believe, and allowed him the courage to continue some research interests and brought him back to New York in good time to carry out the great work that is the important focus of this volume.

Enigmas of Culturing Cells

Tragedy struck soon after Porter returned from this adventure. In September 1941, Porter's wife, Elizabeth, was diagnosed with tuberculosis, and she entered the New York State Hospital for Incipient Tuberculosis in Ray Brook. Keith remained in New York City with their six-month-old son, who died six months later from tuberculosis. In October 1942, Keith himself was diagnosed with tuberculosis and joined his wife in the sanatorium. While his disease abated sufficiently for him to return to New York after eight months, Elizabeth remained a patient there for more than three years.

While at Ray Brook, Porter became so restless and curious that he soon got involved in research in the sanatorium laboratory to learn how to diagnose cases of tuberculosis. Before 1945, when his Rockefeller colleague René Dubos developed a culture medium to produce homogeneous cultures of mycobacteria, there were crude techniques for cultivating these bacilli. The cells grew slowly in large clumps, and they had to be ground up to separate them for analysis, a technique that was not only dangerous for researchers but damaging to bacteria. This enigma suggested a research project to Porter that demanded a closer look at this bacterial cell.

Finding artifacts in cultured tubercle bacilli. Collaborating with Diran Yegian at Ray Brook, Porter examined variants of the tubercle bacilli that had been reported for decades in the literature as "pleomorphic forms." These acid-fast granules, beads, spores, and rods were assumed to be structural units of the mature tubercle bacillus or stages in its reproductive cycle. As soon as Porter tried handling these variants, however, he found discrepancies in different preparations of the same material. He established that these forms were indeed artifacts—crushed or traumatized cells—caused by the techniques used to manipulate the bacilli. The experiments provided a clear-cut demonstration that some forms were caused by rough action of the platinum loop in dispersing the cells on culture medium, other forms by chemical stains that altered their number, size, and distribution, and still other forms by the microtome knife as it sliced against the hard paraffin sections of the cells.[21,22]

This laboratory experience was Porter's first encounter with the vagaries of culture techniques, the microtome, and interpreting artifacts in cultured cells. It instilled a great measure of patience and caution while he learned to distinguish actual from artificially created cell structures.

In vitro transformation of cells. The major disruption of tuberculosis in Porter's life and work coincided with other absences in Murphy's laboratory. Hotchkiss and George Hogeboom were working on projects related to war efforts (see Chapter 4). At the Institute, Murphy and Sturm carried on with leukemia studies, while Claude toiled nearly alone on improving his methods of differential centrifugation. Porter had been reassured by Gasser and Murphy during his illness that his position would be waiting for him after his recovery.[23,24] In later reminiscences, Porter said, "I was fortunate in getting into the laboratory with Murphy, because he was extraordinarily patient; a less patient person would have probably kicked me out of the laboratory after a couple of years of fooling around, but he was not that kind [of person] and it was an experience to be associated with Claude who was a very systematic investigator."[25]

When Porter resumed his post in May 1943, the frog embryo work diminished but did not end for three more years. In its place, Murphy, in his firm but amiable manner, prodded Porter into doing experiments on transformation of normal cells into malignant cells. Porter then turned to applying the techniques acquired with tubercle bacilli to avian and mammalian cells. And, he found this required an entirely different expertise, both in growing cells and in deciphering changes.

The art of tissue culture during the first half of the twentieth century addressed many research questions.[26] Among them were problems in development and growth, malignancy, immunity, propagating viruses and bacteria, and analyzing the effects of nutrition, toxins, and radiation on cells.[27] Ironically, some of Porter's methods for success in growing cells in culture had been developed and highly refined at Rockefeller by two foremost tissue culturists in America, Alexis Carrel and Raymond Parker. But Carrel and Parker left in 1939, the year Porter arrived, and tissue cultivation was abandoned.

Perhaps it was just as well that Porter had no tutors. According to British histologist Edward N. Willmer in his book *Cells and Tissues in Culture*, Carrel espoused such exacting techniques and vigorous asepsis that he gave tissue culture "a spurious and unfortunate reputation for difficulty and almost for mysticism."[28] For example, when Walther F. Goebel, an immunochemist in Avery's laboratory with Hotchkiss, wanted to learn how to bleed chickens, the initial step in many tissue culture procedures, he witnessed the opposing approaches of Alexis Carrel and Karl Landsteiner (see box).

The Mystique of Tissue Culture

by Walther F. Goebel

One day I was having lunch at Dr. Carrel's table and was telling him of our experiment and the dilemma in which we found ourselves, when, much to my surprise, Dr. Carrel said, "Dr. Goebel, if you come to my laboratory on Monday morning at nine o'clock I will show you how to bleed a chicken."

I was taken aback and pleased. Never before had he volunteered to share with me any of his superb laboratory techniques. And so the following Monday morning I sauntered from my apartment to his office. I must confess that I was a bit late on this eventful day. His secretary, Miss Krutcher, was expecting me, and with a hint of reprimand in her voice she said, "You are ten minutes late, Dr. Goebel. The doctor is in the operating room." With this she handed me a package wrapped in black muslin.

I was ushered to an anteroom and told to wash my hands, unwrap the package, and don the operating gown that it held. All this I did, but I was quite unprepared to find my surgeon's gown dyed a dead black. The cap, which pulled down over my head with only slits for my eyes, was also black. If all this was startling and a bit puzzling, it was nothing to that which followed. Miss Krutcher opened the door of the operating room. I peaked in and gasped. The walls of the room, the ceiling, the floor, and even the operating table were all painted ebony. The garb of the three technical assistants and that of the great surgeon himself were the same as mine. We looked like black-clad Klu Klux Klanners. There, before me, prone and anesthetized on the black operating table and covered with black towels, was the unfortunate victim, a white Leghorn hen.

The whole dramatic scene was illuminated by a dazzling overhead operating lamp. One of the assistants served as an anesthetist, the other passed the sterile surgical instruments, carefully laid out on a black towel, to Dr. Carrel. The room was heavy with ether as I entered. The only part of the chicken that showed was the underside of one wing. The rest of the bird was covered with black towels. A small cannula had been inserted into the vein of the bird, and the blood (red) was gently flowing into a sterile tube held by a third black-clad technician . . .

The following day I had lunch at Dr. Landsteiner's table, where I regaled them with my experiences of the day before. Landsteiner

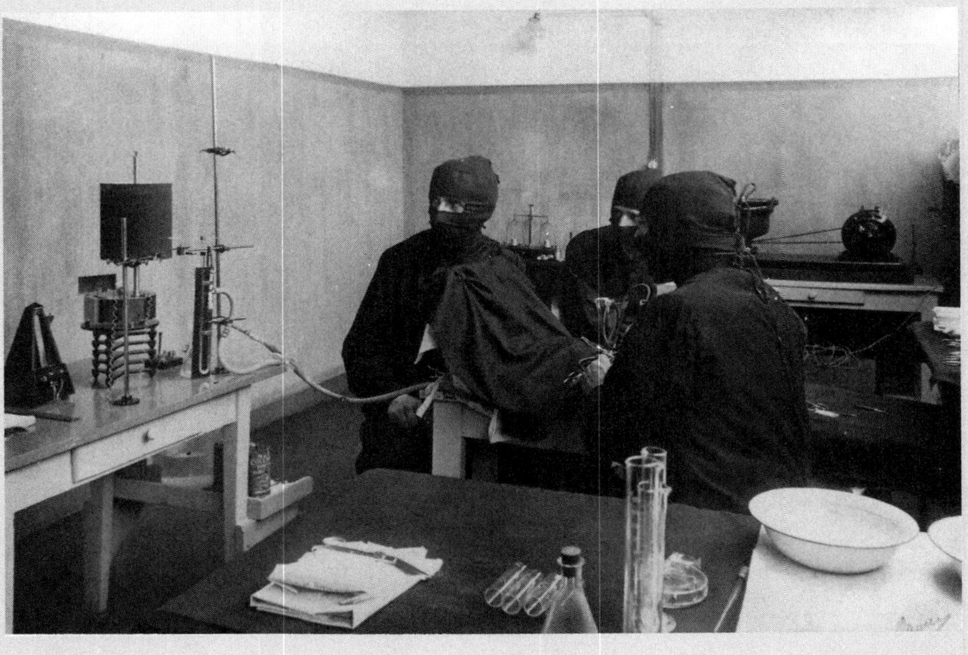

Sterile techniques in Alexis Carrel's operating room. Courtesy of Theodore Malinin.

laughed, an unusual reaction for so serious a man. Then he turned to me and said, "Dr. Goebel, venn you can come to me tomorrow I vill show you how to bleed a schicken." So the following day I went. In his laboratory, he had half a dozen cages piled high, each containing a bird, all to be desanguinated. He reached into the first cage, grasped the bird by the legs, and deftly slung the raucous hen in such a manner that it was firmly held under his left arm. A technical assistant completed the job. It required about a half hour to do the whole lot.

What a difference between the two Nobel laureates!

Excerpted from a manuscript by Walther F. Goebel. ca. 1983. *The Rockefeller Institute: An Irreverent Memoir.* 127–130. Collected papers of Carol L. Moberg.

At the beginning of World War II, Willmer reported, "tissue culture lay becalmed in the doldrums." There were neither uniform populations of cells, akin to the pure strains used by bacteriologists, nor media whose composition could be accurately defined. The individual researcher faced numerous, even esoteric, hurdles in preparing the cells, extended periods of trial and error in manipulating media and cells, and a significant lack of uniformity from one experiment to the next. Fortunately, Parker published a tissue culture manual in 1938, and this helped demystify some of Carrel's methods.[29] However, even while referring to Parker's manual for a few basic techniques, Porter worked alone to adapt growing cells in culture to his singular needs. His essentially self-taught skills also allowed him to be flexible and innovative when the next challenge came.

Porter began by repeating important tissue culture studies of Wilton R. Earle at the National Cancer Institute, who had elaborated an in vitro method to produce tumors in mouse fibroblasts.[30] What puzzled Murphy and Porter was that Earle found untreated cultures became malignant as frequently as the carcinogen-treated cultures. Earle attributed their spontaneous transformation to trace contaminants in the culture; such "transformations" were probably the result of introducing contaminating transformed cells from a continuous cell line propagated in their laboratory. Murphy felt the question of whether transformation was induced or spontaneous could be resolved in his laboratory, where there would be no contact with potential contaminants.[31]

Within a year, Porter prepared a manuscript (never published) in which he reported he could not reproduce Earle's malignant transformation of cultured cells. While the cultures derived from chick muscle began to show many abnormalities, after several transplants, injecting these abnormal cells into chicks produced no tumors at all. He found malformations during the transplantations, specifically, many small nodules but no malignancies. Porter's cautious analyses of these experiments include details of culturing conditions and of cell cytology and emphasize "certain technical problems which make the exploration of in vitro changes in cells very difficult."[32]

In this set of experiments, he learned even more about the problems and vagaries of tissue culture, particularly in identifying which traits indicate malignant or normal cells. For all that anyone knew at the time, transformations may have been induced by an unidentified chemical, hormonal, or microbial contaminant present in the biological culture medium. The terms abnormal, transformed, and malignant were still nebulous since no one had yet been able to study the interior of a normal cell, let alone knew how it differed from a cancerous one.

This unproductive research no doubt put Porter in a quandary about how to proceed. Since Murphy continued to believe the microsomes that Claude was separating from normal and tumor tissues played a role in normal development, he asked Porter to expose the microsomes to cultured cells. This was a challenge, Porter later said, "because there were no packaged media in those days. We had to make our own from scratch. . . . I never had any contact at all at Harvard or anywhere else with cultured cells, and it was exciting for me to see cells grow out of pieces of tissue and adopt a particular form in culture, and I had to examine all of these cultures that were exposed to microsomes to see if I'd achieved any transformation in vitro. Actually some of the cultures did show transformation but I was never able to propagate those cells beyond the primary culture stage."[33]

At this point, the laboratory's routine was interrupted by a serendipitous event. Porter's focus on the microsomes as the potentially elusive sarcoma agent finally brought his talents together with Claude's. In 1943, a new line of investigation of great importance was opened by an instrument that made it possible to study the vast wilderness inside the cell: the electron microscope.

Electron Microscopes and the Nascent Cell Biology

The electron microscope's developmental decade before 1943 lay outside the major centers where working biological scientists flourished. Early models were handmade by physicists more interested in the instrument and its technical aspects. The microscope works by accelerating and focusing a beam of electrons in a magnetic field, similar to the way glass lenses in a light microscope focus beams of light. Because electrons are scattered when they encounter matter in the air, the column inside the microscope must be evacuated during the operation by means of a vacuum pump. The specimen image becomes visible on a fluorescent screen as a result of electrons encountering the specimen under observation, some colliding and being impeded by the structure and others traversing the electron porous components, thereby activating the fluorescent material on the screen. The best resolution attainable with the electron microscope is several hundred times greater than that of the light microscope, thereby revealing fine detail of particulates down to the size of larger molecules.[34–40]

These instruments were expensive, difficult to obtain, highly specialized, and especially difficult to use with biological materials. The microscope demanded extensive training in its operation and careful, time-consuming preparation of the specimens before meaningful images could be produced. Literally hundreds of specimens had to be examined and photographed to find one that had been imaged in exactly the right orientation, like a scene frozen in a tableau. Applications to biological materials were rare, for the all too obvious reason that specimens examined in the microscope are the dead-end products of a series of manipulations (desiccation, damage by electron beams) that alter their appearance from the living state.

In Europe, the first commercial Siemens & Halske (Germany) electron microscope was delivered in 1939. However, German scientists used prototypes from 1936 to 1940 to look at diatoms, bacteria, viruses, and bacteriophages. Some early biological specimens examined with this instrument are listed in the table.[34]

In America, the first commercial model of an electron microscope, the RCA EMB, became available in 1941. Vladimir Zworykin, who was head of electronics research at RCA and a pioneer of television technology, recruited James Hillier from Toronto to build the EMB model. According to John Reisner, an early RCA engineer, the instrument cost $9,600, which represented the average annual salary of a full member position of The Rockefeller Institute or many times the price of a luxury automobile.[38]

Zworykin also initiated a research program under the National Research Council. Thomas Anderson, a postdoctoral physical chemist, was chosen for the express purpose "to investigate the biological usefulness of the electron microscope being built at the RCA laboratories."[41] Anderson operated the microscope under the direction of a committee of eminent biologists, chaired by Stuart Mudd, head of medical microbiology at the University of Pennsylvania. From 1941 to 1943, these scientists worked according to a prearranged calendar at the RCA laboratories with Anderson. Their goal was to advance its development for biology.

Anderson's critical efforts set standards of procedures in taking and interpreting micrographs that excited the scientific world. To ensure micrographs represented what an experimenter claimed, he compared the same specimen prepared in as many ways as possible. "Everything we looked at was novel in those days," he wrote. "The electron microscope had given us the power actually to see as individual objects things that had previously been only mental concepts."[42] Initially, he was skeptical of electron microscopy because in his earliest micrographs "bacteria appeared as big blobs, and viruses as smaller blobs." Until better tissue preparation methods of biological specimens were invented by Porter, the electron micrographs by Anderson and others were essentially images of dried silhouettes.[43]

Nevertheless, after a few weeks of using the new instrument, Anderson found the "simplicity and directness" of obtaining results was "immediately reassuring."[41] For example, he and Wendell Stanley could measure the sizes and shapes of several plant viruses and thus confirm measurements that had been inferred earlier using ultracentrifuges and diffusion.[44] The strength and originality of Anderson's methodology prevented many clashes of electron microscopists over interpretation of what they were seeing. Recognizing that "all the specimens for electron microscopy are artifacts," he commented, "the important thing is to be able to

Some biological specimens seen by electron microscopy before 1945

In Europe	
Diatoms	Krause, F. 1936. *Zeits. f. Physik.* 102:417–422.
Bacteria	von Borries, B., E. Ruska, and H. Ruska. 1938. *Klin. Wochenschr.* 17:921–925.
Viruses	Krause, F. 1938. *Naturwiss.* 26:122.
Tobacco mosaic virus	Kausche, G.A., E. Pfankuch, and H. Ruska. 1939. *Naturwiss.* 27:292–299.
Bacterial flagellae	Piekarski, G., and H. Ruska. 1939. *Klin. Wochenschr.* 18:383–386.
Bacterial membranes and capsules	Frübrodt, E., and H. Ruska. 1940. *Arch. Microbiol.* 11:137–154.
Tubercle bacillus	Lembke, A., H. Ruska, and J. Christophersen. 1940. *Klin. Wochenschr.* 19:217–220.
Bacteriophages	Pfankuch, E., and G.A. Kausche. 1940. *Naturwiss.* 18:46.
In North America	
Tobacco mosaic virus	Stanley, W.M., and T.F. Anderson. 1941. *J. Biol. Chem.* 139:325–338 and 338–344.
Bacteria	Mudd, S., K. Polevitzky, T.F. Anderson, and L.A. Chambers. 1941. *J. Bact.* 42:251–264.
Vaccinia virus	Green, R.H., T.F. Anderson, and J.E. Smadel. 1942. *J. Exp. Med.* 75:651–656.
Papilloma virus	Sharp, D.G., A.R. Taylor, D. Beard, and J.W. Beard. 1942. *Proc. Soc. Exp. Biol. Med.* 50:205–207.
Equine encephalitis virus	Sharp, D.G., A.R. Taylor, D. Beard, and J.W. Beard. 1942. *Proc. Soc. Exp. Biol. Med.* 51:206–207.
Mycobacteria	Rosenblatt, M.B., E.F. Fullam, and A.E. Gessler. 1942. *Amer. Rev. Tuberc.* 46:587–598.
Bacteriophage	Luria, S.E., and T.F. Anderson. 1942. *Proc. Nat. Acad. Sci.* 28:127–130; and Luria, S.E., M. Delbruck, and T.F. Anderson. 1943. *J. Bact.* 46:57.
Muscle and protein fibers	Schmitt, F.O. 1945. *Harvey Lectures.* Series 40. 249–268.

identify the artifact and correct for it either by better specimen preparation or by devising what the original specimen should appear to be."

By late 1942, according to Reisner, only eight of the first thirty EMBs produced were in academic institutions.[38] The few rare biological researchers included Cecil Hall (who tried to develop the instrument for Kodak) and Francis O. Schmitt, both at the Massachusetts Institute of Technology, and Joseph Beard at Duke University, who had worked with Rous at Rockefeller until 1937. This rarity of academic microscopes does not explain why Claude and Porter did not have earlier access to one than they did. On no less than three occasions, important opportunities appeared, but no evidence exists to suggest the Murphy laboratory pursued them.

The first occasion was too early in the development of the microscope, and Claude's tumor material was too undefined to submit to this kind of analysis. During the mid-1930s, he received an illustrated manuscript and letter from Ladislaus Marton, a pioneer in microscope construction and observation of biological specimens.[45] Claude showed Marton's micrographs of bacteria to Rous, who, as editor of *The Journal of Experimental Medicine*, refused to publish the paper because, Claude said, he found the images "were not better than in the light microscope."[46] This comment no doubt also discouraged Claude, for the time being at least, from considering the new instrument for his own work.

The second occasion was closer at hand. Two groups of scientists, who participated in Anderson's project, were more concerned with known viruses than the still nebulous particles resulting from Claude's centrifugation experiments. At the time, Murphy and Claude had not yet accepted the viral nature of the Rous agent, and their ongoing investigations were leading in directions that were still unsuitable for the electron microscope. Had the Murphy laboratory seen Anderson's micrographs of isolated viruses, their interpretation of their own images a few years later might have been less complicated.

One scientist working with Anderson was Kurt Stern from Yale. In April 1940, after Claude had lectured at Yale, Stern informed him "that through my friend in Philadelphia the first photographs obtained from our 4 x sedimented Rous agent preparation with the aid of the electron microscope showed spherical bodies embedded in a net-work of other material."[47] (This friend may have been Stuart Mudd, but neither the friend nor the microscope are identified further.) By late 1941, Stern enrolled in Anderson's program "to undertake the electron-optical study of our Rous agent material and the inhibitor."[48] The first session produced some fifty micrographs of centrifugally purified virus from chicken tumors that showed "slightly ellipsoid dense bodies."[49] Mudd questioned whether Stern's micrographs actually showed "elementary particles of the virus or possible carrier particles from this host tissue, such as mytochondria [*sic*]."[50] Stern was asked to provide control samples for a reevaluation, and he said he would use two techniques for producing them.[51] Virologist Thomas Rivers, who was then physician-in-chief of The Rockefeller Institute's Hospital, advised Anderson that the "one control material that I would absolutely demand is that prepared from normal chick embryos in the manner that Claude and others have used" because it "has all of the usual physical and chemical characteristics of preparations containing the Rous virus." Rivers also doubted that Stern could be "certain that this extra-embryonic fluid does not contain other particles in addition to the Rous sarcoma."[52] Stern had assured Mudd that they would use "appropriate caution in our interpretations" of the electron micrographs; however, since none of these experiments were published, this suggests the second set of micrographs neither clarified the problem of artifacts nor made a convincing visual argument for a virus.[53]

The third occasion involved other scientists working with Anderson from Rivers's own Hospital laboratory. After Stanley reported his work on plant viruses with the electron microscope, Rivers said, "I began to get after my boys to get some pictures of the elementary bodies of vaccinia virus, but, for some reason that I still can't fathom, they were reluctant to do this. Perhaps they just didn't want to take the trip to New Jersey." Finally, in 1942, Anderson, Joseph Smadel, and Robert Green made the pictures he wanted. "The work," he continued, "was highly technical and required great skill . . . [of Anderson]. . . . The pictures those boys got of elementary bodies of vaccinia virus made everybody sit up. Everybody expected those bodies to be spherical—instead the electron micrographs showed that they were roughly brick-shaped . . . [and] had an internal structure and a limiting membrane, and were in fact morphologically closer to bacteria than to plant viruses like tobacco mosaic virus."[54] This time Rous deemed electron micrographs of the isolated virus particles worthy of publication in *The Journal of Experimental Medicine*.[55]

That same year, 1942, after Claude met Francis Schmitt at Bensley's symposium, he sent Schmitt several preparations of chromatin threads he had isolated from nuclei for study in the MIT microscope. Schmitt was interested in adding this substance to his studies of various protein fibers, but he found Claude's preparations so unsatisfactory that he was unable to resolve any structure in the micrographs.[56,57]

Meanwhile, the growing impact of Stanley's work influenced Rockefeller to buy its first electron microscope. On 5 June 1944, the Board approved the $6,000 purchase of a small table model RCA for use by Stanley at the Institute's branch of Animal and Plant Pathology located in Princeton, New Jersey. By this time, however, the nascent cell biologists in New York City had benefited from a full year's collaboration on another singular instrument.

Rockefeller and Interchemical Scientists Examine Three Biological Preparations

Luck or chance, whatever one wishes to call it, intervened for Claude and Porter. Henry Green, head of the physics section at the Interchemical Corporation in Manhattan, contacted Claude on 28 May 1943. This industrial microscopist, who specialized in measuring the size of particulate materials by microscopical methods, had just read Claude's paper in the current issue of *Science*, "The constitution of protoplasm."[10] He sent a letter asking whether "it would be possible to obtain some of the centrifuge portions of protoplasm that you describe. . . . We have in our laboratory at Forty-fifth Street a R.C.A. electron microscope which, we believe, could handle the submicroscopic particles in the cell plasm with great ease and might, therefore, reveal structure of interest. If you would care to let us examine some of your material or at least give us such information that would lead to our obtaining it elsewhere, we would be very glad to let you see the results of our investigation."[58]

Green's invitation launched an unusual eighteen-month collaboration between academic scientists and a commercial research laboratory. Interchemical's RCA EMB microscope was the fourth one produced and the first one in New York City. It had been purchased in 1941 by Albert E. Gessler, Interchemical's director of research. The microscope's supervising technician, Ernest Fullam, used the instrument officially to examine the particle size of industrial finishes of paints, printing inks, and abrasives. After 1942 he used it for research related to the Manhattan Project. However, Mary Schuster, another Interchemical microscopist, recounted how Gessler felt "the microscope had far more potential value for the physical study of animal tissue . . . and that great forward strides could be made as a result of its use in the field of pathology, particularly in regard to cancer."[59]

According to Fullam, Gessler had a close friend who died of cancer, and this experience committed him to finding its cause.[60] Claude's *Science* paper further heightened this collaboration by suggesting that particles in normal and tumor cells were possibly self-duplicating and thus the source of cancer. Consequently, Interchemical became interested in expanding its use of research instruments—electron microscope, spectrophotometer, and an experimental microtome—for biological research. Its stake became even more viable in December 1943, when the company received a grant from the Lillia Babbitt Hyde Foundation to support investigations in cancer research. Claude and Porter interacted directly with Gessler and Fullam on these research projects that took place after regular business hours.

Fullam trained in microscopy at Cornell University and learned to use the electron microscope at the University of Pennsylvania. His course in the theory of electron optics was taught by Ladislaus Marton using the prototype EMA instrument he designed for RCA, which was never marketed because it was too slow, difficult to use, too costly to manufacture, and contaminated easily. However, experience with working models had prepared Fullam for the rough operating conditions that prevailed in its successor, the EMB designed by James Hillier. Fullam called the new model an extremely temperamental instrument: "Alignment was extremely difficult. You had to be a tall person to get good alignment. For a short person,

Ernest Fullam at the Interchemical Corporation RCA EMB electron microscope, 1941.

it was impossible unless you worked with mirrors. . . . Viewing the image as we translated the specimen was impossible in the early days because the hand controls were at the top and the viewing screen at the bottom. . . . At first we also frequently went beyond the travel of the screen so we couldn't tell where we were. Sometimes we had to take the microscope apart and realign the system to find out where we were.

"Another problem involved magnetic fields. Once, I put an electric clock on the microscope and rigged it so it could be used to time exposures. I could see a beautiful image on the microscope, but with the clock on I could never get a good picture. After some experimentation, I found the magnetic field of the clock was imparting a 60 cycle oscillation to the beam, which killed all my resolution. Another time, a magnetized screwdriver lying on the table of the microscope cost me days of trying to find out why I couldn't get alignment.

"There's one other little point. I soon found that kicking the microscope or slapping it sharply on the side when the high voltage wouldn't come on was very effective in returning it to operation."[61]

Claude was very enthusiastic about the prospect of an electron microscope, a trained technician to operate it, and a host of materials that he wanted to examine. For nearly a year, he sent a variety of samples to Fullam, including both normal and tumor rat tissues, protoplasmic granules, and chromatin threads that he was isolating. Although no micrographs of chromatin were ever published, the laboratory's annual report for 1943–1944 contains three images, showing perhaps the first views (at a magnification of 18,000) of what he called "bead-like structures" spaced along a thread.[62]

Interchemical fixed some samples with a freeze-drying apparatus or dried them directly on a nitrocellulose membrane, while Claude sent some fixed in paraffin.[63–65] These fixations caused at best vague images. When time allowed, Fullam took electron micrographs and sent them to Claude and Porter for interpretation. By mid-1944, Porter joined Claude on visits to Interchemical so they could learn how to handle the specimens and to use the microscope themselves. Essentially, three different preparations of biological tissues merited study in the microscope: the isolated subcellular fractions (mitochondria and microsomes), sectioned slices of normal and tumor tissues, and intact cultured cells.

Subcellular fractions. The first publication by Claude and Fullam in January 1945 contained the first electron micrographs of a subcellular component, the mitochondria that had been isolated by differential centrifugation from a rat lymphosarcoma. At exciting new magnifications, from 4,100 to 12,000, the mitochondria ranged from 0.6 to 1.3 micrometers in diameter, which was considerably larger than Claude's original estimate of size but now agreed with Bensley's figures. This diameter was normally too thick to permit penetration of electrons, but since the normally spherical mitochondria flattened out during isolation and purification, the micrographs revealed somewhat murky internal structures. The specimens were prepared using Formvar, a polyvinyl plastic, as the supporting membrane, preserved with potassium dichromate, and then air dried. Claude tested four different fixatives, including osmium tetroxide, which he noted caused a certain amount of clumping. The effect of this treatment led to cautious interpretations of the morphology of mitochondria and suggested the need to examine mitochondria "in their normal environment, i.e., the intact cytoplasm," because with isolated specimens it was impossible to study structural relations within cells. Their micrographs, however, led them to observe a limiting membrane surrounding mitochondria, not only maintaining their general shape but also providing further evidence that these particles were "definite physical entities" in the cytoplasm. The paper also includes a blurry micrograph of "microsomes" to show they were distinct entities from mitochondria.[66]

Cell sections. Claude and Fullam encountered enormous difficulties when they tried to look at mitochondria in the cytoplasm. Some difficulties related to preparing their specimens of tissues from guinea pig liver that were fixed with osmium and embedded in a camphor-naphthalene mixture. Other troubles came when sectioning, or slicing, these specimens with Interchemical's high speed microtome. This technology had been well developed for the needs of light microscopy, and a newer version was built by Gessler and Fullam for electron microscopy that could whirl the knife at speeds of 65,000 rpm. According to Fullam, the microtome was "a fearsome piece of equipment, and I found it could be quite dangerous to operate. . . . It could be worked after a fashion, but it certainly wasn't easy. The worst fault it had was the ear-splitting shriek it produced, which drove everybody out of the room."[61] Although satisfactory sectioning techniques for electron microscopy would take several more years (see Chapter 6), at this early stage Claude and Fullam published micrographs revealing holes in the specimens that were torn by the microtome knife. They observed that

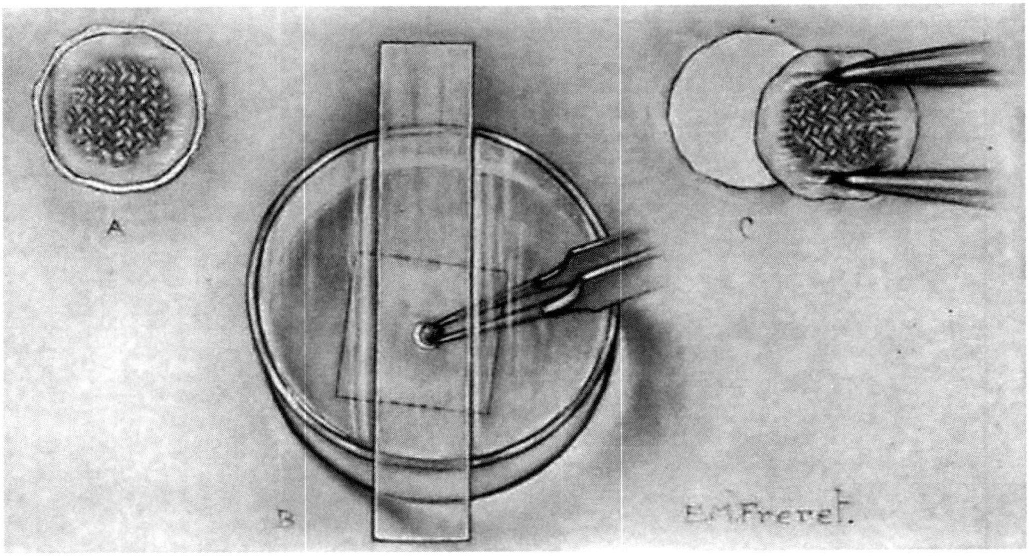

1. Sketch depicting steps to put cultured cells on electron microscope grids. After cells are grown in culture on glass coverslips coated with Formvar and fixed with osmium tetroxide, they are placed underwater, where a small area is cut out of the film with watchmaker's forceps (A and B). The flap of film is peeled from the coverslip and placed on a wire mesh grid that holds the specimen in the electron microscope (C). From Porter, 1953. *J. Exp. Med.* 97:727–750.

mitochondria in their sectioned tissues were completely opaque to the electron beam but were rod shaped in contrast with their more spherical shape when viewed as isolated particles. They also observed small particulate elements, roughly 80–100 nanometers in size, distributed throughout the cytoplasm, which they claimed were the submicroscopic components of the cytoplasm, or microsomes, that Claude had isolated by centrifugation.[67]

Cultured intact cells. The most successful specimens tested during the collaboration on the Interchemical microscope were cells that had been cultured by Porter.[68] Claude's subcellular fractions and cell sections gave no information about structural relationships within the cell. And, since glimpses were expected inside an intact cell, Porter learned it would demand innovative and radical modifications in his tissue culture techniques to meet the rigors of the electron microscope.

On one of Claude's early visits to Interchemical, Porter said, "I tagged along to see this remarkable instrument and learned such basic points as a) the specimen had to be dry and b) it had to be thin. Obviously there was nothing in this to excite a student of cells, so from the outset there was little excitement re[garding] the prospects of looking at cells. . . . Fortunately I had become involved in culturing chick cells because Dr. Murphy, our chief, was certain that if we put Claude's microsomes on cultured chick cells, we would witness a transformation to tumor cells. . . . So I began to dream about ways to get cultured cells into the e[lectron] m[icroscope] (in the hope of detecting the earliest evidence of transformation)."[69]

Porter was aware that anyone "who has grown and observed cells in vitro knows they have a tendency to spread out to extreme thinness, apparently in an attempt to cover as much of the substrate as possible." So he decided "such diaphanous cells might be suitable for electron microscopy, at least in their thinner margins."[70] He devised three refinements to grow cells that spread out in a thin layer in the hope of visualizing a more intricate landscape inside a whole cell than could be seen with a light microscope.

Adapting a hanging drop culture method. Porter began growing cells using the hanging drop procedure. This method was originated by Ross Harrison in 1906–1907,[71] modified in 1911 by Carrel, who cultivated chick fibroblasts and substituted plasma for lymph,[72] and further modified in 1912 with physiological salt solutions or nutrient plasma clots by Margaret and Warren Lewis.[73] Porter's adaptation involved growing the cultures of chicken fibroblasts directly on a supporting Formvar film. This was a vast improvement over Anderson's technique at RCA of placing a droplet of cells on thin collodion films affixed to the microscope grids and "hoping" some cells would remain on the grid after rinsing the specimen in distilled water.[41] Porter used a dissecting microscope and watchmaker's forceps while working under water to transfer the cells, intact with the Formvar film, onto the wire mesh grid that held specimens in the electron microscope. In Porter's method, cells that would naturally be surrounded by salt solutions had little contact with distilled water even during the transfer to the wire grid, and the plastic membrane eliminated surface tension on the cells. These delicate techniques are described briefly in the 1945 paper, but they were later illustrated and elaborated by Porter (figure 1).[74]

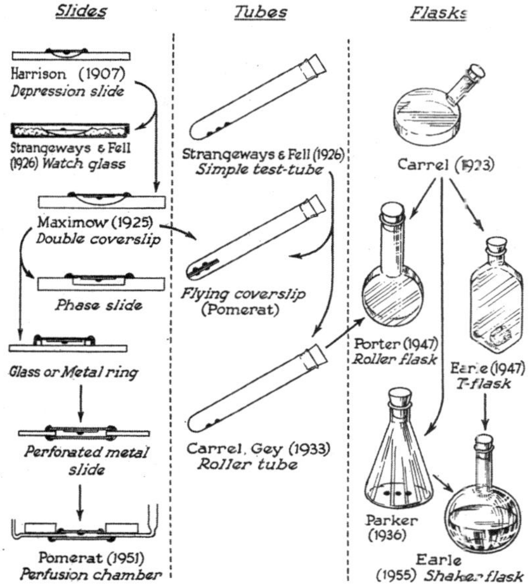

Keith Porter's roller flask for tissue culture. From Paul, 1959. *Cell and Tissue Culture*. E & S Livingstone, Edinburgh, Scotland. ©Elsevier.

Devising a roller flask. Porter found the hanging drop technique too confining. It used so little medium that following changes in the cells was difficult, and the cells had a short life. As early as 1923, Carrel had addressed this problem by devising a sealable flask, now called the Carrel flask, in which pH could be adjusted and culture medium replaced. Ten years later, a roller-tube system of cultivation was developed by George Gey[75] and Warren Lewis,[76] in which rotating racks of stoppered test tubes permitted an easy way to add medium. Porter combined these techniques of growing cultures directly on the glass wall of the tubes and constantly bathing them in liquid media. He created what he called a "roller flask," a long, narrow vessel with two parallel flat sides that was blown from a regular Pyrex test tube, and fit rows of them on rotating racks. He attached a Formvar-coated glass slide inside the flask with a thin plasma clot. The tissue supplying the cells was planted in a small clot on the exposed surface of the slide and was fed nutrient medium on a regular basis until satisfactory cells grew either under or beyond the clot. In this method, the undisturbed cells were removed while still attached to the coated coverglass slide, where they were easily fixed and transferred to wire mesh grids.[77]

Concocting a novel fixation. Porter's experiments with tubercle bacilli had alerted him to the fact that fixatives for light microscopy could easily create bizarre structures. Yet, fixation for electron microscope specimens was even more challenging since the specimens had to be examined in a vacuum and therefore had to withstand removal of liquids without losing their shape. Of the four fixatives tested, Porter found that "osmium tetroxide has been the most satisfactory in that it yields what impresses one as being the most truthful picture of the cell." While other fixatives produced alterations in the cells, osmium tetroxide prevented dehydration effects and served "as a differential stain, for by its action some cellular structures are apparently rendered more electron-scattering than others."[68] He credited English cytologists Thomas Strangeways and Ronald Canti, who showed in 1927 that osmium tetroxide (OsO_4) produced "an almost perfect preservation of the cell and its internal structure" for dark-field microscopy.[78] Porter introduced a way to apply osmic acid as a vapor from a 2% solution. He further prolonged the fixation and subsequent washing because this led to the degradation and subsequent extraction of the proteins of the cytoplasmic matrix, thus reducing the thickness of the cells and enhancing the contrast of membrane-bound subcellular components.

These new techniques resulted in exceptional electron micrographs with high contrast and sharp definition of the structures. Looking back in 1984 on these early micrographs, Porter acknowledged "the first images were not that much better than the light microscope images. There were things that we couldn't see with the light microscope, but we had no idea what they were and so the challenge was there, but I don't think that I projected the electron microscope into the future. The instrument that we used was the old EMB which was a toy compared to what we have nowadays."[33] Nonetheless, with slight modifications, many of the principles he instituted are still followed.[79]

Going beyond Vision: First Glimpses inside a Cultivated Cell

The 1945 article containing the original micrograph of a cultured intact cell, "A study of tissue culture cells by electron microscopy," opened the world of the cell for discovery.[68] In two unpublished lectures, Porter recalled the moments after the first cultured cells appeared under the electron microscope in the summer of 1944. "Within a short time we discovered on one grid—which had only three or four cells on it—*a* cell in *an* opening in the grid, and it became the most micrographed cell in history. It was wonderful, believe me, we had never seen anything like it. Men have visited the moon . . . but we were the first . . . to see particles, to see structures that the light microscope had not been able to resolve. And there were Claude's microsomes, and there was the lacework of [what] eventually came to be known by that horrendous term 'endoplasmic reticulum.'"[80]

Porter also recalled that after putting Fullam on his train to Princeton, he and Claude went to Times Square to eat at an all-night restaurant and to discuss their exciting discovery. It was here, Porter noted, "that I made the shocking discovery that my colleague wanted me to find a teaching job in a small university, somewhere remote from N.Y. His timing with this advice was unfortunate because I too had enough imagination to sense the importance [of] what we were doing and recognize its significance for me especially, a young investigator just starting up the ladder. We managed to live within the constraints of mutual watchfulness . . ."[81]

Murphy's reaction to the micrographs was less sanguine. Writing to laboratory member George Hogeboom, who was on active duty in Florida, he said, "Porter and Claude have been having kittens over the electron microscope pictures, but what they prove God only knows. Perhaps some doubt on Claude's hard and fast classification of particles! In any case Porter thinks it is by far the best thing this lab has produced. I hope he is right, but I just can't get excited over pure morphology . . . [Porter] has kept everything very secret and did not want me to even tell Gasser about it. It is certainly very interesting and the methods are clever."[82,83]

The task of writing the paper took nearly five months and was achieved mostly by letter, with Claude in New York or Woods Hole and Porter spending the summer with Elizabeth at the Ray Brook sanatorium. Their correspondence reveals some of their exhilaration as well as their anxieties while describing objects in the cell that no one had seen before.

On 20 July, Claude sent Porter a complete set of contact prints with notes on the appearance of mitochondria that contained unknown granules. He thought the dark granules around the nucleus were not fat granules but appeared to "represent the Golgi bodies." He postulated the reason the Golgi apparatus had never been seen in living cells "is probably because a reticulum or 'canalicular' system was expected (as in fixed preparations) and the isolated bodies were thought to be fat globules." He thought publishing six illustrations was extravagant but noted that Interchemical was willing to pay for three of them.[84]

Porter's reaction to the contact prints was that they had "scarcely scratched the surface of possibilities." On his first reading, he had visions of a working endoplasmic reticulum while he became absorbed in studying the smallest details: "Is it possible that the reticulum shown in the cells more adequately fixed with osmium may represent channels of motion for the extremely small granules?" He also worried whether artifacts, particularly in the appearance of the ground substance, were being produced by the fixations. Consequently, he suggested they simply "state what we see with the minimum of theoretical considerations." Since Claude was urging that the manuscript reach *The Journal of Experimental Medicine* editor Peyton Rous by 15 August so it would appear in the November issue, Porter suggested they leave for the immediate future other experiments to resolve the fact versus artifact problems.[85]

By early September, Porter was still writing and admitted the lingering effects of his own TB infection were slowing him down. Reacting to Claude's description of electron microscope workers from other institutions who had recently visited Woods Hole—Arnold Lazarow, A. Glenn Richards, and Francis Schmitt—Porter admitted he was anxious that they "not be superseded on the publication in spite of the delay.[86] After two months of revising back and forth, their manuscript was submitted on 11 November and published four months later, in March 1945.[68]

The opening lines of the article herald their success that the cultured cell could be prepared for electron microscopy by relatively simple means and that electron micrographs rendered structural details not previously seen inside cells: "In view of these findings it seems probable that this new technique will make possible the study of the morphology, composition, and behavior of cell components which, because of their small size, are beyond the resolving power of the ordinary light microscope."

This article contains what Porter called the "most micrographed cell." The image is of a fibroblast-like cell cultured from chick embryo tissue. Actually, the printed image is a composite of five electron micrographs at a magnification of 1,600. The field of view in the RCA EMB was not large enough to include the entire cell, so five individual pictures were joined together to form the single cell. To emphasize the light microscope's limited resolving power, a photomicrograph of the same kind of intact cell was taken and printed alongside this electron micrograph, at a magnification of 900 and enlarged to 1,250. The comparative images also made a convincing case that the electron micrograph represented the same entity and had introduced no artifacts (figure 2).

The lone cell in this famous electron micrograph emitted much more structural information than Porter, Claude, and Fullam expected, and they were confronted with how to interpret and describe what they saw. In place of the typical "Results" section in a paper, they provided instead an "Observations" section, noting, "It is not at present possible to do more than point out the more obvious details that appear in the various parts of the cell." Some of their first glimpses inside the subcellular terrain remain cautiously exacting:

> Cell Outline: variety of irregularities . . . associated with migration . . . jagged points under tension . . . finger-like processes not under tension . . . tentacles active in advance of migrating cell borders
>
> Ground Substance: particulate elements . . . small particles, or microsomes . . . lace-like reticulum . . . vesicle-like bodies
>
> Mitochondria: impressive clarity . . . masses of greater density and variable size appear distributed along the body
>
> Golgi Bodies: dense bodies characterized by rather angular outlines
>
> Cell Nuclei: too thick . . . for detailed electron microscope study
>
> Intercellular Relationships: claw-like processes . . . not at present interpretable

The authors conclude optimistically and with keen awareness of a new world in science that had just opened to them: "Micrographs taken at the conservative magnification of 1,600 are so packed with detail that the smaller clearly resolved structural features are not easily seen until the pictures are enlarged two or three times." Based on these spare, careful observations, they predicted that the fidelity of novel cell structures visible at such high power would advance biological electron microscopy.

After a Landmark Paper, a New Microscope for Cancer Research

Once the manuscript for the cultured cell paper was submitted to *The Journal of Experimental Medicine* in late 1944, there was renewed pressure to visualize a malignant cell that could be compared to the normal cell they had just studied. Porter wrote anxiously to his Ray Brook colleague Diran Yegian that "we are obliged because of our connection with a cancer research laboratory to get things going at once on the cancer cells. You know, there may be an agent there. . . . I don't expect to find anything unusual but the search is on and has become the problem of the hour. It has meant a reorganization of mind and work and many headaches. When you get yourself involved in something like this much of the fun in research is replaced by difficulties with people. Someone is usually keeping you stirred up by interfering, rushing, criticism or jealousies."[87]

Fortunately, Porter continued, the laboratory's facilities had improved. The International Health Division (IHD) of The Rockefeller Foundation installed an RCA EMU-1 electron microscope in the Institute's Old North Building (now Flexner Hall). This was the physics laboratory of ultracentrifuge designer Edward Pickels, who was also charged with applying

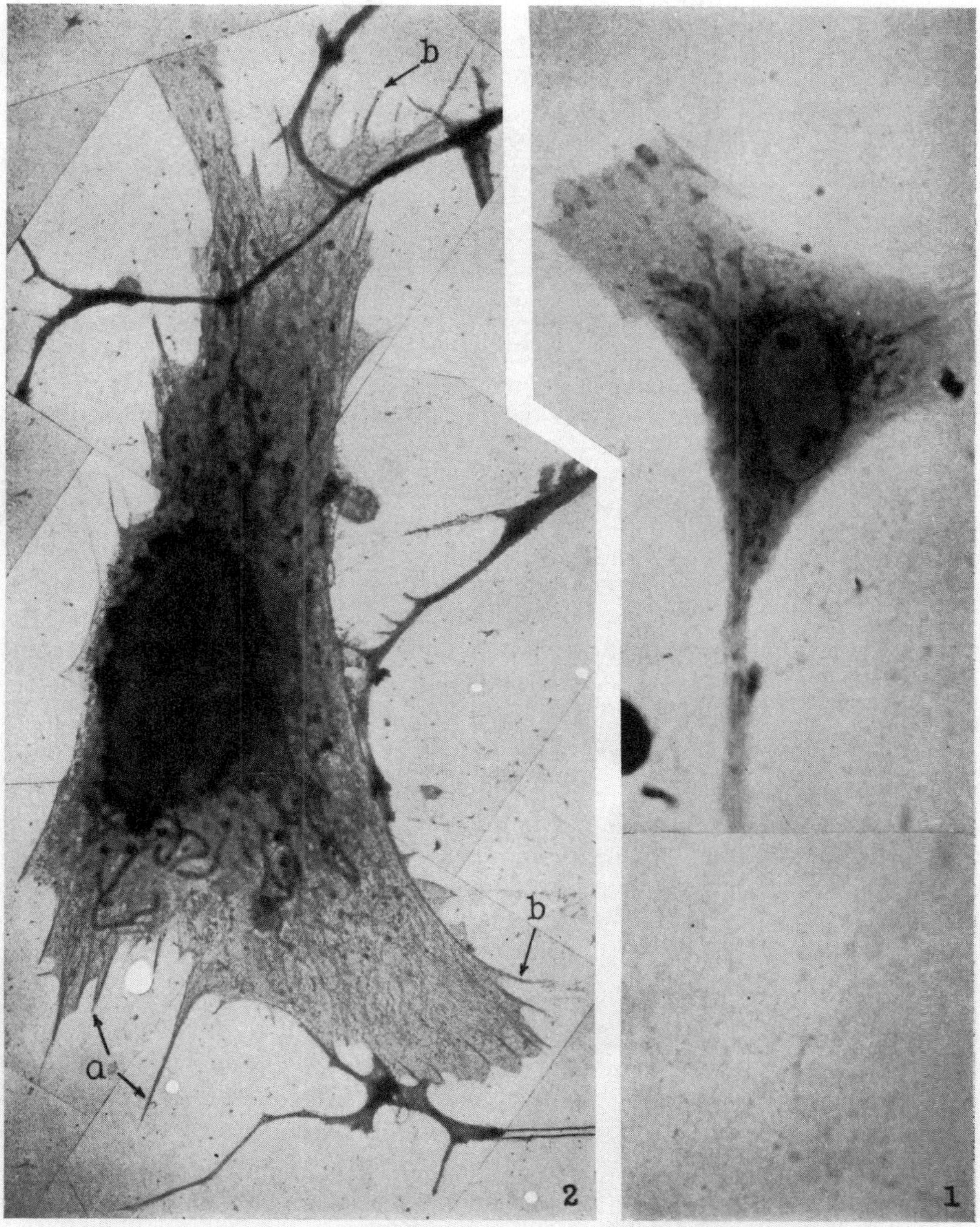

2. Left: first electron micrograph of an intact cell, a fibroblast from a chicken embryo, at a magnification of 1,600. This figure, a composite of five micrographs, was taken on 6 July 1944 with an RCA EMB microscope. Right: light micrograph of an intact cell for comparison with the electron micrograph, at a magnification of 900 and enlarged to 1,250. From Porter et al., 1945. *J. Exp. Med.* 81:233–246.

yet another new tool to virus research.[88] Porter had seen this microscope in operation and told Yegian that "as soon as several little wrinkles are ironed out we shall be given access to it and will probably get something done."

For many years, the IHD had studied physical properties of viruses by indirect means. Discouraged by all the ingenious ways of getting at size and mass, including the ultracentrifuge of Pickels, "the investigators were still in the dark as to the actual structure and appearance of the objects. They wanted to *see* these bodies that they had been filtering through collodion films, whirling out of solutions, and pulling out of mixtures by electrical attraction." For this purpose, they invested in an electron microscope that they hoped would "serve as a weapon against the viruses and other hidden agents of disease."[88]

Almost immediately, the IHD scientists sought the collaboration of Claude and Porter because they were overwhelmed in learning the instrument's "many peculiarities . . . and many refinements made in the matter of preparing mounting films and the specimens themselves." Confronting difficulties that would puzzle virus microscopists for many years, they found it nearly impossible, like Stern and Anderson in their earlier viewing of the Rous Sarcoma Virus, to know how to clearly differentiate particles representing animal virus from those representing cell debris, aggregated protein, and other extraneous materials. Moreover, they were aware that due to "the way in which specimens must usually be prepared, there is great irregularity in the distribution of virus particles on the mounting film and there is no satisfactory method of establishing any direct quantitative relationships for particles of a given type, between the number seen in a particular photograph and the number contained in the fluid specimen."[89]

Although the IHD scientists were not cancer researchers, they announced they wanted to learn "through electron microscopy something about the invasion of cells by virus and the multiplication of the virus within the cells. Since Drs. Albert Claude and Keith Porter of the Institute had already worked out a technique of growing tissue cells directly on the thin mounting films within a culture medium, the invitation to collaborate in the extension of their studies was accepted."[89]

Nothing should have seemed more logical to Porter than to follow up his earlier experiment by looking at a cultured Rous sarcoma cell in the electron microscope. As early as 1911, Rockefeller's Alexis Carrel claimed to have cultivated the Rous sarcoma and found that while tumor cells had no outstanding properties compared with normal cells, they produced tumors when grafted.[72,90] Rous, as mentioned, tried but could not repeat Carrel's experiments. Fifteen years later, Carrel made a systematic attempt to cultivate the Rous virus in cells for long periods and claimed the cells retained their malignancy indefinitely.[91] Nevertheless, like other mysteries of Carrel's experimental procedures, these techniques ended with his retirement in 1939.

Porter disregarded these precedents in his need to establish the cultured cell as a reliable test system for electron microscopy. From late 1944 through 1945, he made and recorded dozens of experiments in laboratory notebooks labeled simply "T.C." with the experiment number and date.[32] He made endless variations on improving methods (clots versus plasma), making comparisons with light microscopy, varying fixations, and trying organic compounds to act as "stains." The process to assemble various media, maintain cultures, estimate cell growth, and interpret artifacts under such scouting conditions was extremely labor intensive and time consuming. Porter also examined a considerable variety of cell types—from fibroblasts, myoblasts, macrophages, and mesothelial cells of chicken, to kidney and skeletal cells of mouse and rat, to corneal cells of rabbit, and to human fibroblasts and a variety of human tumor cells.[92] His culture technique was starting to justify its further development, but innumerable tasks remained.

Beginning in 1946, Porter's work was advanced by Murphy, who obtained a special research grant, the first outside funding of this kind in the history of the Institute. In 1944, the American Society for the Control of Cancer was being reorganized by the influential industrial and advertising leaders on its board and was renamed the American Cancer Society (ACS). Its board outlined a multimillion dollar plan to support cancer research that would be financed from an annual national fund drive. Murphy, an ACS board member who contributed considerable effort to this reorganization, noted in 1945 that "over a million dollars will be available for research this year and several times that from next year's drive."[93] To acquire scientific credibility and to avoid conflict of interest problems, the ACS asked a governmental agency, the National Research Council (NRC), to establish a scientific review group to advise how to allocate their funds in awarding research grants.[4,5] In 1946, Murphy resigned from the ACS after turmoil between the business executives and the scientists, and he joined the NRC's Committee on Abnormal Growth (later changed to Growth, or COG). Shortly thereafter, he sent a research proposal to Cornelius P. Rhoads, chairman of the committee. Rhoads, who succeeded James Ewing as director of Memorial Hospital, had just been appointed director of its newly founded affiliate, the Sloan-Kettering Institute for cancer research. Knowing that Rhoads favored a scientific approach to the cancer problem and buoyed by seeing the first micrographs of tumor cells emerging from his laboratory, Murphy proposed a joint project between Keith Porter and Memorial Hospital to examine "the structure of normal and malignant human cells as revealed by electron microscopy." Memorial physicians would obtain and culture cells using Porter's procedures, while Porter would prepare the grids and examine and interpret the micrographs.[94]

In December 1946, the ACS awarded Murphy a $15,000 grant for "studies of cytoplasmic particles in neoplastic disease." The one-year grant was renewed in 1948 and 1949 (eventually totaling $43,000) and was supplemented with two postdoctoral fellowships in 1949. The grant funds were used for technical staff, accessories, and minor apparatus related to the electron microscope, at first the IHD microscope and then a new instrument purchased in 1948 (the microscope story continues in Chapter 5).[95] For The Rockefeller Institute to be a grant recipient was not that surprising since Murphy could specify only a dozen major "cancer research groups with coordinated programs" in the United States at the time.[96]

Nevertheless, accepting outside funds went against the Institute's founding objective to support its research programs solely from its endowment. This ACS grant caused the administration to take extraordinary measures. At its 18 January 1947 meeting, the Board of Scientific Directors passed a formal resolution of recommendation to the Board of Trustees that the grant be accepted. In response, each trustee on the executive committee had to sign a written statement approving the resolution.[97] The following year, when the grant was about to be renewed, Gasser was obliged to ask John D. Rockefeller Jr., chairman of the board, to "approve of the acceptance" of the grant. Gasser also reassured him, "Experience has afforded no new information negating any of the reasons for the acceptance of the first grant. The fund is used in such a way that its discontinuance at any time would not prove to be embarrassing."[98] For his part, Murphy was pleased that the COG succeeded in developing and directing cancer research on a broad basis. He thanked its chairman Lewis

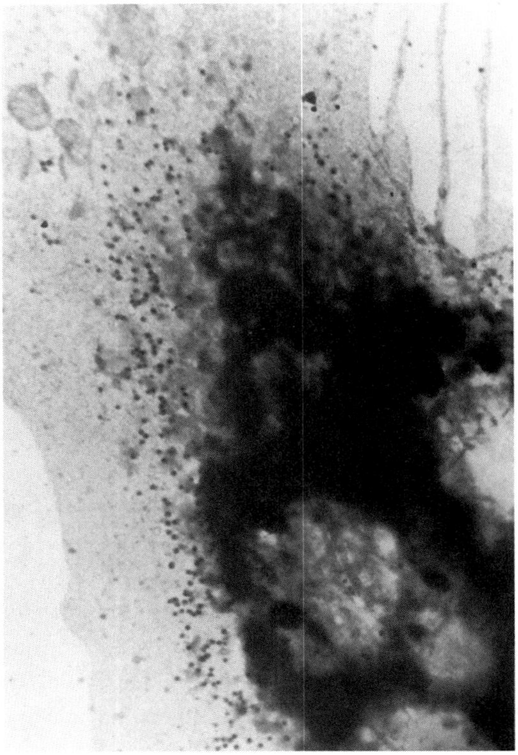

3. First electron micrograph of an intact cell, also a fibroblast from a chicken embryo, infected with chicken Tumor 1, the causative agent of the Rous chicken breast sarcoma (RSV). From Claude et al., 1947. *Cancer Res.* 7:421–430 with permission from the American Association for Cancer Research.

H. Weed "not only for relief of inadequate support but doubly so because it has broken down the unfortunate policy of the Institute about accepting money from outside sources."[99,100]

Amid experiments with the capriciousness of tissue culture, Porter was able to refine his original techniques and produce four publications showing different cultured cancer cells: chicken sarcoma, rat sarcoma, a mammary carcinoma of mice (involving the Bittner milk virus), and a human Hodgkin's tumor. The resulting micrographs of these specimens acknowledge the use of the IHD electron microscope; Pickels participated in the first two studies before leaving the IHD in 1946, while Claude worked only on biochemical analyses of chicken sarcoma cell fractions. When Porter then took over the microscope, he credited Pickels's cooperation for making it "possible to proceed fairly rapidly in our examination of the finest detail of cell structure. Much of this detail has been new and so has created problems of interpretation which could only be solved intelligently by examining a mass of material."[101]

The most important micrographs from this period were those of virus-infected cells from two chicken tumors, Tumor 1 (Rous Sarcoma Virus, RSV) and Tumor 10 (isolated by Murphy in 1931). While they had been well characterized by others in Murphy's laboratory during the previous twenty years of study, this was the first microscopic study. Images of RSV-infected cells revealed discrete particles scattered in doublets or chains throughout the cytoplasm and at the edges; these particles had a diameter of 67–80 nanometers, which corresponded in size to the particles isolated earlier by Claude using graded pores (figure 3). In contrast, cells infected with the slower growing Tumor 10 revealed individual particles 60–70 nanometers in compact crystalline-like arrays, forming patches like bacterial colonies; they appeared to be on the surface of the cell. The particles found in both tumor types were further distinguished from Claude's microsomes by their rough shape, uniform size, and greater density.[102]

In many aspects, micrographs of malignant cells presented more problems of interpretation than those of normal intact cells. It was hoped the images would show something about the identity, localization, and structure of viruses inside a cell. Discouragingly, the numerous virus-like bodies gave enhanced but not convincing evidence of a viral cause for tumors. The infected-cell micrographs were nevertheless notable because Porter used shadow-casting techniques, introduced by Robley Williams and Ralph Wyckoff in 1944.[103] This technique improved their ability to calculate the particles' size and location in the cell and to compare their micrographs with those of Bittner virus particles taken by other electron microscopists.[104,105]

To prove these particles were not artifacts, Porter went on to take micrographs of a variety of other cells, including those from mammalian tumors. After comparing 450 micrographs of normal with malignant rat sarcoma cells, Porter and Helen Thompson found more "endoplasmic granules" and dense bodies in malignant cells than in normal cells and cautiously linked them to a rapidly growing malignancy.[106] They also observed virus-like particles inside carcinoma cells infected with the Bittner virus, although they did not publish comparative images of normal cells because their results were too inconsistent.[107–109] In one last futile effort to study cancerous tissues, Porter and John Jacquez examined tumor nodes from patients with Hodgkin's disease; when they found no virus-like particles in the nodes and no differences between normal and diseased cells, they offered the cautious speculation that the images "favored the nonviral etiology of Hodgkin's disease."[110]

Controversial Interpretations

Despite an improved microscope giving a better resolution of cells, Porter was keenly aware that all the micrographs were fraught with problems imposed by the novel materials and methods. Although his numerous experiments convinced him the virus-like particles were not artifacts, he acknowledged that any interpretation of the differences between normal and malignant cells remained "highly speculative."[104]

From the outset, there was controversy. Within Murphy's laboratory, the scientists created more questions than answers in comparing normal and tumor cells. Their cautious interpretations of so many new details in the micrographs took place in a vacuum. They had no previous knowledge to draw on, no experience with this kind of virus research, and no one to guide or review their scrutiny.

Some issues, which may be difficult to appreciate today, were evident to the laboratory:

- Morphological identification of virus particles was not yet possible, so they could not be reliably differentiated from other still unknown or unidentified components of a normal cell. Even virologists of the time questioned whether the particles were the cause or result of the malignant transformation.
- The action of chemical fixatives on the cells was unknown, so environmental factors of temperature, light, and concentration had to be controlled from one experiment to the next.
- The thickness of sarcoma cultures precluded more descriptive statements about whether the particles were located in the cytoplasm or nucleus.
- Using natural media from chickens introduced the possibility that other intercurrent avian viruses would also infect and appear in the cells.
- Porter and Hogeboom introduced an isotonic solution of calcium ions as a vital stain for mitochondria in 1946, but its effects could not be calculated, and these electron microscopy studies went unpublished.[111]

These uncertainties were compounded by a troublesome incident concerning Claude's handling of publishing their data. He and Porter presented some very preliminary observations and micrographs on both chicken and rat sarcoma cells at the March 1946 meeting of the American Association for Cancer Research in Atlantic City.[112,113] But these findings were not published for another year and a half, in July 1947.[102,106] In the meantime, for unknown reasons, Claude on his own wrote and submitted a manuscript on the chicken sarcoma work to Peyton Rous, editor of *The Journal of Experimental Medicine*, without first showing it to Murphy. This action violated an Institute practice that manuscripts had to be approved by the laboratory chief before submission to a journal.

In brief, Rous completely rewrote the manuscript, penning his comments and corrections in between the double-spaced lines. He removed many of Claude's speculations, and in place of Claude's conclusion that "the tumor agents appear to possess properties commonly attributed to viruses," Rous substituted what for him—despite his unwavering belief in the viral cause of this sarcoma—was a more circumspect sentence: the bodies seen inside the chicken tumor cells, he wrote, "have all the appearance of extraneous bodies."[114]

When Murphy learned what had happened, he further revised the manuscript and submitted it to *Cancer Research*, making Claude, Porter, and Pickels its authors. The day after it was received for publication, Murphy berated Claude in a letter for attempting to "by-pass" him. He expressed anger that the draft had not credited the early work of Rous and Murphy on the tumor agent but simply began with Claude's purification of the agent. "If you did this," Murphy wrote, "on the mistaken idea that I would want to hide the fact that in the past, on the evidence available, I was opposed to the acceptance of the virus theory, you inexcusably misjudged me." He further accused Claude of several instances in which he had "resorted to devious methods to avoid giving references to my papers" and called this one "the culmination of a series which convinces me that you are resentful and disloyal to our group." These omissions, Murphy added, had been taken care of in his revision of the paper.[115]

Claude's reply, two months later, was that the paper dealt with morphology of the cells and should "not be padded with irrelevant matter." He further contended it was not fair for Murphy to have chosen this paper to correct his views some thirty-five years later: "As far as I am concerned, I am afraid I cannot assume the responsability [*sic*]."[116,117] Murphy seems to have reasoned more broadly in what turned out to be his final utterance on the chicken tumor agent: his revision not only incorporated the original sarcoma studies and took most of Rous's suggestions on the manuscript, but he also reinstated the one idea Rous had removed. Going beyond Claude's limited interpretation of shapes in the micrographs, Murphy drew a fuller conclusion (than Rous or Claude) that was based on the entire spectrum of evidence going back to the original 1911 studies. His final sentence, as published, reads "*In all these particulars*, the chicken tumor agents appear to possess properties commonly attributed to viruses" (emphasis added).[102]

Outside the laboratory and the Institute, the initial and lingering reaction to Claude and Porter's electron microscope studies was generally one of disbelief that viruses were visible in cells. At the time, the electron microscope was still not an accepted instrument for biological study. Nor was the cultured cell yet accepted as a reliable test system. Tools to make definitive identification of viruses were still missing. There had been no effort to observe the same type of cell before and after infection, so there were no experimental controls; there was also no way to follow cellular pathogenesis and changes in form of the assembling virus particles. The micrographs were so novel and advanced in cancer science that they were years away from being confirmed by other scientists.[118]

The presence of particles in Rous sarcoma cells was not confirmed by others for six years. Even then, virus particles were scarce, not all the cells contained particles, and the particles were found in control animals, suggesting the viruses might be present in a latent, asymptomatic form in healthy animals. Moreover, in size and morphology, the spherical particles observed in Rous sarcoma cells were nearly indistinguishable from viruses causing other chicken diseases.[119,120]

As for the micrographs of the Bittner virus–infected cells, Robley Williams was still skeptical twelve years later that Porter had identified the cause of the mammary tumor. He observed that questions of identity depended "upon the development of sensitive and reliable bioassay systems and cytochemical techniques that will enhance the discrimination with which morphologic similarities can be established."[121]

As late as 1960, Françoise Haguenau and Leon Dmochowski observed that in electron micrographs the virus-induced tumors all resembled each other and the number of visible particles related to the potency of the extract used in the experiment.[122,123] Despite an unmistakable presence of virus particles in tumor cells, as Porter astutely reported to Murphy in 1946, they were not destroying the cell so that differences between normal and malignant cells were "differences in quantity rather than quality" and expressions of "differences in growth activity."[92] Even at this early stage, his guarded interpretation was that no subcellular structures seen with the electron microscope could be attributed exclusively to tumor cells.

In several significant ways, the rift over the chicken sarcoma paper changed the course of science. Personally, it was the beginning of the end of Claude's tenure at the Institute, as Murphy's angry letter also urged him to find a position in Europe where "you could come nearer to living the way you like to live, you would be your own boss, and would be away from a position in which you are apparently not happy."[115] Perhaps more understandable, the 1947 paper with its micrographs of chicken tumors reported the last experiment on Rous Sarcoma Virus at Rockefeller for nearly three decades, until the arrival of molecular oncologist Hidesaburo Hanafusa in 1973. Scientifically, however, because Murphy believed that tissue culture and electron microscopy would open a new field of investigation, he acquired prestigious outside collegial and financial support to help ensure its future. Above all, the electron microscopy of malignant cells, which ceased on Murphy's retirement in 1950, concluded yet another phase of cancer research at the Institute. Its end became an auspicious beginning of modern cell biology.

A Second Milestone in Cell Biology

The significance of electron microscopy at this formative stage was to provide convincing views of structures inside a cell beyond the realm of biological entities then known to exist. The light microscope had shown a few distinct bodies inside the cell (nucleus, mitochondria, and Golgi apparatus) and suggested the cell was a liquid bag of enzymes with the bodies floating in it. The electron microscope yielded a very different picture of a unit of highly organized semisolid gel holding together a remarkable universe within itself. The discovery of a lacework of strands and vesicles was the first evidence of the "invisible" organization of a cellular ground substance (later known as the endoplasmic reticulum), which was to have important implications for the future course of the nascent cell biology.

In introducing electron microscopy in cell research, Porter more than kept his initial promise to Murphy that he had "the eagerness, determination and sincerity of purpose which makes rapid learning possible."[13] During his five-year evolution from frog embryologist to tissue culturist to electron microscopist, he acquired an advantage over his contemporaries of an in-depth knowledge of the objects of interest. While using the light microscope, he developed keen skills in manipulating live tissues and discerning true artifacts. When the electron microscope challenged these notions, he continued to be guided by the premise that controlling the quality of the picture was directly related to methods used to prepare cells.

There was no other way for these biologists to confront the complexity of cells visualized in the electron microscope until there were standard methods of making and reading electron micrographs. Porter's special contribution reflected a natural sense of biological structures that allowed him to select specimens and devise methods that would measure up to these goals. Like Claude, he exercised gentle, careful handling of biological specimens so there would be a full accounting of the cell. With a minimum of treatments, he produced a specimen having the essential quality of thinness and a near absence of artifacts that could be examined intact and in its entirety.

Despite persistence and inventiveness, Porter's early thinking about electron microscopy was conservative. As he observed at a 1947 Institute Friday staff meeting, "From these very crude beginnings it may be possible eventually to build more accurate concepts of cell morphology and thereby provide a framework around which to construct explanations for the highly integrated cell functions."[124] Although Porter's early experiences with these Lilliputian worlds enabled him to achieve such early success, his manipulation and mastery of cell preparations would improve dramatically during the next decade (and throughout his scientific career).

These early gains were beset by other problems in interpreting the objects and relationships of cell structures in the micrographs. One problem was settled by the Murphy laboratory. Claude's "microsomes" were observed in both normal and malignant cells, and they were clearly distinct from the virus-like particles in infected cells. This observation ended his theory, and Murphy's, that the microsomes were the chicken tumor agent. More problematic for the scientific world, the normal and malignant cells looked just the same except for the presence of virus-like particles. And while the particles pointed to a difference between these cells, there was still more to find. In other words, the *formal* genesis of cancer was still invisible. Seeing structures was not sufficient.

Nonetheless, at this stage, it was not necessary to accept the identification of viral particles to value the more significant discovery of a new world that had just opened. The resulting micrographs set standards of the day for showing more structural details than the corresponding light microscope pictures and, for the first time, the presence of virus particles in infected cells. It was not an overstatement to say these images marked the beginning of the electron microscope era in cell biology. In 1995, fifty years after the landmark 1945 paper with the first electron micrograph of an intact cell, George Palade pronounced it "the birth certificate of a vigorous baby that was baptized a few years later Cell Biology."[125]

During the immediate aftermath of the 1947 tumor cell paper, the question propelled by the electron microscope was what, if anything, could scientists possibly see in a cancer cell when they still did not know what the inside of a normal cell looked like. The new morphological studies needed additional bridges to understand cell functions. Fortunately, they were right at hand in Murphy's laboratory. The first milestone was Claude's isolation of subcellular components by cell fractionation. The second was electron microscopy that could map the cell's natural landscape. The third milestone was determining biochemical composition and measuring biological activities of cells; this phase, which had been initiated by Claude, was energized by the arrival of George Palade in 1946 and is described in the following chapter.

Even before Murphy retired in 1950, the laboratory once again renewed his persistent vision that any attack on cancer had to come from first understanding the anatomy and chemistry of normal cells. The new science was barely recognizable, unnamed, and amorphous, but it was on its way.

Chapter 4

Stalking Chemicals inside Cells

Carol L. Moberg

> It does not matter who is right; the only important question is, what is true.
>
> Robert Russell Bensley

Anatomist Robert Russell Bensley, who began studying cells in the early 1900s, was known for warning that subjecting cells to harsh treatments could disorganize or distort them. According to colleagues at his seventy-fifth birthday symposium in 1942 at the University of Chicago, Bensley liked to compare the cell to a watch. A watch, he said, cannot be understood by looking at its closed case, observing the hands go round, or grinding it up to analyze the various amounts of metals inside. However, as Arnold Lazarow commented, "the watchmaker knows each part separately; he knows that the proper combination of copper and beryllium will give the mainspring its necessary resilience, and he knows how each of the pieces fits into the whole."[1] Furthermore, according to E.S. Guzman Barron, "Did not Professor Bensley do some watch-grinding himself when he proceeded to isolate mitochondria from the liver" to understand the structure, composition, and function of a cell's individual parts? At the time of the symposium, Bensley's laboratory was actively engaged in studying an oxidative enzyme system in the mitochondria.[2]

James Murphy's objective in the laboratory since Albert Claude's arrival in 1929 had been to find a chemical difference between normal and cancer cells. This intent was rekindled by Claude's identification of similar particles in tumor and normal cells in 1937. By 1941, two of the three approaches to studying cancer advocated by Murphy, in response to the National Cancer Act of 1937, were under way in his own laboratory. Claude was preoccupied with the mechanical separation of cells into fractions that could be chemically analyzed. Keith Porter was enmeshed in methods with growth and differentiation in frog embryos and transformations of normal cells in tissue culture. The third approach was initiated in 1941 when Murphy hired George Hogeboom for a series of comparative biochemical tests.

Chemical investigations of enzyme systems in the cell exemplified a trend that was just gathering momentum among biologists. This approach seemed like a logical extension of Murphy's earlier enzyme and transmissible mutagen theories in which a harmless ubiquitous cellular chemical becomes transformed into a carcinogen. His vague theories of enzyme systems were invigorated by the major discoveries of Rockefeller colleagues John Northrop, who characterized pepsin and trypsin, and Wendell Stanley, who was working with tobacco mosaic virus. Both devised biochemical methods for isolating and crystallizing proteins and enzymes. Their findings suggested an intriguing analogy between viruses and enzymes due to their ability to act as chemically transforming principles within the cell.

Not surprisingly, about this time, Peyton Rous was also becoming attracted to the concept that viruses seemed related to enzymes. In 1940, he wrote to Herbert Gasser about a manuscript by Northrop called "The quick and the dead" and remarked that "if viruses are enzymes, existing in the tissues as peculiar, inert proteins, like the precursors of pepsin and trypsin, then certainly the place to look for these peculiar proteins is in tumor cells. Ordinary infective viruses cannot be evoked from cells by any means now known, nor is there any evidence that they are fabricated by them under natural conditions. Yet by an immense diversity of methods cells can be rendered neoplastic, and under conditions which completely rule out the immediate introduction of an infective agent from outside."[3] (Northrop's essay was not published until 1946.[4])

Murphy was also influenced by the work of German biochemist Otto Warburg, a Nobel laureate in 1931 who brought a flowering of research on enzymes in cellular metabolism and respiration of tissues. The manometric apparatus for respiratory studies, which bears his name, was becoming a common instrument of biochemistry. Warburg extrapolated from his work on anaerobic glycolysis of cancer tissue to state that cancer was caused by insufficient oxygen in the cell. This controversial theory, which Murphy had dismissed in 1925 after doing experiments on the chicken tumor agent, nonetheless stimulated many scientists in the 1930s to take up the biochemical problem of respiration as the basic change in the malignant transformation of cells. A primary achievement of anatomist Bensley in 1935 was isolating mitochondria to measure their oxidative enzymes. It became evident that Claude's newly isolated fractions needed similar analyses.

A more direct influence came from a group of physician-scientists in the Institute's Hospital who had become fascinated with the possibilities of enzymology. In 1940, organic chemist Walther Goebel brought biochemist Mark Adams from Columbia University. The laboratory organized a seminar series to explore the literature in this field, using as a text the monograph by British biochemist David Green, *Mechanisms of Biological Oxidations*.[5]

Energized by these research initiatives around him, Murphy hired George Hogeboom as his biochemist. Hogeboom trained as a chemist before getting his MD in 1939. When Murphy first met him at Johns Hopkins, he was surgical resident on the Halsted Fellowship. About this time, Hogeboom's father died of cancer, an event that led him to write Murphy on 13 September 1940 reporting he had become "primarily interested in research, particularly research in cancer."[6] He joined the laboratory in September 1941 to work directly with Murphy and Sturm, and his first chemical experiments were respiratory and permeability assays on normal and leukemic cells. He also collaborated with the Hospital's hematologist Cornelius P. Rhoads on in vitro chemotherapy assays of tissues from patients with lymphosarcoma. Then he and Mark Adams studied two enzymes involved in melanin production. Their work on mammalian tyrosinase and dopa oxidase used Claude's fractionation techniques and indicated that tyrosinase was associated with the newly discovered "microsome" fraction.[7–9]*

Only ten days after Claude returned from the Bensley symposium in 1942, he embarked on a search for enzyme activities (not single chemicals) in the cytoplasmic fractions he was isolating. The meticulousness that he brought to developing and interpreting his

*By late 1941, David Green, then at Columbia University, Charles Hoagland at Rockefeller, and Adams (who had moved to New York University) formed the Enzyme Club of New York City, a select group of biochemists who met monthly for fifty years. Adams gave the club's first talk, 14 January 1942, on the tyrosinase work with Hogeboom.

George Hogeboom, 1942.
Courtesy of The Rockefeller Archive Center.

Walter Schneider, 1946.
Courtesy of The Rockefeller Archive Center.

centrifuge experiments applied equally to his new quest to understand the activities of cell fractions. Even though Claude was conversant with many chemical methods of staining and identifying various elements and molecules, he knew little biochemistry and nothing about enzymatic assays. He initially enlisted the aid of his colleague Rollin D. Hotchkiss. A few months later, Hogeboom joined this search and began playing a key role in yet another scientific beginning in cell biology.

Looking for Functions in Cell Parts and Particles, 1942–1944

by Rollin D. Hotchkiss

A biochemist at The Rockefeller Institute, surrounded daily by physicians, microbiologists, and immunologists, I was likely to go alone to lunch in the staff dining room and at in-between times, determined by some timed experiment in which I was involved. I enjoyed joining some subgroup of the Michaelis-Landsteiner-Levene-Bergmann departments and often came into an ongoing discussion of a current experiment or publication from which I could learn a lot while trying to sense what had gone on before I got there. Medical questions, viruses, bacteria, and physical chemistry could be met at other tables. Science, politics, and administration were a frequent topic of conversation of James B. Murphy and his small group, which included my friend Keith Porter, whom I often joined. It was there that I first began to talk with Albert Claude. He often would ask politely about my work and, in a turn about, talked modestly but not evasively of his work with tumor particles.

It was thrilling-chilling to hear Claude tell about particles—indistinguishable from similar ones in normal chicken tissue cells—that would transmit and reestablish a classical tumor when introduced into normal chickens. I knew that others, including Peyton Rous, held the view that viruses (not identified as "particles") were the typical agent transmitting other tumors. Biologists were well aware that different tumors presented different phenomena. But nevertheless, in the 1940s, many scientists expected an eventual unity to show up in what was widely called "the cancer problem" and some kind of tumor factor to underlie all the types of unregulated growth.

One difficulty was that virus (filterable) and particle (centrifugable) seemed then to be alternative definitions requiring an either/or identification for the different tumors. Actually, laboratories using bacterial filters to achieve sterility with viruses considered pathogenic were unlikely to use the reckless ultracentrifuge techniques that were beginning to be employed in work with ones less dangerous for humans. Some plant viruses had just been isolated as crystallizable proteins in the 1930s, but most medical researchers considered viruses to be bacteria, tiny enough to be filterable. That was mysterious enough, although at this time medical researchers had not widely recognized the elegantly differentiated metabolic capabilities that bacteria enjoyed. Indeed, René Dubos's *The Bacterial Cell* in 1945 was for some a more ground-breaking revelation to a wider audience perhaps than Marjory Stephenson's *Bacterial Metabolism*.[10,11]

This period was a prime time to begin looking for function in the formed elements of cells. While microscopy was continuing to probe into ever finer details of structure, some biologists had begun to realize that merely describing a growing catalog of membranes, walls, granules, and fibers was not enough to explain the dynamic processes of cell physiology. At the Woods Hole Marine Biology Laboratory in 1935, I had the good fortune to meet visiting scientist Heinz Holter, and for a summer experience I volunteered to help him determine aminopeptidase of sea urchin eggs in the successive layers produced by centrifuging the intact eggs. Certainly an amateur then, I was happy to be thanked in the eventual publication for my assistance.[12] It resulted in my being invited to Holter's home laboratory—the Carlsberg Laboratory in Copenhagen, Denmark—two years later. There I learned more about such new approaches to this biochemical interface and hence was conceptually prepared to find functions associated with structures wherever I looked.

This rapprochement in biology was still very new at the macromolecular level among the biochemists. Preferring the advantages of homogeneous catalysis, they sought the soluble dispersed enzymes, working in an uncomplicated environment. Even so enterprising a biochemist as David Green, ready to engage any cell and face any biologist, classified a large group of enzymes as associated with structure and (by implication) unavailable for study.[5]

Into this situation I willfully plunged one day in November 1942. At that time, I had been left in the custody of Director Thomas M. Rivers, when my chief, René Dubos, moved for a time to Harvard University. On the third floor of Theobald Smith Hall (to which I had moved with Dubos in 1940), in the vestibule before my laboratory, were three large refrigerated compartments. Albert Claude from the floor below had some time before mounted in two of them a standard International centrifuge with high speed attachment. He would arrive there quietly on crepe-soled shoes, with or without a friendly greeting, busily start centrifuging his cell homogenates, and usually soon disappear. On one of the intervals, however, he wistfully asked me, "Do you know how to analyze for enzymes?" I replied that there are many enzymes; which ones interested him? He said, "Well, such things as trypsin, or pepsin . . . ?"

But his question and answer were not as innocent as he let them sound; he surely was aware that I was busy exploring the effects of antibiotics on bacterial metabolism. More completely presented, his postulate was that the subcellular particles he was isolating were the "factories of the cells" (his very words), and he was willing to cooperate on any enzymological analyses a colleague would help him with. Much later I found out that he'd recently made inquiries of Van Potter of the University of Wisconsin regarding enzymes. And he'd already arranged elsewhere for quantitative analysis of the elements (carbon, nitrogen, phosphorus, etc.) of cell particles and some search for the nucleic acids, RNA and DNA.[13]

As a chemist, I'd often been approached by the biologists for help or advice. In the free environment of Rockefeller, this was the historic way collaborative projects began, although it could also be a first step down the garden path of over-involvement. But Claude's open inquiry seemed to be a real opportunity for me as well, and I soon agreed to do some exploration. This of course was in the middle of World War II. Already in the uniform of the Naval Reserve since the Pearl Harbor attack of 1941, I felt obligated to continue my own antibiotics work and another small war research project I'd generated.

The daily pattern of Claude's particle purifications allowed me at first to cooperate by merely extending occasional work days into the late night hours. I started looking in cell particles, not systematically, but in limited fashion, for those enzymes and components that I felt able to handle, at first using mainly Warburg respirometry as the tool.

My first experiment was on 23 November 1942. About 4 p.m., Claude brought me a suspension of what he called large granules freshly isolated from guinea pig liver cells. And by 11 p.m., I'd shown them to have a significant ability to oxidize the unnatural D-leucine and D-phenylalanine isomers, but not the L- or natural ones. Centrifuging out the small particles took longer, and that time I tested them only the next day, finding no D–amino acid oxidase. I began a series of experiments, thereafter bringing the small particle fraction into a late run starting around midnight and running on into the early morning of the second day. These assays and others on the same pattern ran around the clock from about 4 p.m. to 4 a.m.—all after a regular day's work. The pace was hard on family life but not atypical of wartime projects.

My records show at least nine successful repeats of this finding on rat and guinea pig livers, with additional evidence that the D–amino acid oxidase as expected produced ammonia, using flavine-adenine cofactor that seemed to be optimally adequate in the cell fractions. Only D–amino acid isomers were oxidized. Moving on to other enzyme assays, I found active oxidase activity on succinate and malate but no sign of components of the Embden-Meyerhof carbohydrate-using system. Considerable catalase activity was demonstrated in the large granules, but to a much lesser extent in small particles. Claude was delighted by these things and made preliminary statements about them at scientific meetings.[14,15]

I was vaguely aware that Kurt Stern, then at Yale, had reported some oxidase activities in sedimented fractions from muscle tissue, but I was practically ignoring the literature.[16] I was too occupied with my own program of investigating effects of antibiotics on bacterial metabolism to study also the literature relevant to cell particulates. Therefore, while I certainly claim an interest in this field, I can admit that I more nearly entertained it than fully entered it!

Particularly interested in phosphorylative metabolism (which I'd found blocked by gramicidin and by dinitrophenol), I found there were some ATPase and simple phosphatases in the liver particles. And in February 1943, I was excited to find that the large granules had a compartmental organization (like my bacteria!), probably indicating a limiting membrane and some kind of endoplasm. Simple heating or classical cold acid extraction would liberate from these particles such standard soluble metabolites as inorganic phosphate, labile and stable organic phosphate esters (seemingly including ATP), and amino acids. So far as I know, this was the first explicit demonstration of this structural makeup, one we never published as such. It did become implicit later in Claude's eventual description of the particle preparation.

A few months after I began, we were joined by George Hogeboom, a gifted young worker especially interested in cancer research invited into the Murphy department and glad to take part in the program. He was an MD from Johns Hopkins, well grounded in biochemistry, friendly, and delightful. I introduced him into the experiments sometime in early 1943, and he joined enthusiastically. Did I have to teach him anything? I remember clarifying a few simple points: to assay an enzyme activity, one had to expose the cell particles to an ample excess of pure substrate; and to measure a potential substrate content, one had to supply an excess of pure enzyme. In both cases, one then measured the formation of product (or uptake of oxygen) over a series of time intervals. But he needed little help—perhaps not even that much—and soon we were planning together the execution of a broader range of enzyme searches.

At that time, some enterprising modern biochemists were using crude tissue "brei" or cell-free homogenates to search for the enzymatic capacities available to the tissue. This approach made use of the most unmanipulated, undisturbed, native brei they could achieve. Claude was surely making one of the cleanest first steps of fractionation imaginable by separating heavy and light particles and supernate from the cruder homogenate. Since it was Fritz Lipmann who had first acquainted me with D–amino acid oxidase (in some joint work on gramicidin), at some point I told him about the new findings with that enzyme. How depressing it was to find that my short and apparently clumsy oral description of the work totally failed to impress him! In fact, he exclaimed almost angrily that I was doing "dirty biochemistry" that he didn't like to hear about. I tried diplomatically to counter that we were purifying the cruder cell extracts he so often used. How unlike every other scientific discussion I can remember—warm friend and counselor as he was to me over very nearly fifty years! Since then I have realized that this magnificent biochemist knew everything he needed to know about D–amino acid oxidase but in 1943 failed to grasp that we and Claude were concerned with the *enzymology of tissues* and not a study of the already known enzymes. It had become clearer to him after our first paper came out, and Lipmann remarked admiringly that Hogeboom must be a brilliant scientist.[17]

Certainly Hogeboom's grasp was brilliant, and his work skillful; we collaborated well and things moved sharply along. Once in a while he was needed on a special (chemical warfare) project, but he usually could give his entire time to reading and working on our program. He took up and energetically investigated other aspects of the oxidase systems. We readily demonstrated a cytochrome respiratory system that oxidized the components of the dicarboxylic acid system. We also designed an assay for transaminase activity and found it to be present, again chiefly in the large granules. In fact, it was becoming apparent that the small particles had no substantial enzymatic activity of any of the types we were detecting, and at most only what might be contamination with those present in the large granule fraction.

Hogeboom also found the time program a bit heroic. After completing one of these late night ventures, he left me a note. Following a summary of the respiration results of that night's assay, his note concluded with "I am planning to show a great increase in O_2 uptake and general well-being as soon as I get to bed, if ever—Yours, G." Claude was sympathetic enough about the late shifts, which involved our working on for hours after he himself went home. He often faithfully promised to come in earlier next time, but he found it somehow hard to bring himself into the lab much before 11 a.m. most days. But, even a 9 a.m. start would have saved us only a small part of our late-running efforts. As time went on, other assays were shorter, though "fresh particles" at 4 and 6:30 p.m. always required some night work.

Although the literature contained reports that other workers had found enzyme activities in cell particulates, I think Claude was in some sense a true pioneer. Not on the trail of a favorite single enzyme, and not even searching and performing the specific enzyme assays himself, he set a standard of estimating the enzyme distribution (e.g., in particles A, particles B, and supernate) *in relation to the whole*. This was a kind of accountability not common with biochemists more likely to be looking for a pure and convenient enzyme source. And by the time of his 1944 conference discussion, he had a fairly sophisticated view of some of the different biochemical systems being revealed and of what prior investigations had shown.[15]

But concerns had also grown in our minds that the methodology and concepts identifying the subcellular fractions were incomplete or far from ideal. Normally, I was at my own work upstairs while the particles were being prepared. But once or twice I had come upon Claude working at zero degrees, with a piece of liver and a beaker of homogenate sitting in a puddle of water with a few expiring pieces of ice five or six inches away at the edges of a large dish; I mildly scolded him about temperature control. Also, though he'd monitored the particles microscopically in separate small experiments, he trusted the naturalness of alkaline water as a primary agent in the cell dissolution. Eventually particles were transferred into a lightly buffered isotonic sodium chloride solution, but not a canonically balanced one such as Ringer's or Zenker's solution. His recipe, finally published in 1946,[18,19] mentions osmotic instability of large granules that is in keeping with my findings of their compartmental nature and extractable "soluble components." I had been usually reporting the proportion of enzymes in the supernates, but by then it was necessary to suppose that much of the supernate enzyme may have leaked out from fragile particles.

Hogeboom began sometimes to prepare the cell particles himself and make his own choice of the particle environment. In 1946 he reported in the department records some complexity of sizes and osmotic sensitivity of the enzyme-bearing particles. By then, the large granule preparation was admitted to be a mixture, part of which was described as being mitochondria. Such refinements eventually led me to withdraw from a collaboration in which I had limited opportunity and ability for adequate control.

The success of Claude and Porter in 1945 in obtaining electron micrographs of tissue cells had drawn them into a different productive endeavor, so the particle work began to pass over into other hands. In 1946, Walter Schneider, from Van Potter's laboratory, had come to the Murphy department for a year, during which his enthusiasm and ideas were enjoined. In 1948, Hogeboom moved to the National Institutes of Health, and Schneider soon joined him there. They continued to add definition to the subject until George Hogeboom's tragic early death in 1956, after which Schneider continued developing the methodology.

An important prospect opened up for me in 1946. Maclyn McCarty had agreed to form a department devoted to rheumatic fever and streptococcal disease and was essentially leaving the DNA-induced "transformation" area he opened. I'd been entranced by that problem since 1938, and now it was conveyed to me by Dubos that I would be welcomed if I rejoined Oswald T. Avery in that work before he retired. I proceeded to do so, totally abandoning the particle work, and have happily been able to make contributions among the broad opportunities that DNA has afforded.

I believe that we contributed an early stimulus to analyzing enzyme localization in cell particles. Had I been working under grant support, I suppose there would have been the usual obligation to credit the time and effort spent by issuing publications, even inconclusive ones. Fortunately, I could withdraw after I appreciated the complexity of the material I had been using. There was a joint first article.[17] Later publications reflected some of the developing sophistication to which I did not contribute. While my measurements were accurate and "correct," they were not definitive, although I hope and believe that such unrecorded steps do have a place in the history of pathfinding themes and motivations in cell biochemistry.

These slender threads have grown into the long warp lines in the fabric of enzyme localization—even now, still in the loom—in the productive environment at Rockefeller. Ultimately, the building up included much development in other institutions, making this topic one of the important ones in modern cell biology. In the line that includes the cell small particles, the ribosomes, and protein synthesis, the ultimate world contributors make up a huge group that is dwarfed only by its beneficiaries, which include all of us.

All science is really a continuous collaborative growth. Still, scholars tend to look for discontinuities—new beginnings or significant advances to satisfy their sense that at certain times particular obstacles had to be overcome by discovery or insight. Taking this view, I would say that in the grand and broad movement of localization of functions in structures, Albert Claude was the one who inspired the description of the enzymatic makeup of subcellular particles. For me, it emerged to the surface in that "innocent" question about his urge to look into "the factories of the cells" on a certain evening in November 1942.

Passing Enzyme Work into New Hands

The experiments described above by Rollin Hotchkiss lay dormant for three years before they were published.[17] This hiatus could be attributed to such obvious factors as war efforts: Hotchkiss had research obligations related to Navy projects, and Hogeboom left in 1943 to spend two years with the Chemical Warfare Service in Florida on a team directed by Rockefeller's scientific director Gasser. Hogeboom apologized in regular letters to Murphy and Claude that he was so absorbed in writing six papers on mustard gases that he had little time to write up the experiments. The paper awaited Hogeboom's return to Murphy's laboratory in 1945, but by that time many things had changed.

Other scientific hurdles contributed to this delay. Claude was preoccupied with perfecting his cell fractionation techniques to get a purer isolation of mitochondria and trying to remedy the misidentification of large and small granule fractions. By expending effort to obtain purer fractions, he was also ensuring the enzyme assays would be more reliable. When the electron microscope was made available in 1944, he and Porter turned their attention to its novel possibilities in studying cells. Thus, by 1944, according to his annual report to Murphy, Claude did no further oxidative enzyme essays and instead conducted only a few physical and chemical analyses for nucleic acids.[13,20,21]

Fortunately, in the absence of Hogeboom and Hotchkiss, Claude became acquainted with another valuable biochemist, Van R. Potter of the McArdle Memorial Laboratory at the University of Wisconsin. Potter was a proponent of Warburg's theory that cancer results from deficient respiratory enzymes in cells. Potter and Conrad A. Elvehjem had assayed many cell types for components of the respiratory or succinoxidase system: succinic dehydrogenase, cytochrome c, and cytochrome oxidase. They used the newly devised Elvehjem homogenizer, a rapidly rotating glass pestle in a test tube, to mince tissues against the sides before measuring quantities of enzymes in the total homogenate.[22] Potter had speculated in 1941 that the succinoxidase system resided in a "smaller architectural entity which is a functional unit in the system, and many of the properties seem to coincide with those which have been ascribed to mitochondria by Bensley."[23]

When Potter met Claude at a March 1942 meeting in Chicago, he learned that entire cells were unnecessary for the effective function of these enzymes and, more intriguingly, that enzymes might be bound to the particles in the fractions. He immediately asked for Claude's help in separating the liver into nuclei and three fractions of cytoplasm, saying he wanted to "*quantitatively* analyze the four fractions . . . to see if the sum of the various parts [enzymes] is equal to the whole."[24]

Claude sent Potter detailed instructions for separating liver cells into components using the multi-speed attachment on the International centrifuge along with preliminary biochemical data on liver tumor cells that showed no significant differences in enzymes from normal cells. However, he admitted that in the new studies he had not determined the distribution of enzymes in each fraction in terms of the total activity: "I agree with you that, to have any value, studies of this sort should be quantitative and should be considered in the function of the original activity of the whole cell."[25]

Unfortunately, Potter did not have a priority rating during the early months of the war that would allow him to buy the multi-speed attachment, so he wrote Claude again asking how to separate nuclei from cytoplasm using his Beams ultracentrifuge. Armed with this methodology, he and graduate student Walter Schneider measured the enzyme distribution of RNA and DNA and found DNA in the nuclear fraction.[26] An important result of Schneider's dissertation was introducing the method (still used) for extracting nucleic acids from tissues with heated trichloroacetic acid. In one of those amusing, serendipitous moments in science, Schneider's method succeeded after he mistranslated a German instruction and substituted the word simmer for *zimmer* (room) temperature.[27]

Even before Hogeboom returned to Murphy's laboratory, Claude was under pressure to publish the early enzyme results. Without performing any further enzyme experiments, he presented the same data at two conferences. In July 1944, at the AAAS Research Conference on Cancer on Gibson Island, Maryland, Claude reported that oxidative enzymes found in *normal* cells by Hogeboom and Hotchkiss were associated with the large granule or mitochondrial fraction. Claude also speculated that mitochondria, like chromosomes, retain a "considerable degree of independence and stability" and that during mitosis the mitochondria are "transmitted as such to the daughter cells;" in other words, "they are capable of perpetuating themselves by the mechanism of self duplication."[15] Then, in 1945 at the Hershey (Pennsylvania) Conference on Intracellular Enzymes in Normal and Malignant Tissues, he was no doubt caught by surprise on learning from Potter that he and Schneider had repeated the Hotchkiss-Hogeboom assays and confirmed that the large granules yielded most of the succinic dehydrogenase and cytochrome oxidase.[28] While this news may have increased pressure on Claude to publish his research, the paper was still a year away.

In the meantime, Potter inquired on 2 February 1945 whether Schneider could join Claude to learn "the centrifugal methods which you employ in the separation of the cytoplasmic granules." Schneider's analytical methods for determining nucleic acids in normal and tumor tissues, Potter wrote, made him "very well qualified to do enzyme studies."[29] (Schneider was the first of three young scientists from Potter's laboratory to become cell biologists at Rockefeller; the arrivals of Philip Siekevitz and Günter Blobel are described later.)

Hogeboom Returns

Anticipating a return to the Institute in July 1945, Hogeboom admitted to Murphy that "a further exploration of the particles . . . doesn't sound like cancer research." However, he continued, "I believe I'm a chemist at heart—with an intense desire to find out what goes on within the cell. Do you think you can put up with a long-haired enzymologist around your department . . . I am firmly convinced that something eventually will come of that approach to the cancer problem."[30] Murphy's reply reflected his continuing control of the lab's direction. "Your hope to develop into an enzymologist is OK by me. I may bring a little pressure to have you keep cancer in the picture for a very definite reason I will explain to you later."[31]

The paper was finally written by Hogeboom, but only after he made new enzymatic assays on tumor cells during the next year. At the same time that Murphy was pressing Claude and Porter to visualize cells infected with the chicken tumor agent under the electron microscope, he was pressing Hogeboom to make comparative studies of oxidative enzymes in precancerous livers of rats that had been fed a diet containing the carcinogen butter yellow. To Murphy's disappointment, the tumor cells showed no significant change in the distribution of cytochrome oxidase, although there was some lowering in succinoxidase. To explain this puzzling fact, Murphy assigned Hogeboom to study the obstacles in unraveling this enzyme system, thus ushering in Hogeboom's fruitful career in the field of cytochemistry.[32]

Almost anticlimactically in October 1946, two articles on oxidative enzymes in particulate components appeared in the very same issue of *The Journal of Biological Chemistry*.[17,33] The one by Schneider was received for publication on 26 June. The other by Hogeboom, Claude, and Hotchkiss was received on 22 July, the same day that Claude sent Schneider a critique of his manuscript, advising, "it would be nice if you could mention [Hogeboom and Hotchkiss] in your discussion," since Schneider referred only to the preliminary data that Claude reported orally in the two 1944 conferences.[34]

The significance of both papers is that there was a preponderant amount of activity of a certain enzyme associated with a particular subcellular fraction. Despite similar results, Schneider used the homogenizer that yielded better mitochondrial fractions, but he had to fractionate liver homogenates using a low speed centrifuge (2,400 times gravity, rpm not given). The Rockefeller scientists, in contrast, used a fine mesh to break apart the cells that resulted in more broken particles, but they could spin down the homogenates with three times more force than Schneider. They also provided comparative analyses of enzymes in tumor cells. Both papers agreed that sixty to seventy percent of the respiratory enzymes cytochrome oxidase, succinoxidase, and succinic dehydrogenase resided largely in the large granule or mitochondrial fraction. Another important aspect of both papers is the use of tables showing quantitative recovery; that is, the distribution of enzymes in each fraction. These were powerful indications that differential centrifugation could link biochemical activity to biological structure.

Linking Cell Energy, Enzymes, and Mitochondria: Enter Walter Schneider and George Palade

Even with these 1946 papers, Claude remained hesitant about renaming the large fraction mitochondria. He felt this fraction was still a mixture of mitochondria and secretion granules that needed better cytological identification. New methods and clearer results came with the arrival in the laboratory of two individuals, Walter Schneider and George Palade.

Schneider arrived in New York in August 1946, a year after receiving a PhD in physiological chemistry as well as a cancer research fellowship from the Jane Coffin Childs Memorial Fund. He had not previously met Claude, Hogeboom, or Hotchkiss. "Nor was I aware that I would be in the Laboratory of Dr. James B. Murphy," he said. "Claude had informed me before I came that a mature Rumanian pathologist, George Pallade [*sic*], was also coming on a fellowship. . . . Claude and I conferred on what I should do and I suggested measuring cytochrome c, the third component of the succinoxidase system, in liver fractions."[27]

Since two enzymes in the succinoxidase system had been linked with the large granule fraction, it followed that the third one should also reside in the same structure. According to Schneider it was agreed that "Claude would do the fractionations, I would analyze for cytochrome c, and Hogeboom would do succinoxidase assays. Before we could start, I had to learn how to extract and measure cytochrome c using a method developed by my colleague at McArdle, Ken DuBois. This took quite a few weeks, during which I learned that Claude's work times were unusual—he would come in at about 4 p.m. and work late into the night."[27]

During Schneider's fellowship year, two new techniques were introduced that greatly improved the quality of the preparations. Separation of fractions was improved because they used the new refrigerated centrifuge run under a vacuum, the one designed and constructed by Josef Blum (described in Chapter 2), that reached much higher speeds than either Schneider or Claude had used before. More importantly, they found the biochemical properties of the mitochondria differed depending on the solutions used to prepare the fractions. If the large granule fraction was prepared in water, it lost most of its cytochrome c activity. If this fraction was prepared in an isotonic saline, the granules retained cytochrome c, which could, however, be removed with a water wash. Conversely, if the other fraction, the microsome fraction, was prepared in water, it contained cytochrome c, whereas the enzyme was lost when isotonic saline was used. Their important conclusion was that "the study of the media to be used for the isolation of the components of the cell would seem to be of paramount importance if a true picture of intracellular processes is to be obtained."[35]

A few weeks after the arrival of Schneider, George Pallade (the original spelling of his last name) joined Murphy's laboratory. He was born in 1912 in the Romanian city of Jassy, the old capital of Moldavia. In 1930, despite his father's hope that he would follow his career as a philosopher, a cousin in medicine convinced the eighteen year old about the possibilities of doing experimental science. This idea appealed to him as something concrete "in terms of working with your mind and your hands rather than your mind only" as in philosophy. He entered the ten-year curriculum in the medical school at the University of Bucharest, where his growing interest in science attracted him to anatomy professor Francisc Rainer and biochemistry professor André Boivin. As a research assistant under Rainer, Palade embarked on the required doctoral thesis to examine the microscopic anatomy of the kidney and urinary tubule of the dolphin. His study, which included working with fishermen in the Black Sea, was akin to taking the kidney apart and putting it back together. By cutting thin sections through the kidney, he greatly improved the microscopic resolution, and by looking at the sections in sequence, he could extrapolate this anatomy into three dimensions. Thus, he was able to correlate the kidney's glomerulus and nephrons with the special needs of this mammal in its salt water environment. In his initial scientific experience, he was able to link structure to function at the light microscope level.[36,37]

After receiving his MD in 1940, he became a teaching assistant at the medical school in human anatomy and embryology while serving in the Romanian army during World War II. After the war he felt "uncomfortable" practicing medicine at a time when the patient looked "at you as a fellow with extraordinary powers . . . and in fact, you realized you didn't know" enough. His decision to pursue a career in research was encouraged by Rainer's successor Grigore T. Popa, who had studied in America as a Rockefeller Foundation fellow in 1925–26 (not related to The Rockefeller Institute). In his letter of introduction to the Foundation in October 1945, Popa announced that Palade was "coming to America to study problems of heredity and the structure of living matter. He has money and does not ask material help to the Foundation, but he needs a moral help and introduction letters [*sic*] to some American scientists," and he specifically suggested professors Thomas H. Morgan and Robert Chambers.[38]

In March 1946, Palade joined the New York University laboratory of Chambers, who was known for developing microsurgical techniques to study cellular structure. This was a very brief experience since Chambers retired that July. Fortunately during this time, Claude showed electron micrographs of cultured cells during a seminar at NYU, and Palade became immediately "fascinated by the possibilities." After talking with Claude over lunch that day, he sent him a formal letter of application, requesting "to work as a research fellow under your supervision on subjects to be indicated by you."[39] After Claude made a critical study of Palade's publication on the dolphin kidney, he advised Murphy the work "is a classical one, and the way in which the work was carried out and reported suggests that Dr. Palade is a mature and thorough investigator."[40] On Murphy's recommendation, Palade was invited to be an unpaid visiting investigator, and two years later he was appointed to the scientific staff as a salaried assistant. (He remained at Rockefeller for twenty-eight years.)[41,42]

Palade's first experiments dealt with cell fractionation. At first, he was guided in every step by Claude. "The routine required that everything should be set in the cold room the evening in advance, absolutely everything," Palade said. "Down to the last detail every piece of instrument, every piece of glassware had to be there, checked in advance. . . . This business of very careful resuspension [of the fractions] was stressed and repeatedly stressed, but once I mastered the technique under his direct supervision—with him looking not only over my shoulders but at my hands—he left me alone. Then after that I was on my own. I could do whatever I intended to do."[43]

Palade began by trying to settle Claude's concern about the purity of fractions. According to Schneider, due to Claude's daytime absences in the laboratory, "Hogeboom, Palade and I found ample time to discuss the large granule problem—namely, what cellular component were they?"[27] Their collaboration went quite far and in fact led to the advent of combining centrifugal, biochemical, and electron microscope studies. The basic question they pursued in this now classic 1948 study was how the large granules could really be recognized as mitochondria when they were separated and studied outside the cell.[44]

At the time, mitochondria could be identified by cytological criteria: they stained with Janus Green dye and could be visualized as elongated structures inside cells using a light microscope. In their new efforts to get pure fractions of this subcellular component, the three scientists used a series of techniques. They adopted the Potter-Elvehjem homogenizer. Introducing a more intricate centrifuge technique, they began with the International at low speeds to remove the nuclei and then moved to the new Blum prototype centrifuge, using two speeds: 24,000 times gravity to sediment mitochondria and 41,000 times gravity (no rpm given) to sediment the submicroscopic particles.

Analyses of and changes in the medium used for centrifuging led to a major breakthrough in isolating mitochondria in a recognizable form. As Hogeboom, Claude, and Hotchkiss had suggested in 1946, injuries to the mitochondria during fractionation caused their shape to change from spherical to elongated. But Porter remembered how meticulous Claude was in resuspending his fractions from the centrifuge tubes. "They'd be plastered into the bottom of the tube, of course, and he'd go in as though he were making mayonnaise. . . . He would add saline a drop at a time and just move his wand or glass rod, whatever he had, very carefully so as not to denature any of the components that he was trying to suspend."[45]

The new experiments revealed that the distortion was instead due to the medium in which the cells were suspended. In 1943, Claude used an alkaline medium that Lazarow criticized as promoting the disintegration of the particles. So in 1946, Claude switched to a buffered saline medium, and this caused the particles to aggregate. The saline solution also caused the mitochondria to swell and become spherical before eventually disintegrating.

In efforts to eliminate aggregates of mitochondria, Hogeboom, Schneider, and Palade tested homogenates of liver in different aqueous media. Schneider gave Palade credit for "the inspiration of using sucrose solutions to prevent aggregation and to provide a suitable osmotic environment for the mitochondria. These findings . . . were the basis for a revolution in biochemical research that led to the definition of the functions of many subcellular components."[27]

Sucrose was the molecule of choice due to the high solubility and hypertonicity that can be reached in preserving mitochondrial structure. The 1945 micrographs by Claude and Fullam of mitochondria isolated in saline had revealed spherical shapes in small clumps. Porter then took electron micrographs to compare isolated mitochondria in saline with those isolated in sucrose solutions. The mitochondria suspended in sucrose showed distinctly elongated or rod-shaped forms, the shape seen when mitochondria were viewed inside cells under the light microscope as well as the shape seen by Claude and Fullam in 1946 in sectioned tissues.

Porter, not Claude, was credited with taking the electron micrographs for these experiments. At the time, he and Hogeboom were trying to find a vital stain for electron microscopy that would be specific for mitochondria. In unpublished observations, they reported that calcium ions "trap as an insoluble salt a substance which is slowly (and perhaps enzymatically) liberated within the mitochondria."[46]

The sucrose medium not only eliminated the aggregates but also allowed mitochondria to retain their stain with Janus Green, something not possible in saline solutions. Further, there were more rod-shaped forms in 0.88 molarity (a unit of concentration) sucrose than in 0.25 molarity sucrose. While the 0.88 molarity was best for morphological reasons, it proved no good for biochemical reasons because it inhibited oxidative phosphorylation. A compromise was eventually reached to use 0.44 molarity sucrose that did not affect phosphorylation yet preserved their intracellular structures almost as well.

Palade worried what Claude would say when he learned about this switch to sucrose. "When we came with the sucrose method, it was, if you like, something done against what he had established before. More importantly, it was done without consulting him, and it was done in his laboratory. Well, I was expecting that we were going to be reprimanded. . . . The very interesting thing is that we were not reprimanded and not fired. . . . He was annoyed but somehow he realized that was a significant contribution. Of course it would have been much better to be done with his knowledge and, if possible, with his collaboration."[45]

Turning to biochemical assays on the mitochondria, they found their newly improved and purified fraction had about seventy percent of the succinoxidase over the original homogenate, with about a three-fold concentration over the whole homogenate. Some pentose nucleic acid was found, but not as much as in the submicroscopic particles. Their quantitative recovery indicated that not much material was lost during the fractionation procedure. Hogeboom, Schneider, and Palade also explored Claude's worry that "secretory granules" were contaminating the large granules. They applied a neutral red stain and found among the Janus Green–stained mitochondria a small number of spherical granules and lipid droplets. Using various washing procedures, they removed both this material and the small amount of pentose nucleic acid. These final procedures ensured a relatively intact and pure mitochondrial fraction.

At the end of this collaboration, Palade was still puzzled about the ways of American science. While "I started with the idea [of the sucrose procedure]," he said, "evidently I didn't have the technology, so they were doing the work. And we started discussing something that for me appeared funny: what is going to be the order of the authors? . . . George came and told me that it's going to be Hogeboom, me, and Schneider. So I said, 'OK, no problem.'" When Hogeboom returned a few days later "and asked 'what about putting you at the end, because Schneider is at the beginning of his career and publications count?' I wasn't operating under the idea that I have to use every opportunity to promote my future career. . . . Little by little I discovered that if you are at the end of the list of authors, provided the list is not very long . . . it's really a good position."[47]

After the Classic 1948 Paper

Fuller biochemical stories of mitochondria were pursued in many places outside Rockefeller and are not discussed here except for three: the confirming mitochondrial investigations by Hogeboom and Schneider (described briefly below); the further evidence for the individuality of mitochondrial structure and function from the electron microscopy of Palade (see Chapter 6); and the greatly refined methods and discovery by Christian de Duve that mitochondrial preparations also contain lysosomes (see Chapter 11A).

Once the Rockefeller scientists adopted the sucrose method, many scientists also found the same localization of oxidative enzymes in mitochondria. Almost immediately, biochemists Eugene P. Kennedy and Albert Lehninger at the University of Chicago applied the new method of isolation and demonstrated that the mitochondrial fraction oxidized fatty acids, and they later determined this organelle carried out all the citric acid cycle (Krebs cycle) intermediates and ultimately generated ATP, the cell's nearly universal energy source. A few months later, Schneider (who had returned to Wisconsin after his fellowship year) and Van Potter showed that mitochondria oxidize other enzymes. At about the same time, David Green and his colleagues at Columbia confirmed these observations. Meanwhile, Hogeboom's final experiments at Rockefeller led to finding an enzyme (DPN–cytochrome c reductase) in both the mitochondrial and microsome fractions that links the dehydrogenase cofactors to the cytochromes; this was the first finding of a double localization of an enzyme and the first enzyme identified in the microsome fraction. Isolated mitochondria quickly became a standard preparation to begin examining their complex respiratory systems.[48–50]

Then, in 1948, Hogeboom and Schneider were reunited at the National Cancer Institute (NCI). According to Schneider, Harry Eagle, the NCI's new director, had "learned of our work on isolation of mitochondria, [and] offered George a free hand in setting up his program in Bethesda. There was plenty of money for equipment and salaries, good lab space, good prospects for promotion, and the lure of suburban living." The two scientists tried to recruit George Palade to join them, without success. Schneider said that Palade "preferred New York life to that in Washington and he may have had an antipathy to working for the government, having just escaped from communist Rumania [*sic*]."[51] Like many people coming from central Europe at that time, Palade said he was given "a special status called displaced person," and even though Eagle "managed to get some kind of waiver in my case" to work at the NCI, he decided to stay at Rockefeller.[47,52]

For the next eight years, the collaboration between Schneider and Hogeboom produced significant evidence of the intracellular distribution of enzymes. They produced fifteen major publications that set standards for work in this field of cytochemistry. Hogeboom mastered well the mentoring of Hotchkiss on enzymes and Claude on fractionation techniques, while

Schneider's knowledge of biochemistry showed him to be an accomplished student of Van Potter. By 1953, they had described a dozen components that should be looked for in the nuclear fraction, thirty enzymes in mitochondria, and eight components of the microsomal fraction.[53] Their prodigious efforts seem more remarkable considering that Hogeboom contracted tuberculosis from one of his children in 1952 and spent more than a year in a sanatorium.

Finding more intracellular enzymes. Hogeboom and Schneider worked from the premise, echoing Claude's concern, that "a biochemical property can be ascribed to a given structural component of the cell only when at least three criteria have been satisfied: 1) the cell fraction must be cytologically defined in terms of the intracellular component it represents; 2) the validity of the method of assay for the biochemical property must be established; and 3) the biochemical property must be concentrated to a greater (preferably, considerably greater) extent in the cell fraction than in the original whole tissue."[54]

Consequently, they found cytochrome c was associated with mitochondria as much as succinoxidase, thus linking the oxidative cycle with the electron transfer chain for creating energy.[55] They also standardized methods for recovering enzymes so that they could differentiate mitochondrial enzymes from those in the soluble fraction.[54] In addition, Hogeboom used the counter-current distribution techniques that he learned at Rockefeller from Lyman Craig in 1946 to purify the necessary but unacceptable commercial preparations of diphosphopyridine nucleotide (DPN) needed for the enzymatic assays.[56]

As part of the NIH devoted to cancer research, Hogeboom and Schneider compared distributions of enzymes, nucleic acids, and solid proteins in tumor versus normal cells. Not too surprisingly, their inconclusive results showed reduced enzymes and mitochondria with altered biochemical properties in cancerous cells. Returning to normal tissues, they further localized two enzymes within mitochondria, cytochrome oxidase and DPN–cytochrome c reductase, that ultimately lead to the production of ATP, the energy molecule.[57]

Refining methods. Their work led to a new standard procedure for isolating the nuclei from cells to obtain as homogeneous a fraction as possible that, like the mitochondria, would retain their intracellular appearance. They modified the ionic medium containing citric acid (used by Alexander L. Dounce) by adding calcium to the hypertonic sucrose solutions; as a result, many enzymatic activities previously found in isolated nuclei were greatly reduced or entirely lost by eliminating the cytoplasmic contaminants. Moreover, they found the isolated nuclei contained a soluble enzyme absent in the cytoplasm that could synthesize the respiratory coenzyme DPN.[58,59]

Hogeboom's final experiments involved refining Claude's fractionation process as a way to look for additional fractions and other intracellular components. In collaboration with centrifugation expert Edward L. Kuff and electron microscopist Albert J. Dalton, they used a hand-crafted mixing machine to create a continuous sucrose density gradient. After removing the cell debris and nuclei, the cytoplasmic extract was placed on top of the sucrose and centrifuged for various lengths of time and at various speeds, which resulted in a gradient of multiple layers (fractions) that were subjected to enzymatic assays and electron microscope examination.[60] In addition to the mitochondrial and microsome fractions, they detected a third fraction. As Hogeboom noted in a footnote to their paper, Christian de Duve in Belgium had also just separated the cytoplasm into three fractions, thus supporting their findings, although he used a refined technique of differential centrifugation.[61] These two papers were the first to indicate the presence of previously unknown vesicular structures later named lysosomes and peroxisomes (see de Duve, Chapter 11A).

As early as 1952, Hogeboom revealed to Porter his discouragement with the bureaucracy at NCI. Porter responded that Gasser was "definitely interested" in bringing Hogeboom back but that "it was now too late in his career as Director to open any new departments."[62] Not long afterward, Claude arranged for Hogeboom to spend a sabbatical year as visiting professor at the Bordet Institute in Brussels to open a new research program on "the enzymatic constitution of the cell."[63] A month before he was to join Claude, Hogeboom had a minor operation which led to thrombosis and sudden death.[64]

In one of their last papers, after a decade of working together, Hogeboom and Schneider felt their studies were "still in their infancy." They concluded the mitochondrion was a "highly organized structure in which the various enzymes are so localized that the products of one reaction are formed at an appropriate site for the next enzymatic transformation." Yet they remained cautious that the isolation methods they developed were not ideal tools for the study of subcellular particles because of the need to expose them to foreign environments. Despite their misgivings of the cell fractionation technique, they were able to confirm their biochemical results with independent and less generally applicable methods.[65]

A Third Milestone in Modern Cell Biology

Cytologists had known about the existence of mitochondria for a long time. They had seen these microscopically small bodies in most cells of the body, could stain them with certain vital dyes, and learned they contained enzymes that could react with these dyes.

Once Murphy's laboratory could isolate a structurally well-preserved mitochondria, they made two contributions advancing the study of this organelle. In 1946, Hogeboom, Claude, and Hotchkiss concluded that succinoxidase and cytochrome oxidase were localized exclusively in the mitochondria.[17] In 1948, Schneider, Hogeboom, and Palade established an experimental model linking evidence from cell fractionation, electron microscopy, and enzyme studies.[44] Importantly, what these studies demonstrated was that enzymes and other compounds of biochemical interest are localized in specific structures and not distributed haphazardly throughout the cell.

During the following dozen years, substantial chemical knowledge about the mitochondria was accumulated by many other scientists to show they supply practically all the necessary biological energy, primarily from the enzymatic oxidation of chemical compounds in foodstuffs. After a long period of speculations about the metabolic individuality of this organelle, it became clear that mitochondria justified Claude's designation as "power plants" of the cell.

Palade recalled these early years were somewhat tentative. "When this work was started there were some stepping stones in view in this poorly explored border between cytology and biochemistry. Now I believe that we have a more or less continuous path. In due time it may become a good, open road."[66] Within a short time, their discoveries added another milestone in modern cell biology.

Chapter 5

The New Science Emerges

Carol L. Moberg

Herbert Gasser, the second scientific director of Rockefeller from 1935 to 1953, became a champion of cellular physiology. During this period, he gave substantial thought to what scientific directions should be followed, changed, or created. Of the three fields he considered in 1940—biophysics, genetics, and chemotherapy—it was his support of biophysics that became a significant force in fostering the venture into modern cell biology.

Trained as a physiologist, Gasser had worked with Joseph Erlanger at Washington University in St. Louis, where they devised methods to record and study electric currents of nerve fibers, work for which they shared the 1944 Nobel Prize. Gasser's medical training and grasp of mathematical and physical sciences equipped him to lead a transition from Simon Flexner's focus on diseases to a broader approach to human biology. His tenure encompassed the Great Depression that depleted the Institute's endowment and World War II, when most Institute scientists conducted research relevant to military needs. Gasser devoted much time to analyzing the staff's ongoing research as he worked to foster optimal conditions for independent scientific inquiry. He espoused the interesting principle that there are two times for working on a problem—before anyone has thought of it and after everyone else has left it.[1]

In 1940, Gasser proposed a new line of research involving biophysics. For reasons that seem puzzling, only three years earlier he had closed the biophysics laboratory headed by Ralph Wyckoff, who was studying the biological (although not medical) applications of x-ray crystallography and working with Edward Pickels of the International Health Division (IHD) to help design and use the air-driven ultracentrifuge to purify viruses (discussed in Chapter 3). Gasser wanted to reestablish biophysics "under more favorable conditions" by placing it within physiology, a field akin to his own science of electrophysiology of the nerve impulse. The principal reason for a different kind of biophysics laboratory, he noted, "is to be able to penetrate to the fine structure of cells and their constituents" and to find a laboratory head who was "a biologist as well as a physicist—a rare combination and one difficult to find;" in other words, someone who could "judge whether or not a biological problem had significance."[2]

Four years later, in June 1944, with scientists still engaged in war efforts, Gasser was anticipating a presentation to the Board of Scientific Directors. This was the very month that Keith Porter and Albert Claude were taking their first electron micrographs of cultured cells. He applied metaphors of war to assess the current research trends, noting, "We are a very small group in the army of investigators in this country. To make ourselves effective we will have to apply our potentialities to the opening up of new fields." He also opposed entering well-established fields that would "merely add a small increment to the total." Weighing the prospect of a completely new field of science, he suggested one that would utilize biophysical techniques to analyze "cellular organization, the way chemical molecules are fitted together

Herbert Gasser, 1942. Courtesy of The Rockefeller Archive Center.

into a coherent structure that makes function possible. The idea that cellular activity depends upon the detailed structure of the cell has been a definitely traceable thread in biological thinking for over a century." He provided his own definition of the term "ultrastructure" (or fine structure) that had recently come into existence as "structure in molecular dimensions."[3]

Gasser's interest in the architecture of the cell was enhanced by Claude, George Hogeboom, and Rollin Hotchkiss, who in 1942 had just begun to separate the cell into its component parts and to characterize them chemically. In singling out this work for praise, he noted the limited physical methods available at Rockefeller for the investigators to continue this exploration of the cell. "We have not got either of the two most important direct methods," namely, x-ray diffraction and electron microscopy. Moreover, the scientists being drawn to these fields were "going to other institutions to get their pictures taken." This statement referred directly to two electron microscopes, one being used by Porter and Claude at the Interchemical Corporation and the other by Thomas Rivers, Joseph Smadel, and Wendell Stanley at RCA in Camden, New Jersey.

To introduce biophysics into the midst of medical researchers was not that simple, Gasser noted in the report. The Institute did not need someone who knew just physical techniques but rather "someone who knows biology and who knows the whole gamut of the methods and where they can be applied. We need someone with weight enough to tie the whole thing together and a fund of ideas that will catalyze the effort in all the special instances where the type of research is available." Among several American and European scientists in various fields whom he thought could head such an effort, he suggested Francis O. Schmitt, head of the Department of Biology and Biological Engineering at the Massachusetts Institute of Technology. Reviewing the work of Schmitt, a former student of Erlanger, Gasser remarked in detail on his recent observations on collagen using both x-ray diffraction and electron microscopy, the two instruments lacking at Rockefeller. He also conveyed advice from René Dubos, a premier biologist and a full member of the Institute, who urged approaching Schmitt quickly because MIT was planning a greatly expanded program in biology.

The Board authorized Gasser to consult with Schmitt about individuals who might be available to direct such a laboratory and, after Schmitt showed interest, to offer him membership in the Institute and a supporting staff to create a biophysics laboratory. However, Gasser believed the odds of Schmitt accepting were low, due to such perennial hiring problems as living conditions in New York, commuting expenses, and the unlikely prospect that Schmitt's colleagues Cecil Hall and Richard Bear would be appointed as full or even associate members of the scientific staff. Schmitt was apparently happy at MIT because he was reducing contacts with biological engineering and increasing those with chemistry and genetics. So it was no surprise that in early 1945, Schmitt declined the invitation with the cryptic comment to Gasser that "R.I., like Stalingrad, is a pretty formidable citadel. While the front seems pretty far down the Hudson, battles have a way of see-sawing, and I suspect that enemy patrols have been reconnoitering as far behind the lines as the Charles."[4]

A year later, in 1946, Gasser began organizing ways to integrate biophysics into the ongoing science, believing that new funds would become available after the war ended. In a report entitled "Future Needs," he debated whether physics should be brought to biology or biology taken to physics. While renewing an emphasis on biology, he doubted whether the $100,000 proposed for biophysics could pay for physicists with specialized training, very expensive instruments, and a complement of shopmen and technicians to run them. On the other hand, he noted many laboratories were using sophisticated physical technologies as a dominant method of investigation. Gasser then looked more closely at the current work in cytology and reasoned how biophysics might contribute to what was already in place, because, in his words, "one approach to a biological problem radiates to another and . . . a number of techniques must be brought to bear on the solution of a single fundamental question." Among his recommendations to the scientific directors in this report was establishing a unit devoted to "cellular physiology, included because of its underlying importance for all of the special branches."[5]

Unraveling the nature of cancer was the kind of "fundamental question" Gasser had in mind. While his views on cancer causation are unknown, he nonetheless sensed a great opportunity in Claude and Porter's research to make potential contributions to the cancer problem. "Work in Murphy's laboratory," he wrote, "is integrated by the general proposition that whatever contributes to an understanding of intracellular mechanisms also contributes to the cancer problem. Porter's studies of the cytology of cells in tissue culture are leading progressively to clearer pictures of the intracellular morphology."[6]

Before Gasser invested in a full-fledged laboratory devoted to cellular studies, he foresaw some imposing challenges, particularly improving the instruments, techniques, and interpretations. As an astute observer of work in James Murphy's laboratory, he noted that research on the enzymatic activity of intracellular fractions, which Gasser referred to as granules, needed "better methods of isolating and preserving." While encouraged that the "location of constituents in a cell has come to life again," he remarked that the "microtome is far from an instrument and is not ready for mention at this time." And, he pondered whether the earliest micrographs of chicken tumor cells "look like malignant cells, though there is no proof," and disparaged Porter's claims for not being "as conservative as one would like."[7]

Solutions to these problems became more urgent when the electron micrographs of virus-infected cells by Claude and Porter began to appear in late 1945. Gasser sensed how they opened an exploration of a completely unknown and previously unseen territory. Within two years, all three needed improvements were under way: the Institute supported development of a new microtome and purchased a new electron microscope, and Porter obtained an outside grant from the American Cancer Society to help found the Tissue Culture Association (later in this chapter). Gasser's early reflections evolved into a commitment to establish what would eventually become cell biology.

Early Microtomes for Electron Microscopy

Initial efforts to create thin biological sections for viewing in the electron microscope were far from satisfactory. Even though it may have seemed logical to adapt methods and microtomes that had been successful with specimens using light microscopy, it took from 1943 until the appearance of the Porter-Blum microtome in 1953 before sufficiently thin, high quality specimens could be prepared with some degree of regularity for viewing in the electron microscope. Some unique problems included creating thin enough sections for the electron beam to penetrate, preserving a natural form that retained spatial relationships, and replacing water with a solid enough matrix that could be sliced. The latter problem meant finding a cutting edge both sharp and hard enough to shave sections of 0.1 micrometer, which was up to a hundred times thinner than sections for light microscopy.[8]

Early microtomes by A. Glenn Richards, Thomas Anderson, and Robert Hance[9] in the United States, Manfred von Ardenne in Germany,[10] and Fritiof Sjöstrand in Sweden[11] were based on the Cambridge rocking microtome designed by Charles Darwin's son Horace around 1885.[12] In 1943, Harold C. O'Brien and G. Murray McKinley at the University of Pittsburgh devised a "cyclone knife" microtome that worked by driving a steel knife edge whirling on a centrifuge rotor at a speed of 12,500 rpm and had a cutting speed of 140 feet per second that was calculated to produce a section 0.1 micrometers thick. They hypothesized that the knife chipped off rather than sliced because it blitzed past the fixed specimen at a speed greater than the capacity of the block to transmit the compression, thus reducing distortions, much like a rifle produces a small hole in a pane of glass compared with a hammer that delivers a shattering blow.[13] In 1944, they presented pictures of sections made with this high speed microtome, which Porter later described as excellent in theory, but "in practice the hazards and difficulties involved in harvesting the sections made the procedure impractical . . . it would be correct to say that in sectioning, the block was in a flash transformed into a cloud of sections."[14]

The Interchemical microtome. In 1944, coincident with Claude and Fullam's collaboration on the electron microscope, Interchemical was developing its own microtome. This instrument, according to Mary Schuster, had been "re-designed and rebuilt several times" and was in daily use at Interchemical.[15] Its construction was based on the same principle as the one of O'Brien and McKinley, although its knife ran at even higher speeds (49,000 and 57,000 rpm), used a variety of steel knives and razor blades, relied on special embedding media, and could cut sections 0.1–0.8 micrometers thick, depending on the material.[16,17] As mentioned in Chapter 3, Fullam found this microtome "a fearsome piece of equipment" that was dangerous to operate and emitted an ear-splitting shriek that drove everybody from the room.[18]

Claude provided Fullam with tissue samples and valuable advice for their microtome experiments and development, which likely accounts for two methods that were introduced to deal with the specimens: they were collected on Formvar-coated slides and fixed with

Claude-Blum microtome, 1946–1947. This prototype hand-turned instrument was the first to allow continuous motion of the tissue block and collection of sections on water for easy transfer to electron microscope grids. Courtesy of Philippa Claude.

osmium tetroxide. As Claude and Fullam noted in their 1946 *Journal of Experimental Medicine* article, they used an Interchemical microtome to make sectioned cells from 0.3 to 0.6 micrometers thick of the guinea pig liver and looked at them with the electron microscope. Despite their introduction of another embedding medium, a eutectic mixture of camphor and naphthalene, their micrographs were unsatisfactory because the microtome created visible holes in the sectioned slices. They concluded that suitable sections for electron microscopy depended not only on an adequate microtome, "but, in a considerable degree, on the type of fixation and embedding to which the tissue must be subjected prior to sectioning."[19]

Claude's experiments with fixation techniques became even more problematic and eventually regrettable when he and Palade began in 1946 to isolate and identify the cell organelle known as the Golgi apparatus. This structure had first been visualized by Camillo Golgi in 1898 while studying nerve cells. After he devised a metal staining method, called the black reaction, he discovered a basket-like network or internal reticular apparatus surrounding the nucleus, which the stain put in sharp contrast with other parts of the cell.[20,21] Yet, for fifty years during the era of light microscopy, a controversy persisted as to whether the Golgi apparatus was a true cell organelle. Scientists argued whether the fixation methods revealed a common cell structure with variable forms, a cytoplasmic substance, or an artifact.[22–24] Claude and Palade tried to extract the Golgi from a variety of cell types using the Elvehjem-Potter tissue homogenizer (later differential centrifugation of homogenates in sucrose) and testing several fixatives. After three years of study, they presented evidence that "the Golgi apparatus is a gross artifact, namely a myelin figure, or a complex of myelin figures, which develop in cells during fixation."[25,26] Although their observations were made using only a light microscope, Porter's earliest exploratory studies of the cell with the IHD electron microscope no doubt enhanced their hypothesis. In 1946,

Porter reported to Pickels that "the Golgi apparatus is composed of small osmophilic vacuoles and is not constructed of strands or canaliculi as heretofore contended. The latter structure is an artifact of light microscopy."[27]*

The Claude-Blum microtome. In 1946, Murphy noted in his annual report to the Board of Scientific Directors that because the sectioned specimens were unsatisfactory, "an experimental microtome had been built which was designed to give sections of 0.1μ [micrometer] . . . [but] an entirely satisfactory blade has not yet been found and further investigations of methods of fixation and embedding of tissue to render them suitable for thin sections" were in progress.[31] A high speed microtome had been under development after the Board voted a year earlier to appropriate $500 from the Contingent Fund.[32]

Within a few months, Josef Blum, who was simultaneously designing and producing centrifuges in the Instrument Shop, built a low speed (hand turned) microtome for Claude. This instrument had a revolutionary principle of reduction gears, so that the advance of the embedded specimen block was controlled by pulleys and belts that could be altered to make sections of various thicknesses. A single section of the right thickness was more desirable than a cloud of sections spewed into the air. Since the passage of the block over the knife edge on the return stroke damaged both the surface of the block and the knife edge, a "single pass" or "bypass" mechanism was also devised that caused the block to move in a circular path away from the knife edge after each downward stroke, an important feature of successor microtomes. Another feature was a water trough for collecting the cut sections, and this permitted an easy transfer of the floating sections to the support grids.[33]

Sanford Palay, a National Research Council fellow who came to work with Claude in 1948, had occasion to test and use the Claude-Blum microtome. He found it "clumsy and very inconvenient."[34] The microtome had to be used in a cold room because the embedding mixture of naphthalene and camphor became very hard at 4°C yet conveniently vaporized at room temperature after the sections were placed on the grids. Moreover, the embedding mixture would melt when the block passed over the knife edge and refreeze on the upward pass.

Another problem was Claude's unusual choice of a knife blade. The cutting tools for earlier microtomes were either razor blades that were too flexible or steel knives that needed to be resharpened after nearly every use. Fullam and Claude used steel knives in their study of sectioned guinea pig liver cells. However, Claude had been impressed on his trips to Mexico, many years earlier, by the knives and arrowheads that were fashioned from an extremely sharp and amorphous form of glass called obsidian, and their sharpness suggested their crystal structure would not be the same as well-tempered metal.[35] In 1945, he asked Bausch & Lomb to produce glass knives that would be suitable for sectioning materials at least 0.2 micrometers thick.

*As a postscript to Porter's observation, Palade and Claude did not discard their mistaken interpretation for another six years. In 1952 there were still many adherents of the view that the Golgi apparatus was real, and Claude wanted to publish their evidence that it was an artifact. Although a paper never appeared, he drafted a manuscript reviewing arguments for and against their view while maintaining that "the Golgi apparatus of classical cytology is an artifact, but to be sure, it is a highly organized artifact."[28] He then suggested to Palade that perhaps the matter should be dropped until some biochemical evidence and new tools and methods became available rather than to continue arguing fruitlessly on the basis of observations provided by nearly obsolete techniques. In response, Palade speculated whether the electron microscope might resolve this issue and agreed that "we can very well keep still and let the others continue the polemic."[29] When electron microscopy studies did appear in 1954, especially those of Albert J. Dalton and Marie D. Felix from the National Cancer Institute, combined with refined fixation techniques, some instituted by Palade, the Golgi apparatus was at last recognized as a stack of flattened membranous sacs; it soon became accepted as a respectable, ubiquitous cellular organelle.[30] During the 1960s and 1970s, Palade and his students became foremost interpreters of the Golgi's role in protein transport, and in 1981 he and Marilyn Farquhar provided a history of the Golgi "from artifact to center stage."[24]

When the company sent Claude some experimental crown glass knives, 10 millimeters wide and ground to an angle of 30°, they expressed doubt, even pessimism, whether these knives could stand much shock.[36,37] After Claude's trials failed with these glass knives, in 1947 he tried Eversharp Schick ordinary safety razor blades that had been rejected as "not suitable for shaving purposes" because he had seen under the microscope that their cutting edge was sharp and clean.[38,39] And when these trials failed, according to Palay, he tried hard lumps of optical glass from Bausch & Lomb that were fractured "with blows from a hammer or dropped on the floor, and the pieces were selected for the smoothest, curving or nearly straight, thin edges. The fragment was fitted into the microtome knife holder and could be expected to cut only a few sections before the edge crumbled."[33]

Palay found it very difficult to secure the glass fragments in the knife holder and to cut two sections in a row having equal thickness. As a result, he commented, "the images we obtained of them in the electron microscope (an RCA EMU B) were only charred remnants of cells." No paper ever appeared using the Claude-Blum microtome, although Claude described and illustrated it in his 1948 Harvey Lecture without mentioning either his trials with glass knives or identifying the knife blade material.[40] Two years later, Harrison Latta and J. Francis Hartmann at the Massachusetts Institute of Technology reported success using glass knives.[41] In retrospect, Claude's failure may have been due to using a hard optical glass rather than a softer plate glass. His experimental microtome nevertheless became a prototype for a highly successful instrument for electron microscopy, the Porter-Blum microtome (see Chapter 6).

When Palay did not make any significant progress in producing acceptable sections, he and Claude turned to using the method of replicas to examine biological specimens with the electron microscope. This method was used in the early 1940s by physicists to study the crystalline structures of metal surfaces and was adapted by James Hillier and Richard F. Baker to examine cell surfaces.[42]

The Method of Replicas

by Sanford L. Palay

Claude had invented a procedure for making Formvar replicas of red blood cells. Blood from a chicken was spread on a glass slide and allowed to dry. The slide was dipped into a dilute solution of Formvar. When the resultant film had dried, it could be floated off onto a water surface, grids could be deposited onto it, and then the film could be picked up again on a clean glass slide, trapping the grids underneath. When the grids were picked off, they carried a piece of the film, which was thin enough to be examined in the electron microscope. The Formvar film faithfully replicated the protrusions and depression in the surface of the blood cells in the original specimen. . . . At this time, Claude had a close association with the geneticists at the Cold Spring Harbor Laboratory on Long Island. Among them were Barbara McClintock, studying the genetics of maize, and Bruce Wallace, studying *Drosophila*. Claude thought that his replica method could be adapted for studying the chromosomes of fruit flies, which for decades had been examined in flattened spreads on glass slides. Claude persuaded Wallace to bring us a couple jars of fruit fly larvae and show us how to tease out the salivary glands in a drop of acetic acid and prepare squashed preparations of the giant chromosomes in them. It was not difficult to dry them down and coat them with Formvar. In a short time, we had examined dozens of chromosome preparations and had developed the idea that the polytene chromosome consists of parallel threads, each bearing small globules arranged in register and corresponding to the cross-banding that was so characteristic in stained preparations. In order to discover what the globules consisted of, we secured a preparation of purified deoxyribonuclease from Maclyn McCarty (whom we met in the lunchroom) and treated the spreads with it. As the globules became smaller, we concluded that they contained DNA.

It was crude and indirect evidence, but we decided that the two papers, one by Claude on the chicken erythrocyte and one by me together with Claude on the chromosomes, should be written up and submitted to *The Journal of Experimental Medicine*, which at that time was the only journal with adequate illustrations. Many of the early papers using electron microscopy had been published in this journal, including the classic paper by Porter, Claude, and Fullam (1945) on the microscopy of cells in culture. After the papers were submitted, Claude somehow managed to persuade the editor, Peyton Rous, that they should be accepted, probably on the basis that the technique was novel and opened a new field of investigation to the electron microscope.[43,44] Replica techniques have flourished in electron microscopy since that time, but no one followed up our work on red cells and chromosomes. Much better methods were developed subsequently. . . .

Claude's laboratory was not an independent administrative unit. Formally, it belonged to the laboratory of James B. Murphy. . . . The star . . . was Keith Porter, who had earlier collaborated with Claude but was at this time in active competition with him to succeed Murphy as head of the laboratory. Porter carried on his campaign in a mischievous fashion, deprecating Claude sometimes within his hearing, and often causing pain. As a result, there was considerable coolness between the two, even though they worked in neighboring rooms and often shared equipment, such as the electron microscope. . . .

Working with Claude was an absorbing experience . . . arriving at about 9:30 each morning, I would busy myself with laboratory work or reading until Claude arrived about two hours later. He would make telephone calls and contend with his correspondence until 12:30, when the whole laboratory went to lunch in Founder's Hall. Lunch was the one time in the day when the entire staff of The Rockefeller Institute gathered in one place. Generally, laboratory groups sat together at long tables arranged in ranks across the room. As the tables were larger than most groups, there was perforce some mixing of members. In addition, some people insisted on sitting with different groups each day, and so there was a good deal of interchange and an enviable opportunity to converse with scientists from all fields from day to day. . . .

The afternoons were spent in laboratory work, and often Claude and I would examine the replicas in the electron microscope together. The microscope was installed in a large room in the basement of Theobald Smith Hall [and] was surrounded by a circular black drape, which we closed around us when we were ready to take pictures. In a neighboring alcove, the plates were developed and prints were made. It was all rather makeshift.

Around 6:00 we would go to dinner. Often, I simply went alone to the dining room of the New York Hospital. . . . Almost as often, Claude and I would have dinner together. We usually went to a Slovak hall in the mid-seventies, where we could have a nourishing dinner for little money. Then we would return to the laboratory and continue our work until around midnight. Claude would buy a *New York Sun* and take a taxi to the West Side, where he boarded the Broadway Seventh Avenue subway for his ride up to Riverdale, where he lived with his sister and his daughter, Philippa. I accompanied him in the taxi and in the subway until my stop at 125th Street. . . . It was in these evenings, the shared dinners, and the shared rides home that I got to know and love Albert Claude. . . . He was interested in everything—scientific, cultural, human. He could discourse on modern music, especially the work of his friend Edgar Varese, as well as on the role of mitochondria in the cell. He had a clear vision of what electron microscopy should be good for in biology. I remember and have often quoted his injunction that we should see more with the electron microscope than with the light microscope, not less. This particular adage is still pertinent when skill and flair in the use of the microscope are neglected for "practical" results in immunochemistry.

Excerpted by Palay from his reminiscences in "A concatenation of accidents," *The Neurosciences: Paths of Discovery II*. 1992. F. Samson and G. Adelman, editors. Birkhäuser, Boston. 191–212.

Porter and the Founding of the Tissue Culture Commission, 1946

Concurrent with Claude's efforts to produce better specimens of cells with a microtome and replica methods, Porter was engaged in finding ways to improve methods for cultured cells. His impetus to create a new organization devoted to tissue culture arose from many initial frustrations. It was, he said, "like being dropped into completely unexamined territory . . . to obtain an over-all picture of cytoplasmic organization as depicted by the electron microscope."[45,46] Further, his initial successes attracted so many postdoctoral fellows to the laboratory who needed to learn these new techniques that he grew impatient. He told Raymond Parker, Alexis Carrel's former associate at the Institute who had moved to the Connaught Laboratories at the University of Toronto, that he wanted to "get tied up with a good tissue culture lab so that so much of my time would not be taken up with tissue culture problems. The electron microscope end of things is much more my meat. If only some of those boys who control the cash had enough intelligence to see the possibilities, they would really get a good crowd of specialists together and in a few years shake some information out of the cell. . . . The quality of one's electron micrographs is very largely determined by the methods used in preparing the material. This is probably more true of these tissue cells than of most other materials."[47]

Getting specialists together in fact began in late 1945, when Charles Pomerat, an anatomist and pioneer in tissue culture from the University of Texas, visited Porter's laboratory. He then traveled to laboratories of well-known cytologists who were using in vitro research on living cells to solve other kinds of problems: in particular, George Gey and John Hanks at Johns Hopkins and Wilton Earle and Jack Dalton at the National Cancer Institute.[48]

As a result, Pomerat and Porter decided these scientists should discuss their problems and results at special session of the April 1946 meeting of the American Association of Anatomists.[49] Among the nine papers dealing with tissue culture, according to Sergey Fedoroff, the one by Keith Porter stood out. His electron micrographs with details of mitochondria, endoplasm, and Golgi were of "exceptional clarity . . . [and revealed] new structural detail in the cytoplasm, including the netlike pattern of extremely fine canals or chains of vesicles now known as endoplasmic reticulum."[50]

A month later, Gey was so excited by another presentation by Porter at the American Association for the Advancement of Science that he wrote to Pomerat urging the formation of a new panel on tissue culture under the Committee on Growth of the National Research Council.[51] But Porter advised Pomerat against this route because all "plans and promotions will be subject to review by the executive of the Committee and . . . much of what is planned may be nipped in the bud." He proposed instead seeking an outright grant from the Committee so they could enjoy greater freedom as a "service society" that would provide "information, techniques, bibliographies, and even prepared salt solutions and frozen dried plasma and serum."[52] As a model, he suggested the experience of the Commission on Biological Stains, whose chairman, Harold J. Conn, had advised him to create a self-supporting agency independent from government control and to develop a standardization of materials, but never a standardization of methods.[53]

As mentioned in Chapter 3, Murphy was a member of the Committee on Growth and was instrumental in obtaining the Institute's first outside funds in late 1946 to study "the structure of normal and malignant human cells as revealed by electron microscopy." Also through Murphy's intervention, Claude was appointed to the Committee's Panel on Cellular

Biology, chaired by George J. Streeter, whose mission was to advise on research support for the general problem of "determining the precise cellular structures that are involved in the onset of cancer." In November 1946, Claude organized the "Conference on Tissue Culture in Relation to Cancer" in Hershey, Pennsylvania. The conference was attended by a small group of leading tissue culturists, including Albert Fischer and Honor B. Fell from Europe, who discussed physical factors and nutritional requirements affecting animal cells in vitro, the effect of electrolytes, and the cultivation of various tissues and cells; they also showed motion pictures of living cells. Ross Harrison gave the keynote address on "The Historical Development of Tissue Culture."

A month before this conference, Porter wrote to several individuals invited to Hershey and included a draft for "The organization of a commission on tissue culture media and procedures." He suggested five aims: to prepare a few of the most common constituents of culture media for distribution to tissue culture laboratories; to examine reported procedures that would simplify and standardize tissue culture techniques; to compile a bibliography of studies that utilized cultured cells; to organize a library of pictures of cultured cells as they appear when grown under certain conditions; and to produce regular progress reports on these aims. He further proposed that a group of five commission officers (drawn from the American Association of Anatomists, the Society for the Study of Development and Growth, and the Committee on Growth) would carry out all the plans, outline policies, raise funds, and prepare the written reports.[54]

The Tissue Culture Commission (TCC) was officially created on 13 November after the final session of the conference in Hershey. Porter was chosen chairman of the TCC. Margaret R. Murray, a professor of anatomy at Columbia, was chosen secretary, and she immediately assumed the enormous responsibility of preparing a bibliography of tissue culture literature, amassing some 15,000 citations to original articles written before 1950.[50,55] The executive committee consisted of Duncan C. Hetherington from Duke University, Honor B. Fell from Cambridge University, George O. Gey, and Charles M. Pomerat. According to the commission's first official circular, other founding members were Austin M. Brues, Gladys Cameron, Dale R. Coman, Wilton R. Earle, Cutting B. Favour, George H. Paff, Raymond C. Parker, Henry S. Simms, Paul A. Weiss, and Philip R. White.[56] Membership was open to anyone engaged in research using tissue culture methods, and within a year, more than a hundred scientists from the United States and Europe had joined.

In its initial meeting, Porter recalled there was an element of "impatience with the fact that tissue culture was an art practiced by remarkably few laboratories."[57] The TCC's view was that more rapid development and greater use of tissue culture would lead toward standardization of materials, increased laboratory efficiency, and better reproducibility of experiments. Their temporary wish list of media included chicken plasma, placental cord serum, horse serum, serum ultrafiltrate, embryo extract, and even buffered saline solutions and triple-distilled water.[56]

A laboratory for testing culture media had top priority for Porter, and it took two years of intense effort to encourage the manufacturers of biologicals to produce and sell media components for tissue culture. His proposal for a testing laboratory was rejected at two seemingly obvious places. At Rockefeller, Gasser's rejection was based on the Institute's strict policy of not-for-profit research. Testing a commercial product meant there would be "adverse comment on the merits of taking up the time of a member of the staff that could be utilized more profitably in his own interests," Gasser noted, and "in the end there is involved the certification of a product by the Institute. This is a situation which we have always carefully avoided."[58] At Johns Hopkins, Warfield Firor and George Gey rejected the idea of a combined media and training laboratory for lack of funds and space.

By early 1948, Porter found laboratory space at the Sloan-Kettering Institute, located across York Avenue from the Institute; its director Cornelius Rhoads was also chairman of the Committee on Growth. Funds for two technicians were provided by the American Cancer Society via a (separate) grant from the Committee on Growth. Porter assumed full supervision of the testing that began in July 1948 and continued until July 1950.

Porter also enlisted the enthusiasm and complete participation of two commercial labs, Difco in Detroit, Michigan, and Microbiological Associates, then in Coral Gables, Florida. "There was a reasonable hesitancy on their part at first because of the limited and uncertain market and because extremely little experimental work was available to guide in devising production methods. They recognized, however, that in the next few years there would probably be a growth in the popularity of cultured material for experimental work and that a commercial supply of media would stimulate this development."[59]

Initially, Porter's testing laboratory advised the companies on how to prepare the materials, and then they examined and certified their products. Some test objectives were to detect frank toxicity or substandard properties; to analyze various components of the media, including lipid, serum phosphorus, calcium, and nucleic acid phosphorus; and to study problems associated with using and preparing media. To this end, each commercial sample of media was tested against an equivalent sample that was freshly prepared in the laboratory. During the first year, almost half of some materials submitted for scrutiny were rejected. The test object was the cultured cell, and reliance was placed chiefly on behavior and growth of cells subjected to different media. Growth was measured in terms of colony diameter. Using a particular strain of cells proved to be impractical, so fresh tissue explants from fourteen-day-old chick embryos were used instead. While Porter admitted this test procedure seemed crude, he found that careful controls followed by a single operator gave excellent uniformity. The results of his efforts showed that culture media of suitable quality could be processed, packaged, and distributed commercially. The first media, referred to by TCC committee members as Difco's Delights, went on the market in January 1949.[60] A few of the commercial media are listed in the table.[61]

Predictably, many potential customers were discouraged by the initial high prices. Porter himself was one of the first to protest, and Difco responded they were "not considering recapturing developmental and investigative costs but . . . only interested at this time in pricing materials on the basis of costs of preparation."[62] Nevertheless, Porter acknowledged the superiority of the commercial preparations that were stable for long periods, entailed little waste, and would in time allow him to dispense with one laboratory helper and simultaneously expand his experimental work.[63] Other customers were discouraged by difficulties in learning to match the new media to culturing specific types of cells. Porter expected this reaction, noting that "in general, the younger, newer, and less demanding workers have adapted the materials to their needs with greater success than their older contemporaries."[59] He further predicted that once there was widespread success in the mass culturing of viruses, there would be increased interest and demand by researchers using commercial media for many other scientific investigations.[64]

Tissue culture media certified by the Tissue Culture Commission and commercially available in March 1949

From Difco Labs, Detroit, Michigan	
Fowl plasma	Desiccated, from Plymouth Rock cockerels, 6–9 months old
Chick embryo extract	Dried or frozen from 11-day-old embryos
Human placental serum	Filtered and freeze dried from fetal cord blood
Horse serum	Dried
Distilled H_2O	Triple distilled from an all-glass Pyrex still
Balanced salt solution	Developed by George Gey. Low calcium ion content, a pH of 7.4, and equilibrated with room air
Phenol red solution	Sterile, 0.5% in 0.85% NaCl, non-toxic pH indicator
From Microbiological Associates, Flemington, New Jersey	
Balanced salt solutions	X6 and X7 vary in phenol red content, Z contains less $NaHCO_3$. From H.S. Simms. Used with ultrafiltrates
Serum ultrafiltrates	Ox, horse, sheep, or human. A supplementary nutrient to prevent accumulation of fat in tissues. Recommended for virus cultivation
Fischer V-614	A partially complete synthetic medium developed by Albert Fischer. Recommended for studies in cell nutrition

On the fortieth anniversary of the founding of the TCC in 1986, Porter noted how huge and flourishing the tissue culture enterprise had become. In contrast, he recalled its meager and inauspicious beginning with the first autoclaved serum that looked like molasses. "Many of you cannot believe," he said, "that there was ever a time when one could not order media for tissue culture. But, there was such a time, and some of us invested days, yes, many days, in collecting serum and plasma and embryo extract, not to mention washing glassware."[57]

Porter also addressed the lack of personnel trained in tissue culture. He enlisted support from Harry Eagle, scientific director of the National Cancer Institute, who in 1955 devised a synthetic medium to replace whole serum. Eagle was immediately enthusiastic about Porter's proposal because he was also being hindered by a dearth of trained people that kept him from starting research projects at Bethesda. In October 1947, the TCC submitted a proposal to Eagle describing a laboratory apprenticeship program, based in part on Murray's graduate course at Columbia, that would train MDs and PhDs for two years and technicians for one year in the techniques and values of tissue culture. When Eagle deemed this proposal "excessively elaborate," Porter suggested a month-long tissue culture course during the summer, and this program was funded by the NCI.

The first TCC course was given in July 1948 by Raymond Parker at Connaught Laboratories. Among the first sixteen students were Eagle and Don Fawcett, an anatomist from Harvard who would join Porter's Rockefeller laboratory a few years later. In 1949, the course moved to Cooperstown, New York, at the Mary Imogene Bassett Hospital, affiliated with Columbia University, where it was directed by John Hanks, professor of microbiology at Harvard. The course continued in other locations until 1982.

Until 1953, Porter helped to equip the training laboratories, and he compiled films and slides for demonstrations for these summer programs. He also gave the lecture-demonstration on his unique use of cultured cells for electron microscopy. By 1955, Porter believed the

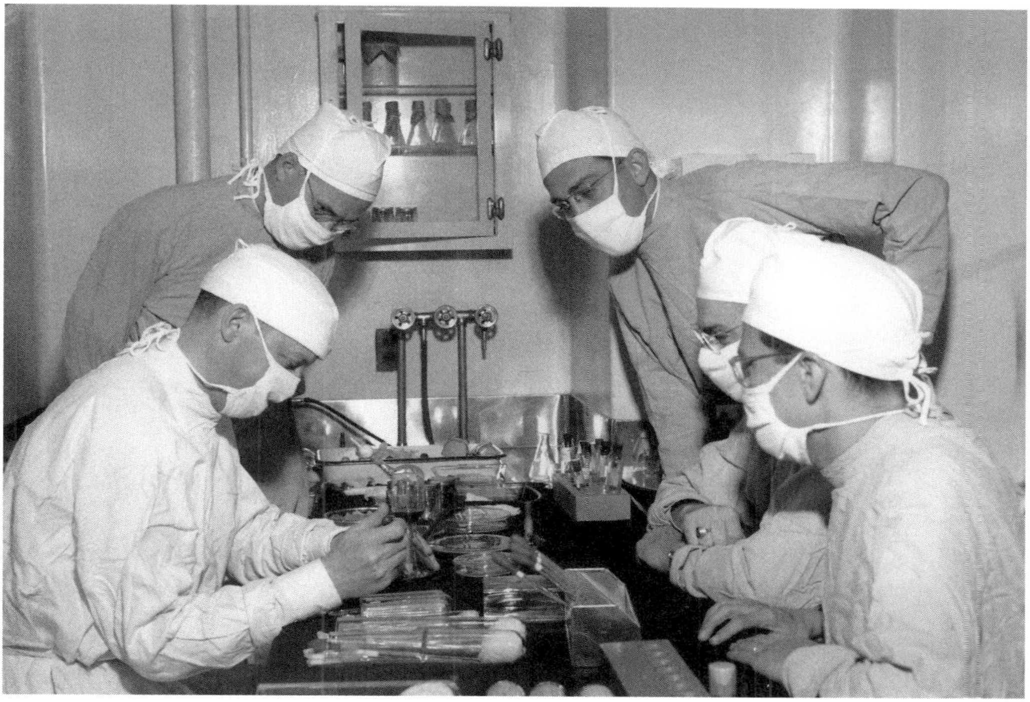

Keith Porter, left front, teaching cell culture methods at the first Tissue Culture Commission course in the laboratory of Raymond Parker at the Connaught Laboratory, Toronto, Canada, July 1948. Printed with permission from the City of Toronto Archives, Fonds 1266, Item 126816.

training program was achieving its goals since techniques once considered the specialty of a few laboratories were now in the service of many. Further, the sanctity of black gowns and masks perpetuated by Carrel had been replaced by standard aseptic methods. He noted in his introduction to the training manual that "the cultivation of tissue cells could be accomplished with modest equipment and easily acquired skills and further, and perhaps of greatest importance, a laboratory and lecture course in the subject would be both feasible and desirable for a college curriculum."[65]

Through his efforts on behalf of the TCC, Porter's talents as a tissue culturist also improved. At its initial meeting, the TCC discussed the viability of finding a chemically inert substitute for the plasma clot, thus taking a preliminary step toward the development of a synthetic nutrient medium. For years, the plasma clot was accepted as the most satisfactory support for the migration and multiplication of cells. However, due to its derivation from natural sources, its composition was unknown, variable, and uncontrollable. The plasma could also transmit avian bacteria and viruses. Further, plasma was difficult to collect and store. On his own initiative, Porter thought that purified fibrinogen clotted with thrombin might provide a more suitable support because its composition could be determined precisely, and its standardized nature could provide the same source throughout the experiments. With Clinton Hawn of Harvard Medical School, Porter examined clots in tissues formed by purified fibrinogen and thrombin. They used chicken plasma fractions to isolate concentrates of fibrinogen and thrombin by the method of the Edwin Cohn group at Harvard. These fixations were incubated in flasks with various tissues such as chick muscle explants, and they observed clot formations in the flasks. Although several publications appeared from 1947 to 1949 with striking electron micrographs of cultured cells, the results with this culturing technique were not entirely successful because fibrinogen in purified clots could not resist lysis as completely as in plasma clots.[66–68]

Another innovative effort by Porter and George Hogeboom during 1946 was to devise what was perhaps the first example of a cytochemical reaction that could be used in electron microscopy. When myoblasts from rat hearts were cultured in a neutral isotonic solution containing calcium chloride, the reaction produced prominent granules inside the mitochondria that were visible in the micrographs.[69] They went as far as looking at these cells in one of the first high voltage microscopes being set up by James Hillier in the University of Pennsylvania laboratory of Stuart Mudd.[70] The fact that the phenomenon was specific for one component of a cell, Porter told Edward Pickels of the IHD, meant it "should contribute greatly to our knowledge of intracellular organization and processes."[27] Investigation of this topic apparently was not pursued because Hogeboom quickly returned to chemical analyses of enzymes in fractions of mitochondria.

The TCC changed its name to the Tissue Culture Association (TCA) in 1949 upon the adoption of a constitution. Within its first three years, it had become a ready source of supplies, information, and new investigators. What emerged in the TCA's assessments of its achievements were some philosophical underpinnings of modern cell biology. The committee was generally in favor of changing this "clever specialty" of tissue cultivation from a service into a science of importance. In particular, a 1948 essay by John Hanks captured this spirit. In "A menagerie: The chiropractor, the pugilist, the pet dog and the tissue cultivator," Hanks argued that the short summer courses were comparable to the task of training pugilists and dogs to do a few tricks, and he predicted the art of tissue culture would progress in many sciences of biology and medicine only when those few who were making practical applications of tissue cultivation began to study fundamental problems and train students. He proposed instead creating full-time departments called Cell Biology where students could earn a graduate degree and professional status.[71]

From the beginning, Porter's goals were clear: he wanted to "shake some information out of the cell." In 1950, Pomerat praised Porter's revival of tissue culture, noting he was "a quiet person, yet one who had sheltered and reanimated the research spirit in nearly two-score of displaced investigators. . . . The history of the arts and the sciences has recorded all too inadequately how important have been the contributions of public-spirited citizens who have quietly sponsored, in all times and in all places, the pabulum essential for discovery."[72] For Porter, tissue culture, like electron microscopy, cell fractionation, and biochemistry, provided a much-needed bridge to the fundamental study of cell life.

A New Electron Microscope in 1948

Even though Gasser was investing money from the Institute's Contingent Fund to develop an experimental microtome, he exercised great caution before purchasing a full-sized electron microscope. It took four years, from the $6,000 purchase of the small RCA EMC console microscope in 1944, for the use of Wendell Stanley to visualize viruses at the Institute's branch in Princeton, until 1948, when a full-sized model was acquired for the Institute in New York. A substantial reason was the enormous sum of money to be invested. The cost was $16,000 for the instrument plus nearly an equal sum to construct housing for the delicate scope, which ideally had to rest on the solid rock foundation of New York City, and to hire one skilled technician to operate and keep the instrument in working order.

A more consequential hesitation on Gasser's part was that electron microscopy of biological materials was truly in its infancy. He needed to be convinced. Gasser may have worried, in George Corner's words, that the instrument was only a "formidable scientific toy."[73] As editor of *The Journal of Experimental Medicine*, Gasser had personally approved for publication several papers illustrated with electron micrographs taken by Thomas Anderson that he found exciting but far from intelligible. For example, a 1942 paper included a micrograph of the vaccinia virus from the laboratory of his close advisor and colleague Thomas Rivers, Rockefeller Hospital's physician in chief.[74]

Some of Gasser's skepticism dissipated with the reception of the 1945 Porter, Claude, and Fullam paper in *The Journal of Experimental Medicine* because it revealed considerably more information about cells than the earlier micrographs showing only profiles of microorganisms. The Board of Scientific Directors was treated to a demonstration of the electron microscope by Porter and Claude in October 1945, and this session further enhanced the value of the new instrument.

After little more than a year of shared access to the IHD microscope, Murphy and Porter began to agitate to acquire a microscope of their own. Several factors came into play during the critical year 1946. On Murphy's vigorous recommendation, Gasser promoted Porter from the position of assistant to associate by singling out how he "opened up a new field of investigation" and showed how the electron microscope can be used to investigate the intact cell. He also noted Porter's discoveries of a "heretofore undescribed cell system, the endoplasm," and a "clear-cut difference between normal and malignant cells." Murphy further informed Gasser that Porter had received outside offers of positions from Harvard and the University of Colorado, at twice the Rockefeller salary and with superior research facilities.[75] This was Murphy's hint that the IHD microscope could not continue to serve the laboratory's expanding needs. As it turned out, all papers published from Murphy's laboratory from 1945 to 1948 acknowledged the use of the IHD microscope, but after Pickels left the IHD in mid-1946, maintenance of this microscope deteriorated considerably. Despite Murphy's complaint to Gasser that the IHD arrangement was unsatisfactory, Gasser replied that its use "is on a personal basis, it does not take place at the official level." He therefore would not approach the IHD but offered to deliberate on another solution.[76,77]

Murphy then suggested that Gasser offer to buy the IHD microscope, but the IHD refused.[78] Meanwhile, with Murphy's help, Porter and Hogeboom were nourishing other leads to set up cancer research laboratories using tissue culture and electron microscopy at other institutions. As part of Murphy's role in obtaining a grant from the Committee on Growth for electron microscopy (see Chapter 3), there were also negotiations at Johns Hopkins with Firor and Gey and at Sloan-Kettering with its director, Rhoads. If the Institute did not consider Porter "a career man," Murphy prompted Gasser, then Rhoads was preparing to take Porter away and set aside space in the new hospital for a department of electron microscopy.[79,80]

The decision to buy a microscope became more compelling in 1948 after Rivers, who worked also on poliomyelitis, saw unpublished micrographs by Francis Schmitt and Eduardo De Robertis that seemed to show the polio virus traveling along the inside of a hollow nerve axon. This evidence was an exciting confirmation of what had previously been only conjecture.[81] Gasser, as a neurophysiologist and physician, was certainly impressed by Rivers's reports about clear views of nerve fibers since it lent increasing value of an electron microscope to his own field of research. In addition, Porter and Claude continued to provide good micrographs in a few publications comparing malignant and normal cells that, while controversial in interpretation, continued to show improved resolution. This is because as the tissue culture techniques improved, so did the micrographs and with them the value of using this new instrument in biology.

Finally, on 30 April 1948, the Board of Scientific Directors voted to appropriate from the Institute's Contingent Fund for "Miscellaneous and Special Items" the budgeted sum of $16,375 for a microscope. The actual cost of the RCA model EMU-2 was $16,291.13.[82] In October, the Board of Trustees voted to accept another grant of $18,000 from the American Cancer Society to set up a laboratory for the electron microscope that included a darkroom and air conditioning unit. This laboratory, located in the third basement of the New North Building (now Theobald Smith Hall), also housed the older IHD electron microscope.

Ironically, after this long decision process, Gasser was one of the first scientists outside the Murphy laboratory to use the new microscope. According to George Palade, Gasser "was very interested in finding by electron microscopy something that was dear to his heart—the dimensions of the C fibers, the pain conductors in the central nervous system which were too small to be seen by light microscopy. He devoted tremendous amounts of time trying to find those fibers. Working by himself with the microscope, he sometimes got furious because the microscope was not behaving and kicked it into shape. Actually he insisted in doing the work himself with a very good technician, Lucien Caro."[83] Although Gasser had electron micrographs of nerve preparations in 1948, his first published electron micrograph did not appear until two years later. He eventually used the microscope to confirm findings he had made twenty years earlier with electrophysiological instruments and was able to measure accurately the size of unmyelinated axons and to show their relationship to the Schwann cells in which they are embedded.[84,85]

Creating a Laboratory Called Cytology

The year 1949 was eventful in several other respects for the origins of modern cell biology. After a decade of watching, commenting on, and nurturing the research in the Murphy laboratory, Gasser was preparing to create a new field of science devoted to cellular studies.

Perhaps due to his enthusiasm for what the microscope could do for his own science, Gasser decided to purchase another one just six months after the RCA was installed. There were two reasons.

The RCA microscopes, even after a decade of production and improvements, had deficiencies, lacking such features as stable performance, astigmatic lenses, and excellent contrast.[8,86,87] Before ordering a new instrument in 1949, Porter sought advice from the major American architects and users of the RCA instruments, James Hillier, Cecil Hall, Ralph Wyckoff, and Robley Williams. The choice this time was a Philips 100 made in the Netherlands. This microscope featured an intermediate lens to provide superior continuous magnifications, a higher voltage whose beam was expected to penetrate without destroying thicker specimens, and an electrical focusing device; it also used film instead of plates. Although Porter found the RCA gave a slightly greater resolution, he claimed he could do "at least twice as much with [the Philips] in a given length of time and with far less fatigue."[88]

While anticipating the arrival of the Philips microscope, the Institute also initiated an auxiliary service to make it available to Institute scientists and provided support for a technician, supplies, and other accessories. The new instrument would serve, Porter told the Philips sales manager, as "a general service instrument for the Hospital and the Laboratories of the Institute. For the staff of the hospital it will provide a means for examining specimens from patients for the presence or absence of infectious agents. It will thus become of value in the diagnosis of disease." For the laboratories, the microscope would be used for medical research "to distinguish between normal and cancerous cells in terms of their fine structure . . . [in which] the greater power of penetration [of the Philips] will be a general advantage for the examination of thick specimens."[89] Since staff scientists had to be instructed and personally assisted, usually by Porter or Palade, this task demanded considerable time and effort. Porter once complained of such heavy usage by others that he found himself getting too involved in their scientific research programs.

By 1949, the Murphy laboratory was moving in several directions. The controversial micrographs of virus particles in malignant cells had ended one phase of cancer research in Murphy's laboratory. A general attitude to change research direction was emerging among the laboratory members, one often discussed with Murphy during their lunch together in Welch Hall. The key question was why they should study malignant cells when no one knew what constituted a normal cell.[90] In addition, Hogeboom and Schneider continued their chemical analyses of the mitochondrial fractions when they moved to the National Cancer Institute; they tried, in vain, to hire Palade, but after a visit to the NCI, he determined there were "several risks for an alien" and decided to keep company with Porter and Murphy.[91] Porter took over the new microscope and began exploring cells in collaboration with numerous postdoctoral fellows. Palade was just beginning to find his way with all these methods.

Claude resigned from Rockefeller in April 1949 and, unlike the others in the laboratory, remained focused on doing cancer research. After Murphy's rebuke concerning the writing of the paper on malignant versus normal cells in the summer of 1947, Claude became more isolated from active research in the laboratory. He had no success producing a microtome, a clear description of the Golgi apparatus, or a useful application of the replica methods. In 1947, he had attended the Fourth International Congress of Experimental Cytology in Stockholm and visited cancer research laboratories in Belgium, France, Denmark, and Sweden; within a few months he turned down offers to work in Liège and at the Queen Elizabeth Institute in Brussels. A year later, he accepted an offer from the Free University in Brussels to assume the scientific directorship of their Jules Bordet Institute. Established in 1940 as a cancer research and treatment center, the Bordet Institute was taken over by the German army during the war, then was occupied by British troops, and soon after became a camp for displaced persons. By 1947, the Jules Bordet was being rebuilt as a cancer hospital with extensive space for cancer research. In response to a formal welcome at the Bordet Institute on 7 July 1949, Claude announced that "five or six years from now we ought to be able to call cancer by its name, namely a viral disease, and in this large sense, by consequence, contagious. Philosophically, I have never been attracted by the theory which generates an anarchic cell from a cancerous cell. . . . Without a doubt, the viruses of cancer, for they are certainly multiple, will one day find their master."[92]

Also in 1949, Murphy was a year away from the Institute's official retirement age. Gasser had changed Simon Flexner's policy of abolishing a laboratory and disbanding its staff when a full member retired or died. Instead, on the member's departure, he preferred to judge "everyone in that laboratory individually and retain those whom he judged would offer productive years in research." In this way, he advanced the careers of several junior staff members.[1] Since Gasser had given a decade of thought about creating an independent laboratory devoted to cellular fine structure and had maintained a thoroughly critical evaluation of research by Claude and Porter during these years, it is not surprising that he was prepared to turn his observations into action.

Lastly, in 1949, Gasser assessed the Institute's future for John D. Rockefeller Jr., chairman of the board. A major section of his report focused on preparing the Institute "to intensify investigation of fundamental life processes at the level of the cell and its constituents." He was committed to a goal of producing "scientific capital" or the production of new knowledge, which he explained could neither be directly planned nor managed: "No one knows how. But the conditions for it can be maintained as they are now and always have been. The quintessence of the conditions is reliance on originality. That means fostering individuals and allowing them freedom."[93]

Among many new trends he predicted for medical research, his most elaborate prediction concerned the investigation of the structure and function of the units within the cell: "Thanks to the newer instruments, which permit greater resolution, the field of fine structure is just opening up; and it has brought into view the possibilities associated with a large extension of the range of vision." While he acknowledged that penetration to the molecular dimension of cells was not in sight, he saw that cell biology would foster its advent. "What is going on inside of the cell occurs in a microcosm of complexity. The cytoplasm is filled with particles, and something about the nature of the particles is now emerging. The particles may be thought of as intracellular organs. As the particles are defined, problems are set for a long pull ahead. One wants to know, step by step, the chain of chemical transformations peculiar to each particle, and then the relationships of each chemical system within the cell to the others in the cell."

Among his concerns, he worried that the undeveloped state of morphology was too often following consultation of a handbook and not keeping pace with other fronts in biology, but he argued that the electron microscope was changing this trend: "This enterprise belongs to a sector of investigation in which modern physical methods have one of their greatest opportunities for contribution to contemporary biology. In order to understand function it is necessary to have knowledge of the structure of the mechanism." He was also concerned that techniques might become dictators of the problem, rather than servants of it, so that what the Institute needed was someone who could fulfill the exacting conditions of the instruments yet be "temperamentally amenable to cooperation with associates having the same broad end in view." He concluded the likelihood of fulfilling such conditions must be "an outgrowth of what we already have."[94]

Accordingly, in July 1950 on Murphy's retirement, Gasser established a new laboratory of cell structure, called simply Cytology, that set up Porter and Palade as a separate unit within his own Physiology Laboratory. A newly created space in the third basement of Smith Hall housed the electron microscopes in dust-free rooms, biochemical and general laboratories, and an office. This unusual assignment meant, among other things, that Gasser supervised their research, approved their manuscripts, and oversaw their annual reports until the time when they might be promoted to full members and could establish an independent laboratory. He also secured the appointments of Porter as an associate member and, a year later, Palade as an associate. This development very likely reassured Palade and Porter, who had worried that they were about to become "orphans" on Murphy's retirement. When Murphy died suddenly just two months later, they were truly on their own, although still under Gasser's careful watch for the next three years.

In retrospect, Gasser invested in Porter and Palade because he had seen first hand that they were capable of succeeding beyond what they had already put into place. Gasser "made it easy for us in many ways," Porter recalled. "He was patient, he was considerate, he was interested in what we were doing, and he knew what we were doing. I remember one occasion when I had to go to his office with the annual report that I had written and he took it apart word by word. This was his attention to detail.

"I remember another instance when I had investigated fibrinogen or its transformation into fibrin. We were doing this with the EM and we were a little tardy in getting it published. Gasser came into the lab one day and said, 'Porter, what have you done with that study you made of fibrin.' 'Nothing, Dr. Gasser,' I answered. I was rather worried. He said, 'Well, you had better get in print,' he said, 'because I was at the academy last week and I heard somebody down there give a talk and he illustrated it with pictures that looked very much like the ones you've taken.' I wondered who could that could have been? So we hurriedly got the manuscript ready for publication. Then it developed that a fellow from Harvard had described this fine structure of fibrin without mentioning the source of his pictures. The pictures had come from us! We had loaned them to him because he had loaned us the fibrinogen from his laboratory. . . .

"Gasser was really interested in everybody, not just us, in our generation. . . . He elevated us from a position of relative insecurity to a position of security in our particular effort."[83]

From this point forward, Porter and Palade became champions of the electron microscope. Their quest was to establish norms where none yet existed and to map the previously unseen fine structures of cells. Throughout long careers, they constantly sought better microscopes, higher resolution, and sharper micrographs in order to make observations at the pioneering edge of vision.

Part 2

New Worlds inside the Cell
1950–1960

Chapter 6

Cytology: The Tentative Beginnings

Carol L. Moberg

Keith Porter and George Palade had clear ideas about the investigations that their "new" cytology unit could follow and the equipment needed to accomplish them. Anticipating their future work, Palade wrote a formal memo to Porter in October 1949 and proposed two projects. He suggested that the most "favorable and attractive object" was to study the nucleus to elucidate "an old and deceiving problem, namely the fate of the chromosomes in the interkinetic phase." He saw a number of technical problems in accomplishing this, particularly those related to fixation and better handling of cell sections. Thus, the second project he proposed was rebuilding Albert Claude's microtome to add an electric motor and finding a steady microtome knife. He introduced the intriguing hope that they could clear up the black spots that marked the place of nuclei in previous micrographs. And, most optimistically, he predicted, "We will have a complete picture of the fine structure of a normal cell on the electron microscope scale."[1]

In notes prepared for discussions with Herbert Gasser, Porter outlined the kind of experiments he had in mind. Curious among these was his lingering youthful love of work with frogs and salamanders on nuclear-cytoplasmic interactions, a project that remained in his dreams while cell structure occupied his full-time energies. His initial approach to organizing a cytology laboratory can be found in the advice he gave to others outside the Institute, which reflected a continuing evolution of his own staff and equipment. Initially, he recommended morphological and structural studies, cytochemical studies, cytogenesis or histogenesis, and biochemical and physiological studies, all comparing normal with malignant cells.[2] By the mid-1950s, Porter modified this advice and eliminated any suggestion of work on cancer cells. He instead recommended establishing three separate disciplines focused on the normal cell, physical chemistry, biochemistry, and cytology, each headed by a PhD who would be aided by a postdoctoral fellow and one technical assistant. Reflecting the Institute's organization, he emphasized that bigness was not a virtue, that working scientists should be free of administrative duties to focus on research, and that the laboratories "should not be permitted to drown themselves with apparatus and technical aides acquired through outside grants." He concluded with what he called his "wild idea" that he had been promoting in many places and for several years: creating a National Institute of Cell Biology whose emphasis on normal biology would counterbalance other institutes devoted to various diseases.[3]

For its first three transitional, even probationary, years, the Cytology laboratory was small. Except for Porter and Palade, the scientists were either guest investigators or postdoctoral fellows who wanted to learn the new techniques of electron microscopy, tissue culture, and cell fractionation and to apply them to a variety of scientific interests. There was a need for a biochemist to replace George Hogeboom and Walter Schneider. On 19 January 1950, Porter sought the help of Van R. Potter of the MacArdle Memorial Laboratory at the University of Wisconsin. Whereas four years earlier Potter sent his new PhD, Schneider, to work with Claude on the biochemistry of isolated subcellular fractions, this time Potter had no one in mind, and the post of biochemist went unfilled until 1954, when he recommended Philip Siekevitz (see Chapter 7). Where there had been twelve researchers in James Murphy's laboratory in 1948–1949 and eight in 1949–1950, there were only two others in 1950, Parker Vanamee, a physician who wanted to study connective tissue, and Hertha Meyer, a German-born Brazilian protozoologist who spent a year learning how to culture the one-celled organisms and to take electron micrographs of *Trypanosoma cruzi*. Two others came when Meyer left, virologist Francis Kallman and Jerome Wolken, a biophysicist interested in chloroplasts and photosynthesis. During 1952–1953, Kallman and Vanamee were succeeded by George Pappas, Carl Robinow, Maria Rudzinska, and Don Fawcett. A year later, the group expanded with the arrivals of Sanford Palay and Albert Sedar (see Chapter 9).

Two Cautionary Experiences

The matter of Peyton Rous. Just as the laboratory was taking shape, Peyton Rous managed to create a major problem for the new cytologists. According to Porter, Rous "tried to take us over with the thought of adding us to his stable of slaves."[4] Admittedly, Rous as editor of *The Journal of Experimental Medicine* knew a great deal about the cell research going on because, according to Porter, "he had come around to see the pictures and he had accepted the manuscripts without question." Rous remained convinced that viruses cause the generality of tumors, and Porter's micrographs showing virus-laden cells no doubt further encouraged his belief that neoplastic viruses could be present in cells, albeit in a latent form.[5]

After Gasser established an auxiliary electron microscope service for use by Institute scientists, Porter and Palade complained they were spending too much time instructing the scientists about how to use the electron microscope and getting involved in too many other research programs. Rous took advantage of this service to ask Porter and Palade to compare micrographs of two Shope virus-induced rabbit papillomas. When the fixed tissue sections gave encouraging but inconclusive results, Rous, collaborating with Stanfield Rogers, resorted to trying to culture the tumors, a task of great difficulty that resulted in an overwhelming number and variety of specimens. Rous mistakenly believed Porter and Palade would do all the electron microscope work if he provided the specimens, and he became annoyed when only a few micrographs were being taken of the numerous samples he was preparing. Porter told Rous his preparations were unsatisfactory for microscopy and that "it would require many weeks or even months of concentrated effort on my part to obtain convincing pictorial evidence of viruses in such material. This I cannot afford." The situation ended somewhat acrimoniously in May 1952 after Porter told Rous that he and Palade were never more than "technicians in the solution of Dr. Rous's problem."[6,7] "Fortunately," Porter later reminisced, "Gasser came to our rescue" and allowed the two cytologists to continue transforming electron microscopy of cells into an independent discipline and to minimize the personal assistance to others.[5]

A search for acid phosphatase. The laboratory's studies of the nucleus were short lived. Pursuing work on the intracellular distribution of enzymes, which Hogeboom and Schneider began while in Murphy's laboratory and were continuing at the National Cancer Institute, Palade began by looking for the intracellular localization of the enzyme acid phosphatase in liver cells. (About this same time in Belgium, Christian de Duve was also trying to unravel riddles of acid phosphatase; his story appears in Chapter 11A.) Palade made a comparative study of two techniques for analyzing the same material: a biochemical analysis of the nuclear and cytoplasmic fractions isolated by differential centrifugation and a histochemical analysis using a Gomori staining technique. Surprisingly, he found no correlation between the results: the biochemical analysis showed acid phosphatase was concentrated in the cytoplasm (almost half in the mitochondria), whereas the histochemical method showed it concentrated in the nucleus. He then learned the histochemical test did not reveal differences in enzyme concentration among the various cell structures but instead showed the differences in their affinity for another substance, lead phosphate. This test also caused considerable fixation damage and the formation of artifacts. He concluded a biochemical approach to study intracellular compounds would be superior because the material could be better preserved and there were more reliable methods for demonstrating enzyme activity. In this brief set of experiments, he learned more about techniques than the nucleus.[8,9]

Further work to identify and possibly clear up the black spots in the nucleus was halted by a different problem. Among the sixteen Institute members who headed laboratories in 1950, Alfred Mirsky had been studying the nucleus for nearly a decade and at this time was also making a biochemical study of the nucleus (see Chapter 14). Just as Claude had earlier tread on Mirsky's scientific territory by isolating chromatin threads from nuclei, Palade was now making incursions on Mirsky's research program by looking for enzymatic activities in the nucleus. The issue was subsequently resolved when Palade hired biochemist Philip Siekevitz and they began looking at cell components other than the nucleus. As Porter wrote to Thomas Rivers, interim director of the Institute after Gasser's retirement in 1953, he had persuaded Palade "to consider alternative approaches to the same problem, which is to define the function of a certain small granular component of the cytoplasm."[10] Within two years, their discoveries about this granule added another milestone in modern cell biology (described in Chapter 7).

During these trying experiences, the laboratory was juggling several experimental systems to improve procedures for electron microscopy. It was becoming necessary, in particular, to overcome the limitations imposed by cultured cells. Within three years, the laboratory had its first successes: Palade vastly improved the fixation method for specimens and Porter developed a new microtome—techniques that made it possible not only to obtain a more realistic view of the cell but also to see what the organelles looked like inside.

New Techniques for Preparing Specimens

Whatever procedures they used—culturing cells or tissues, fixing, embedding, and slicing the specimens—each step to render the specimen suitable for visualization in the electron microscope meant trying to preserve cells in as near life-like form as possible. As often as not, each of these procedures could alter the original living material and produce artifacts in the final image. Thomas Anderson, a pioneer in biological electron microscopy, understood that "artifacts are always with us in biology, but the important thing is to recognize the nature of the artifacts, so that they can be corrected . . . either by better specimen preparation or by devising what the original specimen should appear to be."[11]

Fixatives. In their first annual report to Gasser in 1951, Porter and Palade took the position that an intelligent interpretation of artifacts could be made if the action of the fixative was understood.[12]

Prior to electron microscopy, most fixatives for *cultured* cells were acids or contained protein precipitants that caused the cytoplasm to draw up into dense clumps so that preexisting fiber patterns were retained.[13] Porter had learned, for example, that formaldehyde, applied in isotonic saline and buffered at pH 7.0, gave a more faithful fixation and hardened the specimen so that subsequent drying did not shrink it. On the other hand, osmium tetroxide, introduced by Thomas Strangeways and Ronald Canti in 1927, was more valuable because it preserved the normal cell architecture and simultaneously produced density differences that aided in segregating structures. As described in Chapter 3, Porter adopted and modified the fixative's application as a vapor to preserve cells for electron microscopy, and he reported the technique in the classic 1945 paper.

Finding new fixatives of *whole tissues* for thin sectioning became more imperative after Sanford Newman, Emil Borysko, and Max Swerdlow at the National Bureau of Standards reported success in 1949 using methacrylate to replace the conventional paraffin or hardened waxes as an embedding medium. This plasticized resin gave an optically clear matrix from which the sections could be cut.[14,15] In 1951, when Palade set out to find a fixative to use with methacrylate, he was guided by Porter's pioneering use of osmium tetroxide.[16]

Although osmium could fix cultured cells quite well because they are only one or two micrometers thick, it did not perfuse well into the thicker tissue cells. As Palade described in a letter to Claude in Belgium, he discovered the "morphologic changes sustained by these cells indicate that fixation is preceded by a rather intense acidification of the tissue. It is for this reason that I have tried two things at the same time: adding, on one hand, a fixative of inhibitors capable of blocking the metabolic reactions generated by acid . . . and, on the other hand, buffering the solution with osmium tetroxide with a pH in series. The first attempt gave no results; the fixation was even worse than that obtained with tetroxide alone. The second showed that the quality of the fixation is very clearly influenced by the pH and that the best fixation can be obtained, for a long list of tissues and organs, in buffering the fixative (osmium) with the buffer solution of Michaelis at pH 7.5 (the physiological pH that produced a nearly normal appearance)." Palade bragged to Claude, "The results are very impressive: one is tempted to believe that the entire microscopic anatomy can be done on this new scale."[17]

The mixture containing one or two percent osmium tetroxide buffered with veronal-acetate at pH 7.3–7.5 was affectionately referred to as "Palade's Pickle." It was rapidly adopted by others, particularly when it was recognized that refined tissue mincing could be used to produce small blocks of tissue for immersion in the fixative.

Other investigators tried to control the osmium fixation process, notably Fritiof Sjöstrand and his students Johannes Rhodin and Henrik Zetterqvist at the Karolinska Institute in 1954–55. They used a balanced salt solution with little buffering capacity that produced good samples, but their method was not widely adopted.[18–20]

In 1953, Porter and Kallman produced a definitive paper concerning the chemistry of how osmium tetroxide worked as a fixative and established the criteria that "the cell should not detectably change its shape under the action of the fixative and that its cytoplasm should not, under high resolution, show discontinuities and lacunae of irregular size and angular form." Earlier papers on the fixation properties suggested it reacted with lipid double bonds and perhaps also with proteins. Porter and Kallman extended the work on proteins by altering the concentrations of the fixative, the length of the fixation process, and the protein concentrations using purified serum albumin, serum globulin, and fibrinogen. They learned that osmium tetroxide worked by decomposing and leaching out the proteinaceous matrix surrounding the formed elements of the cell, such as the mitochondria and endoplasmic vesicles, thereby leaving behind a membrane skeleton that was of excellent quality for electron microscopy.[21]

Still, all was not perfect with osmium tetroxide; not all the cytoplasmic macromolecules were rendered sufficiently insoluble to withstand leaching in subsequent processing steps. This fixative also destroyed all enzymatic activity so that cytochemical reactions could not be demonstrated. The problem was not solved until ten years later, in 1963, when David Sabatini, Klaus Bensch, and Russell Barrnett at Yale discovered that glutaraldehyde gave the best general preservation of cellular fine structure.[22] (Barrnett had been a member of the Porter/Palade laboratory in 1955–1956; Sabatini, an MD and a postdoctoral fellow in Barrnett's lab, would earn a PhD under Palade in 1966. A description of the 1963 discovery is included in Sabatini's article in Chapter 8A.)

Staining. Contrast was long recognized as essential for distinguishing and evaluating various components in micrographs of cells. Much earlier, many laboratories experimented with heavy metal stains, such as uranyl acetate and phosphotungstic acid, with limited success. In 1945, Francis Schmitt at the Massachusetts Institute of Technology (MIT) showed that phosphotungstic acid could be used as a stain after osmium tetroxide fixation without having to extract the methacrylate embedding medium.[23] Ten years later, Cecil Hall found this stain also worked very well as a "negative" enhancer of the electron microscope image.[24] However, after epoxy resins replaced the polymethacrylates, the heavy metal stains were inadequate for providing contrast because the epoxies were so dense that there was no sublimation of material during electron bombardment. Epoxy resins are also quite hydrophobic and interfere with the penetration of heavy metal stains.

In 1958, Michael Watson, who spent 1954–1956 in the Porter/Palade laboratory, made a detailed comparison of the heavy stains uranyl acetate, phosphotungstic acid, and phospho-molybdic acid to test which were best for certain cell structures. He used "Palade's Pickle" and polymethacrylate embedding and added isotonic sucrose (the latter suggested by postdoctoral fellow James Caulfield).[25] Watson emphasized the importance of applying the stains after embedding and recognized two types of staining: one produced a marked change in density and the other a change in morphology. Both changes depended on the differential binding of heavy metal groups to specific intracellular structures.[26] Watson also introduced barium and lead salts as stains and found these to be highly effective with methacrylate resins and, somewhat later, with the epoxy resins.[27] While the two salts differed in their ability to stain structures, Watson concluded the alkaline lead stain was preferable because of its affinity for many cellular structures and its high density. This work allowed him to produce the first good electron micrographs of the ribonucleoprotein particles on the endoplasmic reticulum membranes, thus making lead hydroxide the stain of choice for these particles.[28]

The Porter-Blum Microtome

In 1950, Gasser commented in a paper containing his first electron micrographs of nerve fibers that the microtome was still not perfected and that the "effectiveness in the use of the electron microscope awaits a considerable advance in the techniques of section cutting."[29] By this time, Porter and Palade were prepared to take advantage of developments in low speed microtomy that had occurred during the two years since Claude called attention to his prototype hand-turned microtome in his 1948 Harvey Lecture (described in Chapter 5).[30]

Among the important developments, in 1948 Daniel Pease and Richard Baker at the University of Southern California introduced a low speed instrument using a Spencer rotary microtome,[31] and in 1950 Ludwig H. Bretschneider in the Netherlands produced one based on the Cambridge rocking microtome.[32] While both groups could cut sections in the submicrometer range, the resulting micrographs showed artifacts that they concluded were due to the fixation and embedding procedures. Further progress came in 1949 when Newman, Borysko, and Swerdlow, who, in addition to introducing the methacrylate embedding medium, produced a microtome having a simple desk lamp over the specimen that created a thermal advance system for the knife.[14,15]

In 1950, James Hillier and Mark Gettner elaborated on Pease and Baker's technique and independently developed a trough for collecting floating sections on a water surface to permit their easy transfer to support grids.[33] The Claude-Blum microtome had this feature, but its description was not published. Writing from Brussels in 1951, Claude complained to both Palade and Bretschneider that Hillier and Gettner had not given him credit for designing the liquid reservoir feature on his prototype microtome: "Gettner visited my laboratory in 1947 and saw the microtome which was described to him, including the well in front of the blade where sections can expand directly on liquid. If you write about the subject," Claude told Bretschneider, "I hope that you will have the opportunity to correct this 'oversight.' "[34,35] Bretschneider replied, "The Gettner error will be corrected in my article."[36,37]

Another important advance in 1950 was the introduction by Harrison Latta and J. Francis Hartmann, in Schmitt's laboratory at MIT, of glass knives to replace the unsatisfactory razor blades or heavy knives. They described how to crack off strips of plate glass, striving for a 45° angle, and to insert them into the chuck of a Spencer microtome. The result, as revealed by electron micrographs, was very satisfactory.[38]

During Palade's search for a better fixative, he was also upgrading the prototype Claude-Blum microtome as he had suggested to Porter in the 1949 memo: he attached an electric motor to make it a low speed instrument and used the recently introduced glass knives to cut the sections.[16] In August 1951, when Claude wrote to Palade asking whether Blum would build him a replica of his microtome, Blum refused, saying he had neither the time nor the men needed. Palade nevertheless reassured Claude that "it was very likely that this fall we are going to modify once again the old model to reduce even more (and to better regulate) the advance. I am nearly sure that this time Blum will occupy himself with this problem with priority, since Papa Gasser himself has become in the meantime an electron microscopist, and, in this capacity, seems to be very eager to have your model of microtome."[17]

A year later, in the autumn of 1952, after Porter received funds of $783 from the Institute, Josef Blum, the instrument maker who had designed Claude's microtome, built a new prototype. Palade informed Claude that it was "a happy combination of your old microtome with a 'gadget' of Porter's and several new ideas of Blum. Despite this rather charged heredity, the prototype we use works very well and Blum is ready to construct another for you. He already has the majority of the pieces and says that he can have the microtome in two months ready to ship to Brussels." The cost was $400.[39]

1. Side view of the early Porter-Blum microtome with microscope attached, 1953. The horizontal bar has a specimen holder at one end that is drawn vertically past the knife. The instrument, like the Claude-Blum microtome, has a single-pass motion to guide the specimen over the hand-turned knife on the cutting stroke and bypass it on the return stroke. The specimen was moved toward the knife by thermal expansion of the horizontal bar heated by a reading lamp. Thumb screws adjusted the thickness of sections, and a water bath collected the cut sections. Courtesy of Keith R. Porter.

The Porter-Blum microtome was a vast improvement over earlier attempts.[40–42] It was evident to Porter that the new instrument would have to use a "single-pass" mechanism of the knife in order to take advantage of the methacrylate embedding medium, glass knives, and the collection of sections on fluid surfaces. This mechanism was a special feature of Claude's microtome that influenced all subsequent designs of instruments for ultramicrotomy.

In the new prototype, Blum introduced a different single-pass mechanism that consisted of a horizontal bar with a collet-type chuck on one end to hold the embedded specimen and two sets of pivots (constituting a gimbal) at the other end to hold the bar and provide for its universal motion. In operation, a parallelogram movement guided the specimen over the knife on the cutting stroke and bypassed the knife on its return stroke. The specimen was moved toward the knife by thermal expansion of the horizontal bar heated by a reading lamp. "In its simplicity," Porter recalled, this model was "a charming and reliable instrument."[28] One pass of the specimen block against the knife severed a section of plastic, which was examined by a dissecting microscope mounted on a flat movable plate. If color and thickness were satisfactory, a ribbon of consecutive sections was cut and fell into a water trough below; individual sections could then be retrieved and mounted on grids for examination in the electron microscope (figure 1).

For a later model, Blum designed and built a hand-driven advancing mechanism to produce precision serial sections of uniform thickness. The commercial production

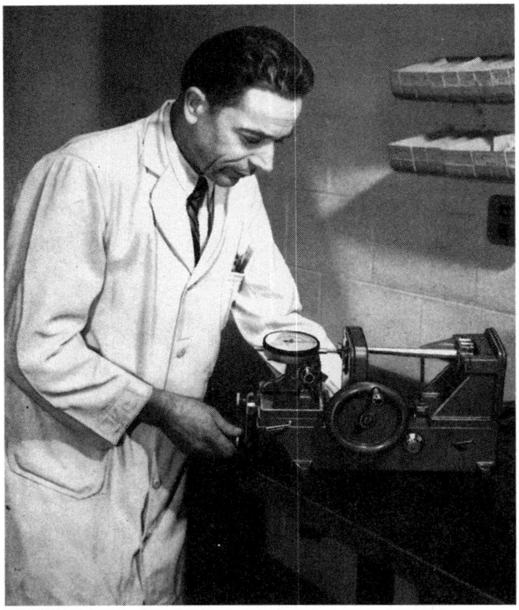

Josef Blum at Sorvall Porter-Blum microtome, 1953. The MT-1 was the first commercial version of the microtome shown in figure 1. In this version, the tissue block is advanced to the knife by a mechanical device controlled by a knob that can vary section thickness from 25 to 500 nanometers. Courtesy of Boeckeler Instruments Inc., RMC Products Group.

of this model, designated MT-1, increased its success, particularly when Blum left Rockefeller in 1953 to join Ivan Sorvall, Inc. of Norwalk, Connecticut, where he took charge of the commercial manufacture of the instrument. On this model, the knife was advanced toward the specimen by coarse and fine screws, and the section thickness could be varied from as little as 25 up to 500 nanometers per section. Other important features included unlubricated pivotal bearings for all moving parts that required little adjustment and a metal hood enclosing the bar and advancing mechanism to protect them from dust and effects of ambient temperature changes. Blum had initially discarded the idea of the motor-driven microtome that Porter and Palade tried on Claude's instrument; problems arose due to the high frequency vibration, low cutting speed, and longer cycle that led to thermal fluctuations.[43] By the early 1960s, when Sorvall produced a second model, the MT-2, having a similar suspension and feed mechanism, a motor drive was incorporated that could provide a rhythmic motion of variable speed and produce a greater range of section thickness.[42]

The Porter-Blum microtome was originally designed to use glass knives purchased from the Acme Glass Company in Cambridge, Massachusetts. Later the laboratory made their own, considerably less expensive glass knives cut from one-fourth–inch plate glass with pliers; if imperfections were found after use, the knives were discarded. When Sorvall began producing the MT-1, it came with six glass knives and an auxiliary upper stage so that investigators could use the metal knife holder and knife supplied with other microtomes.*

Even before a definitive description of the Porter-Blum instrument was published in 1953, word had gotten around about this superior, reliable, and easy-to-use microtome. Porter was receiving countless inquiries from scientists outside the Institute about how to obtain one. He told the eager investigators that Blum and his co-workers in the Institute's Instrument Shop were available to make them for about $400–$500. When Sorvall began

*An important climax to cutting edges came with the introduction of diamond knives by Humberto Fernández-Morán in 1953.[44] Although an improvement, one drawback was the high cost of both the initial purchase and the resharpening. (In the United States, diamond knives were produced and resharpened by DuPont.) Therefore, many laboratories used glass knives for routine work and saved the diamond knife for the final, publication-quality specimens. By 1956, Porter began using diamond knives obtained from Ivan Sorvall, while the Rockefeller Instrument Shop made diamond knife holders for him and outsiders who inquired. By the early 1960s, he switched to DuPont knives. In a memo to himself and those who inquired about its use, he referred to a later paper by Humberto Fernández-Morán[45] and concluded the knife is "especially useful for cutting materials and objects that cannot be sectioned with glass or steel knives . . . it has been used successfully to cut cone, teeth, hair and harder inclusions of cells such as yolk and melanin granules, as well as wood and plant products of many types."[46]

producing them for about $700, the Rockefeller machinists had to stop their production during regular working hours because it was interfering with other duties. John Doherty, a later head of the Instrument Shop, was a new employee in the mid-1950s and recalled how he and his machinist colleagues then built Porter-Blum microtomes during their lunch hours and on weekends. Each instrument took nearly 100 hours to build, since all parts were made by hand, including hand-engraved knobs with increment measures. By 1960 the price was nearly $1,000, which brought the machinists considerably more than their $3.50–$4.00 hourly wage.[47]*

Of historical interest, no patents were taken out on this instrument because, as Porter later explained, the Institute "was operated for the benefit of mankind." Its policy, in keeping with its philanthropic purposes, was to make all its discoveries freely available to the public. In discoveries involving therapeutic drugs or new instrumentation, the Institute filed patents to insure purity and precision and to avoid exploitation, but neither participated in their commercial preparation and sale nor received royalties or other benefits from the patented items. Porter added that even though Sorvall took out a patent on their microtome, "no royalties were distributed to the designers."[48]

Three other microtomes with single-pass mechanisms for ultrathin sectioning were also reported in 1953. Two were American made, one by Michael Watson at the University of Rochester (who was about to join the Porter/Palade laboratory)[49] and the other by Mark Gettner and Leonard Ornstein in New York City,[50] but the one that dominated the European market was the "Ultratome" designed by Fritiof Sjöstrand and built by LKB in Sweden.[51] Within a few years, microtomes from other companies appeared on the market: in the Netherlands, the N.V. Philips Company built one in 1955 designed by Hendrik B. Haanstra; in England, the Cambridge Instrument Company made one designed by Andrew F. Huxley; in Germany, Leitz produced one designed by Humberto Fernández-Morán; and still later, Hitachi of Japan marketed one along with their own model of an electron microscope.

As early as 1954, a microtomy workshop at the New York Academy of Sciences showcased a dozen new microtomes from various East Coast designers and manufacturers. Irene Manton, a pioneer electron microscopist in England, recalled that some were crude and "others almost comically complex, but only the Porter-Blum behaved perfectly."[52] Although many microtomes were similar to the Porter-Blum instrument, it was the simplicity and reliability of the MT-1 Porter-Blum instrument that made it so successful. It remained in production well into the 1980s.[28,42]

The Organelle Status of Mitochondria

These improvements in techniques for electron microscopy opened many paths to a systematic exploration of cells for structural patterns of general significance and to intensify the correlation of fine structure of cells with their function. While other members of the laboratory gained amazing views inside various cell types (see Chapter 9), Porter and Palade pushed ahead with mapping the structures of the two organelles that had preoccupied the laboratory since 1945—mitochondria discussed in this chapter and the endoplasmic reticulum (discussed in Chapter 7). In a very short time, their approaches became a discipline-building strategy

*The Claude-Blum microtome (1947), an original Porter-Blum microtome with a mechanical advance, and a motor-driven MT-2 Porter-Blum microtome belong to the Merrill Chase Historic Scientific Instrument Collection at Rockefeller.

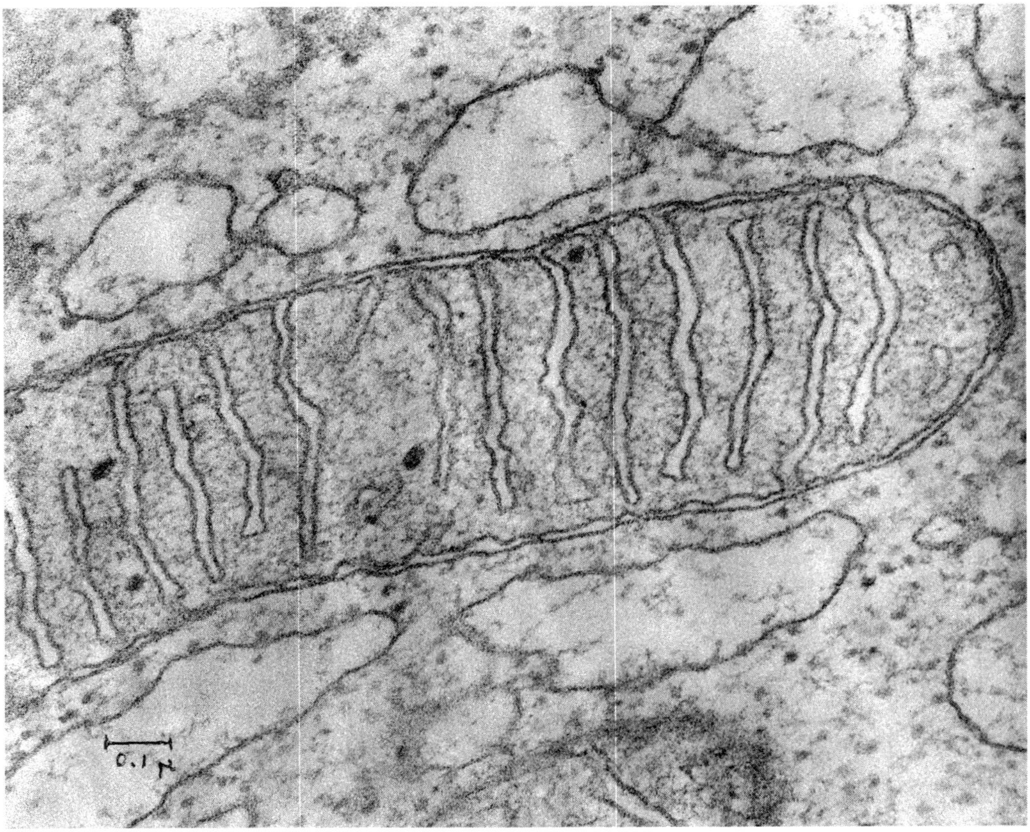

2. Electron micrograph of a mitochondrion within an axon of a mouse neuron in tissue culture, 1968. Courtesy of James Jamieson.

as they linked better electron micrographic images with evidence from cell fractionation and biochemistry.

By 1953, more was known about metabolic pathways within a cell than about which corresponding cell organelles determine them. In the case of mitochondria, the light and electron microscopes had revealed their existence, and there was a large, separate body of biochemical information about enzymes as a source of cellular energy. Nevertheless, for a long time this energy was thought to be produced by a structureless entity.

In 1947 and again in 1953, Robert R. Bensley, who had given critical help to Claude a decade earlier, reiterated his belief that the "mitochondrial structure is expendable" and that "to erect the mitochondria to the dignity of an organelle or cell organ . . . is a blunder." He called mitochondria "simply temporary aggregates of complex composition" and claimed that he had seen with the light microscope how they "disappear" after feeding, observations that were possibly due either to variations in refractive index or the limit of resolution of the light microscope.[53,54] Palade noted that just because Bensley demonstrated that mitochondria lose or gain materials, this fact "does not preclude their being recognized as cell organs, because such changes are true for any organ on any scale. What is stable in a given organ is only a certain pattern of organization in which the various compounds fit while continuously turning over." Such a situation, he asserted, "should be reexamined with the electron microscope."[55]

Consequently, after finishing his studies on fixation, Palade turned to microscopy and became the first to report the new elements he discovered inside mitochondria.[56] The mitochondria, he reported in 1952, are bounded by an outer membrane that separates them from the rest of the cytoplasm. There was also an internal membrane system of parallel, regularly spaced ridges—each ridge having a central transparent layer covered by a dense, thinner layer on each side—that protrude from the inside surface of the membrane toward the interior. He named these internal ridges *cristae mitochondriales*, based on observations he had made on cell sections from a variety of animals and cell types. His experiments utilized the fixative he had developed, a prototype Porter-Blum microtome with a glass knife, and a newly acquired RCA EMU 2a microscope (figure 2).

In all cases, mitochondria were seen to possess a limiting membrane, a system of internal ridges, and a matrix between them that occasionally contained granules. The outer membrane, whose existence had long been postulated after studies of osmotic pressure and permeability (by Hogeboom, Schneider, Albert Lehninger, and David Green, among others), appeared uniform in structure. Depending on the direction in which a section was cut, the various contours of this membrane in the electron microscope suggested mitochondria have random shapes, while under the light microscope their normal shape appeared elliptical.

The uniform and universal appearance of the outer and inner membranes prompted Palade to state they "may be used in the future as a criterion for the identification of mitochondria in electron microscopy." Based on established biochemical findings that enzymes in mitochondria in vitro carry out coordinated and complicated oxidative processes, Hogeboom and Schneider claimed these processes should be retained in a "definite spatial relationship."[57] Palade postulated that these enzymes are built into the solid framework of the inner membrane and "arranged in the proper order in linear series or chains—a disposition comparable in design and efficiency to an industrial assembly line" (figure 3).

Not everyone agreed, particularly not Fritiof Sjöstrand of the Karolinska Institute. As the history of scientific research amply illustrates, there is a stimulating value in honest scientific controversy. Especially in biological research, vital processes are difficult to observe accurately, experimental conditions are hard to control, and reaching conclusions from two-dimensional images presents obvious difficulties for interpreting three-dimensional living structures. The controversy between Palade and Sjöstrand involved interpretations of electron micrographs and centered around the organization of the infoldings of the inner mitochondrial membrane (cristae) and the connectivity of the compartments they enclose. A few months after Palade reported his observations, Sjöstrand published his first reports on the fine structure of mitochondria with no mention (and perhaps no awareness) of Palade's work.[57,58]

Sjöstrand, a prominent electron microscopist in Europe, had been a postdoctoral fellow in Schmitt's laboratory at MIT during 1947–1948. He joined their studies to measure macromolecular assemblies directly from micrographs and to compare these findings with other biophysical methods such as x-ray crystallography. He had also visited Murphy's laboratory at Rockefeller in 1949 to discuss Porter's methods of preparing tissue culture cells as well as the design of microtomes. A few years later, he learned about Palade's new method of fixation through personal communication with him. Thus, it is no surprise that, while working independently, Sjöstrand used techniques very similar to those of Palade and Porter to obtain sharp micrographs of mitochondria. Even so, his interpretation differed from theirs.[59]

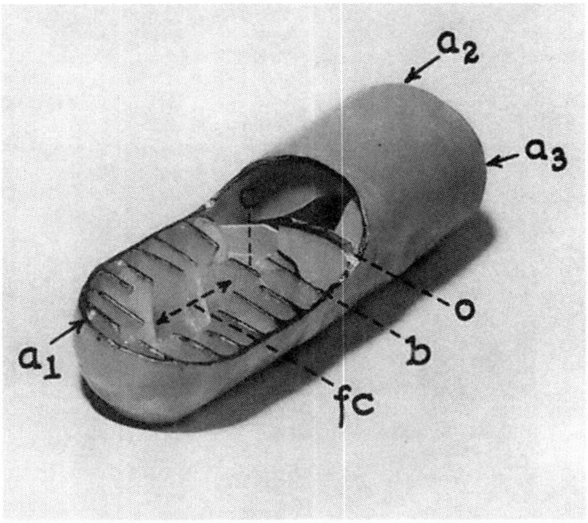

3. Palade's interpretation of inner mitochondrial membranes. The wax model shows the inner membrane folding into the interior to form cristae and leaving an open channel in the middle. From Palade, 1953. *J. Histochem. Cytochem.* 1:188–211 with permission from SAGE Publications. ©The Histochemical Society.

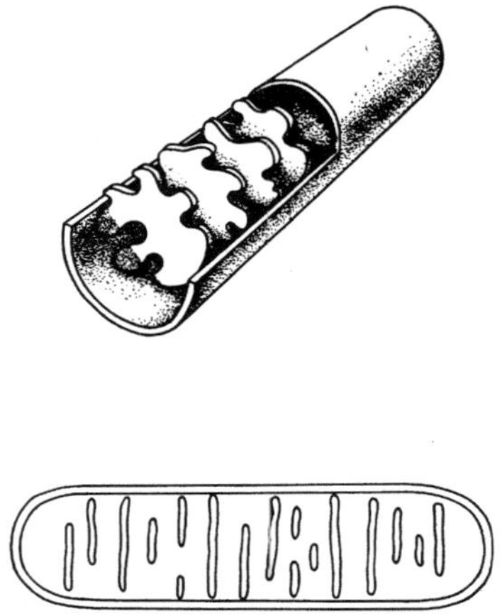

4. Sjöstrand's interpretative diagram shows inner mitochondrial membranes as separate structures and not contiguous with the outer membrane. ©Academic Press, G. Oster and A.W. Pollister, editors, 1956. *Physical Techniques in Biological Research.* 284.

Sjöstrand insisted that only longitudinal sections—those cut on the long axis of the mitochondrial shape—could produce a true (normal) picture of their topography. Because his micrographs had a higher resolution than Palade's, he was able to describe for the first time an outer double membrane in mitochondria "of mainly protein character separated by a double layer of lipid molecules." He also observed that mitochondria have a system of internal double membranes that looked to him like a stack of plates. In an attempt to address the size, shape, and distribution of macromolecules (enzymes and proteins) in this architecture, Sjöstrand and his student Johannes Rhodin also measured the dimensions of inner and outer membranes and found them to be identical. A few months later, after further experiments comparing micrographs of mitochondria from kidney and pancreas cells, he strengthened his argument that the inner membranes were enclosed, separately walled vesicles that he called *septa.* "No support has been found for the opinion of Palade," he wrote, "that the inner membranes represent folds of a single-edged surface membrane." Instead, the inner membranes "are individual structures with only topographic relations to the outer membrane." He did not provide any biochemical evidence to bolster his ideas about how he viewed the role of these sacs in segregating enzymes (figure 4).[60,61]

Meanwhile, during an April 1953 symposium in Chicago organized by the Histochemical Society and devoted to the structure and biochemistry of mitochondria (not attended by Sjöstrand), Palade was the only participant to present electron micrographs. He showed more recent images of a large variety of cells taken at a higher resolution, this time including protozoans and plant cells. He now agreed with Sjöstrand's observation that the outer membrane is double. Expanding on his original observations of a structural pattern of organization characteristic of all mitochondria, he argued that its ubiquitous nature constituted a basic, fundamental pattern in the organization of the cell, thus giving it a clear status as a cellular organelle. In particular, he restated his view that the cristae or ridges of the inner membrane are shaped like baffles that communicate freely with the peripheral compartment. To clarify this conclusion, he presented serial sections showing cristae anchored to the inner membrane and protruding various depths toward the interior chamber. Then, bolstering his argument with diagrams, he showed that when sections were cut either longitudinally or crosswise to their axis, these infoldings did not extend completely across the mitochondria, leaving a channel open in the central interior. Finally, he provided two further links between the enzymes and the membranous ridges: the cristae are most clearly packed in mitochondria of heart muscle cells known to contain the greatest concentration of oxidative enzymes; and these ridges contained most of the oxidative enzymes even after he used centrifugation procedures to isolate mitochondria from whole cells.[62,63]

The contrasting interpretations of mitochondrial structures surfaced when Porter and Sjöstrand participated in the Third International Conference on Electron Microscopy in July 1954 in London. This time, the clash was over techniques. Sjöstrand focused on ways to produce the highest resolution micrographs to obtain a more precise measurement and identification of molecular architecture, or what he now called *ultra*structure. Because he believed there is more order in a living than in a fixed specimen, he opposed methods of disturbing native structures with cell fractionation and criticized Palade's fixative for disorganizing the cytoplasm. Instead, he advocated that the tissues be injected, cut, and fixed quickly with a fixative he devised that would maintain physiological conditions.[64]

Porter presented two of Palade's papers (Palade remained in the United States), one on techniques and the other on the "small particulate component of the cytoplasm" (discussed in Chapter 7). The paper on techniques countered Sjöstrand's arguments; it dismissed as impossible the ability to obtain precise measurements of membranes because the fixed material "cannot be sufficiently true to nature to permit us to deduce . . . which particular kind of molecule would best fit a given spacing." It also objected to Sjöstrand's emphasis on a physiological fixative as not important since water concentration of living cells can easily differ from one type to another. The best fixative, Palade wrote, "leaves the specimens with a maximum of organization." Admitting that fixation for electron microscopy had not yet reached its final solution (see Chapter 8A), he advocated integrating techniques having more reliable data, namely, those "derived from studies in cell chemistry and cell physiology," as a way of accounting for activities of the cell.[65]

Porter devoted the presentation of his own work at the meeting to describing the fine structure of the outer cell wall. Afterwards, he wrote Palade about the "confrontations" with Sjöstrand. Palade replied, "Our friend Sjöstrand is made up of worse material than I suspected—worse even than the one implied in your comment about paranoia. I will take care of his desire in the coming paper but most probably he will not like it."[66]

Palade attended a meeting in Maine a month later and wrote again to Porter, who was still in Europe visiting electron microscopists, including Albert Claude. Palade complained the electron microscope presentations in Maine were generally poor, except for the one by Sjöstrand's student Rhodin: "I still believe that the Swedes have an advantage in resolution—their morphological interpretations are improving—but their ability to integrate their results in general cytology and in physiology remains nil. This is our main asset and it will be wonderful if, in addition, we could regain the lead in high resolution. Fly therefore to Berlin, don't get mixed up with the Russians, and bring back an 'Ubermikroskop' that will show even sodium ions crossing through the membranes."[67] Porter did visit Siemens, where he had first-hand experience with their newest microscope and found it had much better resolution and sturdiness than the RCA microscopes at the Institute. A few months later, he purchased a Siemens ultramicroscope for the laboratory.

The "coming paper" referred to by Palade was to be his last on mitochondria. He and Porter had by then turned to unraveling the structure of Porter's newly found organelle, the endoplasmic reticulum (described in Chapter 7). At the 1955 International Symposium on Enzymes in Detroit, devoted to enzymes as units of biological structure and function, Palade used micrographs to map the known biochemical reactions onto the spatial relationships that could be seen in mitochondria. By linking biochemical fractionation studies made in tandem with microscopy, he hypothesized the mitochondrion is a "two-phase system consisting of a solid and liquid phase" with the solid phase corresponding to the membranes (as molecular films) and the liquid phase corresponding to the contents of the chambers. He also postulated that the repeated enzymatic chains are built in, or "woven into, the texture of the mitochondrial membranes in the same manner as repeated decorative patterns are woven into a sheet of damask."[68]

The dichotomy persisted into 1956. Porter organized an American conference on fine structure in which all papers on mitochondria used Palade's term cristae, whereas a few months later Sjöstrand organized a European conference on electron microscopy in which participants referred to the mitochondria's inner double membranes. Sjöstrand's ambition to observe the microscopical molecular basis of life was given more weight when he helped found the *Journal of Ultrastructure Research* in 1957.

There was no convincing biochemical evidence to decide these competing views until 1961, when chemist Peter Mitchell at Edinburgh University proposed a theory of cell energy production, for which he won a Nobel Prize in 1978.[69] According to Palade, "It was difficult to understand the existence of two compartments and to understand what was the meaning of the internal mitochondrial membrane. It took a long time to understand the biological rationale for this structure. We had to wait until finally the chemiosmotic theory of oxidation-phosphorylation was worked out because that particular concept required a tight membrane."[70] Mitchell's concept demonstrated that it is the enzymes in the membrane, rather than in granules or in solution in the matrix, that form the common link: the coupling of energy from oxidative to phosphorylating enzymes (from respiration to ATP synthesis) is not a chemical one but a physical one made possible by a specific orientation of the enzyme molecules in the membrane. In essence, the functional units are the same as the structural units.

Eventually, most researchers accepted Palade and Porter's interpretation of mitochondrial structure, which has been the standard definition in textbooks for nearly four decades. As late as the 1990s, however, Sjöstrand still had not accepted Palade's view of cristae as folds, according to science historian Nicholas Rasmussen, and went so far as to propose an alternative structural model that also refuted Mitchell's biochemical theory of energy production.[59,71] Interestingly, new views are currently emerging about the structure of mitochondria. The transmission electron micrographs of Palade and Sjöstrand provided only two-dimensional views, but the shift to direct molecular approaches has made it possible to reconstruct mitochondria in three dimensions. With the advent of electron microscopic tomography (similar to CAT scans), even using the traditional fixation and embedding techniques of Palade and Sjöstrand, scientists now find that Palade's baffle-like cristae is not quite an accurate representation: the cristae are now seen connected to the inner membrane not by shelf-like ridges but by narrow tubules. This finding is similar to one made in 1966 by W.Th. Daems and Eddie Wisse in the Netherlands, who named these tubular structures *pediculi cristae*.[72] The significance of tubules, with their greatly increased surface area, is presumed to correlate better with Mitchell's theory of how cell energy is produced and regulated.[73,74] In the early 1960s, Efraim Racker and his colleagues at the Public Health Research Institute in New York City were able to identify and purify factors of the ATP synthase complex and to demonstrate their attachment to the inner mitochondrial membranes, thus proving Mitchell's hypothesis.[75,76]

Nevertheless, at the time, the controversy between Sjöstrand and the Rockefeller scientists was a pivotal event in directing modern cell biology toward even firmer ground. In the end, neither determined the exact structure nor elucidated the mechanisms of mitochondrial energy production. As Palade noted, "the arguments were essentially arguments of over-interpreted morphology." He and Porter were working at the level of fine structure, while Sjöstrand preferred speaking of *ultrastructure*. But, Palade added, "we didn't become dogmatic, and this was in the long run something extremely important."[70]

At the time, Porter and Palade were not yet leaders or authorities in this kind of science. They were still finding their way through the challenges of looking at dead specimens that had been fixed with harsh chemicals and shaved into pieces. By 1953, when they had generated techniques for producing consistent micrographs, the plausibility of their interpretation of structures then rested on evidence generated by other, more traditional methods. They did not begin with a theory; rather, by analyzing a body of observations, they reached a plausible explanation about how processes occur in living organisms. When cell fractionation, biochemical analyses, and sharp micrographs provided data related to a particular cellular function and could be traced to the same organelle, an overall interpretation could be justified. The long trials with mitochondria provided the penultimate step toward mapping the structure and function of the endoplasmic reticulum (see Chapter 7).

Cytology Becomes an Independent Laboratory

In late 1952, Palade wrote to Claude describing "a rather agitated period in the history of our Institute. Papa Gasser is preparing to take his retirement in the month of June and we are getting ready to learn shortly the name of our future father. It is very possible that it might be a good father or even an ogre and, in this case, it will be necessary for me to find a place somewhere else."[39]

In one of his last acts as scientific director, Herbert Gasser put the fledging Cytology laboratory on an independent basis and no longer under his close supervision. He was effusive in his praise for Porter and Palade, who "have built up a cytology laboratory that has individuality on a world scale . . . [based on] the excellence of their techniques and the fineness of their workmanship. They are capable of going places and are going. . . . Their principal tool is the e.m. But they are much more than picture takers . . . they view what they are doing through eyes of trained biologists."[77]

Gasser singled out Porter's contributions to tissue culture and microscopy, noting with some pride after all these years that "his work has the respect of authorities like F.O. Schmitt . . . [and] he is emerging as a prominent figure among those working this field, as far as I know he is no. 2." Singling out Palade's fixation technique that he called "Palade's Pickle," he observed how it had contributed to the outstanding resolution of their micrographs. He further remarked that the Porter-Blum microtome had now given the Institute the best sectioning laboratory in the country as well as "the lead in the beauty of detail revealed in sectioned material." Cytology, he added, had become the most visited laboratory at the Institute.

He then recommended that Porter be promoted to member and Palade to associate member. The Porter recommendation lay fallow while the new president, Detlev Bronk, made his own review of promotions. After Palade was named associate member and made a co-head of the laboratory, he wrote Claude that this appointment "makes me, from the point of view of the title, the equal of Porter. He is even so my elder in this rank and is the one who is head of our laboratory, a situation which agrees with me since I must not waste my time with the difficult affairs of the administration. Papa Gasser is now retired—he has become as you know a converted electron microscopist. . . . It is because of his interest in electron microscopy that we owe a large part of our present fortune. His successor . . . seems much less enthusiastic by the accomplishments and promises of electron microscopy. Our star will shine less brightly during his reign."[78] Despite Palade's misgivings, in 1956, Bronk promoted Porter and Palade to members and professors (an added designation when the Institute became a university). They continued to share leadership of the laboratory until 1960.

On Gasser's departure in 1953, the stage was set for the laboratory—challenged by instruments, bold in their experiments, yet cautious about interpreting micrographs—to tackle the endoplasmic reticulum, an organelle that was obscure before electron microscopy.

If Palade's prophecy was incorrect, Gasser's was not. Concluding his long commitment to cell biology at Rockefeller, Gasser noted, "The work of putting these structures on the map I venture to predict will turn out to be a classic."

Chapter 7

The Endoplasmic Reticulum Discoveries

. . . . .

7A. Putting the Endoplasmic Reticulum on the Cellular Map

Carol L. Moberg

The first decade of modern cell biology began with Keith Porter and George Palade's joint and separate contributions to elucidate the structure and function of the endoplasmic reticulum (ER). This organelle was the first subcomponent of the cell found by Porter using the electron microscope, and it occupied the thoughts and work of the Rockefeller group for nearly three decades—from the discoveries of microsomes in the early 1940s, to their identification as membrane-associated groups of ribosomes in the mid-1950s, to their role in protein synthesis and secretion in the 1960s. This successful research laid the cornerstone of the new science.

As far back as the mid-nineteenth century, there had been interest in what might constitute the fine structure of protoplasm surrounding the nucleus. Early cytologists hypothesized that even though the cytoplasm appeared empty, it might contain intracellular structures. Yet, the nature of protoplasm and its possible constituents remained unanswerable with light microscopy. In addition to the barely visible mitochondria, some terms used to refer to other possible structures in the cytoplasm were also terms used initially by Albert Claude and Porter.

One important word was "ergastoplasm" (literally, work plasm), coined by Charles Garnier in French publications between 1897 and 1900 to describe a component staining with basophilic dye in the cytoplasm of glandular cells.[1] This observation was extended by many others who found this property in additional cell types, so that by the early twentieth century the term "basophilia" referred specifically to an acidic area in the cytoplasm.[2] The next advance came when basophilia was shown to be due to a high ribonucleic acid (RNA) content, first by Jean Brachet in 1933, using ribonuclease;[3] then by Torbjörn Caspersson and Jack Schultz in 1939, using ultraviolet spectrophotometry;[4] and by James N. Davidson and Charity Waymouth in 1943, using chemical methods.[5] These findings, along with the observation that high basophilia was found in cells known to produce vigorous protein synthesis, led Caspersson in 1941[6] and Brachet in 1942[7] to postulate the importance of RNA in protein synthesis. The findings, cited in papers by the Rockefeller scientists, undoubtedly influenced Claude in locating RNA in his microsome fraction; Claude and Porter in equating ER with ergastoplasm (and with basophilia) and speculating on the ER's role in protein synthesis; and Palade in connecting the particles found on the ER with protein synthesis.

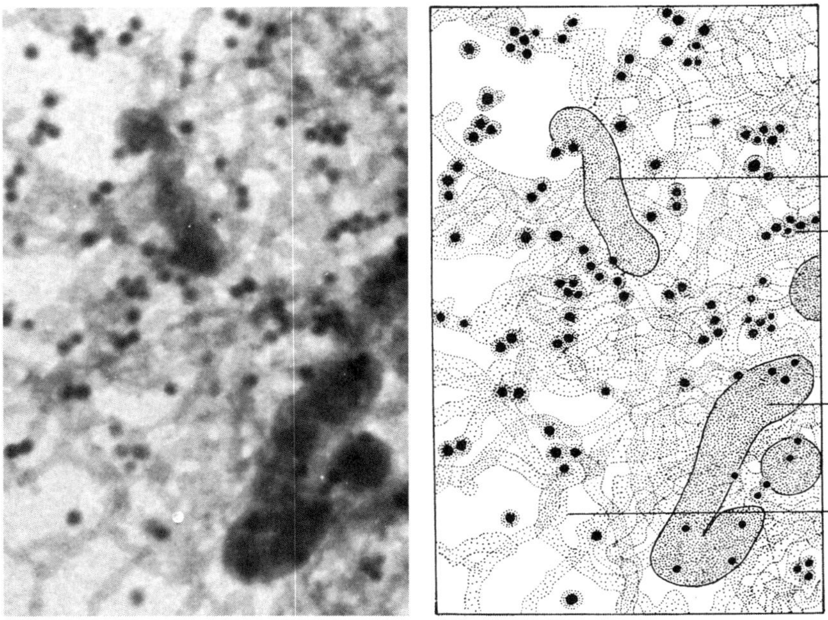

1. Left: electron micrograph by Keith Porter of epithelial tumor cell showing strands of endoplasmic reticulum. Right: Porter's descriptive diagram of the same portion of the epithelial tumor cell showing strands of endoplasmic reticulum. From Porter and Thompson, 1948. *J. Exp. Med.* 88:15–24.

Porter's initial descriptions in 1945 about the ER as "a lace-like reticulum" were general. Yet, a year later, he and Murphy more confidently speculated that the bodies scattered along the finely divided reticulum were "believed to be the same as those isolated by high speed centrifugation and known as microsomes."[8] By 1947, Porter formally equated the endoplasmic material with the "small particle fraction of broken-up cells described by Claude in 1943."[9] He further speculated that the newly found component "is the most active part of the living unit. It is present in one form or another in every cell so far examined. It makes up a fibrillar apparatus concentrated around the central body and sends fine extensions of its substance throughout the cell. The vesicular structure of the small units suggests a secretory activity."[10]

The term "endoplasmic reticulum" appeared in print for the first time in 1948, but only in a label of a published drawing by Porter of a cultured cell.[11] In this case, Porter drew the live cells he was seeing with a phase-contrast microscope. He inferred that the small vesicles and strands forming a reticulum in the electron micrograph were the same elements he was seeing in living cells that were "involved in a streaming motion which carries them about in the cell" (figure 1).[12]

Further investigations into the ER had to await improvements by Porter and others in both cell sectioning and fixation, described in the previous chapter. By 1951, when Porter could look at cultured cells fixed in Palade's newly developed buffered OsO_4, he was able to see arrays of filaments in the cytoplasm that were "in fact fine tubules, or strands of vesicles."[13] Consequently, he and Frances Kallman formally introduced the ER as an organelle in 1952, noting it was "made up of vesicular or canalicular elements which sometimes constitute a complex reticulum . . . [and] is part of the innermost cytoplasm of the cell, the endoplasm." This new observation emphasized the appearance of small granules, 25–200 nanometers in diameter interspersed within the reticulum. Since the electron microscope's resolution was still not that good, the authors tentatively speculated (wrongly, as it turned out) that the granules might have significance in "growth and proliferation in the cell."[14]

Also in 1951, Porter made an accidental discovery when he provided electron micrographs of muscle cells for Charles Ashley at the Presbyterian Hospital in Chicago.[15] This was his first glimpse of the sarcoplasmic reticulum, a variation of the ER found in muscle cells. Again, the micrographs were unsatisfactory due to excessively thick sections. However, the observation led to Porter's working visit during the summer of 1951 with Stanley Bennett in Seattle to look more closely at striated muscle. The tissue specimens were fixed with Palade's buffered solution, embedded in methacrylate, and sectioned with glass knives attached to a Spencer rotary microtome having a thermal advance of the specimen. (All these trials were valuable contributions to the ongoing development at the Institute of the Porter-Blum microtome.) The published images provided a good look at a reticular system in muscle cells,[16] although they were not as clear as those Porter and Palade saw four years later in other cell types (figure 2).[17] The ER-type system in muscle cells was different enough so that Porter coined the name "sarcoplasmic reticulum" to describe an apparent connection of the reticulum to the long, thin contractile fibers.

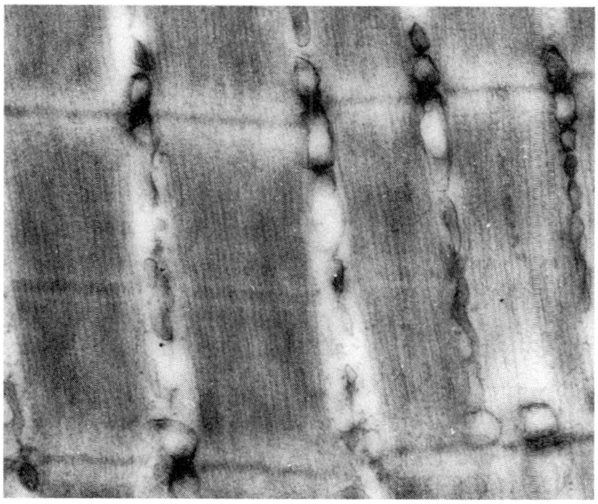

2. Electron micrograph of the sarcoplasmic reticulum, a special type of smooth endoplasmic reticulum found in muscle cells. Bundles of myofibrils from an *Ambystoma* larva that fill most of the field are separated by membrane-limited structures of the sarcoplasmic reticulum. From Porter and Palade, 1957. *J. Biophys. Biochem. Cytol.* 3:269–300.

At the time, Porter was unaware of a paper on this topic published in 1902 by Emilio Veratti in Italy. Veratti used the black reaction stain devised by Camillo Golgi to highlight a reticular apparatus in muscle cells of various animal groups so that it could be seen with the light microscope.[18] When this historical lapse was called to Porter's attention, he was somewhat embarrassed. But wishing to credit Veratti's remarkable paper as well as his own rediscovery of the sarcoplasmic reticulum using the electron microscope, he had Veratti's original paper translated and published as part of a 1961 supplemental issue of *The Journal of Biophysical and Biochemical Cytology* devoted to modern studies of the role of this organelle in controlling muscle contraction.[19]

In 1952, the existence of the ER was still so obscure (even unknown to others) that Porter, in a lecture at Harvard University, introduced biologists to the electron microscopy of cells by noting that "some of my acquaintances who think less highly of the [ER] system than I do facetiously refer to it as porterplasm." He anticipated skepticism and questions about artifacts in such harshly treated cells and therefore presented abundant data along with numerous micrographs. Manuscripts of his many (unpublished) lectures during this period testify to his energy and talent in arguing for the acceptance of the ER as an organelle. He in fact claimed it was unavoidable to accept the ER system as a fundamental component of protoplasm and necessary to consider its possible role in the cell. At this still early stage, he was strenuously promoting the establishment of a new field of science that would replace the old cytology.[20]

The ER in cultured cells. In anticipation of criticism, Porter undertook a massive investigation of many cell types in vitro (whole-mounted cultured cells). He looked at fibroblasts, myoblasts, mesothelial cells, sarcoma cells, macrophages, lymphocytes, osteoblasts, and kidney and mammary epithelial cells from chicken, rat, rabbit, and mouse. These were the most easily cultured cells and, in the absence of a good microtome, provided valuable specimens for observing and comparing the fine structure of protoplasm.

A detailed description with micrographs of the ER was first published in 1953.[21] Porter stressed the quality of electron micrographs depended on a high degree of environmental control of the cells. For this reason, he provided careful diagrams to illustrate cell culturing and fixation for electron microscopy, displaying his methodical thoroughness. The images resulting from these micromethods, he argued, left no doubt that the ER is a universal element in cells. Nonetheless, while images of cells appeared rather uniform, there were degrees of variation from one cell type to another; for example, in some cells the reticulum disappeared and seemed to be replaced by apparently unconnected vesicles.[22–24]

To counteract criticisms about artifacts created by fixation methods, he compared live cells seen with phase-contrast light microscopy with electron micrographs of the same cells. He was able to show that even low magnifications of light microscopy revealed structures of similar size and intracellular distribution and had refractilities similar to those observed with electron microscopy (EM). Moreover, whether he used osmium, increased fixation time, altered fixations of formaldehyde, or simply dried the specimens, all methods produced similar images of the ER distribution in cells and greatly reduced arguments about fixation artifacts. Porter based the identification of the ER on a basophilic component having a similar form both in the hematoxylin- and toluidine blue–stained components, just visible with light microscopy, and in osmium-fixed EM images of the ER. His method of comparing light and electron micrographs of similarly treated cells (initiated in the 1945 report with Claude and Fullam; see Chapter 3) was a key element in the methodological approach he was developing to put the ER on the cellular map.

He further compared the ER in cultured cells with those that had been observed in an isolated microsome fraction by George Hogeboom, Walter Schneider, and George Palade in their 1948 study.[25] The newer EM images of microsomes showed a "highly variable, partly amorphous component which . . . displays enough similarity to vesiculated strands to convince one that it is made up, in part at least, of elements of the reticulum." This observation of a reticular system in microsomes, also made by David Slautterback, strongly indicated the ER ends up in one form or another in the microsomal fraction.[26] "With a few reservations," Porter noted in his 1952 annual report to Gasser, "this finding justifies the transfer of the biochemical data obtained on the microsome fraction to the endoplasmic reticulum."[27]

Porter next presented the ER, in 1953, as a unit system "set off from the matrix of the ground substance by a membrane . . . [that] makes vesicles of the rounded elements and canaliculi of the strands." This statement put him in disagreement with Albert Dalton, who interpreted the submicroscopic strands in the ground substance of the cytoplasm as sectioned lamellae or fixation artifacts.[28,29] Porter also refuted Wilhelm Bernhard's picture of the ER as filaments, blaming poor fixation techniques and arguing this interpretation was no longer valid when using Palade's buffered osmium. To Porter, the ER was a labyrinthian cavity of branching tubules, flattened sacs, and sheets that extended throughout the cytoplasm.[30]

The ER in sectioned cells. This 1953 paper also revealed Porter's dissatisfaction. More work, evidence, and arguments were necessary to convince critics that his interpretation of the ER was the correct one. He therefore continued to study various cell types, this time incorporating the newer methods of thin sectioning of fixed, stained tissues and the newly developed Porter-Blum microtome (see Chapter 6).

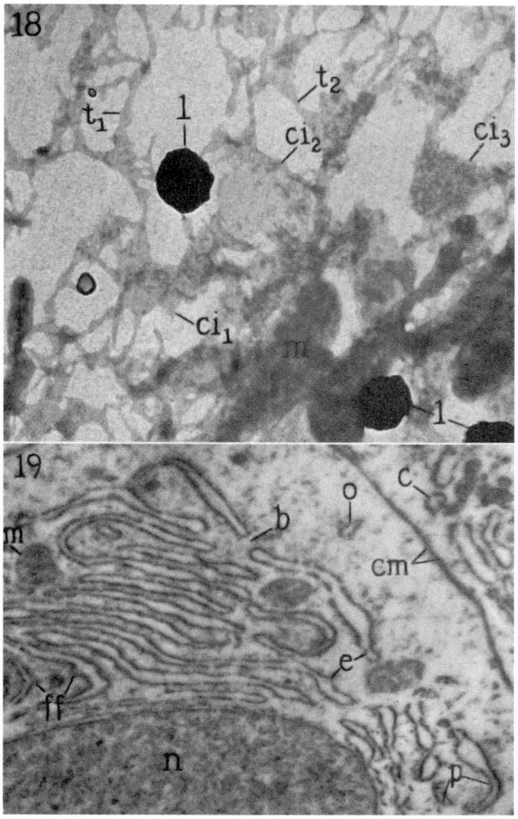

3. Comparison of the appearance of the endoplasmic reticulum (ER) in whole tissue-cultured cells (top) with thin sections of ER in situ tissue cells embedded in methacrylate and thin sectioned (bottom); both are epithelial cells of the parotic gland from the same rat. The comparison shows the ER in sectioned material is similar to that described earlier in cultured cells. The finding ruled out artifacts of tissue culture specimens and suggested the ER is a significant, universal component of all cell types examined in situ as well as in vitro. From Palade and Porter, 1954. *J. Exp. Med.* 100:641–656.

Porter's first observations on the ER using these newer techniques were delivered on 16 April 1954 before the Histochemical Society in Atlantic City, where he supplemented and confirmed his earlier observations on cultured cells.[31] Cultured cells, he said, had "recognized limitations and are invariably subject to the criticism that some artificial condition of culturing is responsible for the observed structures." Nevertheless, there was a great similarity of ER in both cultured and thin-sectioned sarcoma cells that had been fixed and stained with osmium and shown at the same magnification of 12,500. Sections from many other cell types also showed close-layered arrays of membrane-bound vesicles. In some sections, there was fusion of an individual vesicle with an adjacent one, indicating a continuum of the system. The basophilia, located in the same cellular region in toluidine-stained cultured cells and in the thin-sectioned cells, seemed to coincide with "discrete clumps" of basophilic material that had been seen in liver cells by Helen Deane.[32] Finally, at a higher magnification of 63,000 of extensive layered profiles in a pancreatic acinar cell, he noted for the first time that there were particles 100–150 ångströms in diameter both covering the membranes and between profiles in the cytoplasm. He made the prescient observation that the basophilic component, particularly in secretory cells, was a "concentration of vesicular and canalicular elements of the ER with small basophilic granules associated." There was now a correlation between the distribution of the staining properties in the histological sections, the distribution of dense particles seen in electron micrographs, and the RNA found by Claude (figure 3).

The ER in situ. In 1954, Porter and Palade began publishing an extended series of studies on the ER. The first paper, subtitled "Its identification of cells in situ," was based on a new way to look at cells taken from animal tissue rather than from tissue culture. It established a cautious but measured correlation of appearances between sectioned and cultured material. The term "in situ" was new to electron microscopy, so they provided a precise definition: "The expression is used to designate cells found in tissue blocks excised for fixation directly from living animals. The morphology of such cells, which have retained until excision their normal connections within their respective tissues and organs, provides the best possible approximation of the morphology possessed by cells in their normal surroundings; i.e., *in situ*."[33]

Porter and Palade were trying to ascertain whether the ER is a network of connected vesicles. Working with sliced cells, they could see spherical vesicles only when they lay in the plane of the slice; when looked at perpendicular to the plane, the vesicles looked at best like vacuoles. Since the ER network's connections were easily severed by the microtome, they argued that "cell sections constitute a much less favorable material than whole mounted, cultured cells for studying the general morphology and intracellular disposition of a highly dispersed and complicated component such as the ER."[33]

Monocytes and macrophages, mesothelial and endothelial cells, and various types of epithelia were examined in vitro in whole mounted cells, in sections, and in situ after sectioning, this time using the Porter-Blum microtome and the RCA EMU 2E and Philips EM 100 microscopes. Their new micrographs confirmed that all cells except the adult mammalian erythrocyte had an identifiable ER, whether the cells were fixed in situ or cultured, and despite different fixations. This was compelling evidence that the ER was a component of the living animal cell because its profiles had a dense, continuous outline surrounding a content of lighter density, suggesting there were cavitary structures such as tubules and vesicles. These profiles resulted from sectioning through what Palade termed "cisternae" to indicate this continuity of membranes and contents. In particular, the serial sections of endothelial cells showed that what once seemed to be independent elements were in fact part of a "continuous, tridimensional reticulum."[33]

On a different aspect of this work, Palade, working independently, began to define the continuous nature of the ER and its variations in specific cell types. Subtitled "Simple dispositions in cells in situ," he described ER sections from nearly forty mammalian and avian cell types.[34] In general, the ER varied in shape from circular to oblong; it had a small diameter (50–250 nanometers), a dense membrane surrounding a light content, and no internal structure. The ER could be distinguished from the ground substance at the resolving power of the electron microscope, where in some cases it was homogenous and in others granular or fibrillar. Again, the partially verified assumption was made that profiles seen in many micrographs were actually sections that cut through vesicles, tubules, and cisternae. Furthermore, variations in the angles of sections were responsible for some vague images.

Palade noted the limitation of using only morphological criteria to assert a similar nature of the ER in different cell types having a cell-wide system of interconnected vesicles and tubules. In fact, he described two general subtypes of ER. A circular type with smooth membranes is found in leucocytes, mast cells, adipose cells, adrenal cortex cells, and parietal cells of gastric glands. Another predominantly tubular type with particles on its outer surface is found in secretory cells such as liver and pancreas.[35] He further noted that the membrane enveloping the cavities of the ER appeared to be continuous with both the outer cell membrane and the nuclear membrane, an observation also confirmed by Michael Watson[36] and Björn Afzelius.[37] These observations strengthened Palade and

Porter's assertion that the ER is a cell-wide structure whose contents are connected to both extracellular and intranuclear spaces.

In a now classic study called "A small particulate component of the cytoplasm," Palade examined even more cell types. He focused on the granular nature of the ER, the small particles sometimes referred to at this early stage as "Palade bodies" and later renamed ribosomes (see Chapter 7B). Earlier, Porter had suggested that granules from 100 to 150 ångströms, rather than any component of the reticulum, represented the basophilic material.[31] Palade enlarged on this observation by showing that in secretory cells, such as pancreas, salivary, and mammary glands, many granules were soon to be "attached" to the membranes of the ER, while there were others lying freely in the cytoplasm. In other cell types, such as epithelial cells, lymphocytes, blood granulocytes, and spermatocytes, all the granules were sparse and unattached to cytoplasmic membranes.[35]

Palade dismissed the idea that the loose granules were artifacts of fixation because they were seen after fixation with formaldehyde as well as osmium. He referred to the work of Mary Petermann at the Sloan-Ketttering Institute, who had found an RNA component in particles isolated from liver homogenates,[38,39] and Hubert Chantrenne at the University of Brussels.[40]

The question remained whether attached granules were artifacts of fixation or whether they preexisted in the cell. George Pappas, a postdoctoral fellow in the laboratory at that time, recalled a tea-time discussion about the reality of the "Palade particles." Porter believed the ER membrane structure was of two types: one was thick and probably looked rough because its surface was damaged with the electron beam, while the other was thin and appeared smooth. It was becoming clear, Pappas noted, that some chemistry was needed.

Slautterback, also at Sloan-Kettering, using electron microscopy, found some "small microsomes" similar in size and density to Palade's particles at the surface of "large microsomes."[26] This prompted Palade to suggest the ER might be fragmented; in other words, that neither the loose nor attached granules were artifacts but that there might be a chemically diverse grouping of particles. He based the idea of a class of cytoplasmic components on ultracentrifuge findings by Petermann, whose various sedimentation values for particles were later attributed to the dimerization of single particles (or the absence of magnesium), causing them to break into large and small subunits.

As in his research on mitochondria, Palade was further persuaded that using only electron microscopy to study the ER was not enough. After attending a meeting in Maine during the summer of 1954, he wrote Porter, then in Europe visiting electron microscopists, that the few studies presented were "strictly and narrowly morphological, with no attempt at integration in general cytology and in cell physiology. As a result, everybody was slightly disappointed. The physiologists—the two Coris among them—could not grasp the significance of the new structure, and the morphologists felt bad because of the lack of response of the physiologists."[41] To make a positive identification of the particles, Palade proposed that the next phase of studying the ER should be "an integrated electron microscopical and biochemical study of cytoplasmic fractions."[35]

Clashing visions over the endoplasmic reticulum. Before turning to this phase, however, a brief description is needed about the continuing challenge by Swedish scientist Fritiof Sjöstrand over interpretations of ER structure. Like their controversy over mitochondrial structure, described in Chapter 6, the Swedish and American scientists used equivalent

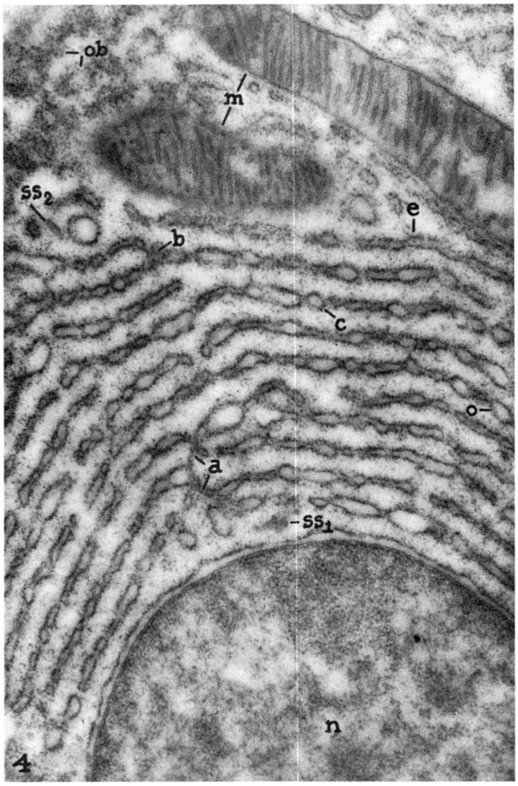

4. Electron micrograph of numerous rough-surfaced profiles of endoplasmic reticulum (ER) in a pancreatic cell of a guinea pig. Along with the nucleus (n) and two mitochondria (m), the ER profiles appear aligned in parallel rows and show their lumen, limiting membrane, and attached granules. Most profiles are elongated (e), but circular (c) and oval (o) profiles are also present. From Palade, 1956. *J. Biophys. Biochem. Cytol.* 2(Supplement):85–98.

instruments, techniques, and biological objects, yet differed sharply as to what their images actually revealed about the cell.

The electron microscope once again allowed Porter and Palade to produce considerable evidence that the ER was an organelle, not an artifact, and to show that the reticular network they visualized with the electron microscope represented the in situ tissue (figure 4). They were careful in their publications to respond to criticisms and varying interpretations by others. As mentioned previously, Porter refuted Dalton's and Bernhard's pictures of the ER as sectioned lamellae or filaments due to their poor techniques. However, the new argument advanced by Sjöstrand had to be taken more seriously.[42–44]

Sjöstrand's controversy with the Rockefeller scientists again emanated from his reliance on strict morphological interpretations based on criteria found in the micrograph and the specimen. His main objection was that even though they all seemed to use similar techniques to look at exocrine pancreatic tissue, he did not visualize a structure that was either endoplasmic or reticular and dismissed Porter's designation of ER as "unspecific and useless." Instead, he referred to paired cytomembranes, with irregularly shaped particles attached to one side, surrounding irregularly formed, closed, and, in most cases, very flat spaces. The point of controversy was whether the membrane of the cisternae, or flattened vesicles, was continuous within a larger membrane. Sjöstrand saw separate parallel membrane systems.[45]

Porter and Palade, on the other hand, saw these membranes as forming a closed compartment of a finely divided vacuolar system. To answer Sjöstrand's argument, they had to make the case that paired membranes seen in tissue slice preparations were parts of a more continuous structure (like the ones seen in tissue culture preparations), and therefore they showed intermediate specimens of sliced tissue culture cells. In their serial sectioning of cells, they reconstructed a three-dimensional picture from micrographs to show that apparently isolated components were part of a continuum. These sections regularly showed that an enclosed tubular space in one section could be fused with an adjacent tubular space in the next section. They also overcame Sjöstrand's criticisms by examining and comparing whole cultured cells, slices of cultured cells, and slices of ordinarily prepared fixed cells, procedures which Sjöstrand did not do. Sjöstrand's ideas of isolated membrane-bound structures, they concluded, had resulted from his analyses of single rather than serial sections.[33]

Palade further advanced the view that the presence of "intracisternal granules" in exocrine pancreas cells indicated the ER is a "cavitary system."[46,47] As early as 1902, Alexander Mankowski had noted such granules in cavities with the light microscope.[48] However, Sjöstrand felt the paired membranes were so closely parallel and tightly spaced that they could not be enlarged to enclose large bodies.

One difference between the Rockefeller and the Swedish researchers can be attributed to their interpretative techniques.[49,50] Porter and Palade based their hypothesis of ER structure on its possible function, comparing morphological connections in situ with those in vivo. Their effort was a significant outgrowth of the earlier work of Claude and Hogeboom because it incorporated biochemical studies that linked the ER structure to its basophilia and RNA content, all of which eventually led to determining the ER's function in protein synthesis. Sjöstrand, on the other hand, based his interpretations on a visual and geometrical approach to cell structure in high resolution micrographs and deemphasized any connection to ergastoplasm or basophilia.[44] His strictly morphological argument was reasonable for a time because his micrographs provided a greater degree of order that appeared more representative of the living state of the cells. However, support for his view declined by 1960, while Porter and Palade's speculation was eventually confirmed by the work of Philip Siekevitz, James Jamieson, and Palade on protein synthesis and secretion in the pancreas showing a continuity of the intracisternal space with the outside of the cell (described in Chapter 7C).

Another difference relates to the persistence and thoroughness of the Palade and Porter experiments, which brought about an acceptance of their view of the ER that endures today. Using multiple means of identification, they established the structure of the ER as a continuous system of intracellular membranes lying closer to the nucleus than to the plasma membrane. Some cells have visible granules, and because some of them are attached to the outside of membranes, the ER system is responsible for basophilia or cytoplasmic RNA content. By 1955, even though Porter and Palade knew its membranes enclose substances of a yet unknown nature, they were able to hypothesize that because the granules contained cytoplasmic RNA, the ER system was likely to be found responsible for cellular protein synthesis. At the conclusion of this structural phase of the research, Palade could comment to Porter that their work definitely showed "why we didn't become ergastoplasmologists."[51]

In retrospect, by 1955, the exploration of the ER, from initial papers to many substantial reviews, had evolved into a more reliable discipline-building strategy. Even though both Porter and Palade regarded the electron microscope as bridging the gap between the limits of light optics and the evidence from physical and biological chemistry, there is a noticeable difference in their approaches toward studying cells, one that strengthens the emerging discipline of cell biology.

Porter regarded the electron microscope as spurring his scientific curiosity to resolve what was previously assumed were empty parts of protoplasm. In particular at this stage, he was beginning to contemplate how the ER might be an elastic framework that could be regarded as "cytoskeletal." His seminal Harvey Lecture in March 1956 contained only a few comments regarding ER function, work that was just beginning to emerge from Palade and Siekevitz. He questioned instead the form of the ER and factors controlling it: "Are these inherent in the system itself; does it hold the initiative in determining not only its own shape and organization but also that of the cells of which it is a part?"[52]

He also countered an attack by "an eminent physicist among electron microscopists," without naming Sjöstrand, who criticized biologists for indulging in a freewheeling approach to studying cells: "Investigators still appear on the fringes of our society who question the authenticity of what the rest of us accept as factual," Porter said, "but such terrorists are neither numerous nor long-lived. They soon join us in extolling the virtues of our OsO_4 and our new gadgets." He defended his role as searching out "similarities and repeating designs in the fine structure of biological materials even as complicated structurally as tissue cells. In recognizing these, and more especially the variations related to function, we may hope to gather some clues to the possible roles of these new structures." While he elaborated on facts about these structures found in micrographs, he cautiously concluded that "the fabric of speculation against which they have been projected is thin indeed and will have to be rewoven many times before it will stand much wear."[52]

Palade, in contrast to Porter's emphasis on submicroscopic morphology of protoplasm, projected in 1956 "a new morphological background against which old and new physiological data will be interpreted." Not content with simple morphological criteria, he sought correlations, at the corresponding dimensional level of structure, with histochemical and cytochemical data. He questioned whether patterns "have anything to do with the type of cell or the type of cell products." He compared himself to the French explorers during Napoleon's expedition to Egypt who came face to face with hieroglyphs. Like those explorers, Palade said, "I am recording the patterns, and I am waiting hopefully for a biochemical [Jean François] Champollion to decipher their meaning."[53]

With work described in the following pages, the stepping stones toward putting the ER on the cellular map soon turned into a more or less continuous path. Porter and Palade were assuming the complex structures they were seeing with the electron microscope were there to carry out precise, efficient functions.* Up to this point, the structural mapping of the ER, onto which its function could now be plotted, was drawn almost entirely from thin-section electron micrographs. As described in Chapter 9, Porter retained a strong interest in the fine structure of different cell types. And, as this chapter continues, Palade concentrated on the ER in the acinar cell, the digestive enzyme-secreting cell type of the pancreas, to develop a model system for understanding the synthesis and processing of proteins as a universal function of cells. He later recalled the appeal of the acinar cell for its amazing organization "whose cytoplasm is packed with stacked endoplasmic reticulum cisternae studded with ribosomes. Its pictures had for me the effect of the song of a mermaid: irresistible and half transparent. Its meaning seemed to be buried under only a few years' work, and reasonable working hypotheses were already suggested by the structural organization itself."[54]

*Even while this work was still in progress, its significance was recognized by three major awards: Keith Porter received the Gairdner Foundation Award in 1964; George Palade received the Albert Lasker Basic Medical Research Award in 1966; and Porter, Palade, and Albert Claude shared the Louisa Gross Horwitz Award in 1970.

7B. The Microsome Fraction and Ribosomes: An Integrated Electron Microscopical and Biochemical Study

Philip Siekevitz

George Palade moved forward in the ER studies to make a positive identification of the particles associated with the endoplasmic reticulum. For this endeavor, in 1954 I became the first biochemist to be associated with this Rockefeller group since the departure of both Walter Schneider and George Hogeboom in 1948–1949 and also the first new staff member hired by the laboratory of cytology.

In 1942, I received a degree in biology at the Philadelphia College of Pharmacy and Science, the oldest pharmacy school in the country that had good biology and chemistry courses. Though less expensive than the University of Pennsylvania in Philadelphia, my home town, it was necessary for me to work for two years after high school to earn enough for college. I then spent four years in the army, after which I was able to take advantage of the G.I. Bill to study biochemistry at the University of California in Berkeley, receiving a PhD in 1949 under David Greenberg. My thesis described the metabolism of the amino acids glycine and serine in liver slices by using the newly discovered tracing technique with the isotope carbon-14.[1,2]

The biochemistry department at Berkeley was one of three in the country studying protein synthesis by following the incorporation of radioactive amino acids into proteins of incubated tissues, primarily liver. Before biochemists began using radioactivity to trace protein synthesis, scientists thought there was very little production and breakdown of proteins in adult cells. They assumed that once cellular proteins were formed, they lasted the lifetime of the cell, and that the only proteins being metabolized were those from the breakdown of dead and dying cells. So when radioactive amino acids were incorporated into the protein molecules of cells, there came a distinct surprise. This technique revealed newly synthesized proteins being produced in an adult organism. Actually, the tracer results could not describe protein synthesis because no net synthesis of proteins ever occurred. Instead, the noncommittal term "incorporation into proteins of radioactive amino acids" was used.

While learning how to perform these new experiments in Greenberg's laboratory, I became intrigued by this exciting field. With Greenberg's help and an NIH fellowship, I then joined the laboratory of Paul Zamecnik and Ivan Frantz of Harvard Medical School at the Massachusetts General Hospital. This laboratory and the one of Henry Borsook at the California Institute of Technology were the only two others investigating "protein synthesis" using radioactive amino acids.

Trying to study protein synthesis in whole cells at that time was not very productive. Zamecnik regarded the cell homogenate as a "biochemical bog," so the cell fractionation technique of Albert Claude, refined by Schneider and Hogeboom,[3] offered a new stepping stone to isolate and study components involved in protein synthesis. During my years at

Harvard (1949–1951), I worked out a system that used separated subcellular fractions in incorporation experiments.[4]

My first experiments repeated those from other laboratories and showed that when a whole homogenate was incubated with radioactive alanine and other components such as oxidative substrates and then fractionated into individual subcellular fractions, the proteins of the microsomal fraction had the highest specific radioactivity. When individual subcellular fractions, either separately or combined, were incubated under the same conditions as for a whole homogenate, the highest specific radioactivity was found when mitochondria and microsomes were incubated together. Earlier experiments, reported in abstract form by the Greenberg and Borsook laboratories, had similar findings but less well-defined subcellular fractions.

Then, based on the earlier work of Claude,[5] I learned about ways other than centrifugation to sediment the RNA out of the mitochondrial supernatant fraction. When this fraction was made to pH 5 or treated with magnesium or calcium and then combined with the mitochondrial fraction, these sediments also had the highest specific radioactivity. This evidence strengthened the hypotheses of Brachet and of Caspersson, made a decade earlier, that RNA was important for protein synthesis.

Another key aspect of the work dealt with the mitochondrial energy requirements, and since this experimental field was new to me, I was helped by advice from the nearby Fritz Lipmann laboratory at Harvard. There was a need for oxidative energy in the form of Krebs cycle substrates, the interacting system between carbohydrate, lipid, and amino acid metabolisms; the radioactive uptake paralleled the measured oxidation of ketoglutarate or of succinate. This finding confirmed earlier reports from the Greenberg and Zamecnik laboratories.[6,7] More to the point, there was a need for production of the high energy compound adenosine triphosphate (ATP) by mitochondria, since adding dinitrophenol that uncoupled ATP production from oxidation, or adding hexokinase plus glucose that competed with the microsomal radioactivity uptake system for ATP, resulted in a much diminished uptake. Frantz and Zamecnik had similar findings in a liver slice system.[8]

I remember showing the paper to Fritz Lipmann, who expressed some qualms about the results, particularly concerning the low incorporation rate, a fact that also bothered me. Two years later, however, Zamecnik and Elizabeth Keller modified my system to use a soluble ATP-generating system instead of mitochondria, and they obtained higher rates of incorporation into microsomal proteins.[9] However, my paper was the first to show that well-defined subcellular fractions could be used in in vitro systems to study protein synthesis and that energy in the form of ATP was necessary for incorporation of amino acids.

Early investigations on mitochondria. This work on the role of mitochondria aroused my interest in an increasingly important field that was filled with conflicting ideas as to the biochemical mechanisms by which mitochondria made ATP. I applied to Van R. Potter at the McArdle Cancer Laboratory at the University of Wisconsin in Madison and was offered a fellowship in oncology. This move in 1951 was a personally happy one, since my wife Rebecca, whom I had met in Berkeley, was a native of Wisconsin and a graduate of the Madison campus.

Potter was one of the leaders in oxidative metabolism in mitochondria and was also well known for introducing, with Conrad Elvehjem, the homogenizer for mincing tissues. I quickly began to work on the role of adenine nucleotides in mitochondria. By this time, due to the pioneering studies of Claude, Hogeboom, and Schneider and of Albert Lehninger at Johns Hopkins, it was well established that enzymes of the Krebs cycle were bound to mitochondria. Also, the oxidations generated during reactions of this cycle led to the production of ATP. How ATP was generated or coupled to oxidation and secreted from mitochondria remained a problem. The biochemical mechanisms of this process were intensely disputed by the half dozen "experts" in the field. I remember verbally violent, though respectful, discussions on these questions at annual meetings of the American Society of Biological Chemists. Potter and I never even came close to solving the biochemical mechanism of mitochondria ATP genesis, and neither did any of the select participants in the heated discussions. This discovery was left to an "outsider," Peter Mitchell, of Edinburgh, who in 1961 proposed the chemiosmotic mechanism, which was accepted by the biochemical community after years of arguments and for which he received the Nobel Prize in 1978.[10,11]

Of the six studies that I published while with Van Potter, three were important for my future career. Two concerned the manner in which oxidations within the mitochondria were controlled by the ADP-ATP cycle.[12,13] The third postulated a biochemical mechanism of ATP secretion that attracted some supporters at the time[14] but was later superseded by the secretion of ATP via a specific membrane carrier. It was clear from the focus and contents of these papers that I was interested not only in the biochemistry of the mitochondria, but also in the physiological, or one might say cell biological, role of these organelles; that is, how mitochondrial structure was related to its function.

An invitation from The Rockefeller Institute. Toward the end of my third year at McArdle, I realized I was a competent biochemist and had been elected to the American Society of Biological Chemists. But as the father of two infant girls, I also knew I had to find a more permanent position. On a fateful day in early 1954, Potter burst into the lab, letter in hand, and exclaimed he had a position for me. The letter was from Palade inquiring about my status at McArdle and whether I was available for a position at Rockefeller. This was at least the third inquiry to Potter from The Rockefeller Institute. The first came from Claude, when Walter Schneider was sent, while a second query from Porter in 1950 elicited no candidates. I was aware of the fame of the Institute and knew the excellent work of Hogeboom and Schneider, although very little about the work of Porter and Palade. After I replied to Palade that I was interested, he quickly invited me to visit the Institute in the spring of 1954. On my arrival there, I met Palade, Porter, and other members of the cytology group, postdoctoral fellows George Pappas, Maria Rudzinska, and Albert Sedar. I gave a seminar to the lab group, hoping my biochemical experiments and results were understandable. The laboratory occupied the third basement of Theobald Smith Hall due to its solid foundation for the electron microscopes on New York City bedrock; it was minimally equipped with biochemical apparatus—a high speed centrifuge, a spectrophotometer, a chemical hood, and a cold room, all apparently left from Hogeboom's tenure. During this visit, Palade offered me the position of assistant, and I accepted then and there.

Back in Madison, I concluded I was offered this position because Porter and Palade wanted to learn whether the endoplasmic reticulum had any function in protein synthesis. I also read for the first time some papers on the endoplasmic reticulum and found that Porter's 1953 *Journal of Experimental Medicine* paper referred to my 1952 paper on in vitro amino acid incorporation into microsomal proteins.[15] So they knew about my work. Except for two later contributions on mitochondria, working with Michael Watson at Rockefeller and with Olov Lindberg and Lars Ernster in Stockholm, my future lay in other aspects of cell biology.

Shortly after my arrival at Rockefeller in the summer of 1954, two other faculty appointees appeared in the lab. Montrose Moses, a cell biologist from Duke, was interested in ultraviolet microscopy. Michael Watson, an electron microscopist from Rochester, was well known to Porter and Palade for his innovative microscopy techniques and was to become well known as well for his electron microscopy work on the nuclear pore complex in the nuclear envelope (see Chapter 9H). Palade or Porter ushered the new appointees individually into the presence of the rather brusque Thomas Rivers, interim director of the Institute between Herbert Gasser's retirement and Detlev Bronk's arrival. All I remember of my meeting was that Rivers asked why I wanted to come to Rockefeller, and all I could quickly reply was that I had been invited.

Combining biochemical and morphological studies on mitochondria. My first project with Watson investigated whether the oxidative enzymes of mitochondria were really in the membranes. We were spurred by a recent hypothesis that succinoxidase was closely associated with liver mitochondrial membranes.[16] We conclusively showed that some enzymes, such as those that transfer electrons from succinate to oxygen, are localized in a membrane fraction that was recognizable by EM.[17,18] This finding extended the work of Schneider, Hogeboom, and Palade by identifying these enzymes in electron micrographs of a mitochondrial fraction. A glaring negative finding was the lack of oxidation of the enzyme DPNH–cytochrome c reductase (DPNH) and of the oxidative phosphorylation pathway. We used the detergent deoxycholate (DOC) to solubilize the membranes and then spun down the solution to get a fraction containing both membranes and the bound enzymes. Similar findings were observed by David Keilin and Edward Hartree in 1949 using hypotonicity and mechanical breakage to get a muscle mitochondrial subfraction with DPNH oxidizing activity[19] and by Lehninger and co-workers, who used one percent digitonin to isolate fragments that also had phosphorylating activity and DPNH oxidizing activity.[20–22] However, neither of these studies was accompanied by electron micrographs. Our contribution was finding that succinoxidase is a mitochondrial membrane protein. The lack of oxidizing activity, using other Krebs cycle substrates and phosphorylations, was probably due to the inactivation by DOC that inhibited these enzymes.

An auxiliary finding came from an ingenious idea by Watson to layer the detergent on top of a mitochondrial pellet for a short period and then fix the preparation and section the pellet at different layers. One result, found in time-lapse pictures, showed mitochondrial disruption, from an initial swelling to contraction to solubilization of the matrix contents and then to fragmentation of membranes. Another result, not realized at the time, was the first finding of a reconstitution of the membrane. While DOC inhibited succinoxidase completely, once the treated mitochondrial fraction was spun down and the DOC washed out, the enzyme activity and presumably the membrane structure was regained, a process later termed "reconstitution." Soon thereafter, we repeated the experiments of Lehninger with digitonin and found by EM observation that this phosphorylating preparation was indeed membranous in origin.[23]

In the spring of 1956, Lindberg and Ernster visited our laboratory and invited me to join them for six months at the Wenner-Gren Institute at the University of Stockholm, which was becoming known for its work on mitochondria. During this sojourn with my family amid the dark, cold winter months of Scandinavia, I worked arduously and made two further observations on mitochondria. While working on the "latent" enzyme adenosine triphosphatase (ATPase) of mitochondria, a subject of intense interest at the time, I found that dinitrophenol-activated ATPase was part of the oxidative phosphorylation mechanism, while the Mg^{2+}-activated ATPase obtained by DOC treatment was the result of disrupting the connection. Also, using the Watson-Siekevitz technique, it was found that the Mg^{2+}-activated ATPase resided in the membrane.[24] In another study, I showed that the well-known ATP–inorganic phosphate exchange reaction occurred at only one of the respiration chain sites of coupled oxidation to phosphorylation.[25]

While these investigations ended my work on mitochondria, they were important first steps for my next project with Palade in a combined morphological and biochemical study to localize proteins and enzymes in another cell organelle. Palade had a grand strategy to study protein synthesis and secretion from the standpoint of individual cell components. The first problem to be solved was, what exactly was the microsome fraction that was increasingly being thought of as the cellular site of protein synthesis?

Identifying the microsome fraction. The term microsome is not particularly descriptive in this instance because it only describes what emerges after centrifuging at high speed a homogenate of liver or pancreas cells. The pellet spun to the bottom of the tube was called the microsome or small particle fraction. Early hints from Claude and Porter suggested this sedimentable fraction was the endoplasmic reticulum that fragmented during this process. However, definitive proof was lacking. Now, with improved EM techniques, and with me as the biochemical partner, Palade suggested a new approach. We would use liver and pancreas for the simple reason that these tissues had been seen in situ with the EM having many particles attached to endoplasmic reticulum membranes.[26] The particles could therefore be used as markers for the endoplasmic reticulum. If we examined a microsome fraction by EM and found particles attached to membranes, we could compare them with the earlier micrographs of intact tissue slices to confirm that the microsomes (particles) were derived from the endoplasmic reticulum.

In our first studies on rat liver, both the tissue sections and the isolated microsome fractions showed some ER membranes having particles attached, while others did not.[27] Later, Palade named the particle-free membranes smooth endoplasmic reticulum or smooth microsomes, while those laden with particles were called rough endoplasmic reticulum or rough microsomes. The appearance of particles on the membranes of the microsomal fraction was the first indication that the microsomes were derived from the endoplasmic reticulum by a generalized pinching-off process rather than by mechanical fragmentation. In other words, during the homogenization of the tissue, the membranes were not broken but fragmented into a collection of isolated little vesicles. If they had broken, their contents would have spilled out. These discrete units were confirmed by comparing the microsome fraction with in situ tissue and finding comparable dense contents enclosed by the membranes (figure 1).

Identifying the nature of the particles. The next step was to make a "chemical dissection," a term used by Palade, to separate the microsome fraction into its biochemical constituents in order to relate a structure, such as membranes or particles, to its function. The analyses of the microsome fraction gave a high content of phospholipid, RNA, and protein. There was also a positive assay for the enzyme DPNH that Hogeboom had found earlier in this fraction.[28] An important technique to separate the particles from endoplasmic reticulum

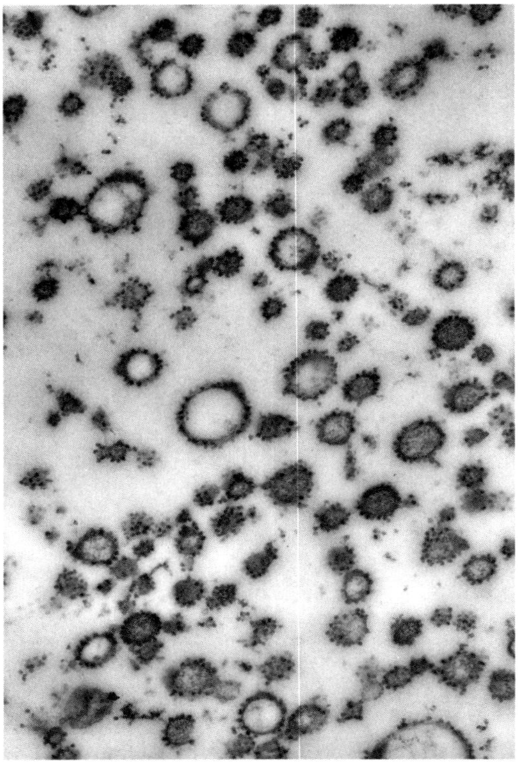

1. Electron micrograph of isolated microsome fraction from a pancreatic cell of a guinea pig. The microsomes appear as spherical or ovoid vesicles having two different components, a limiting membrane and small, dense particles attached to the outer surface. These particles were found to be identical in fine structure and similar in size to the rough-surfaced elements of the endoplasmic reticulum (ER), which suggested the particles derived from the ER by a pinching off process rather than by mechanical fragmentation. Courtesy of Philip Siekevitz.

membranes used DOC to solubilize the phospholipids, the reductase enzyme, and the membranes, but not much of the RNA. This dissection yielded DOC-insoluble particles and DOC-soluble vesicles and membranes. The particles could then be isolated and characterized. We learned that the particles consist of ribonucleoprotein and that the other compounds in the microsome fractions (phospholipid, protein, and enzymes) are part of either the membranes or the content of the vesicles. Since we found the particles to consist of RNA and protein, we initially called them ribonucleoprotein particles (RNP particles).[29]

At the same time, the Zamecnik laboratory at Harvard also used DOC on the microsome fraction and found the detached particles had a high RNA content.[30] But they did not use the electron microscope to look at in situ tissue or particles. What we did was to spin down these treated fractions and examine the pellet by EM, and here we found small particles of 100–150 ångströms having a high RNA to protein content. This supported our biochemical findings that the microsomal RNA is associated with the small particles, whereas the other substances were associated with the membrane or content of the microsomes.

This experiment with liver was the first indication that the microsome fraction was composed of three morphological and biochemical entities: 1) a membrane-enclosed space; 2) membranes having a high phospholipid to protein ratio, and most likely microsomal enzymes such as DPNH; and 3) attached RNA-containing particles. The evidence that was also accumulating on the microsomal fraction's ability to incorporate radioactive amino acids into proteins helped us focus our work on determining whether the RNA-containing particles were the sites of protein synthesis. An attractive premise for us was that protein synthesis is an obligatory and universal function of all cells.

Confirming results in pancreas exocrine cells. We then shifted from rat liver to perform similar experiments on guinea pig pancreas. The endoplasmic reticulum in the predominant cell of the pancreatic gland, the acinar or exocrine cell, occupies a greater proportion of the cytoplasm than liver cells and, as Palade noted, is a champion in protein production. These exocrine cells have an extensive rough endoplasmic reticulum and also a large population of free RNP particles. This meant we had a better chance to obtain a more homogeneous microsomal fraction and to separate free particles. Our results with pancreas confirmed and extended our previous findings on liver that the microsomes are fragments of the endoplasmic reticulum. The EM pictures of in situ pancreatic cells and of isolated particle fractions were more dramatic than liver tissue because they had many parallel arrays of particle-studded membranes.[27]

The free RNP particles, like the attached particles, also proved to consist of ribonucleoprotein. It then became evident that our RNP particles were identical to the structure that had been examined since the early 1940s by biophysical procedures as if they were high molecular weight complexes or macromolecular assemblies.[31–34] What we would learn using cell biology techniques was that instead of a macromolecular structure, the RNP particles were in fact small organelles with a critical function.

There were two other major findings. Pancreatic cells had no DPNH in the microsome fraction, but we obtained another particle fraction, called postmicrosomal particles, when the supernatant from the spun-down microsome fraction was sedimented at high speed. Both types of particles had a high RNA to protein ratio and were about the same size, 15 nanometers, whether obtained with DOC (later named attached particles) or without DOC (the postmicrosomal particles, later named free particles). The question was raised as to what functional differences might exist between the two types, but no answer was forthcoming at that time.

Another especially interesting finding of the pancreatic microsome work was that many free particles seemed to be arranged in strings or clusters. Palade had earlier observed these same formations of attached particles on the membrane surface.[27] We also had no explanation for this observation at the time. However, in 1962 and 1963, others found the particles were attached to a string of the newly discovered messenger RNA, or mRNA, which carries the instructions for making proteins from the gene in the nucleus to the RNP particles.[35,36] Stumbling onto this critical courier in protein synthesis before its cellular, let alone its molecular, pathway was even known reveals the nature of scientific discovery: many early valid observations must await further data for their explanation.

The two papers on liver and pancreatic microsomes provided conclusive evidence for the origin of the microsomes in the endoplasmic reticulum and for the nature of the particles.[27,29] With the approaching publication of these articles, Porter wrote to President Bronk to decry the publicity put out by "the great hungry boys in Berkeley" who were getting credit for ribonucleoprotein particles. He undoubtedly referred to a new work by Fu-Chuan Chao and Howard Schachman on analytical ultracentrifugation studies of particles.[37] Porter recommended that Rockefeller issue a news release about the upcoming Palade-Siekevitz papers, to which Bronk heartily agreed.[38]

During the late 1950s, a number of laboratories were working on microsomes and on RNP particles, and a large part of the first symposium of the Biophysical Society in 1958 at the Massachusetts Institute of Technology was devoted to these subcellular components. It was on this occasion that Richard B. Roberts proposed the name "ribosome" as being less awkward and having "a pleasant sound."[39] This name soon supplanted Petermann's label "macromolecular nucleoprotein particles" as well as our terms "ribonucleoprotein particles" and "Palade bodies." It was also at this meeting that Palade's presentation of our microsome work concluded, with a few qualifications, that the dense particles (15 nanometers) were found in practically all animal and plant cells. Either free or attached to ER membrane, they are the structural substrate of basophilia. While he knew this assumption needed to be tested in more kinds of tissues, he was confident that the "RNP particles are cytoplasmic components of widespread occurrence, probably basic structural elements in the organization of the cell." He ended by pointing to the next path in our research: "What seems to be particularly disturbing at present is the meagerness of our information about the functional role of these particles. Without more knowledge on this aspect, progress is uncertain even if it takes the appearance of elegant physicochemical data."[40]

Once we had identified the nature of microsomes and the chemical composition of the ribosomes, our next objective was to learn their function: were ribosomes involved in in vivo protein synthesis? For this, we conducted a series of experiments on the pancreatic exocrine cell of the guinea pig, proceeding on the assumption that this cell is a biological machine synthesizing a large amount of proteins, mainly digestive enzymes and proenzymes or zymogens, in a process in which each cell component has a special role in the production and handling of these enzymes. Many of these enzymes had already been isolated as crystalline proteins, so there would be a certain ease in manipulating this cell experimentally.[41]

Isolation and enzymatic activities of cell fractions. The work started with an attempt to make a more complete fractionation of the pancreas and to isolate a purified zymogen fraction. In the late nineteenth century, Rudolf Heidenhain in Germany established that secretory enzymes of the pancreas decrease in number after eating, when the pancreas discharges its juices into the intestine, and increase again slowly to the previous concentration.[42] He also postulated the enzymes existed within temporary deposits, which he labeled zymogen granules.[43] Some zymogens were found by John Marshall in 1954 using fluorescent antibodies[44] and a year later by Lowell Hokin in cell fractions.[45] In Palade's first electron microscopy studies of the pancreatic exocrine cell, he noticed a similarity in density and texture of the zymogen granules with the material segregated in the endoplasmic reticulum.[46] This impetus to isolate a zymogen fraction was a rudimentary step in stalking a process that might secrete granules in the ER's cavitary system to the zymogens.

The laboratory's standard cell fractionation procedures were adapted to the pancreatic cells, and the fractions were systematically examined in the electron microscope.[47,48] As a result, we obtained a satisfactory fraction of zymogen granules in addition to the nuclear, microsomal (having both free and attached ribosomes), and mitochondrial fractions. Next, the separate fractions were analyzed biochemically and measured for their total enzymatic activity and protein contents. Earlier experiments on protein formation in the pancreas by the Alfred Mirsky laboratory at Rockefeller had measured total protein in mouse pancreas and found a cyclical nature of synthesis and secretion.[49] But our new experiments focused on accounting for activity and distribution of enzymes in all fractions and indicated that the exportable enzymes being studied, i.e., ribonuclease and trypsin-activated proteases such as chymotrypsinogen, were found in all cell fractions; the highest concentrations were found in the zymogen granule fraction. Our experiments confirmed the old hypothesis of Heidenhain that zymogen granules deserved their name as containers of enzyme precursors: trypsinogen and chymotrypsinogen, among others. Also, this new approach to enzymatic activity looked promising for studying the movement of proteins in the pancreatic cells.

To overcome the disadvantage of measuring cumulative activities of all pancreatic enzymes, we isolated chymotrypsinogen using column chromatography and determined that sixty percent of it resided in the zymogen granules. With this preparatory work, using known enzymes as markers, we were able to analyze various cell fractions in static morphologic and biochemical terms.[48]

In vivo incorporation of radioactive amino acids into proteins. My 1952 paper had showed radioactivity was more actively incorporated into the proteins of microsome fractions than any other cell fractions. While Palade originally used subcellular fractions incubated in vitro, now we modified the technique to do in vivo studies using guinea pigs, a mammal chosen because of its easily dissected discrete pancreas.[50]

The basic technique involved injecting the radioactive amino acid leucine into guinea pigs, which quickly became incorporated into any newly forming protein in the pancreas. On removing the pancreas, we obtained proteins from different fractions and measured radioactivity in each. We could initiate a secretory cycle by feeding starved animals and at selected time intervals examine cell morphology, analyze fractions, integrate the results, and compare them with those from starved animals. We found maximal enzyme activity in both the zymogen granule and microsomal fractions of the cells just after the animals had been fed.

Biochemical results of initial experiments showed that radioactive leucine became concentrated in the proteins of the microsomal fraction. These results mirrored the earlier ones from the Mirsky laboratory in which N^{15}-glycine injection into mice was followed by fractionating the pancreas into a high speed fraction containing RNA and a supernatant fraction.[51] However, our microsomal fraction, which was also identified by EM, gave the highest radioactivity, thus allowing us to link the high occurrence of RNA in this fraction with protein synthesis. Our results agreed with the Mirsky laboratory that there seemed to be a steady level of protein synthesis and degradation in the pancreas.[52]

The in vivo experiment was then refined, and the microsomal fraction was further subfractionated into two fractions: a DOC-insoluble fraction of attached particles and a DOC-soluble fraction containing microsomal vesicles, a procedure based on our earlier work[27] and that by the Zamecnik group.[30] Not surprisingly, the highest protein-specific radioactivity was found in the proteins of the attached ribosomes. This was followed at time intervals when radioactivity declined in the ribosomes and increased in the contents of the endoplasmic reticulum.[50] This indicated a possible transfer of the radioactive proteins from the ribosome particles into the ER contents.

Synthesis of a specific enzyme. The stage was set for the next series of experiments in which the synthesis of a single secretory protein was followed from its first appearance to its final secretion from the cell.[53] The data obtained so far by cell fractionation procedures had not yet answered whether synthesis occurred at one site followed by transport to others or at multiple sites proceeding simultaneously but at various rates.

To solve the problem, we followed the time course of the synthesis and intracellular movement of the zymogen chymotrypsinogen using the basic protocol above of feeding starved animals. We collected pancreatic tissues at various intervals and determined the specific radioactivity of the enzyme in the mixed proteins of cell fractions. By reserving some of the tissue for electron microscopy, we could also identify the site of synthesis of a known protein by finding the first structure or compartment on which the radioactivity appeared. There was also a considerable effort to isolate a purified form of the enzyme, following methods developed by Rockefeller biochemists Werner Hirs, Stanford Moore, and William

Stein to determine the structure of ribonuclease.[54] This involved sulfuric acid extraction, ammonium sulfate fractionation, isoelectric precipitation, and final ion exchange column purification to determine the specific radioactivity of the enzyme purified from the various fractions. Showing the extent of purity of the enzyme and the lack of contamination by free radioactive leucine enhanced the reproducibility of the result.

At the early time points, the highest specific radioactivity appeared in the attached ribosomes. At later time points, the radioactive chymotrypsinogen predominated in the intracisternal (endoplasmic reticulum) content and later still predominated in the zymogen granules. In contrast, the unfed control animals showed no activity in this fraction. An interesting finding was that it took a minute or less to synthesize an exportable protein and an hour for it to reach the zymogen granules. In this way, we produced evidence of movement of the newly synthesized enzyme between various subcellular components. Palade commented later, "at that time, this experiment was a significant technical achievement."[55]

In contrast to the attached ribosomes, the free ribosomes were found to incorporate leucine into their proteins at a considerably lower rate. A reverse picture was obtained when a ribonucleic acid precursor, adenine-8-C^{14}, was injected. This time the most active fraction was in free ribosomes, while the attached particles showed negligible activity.[56] These results provided the basis of our hypothesis for the difference between the free and attached particles: free particles produce proteins for intracellular housekeeping, while attached ribosomes synthesize proteins for export.

The Mirsky laboratory had extracted a semi-purified chymotrypsinogen having higher radioactivity than the proteins in their final supernatant or high speed fraction, which they called a complex, and concluded the enzymes in this complex were not the precursors of the secretory enzymes.[51] Their opposing view could be ascribed to different experimental protocols that led to separate conclusions, specifically due to our use of electron microscopy to identify the cell fractions. Our findings were distinguished because for the first time we found that a highly purified secretory protein was synthesized in the attached ribosomes, secreted into the interior space of the endoplasmic reticulum, and finally, in the case of the pancreas, ended up in the zymogen granules prior to exiting from the cell.

Later experiments in the laboratory using isolated microsomes showed other secretory proteins, both pancreatic amylase[57] and a truncated protein,[58] behaved in a similar manner. Several years later, other Rockefeller investigators showed that also in liver the attached ribosomes synthesized secreted serum proteins, while free ribosomes synthesized endogenous liver proteins.[59] Palade concluded that noticeable changes "affecting either the relationships, or the structure, or the location of a number of cell components have been detected by electron microscopy. Experimental conditions and complementary biochemical information indicate that the changes described are functional modulations of the structure under consideration. In my belief, these findings clearly support the thesis that the electron microscope can be a useful tool in physiological research at the cellular or intracellular level."[60,61]

In vitro incorporation of radioactive amino acids. The in vivo experiments with pancreas cells prompted attempts to verify whether ribosomes alone could synthesize proteins in vitro. We needed a control experiment to show whether these particles in the absence of other microsomal components, but in the presence of an energy-producing system and soluble RNA, were capable of amino acid incorporation. Working with graduate student

Jack Kirsch, we found that incorporation proceeded in tissue slices, homogenate, or an isolated fraction. An in vitro analysis of five different amino acids also established that the ribosome, free or attached, is the site of protein synthesis in the cytoplasm.[62]

Attached ribosomes: Connecting synthesis with transport. Between a protein's first appearance on the ribosomes and its next one inside the endoplasmic reticulum, two important operations were postulated: the release of newly synthesized enzymes from the ribosomes and movement across the membrane from the cytoplasm into the ER's cisternal space.

We explored various means to find a physiological agent responsible for detaching enzymes from the ribosomes. They could not be removed by washing, so we tried several reagents. The enzymes could be released by the magnesium chelating effects of adenosine triphosphate or inorganic pyrophosphate, but this treatment also caused the ribosomes to fall apart.[63] Ribosomes could also be released by treatment with the polyamine spermine, but this caused the ribosomes to aggregate, undoubtedly due to the tight binding of the positively charged spermine to the negatively charged RNA.[64]

These negative results were followed later by an important piece of evidence for the transfer of enzymes across the cisternal membrane. An in vitro experiment followed in time the radioactive appearance of the digestive enzyme amylase. The newly synthesized amylase was found first associated with the attached ribosomes, and then a short time later it was released from the ribosomes and appeared primarily in the vesicles.[65] Other researchers, looking at HeLa cells and liver cells, had found a transfer of proteins from attached ribosomes to the cisternal space. But the EM pictures from our experiments provided convincing evidence, showing an actual connection between the two stages. This would become the first step in a postulated pathway of protein secretion. The ribosome was emerging as a key element in protein synthesis.

The Golgi complex in the formation of secretion granules. While following the time course of radioactive chymotrypsinogen in various subcellular fractions, we noted a mystery. When intracisternal granules were present in the rough ER, there were also large vacuoles filled with a dense material that appeared in the Golgi zone. The material was similar to the zymogen granules on one side and to the intracisternal granules on the other, so we assumed this was the same material in three different locations. The question was what was happening in between. Electron microscopy could not demonstrate movement, direction, or rate of transport of the granules. Cell fractionation, as we practiced it, was imperfect and did not contain a specific fraction of the Golgi complex; its contents were likely dispersed among other fractions. The mystery, centering on the role of the Golgi, was solved by Jamieson and Palade using a different technique—in vitro pancreas slices.

The possibility that the Golgi complex was involved in the secretory pathway was raised by Sjöstrand in 1954,[66] by Marilyn Farquhar and Robert Wellings in 1957,[67] and by Hokin in 1955.[45] What was needed to implicate the Golgi complex in the secretory process was a technique to trace in time and space, using an appropriate labeling technique, the movement from one cell compartment to another all the way from attached ribosomes to zymogen granules.

Use of autoradiography in tracing the secretory pathway. The technique of autoradiography was just being adapted for electron microscopy by Lucien Caro and Robert van Tubergen.[68] Caro had joined the Porter/Palade laboratory in 1959 as a technician and then earned a PhD at Yale. His thesis and great contribution was the application of autoradiographic techniques to biological specimens. When he returned to Rockefeller, he applied this technique to the problem of secretory protein pathways in the pancreas.

Caro injected amino acids (radioactively labeled with tritium) into the guinea pig, following the earlier procedure. Samples of pancreatic tissue were taken at various times after injection of the radioactive amino acid, fixed with osmium, and embedded in plastic, and thin sections were prepared for transmission electron microscopy. The methods for autoradiography are relatively simple: light or electron microscopic sections are overlaid with photographic emulsion (for light microscopy) or with a single layer of silver halide crystals (for electron microscopy). Control experiments were done to ensure that most of the soluble radioactivity was removed during the fixation and subsequent washing and that staining removed only a small percentage of the radioactive proteins. Following exposure and development to produce silver grains visible in the light or electron microscopes, sections were stained to improve contrast. A similar procedure had been tried previously at the light microscope level,[69] but the structural resolution was not good enough to identify positively the structures involved, so they could only be inferred from their distinctive cell location.

In the Caro studies, however, both light and EM images showed a clear pathway: five minutes after injections, most of the radioactive label appeared over the rough ER; at twenty minutes it appeared in the Golgi complex; and at sixty minutes it appeared in the zymogen granules. There was a drawback since the radioautographic film did not show a single source of radioactivity but instead a circular path due to the "spread" of the emitted source over the emulsion or the silver halide crystals. Nonetheless, the data were dramatic in indicating the structures covered by the radioactive grains.[70,71]

This study confirmed our earlier results of three separate locations for the proteins, this time at the in situ level. The Golgi complex was now firmly placed into the secretory pathway. Our interpretation was that as the proteins pass through it, they are progressively concentrated in condensing vacuoles and, when completely filled, become zymogen granules. The details of this process were solved by Jamieson (see Chapter 7C).

Other investigators outside Rockefeller were beginning to trace a similar secretory pathway. Some were strictly biochemical approaches. Anna Laird and A.D. Barton followed a hormone-induced increase in amylase activity in the pancreas over time intervals in various fractions, beginning with the microsomes. They hypothesized that the protein was released in soluble form and recondensed into granules by the Golgi complex. However, they did not isolate ribosomes and did not provide micrographs of their fraction.[72] Allan Morris and Sherman Dickman agreed with Laird and Barton while using a radioactive amino acid protocol, but they did not isolate ribosomes and did not verify the purity of the fractions.[73] On the other hand, Sjöstrand was following a strictly morphological approach, and during the 1961 Ciba Foundation Symposium on the Exocrine Pancreas, he argued that ribosomes were artifacts of fixation and that the Golgi was the site of both zymogen granule membrane formation and protein synthesis.[74] At this same symposium, Palade refuted Sjöstrand once again by basing our interpretations on a combined cytologic and biochemical approach. Our theory was strengthened when Palade presented evidence on isolated ribosomes having a high RNA content and their ability to incorporate radioactive amino acids before moving to the Golgi. It was this combined strategy that showed limitations of either a strictly biochemical or morphological approach.[71]

Isolation of Golgi fractions by centrifugation. Caro's work had shown only an association of proteins with elements of the Golgi complex, but not exactly where they were located within the complex. What was needed was a biochemical analysis of a separate Golgi fraction. Thus, with Canadian postdoctoral fellow John Bergeron and graduate student John Ehrenreich, we isolated a purified Golgi fraction from liver to look at its morphological and biochemical characteristics.[75,76] We instituted a procedure that relied on the injection of alcohol in rats to produce very low density lipoproteins (VLDL) that ended up in the liver cells' Golgi cisternae. This facilitated the isolation by sucrose-density gradient centrifugation of three Golgi subfractions based on differences in density due to their VLDL contents. By comparing the isolated lightest fraction with one seen in situ, we identified it as being derived from the trans-secretory Golgi face. The second, intermediate fraction was a mixture of mostly the lightest elements and lighter material. The third fraction was similar to the lightest one and was identified as being derived from the cis-Golgi opposite face. A short time later, Marilyn Farquhar, John Bergeron, and Palade amplified this work by using enzyme cytochemistry on the same isolated Golgi subfractions (see Chapter 13C). They examined 5′-nucleotidase (presumably a plasma membrane marker), acid phosphatase (for lysosomes), thiamine pyrophosphatase (for trans-Golgi cisternae), and glucose-6-phosphatase (for ER microsomes). They found the nucleotidase in the Golgi fractions as well as in the plasma membrane fractions, acid phosphatase in VLDL-filled vacuoles that was indigenous to the secretory (Golgi) granules and lysosomes, and thiamine pyrophosphatase in the three Golgi subfractions. This cytochemical examination at the EM level allowed them to distinguish between indigenous and contaminating activities in the Golgi fractions.[77]

7C. Establishing the Secretory Pathway of the Pancreatic Exocrine Cell

Philip Siekevitz and James D. Jamieson

What was emerging from this in vivo work was a tentative reconstruction of how newly formed cellular proteins were moving within the cell for secretion to the outside.

From a morphological point of view, George Palade commented in 1961 that it had been somewhat "disappointing to find out that the most elaborate structure found in the exocrine cell, namely its voluminous and orderly disposed endoplasmic reticulum, is not involved in protein synthesis per se." The ER, he conjectured, performs an auxiliary operation such as segregation and intracellular transport of proteins, more like a superstructure involved in the shipping and packing of the cellular plant. Nevertheless, he was intrigued that the ER's spectacular size might be essential for a glandular cell like the pancreas to achieve such large-scale synthesis.[1]

At the same time, it was also becoming clear that the most important operation in protein secretion occurs in the structural components in the cytoplasm. The proteins appear in finished form on the ribosomes attached to the ER membranes. At this point, the laboratory began testing their hypothesized steps in the secretory pathway. The first stage appears in the attached ribosomes, where protein synthesis ends and intracellular transport begins. In the second stage, proteins are segregated in cavities of the rough cisternae of the ER. The third stage takes place in large vacuoles of the Golgi complex, where proteins are concentrated in smooth-surfaced vacuoles. In the fourth stage, proteins are stored in zymogen granules. At this point, the granules are discharged out of the cell into the lumen of the pancreas. Conclusive experimental evidence for stages two through four, described in the following paragraphs, turned this hypothesis into a sound working model.

From the endoplasmic reticulum to the zymogen granules. Transport of secretory proteins from the cisternal space to the condensing vacuoles was investigated using autoradiography and an in vitro system that could give more precise cytochemical data, namely, the tissue slice system. Lowell Hokin had used a pancreas slice system to synthesize amylase as early as 1951.[2] Two members of Palade's laboratory decided to try this system, James Jamieson, an MD graduate student of Palade, and Colvin Redman, a postdoctoral fellow and former graduate student of Hokin.

Jamieson became interested in research during medical school at the University of British Columbia (UBC), which permitted him time to work on experimental hypertension and smooth muscle structure and function. After reading the papers of Palade and Philip Siekevitz, he decided to learn electron microscopy, but since the university had no electron microscope at that time, he realized he needed more training and considered a postdoctoral position at the Karolinska Institute in Stockholm. However, his research mentors at UBC, Sydney and Constance Friedman, who had come from McGill to found a department of anatomy, advised him to apply to the new PhD program at (what was then) The Rockefeller Institute, a curriculum that required few formal courses and was essentially a laboratory apprenticeship. Arriving in 1960 straight from medical school, he was encouraged to work with Arpad Csabo on uterine smooth muscle control. A short time later, after taking the cell biology course given by Keith Porter, Palade, and Siekevitz and viewing their electron microscope setup, he switched laboratories and was put to work on one of Palade's favorite topics, blood vessels and heart muscle.

According to Jamieson, his job was "to characterize the atrial secretory granules that had been seen by others in cardiac muscle cells and whose function was unknown. We knew nothing about the content of these granules at the time so their characterization was by brute force centrifugal isolation combined with electron microscopy of the fractions—a task made all the more difficult by the fact that the larger the animal, the fewer the granules, which became abundantly clear after several 3 a.m. trips to a Brooklyn abattoir to collect atria from cows. Years later it was discovered by Adolfo J. de Bold that the atrial granules contain atrial natriuretic peptide, a potent regulator in controlling the water-electrolyte balance in mammals.[3,4] After four frustrating years, George and I called it quits. And for George to give up anything, it really had to be going bad. What to do?

"The pancreas next beckoned with the question of how secretory proteins get from the rough ER to the Golgi and beyond. Lucien Caro and Robert van Tubergen were at that time finishing their work on the use of EM autoradiography to map out the pathway of secretory protein transport in the guinea pig pancreas in vivo but the results suffered from poor temporal and spatial resolution. Plotting with Colvin Redman, who had extensive experience with pancreas slices in vitro, I planned out a series of pulse-chase experiments in which secretory proteins would be labeled in vitro in guinea pig pancreatic slices with H^3-leucine for short times, then 'chased' with unlabeled leucine to allow the wave of nascent secretory proteins to traverse the secretory pathway and so give the needed temporal resolution. Concurrent EM autoradiography was to provide information on the way stations through which the proteins traveled, while cell fractionation of the slices would give the spatial resolution that was lacking in EM autoradiography. It all sounded good to Colvin and me but George was skeptical that pancreatic proteases and ribonucleases would be the undoing of the plan. Reluctantly he let me go ahead, because otherwise I had nowhere to go in my thesis work: the alternatives were bleak.

"During the summer of 1963, I set up the slice system, verified the pulse chase, and was able to carry out preliminary light and EM autoradiography as well as gradient fractionation of the slices into rough and smooth microsomal fractions. The latter clearly showed that newly synthesized proteins entered this compartment which autoradiography showed corresponded to the smooth surface vesicles in the periphery of the Golgi complex. That was very exciting, since it showed that secretory proteins move from the ER to the Golgi always in the lumen of a membrane-enclosed compartment. Upon returning from his summer vacation, George wondered what this boy from the Wild West had done, and I proudly showed him. His reaction I vividly recall was one of surprise and probably relief which elicited from him one of his few uses of an expletive. I had been fortunate during those four years while I learned patience and perseverance and became proficient in both cell fractionation and electron microscopy, the backbones of cell biology."[5]

By shifting from whole animal to tissue slices, Jamieson could manipulate the pulse-chase radioactive incorporation much better than the in vivo labeling used previously. At

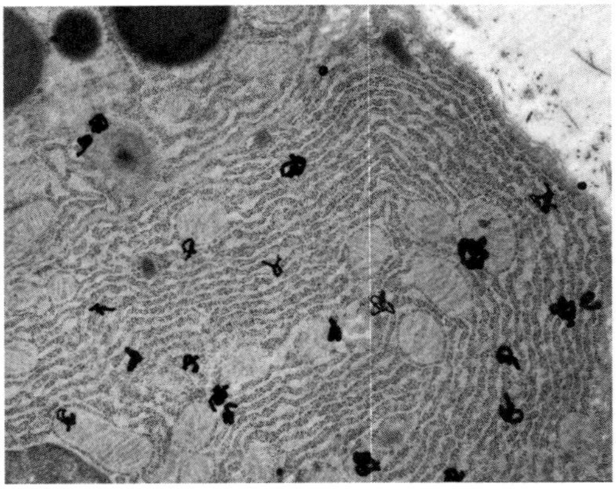

1. The four electron microscopic radiographs in figures 1–4 show the procession of radioactively labeled proteins through intracellular compartments in the secretory pathway in a guinea pig pancreas slice. After a three-minute pulse with radioactive leucine, the label is in the rough endoplasmic reticulum. Courtesy of James Jamieson.

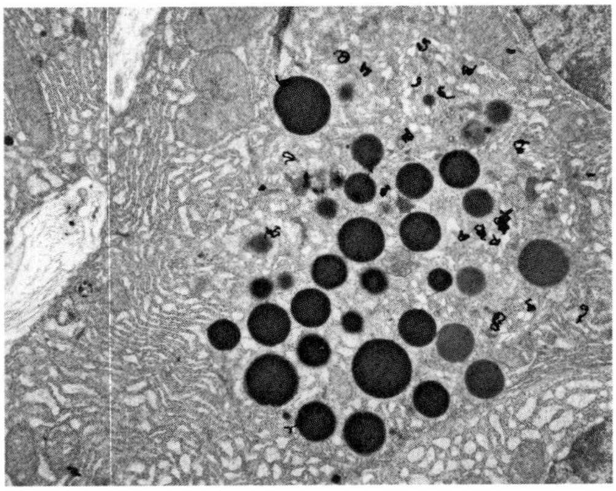

2. After a three-minute pulse followed by a seven-minute chase with non-radioactive leucine, the label is in the Golgi periphery. Courtesy of James Jamieson.

the same time, he fractionated the tissue into various subcellular fractions and determined the specific radioactivity of the proteins in each one. Finally, he and Palade took electron micrographs of the slices to record their structural integrity after various incubation periods.

The first challenge was to identify the subcellular structures in the movement of proteins from the rough ER to the condensing vacuoles of the Golgi complex. Jamieson's kinetic experiments on smooth microsomal fractions, isolated from slices at the end of the pulse and at various times of incubation in the chase medium, provided direct evidence that secretory proteins pass through the small peripheral vesicles of the Golgi complex from their site of synthesis and segregation in the rough ER to their site of concentration in the condensing vacuoles. Cell fractionation data further indicated that, during the time intervals examined, the specific radioactivity of proteins in the cytoplasmic matrix remained constant, thus overturning the assumption that secretory proteins leave the microsomes and pass in soluble form through the cytoplasmic matrix to the zymogen granules.[6]

Jamieson's findings agreed with autoradiographic data by Caro and Palade on intact cells, either in slices or in whole animals. Their data showed that as the time-labeled protein becomes concentrated in the smooth microsomal fraction, the autoradiographic grains are concentrated only over a region of the Golgi complex containing numerous small smooth-surfaced vesicles. The results of cell fractionation and autoradiography were thus compatible and complementary.

The next step was to identify the role of the condensing vacuoles of the Golgi complex that had been observed by electron microscopists. Caro and Palade had previously established that after nascent proteins concentrated in the condensing vacuoles, they are converted to zymogen granules.[7] Jamieson and Palade then confirmed the position of the condensing vacuoles in the secretory cycle by electron microscopic autoradiography and demonstrated that the time taken to accumulate labeled proteins within condensing vacuoles is almost the same in vitro as in the intact cell in situ. Because of the well-defined pulse label, which passes through the cell as a coherent wave, the presence of the label in condensing vacuoles was more strikingly demonstrated than was possible in the whole animal.[8–10]

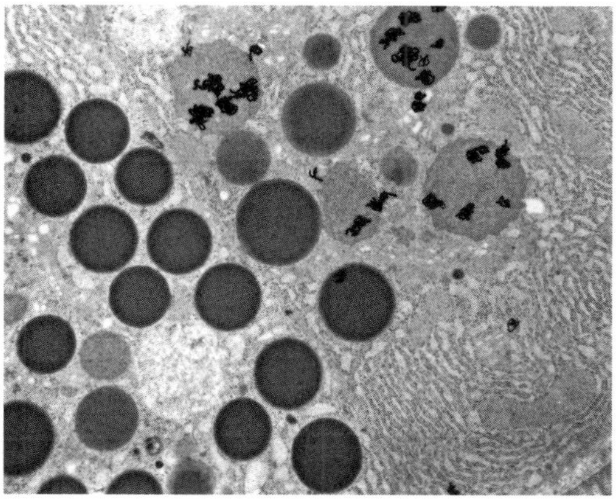

3. After a three-minute pulse followed by a thirty-seven–minute chase with non-radioactive leucine, the label is in the condensing vacuoles of the Golgi complex. Courtesy of James Jamieson.

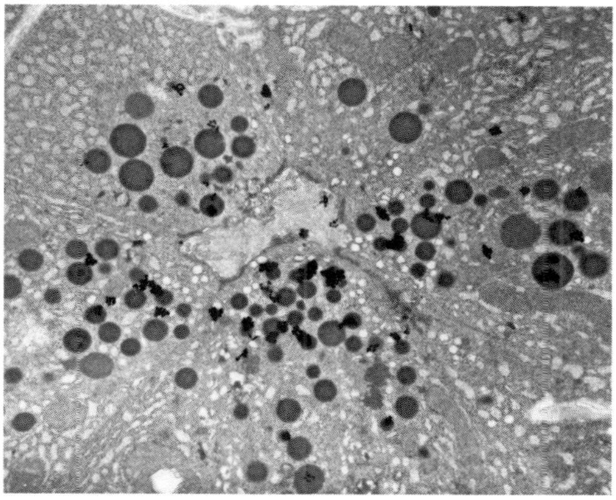

4. After a three-minute pulse followed by a 117-minute chase with non-radioactive leucine, the label is in the zymogen granules. Courtesy of James Jamieson.

The last step was to delineate how long it took for labeled secretory proteins to pass from the rough ER into the zymogen granules. Earlier results using autoradiography or cell fractionation did not agree and ranged from sixty minutes in the former case to ten minutes in the latter. Repeating the experiments, Jamieson found that the sixty-minute interval was correct, since he discovered that the isolated zymogen granule fraction was "contaminated" by condensing vacuoles containing newly synthesized radioactive secretory proteins undergoing concentration to zymogen granules in the trans-Golgi at earlier times.[10]

By 1968, the secretory pathway had been followed from its beginnings at the ribosome/ER level (by Palade and Siekevitz) to its ending at the zymogen granules (by Palade, Jamieson, and Caro). At no time were there movements of the secretory proteins in the soluble cytoplasmic matrix, as had been earlier envisaged by Laird and Barton[11] and by Morris and Dickman.[12] These findings further refuted Sjöstrand's contention that elements of the Golgi were involved in synthesizing secretory proteins.[13] Four years later, in 1972, Palade and graduate student David Castle strengthened this evidence by finding a similar secretory process in the rabbit parotid gland using only autoradiographic studies.[14,15]

Still needed were details to identify the transfer from the Golgi peripheral vesicles to the Golgi condensing vacuoles and then to zymogen granules. This was quickly solved when Jamieson and Palade determined that proteins exit from the ER toward the smooth-surfaced vesicles in the Golgi periphery by budding off transitional elements, part rough-surfaced and part smooth-surfaced evaginations, of the ER. They further suggested that the smooth-surfaced vesicles and vacuoles are functionally continuous but maintain their distinctive membrane compositions despite extensive traffic between them.[9,10]

As Jamieson summarized, immediately after completion of synthesis on ribosomes attached to the membrane of the ER, nascent secretory proteins are segregated in the cisternal space and remain separated from the cytoplasmic matrix by a membrane for the duration of their intracellular existence. The digestive enzymes cross the cytoplasmic membrane only once, immediately upon completion of synthesis. The membranes protect the cytoplasm from the potential hydrolytic activity of the digestive enzymes and enable the cell to direct the transport, concentration, and distribution in quanta of secretory proteins without affecting the resident soluble proteins of the cell (figures 1–5).[6]

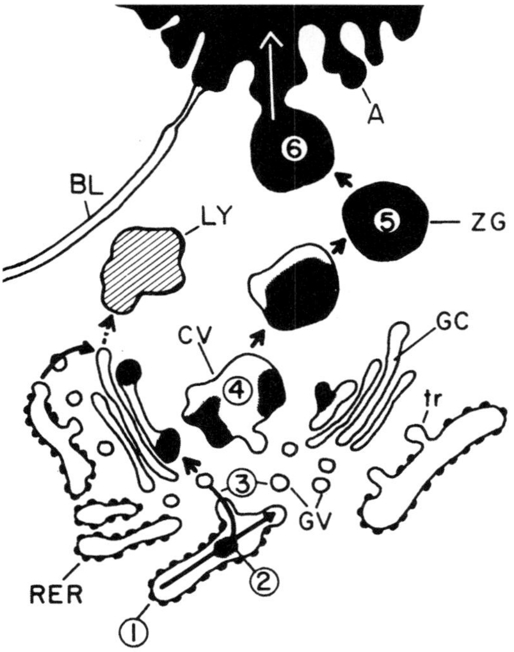

5. Diagram of the (then known) intracellular movement of nascent proteins in the secretory pathway: 1) proteins are synthesized on ribosomes, which attach to the membranes of the rough endoplasmic reticulum (RER) and are transferred across these membranes and segregated 2) in the cisternal space of the RER. They follow two pathways, either to the degradative one of the lysosome (LY) or to the secretory pathway. The latter pathway begins where they are transported 3) by budding processes (tr) from the rough ER to Golgi vesicles (GV), then to the condensing vacuoles (CV) located in the Golgi complex (GC) region 4), where they are concentrated. The concentrated products 5) are stored in zymogen granules (ZG), where they are discharged by exocytosis 6) to the pancreatic ductile system (A). Courtesy of James Jamieson.

To test whether new protein synthesis was necessary for secretion to operate, Jamieson and Palade performed another set of experiments. Following earlier protocols, they used a protein synthesis inhibitor, cycloheximide, to abruptly terminate further input of secretory proteins into the cisternae of the ER.[16] They observed that transport over this link, and over the entire pathway, did not depend on ongoing new synthesis that pushed proteins along the pathway. They also found that the energy for transporting the proteins came from ATP generated by oxidative phosphorylation and not by glycolysis. Without ATP production, there was an abrupt arrest of intracellular transport. This energy requirement, most likely from the oxidation of fatty acids, occurred at the step between the rough ER and the peripheral vesicles or the Golgi complex.[17] Energized by ATP, the cell appeared to operate a lock that regulated the transport of proteins between the two compartments.

Characterizing membrane fractions of the secretory pathway. After Jamieson received his PhD under Palade in 1966, he joined the laboratory staff and turned to study what role the membranes might play in the secretory process. This work was done in collaboration with his first postdoctoral fellow Jacopo Meldolesi, a young MD from the University of Milan, Italy. In essence, they isolated membranes from each subcellular fraction in the pancreatic acinar cell, identified their lipid and protein compositions, and concluded that the membrane of each fraction was distinct and did not mix with the others during the secretory process. They postulated that the distinctive membrane properties were accounted for by recycling of membranes between intracellular compartments, a concept borne out experimentally many years later. In addition, later studies by others demonstrated these differences were typical of all secretory cells.[18–21]

Secretion of proteins from zymogen granules. By 1971, Jamieson and Palade were looking at the final step in the secretory process, the release of proteins from zymogen granules. Twelve years earlier, Palade had observed with electron microscopy how the granules discharge their contents.[1] The membrane of a granule fuses with the cell's outer membrane, leading somehow to fission of the membrane at the point of fusion and release of the granule contents directly into the lumen. Originally called membrane fusion and fission and later exocytosis, this simple process allowed proteins to be released, keeping the cell membrane intact and dangerous digestive enzymes from flowing back into the cell and cytoplasmic components from leaking out of the cell.[21,22] Since the overall surface area of the apical plasma membrane, which is the unique site of exocytosis in regulated secretory cells, did not increase by the calculated fifty- to a hundred-fold due to contribution by zymogen granule membranes under intense secretory stimulation, they also proposed that the excess membrane is recycled for reuse, a concept generally accepted for all exo- and endocytic processes.

Palade and Jamieson then became interested in agents that might activate or inhibit exocytosis. They used the chemical stimulus carbamylcholine and combined it with the pulse-chase technique on slices, quantifying radioactivity in isolated fractions, and autoradiography. They learned that this secretory stimulus affects the discharge of proteins but not their rates of synthesis or their intracellular transport.[23] Redman and Palade also studied the secretory protein release process in liver cells discharging plasma proteins. Using colchicine, which interacts with the fine structures of the microtubules, they found secretion could be inhibited at the level of the Golgi.[24]

Biochemical studies on the release process. A rather important question that had to be solved was whether or not all the secretory proteins were transported and released at the same time after stimulation. The first step was taken by Palade and two graduate students, Alan Tartakoff and Lewis Greene. They separated and semi-purified six secretory proteins from pancreatic zymogen granules and from the soluble secretion discharge using isoelectrical focusing and sodium dodecyl sulfate gel electrophoresis. They found the same proteins in both compartments in relatively similar amounts.[25] Next, postdoctoral fellow George Scheele used an in vitro pancreatic lobule system to study the kinetics of discharge, which was stimulated by carbachol, amylase, lipase, and ribonuclease and four zymogen enzymes. EM pictures showed the architecture of the lobule was preserved during the in vitro incubation. These results led to the conclusion that all the enzymes or pre-enzymes were discharged in synchrony and in constant proportion to one another.[26]

A parallel study, using the same system but adding radioactive tracers, showed that regardless of the secretory stimulant used, the relative composition of the discharged proteins was the same at all time points and that the proteins enter the condensing vacuoles of the Golgi and the zymogen granules in apparent synchrony.[27] Thus, this radioactive labeling experiment reached the same conclusion as the enzyme assay experiments. Studies by Jean-Pierre Kraehenbuhl and Jamieson, using immunological techniques, supported this conclusion since they found four pancreatic enzymes in all cells and in all zymogen granules in a given cell.[28] However, all these results contradicted the argument against parallel discharge by Stephen S. Rothman at the University of California, and particularly his biochemical methods to activate the zymogen into active enzymes.[29]

Six years later, Scheele and Tartakoff joined Palade, who by now had moved to Yale University. They resolved some apparent contradictions between results using subcellular fractionation and those using radioautographic techniques. The problem resided in cross-contamination of the fractions and leakage of proteins from vesicular components during the fractionation process. Once they devised a way to quantitate the contamination and the leakage, the data from both techniques agreed: secretory proteins remain confined inside membrane-bound organelles as they are transported successively from the rough ER to the Golgi complex, to condensing vacuoles, and to zymogen granules.[30]

7D. The Palade Model of Combining Structure with Function

Carol L. Moberg

The detailed investigations establishing the secretory pathway, which some have called the "Palade model," represent the embodiment of modern cell biology. Delineating a central question of how proteins are transported through the cell took many scientists in this Rockefeller laboratory twenty-five years (1950–1975) of combined cytological and biochemical experiments to produce the first coherent picture of this process. By keeping a final goal in mind to correlate structure and function at the subcellular level and by proceeding step by step with their findings, they adapted and adjusted their tools to fit the varied, successive phases of their work.

After countless experiments, these scientists found the endoplasmic reticulum is a receptacle for proteins that were newly synthesized on its attached ribosomes. The proteins are transported through the cell to an extracellular space through a series of connecting membrane-bounded compartments. The Golgi complex was likened to a "bottling station" or intermediary that packaged the secretory products, and the cisternae were believed to proceed along a specified course having no backward movement, as on a conveyor belt in a factory assembly line that packaged granules on the trans (or mature)-face and reformed by new membranes on the cis (or immature)-face. The proteins are carried in discrete containers (vesicles or vacuoles) through locks whose opening requires energy. They cross a membrane only once, between synthesis and segregation in the ER, otherwise transport is associated with fusion and fission of membrane between compartments.[1–4]

The scientific story presented here relates only the initial discoveries of this process. By the 1960s, many other laboratories were following this integrated approach, a strong signal that an all-encompassing science was developing. Aided by the applications of even newer instruments, cytochemical assays, and recombinant DNA technology, scientists continue to build a more dynamic picture of a network whose changes and movements are all highly regulated to secrete proteins. Nonetheless, this model set the stage for gleaning and interpreting a vast amount of knowledge about the cell and has generally withstood another thirty years of accumulated evidence from further experimental work.[5]

In retrospect, at the First International Cell Biology meeting in 1976 in Boston, George Palade referred to this period of "relative scientific affluence. . . . The universality of whatever exists in biological organization between molecules and cells was neither assumed nor predicted. It emerged directly and progressively from the evidence collected over the last twenty-five years. This applies to the universality of ribosomes in all cells as a counterpart to the universality of proteins and protein synthesis, and it also applies to the universality of mitochondria in eukaryotic cells as an expression of the generality of oxidative phosphorylation and ATP utilization."

Palade also realized that this early period of discovery in cell biology was nearly over. Integrating structure, biochemistry, and function for each subcellular component in isolation was no longer sufficient. The interest was already shifting, he noted, "toward regulatory mechanisms operating over the whole domain of cellular organization and integrating the activity of each cell organ within the overall activity of the cell."[6]

Chapter 8

Ribosome Structure and the Synthesis and Movement of Proteins

• • • • •

Introduction

Carol L. Moberg

By the early 1960s, George Palade realized that the isolation of microsomes (fragments of the endoplasmic reticulum [ER]) and ribosomes was opening a remarkably fertile field of research. In a lecture for the centennial of the National Academy of Sciences in 1964, he summarized his views on "The Organization of Living Matter." By then, it was known that the ribosomes consist of equal amounts of RNA and protein, have a particle weight of $\sim 4 \times 10^6$, comprise two unequal subunits, and represent the basic cellular equipment for protein synthesis. The ribosome discoveries had strengthened his view of well-defined functional specializations of subcellular components and "brought forward, more forcefully than before, the concept of necessary integration among various cell organs or subcellular components."

By necessary integration, Palade meant various functions of definite cell organs are carried out by smaller elementary particles at the "subjacent macromolecular level," of which he believed ribosomes were the first relatively well-defined example. Since ribosomal RNA was the only known RNA in cells, Palade thus hypothesized how this integration was working: the ribosomes receive the template of the protein to be synthesized from the nucleus in the form of messenger RNA, accept amino acids activated by enzymes in the cytoplasmic matrix in a soluble form, use a polymerase to link these amino acids by peptide bonds, and finally discharge the ensuing protein either into the cytoplasmic matrix or across the membrane of the ER into the cisternal space (depending on the protein). In this hypothesis, he asserted the "salient aspect is the extent of integration, which is superimposed on functional specialization, and apparently involves practically all important cell organs."

Concurrent with this hypothesis, his laboratory began research on the last ER discoveries presented in this book. This research began while many *final* steps in the secretory pathway were falling into place, but they relate to the *initial* steps in this pathway, namely, how the structure of ribosomes functions to synthesize proteins. These studies relied on earlier knowledge while they dealt with mechanistic questions that lay dormant during those years. The questions concerned how the ribosomes attach to the membrane, how the ribosomes "know" which kind of protein should be synthesized, and which protein factors are involved in moving the newly made proteins from their ribosomal site of synthesis through the ER membrane into the vesicular space of the reticulum. A new era ensued with a new methodology that might be termed molecular cell biology.

Two members of the Palade/Philip Siekevitz laboratory were major players in this work, David Sabatini and Günter Blobel. Each has written his own story of the molecular biology of the ribosomes. In part A, Sabatini describes more about the early background that led to the thinking and the postulation of the "signal" specificity hypothesis. In part B, Blobel delves into the proteins and molecular mechanisms involved in bringing about the fruition of this hypothesis. Of considerable interest in their stories is how many years and how many scientists it took before the questions asked in the early 1960s were finally solved. Blobel's story is excerpted, with permission, from his 1999 Nobel Prize lecture, in which he describes this lengthy process of resolving the questions of how proteins have built-in intrinsic signals that direct them to their correct location within cells and across membranes. The Nobel committee deemed this work "laid the foundation for modern molecular cell biology." In part C of this chapter, James Lake describes his concurrent studies in Palade's laboratory on elucidating the structure of the ribosomes.

8A. Mechanisms that Regulate the Destination of Nascent Proteins

David D. Sabatini

The educational system that prevailed in Argentina in the late 1940s, when I finished high school, required that I then choose a specific university track in which to continue my studies. I found this an extremely difficult decision to make, as I had a wide range of interests both within the sciences and the humanities. I decided to enter medical school, not moved by a passionate yearning to be a physician, but rather because I believed that the study and practice of medicine offered almost limitless opportunities for both scientific exploration and a full appreciation of the human condition.

I graduated with an MD from the University of Litoral in Rosario, Argentina, in 1954, a few months before the forced end of Juan Peron's regime. The beginning of my scientific career was made possible by that event since Argentine universities went into a sort of academic renaissance and teaching positions were again, for a time, opened to all those who qualified without regard for political affiliation.

In medical school I had read several classical textbooks, including those by Giuseppe Levy, Alexander Maximow, and William Bloom on histology and by Herwig Hamperl, Hugo Ribbert, and Gustave Roussy on pathology. They stimulated my interest in the structure and function of cells. At this early stage of my scientific life, I also became acquainted with Edmund B. Wilson through his marvelous introduction to the 1925 edition of his book *The Cell in Development and Heredity*, which I found at the medical school library. I was soon fascinated by questions such as whether the mysterious different cellular organelles illustrated in those books, although cooperating with each other to sustain the activity of the cell, had a certain life of their own. Or, to phrase it in a more scientific manner, by what mechanisms did they perpetuate themselves as distinct structures within the cell and from one cellular generation to another while retaining their individuality? So little was known about these questions that I used to think of them as somewhat akin to those concerning the structure of matter itself. I made some failed attempts to begin a research career in Rosario, and in 1956 I moved to the University of Buenos Aires.

The late 1950s were splendid and exciting times at the University of Buenos Aires, and I was fortunate to become an instructor there in 1956, at the Institute of General Anatomy and Embryology of the School of Medicine. Under the direction of Eduardo De Robertis—a pioneer electron microscopist and the discoverer of synaptic vesicles, as well as the author of the textbook *Citologia General* (in 1946 and translated into English as *Cell Biology* in 1948)—the institute was later to become a celebrated center of cell biology research and training in South America. Although I had little experience in research, I obtained my position through an open competition, involving written and oral examinations, that was announced in the newspapers and intended to recruit aspiring scientists to the basic science departments, to which the professors who had been forced out of the universities in the previous decade were now returning.

Some of the outstanding physiologists and biochemists who left their university positions during the Peronist era were Bernardo Houssay and Luis Leloir (who received Nobel Prizes in 1947 and 1970, respectively) and Eduardo Braum Menendez (the discoverer of hypertensin, now called angiotensin). They had been able to continue their research in Argentina because they were able to establish private laboratories outside the official academic system, using their own financial resources as well as those from private donors. Some of those laboratories prosper to this day.

Other investigators, who had been cast out of the universities, like Eduardo De Robertis, had taken the route of exile. De Robertis had earlier worked in the United States at the University of Chicago from 1939 to 1941 with Robert R. Bensley, a pioneer in cell fractionation. He left Buenos Aires in 1947 to join the laboratory of Francis O. Schmitt, the founder of the biology department at the Massachusetts Institute of Technology (MIT) and creator of one of the earliest electron microscopy facilities devoted to the study of cellular ultrastructure. After gaining international recognition for his contributions to cellular neurobiology, De Robertis settled in Montevideo in 1949 at the institute directed by Clemente Estable, a disciple of the Spanish school of Ramón y Cajal and a patrician figure in Latin American science.

As junior faculty members, we waited anxiously in Buenos Aires for almost a year until De Robertis was able to establish at the institute, with the help of The Rockefeller Foundation, the first electron microscopy facility in the country, to which he transferred a functional laboratory from Montevideo. I owe a great debt to De Robertis, without whom I doubt I would be a scientist today. His keen powers of observation and technical skills set standards for us of world-class caliber for electron microscopy and laboratory experimentation. He was also enterprising, imaginative, and capable of remarkable insights. Most importantly, he rejected the notion that the quality or significance of research carried out in Latin America could be curtailed by environmental limitations. He was stern and highly valued discipline and hard work, but he also cared deeply for those whose commitment to science was demonstrated in the laboratory. He instilled in his students lofty ambitions. A good measure of his expectations was apparent from the caliber of the laboratories abroad he chose to send them to for further research experience. I consider my training with "el maestro," as we all called him affectionately among ourselves, comparable in its rigor and excitement to what I could have had in the best laboratories abroad. In De Robertis's laboratory, I carried out extensive electron microscopy studies on the adrenal gland and began to study secretion in both the adrenal medulla and cortex.[1–4] In fact, as far back as 1960, we proposed a role for the Golgi apparatus in the formation of the catecholamine-containing granules of the adrenal medulla and noted that the granules release their content at the cell surface by an exocytotic membrane fusion event (figure 1). These notions, although a part of the dogma of cell biology today, were not widely accepted at the time.

The Lure of Proteins, Their Origin, and Travails

By the end of the 1950s, as my first educational stage as a physician and a budding scientist was being completed in Buenos Aires, the ground had already been laid in the United States and Europe for the explosive developments that, during the remainder of the century, were to transform the biological sciences. As my turn was coming in De Robertis's group

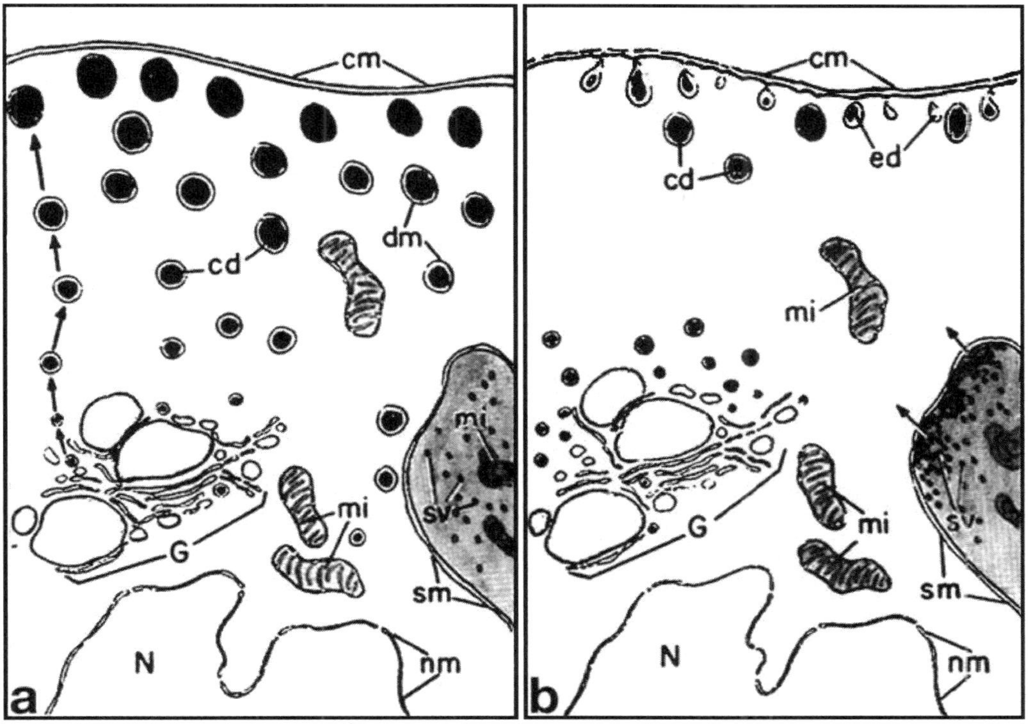

1. A diagrammatic interpretation of the secretory process in a chromaffin cell of the adrenal medulla, 1960. (A) A chromaffin cell in the resting state. Catecholamine-containing granules (cd) bounded by a membrane (dm) form (arrows) from the Golgi apparatus (G). The cell makes a synapse (sm) with a nerve ending containing synaptic vesicles (sv) and mitochondria (mi). (B) A chromaffin cell after electric stimulation of the splanchnic nerve. An exocytic discharge of catecholamines takes place at the intercellular cleft, resulting from fusion of the granule membranes with the plasma membrane (cm). From De Robertis and Sabatini, 1960. *Fed. Proc.* 19:70–78.

to seek training abroad—to which we all looked forward as necessary for the ripening of our scientific careers—I could scan an ample horizon of emerging fields in my search for a laboratory in which to invest the coming years in a foreign land.

A decade was closing that had witnessed a revolution in genetics and the explosive growth of molecular biology that followed the discovery of the double helix in 1953. Proteins were, of course, known to be the final products of gene expression, and their diversity as building blocks and as molecular machines had been recognized as responsible for the myriad of metabolic and physiological processes that sustain life. The first primary sequence of a protein, insulin, had already been determined by Frederick Sanger, who had visited our laboratory in Buenos Aires, and the first glimpses of the three-dimensional structure of some proteins by Linus Pauling, Max Perutz, and John Kendrew were making it clear that much would soon be learned about the particular architectural features that endowed proteins with specific functional activities. I could see that chemistry and crystallography held the key to unlocking these secrets and was attracted to these fields.

On the other hand, electron microscopy studies, initially concerned with macromolecular complexes, had by the mid-1950s already revealed a previously unimaginable complexity in the organization of the cytoplasm of the eukaryotic cell. The notion that the distinct compartments or organelles (whose existence had long intrigued me) have defined chemical compositions and carry out specific cellular functions was also established by a combination of electron microscopy, cell fractionation, and biochemical and enzymatic analysis. This added considerably to the fascination the eukaryotic cell exerted over those like me who had begun to peek at it through the powerful eyes of the electron microscope.

Cell biology pioneers Torbjörn Caspersson and Jean Brachet had already demonstrated a relationship between the protein biosynthetic activity of the cell and its capacity for RNA synthesis, manifested in prominent nucleoli and the intense basophilic staining of regions of the cytoplasm. The process by which proteins themselves were synthesized, however, remained a mystery that was also alluring to me, as it lay at the boundary of biochemistry and cell biology. By 1960, it had been recognized that protein synthesis took place in ribosomes, but the puzzle of how these particles carried out the decoding of genetic information had not been solved.

Of course at that time, few were yet concerned with the even greater mystery of the mechanisms and pathways that insure that proteins, once synthesized, become the building blocks of organelles, or become functional within them. Although our knowledge of the properties of membranes was meager, it was indeed difficult to imagine how proteins, once released from ribosomes, could freely traverse the ubiquitous intracellular membrane barriers to find their correct destination within an organellar lumen or in the membrane that limited a specific compartment. For anyone whose interests straddled the boundaries of cell biology and biochemistry, this should have seemed an insurmountable problem to tackle.

In 1959, I had to make the difficult decision of whether I would get training in protein crystallography, which promised to give the ultimate understanding of biological processes at the molecular, and even the atomic, level, and therefore go to Great Britain, the undisputed mecca of crystallography at that time; or to go to the United States to work with George Palade and Philip Siekevitz, who were beginning to unravel the secretory pathway responsible for the execution of one of the most complex cellular processes.

Cytological studies extending back to the nineteenth century had made it clear that secretion requires the concerted, and probably sequential, action of several organelles, and efforts to increase our understanding of the secretory process were moving toward the vanguard of biological research. During the 1950s, Palade and Siekevitz had pioneered an approach to study the secretory pathway that combined the techniques of cell fractionation with biochemical and morphological analyses, which today is the underpinning for most of the work being carried out in the field of molecular cell biology. I believe it fair to say that it is from those studies that the field of organelle and membrane biogenesis and the broad current concern with protein traffic sprang to life.

Moreover, the work of Palade and Siekevitz had focused foremost on the protein biosynthetic role of the endoplasmic reticulum (ER), an organelle not known before electron microscopy. The discovery that the rough, ribosome-studded portions of the ER corresponded to the highly basophilic ribonucleoprotein-rich areas of the cytoplasm—characteristic of cells with intense protein biosynthetic activity—had placed the ER at center stage in the emerging field of cell biology.[5]

Palade had discovered the ribosome in 1953 as a small particulate component of the cytoplasm visible in his electron micrographs.[6] Siekevitz had a long-standing interest in protein synthesis since his time in Paul Zamecnik's laboratory in Boston, where in 1952 he had pioneered in demonstrating that a microsomal fraction was capable of carrying out the process in vitro.[7] The identification of the ribosome as a molecular machine that carries out protein synthesis had begun with the finding that polypeptide chains, labeled in vivo during short incubations of *Escherichia coli* with radioactive amino acid precursors, were recovered in association with ribosomes when these particles were purified from the remains of disrupted bacteria.[8] Yet, until 1961 (the same year I began to work at Rockefeller), when François Jacob and Jacques Monod proposed the mRNA hypothesis—derived primarily from observations made on ribosome reprogramming in *E. coli* during mating and after bacteriophage infection—the prevalent notion was that individual ribosomes were "congenitally specialized" to synthesize specific proteins.[9]

I decided to apply for a postdoctoral position with Palade, expecting that in Siekevitz's laboratory, where I had asked Palade to place me, I would be able to investigate the function of the membrane-bound ribosomes that characterize the rough ER. Meanwhile, with the support of De Robertis and Houssay, I obtained a Rockefeller Foundation Fellowship. I am still impressed by the fact that I was interviewed twice by officers of the foundation who came to Buenos Aires and devoted a substantial amount of their time to assess my potential as a scientist. Palade agreed to take me into his laboratory, but he wrote that it would be best if I could postpone for six months my arrival to New York, which was planned for January 1961, since Keith Porter and his group were leaving The Rockefeller Institute for Harvard and more space would then become available. Since my Rockefeller Foundation Fellowship had been arranged to begin in January, Palade suggested that I first spend a few months at Yale with Russell Barrnett, a well-known histochemist with whom he had previously collaborated. I knew of Barrnett's work, since he had developed with Arnold Seligman a histochemical technique for the detection of SH groups at the light microscope level, and in my first paper I had used a related method to localize cysteine-rich neuropeptides in the toad hypothalamus.[10]

A Cytochemical Interlude: The Power of Dialdehydes

My stay in Barrnett's lab, short as it was by the usual standards, was immensely productive. I was the most experienced electron microscopist in his group and was coming from a laboratory using cutting-edge technology not yet widely adopted in the U.S. Thus, in Argentina, we had already replaced methacrylate with epoxy resin embedding and had added a new Siemens Elmiskop to our old RCA EMU-2. Barrnett had inherited an RCA EMU-3 from Sanford Palay, a former Yale professor who had trained with Porter and Palade and had recently moved to Harvard, and I was put in charge of that instrument.

The focus of Barrnett's laboratory was on the application of histochemical techniques to electron microscopy, something that was being eagerly attempted in several laboratories, but with little success. Since the 1940s, histochemistry had been a thriving field of research and a powerful tool for histophysiological studies at the light microscope level, but its application to electron microscopy faced considerable challenges. The fine structure of unfixed tissues and cells sadly decayed to almost an unrecognizable state during the harsh conditions of incubation needed for some histochemical reactions. Conversely, prefixation in OsO_4, the universal fixative for electron microscopy, mainly used in a veronal-acetate buffered form introduced by Palade,[11] rapidly and irreversibly inactivated the enzymatic activities that

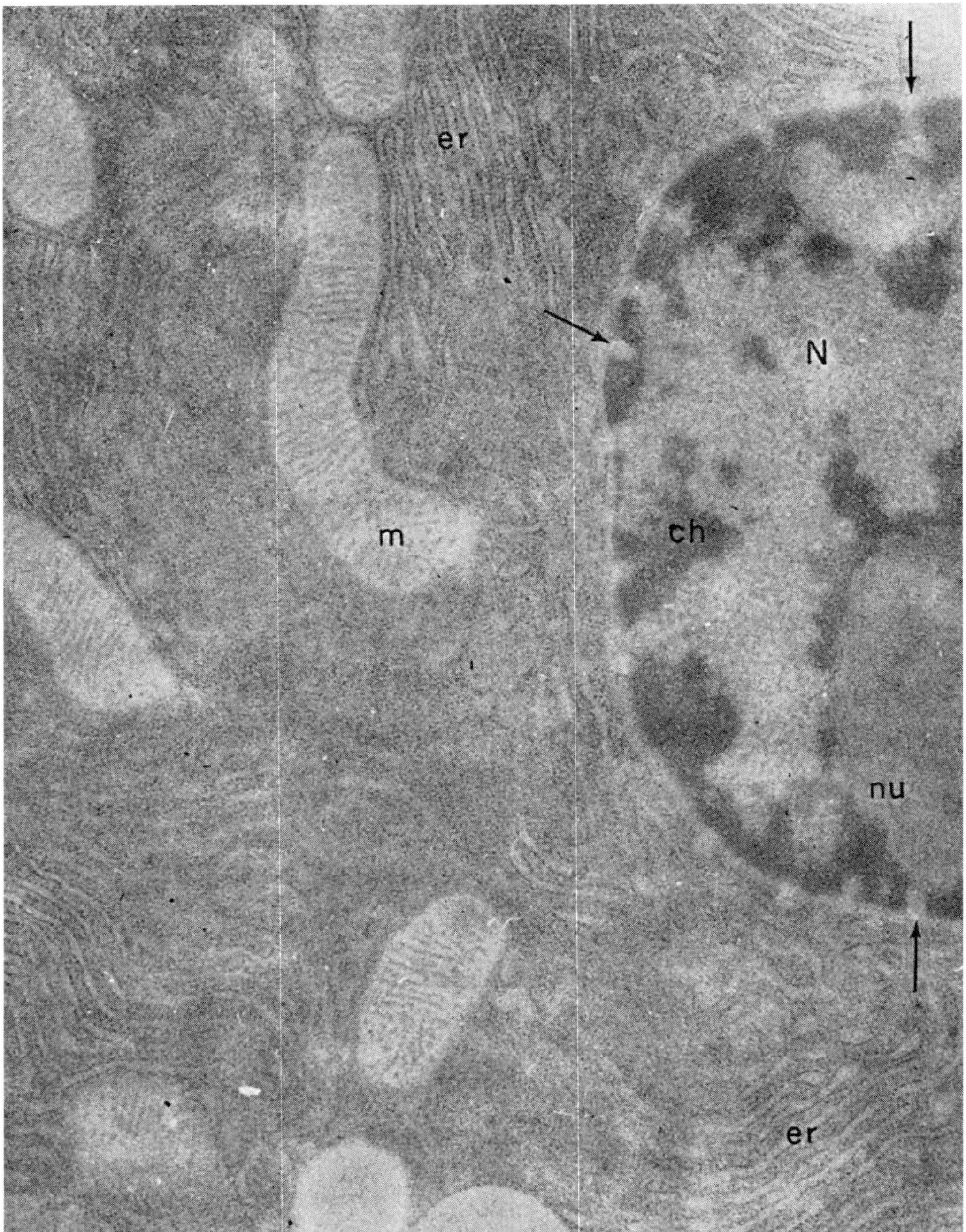

2. Partial view of a pancreatic acinar cell fixed in glutaraldehyde alone, 1963. The structure of the nucleus and of cytoplasmic elements, such as mitochondria, rough endoplasmic reticulum, and zymogen granules, is well preserved. Membranes, however, are not visible unless the sample is postfixed in OsO_4. From Sabatini et al., 1963. *J. Cell Biol.* 17:19–58.

histochemists wanted to relate to subcellular structures. This was also the case with other metal-containing reagents, such as $KMnO_4$ and $K_2Cr_2O_7$ that were used to increase contrast.

At Yale, with my 1956 Fieser and Fieser *Organic Chemistry* textbook opened on my desk,[12] I decided to examine the usefulness as fixatives of a series of dialdehydes, agents that in a single molecule incorporate a double dose of the group responsible for the efficacy of formaldehyde, the preferred fixative for light microscopy. I hoped that some dialdehydes would serve as good cross-linking reagents that, at appropriate concentrations, could prevent the disintegration of subcellular structures while perhaps maintaining some enzymatic or cytochemical activities. By a great coincidence, I found a source of glutaraldehyde nearby. Klaus Bensch, a young resident in pathology working with Don King at Yale and a frequent visitor to Barrnett's lab, had used glutaraldehyde following an industrial recipe to produce gelatin microcapsules containing *E. coli* DNA. They were administering microcapsules to cultured mammalian cells in an attempt to transform them into amino acid prototrophs. My very first experiments showed that glutaraldehyde gave an excellent preservation of cellular structures while allowing the demonstration in situ of several specific enzymatic activities. I soon extended my search for new fixatives to other dialdehydes and related reactive organic compounds, but none was better than glutaraldehyde.

It was a simple matter to choose concentrations and conditions that gave a compromise between acceptable tissue preservation and retention of enzymatic activities. The first paper I published with Bensch and Barrnett had thirty-eight figures and illustrated the cytochemical detection at the electron microscopy level of the products of more than a dozen enzymes.[13] Glutaraldehyde had distinct advantages over OsO_4 as a fixative in that it penetrated rapidly and deeply into tissues and was not nearly as noxious and difficult to handle. However, glutaraldehyde by itself provided little electron contrast, a drawback easily remedied by the introduction of a postfixation step with OsO_4 that could be carried out after the histochemical reaction had been completed and be followed by staining with the usual contrast-enhancing uranyl or lead-containing reagents (figure 2). Postfixation in OsO_4 not only revealed membranes and stained other lipid-containing structures that were seen as negative images after glutaraldehyde fixation alone, but it also reacted with some of the organic deposits generated during the histochemical reactions, making their products more visible under the electron beam. The advantages of the double fixation procedure I developed to preserve fine structural details were well illustrated in a second paper,[14] in which I included work I did at Rockefeller with Fritz Miller, then a visiting scientist in Palade's laboratory, who, upon his return to Europe, assumed the chair of histology in Munich.

I was later particularly gratified that one of the main bonuses of glutaraldehyde was its preservation of cytoplasmic microtubules, which, as components of the cytoskeleton, Keith Porter first described in detail in doubly fixed specimens.[15,16] I believe that the warm and amiable lifelong relationship I had with Keith owed much to his appreciation of the very indirect contribution I made to one of his many important discoveries.

Getting Started at Rockefeller

At Rockefeller, I was assigned a desk and a bench in Siekevitz's laboratory on the fifth floor of the South Laboratory, since renamed the Bronk Laboratory. This was a wonderful place to do science. Siekevitz's enthusiasm was infectious, and he set a marvelous example with his own intense involvement in experiments, readiness to discuss novel ideas, and willingness to give his junior associates considerable freedom in the direction they took.

Although I came to the laboratory as a postdoctoral fellow, after one year I was seduced by the opportunity to join the PhD program at Rockefeller, following the example of other young physicians, who had already done or were doing so and who later developed stellar careers, such as Howard Rasmussen, Gerald Edelman, Charles Stevens, Edward Reich, David Luck, James Jamieson, and Scott Grundy, among others. Palade supported my application and, after an interview with Rockefeller's president Detlev Bronk, I was admitted into the program. This change of status gave me the opportunity to repair some of my educational deficiencies by arranging for some excellent tutorials: in math with Ervand Kogbetliantz, chemistry with Te Piao King and William Agosta, and physical chemistry with David Yphantis. It also provided for the much coveted free photocopying privileges, but did not change the direction of my research. What it did was to greatly enrich my intellectual and personal life as I entered into daily contact with the exceptional group of brilliant aspiring young scientists who were my classmates, including David Baltimore, Robert Barlow, Tony Cerami, Bert Hille, David Hirsch, Bob Klug, Harvey Lodish, Dan Rifkin, and others.

Entering the Rough Endoplasmic Reticulum

In Siekevitz's laboratory, I became a close friend and admirer of postdoctoral fellow Yutaka Tashiro, a very experienced biochemist from whom I had much to learn. He and I shared a profound interest in ribosomes, which he had begun to study in Japan using the analytical ultracentrifuge. Tashiro was interested in the molecular architecture of the ribosome and was using the analytical centrifuge in the laboratory of David Yphantis, a major contributor to technical improvements of that instrument, to characterize rat liver ribosomes. He introduced me to the intricacies of the technique of analytical centrifugation as well as to the use of the sucrose gradient centrifugation methods. These had been recently applied to the study of ribosomes by Richard Roberts at the Carnegie Institution, and with them we could separate and obtain preparative amounts of ribosome monomers and subunits. We frequently visited the laboratory of Mary Petermann at the Sloan-Kettering Institute, who was also studying the physical and chemical properties of eukaryotic ribosomes; and we followed closely the work of James Watson and his colleagues at Harvard, who had shown earlier that *E. coli* ribosomes were composed of two unequal subunits[17] and that the larger of them contained the nascent polypeptide chain.[18]

The workers at Harvard had taken advantage of the fact that *E. coli* ribosomes spontaneously dissociate into functional subunits when Mg^{2+} ions are removed by dialysis. Tashiro, on the other hand, had to use the Mg^{2+} chelating agent EDTA to induce the dissociation of rat liver monomeric ribosomes into the subunits.[19,20] Unfortunately, this treatment also caused the partial unfolding and irreversible inactivation of the subunits. Nevertheless, Tashiro was able to demonstrate that in rat liver ribosomes too nascent polypeptide chains are associated with the large subunits.[21]

I was impressed by the asymmetry and obvious specialization of the two subribosomal domains and asked myself whether the association of the ribosomes with the ER membrane that characterizes the rough portions of the ER involves one or both ribosomal subunits, and how this might relate to the function of the ribosomes. I wondered whether EDTA treatment of rough microsomes would also dissociate the bound ribosomes into subunits and whether, in that case, one or the other subunit would remain bound to the membrane.

The first experiments I did worked like a charm.[22] At low concentrations, EDTA preferentially released small ribosomal subunits, while most large subunits remained associated with the membrane. When microsomes containing radioactive nascent chains labeled after very short pulses in vivo were treated with EDTA, I found that the ribosomal subunits recovered after detergent dissolution of the membranes contained a higher level of radioactive nascent chains than those subunits removed by EDTA. I had also begun an electron microscopy study of free ribosomes and of rough microsomes using the technique of negative staining. It was clear that ribosomal particles contained a cleft that divided them into two unequal regions and that, in general, in membrane-bound ribosomes the cleft lay parallel to the microsomal membrane to which the larger region was adjacent. These observations of course immediately suggested that the nascent chain in the large subunit might play a role in anchoring the ribosome to the membrane. At that time, however, we cautiously stated, "It seems plausible to consider the possibility that strong attachment of 'active ribosomes' is related to passage of the protein through the membrane. However, our work does not indicate which of the following possibilities holds: 1) the ribosomes that are strongly attached are, because of this situation, more active in protein synthesis and, therefore, become labeled in vivo, or 2) the presence of the product of protein synthesis on the ribosomes is what makes them stick to the membrane."[22]

The initial evidence for a role of membrane-bound ribosomes in the synthesis of secretory proteins had been provided by Siekevitz and Palade in 1960.[23] They had reported that very soon after the injection of a radioactive amino acid to an animal, the purified pancreatic secretory enzyme chymotrypsinogen of the highest specific radioactivity was found associated with ribosomes that could be isolated by sedimentation from rough microsomes after detergent treatment to dissolve the membranes. At later times after the injection, however, chymotrypsinogen with the highest specific radioactivity was found in the microsomal subfraction solubilized by detergent, which included the content of the microsomal lumen. Curiously, the route by which the chymotrypsinogen molecules released from the membrane-bound ribosomes found their way to the lumen of the cisterna was not considered, perhaps because it was regarded as a problem to be dealt with later. Indeed, a scheme drawn in 1961 by Palade showed the site of passage of the protein through the membrane at some distance from the ribosomal membrane junction, as if it could be affected by an independent transport mechanism that operates on completed polypeptides released from the ribosome.[24]

Laying the Basis for the Signal Hypothesis

It was left to Colvin Redman, who joined the Siekevitz and Palade laboratory in 1964 after obtaining his PhD in Canada with Lowell Hokins—the scientist who first discovered the involvement of phosphoinositides in controlling secretion—to demonstrate that completed polypeptides, in this case amylase synthesized in vitro by pigeon pancreas microsomes, were not recovered with the medium but remained associated with the microsomes, most likely sequestered in their luminal cavities.[25] These experiments, however, could not exclude that polypeptides discharged from the ribosomes into the medium were rapidly taken into the microsomes by an uptake and transporting mechanism. Therefore, I, in collaboration with Redman, decided to examine the fate of in vitro labeled incomplete polypeptides that were synthesized in a system of liver microsomes and were released from the ribosomes by the action of puromycin. This aminoglycoside antibiotic substitute for aminoacyl tRNA and its incorporation at the C-terminal end of a nascent polypeptide cause premature termination of polypeptide synthesis, purging the ribosomes of nascent chains, which are released in the form of peptidyl-puromycin molecules. We found that the peptidyl-puromycin molecules lost from the ribosomes were not released into the surrounding medium but remained associated with the microsomes, from which they could be recovered in soluble form after dissolution of the membranes by detergent and sedimentation of the ribosomes.[26]

These observations, together with the finding that a significant fraction of the peptidyl-puromycin molecules could also be released from the microsomes by a mild sonication procedure, suggested that such molecules had become sequestered within the microsomal lumen. This was confirmed by the finding that they could also be released using a low concentration of detergent that permeabilized the membranes without causing their extensive disassembly. This method to dissect the microsomes, which worked much better with liver microsomes than with the pigeon pancreas microsomes used by Redman, had been introduced by Lars Ernster, a distinguished Swedish biochemist who had spent a sabbatical year in Siekevitz's laboratory.[27]

We suggested that nascent chains synthesized in bound ribosomes become engaged with the translocation machinery in the membrane before their synthesis is completed and, therefore, that the site of translocation must be in close proximity to the ribosome, most likely at the ribosome-membrane junction.[26] Thus, transport across the membrane appeared to be effected by a non-discriminating mechanism that does not distinguish between complete and artificially terminated incomplete chains when either of these are released from the ribosome. I had become convinced that translocation was a cotranslational phenomenon and that it resulted from the fact that in a membrane-bound ribosome, growth of the nascent chain occurs in an environment that precludes its release into the surrounding medium, which in the cell is represented by the cytosol. We described a model, illustrated in a later paper,[28] that explained the vectorial, unidirectional discharge of the polypeptide into the lumen of the ER as a result of the polypeptide growing in a tunnel or central cavity within the large ribosomal subunit. We suggested that because the ribosome sits on the membrane, this tunnel is made continuous with the ER cisternal space through a permanent or intermittent discontinuity in the microsomal membrane. We wrote, "As visualized at present the transfer mechanism relies primarily on release from the large subunit and on structural restrictions at the ribosome membrane junction and, hence, is nondiscriminatory and possibly passive." In a note to this paper, we indicated, "Unpublished electron microscopic observations by D. Sabatini, Y. Tashiro, and G.E. Palade are the basis of this model. The existence of a channel is suggested by electron microscopy of negatively stained large subunits. Discontinuities in the membrane under the large subunits can be detected in some instances in sectioned specimens."[26]

Electron micrographs were published in 1971 of isolated large ribosome subunits in which a central depression or cavity was apparent because it accumulated negative stain (as expected from the entrance to a tunnel in the subunit).[29] The passive character of the transport of the peptidyl-puromycin molecule into the microsomal lumen was supported by a paper in which Redman showed that ATP was not required for this process.[30]

The two papers representing my work on membrane-bound ribosomes became the basis for my doctoral thesis at Rockefeller, where I received a PhD in 1966. I felt that my work had opened an important avenue of cell biology research, which I very much wanted to follow, hoping to provide a molecular description of the relationship between the ribosome and the translocation apparatus that I presumed existed in the membrane. Of course, very little was known about the basic organization of cellular membranes at that time, but it seemed most likely that the passageway or pore in the membrane through which the nascent polypeptide was transported was of a proteinaceous nature or was constructed of protein subunits, although given our meager knowledge of membrane structure I was hesitant to present such a speculation. Obviously, this was a topic of great interest to Palade, and I was, therefore, delighted when he offered me an assistant professorship and a laboratory being vacated by Sam Dales, who was leaving Rockefeller to take a position at the Public Health Institute of New York.

In my new laboratory, I began to examine more directly the notion that nascent polypeptides indeed grow within a ribosomal tunnel and, in the case of those synthesized on the ER, upon emerging from the ribosome enter into a relationship with the underlying membrane that precludes their release into the medium or its cellular equivalent, the cytosol, as proposed in the model I had generated. I was successful with my first grant application and had started to determine to what extent nascent polypeptides in free ribosomes and in microsomes, which I could label in vivo or in vitro, were protected from the attack of exogenous proteolytic enzymes when a new postdoctoral fellow, Günter Blobel, an MD from Germany who had recently completed his PhD in Wisconsin with Van Potter (Siekevitz's old mentor), joined the Palade and Siekevitz laboratory.

Formulation of the Signal Hypothesis

During his sojourn in Wisconsin, Günter had worked on rough microsomes and was, therefore, already interested in the subject of my research. He was a frequent visitor to my laboratory, and we soon began to hold many animated discussions on this somewhat esoteric subject. I was pleased to find that he had read my two papers on membrane-bound ribosomes carefully and that he was anxious to collaborate with me in experiments in which I had begun to examine the accessibility to exogenous proteases of nascent polypeptide chains in the free and membrane ribosomes. Together, we found that nascent polypeptides labeled in vitro in free ribosomes were easily and completely digested when the ribosomes were incubated with a mixture of trypsin and chymotrypsin at 37°C. But when the incubation was carried out at 0°C, most of the label incorporated in a brief pulse was protected from proteolysis and appeared in protease-resistant fragments of approximately thirty-nine amino acids in length that remained associated with the partially proteolysed ribosomes.[31] Pulse-chase and continuous labeling experiments indicated that the protected segment contained the growing end of the nascent chain, which Howard Dintzis at MIT in 1961 had demonstrated was at its C terminus.[32] Everyone recognized that this end of the nascent chain should be intimately associated with the ribosome since it serves as the substrate to which the peptidyltransferase, an integral component of the ribosome, adds amino acids to the growing chain. Similar results on a ribosome-protected fragment of the growing polypeptide had been obtained in Alex Rich's laboratory at MIT with polyribosomes from rabbit reticulocytes, which produce almost exclusively globin chains that could be labeled in vivo.[33]

All these findings, together with the hydrodynamic properties of large subunits examined by analytical centrifugation in Petermann's laboratory at Memorial Sloan-Kettering,[34] provided strong support for the notion that a cavity or tunnel within the large ribosomal subunit contains the nascent polypeptide, as I had proposed for membrane-bound ribosomes to explain the vectorial discharge of the incomplete polypeptides released by puromycin.

We suggested that the nascent polypeptide chain in the membrane-bound ribosomes would remain protease inaccessible as it grew beyond the thirty-nine–amino acid length sequestered within the ribosome, provided that the exogenous protease did not break the membrane barrier.[26,31]

Tsuneo Omura, Ryo Sato, and their collaborators[35,36] had already studied the effect of proteolytic enzymes on microsomes, showing that they were capable of digesting or dissecting out from the cytoplasmic face of the membranes several sets of proteins—later shown to be anchored in the membrane by C- or N-terminal segments. Their work had also shown that microsomal vesicles incubated with proteolytic enzymes at 0°C remain largely intact and apparently impermeable to the proteases.

In 1970, Günter and I found that during proteolytic digestion of microsomes at 0°C, the bound ribosomes were released from the membranes. Concomitantly, the individual nascent polypeptides that they contained underwent cleavage, generating two sets of segments that remained largely protected from proteolysis. One set of segments remained associated with the detached ribosomes, which, although bearing partially proteolysed ribosomal proteins, still sedimented as intact particles. These segments were of the same length as those protected in free ribosomes and corresponded to the C-terminal portions of the polypeptides. The other set of segments generated by proteolysis corresponded to the N-terminal portions of the growing polypeptides, and these were of variable length, mostly larger than the ribosome-protected ones. The N-terminal segments remained associated with the membranes and were also inaccessible to the proteases as long as the membranes remained intact.[28]

This work was originally submitted to *The Journal of Cell Biology* as a single paper, but the editor found it too long and asked that we split it into two, which we did. One paper dealt specifically with free ribosomes;[31] the second presented evidence for the protection of the N-terminal portions of the nascent polypeptides by the microsomal membrane.[28] In this paper, the model proposing the existence of a tunnel within the ribosome that is continuous with a passageway for the nascent polypeptide through the membrane was graphically illustrated.

Günter and I soon learned that by using puromycin in a medium of relatively high ionic strength, we could dissociate ribosomes within polysomes into subunits that, contrary to those obtained by chelation of Mg^{2+} with EDTA, remained properly folded and were capable of protein synthesis when reprogrammed with mRNA.[37] This allowed Yoshiaki Nonomura, who came to my laboratory as a postdoctoral fellow from Ebashi's group in Tokyo, to undertake an electron microscopy study by negative staining of the structure of individual active ribosomal subunits and the monomers formed by their association and the relationship of the subunits to mRNA within polysomes.[38] This work was followed by a study by Takashi Morimoto[39,40] of the mechanism by which ribosomes assemble into the tetramers that Breck Byers, then a student of Keith Porter at Harvard, had shown form crystalline arrays in chicken embryos upon slow cooling.[41]

Soon after we discovered the utility of puromycin to generate functionally capable ribosomal subunits, Mark Adelman, who had completed his PhD in Chicago working

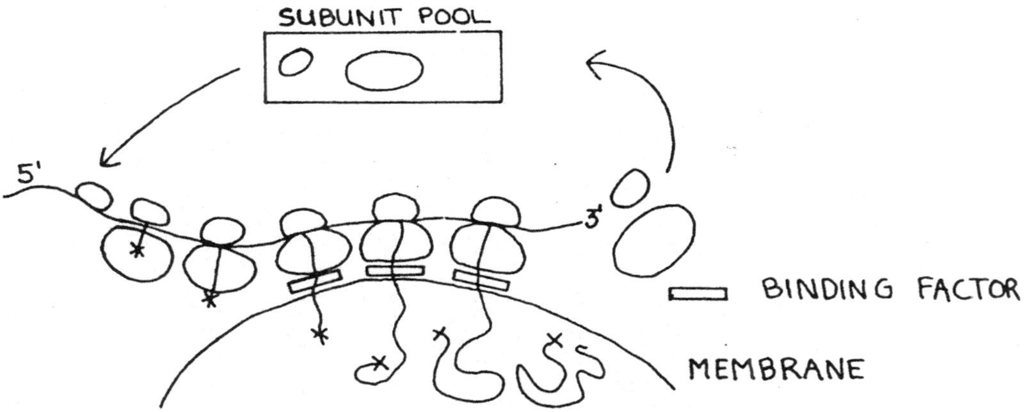

3. The first formulation of what came to be called "the signal hypothesis," 1971. The crosses represent signal peptides at the N termini of the nascent chains, which were proposed to be recognized by a factor that initiates binding of the ribosome–nascent chain complex to the membrane and, therefore, has properties of a signal recognition particle (SRP). From Blobel and Sabatini, 1971. Ribosome-membrane interactions in eukaryotic cells. *Biomembranes*, Volume 2, 193–195 with permission from Springer. ©Plenum Press.

with Ed Taylor in one of the first demonstrations of the presence of actin and myosin in non-muscle cells, joined my laboratory as a postdoctoral fellow to study the role of the nascent chain in the association of the ribosome with the membrane. He developed a new procedure to obtain large amounts of highly purified rough microsomes from rat liver,[42] which he used to study the role of the nascent chain in maintaining the association of ribosomes with the membrane. He showed that, even after the microsomes were treated with puromycin to release the nascent chains, the ribosomes remained associated with binding sites on the microsomal membrane. However, they now could be effectively detached from the membranes simply by raising the ionic strength, which was not possible without previous puromycin treatment.[43,44] We concluded, therefore, that at least two molecular interactions are responsible for maintaining the ribosome-membrane junction: a direct one between the large subunit and a putative receptor in the membrane, which is disrupted by high salt treatment, and a second one that is provided by the nascent chain linking the polypeptide exit site in the large ribosomal subunit to the passageway in the underlying microsomal membrane, which leads to the microsomal lumen.

By 1971 it seemed clear to us that free and membrane-bound ribosomes were structurally identical and functionally interchangeable, something for which Nica Borgese, then a graduate student in my laboratory, was accumulating evidence in subunit exchange experiments.[45] The only difference between free and membrane-bound ribosomes seemed to be that they were translating different classes of mRNAs, with the bound polysomes from secretory glands synthesizing mainly secretory proteins. We had obtained a plethora of evidence implicating the N-terminal portion of the nascent polypeptide in establishing and maintaining the association of a bound ribosome with the membrane. Most salient were 1) the vectorial discharge resulting from puromycin treatment;[26] 2) the role of the membrane in protecting the N-terminal portion of the polypeptide from the proteolytic attack by exogenous enzymes;[28] and 3) the fact that ribosomes containing nascent chains could not be detached by simply raising the ionic strength, which effectively dissociates from the membrane those ribosomes lacking nascent polypeptides.[29,42,43]

The remaining burning question concerned the features in the N-terminal region of a nascent secretory polypeptide that were responsible for the association of the ribosome–nascent chain complex with the membrane. In 1971, it was already established that, as was known for prokaryotes, methionine was the initiating amino acid for eukaryotic proteins. It then seemed reasonable to assume that because no known secretory protein retained an N-terminal methionine, the membrane-binding information that we postulated existed in the N-terminal sequence was removed during or after insertion of the polypeptide into the membrane. We considered the possibility that N-terminal amino acids were modified by acylation with fatty acids or by the addition of cholesterol or other hydrophobic moieties. There are now examples of such modifications in proteins, but none was known at that time other than the formulation of the initiator methionine, which occurs only in prokaryotes. In fact, we suspected that the feature in the nascent polypeptide that served to trigger the association of the ribosome with the membrane could be a stretch of hydrophobic amino acids because we found that polyphenylalanine synthesized in vitro by microsomes programmed with poly-U upon termination with puromycin remained membrane associated.[29]

After many hours of enthralling argumentation in front of my office blackboard, Günter and I arrived at the first formulation of what was later to be known as the signal hypothesis (figure 3). We proposed that "all mRNAs to be translated on bound ribosomes have a common feature, such as several codons near their 5′ end, not present in mRNA's which are to be translated on free ribosomes. The resulting common sequence of amino acids near the N-terminal of the nascent chains or a modification of it (indicated by X) would then be recognized by a factor mediating the binding to the membrane. This binding factor could be a soluble protein, which recognizes both a site on the large ribosomal subunit and a site on the membrane. After release of the chain from the ribosome into the intracisternal space, the ribosomal subunits and perhaps the binding factor dissociate from the membrane and enter their respective soluble pools. The pool of ribosomal subunits would also be fed from subunits generated from free polysomes after chain completion and release and a new cycle would be started."[46] This was a rather prophetic statement that predicted not only the presence of a signal peptide sequence at the N terminus of the nascent chain but also the existence of a signal binding factor, with the behavior later demonstrated by Günter Blobel with his associates and former disciples for the signal recognition particle (SRP)[47,48] and its cognate receptor in the membrane.[49–51]

The Full Flowering of the Initial Model: Molecules and Mechanisms

Before leaving Rockefeller in 1972, my laboratory had nearly completed other studies that established the presence in ER membranes of a finite number of specific ribosome receptors to which ribosomes, even after being purged of their nascent chains, are capable of binding.[52] We deduced that such binding sites contained an essential protein component because the ribosome binding capacity of ER membranes was sensitive to heat treatment and was abolished by mild proteolytic digestion, and we proposed that those sites were spatially close to those through which cotranslational translocation occurs.

With the development of cell-free systems for the translation of natural mRNAs, several laboratories observed that the primary translation products of mRNAs for secretory proteins produced in systems lacking microsomal membranes were somewhat larger than those produced in microsomes.[53–56] It was also shown that the extra segments were at the N termini of the primary translation products. But, of course, it was Günter's laboratory, employing a reconstituted in vitro protein synthesizing system containing microsomal membranes and ribosomal subunits programmed with immunoglobulin mRNA, that provided definitive proof for the signal hypothesis.[57,58] They demonstrated the role of a cleavable signal sequence in initiating insertion of nascent immunoglobulin light chains into the membrane, characterized the signal sequence, and revealed the action of a membrane-associated signal peptidase in its cotranslational removal. In these experiments, the protease protection assay that we had developed was used to demonstrate the sequestration in the microsomal lumen of completed immunoglobulin light chains whose signal sequences were removed cotranslationally.

Over the next decade, in now classical experiments, Günter's laboratory went on to illuminate nearly all aspects of the complex molecular interactions that lead to the insertion of specific polypeptides into ER membranes. Years later, his laboratory provided electrophysiological proof for the existence of a protein-conducting channel in the microsomal membrane that appears to be opened by the signal peptide and to be kept in that state by the bound ribosome.[59]

It is now nearly forty years since Günter and I delineated the basic features of the signal hypothesis. We have now gained enough perspective to view that work, which began with a rather focused concern with the early stages of the secretory process, within a larger frame of reference, one that encompasses more general questions related to membrane and organelle biogenesis. From this we can derive considerable satisfaction, as some of the concepts and ideas we put forward have had a wide impact in the field of protein traffic.

The years that followed the presentation and verification of the signal hypothesis revealed that the cotranslational mechanism first envisaged for the passage of secretory polypeptides across the membrane also applies to resident luminal and integral membrane proteins of the entire endomembrane system, including the ER itself, the Golgi apparatus, the plasma membrane, endosomes, and lysosomes. It is also now clear that a wide variety of signal sequences exist in polypeptides that are inserted in ER membranes, which, however, are all decoded by the same translocation apparatus. Some such signal sequences are transient features of the polypeptide that serve only to initiate its insertion into the membrane, whereas others also serve as permanent transmembrane anchors responsible for specific transmembrane dispositions of the mature polypeptide. Moreover, studies, first on simple transmembrane proteins and later on proteins that traverse the membrane multiple times, brought to light the signaling function of other largely hydrophobic, interior segments of translocating polypeptides that serve to arrest the translocation process and to determine the proper disposition of the polypeptide relative to the phospholipid bilayer in the membrane. Both Günter and I envisaged the complex topology of certain membrane proteins as resulting from the sequential action of insertion and halting the transfer signals operating cotranslationally, before substantial evidence for this mechanism was obtained.[60,61]

The notion we introduced in 1971 of a signal peptide containing information decoded by soluble and membrane receptors that informs the cell of the polypeptide destination was a harbinger of the discovery of signals that determine the posttranslational importation of polypeptides into other membranes and organelles, including chloroplasts, mitochondria, and peroxisomes. Moreover, the role of the signal peptide as the first sorting signal that initiates the journey of a variety of proteins toward diverse destinations as distant from the site of their insertion into the ER as the surface of the cell, or lysosomes, opened the way to the realization that other sorting signals and cognate receptors must exist that mediate the successive sorting events needed to steer each polypeptide along the one pathway that leads to its site of function.

In light of current knowledge, our original model for cotranslational translocation also illustrates, once more, how natural phenomena invariably are more complex than one initially imagines. As the translocation machine whose existence we recognized was dissected to its molecular components, the model, rather stylish at the beginning, grew much more elaborate and ornate, with the inclusion of a host of additional proteins that either assist in the translocation or modify the nascent polypeptide during this process. The Sec61p protein that constitutes the channel consists of at least three polypeptide subunits.[62] Two other ER transmembrane polypeptides, ribophorins I and II, that we along with Gert Kreibich identified as putative ribosome receptors[63,64] are closely associated with the translocation site.[65] They are now known to constitute, with two other subunits, the oligosaccharyl transferase that transfers a glycan moiety to asparagine residues in nascent glycoproteins.[66] Similarly, SRP, the complex factor that targets the ribosomes to the ER membrane, consists of an RNA molecule and six polypeptides,[67] one of which is a GTP-binding protein recognized by the SRP receptor in the membrane.[68] The receptor itself consists of two subunits, both of which are also GTP-binding proteins.[69]

Amazingly, a retrotranslocation machinery is now known to also function in the ER in cooperation with a cytosolic ATPase and to transfer misfolded polypeptides, already segregated in the lumen of the ER or incorporated into the membrane, back to the cytosol for degradation by the proteosome. This machinery is an essential component of a quality control mechanism in the ER that employs luminal chaperones and an elegant transmembrane signaling system that controls levels of protein synthesis and the expression of genes whose products make this organelle a nearly perfect manufacturing plant. Remarkably, certain viruses have learned to evade the immune system by encoding proteins that promote the retrotranslocation and, hence, degradation of MHC class I molecules that otherwise would have presented viral antigens at the cell surface and thus called into action cytotoxic lymphocytes to eliminate the infected cell.[70,71]

It is gratifying for me to think that our initial model provided a framework that facilitated further progress and that, although it has undergone considerable elaboration and fine details have been elucidated, its fundamental aspects are still in place.

Excerpted from Sabatini, 2005. *Annu. Rev. Cell Dev. Biol.* 21:1–33.

8B. Protein Targeting

Günter Blobel

Prologue

I began research in the 1960s, first as a graduate student in the laboratory of Van R. Potter at the McArdle Institute for Cancer Research of the University of Wisconsin in Madison. I continued as a postdoctoral fellow in the laboratory of George E. Palade at The Rockefeller University in New York City. At that time, the intracellular pathway of secretory proteins, from their synthesis to their extrusion from the cell, the so-called "secretory pathway," had already been worked out by George Palade and his co-workers, primarily Philip Siekevitz, Jim Jamieson, and Lucien Caro.[1] Using pulse-chase labeling with radioactive amino acids in tissue slices in conjunction with cell fractionation and autoradiography, Palade and co-workers established that, during or shortly after their synthesis, secretory proteins cross the rough (ribosome studded) endoplasmic reticulum (ER) membrane. It was proposed that translocation across the ER yields segregation of secretory proteins from cytosolic proteins. From the ER, secretory proteins are transported via vesicular carriers through the cisternae of the Golgi apparatus. Finally, vesicular carriers budding from the trans-cisternae of the Golgi complex were observed to fuse with the plasma membrane, yielding externalization or exocytosis of the secretory proteins. The biochemical mechanisms underlying this pathway were unknown at that time.

Several reports in the mid-sixties suggested that messenger ribonucleic acids (mRNAs) for secretory proteins are sequestered in rough microsomes that represent the vesicular remnants of the fractured rough ER. However, it was not clear how the sequestration of these mRNAs was accomplished. One possibility was that an untranslated region that might be common to all mRNAs coding for secretory proteins mediated their attachment to the ER membrane. Other such distinct untranslated regions of other mRNAs might mediate attachment to other organelles. Once attached to their "cognate" organelles, the translation products might then be vectorially transported into these organelles.[2] Another idea was that free and organelle-bound ribosomes might differ in their composition. The distinct organelle-bound ribosomes might function to select cognate mRNAs for the organelles, again via cognate untranslated regions. These and other ideas were frequently discussed in the Palade laboratory in the late 1960s.

The First Version of the Signal Hypothesis

David Sabatini and I tested one of these ideas, namely whether free and ER-bound ribosomes differ in their protein composition. We used one-dimensional sodium dodecyl sulfate (SDS) polyacrylamide gel electrophoresis (PAGE). Although this method did not resolve all ribosomal proteins, we could not detect significant differences in the protein pattern between these two ribosome populations. Therefore, we disfavored the idea that sequestration to the ER of mRNAs coding for secretory proteins occurs via distinct features of ER-bound ribosomes.

Instead, we hypothesized in 1971 that selection of mRNA to the ER membrane is not via direct binding of the mRNA itself, but rather via binding of its nascent translation product.[3]

We postulated that all mRNAs for secretory proteins code for a signatory amino-terminal sequence element that is recognized by a soluble factor that, in turn, binds the nascent chain–ribosome complex to the ER membrane (see figure 3 in previous chapter by David Sabatini). We also postulated that ribosomes, free or ER bound, are indistinguishable and that they cycle between an ER-bound and a cytosolic pool. At this point, there were only a few sequences of secretory proteins established, and there was no evidence that these proteins shared a common amino-terminal sequence element. However, it occurred to us that such an amino-terminal sequence tag might be a transient and not a permanent feature of nascent secretory proteins. Therefore, we had no hesitation to publish this idea.[3] Nevertheless, our proposals at this point were pure speculation without any supporting evidence.

To experimentally test the predictions of this hypothesis, we sought to develop a cell-free system in which protein translation and protein translocation across microsomal membrane vesicles were faithfully recapitulated. We hoped to reconstitute such an in vitro system from defined components. To accomplish this, we began with a series of deconstruction experiments. We started with polysomes. Using puromycin and high concentrations of salt, we isolated functional ribosomal subunits[4] and mRNA still attached to its binding proteins.[5–7] Using the same methods, we succeeded in disassembling rough microsomes. This yielded a virtually mRNA- and ribosome-free membrane fraction.[8] We also isolated "native" small ribosomal subunits that contained translation initiation factors. These initiation factors could be dissociated as multi-subunit complexes.[9,10] With these components in hand, we then attempted to reconstruct a functional protein translation/translocation system from defined components.

While these deconstruction experiments were going on, two important papers appeared in 1972 from the laboratories of Philip Leder[11] and César Milstein.[12] These investigators studied the translation of poly(A) containing mRNAs from myeloma cells (containing primarily mRNA for the light chain of immunoglobulin G [IgG]) in a cell-free system that lacked microsomal vesicles. The translation yielded one major product that was larger by about 2–3 kilodaltons than the mature light chain of IgG, which was secreted from the myeloma cells. By peptide mapping of the radioactively labeled translation product, they could show that it contained an extension located at the amino terminus. A partial sequence of this amino-terminal extension was subsequently established by Israel Schechter and colleagues[13] and revealed a preponderance of hydrophobic residues.

Could this additional peptide function as the determinant for attachment to the ER and subsequent translocation across the ER, as we had postulated in 1971? Or was the larger form the result of an in vitro translation artifact, e.g., perhaps the consequence of an erroneous upstream initiation? Interestingly, in vitro completion of nascent chains in isolated rough microsomes of myeloma cells yielded the mature light chain and not the larger form.[12] Therefore, it was suggested that the microsomal membranes might contain a protease that converts the larger form into the mature form of the light chain of IgG by removing its amino-terminal extension.[12]

Signal Peptidase

Together with Bernhard Dobberstein, we succeeded in reproducing[14] the results of Leder's and Milstein's laboratories. We translated poly(A) containing mRNA from myeloma cells in a cell-free translation system that was reconstructed from heterologous components. We obtained a major polypeptide that was larger by about 3 kilodaltons than the mature form of the light chain that was secreted from these myeloma cells (figure 1).

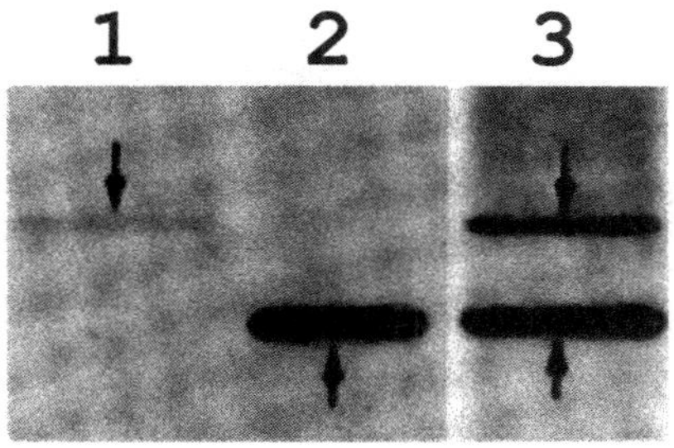

1. Synthesis of the large (pre–light chain [pLi]) and mature (mature light chain [mLi]) forms of the light chain of IgG. [^{35}S]Methionine–labeled proteins were separated by SDS-PAGE and visualized by autoradiography. Lane 1, translation products of poly(A) containing mRNAs isolated from myeloma cells; lane 2, mature light chains secreted from myeloma cells; lane 3, in vitro readout of polysomes ("detached polysomes") prepared by detergent from isolated rough microsomes of myeloma cells. From Blobel and Dobberstein, 1975. *J. Cell Biol.* 67:835–851.

To address the question of whether the larger form of the IgG light chain was a physiological precursor of the mature chain or an in vitro artifact, we were guided by three conjectures. First, the processing enzyme (subsequently termed signal peptidase) is a membrane-associated protease, as had been postulated by Milstein and co-workers.[12] Therefore, this enzyme might be solubilized when isolated rough microsomes are treated with detergent; the "detached" polysomes that could subsequently be sedimented by centrifugation might therefore be free of signal peptidase. Second, as the signal sequence was presumed to mediate the attachment of the ribosome to the cis-side of the membrane, the signal peptidase activity is likely associated with the trans-side of the microsomal membrane. This would assure that signal sequence removal occurs only after signal sequence–mediated attachment on the cis-side of the membrane. Third, after signal sequence–mediated attachment to the membrane, the signal sequence would be translocated across the membrane and thereby gain access to its trans-side. There, it would be cleaved cotranslationally and cotranslocationally, i.e., during but not after translation and translocation were completed. Hence, in detached polysomes, the ribosomes near the 5′ end of the mRNA might contain nascent chains with a signal sequence. In contrast, ribosomes located near the 3′ end of the mRNA should contain nascent chains that already had their signal peptide removed (figure 2).

In agreement with these assumptions, we observed[14] that nascent chains present in detached polysomes can be completed in a cell-free translation system (in the presence of an initiation inhibitor to prevent re-initiation) to yield both the larger and the mature forms of the light chain (figure 2). Moreover, in a time course experiment, we observed that nascent chains lacking a signal peptide were completed first, followed by the completion of chains still containing their signal peptide (figure 3). These results supported the notion that microsomal ribosomes near the 3′ end of mRNA contained nascent chains from which the signal sequence had already been removed, cotranslationally and cotranslocationally. In contrast, ribosomes near the 5′ end still contained their signal sequence. This signal sequence could no longer be cleaved, as the signal peptidase had been removed during the preparation of the detached polysomes. These readout data with detached polysomes suggested that a completed larger form of the light chain of IgG is indeed an in vitro artifact. In vivo, the signal peptide is cleaved off the *nascent* chain but cannot be cleaved off the *completed* chain. Hence, the *completed* larger form is not a physiological precursor of the mature form of the light chain.

Reconstitution of Cotranslational Translocation

The data on signal sequence removal provided us with important clues for our subsequent attempts to reconstitute translation and translocation of the light chain of IgG. In view

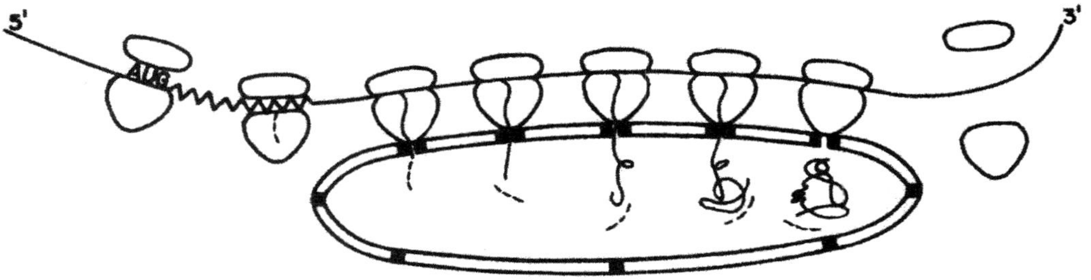

2. The 1975 version of the signal hypothesis. Signal peptidase, not indicated here, was postulated to be associated with the trans-side of the microsomal membrane. When the membrane is solubilized by detergent, signal peptidase would be absent in isolated detached polysomes. The detached polysomes should contain two kinds of chains: those still containing their signal peptide at the N terminus and those with their signal peptide cleaved off by signal peptidase. If signal peptide cleavage occurs during translocation, those ribosomes near the 3′ end of mRNA should contain chains without signal peptide, and those near the 5′ end of mRNA should still contain their signal peptide. In a time course of translation, one should therefore first see the completion of chains that have lost their signal peptide and then the completion of chains that still contain their signal peptide. From Blobel and Dobberstein, 1975. *J. Cell Biol.* 67:835–851.

of the signal hypothesis, it appeared unlikely that the larger form would be translocation competent. That is, cell-free synthesis of the completed larger form in the absence of membranes followed by posttranslational incubation of the reaction with microsomal membrane vesicles was unlikely to yield translocation and conversion of the larger form to the mature form. Indeed, we found that the posttranslational addition of microsomal vesicles to the in vitro translation reaction containing the larger form of the light chain of IgG did not yield conversion to the mature form.

At this point, we experienced an extremely frustrating period in our efforts to reconstitute cotranslational translocation. All microsomal vesicle preparations that we prepared from several tissues of the usual laboratory animals (mice, rats, chicken, guinea pigs, pigeons, and rabbits) severely inhibited the synthesis of the light chain of IgG in the translation reaction. As translation was completely inhibited by the added microsomal membranes, we were not able to detect translation-coupled translocation across these membranes.

A way out of this dilemma came in December of 1974. The chance arose to prepare and to test yet another rough microsome preparation from yet another source: canine pancreas. Surprisingly, addition of canine pancreas microsomal membranes to the translation reaction did not inhibit translation. And, indeed, analysis of the translation products by SDS-PAGE showed that in the cotranslational presence of microsomal vesicles, most of the light chain was synthesized in its mature form.[15] This strongly suggested that the signal sequence of the nascent chain engaged the translocation machinery of the membrane and gained access to signal peptidase on the trans-side of the membrane. The mature chain cosedimented with the vesicles, suggesting that it was segregated within the vesicle lumen.

As a control, we also translated rabbit reticulocyte mRNA.[15] Translation of this mRNA yielded primarily the two globin chains. The two globin chains are cytosolic proteins and therefore should not contain a signal sequence for translocation into the ER lumen. Moreover, there should not be any shortening of the two globin chains as a result of signal peptidase cleavage in the cotranslational presence of pancreas microsomal vesicles. Indeed, we did not observe conversion to smaller forms of the two globin chains in the cotranslational presence of microsomal membranes. Unexpectedly, however, the two globin chains cosedimented with the microsomal vesicles. Could cosedimentation result from cotranslational segregation

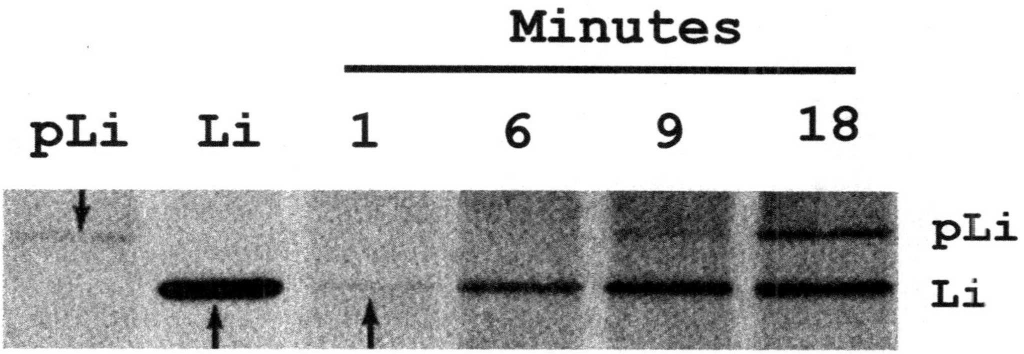

3. Time course of readout of "detached polysomes" (see figures 1 and 2) from myeloma cells in a cell-free translation system. [^{35}S]Methionine-labeled proteins were separated by SDS-PAGE and visualized by radioautography. pLi (pre–light chain), major in vitro translation products of poly(A) containing mRNAs from myeloma cells; Li (light chain), light chains secreted from myeloma cells; lanes 1, 6 , 9, and 18, readout products from detached polysomes after one, six, nine, and eighteen minutes of incubation in a translation system in the presence of an inhibitor of initiation. From Blobel and Dobberstein, 1975. *J. Cell Biol.* 67:835–851.

of the two globin chains into the lumen of the ER? Or was cosedimentation of the two globin chains with the microsomal vesicles the consequence of non-specific binding of the newly synthesized globin chains to the cis-side of the vesicle membrane rather than of translocation to the trans-side of the vesicle membrane (i.e., into the vesicle lumen)? To distinguish between these possibilities, we carried out posttranslational incubation with proteolytic enzymes. We had shown earlier[16] that membranes of microsomal vesicles are sealed in such a way that they protect the content proteins from attack by externally added proteolytic enzymes while allowing proteolysis of surface-bound proteins and even detachment of ribosomes. Hence, posttranslational incubation with proteolytic enzymes should distinguish between proteins bound to the surface of microsomal vesicles and those segregated in the vesicle lumen. We found that the two globin chains were largely degraded by posttranslationally added proteases, regardless of whether translation was carried out in the absence of vesicles or in the presence of vesicles.

Most importantly, posttranslational proteolysis of the translation reaction that was carried out in the presence of microsomal vesicles yielded degradation of the larger form of the light chain but not the mature form of the light chain. These results suggested that even in the presence of microsomal membranes, some of the nascent chains failed to engage the translocation machinery of the membrane, never gained access to signal peptidase on the trans-side of the vesicles, and therefore were completed in the translation reaction as larger forms. Having not been translocated, the larger forms were degraded by added proteolytic enzymes.

Together, these data strongly suggested that we had succeeded in recapitulating the initial events in the secretory pathway in an in vitro system. This opened the way for biochemical analysis of protein translocation across the ER. These experiments marked the beginning of molecular cell biology. Successful in vitro reconstitution was subsequently achieved for many other membrane-coupled reactions.

An Expanded Signal Hypothesis

The results of the reconstitution experiments encouraged us to further elaborate the simple model of 1971 (see figure 3 in previous chapter by David Sabatini). The most significant

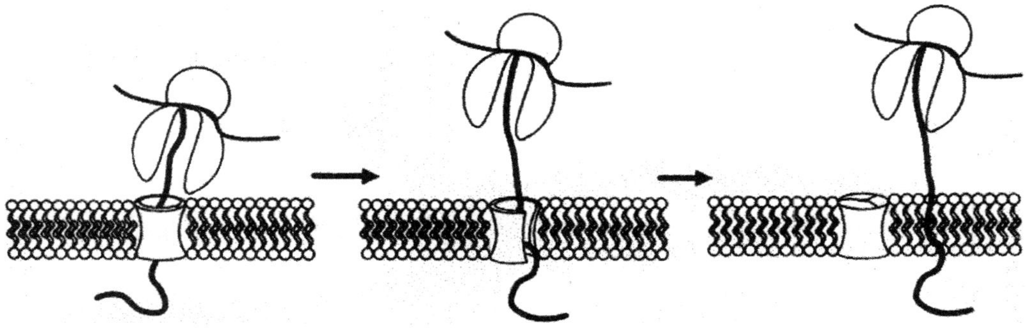

4. Model for catalyzed integration into the lipid bilayer of integral membrane proteins. The proposal that the initial steps in the integration of membrane proteins are the same as those for the translocation of secretory proteins was made already in the 1975 version of the signal hypothesis. Translocation across the protein-conducting channel (PCC) was envisaged to be initiated by a signal sequence, as in the case of secretory proteins. A stop-transfer sequence, corresponding to the transmembrane segment of the translocating chain, would be recognized by the PCC and open it laterally. The transmembrane segment would then be displaced from the aqueous PCC to the lipid bilayer, resulting in the closure of the PCC and the integration of the nascent chain into the lipid bilayer. From Simon and Blobel. A protein-conducting channel in the endoplasmic reticulum. *Cell.* 65:371–380. ©1991, with permission from Elsevier.

additional postulate was that protein transport across the microsomal membrane proceeds through a protein-conducting channel.[14] The protein-conducting channel was envisaged to consist of integral membrane protein subunits. The signal sequence and the ribosome were proposed to bind to these subunits and to assemble them into an aqueous channel allowing transport of the chain in an aqueous conduit across the membrane. The concept of a protein-conducting channel remained controversial for many years until definitive evidence for it was provided by electrophysiological experiments in 1991.[17] In the intervening fifteen years, alternative hypotheses that proposed transfer of secretory proteins across the membrane directly through the lipid bilayer, unassisted by proteins, flourished and became widely accepted.

We also suggested that integral membrane proteins that experience a partial translocation of a segment of the nascent chain to the trans-side of the membrane might use a signal sequence to initiate translocation.

Finally, we predicted that the translocation of proteins across other cellular membranes, e.g., import of cytoplasmically synthesized proteins into mitochondria, is mediated by a signal sequence that would be distinct from that of a secretory protein.

Membrane Protein Assembly

To test whether an integral membrane protein does, in fact, use a signal sequence to translocate a domain to the trans-side of the ER, we studied the glycoprotein of the vesicular stomatitis virus (G protein of VSV).[18,19] This protein is part of the viral membrane and contains an amino-terminal domain exposed on the surface of the viral membrane, a single transmembrane segment, and a carboxy-terminal domain that interacts with the viral capsid. To achieve this topology in the viral membrane, the amino–terminal domain of the G protein must be translocated to the trans-side of the ER, whereas the carboxy-terminal domain needs to remain untranslocated on the cis-side of the ER.

To accomplish this asymmetric integration into the ER membrane, we reasoned that a signal sequence, indistinguishable from those of secretory proteins, might initiate the translocation process in the ER. An additional sequence, which we termed the "stop-transfer" sequence, would then terminate the translocation process. We speculated that the stop-transfer sequence would open the channel laterally, thereby allowing displacement of the stop-transfer sequence from the aqueous protein-conducting channel to the lipid bilayer. The stop-transfer sequence would then become the transmembrane segment that is embedded in the lipid bilayer. The carboxy-terminal remainder of the protein would not be translocated and would remain on the cis-side of the membrane (figure 4).

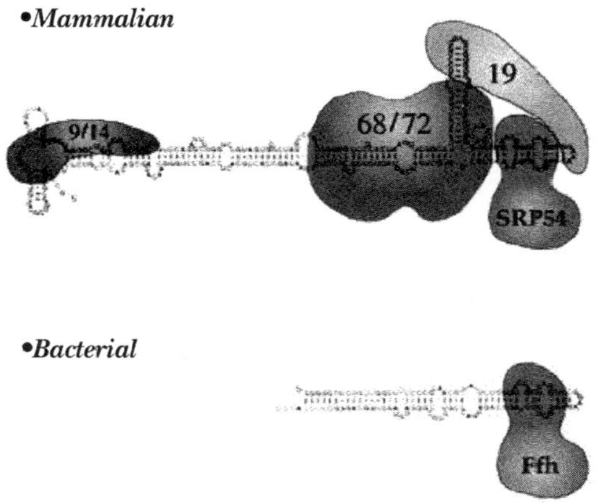

5. Signal recognition particle (SRP) in mammalian cells and in bacteria. Both SRPs contain an RNA and proteins. The mammalian SRP contains six proteins of indicated molecular mass in kilodaltons. The bacterial SRPs contain a shorter RNA and a single protein (Ffh, fifty-four homologue). From Walter and Johnson, 1994. Signal sequence recognition and protein targeting to the endoplasmic reticulum membrane. *Annu. Rev. Cell Biol.* 10:87–119 with permission via Copyright Clearance Center.

It should be remembered that in those days there were still no pure mRNAs available that could be readily obtained by in vitro transcription from recombinant DNA. All we could do was to isolate total mRNA from cells or tissues that contained major species of mRNA whose translation would give rise to major and readily detectable translation products. The VSV-infected cell was a good model system, as most of the total mRNA from these infected cells coded for VSV proteins. Of these, the G protein is the largest and is readily identifiable on the basis of its migration in SDS-PAGE. These experiments were done in collaboration with Harvey Lodish's laboratory at MIT, particularly by his graduate student Flora Katz, and by a graduate student in our laboratory, Vishwanath ("Vishu") Lingappa.

As expected, translation of mRNA from VSV-infected cells yielded the G protein and other VSV proteins. However, contrary to our expectation, in the cotranslational presence of the microsomal vesicles, at least half of the G protein molecules appeared to be synthesized as a *larger* form. Rather than having cleaved off a putative signal sequence, the membranes appeared to have added something to the G protein. We were very puzzled by this result until it occurred to us that the canine microsomal vesicles might contain activated oligosaccharides and oligosaccharidyl transferase to carry out cotranslocational core glycosylation of the G protein. The added core sugars might more than compensate in mass for the loss of a signal sequence and therefore result in a slower migration in SDS-PAGE. Biochemical characterization of both the G_0 and the G_1 species indeed confirmed that G_0 contained an amino-terminal signal sequence that was structurally and functionally equivalent to those of secretory proteins.[18,19] The G_1 species had lost its signal sequence and had acquired core sugars. Moreover, posttranslational proteolysis experiments showed complete digestion of G_0, whereas G_1 was converted to a slightly smaller form. This smaller form represented G_1 that had lost its carboxy-terminal tail. These posttranslational proteolysis data suggested that the G protein had been correctly integrated into the microsomal vesicles! As in vivo, the bulky amino-terminal domain was translocated to the trans-side of the membrane, whereas the smaller carboxy-terminal domain of about forty amino acid residues remained

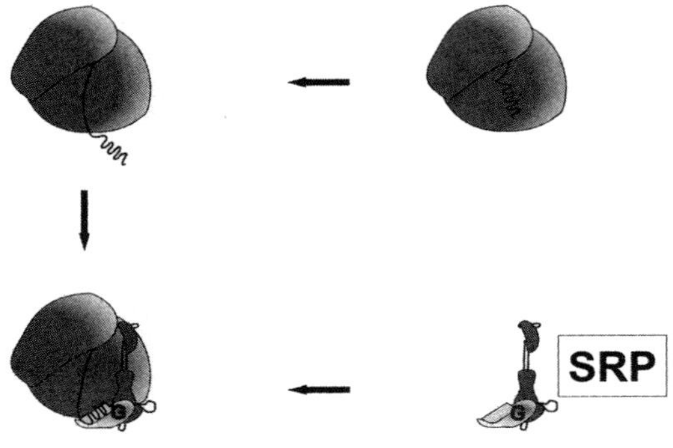

6. Recognition of the signal sequence by the signal recognition particle (SRP). As soon as the signal sequence has emerged from the large ribosomal subunit, SRP can bind to it. Courtesy of Günter Blobel.

untranslocated on the cis-side of the membrane.

Microsomal vesicles were not only able to translocate secretory proteins but also were competent to asymmetrically integrate membrane proteins and to core glycosylate them. These results demonstrated that the asymmetric integration of membrane proteins is not a spontaneous event but is catalyzed in the ER, presumably by the same machinery that mediates the translocation of secretory proteins.

The Signal Recognition Particle

Our next goal was to deconstruct the translocation machinery of the microsomes into functional components. It had been shown that a salt wash of microsomal vesicles removes an activity that is required for translocation of secretory proteins. Such salt-washed microsomal vesicles were incapable of cotranslational translocation of secretory proteins that were synthesized in a wheat germ cell-free translation system. Addition of the salt wash to the in vitro translocation system containing salt-washed microsomal vesicles restored their translocation activity.[20] These data suggested that salt extracted a membrane-associated activity that is required for protein translocation. Peter Walter, a graduate student in the laboratory, discovered that the activity of the salt wash vanished after only a few hours of storage at 0°C. This frustrated all his attempts to purify this activity until he made the important discovery that minute concentrations of the non-ionic detergent Nikkol stabilized the activity. We reasoned that Nikkol protected a hydrophobic site of the activity, perhaps the binding site for the hydrophobic signal sequence. Incorporating the hydrophobic site idea into his purification protocol, he was able to rapidly purify the activity. The activity sedimented at about 11 Svedberg units (S) and by SDS-PAGE consisted of six proteins.[21] Based on an extensive analysis of its function,[22–24] we termed the protein complex "signal recognition protein" (SRP).

It took us another two years before we realized that the purified activity also contained a 7S RNA![25] We then changed the name to signal recognition particle, leaving the acronym SRP intact. A considerable amount of work in both Peter Walter's and Bernhard Dobberstein's laboratories has led to a functional and structural model for the SRP and to the discovery of an interesting homologue in bacteria (figure 5).[26]

Functional analyses[22–24] revealed that canine pancreas SRP bound with low affinity to wheat germ ribosomes. However, SRP bound with high affinity when wheat germ ribosomes were programmed with an mRNA coding for a secretory protein (bovine preprolactin), but not when programmed with an mRNA coding for cytosolic proteins, such as the globin chains. Also, if translation was carried out in the presence of hydroxy-leucine instead of leucine, there was greatly reduced binding of SRP. As the signal sequence for preprolactin is rich in leucine, this result suggested that SRP interacted directly with the signal sequence. Moreover, the interaction of SRP caused a translation arrest when ribosomes were programmed with an mRNA for a secretory protein, but not when programmed with mRNA for cytosolic proteins. The size of the arrested nascent secretory protein corresponded to about seventy amino acid residues, suggesting that binding of the signal sequence occurred after the signal sequence was fully exposed following exit from the tunnel in the large ribosomal subunit (figure 6). When microsomal membranes were added, there was release of the translation arrest, suggesting that the ER membrane contains an SRP receptor that is capable of releasing SRP-mediated translation arrest.

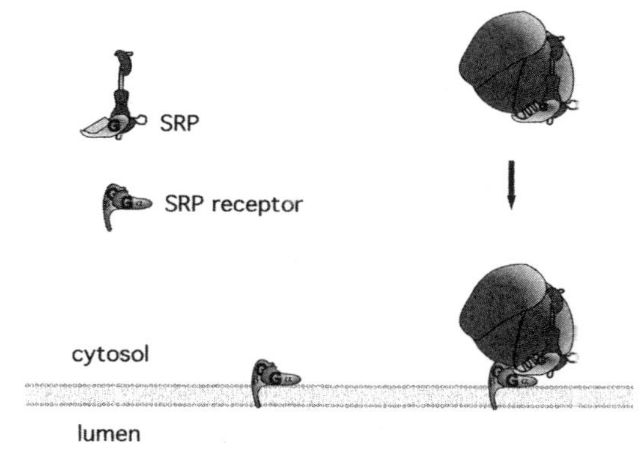

7. The ribosome/nascent chain/SRP complex binds to the SRP receptor in the ER. Courtesy of Günter Blobel.

We found that the in vitro assembled polysomes synthesizing secretory proteins, but not those synthesizing cytosolic proteins, bound to SRP-depleted (salt washed) microsomes, but only in the presence of SRP. Binding was abolished when SRP was pretreated with *N*-ethyl-maleimide. These data strongly supported the notion that protein translocation across the ER is a receptor-mediated event and ruled out proposals that chain translocation occurs spontaneously and unassisted by proteins.

The discovery of the signal recognition particle was an important milestone in the molecular analysis of protein translocation across the ER membrane. Its function corresponded to that of the "binding factor" whose existence had been predicted in 1971 (see figure 3 in previous chapter by David Sabatini). It was the first component of any of the cellular membrane translocation systems that was isolated and characterized. It provided strong support for the predictions made in the signal hypothesis.

SRP Receptor

Reid Gilmore, a postdoctoral fellow in our laboratory, used the ability of salt-washed microsomal membranes to release the SRP-mediated elongation arrest of the synthesis of secretory proteins as an assay to follow the purification of this activity. Together with the membrane's signal peptidase (see next section), the arrest-releasing activity could be solubilized by treatment of the salt-washed microsomal membranes with non-ionic detergent and moderate salt concentrations. The purified SRP receptor consisted of two subunits, a larger subunit and a smaller subunit.[27–31] The larger subunit is a peripheral membrane protein, whereas the smaller subunit is an integral membrane protein with one transmembrane segment. Both proteins were shown to be GTP-binding proteins.[30,31] SRP and SRP receptor interact with each other (figure 7).[32] The SRP receptor is associated exclusively with the ER (including the outer nuclear envelope) and is present in the ER in sub-stoichiometric quantities relative to membrane-bound ribosomes.[27] This suggested that the SRP–SRP receptor interaction is transient in nature. It serves in the targeting but not in the subsequent translocation of the chain.

With the discovery of the SRP and its cognate SRP receptor, the components involved in signal sequence recognition and targeting to the ER had been isolated.

Signal Peptidase

The signal peptidase cleavage site of nascent secretory proteins is accessible to signal peptidase only on the trans-side of the membrane (see previous section). Therefore, signal sequence cleavage is dependent on translocation. Robert Jackson in our laboratory developed a translocation-independent assay for signal peptide cleavage. He solubilized the microsomal signal peptidase with detergent. Using an in vitro synthesized presecretory protein as a substrate, he demonstrated correct endoproteolytic removal of the signal sequence by the detergent form of the enzyme.[33]

The solubilized microsomal signal peptidase was purified by Emily Evans, a graduate student in our laboratory. The purification of this enzyme was a major challenge, as its activity required the presence of lipids and detergents throughout all purification procedures. Omission of lipids led to rapid inactivation of the enzyme. Surprisingly, the purified microsomal signal peptidase turned out to be a complex of five non-identical subunits,[34] all of which are integral membrane proteins.[35–37]

With the SRP receptor and signal peptidase, two components of the microsomal membrane that could be solubilized by detergents and assayed independently of intact vesicles and of translocation had been purified. Our next goal was to purify the putative protein-conducting channel. The proteins making up such a channel were likely to be integral membrane proteins. To purify them from other membrane proteins required detergent solubilization. However, unlike SRP receptor and signal peptidase, channel proteins require reconstitution into proteoliposomes to unambiguously assay their activity.

Reconstitution of Translocation-Active Proteoliposomes

The reconstitution of translocation-active proteoliposomes had been a high priority on our agenda for quite some time before Christopher Nicchitta joined our laboratory as a postdoctoral fellow. However, many attempts by the most talented people in our laboratory had failed to achieve this goal. We had arrived at the conclusion that reconstitution, if at all possible, would probably occur with very low efficiency. Presumably, after detergent solubilization and removal of the detergent to form proteoliposomes, the components that are required for cotranslational translocation are not necessarily reconstituted in the asymmetric orientation and in the stoichiometry in which they exist in the microsomal vesicles. Instead, reconstituted proteoliposomes may contain these proteins in a scrambled fashion and in non-stoichiometric amounts that render them inefficient or inactive in translocation. The SRP receptor, the signal peptidase complex, or the constituents of the protein-conducting channel and perhaps other components that participate in protein translocation may not regain their asymmetric orientation in the reconstituted proteoliposome vesicles. Signal peptidase and the SRP receptor, for example, exhibit their active sites in opposite orientations, on the trans-side and on the cis-side of the vesicles, respectively. Through careful choice of detergents and of other conditions (pH as well as salt and detergent concentrations), we were able to obtain translocation-competent proteoliposomes.[38]

These seminal experiments showed that after complete detergent solubilization of microsomal membranes, the detergent forms of integral membrane proteins required for protein translocation, including a putative protein-conducting channel, could be reconstituted in an active form. This paved the way for the isolation of the protein-conducting channel and allowed the subsequent reconstitution of translocation-active proteoliposomes from purified components.

Electrophysiological Detection of the Protein-Conducting Channel

We first proposed the concept of a protein-conducting channel (PCC) made up of integral membrane proteins in 1975.[14] However, it remained a highly controversial idea, particularly as there was no direct evidence for it. Alternative models of protein translocation directly through the lipid bilayer flourished.[39,40] In 1985, we showed that nascent chains in the process of translocation are accessible to aqueous perturbants.[41] Although these data were consistent with an aqueous PCC, they could not rule out that an aqueous environment was created transiently by the hydrophilic head groups of the lipids rather than proteins forming such an aqueous channel.

Although electrophysiology was a standard approach for identifying and characterizing ion-conducting channels, it had never been applied to the detection and characterization of any protein-conducting channels. It was difficult to predict how such channels would behave electrophysiologically. We had postulated that they would be opened and closed for each protein translocation event.[14] Moreover, we reasoned that the conductance of a protein across the channel is unlikely to be accompanied by a significant co-conductance of ions or small molecules. If this were the case, it would be difficult to maintain the distinct composition of ions (particularly calcium) and small molecules in the ER lumen, the "reticuloplasm," as compared with that in the cytoplasm. Thus, PCCs are likely to be closed when they are not conducting a polypeptide chain. Moreover, they are most likely designed to prevent the co-conduction of ions when they are opened for and during translocation of a protein. Therefore, it was not clear whether PCCs could be revealed at all by electrophysiological measurements. With these caveats in mind, Sanford Simon, a post-doctoral fellow in our laboratory, set out on a most remarkable and exciting journey into completely uncharted territory: to explore PCCs by electrophysiological approaches. These experiments were begun in collaboration with Joshua Zimmerberg at the National Institutes of Health.[42]

Because rough microsomes were too small for patch-clamp experiments, we decided to use the planar lipid bilayer system developed by Paul Mueller et al. in 1962.[43] In this system, vesicles are fused to a planar bilayer that is formed in a hole of a partition separating two chambers. Addition of microsomes to the cis-chamber occasionally yielded fusion of a single rough microsomal vesicle as evidenced by an increase in conductance. Some preparations yielded unitary conductances of 20, 55, 80, and 115 pico-Siemens in 45-millimolar potassium glutamate.[42] By electron microscopy, it can be estimated that each of the rough microsomal vesicles contains about a hundred membrane-bound ribosomes. Many of these membrane-bound ribosomes are potentially in the process of conducting a chain across the membrane. Therefore, it was unlikely that the few unitary conductances that we observed were related to PCCs. Nevertheless, the fact that we could detect conductances at all suggested that fusion of a rough microsomal vesicle with the planar bilayer had taken place.

If the PCCs are indeed either closed or electrically silent when occupied by nascent chains, would it be possible to reveal these channels by releasing the nascent chain without closing them? Could puromycin, an analogue of amino acyl tRNA, achieve this? The peptidyl transferase activity of ribosomes couples puromycin to the carboxy-terminal end of the nascent chain. This causes chain release from the ribosome. In the case of rough microsomes, it had been established some time previously that the chain is vectorially discharged into the lumen of the vesicle.[2] When the puromycin reaction is carried out at low salt concentrations, the chains are vectorially discharged, but the ribosomes remain attached to the membrane.[8] Could the attached ribosome keep a cleared PCC in an open configuration? And could subsequent treatment with high salt release the ribosome and close the PCC?[8]

Based on these considerations, we added puromycin to the cis-chamber of the bilayer that contained a fused rough microsome vesicle. We observed a huge increase in conductance.[17] This effect was specific, as it was observed only when puromycin was added to the cis-chamber, not when it was added to the trans-chamber.[17] The cis-chamber, but not the trans-chamber, contained the exposed ribosomes of the fused rough microsomal membrane. Puromycin, when added to the trans-chamber, could not access the ribosomes and be coupled to the nascent chain. But when subsequently added to the cis-chamber, it again caused a huge increase in conductance. These side-specific data strongly suggested that puromycin indeed cleared the nascent chain and that at low salt the PCCs stayed in an open configuration and were able to conduct ions.

How many PCCs were cleared by the addition of puromycin? Could very low concentrations of puromycin make it possible to demonstrate a time-resolved clearance of individual PCCs? Indeed, when much lower concentrations of puromycin were added to the cis-chamber, we observed clearance, one at a time, of individual PCCs[17] of 220 picoSiemens. Hence, the PCC contains a very large aqueous pore that conducts about ten times more ions than an ordinary ion-conducting channel.[17]

Could one close the puromycin-revealed PCC by dissociating the ribosome by high salt concentrations? Indeed, when a single PCC was revealed at low concentrations of puromycin and at low salt, it was closed by a subsequent increase in salt concentration. This result was consistent with the idea that the puromycin-revealed PCCs are kept open by attached ribosomes but are closed when ribosomes are dissociated.

Can PCCs be demonstrated also in other membranes and by methods other than puromycin-induced clearance of translocating chains? Does the signal sequence serve as a ligand to open (or assemble) the PCC as was postulated in the signal hypothesis?[14] To answer these questions, we turned to the prokaryotic plasma membrane. It had been suggested that the eukaryotic ER evolved by invagination of the prokaryotic plasma membrane.[44] In fact, signal sequences for bacterial secretory proteins function in translocation across the ER of eukaryotic cells, and, vice versa, signal sequences addressed to the eukaryotic ER function in translocation across the prokaryotic plasma membrane. It was therefore conceivable that the PCC of the prokaryotic plasma membrane resembles that of the eukaryotic ER.

When protoplasts of *Escherichia coli* are fused from the cis-chamber to a planar bilayer, the putative signal sequence–binding site of the PCC will be exposed on the trans-chamber. The addition of low concentrations of synthetic signal peptide of the bacterial LamB protein to the trans-chamber, but not to the cis-chamber, caused a stepwise increase in conductance. The stepwise increase in conductance was observed only in the presence of high salt. Each step is likely to represent the binding of one signal peptide to one signal peptide–binding site of the PCC. High salt would reinforce the largely hydrophobic interactions between the hydrophobic signal peptide and the PCC and thus keep the PCC in an open configuration for long periods of time.[45] The conductance properties of the prokaryotic plasma membrane PCC were found to be similar to those of the PCC in the ER, in support of the notion that the two channels share a common history and are conserved.

Taken together, these data strongly supported the existence of aqueous protein-conducting channels that, like ion-conducting channels, were composed of proteins and supported the idea that the signal sequence indeed served as one of the ligands to open the channel. These results led to a paradigm shift in cell biology and strongly argued against models that had reigned supreme for many years and that had proposed transfer of proteins directly through the hydrophobic core of the lipid bilayer.

Visualization of the Protein-Conducting Channel

By genetic approaches, Randy Schekman and co-workers[46] had identified Sec61 as a candidate for the protein-conducting channel in yeast. Tom Rapoport and co-workers used subfractionation and reconstitution into proteoliposomes (see previous section) to isolate the mammalian counterparts of the yeast PCC.[47,48] Their work established that the PCC consisted of a heterotrimer, termed the alpha, beta, and gamma subunits of Sec61.

To visualize the PCC attached to the ribosome, Roland Beckmann, a postdoctoral fellow in our laboratory, was able to isolate the Sec61 complex by detergent solubilization of yeast microsomes and to bind it to isolated yeast monomeric ribosomes.[48] Binding was saturable and suggested that two Sec61 trimers bound per ribosome. Cryo-electron microscopy of these complexes and three-dimensional image reconstruction, done in collaboration with Joachim Frank's laboratory, revealed that the Sec61 trimer bound to a region of the large ribosomal subunit that was near to the previously identified exit site of the nascent chain. The Sec61 complex formed a funnel-shaped structure with a diameter of about 35 ångströms at the side facing the ribosome and a diameter of 15–20 ångströms at the side facing the ER lumen. The Sec61 complex was attached to the large ribosomal subunit at one site. Most interestingly, a section across the center of the ribosome revealed that the tunnel in the large ribosomal subunit aligned with the Sec61 channel. This arrangement suggests a likely conduit of the nascent chain within the large ribosomal subunit and across the Sec61 channel into the lumen of the ER.

The visualization of the PCC attached to the large ribosome and the alignment of the large ribosomal subunit tunnel with the PCC were extremely gratifying confirmations of the predictions that had been made in 1975.[14]

The PCC is clearly one of the marvels of nature. Unlike an ion-conducting channel that opens and closes in only one dimension, the PCC opens and closes in two dimensions, across the lipid bilayer and in the plane of the bilayer. The PCC is not merely a passive conduit; it scans the unfolded nascent chain as it passes across it, responding to a passing stop-transfer sequence by lateral opening. Moreover, the PCC is constructed in such a way that it does not leak significant amounts of small molecules. More recently, evidence has accumulated for reverse translocation and disintegration of membrane proteins. In order to allow degradation of secretory and integral membrane proteins by the cytoplasmic proteasomes, proteins can be reverse translocated from the lumen of the ER or disintegrated (membrane proteins) from the ER membrane.

Posttranslational Translocation across the ER

Protein translocation across the ER can also occur posttranslationally. This was first discovered in yeast.[49,50] Instead of SRP and SRP receptor, this system uses other factors, primarily heat shock proteins, to keep the protein in an unfolded and translocation-competent configuration.[51,52] Moreover, additional membrane proteins interacting with the Sec61 complex are required to facilitate presentation of the signal sequence for opening the Sec61 PCC.

Other Protein Translocation Systems

Work on the ER established the basic principles. Other systems for protein translocation across distinct cellular membranes, for example, the inner mitochondrial membrane or the thylakoid membrane of chloroplasts, were subsequently shown to work on similar principles. Entry of a protein into any of these other "public" translocation systems requires a cognate signal sequence. As in the case of a signal sequence addressed to the ER, each signal sequence is recognized by a cognate signal recognition factor that, in turn, targets the complex to a cognate receptor. These receptors are restricted in their localization to distinct cellular membranes. Translocation then occurs through channels to the other side of the membrane. There are fascinating similarities and differences in the construction of these various translocation systems.

The translocation system of the ER has evolved from the major protein translocation system in the plasma membrane of prokaryotes. Most of the eukaryotic cellular membranes arose by invagination of the prokaryotic plasma membrane and by relegation of specific membrane protein functions of the prokaryotic plasma membrane to these intracellular membranes.[44] The signal sequence of the ER is related to the signal sequence addressed to the major prokaryotic plasma membrane translocation system. Moreover, homologues of SRP (see previous section) and SRP receptor exist in prokaryotes, and the SecY complex of bacteria is the homologue of the Sec61 complex of the ER. Thus, protein translocation across membranes and protein integration into membranes are very ancient and highly conserved membrane functions.

The protein import system into peroxisomes operates with at least two distinct signal sequences and cognate signal recognition factors and receptors. However, it appears that *integral membrane proteins* of the peroxisomal membrane are integrated using the ER translocation system and are then sorted to the peroxisomal membrane, as was first suggested in 1978.[53] It appears that peroxisomal *content proteins* can be transported across the peroxisomal membrane posttranslationally and in a folded and even oligomeric form. A putative peroxisomal membrane pore may not be able to open laterally to the lipid bilayer and, therefore, may not be competent for the integration of membrane proteins. This would explain why integral membrane proteins of the peroxisomal membrane need to be integrated by another translocation system, namely that of the ER.

Import and export of macromolecules into and from the nucleus is not restricted to proteins but includes ribonucleoproteins and deoxyribonucleoproteins (viruses). The nuclear pore complex (NPC) is the common conduit for both import and export of all these molecules and molecular complexes. It is a huge organelle, at least twenty times the mass of the ribosome, with a central pore of about 25 nanometers in diameter. Again, protein import and export are mediated by distinct signal sequences. These signal sequences are recognized by cognate signal recognition factors (karyopherins, or kaps) that ferry between the nucleus and the cytoplasm. The kaps target the import and export substrates to receptors that are represented by a subset of nuclear pore complex proteins (collectively termed nucleoporins, or nups). The NPC is connected on the nuclear side to tubes that extend all the way to the nucleolus. These tubes form a chromatin-free zone and are likely to facilitate intranuclear diffusion of macromolecules.

Other Topogenic Sequences

It is clear that *signal* sequences for protein translocation across distinct cellular membranes and a combination of *signal-* and *stop-transfer* sequences for the asymmetric integration of proteins into membranes are not the only "topogenic" sequences that serve as determinants for protein localization.[44] Following translocation across or asymmetric integration into membranes, many proteins undergo further traffic. For example, lysosomal proteins contain signal sequences indistinguishable from those of secretory proteins,[54,55] and, like secretory proteins, they are first segregated within the lumen of the ER. Subsequently, they have to be sorted to the lysosomes. *Sorting* sequence elements are required to achieve sorting to the lysosome.

Epilogue

What began as an inquiry into how secretory proteins are translocated across the ER proceeded into an exciting voyage that revealed the principles by which cells organize themselves into distinct membranes and compartments. Most cellular proteins, including integral membrane proteins, contain intrinsic sequence elements. These sequence elements are decoded by cognate recognition factors. Cognate receptors and effectors localize the protein to their proper location.

8C. Structural Studies on Ribosomes

James A. Lake

I first heard of cell biology at Rockefeller when I was a postdoctoral fellow at Harvard Medical School from 1968 to 1970, where I had gone to work with Henry Slayter of the Children's Cancer Research Foundation (the Jimmy Fund). Henry had obtained some impressive electron micrographs of ribosome crystals from the amoeba *Entamoeba invadens*, and these looked like a promising source for starting computer diffraction studies on ribosome structure.

The ribosome is, in essence, a molecular factory designed to make proteins according to the genetic message contained in deoxyribonucleic acid (DNA). Before a protein can be made, the necessary information is first copied from DNA into a form that can be read by the ribosomes. This intermediate form is known as messenger ribonucleic acid, or mRNA, a molecule in which the linear order of nucleotides (U, C, A, and G) acts like "bits" of computer code that specify the order of amino acids that are to be incorporated into a protein. The ribosome is the computer that reads this mRNA code, much like computers used to read tapes, and manufactures the corresponding protein. Since all proteins are made on ribosomes, understanding their three-dimensional structure and function was an important biological pursuit. Because the ribosome is so large, containing as many as seventy individual proteins and RNAs, and because it had not yet been crystallized in vitro, it was difficult to study by conventional x-ray diffraction methods. It would take nearly a decade of research until the first ribosomal subunit was crystallized in vitro.[1]

Don Caspar, also at the Jimmy Fund, had developed the theory of quasi-equivalent bonding with Aaron Klug and advised me on diffraction and structure analysis. Don was instrumental in getting me to Rockefeller. After he returned from visiting George Palade at Rockefeller in the spring of 1970, he raved about the electron microscopy studies on ribosomes being done by David Sabatini and Günter Blobel. While there, he told them about the work that I was doing and arranged for me to be interviewed. During my visit, it didn't take long for me, too, to be impressed. Sabatini took me carefully through their analyses of electron micrographs of ribosomes. Over lunch, I gave a seminar on my work at Harvard on the structure of the helical assemblies of ribosomes from the amoeba *Entamoeba invadens*.[2] Two years earlier, David DeRosier and Aaron Klug[3] had shown how it was possible to reconstruct three-dimensional density maps from helical arrays of molecules, and at Harvard I had used their new methods to work out the details of the structure of the eukaryotic ribosome. At the end of the day when the interviews were finished, George Palade offered me a position as a research associate, explaining that an assistant professorship would be difficult, if not impossible, to obtain. A few weeks later I called to accept, but mentioned again that I thought an assistant professorship was reasonable considering my experience and publication record. So, imagine my surprise when on my first day at work that September, as I walked up the marble path to South Laboratory, George Palade, going the opposite direction, greeted me, "Good morning, Professor Lake." Still thinking about this salutation, I walked into the laboratory, found my new desk, and saw an envelope lying on top. It was my appointment letter as an assistant professor.

I quickly settled into a research routine at Rockefeller. During the mornings, I would work on writing a review, and the rest of the day I would continue my work on ribosome structure. The new method of three-dimensional reconstruction made it possible to study ribosome structure by computing the equivalent of x-ray diffraction patterns directly from electron micrographs of ribosome microcrystals. The most favorable types of arrays at the time were helices from the eukaryotic amoeba *Entamoeba invadens*. Using these helices, I had successfully determined a low resolution ribosome map at Harvard, but I wanted to do more because the data were theoretically capable of much higher resolution.

In Boston, I had used the Harvard computer center, which then had an IBM computer with 256 kilobytes of central processor memory, a large amount in those days. In contrast, Rockefeller didn't even have a university computer. So my first job was to find a computer in New York that I could use. My wife, Laura, and I lived near the Brooklyn Polytechnic Institute, where her father had been a professor of mechanical engineering. By good luck, I was able to use their IBM 360 computer, which rivaled Harvard's. Thus, I would work at Rockefeller in the morning and early afternoon, compute at Brooklyn Poly late in the afternoon, have dinner, and then Laura and I would return to Rockefeller in the evenings, where she would continue writing her PhD thesis and I'd do my research. Later, when we moved to Manhattan, the neurophysiologists at Rockefeller obtained a new PDP16 computer and let me use it in the evenings. In fact, I was almost the only user every night. The main problem was that it had only 32 kilobytes of CPU memory, so that the large Fourier transforms that I calculated had to be constantly moved into and out of disk memory and therefore required hours to compute, even using the newly discovered fast Fourier transform algorithm.[4]

Throughout my time at Rockefeller, my science continued to prosper. Although our initial ribosomal structure had been published in *Nature*, a highly visible journal, I wasn't totally satisfied because the electron micrographs we used had not been of complete helices, but from arrays of helices that had been sectioned for electron microscopy. Thus, the helices we analyzed had part of their exterior structure removed by sectioning. This meant that the parts of the ribosome nearest the outer edges of the helix would have been incorrectly reconstructed in our initial studies. In order to calculate a more accurate structure, it was necessary to extend the methods of three-dimensional reconstruction to allow for cases in which information had been partially removed by sectioning. Thus, I spent several months developing the mathematics to separate (deconvolute) the Fourier transform of a helix (representing the desired structure) from the Fourier transform of a rectangular slab (representing the transform of the section). The key to this was provided by some clever work on eigenfunctions done by Tony Crowther in Aaron Klug's laboratory at the Medical Research Council in Cambridge.[5] When all of these parts came together at Rockefeller, I published the first good view of the three-dimensional distribution of density in the eukaryotic ribosome in the *Journal of Molecular Biology* (figure 1).[6]

Just the year before, Yoshiaki Nonomura, Günter Blobel, and David Sabatini had published their studies on ribosomes in the same journal using a visual interpretation of electron micrographs of negatively stained subunits that used model building.[7] Both of our structures revealed the general boundaries of the two parts of the ribosome, called the small and large ribosomal subunits. Given their results and mine, I was convinced that it was possible to develop a good working model of the ribosome if better ribosome crystals were available.

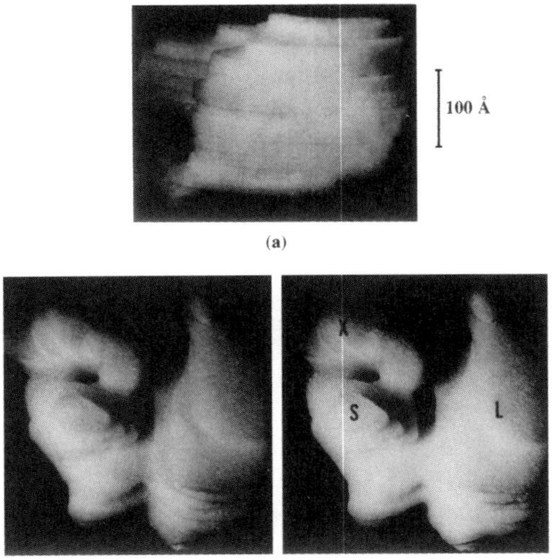

1. The ribosome structure calculated at Rockefeller. In figure a, the large ribosomal subunit is seen from the back. In the stereo image shown in figure b, both the small (S) and large (L) ribosomal subunits, as well as a particle (X) possibly containing eukaryotic protein synthesis initiation factors, are visible. From Lake and Slayter, 1972. *J. Mol. Biol.* 66:271–282 with permission from Academic Press via Copyright Clearance Center.

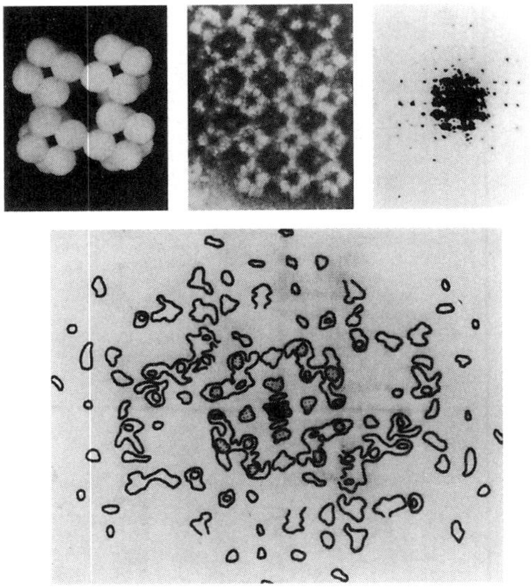

2. An electron micrograph of a small three-dimensional crystal of chicken ribosomes (top center), optical and computer-calculated diffraction patterns of the crystal (top right and bottom, respectively), and a model of ribosome packing in the crystal (top left). From Lake et al., 1974. M. Nomura, A. Tissières, and P. Lengyel, editors. *Ribosomes*. 556. ©Cold Spring Harbor Laboratory Press.

My earlier studies at Harvard had followed the work of Breck Byers on in vivo crystals of ribosomes that can be induced by chilling chicken embryos.[8] Breck was a student of Keith Porter, one of the founding fathers of cell biology at Rockefeller, who had moved to Harvard in 1961. Everyone in Palade's laboratory knew who Keith Porter was, since a picture of him hung at the entrance to our conference room, and underneath a label proclaimed, "Our Father who art at Harvard." Now, continuing this work at Rockefeller, I spent long hours in the cold room, isolating ribosome crystals from chilled chick embryos and learning how to purify the crystals and how to negatively stain them. Sabatini was very helpful and encouraging with the negative staining and procedures for isolating the crystals (figure 2).[9] Our resident electron microscopist, Ivan, helped me learn microscopy. This was not always easy. We had frequent interruptions—the electron microscope images would sway whenever dynamite explosions were set off during the construction of a new subway tunnel to Roosevelt Island. I used to joke that doing electron microscopy at Rockefeller must be like doing electron microscopy in California during an earthquake. Little did I know that in a few years I would actually be able to compare the two situations.

Even after leaving Rockefeller, ribosome crystallization remained our goal. In the late 1970s, our laboratory at the University of California in Los Angeles (UCLA) obtained the first in vitro crystals and crystalline arrays of *Escherichia coli* ribosomes, including helices of small subunits,[10] large subunit two-dimensional crystalline arrays,[11] and three-dimensional crystals of large subunits.[1] This was important because eukaryotic ribosomes are considerably larger, more complex, and harder to work with. These studies paved the way for complete three-dimensional structures of the ribosome that are currently being solved by x-ray diffraction.[12–14]

At Rockefeller I met many interesting people, several of whom became lifelong friends. In particular, I developed a close scientific collaboration with Kevin Leonard, who later moved to the EMBL laboratories in Heidelberg. After Kevin received his PhD at Oxford, he became a postdoctoral fellow at New York University Medical School with Al Kleinschmidt, working on the bacteriophage phi-CbK (a bacterial virus). Since it had an interesting capsid structure and Kevin had wonderful electron micrographs of it, we worked together to decipher its three-dimensional structure. Later, he and I collaborated on using tilting electron microscopy of heavy metal shadowed ribosomes to determine the orientation, left or right, of the structure.[15]

Kevin also developed a new method known as double-carbon-layer negative staining for the electron microscopy of large molecules. This method was ideally suited for ribosome structural studies and was a critical discovery that allowed us to localize specific proteins on the surface of the ribosome.[16] Those studies ultimately allowed us to map correctly the locations of all twenty ribosomal small subunit proteins. Antibody labeling also let us determine the structure, name the features, and map many of the functional regions of the ribosomes. These included the head, platform, base, and cleft of the small subunit and the central protuberance, L1 ridge, L7/L12 stalk, and nascent chain exit site of the large ribosomal subunit.[17]

Around 1972, we began to hear stories that George Palade was leaving Rockefeller and going to Yale University. I had just been reappointed for a second two-year term, which was to start in September, so I didn't pay too much attention to the rumors until they became very frequent. After a while, David Sabatini asked me to join the faculty at New York University Medical School, where he was to become the chair of cell biology. I joined his department in 1973, and three years later I moved to the Molecular Biology Institute at UCLA, where my studies of ribosome structure continued.

My subsequent research on ribosome structure was influenced strongly by my time at Rockefeller: the three-dimensional immunoelectron microscopy mapping of ribosomal proteins with Masayasu Nomura, Larry Kahan, and others; the in vitro crystallization of large ribosomal subunits; and the identification and description of the basic functional sites on ribosomes. The 2006 publication of the complete small and large subunit x-ray structures thrilled me, for it is now possible to see these structures at atomic resolution.[18] I am also proud to say that all of the protein and functional sites mapped by our laboratory were confirmed by the x-ray studies. Also, our subsequent work sequencing and analyzing ribosomal RNAs was key to our discovery of the new animal phylogeny, for which I received the Darwin-Wallace Medal of the Linnean Society of London in 2011. Without my Rockefeller experience, I could not have achieved this.

Chapter 9

Early Explorations into the Fine Structure of Cells

Introduction

Carol L. Moberg

This chapter samples the first decade of an underappreciated path in emerging cell biology, mapping the cell's unknown terrain. Keith Porter, its first scout and persevering guide, once noted, "I am haunted by the question: How do organelles and other entities of cell structure find their 'correct' place in the organization required for normal function?"[1]

The chapters that follow are contributions made during the 1950s by Porter's collaborators and students in this initial prospecting. When he set to work on the basic organization of cells, "this new field had virtually no tradition," said George Palade, and "everybody working in it came from some other province in natural sciences," yet, within a very few years, it "became an area of spectacular growth."[2]

Once biologists gained access to electron microscopes, they hastened to get their pictures into publications. In their eagerness to show all the novel structures they were finding inside cells, they inundated journals with illustrations of minuscule objects at enormous magnifications. There were considerable disadvantages to this haste. The earliest microscopes rendered poor images, and the newly observed objects were not well identified. Scientists vying to write a new lexicon were cautious and prosaic in naming objects according to a taxonomy of shape. After all, they did not yet know what these objects did. As a result, shapes with names like fibrils, particles, double membranes, vesicles, reticula, cristae, or cisternae could refer to diverse objects in a cell. Given the preponderance of micrographs, equivocal terms, and conflicting scientific viewpoints, this initial rush to display cellular particles was, in Porter's view, "perplexing for many readers of biological literature."[3]

An even more serious disadvantage, according to Porter, was that electron micrographs were appearing without regard to light microscope pictures of cells, a practice he had advocated since publishing his first electron micrograph of an intact cell alongside one of a similar cell taken with the light microscope. Like the great naturalists, he took pleasure in observation and searched for a sense of form or wholeness about the cell with a view toward its overall intricate design. While careful to avoid theorizing or generalizing, he sought to learn whether a cell contained a frame or matrix from which all biologists could derive similar information. A continuing thread of his research, extending over his entire career, was to account for a cell's organization.[1]

Pursuing this path, Porter encountered two major hurdles. In contrast with Palade, his persistence in making observations meant he resisted another burgeoning field of science. In the 1940s, biological materials were being described in terms of their chemical composition. To him, this diverted attention from their spatial arrangement in biological systems, which he regarded as unfortunate "because we cannot hope to comprehend the activities of the living cell by analysis merely of its chemical composition or the properties of its component molecules." Porter believed electron microscopy would furnish the atlas that would then attract "devoted users who will seek to relate the form and function of cells at the macromolecular level."[4]

The microscope itself was a formidable hurdle for biologists accustomed to inquiring into life processes. Despite being forced to study dead cells—their vital fluid replaced with harsh chemical fixatives and then whittled into diaphanous slices—Porter used the preserved specimens to focus on how spatial relations might suggest movements and working mechanisms. He first gathered, seemingly at random, still images from the electron microscope to become acquainted with submicroscopic elements. The more cells he photographed, the more he found the same or similar elements in plants and animals and in patterns that appeared universal. While gathering an overall sense of their organization, he also worked toward getting a sense of living cells in perpetual movement that were continuously developing and responding to environments. Some of the cellular processes that attracted him were fibrogenesis, mitosis, tumorigenesis, glycogenesis, and chondrogenesis.

While Porter and Palade were making extensive studies on the "lace-work of densities" in the endoplasmic reticulum (see Chapter 7), the laboratory was also performing many original studies on other cell structures. They included chloroplasts (first with Sam Granick and followed by many others in the laboratory), striated muscle (with H. Stanley Bennett), cilia (with Don W. Fawcett and a little later with Peter Satir), protozoa (with Maria Rudzinska and Albert Sedar), fibroblasts (with George Pappas), neurons (with Sanford Palay), the nuclear pore complex (with Michael Watson), chromosomal structures (with Montrose Moses), the special form of the endoplasmic reticulum in muscle cells (with Lee Peachey), rat ovum (with J. Roberto Sotelo), seminal vesicles (with Helen Deane), and viruses (with Sam Dales).

During this intense period of cell structure observation, Porter organized a conference on tissue fine structure; in January 1956 more than a hundred participants met at Arden House in Harriman, New York, under the auspices of the Morphology and Genetics Study Section of the National Institutes of Health. This was the first conference on the new cell biology based on electron microscopy, and it was arranged in part to counter the criticism that biologists were indulging in a freewheeling approach to this field. So many miscellaneous observations were beginning to appear on such a variety of materials that Porter felt it necessary to examine collectively "standards of quality, of terminology, and of the significances of micrographic evidence." By sorting out the similarities and finding patterns in these materials, he expressed the hope the meeting would "lead to the formulation of several new 'laws' of cell morphology which may help us grasp more readily the significance for the living unit of these various, newly defined structural elements."[5]

In essence, Porter was working toward a day, still in the future, when higher resolution microscopes could display living cells at the molecular level. In the interim, he invested a special interest in cataloging elements of fine structure and helping to prepare detailed cellular maps. He collaborated with many physicians, microbiologists, and anatomists who were eager to learn techniques of electron microscopy and curious to garner interior configurations in a variety of cell types. As the following stories in this chapter relate, knowledge of the cell expanded with every undertaking during a momentous decade of modern cell biology.

9A. Chloroplasts

Philip Siekevitz

Although a biochemist, Sam Granick is included in this story of cell biology because of his work on chloroplast structure and function in photosynthesis. The focus is not on his more extensive chemical and biochemical research on pigments found in plant and blood cells but on his studies of chloroplasts, or plastids, the plant organelle containing chlorophyll.

Granick received his PhD in 1936 from the department of botany of the University of Michigan, where, as a teaching fellow in 1938, he published a milestone paper on the isolation of chloroplasts from green leaves.[1] By grinding leaves with sand in a hypertonic sucrose solution, he could isolate chloroplasts by a process of several low speed differential centrifugations. This was the initial step in determining their chemical constitution. Also, by measuring spectrophotometrically the amount of chlorophyll extracted by acetone from this preparation, he could estimate the recovery of the chloroplasts from leaves by comparing this value to that extracted from the starting leaf preparation. It is interesting that his use of hypertonic sucrose was used a decade later by George Hogeboom, Walter Schneider, and George Palade in their isolation of mammalian mitochondria (see Chapter 4).

That same year, 1938, Granick became a volunteer in The Rockefeller Institute laboratory of physical chemist Leonor Michaelis. Granick had inquired whether he could study the probable chlorophyll protein complex and its similarity to the Warburg yellow respiratory enzyme. He stated his willingness to work along the lines of Michaelis's ongoing investigations on free radicals in oxidation-reduction reactions, since they would be of value to his contemplated investigation of the chlorophyll complex.[2] In his reply, Michaelis wondered whether his "restricted space would allow the starting of a new branch of research within the limited facilities. However, good will and cooperation may make up for other deficiencies. So, I shall be glad to have you in my laboratory."[3] After Michaelis retired in 1940, Granick began to investigate the chemistry and biosynthesis of heme, the red pigment in blood, and chlorophyll, the green pigment in plants. Because plant tissue is easier to obtain in quantity than bone marrow, he studied chloroplasts, using mutants of the green alga *Chlorella vulgaris*, and eventually was able to demonstrate the relation between biological oxidation in animals and photosynthesis in plants.

With the advent of the electron microscope at the Institute, Granick entered modern cell biology at its infant stage. In 1947, together with Keith Porter, he examined the inner structure of the chloroplast, with a particular focus on the "grana" that had been vaguely visualized as tiny dark specks by several investigators using the light microscope.

The Granick/Porter methodology consisted of isolating the grana by first macerating leaves in a blender in either hypotonic NaCl or phosphate buffer (pH 6.5) solutions. This procedure effectively broke up the chloroplasts, freeing the inner constituents from the enveloping chloroplast membrane. The suspension was spun down at low speed to remove debris, and the supernatant was then spun at higher speeds. Phosphate buffer was added to the pellet, and the resultant suspension was placed on a plastic film on top of a wire mesh screen and allowed to dry. The specimen was then examined in an EMU RCA electron

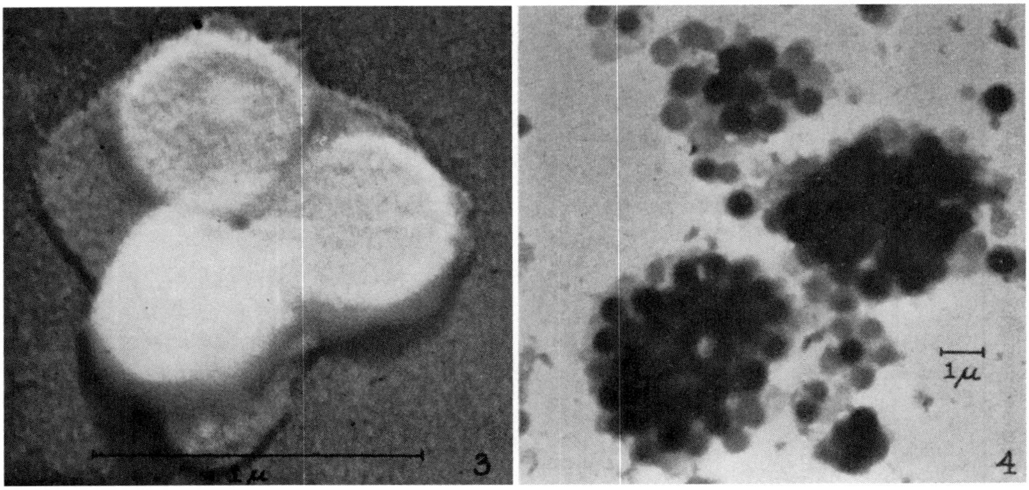

1. Left: electron micrograph of a cluster of three grana shadowed with gold. The grana are held together by a film of matrix and have a diameter of about 6,000 ångströms. ×48,000. Right: electron micrograph of spinach chloroplasts and chloroplast fragments showing the uniformity of the grana in a chloroplast and the variation in density of individual grana in the same chloroplast. ×7,800. From Granick and Porter, *Am. J. Bot.* 1947. 34:545–550. ©*American Journal of Botany.*

microscope belonging to the International Health Division of The Rockefeller Foundation (described in Chapter 3) at magnifications from 8,000 to 48,000 (figure 1).[4]

They questioned whether the grana were definite morphological units or precipitation artifacts; whether the fragments isolated by differential centrifugation were still capable of releasing oxygen, i.e., photosynthesis; and whether a chlorophyll protein compound existed in the chloroplast.

The grana were quite uniform, circular, about 6,000 ångströms in diameter and 800 ångströms thick, and appeared to be embedded in a matrix. Granick and Porter estimated that each chloroplast contained forty to sixty grana stacks. They investigated the results of various treatments of the preparation, such as the extraction of the fat-soluble colored pigments by methanol, and found that the grana were much less thick, having lost about half of their weight, confirming an earlier observation with the light microscope. It was evident that the pigments and the proteins each contributed about one half the weight of the grana. They tried to ascertain whether the chlorophyll was contained in the grana or in the surrounding embedding matrix, but no definitive answer was forthcoming either from their own or other people's work.

Although this 1947 paper left open many questions, it was notable for being the first to present biochemical data linked to a plant biological structure visualized by electron microscopy. It took many years, and the improved methods by Porter and Palade of fixation and thin sectioning, until investigators verified that grana are real entities composed of stacked membrane-bound vesicles with their outer membranes fused together; they were also found to contain chlorophyll bound to proteins, thus having the apparatus for the light reactions of photosynthesis.

Two years after the 1947 paper, Granick twice reported on the morphology of the chloroplast while again showing electron micrographs. In addition to listing the chemical constituents of the plastid then known, he used his interpretation of the micrographs

Sam Granick, 1952. Courtesy of The Rockefeller Archive Center.

to speculate on a specific chlorophyll-protein complex involved in the initial steps of photosynthesis and on the non-Mendelian characteristic of chloroplast "heredity." His gift of assembling a wide range of information and the successful reception of his perspectives by the plant physiologists were very likely what prompted him to write many reviews in subsequent years on chloroplast structure and function. Even though the pigments were his primary laboratory focus, he retained a desire to educate the plant people on the importance of relating structure to function.[5]

For many years, scientists had been finding chemical similarities between hemoglobin in animals and chlorophyll in plants. In 1948, Granick confirmed this resemblance when he demonstrated that protoporphyrin is a precursor of chlorophyll synthesis.[6] Based on this finding, he began in his 1949 Harvey Lecture to build a hypothesis concerning the evolutionary relationship between biological oxidation in animals and photosynthesis in plants.[7] According to David Mauzerall, his long-time colleague, evolution was Granick's favorite topic during the laboratory's daily four o'clock tea. There, he described how all biological observations could be interpreted by evolution, including the biosynthetic pathway of heme and chlorophyll. His statement that "bio-synthesis recapitulates biogenesis" eventually led him to the concept that chloroplasts of plants were inherited by an invasion of a photosynthetic organism into a eukaryotic cell, a concept now widely accepted.

Because of the later use of the green alga *Chlamydomonas* by Rockefeller cell biologists George Palade, David Luck, Günter Blobel, Nam-Hai Chua, and me, it is important to recognize the early work of Granick and Ruth Sager on this model organism. Granick's interest in establishing possible non-Mendelian inheritance of chloroplast traits meant he needed to obtain pigment mutants. Ruth Sager had received a PhD degree from Columbia after studying corn genetics under Marcus Rhoades. This was followed by a National Research Council fellowship to work with Granick in 1949. She spent the first three months taking Cornelius B. Van Niel's famous microbiology and photosynthesis course at the Hopkins Marine Station in Pacific Grove, California. There, she became interested in the unicellular alga *Chlamydomonas reinhardi*, especially because of the ease in obtaining mutants that are defective in various ways. The nutritional studies of Granick and Sager with this organism set the methods for getting this *Chlamydomonas* system working.[8,9] During this time, Sager also collaborated with Palade in making an electron microscopy study of the structure of the wild-type and mutant yellow strains of these organisms.[10]

In 1955, eight years after the first electron micrographs of chloroplasts, Granick brought together the results of improved techniques by Porter and Palade in electron microscopy, including newer micrographs by Palade and Sager, and was able to update his interpretation of chloroplast biology. The images clearly showed the inner membranous nature of the chloroplast, the grana, and the size of the membranes and of the spaces between them. Granick then inferred from evidence of chloroplast development that "plastids are derived only from preexisting plastids" and that "plastids are generally inherited only from the cytoplasm of the egg cell; i.e., maternal inheritance is the rule." Citing other evidence for the presence of chloroplast genes, he claimed that both DNA and RNA are present but cautioned that more biochemical evidence was needed, in addition to the histochemical evidence, to firmly ascertain the presence of DNA.[11,12]

During the 1960s, methods were perfected for isolating purified chloroplasts. Biochemical and histochemical analyses by others confirmed Granick's conclusions that chloroplasts are contained in a semipermeable double membrane and that plastids contain chlorophyll and are active in photosynthesis.[13,14]

In his last experimental cell biological studies, Granick and associate Aharon Gibor demonstrated that the plastids and mitochondria contain DNA and RNA, that these organelles are self-duplicating bodies that do not arise de novo, that the DNA represents a multi-genetic hereditary system not derived from the nucleus, and that these organelles are controlled by an adaptive mechanism that in the plastid is responsive to light and in mitochondria is responsive to oxygen.[15,16]

Granick devoted the final ten years of his life to the biosynthesis of heme and chlorophyll, becoming a leader in this field. His important work in cell biology, however, anticipated the present day thought that both organelles are semiautonomous in that DNA is responsible for the synthesis of only some of their proteins.

9B. Protozoa

William Trager

Maria A. Rudzinska was among the first to apply the techniques of electron microscopy to study the cell biology of parasitic protozoa and of parasitism. Her elegant structural studies reflect a valuable collaboration with Keith Porter in the earliest days of electron microscopy at The Rockefeller Institute.

Rudzinska was born in Dabrowa, Poland, and in 1929 received a PhD in zoology from Jagiellonian University, in Cracow. She had begun studies in Lvov on her favorite research animal, the unicellular protozoa *Tokophrya infusionum*, when the Nazis invaded. All her laboratory records were lost. Although both she and her husband, Aleksander Rudzinski, a scholar in international law, were Jewish, they managed somehow to elude the Nazis, and later the Russians, throughout World War II while they participated in underground teaching. In 1945, with the establishment of a democratic government in Poland, Aleksander was sent to New York as a delegate to the United Nations, and Maria went with him. When the Communists took control of Poland a short time later, Rudzinski refused to vote as they instructed him to. He resigned, and he and Maria asked for and were granted asylum in the United States. During this period, Maria worked with Robert Chambers in the biology department of New York University.

In 1952, Rudzinska received a grant from the National Heart Institute to resume work on the life span and aging of the *Tokophrya*. She wanted to do this research with Porter at Rockefeller after hearing him speak about how the electron microscope was the essential tool for looking at the fine structure of cytoplasm. She wrote Porter that she wanted to use his microscope to learn what "differences exist in the fine cytological structure between young and old, overfed and starved individuals, and those kept on a meager or deficient diet."[1]

Unlike most protozoa, each individual *Tokophrya* has a defined life cycle consisting of a period of infancy, a period of youth, a mature reproductive period, and finally senescence. Reproduction is by endogenous budding, resulting in the formation and release of a free-swimming ciliated form. This swims about for a short time and soon metamorphoses into an ovoid body with a stalk attached by a fibrous disc to a solid substrate. From the body grow several bundles of fine straight tentacles. These are the organelles that catch free-swimming ciliated protozoa, the specific prey of *Tokophrya*. They will not catch flagellates, only ciliates. As Rudzinska once wryly remarked, *Tokophrya* can distinguish between cilia and flagella much better than can biologists. The tentacles not only snare the ciliate, they also paralyze it, and it is through the tentacles that the contents of the ciliate are taken up into the body of *Tokophrya*.

During Rudzinska's three years with Porter, the two obtained some new details on the organism's submicroscopic structure. She cultured the *Tokophrya* directly on Formvar-coated microscope grids and fixed them with osmium tetroxide, the same procedure Porter used for culturing tissue cells. The cytoplasm was found to have an endoplasmic reticulum, mitochondria both inside food vacuoles and in the cytoplasmic matrix, and a honeycombed

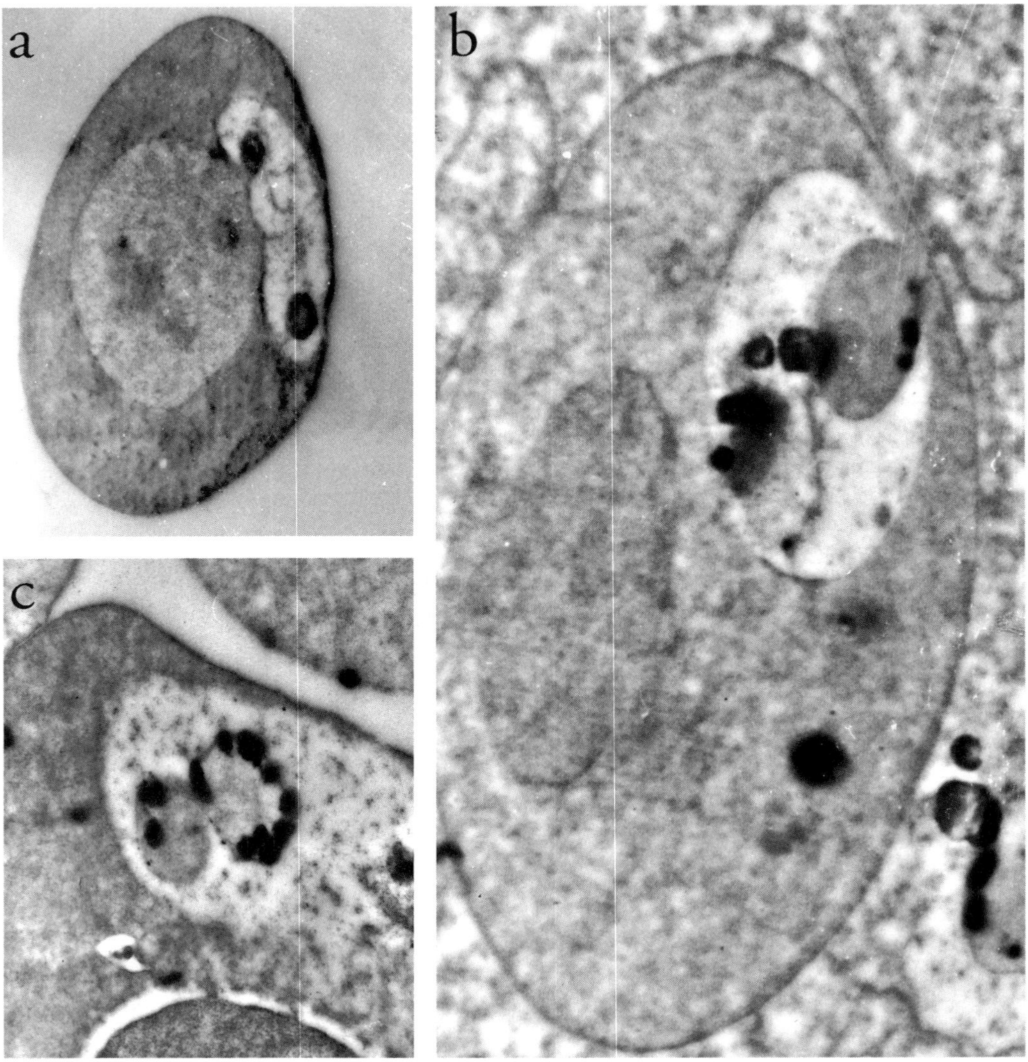

1. Intracellular phagotrophy by the avian malaria parasite *Plasmodium lophurae* in duck erythrocytes as seen by transmission electron microscopy of *Amblystoma* larva, all at ×20,000. a) A young parasite, adjacent to the nucleus of the host erythrocyte, showing the beginning of formation of a food vacuole as well as another small food vacuole. Note that both show the same density as the erythrocyte cytoplasm. b) A large parasite with a food vacuole already having a few small pigment grains but still connected to the erythrocyte cytoplasm. c) Two adjacent food vacuoles, one with few pigment grains and a matrix of about the same density as the erythrocyte cytoplasm, and the other with numerous pigment grains and a matrix of lighter density resulting from the digestion of the hemoglobin. Figures a and c: From Rudzinska and Trager, ©1957. *Journal of Protozoology*. 4:190–199. Figure b: from Rudzinska and Trager, 1959. *Triangle*. 4:97–104.

collection of chromatin granules in a macronucleus with a micronucleus attached to its surface. The electron microscope view of the feeding tentacles showed a sheathed system enveloping a system of long fibers. Using the newly designed Porter-Blum microtome, they were able to look at cross sections of these tentacles and found over forty fine filaments at the distal end of each one, for which they concluded the filaments created a peristaltic action as they carried food from prey into *Tokophrya*. In regard to the aging process, they found an accumulation of lipid particles and some structural abnormalities in mitochondria.[2–4]

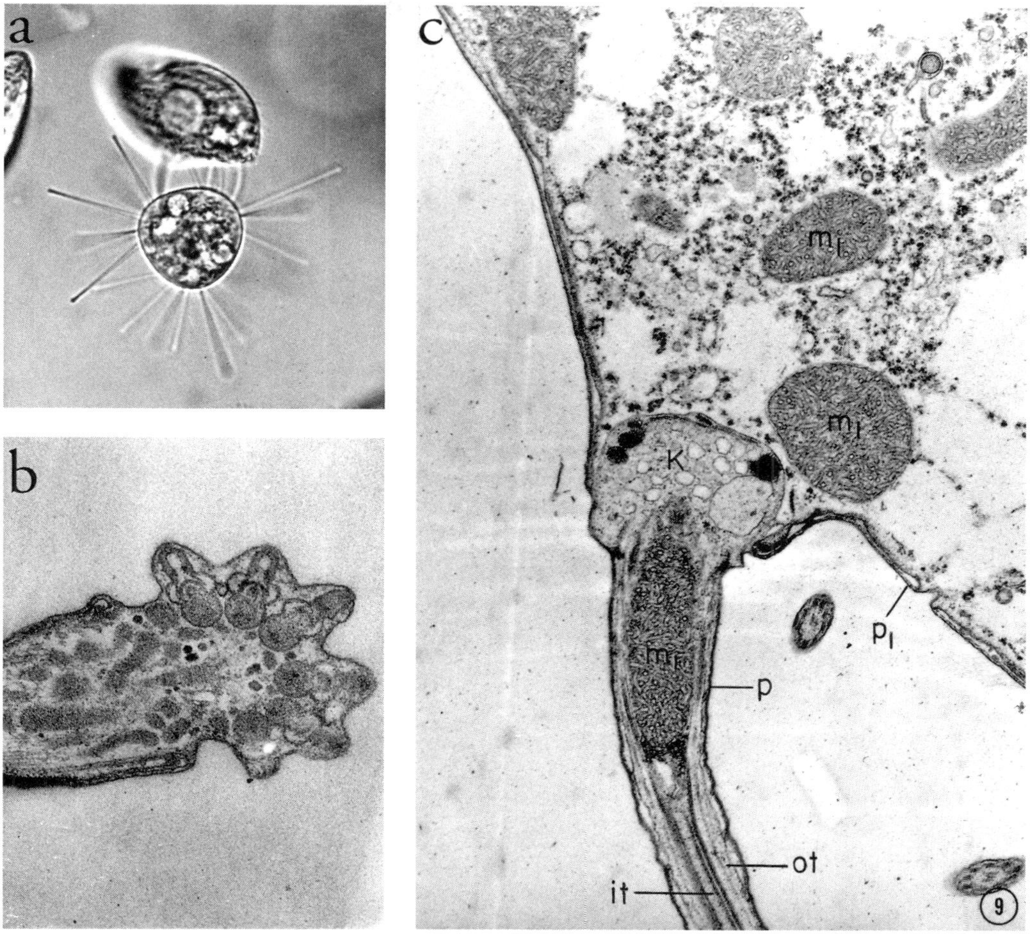

2. The feeding mechanism of *Tokophrya infusionum.* a) Photomicrograph of a living *Tokophrya* (bottom organism) that has just captured a ciliate (*Tetrahymena*) that is held by several tentacles (×500). b) Electron micrograph of a section through the knob at the tip of a tentacle of a non-feeding *Tokophrya* to show the array of the missile-like bodies (×51,800). c) Electron micrograph of a section to show the knob (K) of a tentacle of a feeding *Tokophrya*. Note that the pellicle of the tentacle shaft (p) is continuous with the pellicle of *Tetrahymena* (p1). Several mitochondria (m1) can be seen in *Tetrahymena*, and one is in the inner tube (it) of the tentacle on its way into the body of *Tokophrya* (×19,500). (ot, outer tube of tentacle). Figures a and b: from Rudzinska, 1973. Do *Suctoria* really feed by suction? *BioScience*. 23:87–94 with permission from the University of California Press. ©American Institute of Biological Sciences. Figure c: from Rudzinska, 1967. *Trans. NY Acad. Sci.* 29:512–525 with permission from the New York Academy of Sciences via Copyright Clearance Center. ©New York Academy of Sciences.

In the 1950s it was still the custom for most of the Rockefeller faculty to take lunch in the Welch Hall dining room, a place famous for its scientifically productive lunchtime conversations. It was one of these conversations that brought about my long-term association with Maria Rudzinska, which helped to shape both our careers. One day in 1955, we got into a discussion with Palade about how malaria parasites can enter a red blood cell without disrupting it. They soon realized that electron microscopy, still a new approach to any parasitic organism, might provide some answers. With support from the National Institutes of Health, Maria joined my laboratory, where malaria and other parasitic protozoa were being studied. Her first electron micrographs of the bird malaria *Plasmodium lophurae* revealed a

Maria Rudzinska, 1980.
Courtesy of The Rockefeller Archive Center.

surprising discovery. They showed clearly how malaria parasites within their host erythrocyte feed by a process of intracellular phagotrophy (figure 1). They ingest erythrocyte cytoplasm into food vacuoles, within which the hemoglobin is digested with the formation of the malarial pigment hemozoin. The enigma of how a large molecule like hemoglobin was used by the malaria parasite was easily explained by a few good pictures. Rudzinska, and later others, went on to elaborate on this discovery and extend it to other species of malaria parasites, including those causing human malaria.

Rudzinska and I continued to collaborate on studies of the fine structure not only of malaria parasites but also of other parasitic protozoa, including the intraerythrocytic parasite *Babesia* and kinetoplastid flagellates of the genus *Leishmania*. Among findings of special interest were the fine structure of the dyskinetoplastic state of hemoflagellates and a description of the invasion of hamster erythrocytes by *Babesia microti*. This closely resembles the invasion of erythrocytes by malaria parasites in its early stages but later differs sharply in that the invaginated erythrocyte membrane disappears. Thus, the babesial parasite lies directly in the cytoplasm of its host cell, separated only by its own plasma membrane, whereas the malarial parasite lies in a parasitophorous vacuole and is separated from the host cytoplasm by the parasitophorous vacuolar membrane as well as its own plasma membrane.

The work on *Babesia* led to a collaborative study with Andrew Spielman and his colleagues at Harvard on the developmental stages of *B. microti* in its vector, the deer tick *Ixodes dammini*. The study revealed the gametes of *B. microti* and a remarkable arrowhead-shaped organelle that clearly facilitated penetration of the peritrophic membrane of the tick alimentary tract by the *Babesia* zygote. This work, in turn, led to a further collaboration with Spielman's group on development of the spirochete of Lyme disease in the same vector. Here they demonstrated spirochetes in the salivary glands of the tick.

Throughout all this work with parasites, Rudzinska never abandoned her first love, *Tokophrya*. She found a source of the predaceous suctorian *Tokophrya infusionum* in a fountain pond on the Rockefeller campus. She established it in pure culture, feeding it an axenic culture of the ciliate *Tetrahymena*, and she and her students carried out important basic studies on all aspects of the biology of this organism. Rudzinska's beautiful electron micrographs of this remarkable feeding process go a long way toward explaining what goes on. Of special interest was her discovery of what she called the missile-like bodies. These unusual structures are positioned in the knob-like tip of each tentacle (figure 2). They are oriented with their narrower pointed end directed outward. When a ciliate chances to make contact with the tentacles of *Tokophrya*, it is caught by the tips; the missile-like bodies are clearly involved in this. The knob of the tentacle quickly enters into the body of the prey, and the ruptured membrane of the prey fuses with the membrane of the shaft of the tentacle. The cilia of the prey soon cease to beat, as if paralyzed. Perhaps the missiles are also responsible for this. Besides the missiles, dense bodies and vesicles also enter the prey. All of these structures move through an outer cylinder of the tentacle shaft. Meanwhile, the tentacle has shortened and thickened with the formation of an inner cylinder. Within this, membrane-lined food vacuoles move to carry in all the cytoplasmic contents of the prey, including mitochondria and other organelles (figure 2). It is the missiles that make first contact with the prey, and their complex compartmentalized structure suggests they probably carry out several functions. It will be great fun to understand the molecular biology of these structures and to know what substances they contain; they will surely be interesting, and they might even be useful.

9C. Collagen

George D. Pappas

By the mid 1920s, improvements in cytological techniques had plateaued at the point where light microscopy could function essentially at its theoretical limit of resolution and still provide a good deal of information about the inner structure of the cell. Other physical methods, especially x-ray diffraction, were able to reveal further structures beyond the resolution of the light microscope. With the advent of the electron microscope, it became possible and very exciting, as Francis O. Schmitt characterized it to me, to actually verify the accuracy of x-ray diffraction in real images such as the periodicity of collagen. The structure of teased collagen fibers was first described in the electron microscope by Francis Schmitt, Cecil Hall, and Marie E. Jakus in their classic 1942 publication.[1]

A major contribution to the birth of cell biology was the development of techniques to utilize the high resolution of the electron microscope in studies of the inner geography of the cell itself. Before techniques of embedding and sectioning cells and tissue were developed, the approach was to look at whole cells dried down onto a substrate. These early attempts were unsuccessful, as the cells were too thick to reveal any information about their contents. For example, when Keith Porter and his co-workers turned their attention to finding the virus that causes cells to become malignant, they were not entirely sure they had found the virus particles or the cause of the rat sarcoma. And not believing that he could get any better resolution of the cytoplasmic structure within the whole cells, Porter turned his attention to the structure of fibrin precipitates of fibrinogen.

Fibrin Fine Structure

As part of their search to develop a synthetic medium for the Tissue Culture Association (see Chapter 5), Porter and his postdoctoral fellow Clinton Hawn also examined the fine structure of the fibrinogen-to-fibrin clot formation with the electron microscope. What they saw were clots in a meshwork of single and compound fibers with an easily observed cross-striation of the unit fibers having a constant periodicity. They concluded there was a precise coincidence of the striations of the individual fibers when these are associated side by side to form compound fibers.[2,3]

Thus, they followed up observations made in 1940 by Helmut Ruska and Carlheinrich Wolpers at the light microscope level on the conversion of fibrinogen to fibrin.[4] With the electron microscope, it could now be clearly seen that fibrinogen became polymerized by thrombin to form what Porter described as "needle-shaped, crystal-like protofibrils, which then become aligned into fiber strands by lateral association." He even proposed a model of how fibrin formed—a theory that was closely compatible with the reported physical constants and data from x-ray diffraction studies. The resolving power of the electron microscope was an obvious advantage for studying these issues.[5]

Collagen Fine Structure

Following up on this work, Porter and Parker Vanamee, a postdoctoral fellow of the American Cancer Society, began to look at collagen solutions. They found that fibrin clots seemed to be branching to form interconnected filaments without the direct intermediation of other cellular elements. Porter, however, became more interested in the formation of the non-branching collagen fibers. While fibroblasts synthesize the protein collagen, he wanted to determine what role they play in how the fibers are organized in the extracellular space. He had observed earlier that collagen fibers are not interwoven but rather are layered in a mat-like formation.

Porter's group examined with the electron microscope the solvation and reconstitution of collagen filaments in various collagen-forming tissues, providing more details of these processes. Their controlled conditions led them to conclude that the filaments form fibers, which then associate laterally and longitudinally to form the larger fibers with well-defined periodicity, a finding similar to the one found by Gunter F. Bahr in 1950.[6–8]

When I came to Rockefeller in 1952, I had just received my PhD in zoology from Ohio State University for structural and cytochemical studies of the cytoplasm in the family *Amoebidae*. After I spent a few weeks that summer at the Marine Biological Laboratory in Woods Hole, Massachusetts, Helen Deane, who had worked with Porter a few years earlier, recommended that I contact him for a position. Keith was interested in my graduate experience in culturing amoebae, but he decided I would have to learn about electron microscopy and pursue the problem of collagen morphogenesis in fibroblasts before turning to the fine structure of amoebae. For this work, he helped me obtain a postdoctoral fellowship from Eli Lilly.

Porter's approach to the laboratory was such that he did not turn me over to a technician to show me things. He wanted to show me himself, and he had a marvelous pair of hands. He could do virtually anything with his hands. He showed me how to cut up embryonic chick skin and isolate the areas of the dermis that had fibroblasts and how to embed the primary culture pieces of connective tissue in a clot of rooster plasma, a very interesting procedure. He also showed me how to bleed a rooster wing vein. One day, the rooster got loose and Keith, upset, turned to me and said, "I did that to show you what happens if you don't hold the rooster right." I got the message.

Keith then had me set up primary cultures of embryonic chick dermis fibroblasts that I grew on one side of a porcelain filter in order to determine whether a soluble unprecipitated collagen could be formed on the side away from the cells in culture. I could not find any formed fibers on the noncellular side of the filter. We concluded that the formed fibril of collagen (which is about 25 nanometers in diameter) must be somehow precipitated and formed on the cell surface of the fibroblast. The orientation of the unit fibril was very precise in that we found that the newly precipitated fibril just outside the cell was parallel along the inner "stress fibers" or fibroglial fibers of the cytoplasm. (These are what we have now identified as the intermediate filaments of the cytoskeleton.) Once the unit fibril has been formed, apparently unprecipitated soluble collagen may then precipitate onto the formed unit fibril and thus increase the diameter of the fibril. We measured fibrils "growing" from 25 to 80 nanometers in width, and I showed that the fibroblast cytoplasm contains granules

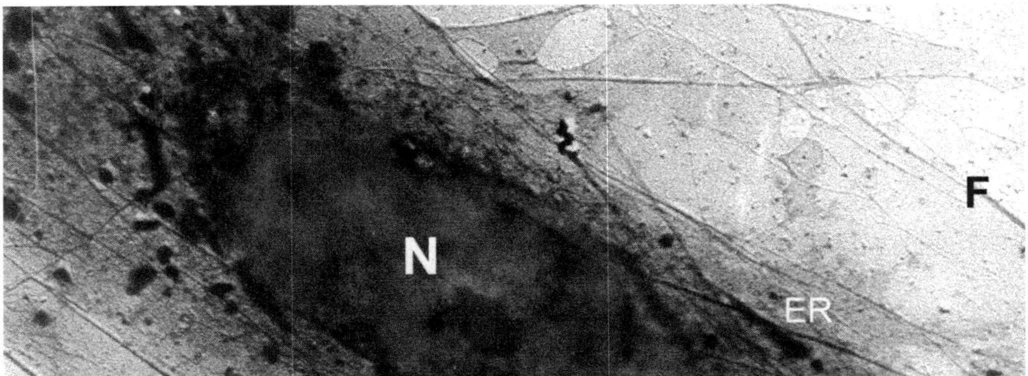

1. Electron micrograph of portions of three fibroblasts (1, 2, and 3) from a seventy-two–hour-old culture of a skin explant from a nine-day-old chick embryo grown on a Formvar-coated coverglass. The cells were fixed for ten minutes with vapors from a two percent aqueous solution of OsO_4. The preparation was then washed in H_2O, transferred to an uncoated electron microscope grid, dried, and then lightly shadowed with metallic chromium to enhance the contrast of the fine fibrils on the cell surface or on the Formvar support substrate. The details of this technique were described by Keith Porter in 1953.[12] Note that the fine collagen fibers (F) run in parallel array along the surface of the fibroblasts. The parallel array of collagen fibers on the fibroblast surfaces is continuous from cell to cell. N, nucleus; ER, endoplasmic reticulum; M, mitochondrion. Courtesy of George Pappas.

that have the tinctural properties of collagen (i.e., staining green with Masson's trichrome stain and blue with Mallory's connective tissue stain).

When we examined the fibroblasts with the electron microscope, we found that these were not discrete granules but rather concentrated portions of the endoplasmic reticulum (ER). We were not able to find any formed fibrils in the lumen of the ER. Obviously, fibrils as such only formed or precipitated on the surface of the fibroblast plasma membrane and not within the cytoplasm. We did not know how the soluble collagen was released or secreted by the fibroblast. Porter suggested that a kind of "shedding" or ecdysis occurs at the cell surface.[9] At that time, we did not know the secretory pathway of proteins that were synthesized by the endoplasmic reticulum into the milieu. It was not until 1963 that

Elizabeth Hay and Jean Paul Revel used tritiated proline and hydroxylproline as precursors for collagen, as these amino acids make up more than one third of the protein collagen. They clearly demonstrated the presence of these amino acids in the lumen of the ER. They followed this labeling from the rough ER to the Golgi apparatus in chondrocytes in the same way that Philip Siekevitz demonstrated other secretory proteins formed in pancreatic acinar cells.[10]

It is clear, however, that formal collagen fibers per se are never seen within the cell, but only after they are secreted and then precipitated in the extracellular matrix. The first unit fibers are seen in very close proximity, if not actually up against, the plasma membrane. After a unit fiber has precipitated, the soluble protein, collagen, called procollagen, can accrete or stick on the already precipitated fiber some distance away from the cell. The important aspect of this is that the orientation of the formed mats of collagen fibers is dictated by the form of the cell itself. In other words, the formed fibers reflect the disposition of the intermediate filaments within the cell, i.e., the cell cytoskeleton. These were originally called the stress or fibroglial fibers. Thus, the cell and the extracellular matrix are in reality a well-organized unit or a continuum. Soon after the Porter-Blum microtome was developed and the use of methacrylate embedding was employed, I noted shrinkage of the collagen fibers and a consequent shortening of the normal 640-ångström spacing. Fiber diameters do increase in time because, as I mentioned earlier, a precipitated fiber can increase in girth by the precipitation of newly secreted soluble procollagen. This can take place at some distance from the cell in the extracellular matrix. However, the pattern of the matting of the collagen is initially determined by the disposition of the fibroblast itself.

Keith was a stickler for writing and wanted things to be absolutely perfect. I thought he really delayed writing up this work. It was finished in the spring of 1954 but not published until 1959 just because he wanted to write the very best prose.[9] When it came to editing, he would ask who wrote a certain part; if I said he did, then he did not find too many things wrong with it; but if I said that I did, he found a few more things to change. Sometimes I would change things around to have a little fun with him. In the meantime, we studied the effects of anticollagen serum on collagen formation in tissue culture with Bill Robbins at Cornell, across the street from the Institute. This was the first time that antibodies to collagen had been found.[11]

By the late 1950s, it appeared to us as an important observation that the intracellular elements of the fibroblasts and the extracellular matrix surrounding these cells were functionally a single entity. The organization of the shape of the cell as determined by the array of the cytoskeletal intermediate filaments is congruent with the arrangement of the precipitated collagen fibers of the extracellular matrix. Of course, these fibrils do not penetrate the plasma membrane, which clearly defines the boundary of the cell (see the electron micrograph in figure 1, which was taken during that period of chick fibroblasts in culture). Parallel arrays of collagen fibers can be seen crossing portions of three cells.

My interest in amoebae lingered, but I was kept so busy learning good microscopy that there was no chance to study them. Of course, the laboratory had other protozoologists: Maria Rudzinska, who worked with *Tokophrya*, Al Sedar with *Paramecium*, and Jerry Wolken with still other protozoa. Soon after I left the laboratory and went to New York University, I was able to start my work on the fine structure of amoebae. Even after I moved to Columbia

University, I would visit Keith once a week during the laboratory's ritual teatime, bringing new electron micrographs. Keith would examine each one carefully, making no remarks at first. After a while he would ask me questions about what I thought of a certain structure or what technique I had used to prepare the cells. He did not elaborate or make declarations, just ask questions. In that way, he would have me see what he thought was significant in a micrograph.

What seemed to me more important at the time was what was going on in the laboratory as a whole, and not that I was working on fibroblasts and collagen. We all sensed that this was a really critical period to find out the fine structure of cells for the first time. Every tissue brought new findings, and this was so exciting.

9D. Neurons

Sanford L. Palay

The spring of 1953 was an exciting time in the history of cytology, and the Porter/Palade laboratory at The Rockefeller Institute was at the center of events. Keith Porter was in the midst of correcting the proofs of a paper describing the endoplasmic reticulum. Don Fawcett and Porter had discovered the microtubules in cilia (they called them "fibrils") but did not get the numbers right at first. George Palade had described the architecture of mitochondria but was engaged in a running battle with Fritiof Sjöstrand about the configuration of mitochondrial cristae and their relation to the internal membrane of the organelle.

The laboratory occupied a large suite of rooms in the basement of Theobald Smith Hall, where Albert Claude's electron microscope and darkroom had been five years before. The space had been converted into spacious and well-designed laboratories. Porter and Palade each had a tiny cubicle for an office. However, there was no space for visiting investigators, which was the normal situation at Rockefeller. We shared a row of desks in a narrow internal passage near the electron microscopes and the photographic darkroom. At this time, other resident visitors were George Pappas and Maria Rudzinska. Don Fawcett came for days at a time but had no fixed abode. Irene Manton came for one of her periodic visits, and Audrey Glauert stayed for a couple of weeks, along with her director, Ernest M. Brieger. Herbert Gasser was frequently in attendance as he worked on the structure of unmyelinated fibers in autonomic nerves and olfactory nerves.

The laboratory was the center of technical advance and the source of new thinking, as well as discovery. Palade invited me to return for six months in order to work with him and learn the new techniques for fixation and embedding. As William Gardner had welcomed the new initiatives for electron microscopy in the department of anatomy at Yale, he readily gave his consent for me to take a leave of absence of six months with pay (but provided no money to support the additional expenses of living in New York City).

Palade and I started with the technical problems of preserving nervous tissue for electron microscopy. As he had demonstrated the architecture of the basophilic cytoplasm in gland cells such as hepatic and pancreatic cells, it was natural to extend this study to the basophilic cytoplasm in nerve cells, the classic Nissl substance. The standard method of fixing solid tissues such as liver for electron microscopy was to extract a small piece of an organ from the living anesthetized or freshly killed animal and dice it with a clean razor blade into small cubes in a drop of fresh, cold, one percent osmium tetroxide. Tissues like the liver or other abdominal viscera could be reached without excessive trauma and fixed relatively promptly after death. But the brain, encased in a bony capsule and meninges, took longer to reach, so the well-known vulnerability to hypoxia had time to alter the morphology of its cells. Besides, cutting it up into small blocks resulted in damage to large numbers of cells with long processes that were severed by the razor blade. The slow penetration of the chemical fixatives allowed plenty of time for artifacts to develop before fixation. Thus, the great advantages of osmic acid as a fixative were subverted by the characteristics of the tissue itself.

In view of these conditions, we began our attempts at fixation with the ganglia of the peripheral nervous system. The superior cervical and dorsal root ganglia were readily accessible in the living, anesthetized animal. They were enclosed, however, in rather thick connective tissue sheaths, which impeded the penetration of the fixative. It proved very difficult to obtain adequate fixation, and we finally resorted to dripping buffered osmium tetroxide over their exposed surfaces in situ while we snipped the ganglia out of the living animal, with its circulation intact. This method yielded at least some reasonably well-fixed neurons.

Neurons in the peripheral ganglia contain only small Nissl bodies and widely dispersed basophilic substance. In order to demonstrate the composition and architecture of the basophilic cytoplasm in nerve cells, it was necessary to examine the classic structures described by Nissl in motor neurons either in the spinal cord or in the medulla oblongata. Then there could be no doubt concerning the identity of the structure described by the new method. But the brain and spinal cord were much more difficult to fix than the peripheral ganglia. Discouraged by this difficulty, I did not pursue the method again until four years later, when the need for a better method was more pressing.[1]

It proved impossible to obtain satisfactory fixation of anterior horn motor cells by immersing the spinal cord in buffered osmic acid in the living animal. I carefully removed parts of the occiput and the atlas overlying the hindbrain of a rat, gingerly inserted a fine needle between the cerebellum and the medulla, and then removed some of the ventricular fluid and replaced it with three percent buffered osmic acid delivered from a 1-milliliter syringe. The method was successful in fixing the floor of the ventricle and the overlying cerebellar lobules. In a few animals treated in this way, I obtained blocks of tissue containing well-preserved large motor cells with typical large Nissl bodies. Purkinje cells could also be studied in the fragments of cerebellar cortex obtained in the same way.

These specimens from a variety of nerve cell types allowed me to construct a generalized portrait of the neuronal cytoplasm and its constituent organelles. Palade and I reported our findings first at the annual meeting of the Electron Microscope Society of America (as it was then called) in November 1953 and then again at the annual meeting of the American Association of Anatomists, in Galveston, in April 1954.[2] The full paper was published in the first number of *The Journal of Biophysical and Biochemical Cytology* in January 1955.[3] In this pioneering study, we showed that the Nissl substance consists of arrays of cisternae of the endoplasmic reticulum with ribosomes both attached to the outer surfaces of the membranes in complex polysomal formations and suspended as rosettes between the cisternae. We identified characteristic mitochondrial features, such as longitudinal cristae and dense granules, and we detected long, fine filaments about 10 nanometers in diameter, the first recognition of neurofilaments in situ in intact nerve cells (figure 1).

We also distinguished a second variety of membranes, which was not associated with ribosomal granules and which we called the agranular reticulum. Although some elements of the agranular reticulum were loosely and randomly dispersed in the cytoplasm, others were tightly packed into complex curved arrays similar to those identified in gland cells by Albert Dalton and Marie Felix at about the same time as fragments of the Golgi apparatus.[4] When we wrote our paper, we took the cautious approach that more electron microscope investigation was needed before these structures could be definitively identified. Our reluctance to admit the existence of the Golgi apparatus at that time is understandable, although

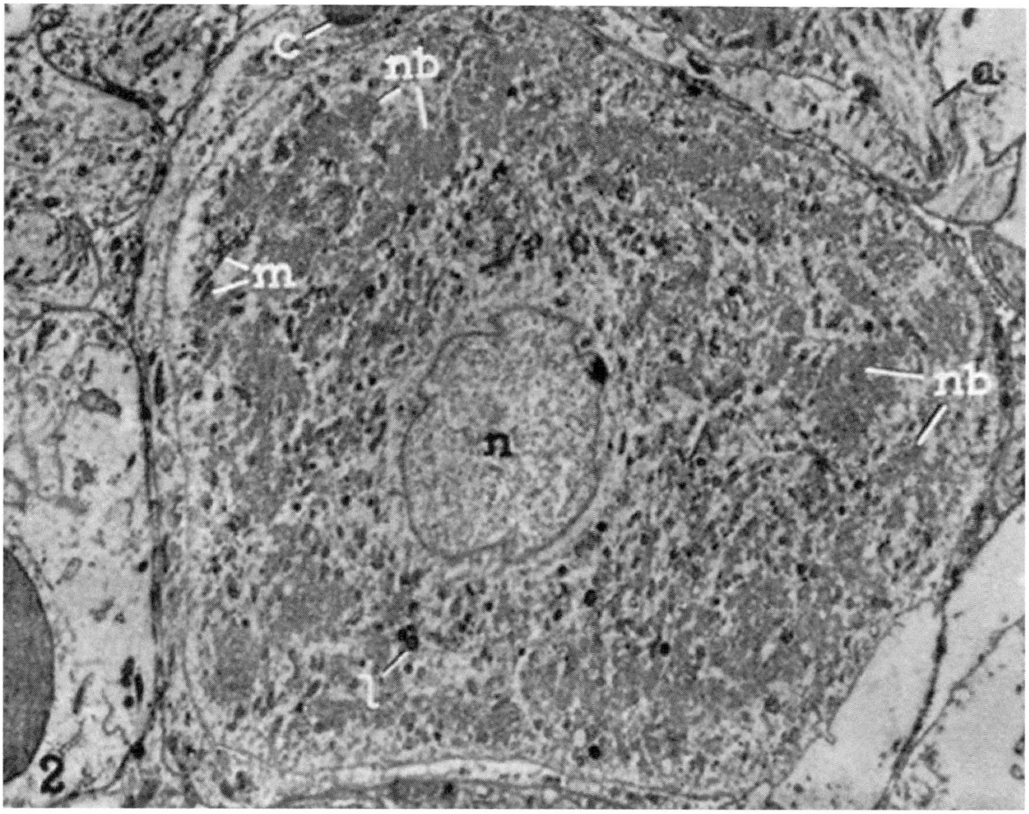

1. Electron micrograph of a section through a dorsal root ganglion cell. The section passes through the nucleus (n) without touching the nucleolus. Small, irregular Nissl bodies (nb), mitochondria (m), and lipid droplets (l) can be seen in the cytoplasm. At c is the nucleus of a capsule cell, and at a, an axon leaves a neighboring perikaryon. ×4,600. From Palay and Palade, 1955. *J. Biophys. Biochem. Cytol.* 1:69–88.

unfortunate. Within a year or so this error was corrected, and the electron microscopic appearance of the Golgi apparatus in nerve cells, as in other cells, was generally agreed on. But a more subtle error was made in not clearly separating the agranular reticulum from the condensed membranous arrays of the Golgi apparatus. After further study with improved methods, the agranular reticulum became a well-recognized entity, pervading the neuronal cytoplasm. It is especially easily noticed in the axonal and dendritic cytoplasm, where ribosomes are either absent or few. In many cells, it furnishes a characteristic cisternal system immediately beneath the plasmalemma, the hypolemmal cistern.

The motor cells in the abducens nucleus provided another important opportunity. Like most motor cells, these neurons are covered with synaptic terminals. If one is to find the site where the nerve impulse is transmitted from one neuron to another, this is the place to look for it. Numerous investigations by light microscopy over the previous six decades had given us a set of features that should be associated with the nerve endings: close proximity to neuronal or dendritic surfaces, a collection of mitochondria, and a more or less bulbous shape, an expansion continuous with a thin nerve process. On the surface of nerve cell bodies and dendrites in the abducens nucleus, I found palisades of nerve endings, exactly as expected: thin nerve fibers expanding into club-shaped terminals and containing many profiles of mitochondria. In addition, collections of small vesicles were aggregated against the presynaptic membrane, and both the presynaptic and postsynaptic membranes were thickened in patches along the shared interface. Most importantly, the confronted membranes were separated from each other by a thin intercellular space. Thus, direct observation confirmed Santiago Ramón y Cajal's inference derived from necessarily inconclusive evidence more than sixty years earlier.

I well remember the moment when I discovered synaptic terminals in my preparations. Palade had gone away on his annual vacation. It was a Saturday early in August, and I was alone in the laboratory. The first field that came into view when the microscope beam illuminated the screen contained collections of mitochondria. On close examination with the microscope, they could be recognized as lying inside terminals on the surface of a motor cell. I became very excited, and having no one with whom to share this great news directly, I telephoned George Pappas at home to convey my exhilaration. I was extraordinarily privileged to be the first one to see the synaptic junction, and I recognized my good fortune.

Naturally, these early discoveries were followed up by much repetition during the next months and years. When I submitted my abstract on the neuronal cytoplasm for the November 1953 meeting of the Electron Microscope Society, I had not yet found the synapses. But after Palade returned from vacation, we decided to alter my planned presentation in order to include a slide or two on the synapse. I devoted three minutes at the end of my paper to a description of the synapse. It fell on deaf ears. No one noticed. Eduardo De Robertis was in the audience and gave a paper later in the same session in which he described completely erroneously supposed synaptic junctions in the acoustic nuclei that were, in fact, merely thinly myelinated fibers adjacent to neuronal perikarya. Francis O. Schmitt, who presided over the session, commented at the end of De Robertis's talk that we now knew the structure of the synapse, completely ignoring what he and everybody else had heard a couple of hours earlier.

For the anatomists' meeting in Galveston in April 1954, Palade and I submitted two abstracts. One was on the fine structure of neurons, with me as first author, and a second, on neuromuscular and interneuronal synapses, had Palade as first author. The American Association of Anatomists had a rule that no one could be an author of more than one presented paper. When we submitted the abstracts, we asked Normand Hoerr, the secretary of the association, to allow us to present these two papers by the same two authors; thus we would not violate the spirit of the rule, as there would be only two papers and two authors. But Hoerr, who had been my anatomy professor at medical school, resisted our pleas and offered the compromise that although only one of us could be listed as the author of a paper, the other could be listed as a technical assistant. Thus it happened that I was accorded the extraordinary privilege of having a [future] Nobel laureate as my technician![5] When the study of the synapse was finally published as a full paper, Palade did not want to be coauthor, and my name appeared alone on it.[6] His interests lay elsewhere.

Both papers were well received by the anatomists, who had no trouble recognizing new and important findings on the nervous system. Probably the reason for the lack of response at the meeting of the electron microscopists is that no one in attendance besides us was interested in the specific cytology of the nerve cell.

These events began the modern study of neurocytology. In a relatively short time, all of the parts of the nerve cell and its satellite or supporting cells were distinguished and described in detail. At the same time, all of us were concerned with correlating structure with function. The composition of the Nissl substance agreed with the structure of basophilic cytoplasm in protein-secreting cells, such as those of the liver and the pancreas, and correlated easily with the high protein synthesizing activity of nerve cell somata. The discovery of a new organelle, the synaptic vesicles, called for a new function that had not been understood before. This new function was supplied by the quantal release of transmitter substances, which had been discovered during the previous year or two by Bernard Katz and his co-workers. This correlation gave rise to the "vesicle hypothesis," which states that neurotransmitter is stored in the synaptic vesicles and that the arrival of the action potential at the nerve ending induces the release of neurotransmitter from the nerve terminal by exocytosis. Discussion of the evidence for and against this hypothesis, which is still the generally accepted one today, would take me too far from the purpose of this article.

Chance played a large role in directing me to electron microscopy when I joined Albert Claude's laboratory. And it was a global upheaval that brought George Palade to Claude's laboratory at the same time that I was there. But once I had learned what was interesting to me, it was not chance that brought me along the path of discovery, but hard work and careful thinking. As Louis Pasteur said, chance favors the prepared mind.

Excerpted by Palay from his reminiscences in "A concatenation of accidents," *The Neurosciences: Paths of Discovery II.* 1992. F. Samson and G. Adelman, editors. Birkhäuser, Boston. 191–212. Printed with permission from Springer Science+Business Media B.V.

9E. Viruses

Samuel Dales

To complete the requirements for an honors degree at the University of British Columbia, I carried out a research project concerning the influence of the thyroid hormone on the embryonic development of the Pacific salmon genus *Oncorhynchus*. It turned out that exposure of the fertilized eggs to the hormone during development resulted in hatched larvae having deformed skeletal parts and eyes bulging from their sockets, reminiscent of the human hyperthyroid condition of exophthalmos. The mentor of this study, the internationally recognized physiologist William S. Hoar, deemed this result to be sufficiently original to merit a publication.[1]

This early experience of an independent research project encouraged me to pursue a long-term research career. Therefore, upon Hoar's recommendation, I enrolled as a predoctoral student at the University of Toronto, joining the laboratory of the physiologist Kenneth C. Fisher, whose major interest was the influence of narcotic compounds upon the oxidative metabolism of cells and tissues. This problem had been researched over the years using invertebrate material and brain and muscle from vertebrates. Upon my arrival in Toronto, a more defined object for analysis, immortalized mammalian cells in continuous culture, became available. The nutritional and physiological requirements for maintaining such cultures were defined by the pioneering studies of Raymond C. Parker and colleagues at the Connaught Laboratories on the University of Toronto campus.

Use of the continuous murine line of L cells enabled me to investigate in a controlled manner the influence of anaerobiosis induced by exposure to carbon monoxide and barbiturates and by other means upon the oxidative glucose metabolism and ancillary phenomena. It turned out that profound suppression of DNA, RNA, and protein synthesis under anaerobiosis was accompanied by a defined, dose-responsive fragmentation of the mitochondria, evident by phase-contrast microscopy.[2,3]

To demonstrate that the mitochondria had actually been fragmented, the treated cells were examined by means of thin sections in the electron microscope. For this purpose, I familiarized myself with the techniques of electron microscopy under the tutelage of Professor Allan F. Howatson in the department of anatomy at the University of Toronto Medical School. Howatson had trained in Britain as a physicist specializing in electron optics and had become interested in cancer research, specifically the application of electron microscopy for observing viruses and virus-infected cells related to malignancy. Combining my experience with cell cultures and knowledge about electron microscopy, I undertook a study on the life cycle of vaccinia, the virus strain used as a vaccine for smallpox virus. The virus-cell system enabled one to elucidate unambiguously individual phases from attachment of the inoculum to release. Howatson was acquainted with Keith Porter, who was also interested in cancer-related viruses appearing in cultured cells as evident in the electron microscope.

In 1960, The Rockefeller University wished to recruit an individual who would become the coordinator of the cell biology course and simultaneously represent a new area relevant to the interests of the department. Upon Howatson's recommendation, I was invited for an interview with Porter and the other head of laboratory, George Palade. I recall my interview with Porter as a long and grueling experience lasting maybe three hours. Toward the end of this encounter, he asked me, "Do you think you are (scientifically) any good?" to which the only response deemed modestly appropriate was to say that I had been "lucky" in some of the discoveries I had chanced upon up to that time. The interview with George was much shorter, mostly over lunch in the dining room in Welch Hall (it is sweet to remember the civilized habits then prevailing at this institution). Evidently I "passed muster" with George, for he made some positive gesture toward Keith about our interview. Following a discussion of a salary scale and conditions of appointment, my visit was concluded with the assumption that a formal notification would be sent to me in Toronto. I surmised that Porter's intention to include animal virology within cell biology stemmed from his long-term involvement with the Rous sarcoma agent, which he had recognized in whole mounts of cultured cells.[4] Palade's interest in animal virology was probably connected with his appreciation that viruses could be used as specific probes for elucidation of certain cellular structures and functions.

After some weeks had passed without receiving further news, I contacted Keith, who informed me that he had assumed that the appointment offered verbally was all that was required for the present. Naturally, appointment as a research associate was very welcome for me. However, documentation attesting this fact was essential for me to obtain an immigration visa from the consulate. It should be remembered that in the 1960s, due to the launching of Sputnik and related events, the United States was very concerned about being outperformed scientifically by the Soviet Union. As a consequence, the issuance of a permanent residency visa to foreign individuals in the sciences, like me, was a short and straightforward process.

My family and I arrived in New York in time for me to start work by 1 July 1960. At that time, the city was in the midst of a torrid humidity-laden heat wave with temperatures soaring into the 90s, climatic conditions to which neither I nor my family members had been accustomed. Despite the unbearable climate, I dutifully made my appearance in the basement of Smith Hall to commence work on that first morning of July only to discover that, with the exception of one or two junior investigators, I was the only person around. The department's complement would not return for some weeks, as it was their habit to escape the stresses of the summer weather in New York and the absence of adequate air conditioning at the work place.

By late August, when the staff had reassembled, we were surprised to learn that Keith was planning a move to Harvard and George would become the exclusive head of the laboratory. Active planning was under way to occupy new quarters on the fifth floor of the South Laboratory, now the Bronk Building, where I was provided with an office and laboratory facilities for propagating cells in tissue culture and for ancillary biochemical studies with viruses. My facility was also used for conducting laboratory exercises by students in the cell biology course.

A series of investigations that I conducted during 1960–1966 were concerned with interactions of several types of viruses with their host cells. The program of research with each virus type was carried out either independently or in collaboration with individuals in other laboratories, especially the animal virology laboratory of Igor Tamm. In each case, the

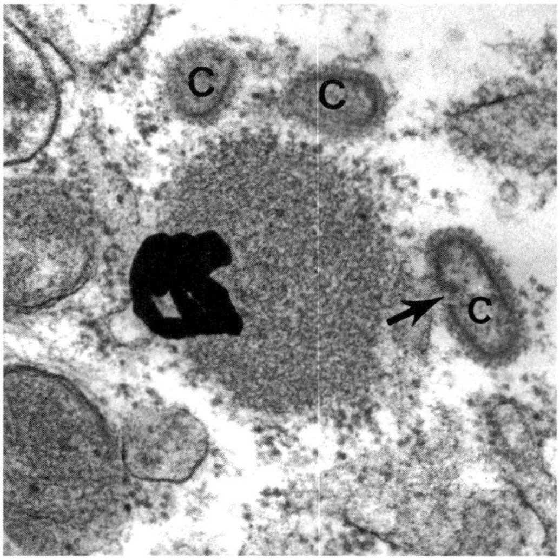

1. Electron microscope autoradiograph demonstrating the uncoating of inoculum vaccinia virus. Cells were infected with purified virus containing tritium-labeled DNA. Following release of the DNA from cores (C) through defined breaks (arrow), the genomic DNA starts to replicate in the cytosol, within defined "factories" containing nucleic acids and proteins. Factories, where the presence of genomic DNA is marked by the silver grain in this autoradiogram, are also sites for virus assembly. ×50,000. Courtesy of Samuel Dales.

quantitative and biochemical investigations were combined with morphological observations. The early events in the virus life cycle, about which little was then known, became one area of primary interest to me. My collaborators and I also explored certain aspects of virus assembly and dissemination of progeny particles involving the modification of cell membranes. These investigations are briefly described below under separate sections.

Vaccinia, a Highly Complex DNA Virus

I pursued my curiosity about the replication of vaccinia, a member of the poxviruses. The DNA of poxviruses is protected by two coats so that release of the genome into the cytosol depends on removal of the outer envelope followed by rupture of the inner coat. Until 1960, the phenomenon of uncoating remained a mystery. Coincident with my studies, Wolfgang K. Joklik, then at the Albert Einstein Medical School, published biochemical data suggesting that after vaccinia virus was divested of its outer lipoprotein envelope, the inner coat somehow became porous, allowing release of the DNA which, thereupon, was accessible to DNase hydrolysis.[5,6] To follow uncoating in my studies, the viral genome was radioactively labeled during replication with tritiated thymidine, and then the isolated progeny virus particles were purified and used to inoculate continuous cell cultures so as to initiate a synchronous infection. With this material, it was possible to trace the fate of the incoming viral genomes in the electron microscope by means of autoradiography.

This technique revealed that vaccinia DNA was extruded into the cytosol, where the virus is replicated, through well-defined ruptures in the coat around the core.[7] When correlated with the biochemical results of Joklik, this information clarified the uncoating phenomenon among poxviruses. My study thus became conceivable in 1962 as a consequence of the efforts by Lucien Caro and Robert van Tubergen, working at that time in Palade's group, who perfected high resolution autoradiography.[8] An example of uncoating of the vaccinia genome is illustrated in figure 1.

The Replication of REO Virus

REO (the acronym for respiratory, enteric, orphan) viruses are agents with a segmented, double-stranded RNA genome of which each segment encodes one protein. The genomes of these cytoplasmic viruses are protected by an inner protein coat, which, in turn, is enclosed within an icosahedral capsid. My interest in early interactions between viruses and host cells led me to follow the ingress of REO virus type 3 into L cells. Initial collaboration with

Peter Gomatos, an MD graduate fellow in Igor Tamm's laboratory at Rockefeller, generated preliminary evidence indicating that inoculum particles of this agent penetrate rapidly into lysosomes of the host, where virions lose the outer capsid, causing the genome to become active. Although this result appeared as an anomaly because the RNA of the invading virus, by coming into contact with the battery of lysosomal hydrolases, including RNases, was vulnerable, we hypothesized that perhaps the double strandedness of the RNA genome and associated proteins ensured the survivability of this genome.[9]

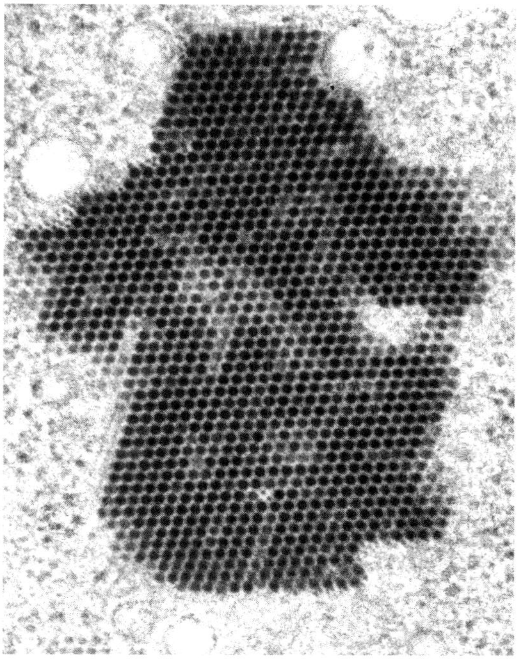

2. Poliovirus assembly; the L cells in this example were sampled at three hours after infection. ×70,000. From Dales et al. *Virology.* 26:379–389. ©1965, with permission from Elsevier.

Our conjecture was confirmed by subsequent studies carried out from 1963 to 1966 with Samuel Silverstein, who at that time was a postdoctoral fellow in our department. Our study using biochemistry, molecular biology, and electron microscopy revealed that inoculum REO virus particles were, indeed, sequestered within lysosomes, out of which their double-stranded RNA genomes escaped intact.[10] Subsequently, Silverstein, after pursuing this problem independently, demonstrated that the uncoated genomic RNA is retained within the inner protein capsule, where it undergoes conservative replication.[11,12] Another unusual feature of the REO virus life cycle is the development of this agent in intimate association with the cell's cytoskeletal elements, the microtubules, and intermediate filaments. I documented this observation as exemplifying one evolutionary adaptation by a virus parasite for exploiting the host cell's structure and function for self-perpetuation.[13]

Intracellular Assembly of Poliovirus

The collaborative study with Richard M. Franklin, an assistant professor in Tamm's virology department, on Mengo virus assembly within L cells led me to investigate the replication of poliovirus, another member of the picornaviruses.[14] I became especially intrigued with this subject from conversations I had with Hans Eggers, a German postdoctoral fellow working with Tamm, who described to me their biochemical work on the synthesis of poliovirus protein using subcellular fractions. I anticipated that examination of the infectious cycle might provide us with complementary data regarding the precise sites of replication. Published work at that time by highly reputable investigators claimed that poliovirus was synthesized inside membranous particles termed U bodies, a concept inconsistent with our knowledge about the synthesis of proteins, the very area that Philip Siekevitz and Palade had pioneered.[15,16] In my study of poliovirus, I collaborated with Eggers and enlisted the direct interest of Palade and Tamm. We demonstrated that the biochemical evidence pointing to synthesis of the viral protein on massive polyribosomes connected to the smooth endoplasmic reticulum was precisely matched by information from the electron microscope.[17] Individual polio virions were detectable as free particles during earlier times. Later on, it became evident that the large pools of viral products accumulating in the cytosol have a propensity to assemble into large, crystalline arrays of individual virions (figure 2). The infectious process also resulted in an accumulation of masses of membranous vesicles that are likely to be formed from the Golgi cisternae coupled with rapid net synthesis of membrane lipids. This occurs coincidentally with leakage of hydrolases out of the lysosomes induced by one of the viral proteins. The cytopathological response to the infection occurs with other picornaviruses, which was documented in my work with Franklin and a postdoctoral fellow from Japan, Kazunobu Amako.[14,18] In retrospect, it appears likely that the U bodies, structures assumed to be sites of poliovirus synthesis, were instead the vesicles of the cytopathological response that had been sedimented in a subcellular fraction.[15,16]

Ingress and Assembly of Influenza Virus

My interest in elucidating how different virus types penetrate host cells led me into a 1962 collaboration with Purnell Choppin, a specialist in this area of virology, on the attachment and penetration of type A influenza viruses. In this study, we used both tissue cultures and the chorioallantoic membrane of chicken eggs, a classic material for propagating influenza viruses in vitro. We demonstrated that inoculum particles become attached to the cell surface by means of peplomers or spikes of hemagglutinin and then are rapidly engulfed within endocytic vesicles by host cells.[19] Although the relevance of this observation to the infectious process was vigorously disputed until the 1980s,[20] subsequent studies by investigators using the powerful techniques of molecular biology and crystallography confirmed our original findings and explained the significance of the acidic milieu within phagosomes for the uncoating of influenza virus genomes.[21,22]

Based on my independent and collaborative studies with this and other viruses, one could conclude that these agents can effect their entry by subverting a normal cellular function for sequestering particulates from the milieu, thereby traversing the cell membrane barrier. The intracellular route traversed by different viruses after entering phagosomes becomes highly variable and unique for each type of agent.

The Rockefeller Cell Biology Course

As the coordinator of the cell biology course from 1960 to 1965, my function was to provide invited guest lecturers with teaching resources, including laboratory materials, to guide the participating graduate students in carrying out their laboratory projects, and to discuss their data with them. At the end of the course, I appraised each student's performance in written reports. I also gave two lectures and designed two laboratory exercises. Up to twenty students were enrolled annually. For me, the most memorable lectures were the introductory two or three in which George Palade provided a tightly organized, elegant general exposition of cellular organization and function, especially as they pertained to protein synthesis. As always, he demonstrated an unrivalled mastery of didactic discourse and showed impeccable illustrations to support and consolidate his statements.

In addition to the innovative, up-to-date subjects covered by lectures from our own group, there were presentations by experts in various topics on cell biology, including some concerning the techniques then in use. It was rewarding for me and, I presume, also for students and faculty in the audience to listen to Cecil Hall of the Massachusetts Institute of Technology tell us about the latest applications of electron microscopy for defining the macromolecular organization of viruses and cellular organelles. The subject of prokaryotes was covered in both lectures and laboratory exercises by the renowned Canadian microbiologist Carl Robinow, whose teaching visits to Rockefeller were for me unforgettable, harrowing experiences. In my role as coordinator, I became enmeshed within the web of this master teacher. A perfectionist in whom the anachronistic habits of a classical European professor were ingrained, he undoubtedly regarded my function in the course as that of an obedient flunky. Ironically, as often happens in academia, some time later I became the chairman of Robinow's department at the University of Western Ontario, where in time he and I became good friends and for whom I was supportive throughout his many productive years as professor emeritus.

The high intellectual caliber and exceptional aptitude for research among the graduate students in the cell biology course were clearly evident in the manner with which they executed investigations into the special projects they had chosen. Most studies included the application of electron microscopy, a technique in which our department excelled. Thus I had both the responsibility and pleasure of supervising and advising the participants in this endeavor. Among the projects that were carried to completion and published in bona fide journals were those of David Baltimore on poliovirus; Nick Acheson, Richard Compans, and Kathryn Holmes on cell-virus interactions; and Fran Messic on the assembly of an RNA bacteriophage within its bacterial host. These and numerous other exceptional students in our 1960–1965 classes included many who developed international reputations, among them Anthony Cerami, Harvey Lodish, James Jamieson, David Sabatini, Carolyn Slayman, Paul Englund, and Samuel Cushman.

A Memorable Laboratory Environment

After such a long interval, any attempt on my part to recall too factually the overall camaraderie and fun prevailing during those days in the 1960s, or the sense among us that we were participating in a unique adventure, might be considered as specious. We certainly were not working at Rockefeller because of any financial incentives, considering the low salary scale then prevailing for junior faculty, fellows, and students. I credit George Palade for being our role model by his style, demeanor, civility in our interpersonal dealings, and a complete commitment to pursuit of the science. Our individual triumphs, whether significant or quite minor, were frequently celebrated by popping champagne, as became evident over time in the ever lengthening rows of unmistakably shaped corks decorating the upper bookshelf of the seminar room.

Occasionally, the laboratory decided to have an evening out at the theater, followed by a dinner at a restaurant selected by one of several gourmand cognoscenti then among us. The daily lunches consumed in the clubby ambience of the dining room in Welch Hall were occasions for free-flowing conversations, sometimes about science, but just as often concerning the arts, politics, or whatever happened to be newsworthy that day. The congenial, relaxing atmosphere of the luncheon, where guests were often hosted, became the informal setting for leisurely dialogues and an opportunity to become acquainted in some depth with these individuals as scientific personalities.

Our single-minded individual research pursuits did not preclude our engaging in much banter and pranksterism. Keith Porter's infectious laugh and his enjoyment of practical jokes incited us into appropriate ripostes. One memorable incident occurred in 1961, soon after we moved into the quarters in South Laboratory. Before completely settling into his new office, Keith departed for a trip abroad, whereupon several of us took this opportunity to astonish the traveler upon his return. We made his new office inaccessible by inflating inside it a gigantic, authentic weather balloon. As this gas bag became inflated, it choked off all available inner space and completely obstructed access through the door. Furthermore, by pushing the intake valve inward so that it became virtually inaccessible, the perpetrators ensured that deflating the balloon would require time and ingenuity. So as to remain evenhanded in our dutiful attention towards our lab co-chiefs, George Palade was equally rewarded during one of his foreign trips. Based on rumors that George was having some administrative conflicts with President Bronk, we noted this disaccord by decorating the access passage to George's office as a simulation of the subway entrance to the Bronx Zoo. I recall that Palade really enjoyed this prank.

Epilogue

It is now quite obvious to me that exposure to the novel concepts about the organization and function of cells then prevailing among investigators in the cell biology department influenced me fundamentally in thinking about mechanisms by which viruses operated within host cells. The presence on campus of many virologists with a sympathetic outlook and common interests made it possible to carry out studies on a variety of cell-virus systems that would not have been possible for me to conduct independently. The energizing experience of the years at Rockefeller unquestionably influenced for the remainder of my productive career both the selection of the topics and the methodologies by which I carried out research.

Maybe we assuage our egos, thereby confirming the significance of our existence, by evoking the most gratifying memories about the formative years in our lives. I invoke this caveat in recollecting these seemingly idealized events of my time in the laboratory. My experiences during those intoxicating years at Rockefeller was a period that was both extraordinary and exhilarating, when a torrent of fundamental discoveries in the all-encompassing field of biology came cascading down upon us, and in which I was a very privileged participant among colleagues in a fabulous department.

9F. Cilia and the Cytoskeleton

Peter Satir

In the summer of 1956 at Woods Hole, Keith Porter repeated his famous Harvey Lecture on "The submicroscopic morphology of protoplasm."[1] For the intent listener, of whom I was one, this lecture summarized the early work on cilia and the cytoskeleton of the Rockefeller group and encapsulated the approach that we may call comparative cell biology. It was not very obvious to everyone at that time that the organelles seen with the transmission electron microscope were the same in every cell and certainly was less obvious that common organelle structure reflected common molecular construction, but Porter presented the case with some clarity, noting that "one may regard the advent of the electron microscope as extremely fortunate for, in its powers of resolution, it bridges rather well the gap between the limits of light optics and the resolutions of physical and biological chemistry."

The modern transformation of cell biology into molecular cell biology, with the understanding of deep homologies in protein structure and cellular organization produced by evolution from a common eukaryotic cell ancestor, was about a quarter of a century in the future, but Porter's lecture was its beginning. He chose two examples of common cellular design to illustrate his thesis of the common construction of organelles. One was the endoplasmic reticulum, which I discussed briefly elsewhere;[2] the second was the cilium. Porter was drawn to the study of cilia in his search for general principles of cellular organization. As he said, "There appears to exist . . . beyond the limits of optical resolutions an 'elastic' framework, undefinable except in terms of its apparent influence on the form and functional properties of the cell and organelles. . . . It is probably not too early to contemplate . . . the structures we have defined, to see if any of them could be regarded as 'cytoskeletal' in function."[1]

Two years before, Don Fawcett and Porter had published a study of a cytoskeletal framework that determined the form and function of the cilium,[3] and a detailed description of *Paramecium* cilia was published with Albert Sedar.[4] The internal cytoskeleton of the cilium, the so-called 9+2 axoneme, was found to be identical in protozoan, frog, and human cilia (figure 1). This finding extended the elegant studies of the English botanist Irene Manton on splayed whole mounted plant cilia.[5] Where others equivocated, from her preparations Manton had boldly proposed the 9+2 structure with some detail as a universal construction of plant cilia. In the late 1950s, continuing her studies, Manton came several times to Rockefeller to work with Porter at Christmastime, when she would use the latest Rockefeller technology and have easy and long access to the electron microscope.

Now Porter realized that from the ubiquity and regularity of the 9+2 structure "one gains the impression that there must be something fundamental in this organization," probably as the basis of ciliary motility that was fixed by evolution. Moreover, when he looked for the framework or cytoskeleton that might be involved in maintaining the cylindrical form of the ciliary axoneme, he settled on the 9+2 filaments—or microtubules—as

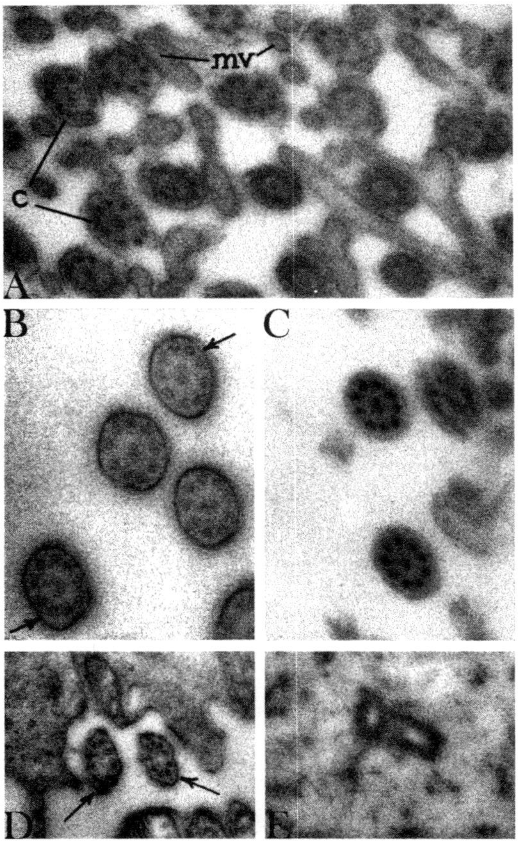

1. Images of cilia, basal bodies, and centrioles. In all figures, the cilium is a natural magnification marker of 0.25-micrometer diameter. Frog (A), *Paramecium* (B), and human (C) cilia show the 9+2 pattern of axonemal microtubules surrounded by the ciliary membrane (arrows). c, cilia; mv, microvilli. (D) Connecting cilium (arrows) of rod photoreceptors show a 9+0 pattern of microtubules. (E) Centrioles are at the pole of a mitotic spindle. From Porter, 1957. *Harvey Lectures*. Series 51. 175–228 with permission from John Wiley & Sons, Inc.

"an obvious candidate." Because the 9+2 pattern was so constant and so striking, and this construction was so difficult to explain except by invoking a common molecular mechanism for the purpose of morphogenesis and motility, the cilium was one of the key organelles in establishing the principle that constant electron microscopic structure reflects its common molecular composition, common mechanistic or physiological principles, and common evolutionary history—which I have elsewhere called "the organelle doctrine." In Porter's Harvey Lecture, particularly with the cilium, arguably the most convincing example, this doctrine was enunciated and demonstrated definitively for the first time.

Porter was not unaware that the ciliary microtubule array was also to be found as the developmental basis of other cell projections that were nonmotile. Perhaps the most notable of these projections was the outer segment of the photoreceptors of the vertebrate eye. The Rockefeller laboratory was not the first to examine these segments; in fact, Fritof Sjöstrand in 1953 had studied the outer segment disc systems as examples of biological membranes,[6] but Sjöstrand did not note the homology of the basal portion of the segment to a cilium. Eduardo De Robertis noted the resemblance,[7] and Porter (later with Eichi Yamada, who visited from Kyushu) confirmed the homology during outer segment growth. Unlike motile cilia, these modified or primary cilia and others like them were missing the central pair of microtubules; in modern terminology they were 9+0 axonemes (figure 1). Moreover, the pattern of nine microtubules began in the basal body. By showing that the repetitive nine-fold symmetry of the basal body matched that of cell centrioles, Porter confirmed the hypothesis of early light microscopic cytologists that centrioles and basal bodies were equivalent organelles.

Porter then asked whether the centrioles also organized filaments (microtubules) within the cell cytoplasm to give it form. Here, he made a mistake because osmium tetroxide, the fixative used for electron microscopy in those early days, was applied at 4°C, which depolymerized cytoplasmic microtubules. He mistook the thin, elongated cisternae of endoplasmic reticulum, which lay parallel to the depolymerized microtubules, for the microtubules themselves.

When, in the 1960s, microtubules were visualized using glutaraldehyde fixation, Porter, then at Harvard, continued the study of their distribution as organizing principles of the cell cytoplasm, contributing largely to our present understanding of cytoskeletal organization.

The Fawcett and Porter paper[3] contained many beautiful and novel micrographs of ciliary and ciliated cell structure, including a striking example of the thick and thin section technique, a staple of the laboratory at that time, which showed how histological and electron microscopic images of ciliated cells could be compared (figure 2). The paper was submitted to *The Journal of Experimental Medicine*, a Rockefeller journal, which declined to publish such a basic study—a fortuitous misjudgment. The study was published in the *Journal of Morphology*, but Porter was dismayed over the quality of the reproduction of the micrographs. This dismay was conveyed to Herbert Gasser, the director of the Institute and an editor of *The Journal of Experimental Medicine*, and in short order, a new Rockefeller journal, *The Journal of Biophysical and Biochemical Cytology*, made its appearance, with stringent standards for the reproduction of electron micrographs and an editorial board of prominent cell biologists, including George Palade, headed by Porter.[8]

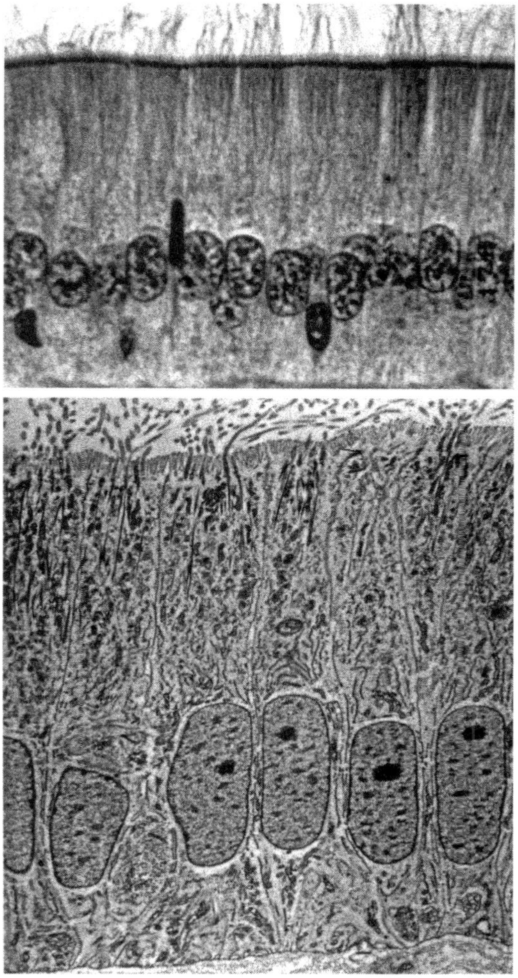

2. Serial thick and thin sections of ciliated cells from *Elliptio* intestine. Note that the structures seen in the thick section light micrograph are readily identified with higher resolution in the electron micrograph. Such images convinced most people that the structures seen with the electron microscope represented known cytological and histological features of cells and tissues. From Fawcett and Porter, 1954. *J. Morphol.* 94:221–281. ©Wiley Periodicals, Inc.

In the fall of 1956, I joined the second entering class of graduate students at Rockefeller. I knew of Porter's work on the electron microscopy of cilia, and, inspired by his Harvey Lecture, I wished to join his laboratory with the ambitious youthful goal to study how the 9+2 pattern of the axoneme produced the beat of the organelle. Before I did this, I needed to learn the basics of electron microscope technique. I thought it would be interesting to learn electron microscopy while pursuing an evolutionary problem that was related to the structure of the modified cilia that Porter had described. The problem was the origin of the vertebrate eye that seems to spring full blown into existence in the cyclostomes; it is peculiar because light strikes the outer segments, the modified cilia, of the eye only after passing through many layers of neural connections and even the inner segments, the cell bodies to which

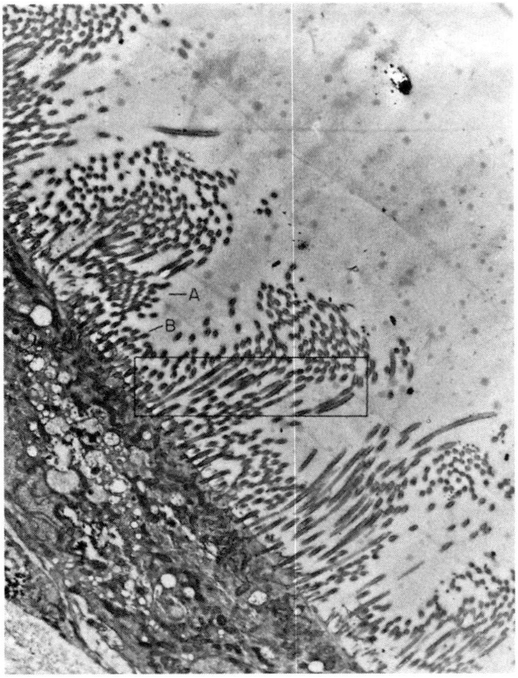

3. The metachronal wave of *Elliptio* lateral gill cilia captured for electron microscopy by quick fixation. The box outlines successive cilia showing systematic displacement because of phase differences resulting in progression of the recovery stroke along the axoneme. A and B: cilia in the effective stroke. From Satir, 1963. *J. Cell Biol.* 18:345–365.

the modified cilia belong. My idea was to look for modified cilia among the many proposed candidates for the origin of the eye among the cells of the cerebral vesicle of amphioxus, which was hypothesized by some to be the undeveloped precursor of the diencephalon. They are there, in modified ependymal cells. But amphioxus proved a difficult organism for a novice to capture for electron microscopy, and my study,[9] although cited,[10,11] remained unpublished.

One of the most elegant electron micrographs in the Fawcett and Porter article, reproduced here as part of figure 2, showed ciliated cells of the freshwater mussel *Elliptio*. Because the fixation and embedding of this organism by osmium tetroxide and methacrylate methods seemed to produce excellent results, Porter suggested that I use this organism when I began my work on motile cilia. While figure 2 shows epithelia from the intestine, much more was known about the ciliary motility of the gill epithelium of lamellibranchs from many years of careful study by James Gray in Cambridge, England.[12] Gill filaments of *Mya* were examined by Fawcett and Porter. *Elliptio* and related species of freshwater mussels were easily obtainable, and my hope was to combine a physiological and an electron microscopic study of the gill cilia based on Gray's work on one hand and Fawcett and Porter's images on the other. I soon settled on the lateral gill cilia as the objects of primary study when I was able to confirm that the metachronal wave—beating cilia with a phase difference between them, first photographed by Gray—could be activated in extirpated small pieces of living gill tissue.

Sometime in 1960, the Porter and Palade laboratory moved from the basement of Theobald Smith Hall to new quarters in South Laboratory (now Bronk Laboratory). At about this time, I had the idea that was to become the key feature of my thesis to capture the metachronal wave by quick fixation with osmium tetroxide. Since the metachronal wave shows all stages of a ciliary beat displayed in sequence along a ciliated cell, if the ultrastructural information in the wave could be read properly, you could see what changed from stage to stage in the beat, and the mechanism of motility would supposedly become apparent.

I tried and succeeded in fixing the metachronal wave with a few millisecond phase differences between the successive cilia (figure 3). Eventually Maria Rudzinska, who was working on suctorian ultrastructure with Porter and William Trager, told me that a Hungarian scientist, Béla Parducz, seemed to have done something similar. He in fact had done this at the light microscope level with *Paramecium*, and his articles were published in German;[13] moreover, he was not the originator of the technique; that was Jōzsef von Gelei in 1927.[14] Of course, neither Parducz nor von Gelei had the electron microscope to look at ciliary ultrastructure, but knowing where and what to look at were not obvious. It took a number of years after I graduated before key features could be properly understood, but eventually the fixed metachronal wave would provide the pictures that led to the first proof of the now accepted sliding microtubule mechanism of ciliary motility.[15]

At this time, Porter and I had a weekly meeting to discuss my progress. When I had produced the image reproduced here as figure 3, I took it into the meeting for him to see. He was quite complimentary; after all, the picture was in reasonable focus, there were no major artifacts, and it showed something interesting. The next week, with a new pile of micrographs to discuss, I began by again showing the image to Porter. We never got further; he had forgotten it, but again he was excited. The next week, the story repeated itself. By this time, I realized I had an image I could use in my thesis.

Dennis Flanagan of *Scientific American* wanted Porter to write an article on cilia for the magazine. Porter, being busy, turned the assignment over to me. Porter, being Porter, refused coauthorship, and so I was able to reproduce many of my thesis pictures, including figure 3, in *Scientific American* in 1961, along with illustrations from many of our colleagues at Rockefeller.[16] The comparative cytology of the 9+2 and 9+0 axonemes could be illustrated rather well through the language of *Scientific American*. However, the article focused on the absence of the central pair in 9+0 cilia and not on the absence of the arms, a new feature of the axoneme described for 9+2 sperm tails in *The Journal of Biophysical and Biochemical Cytology* by Bjorn Afzelius[17] and later shown by Ian Gibbons[18] to be the critical dynein ATPases responsible for ciliary movement. Rather than the central pair, the absence of dynein arms is the explanation of why most 9+0 cilia are nonmotile.

Visitors to Rockefeller at this time included, for several months, John Randall and Sylvia Friton-Jackson, who worked on *Stentor* cilia. Randall, having played a role in the development of radar and in the story of DNA (he was professor and department head at King's College, London, where Maurice Wilkins and Rosalind Franklin worked), now became interested in solving the mechanism of ciliary motility. When he went back to London, he assembled a group that continued working on the ultrastructure of cilia in *Tetrahymena* and *Chlamydomonas*.

Ciliated cells illustrate two additional aspects of cytoskeletal structure commented upon by Fawcett and Porter. One is the striated rootlet fiber attached to the basal body of each cilium, as seen in figure 2. The composition of these particular fibers is still unknown; similar fibers in some algal flagellates are now known to be composed, at least in large measure, of centrin, a Ca^{2+}-binding contractile protein. A second feature is the striated border of short microvilli separating the cilia. Fixation destroyed the actin cytoskeleton of the microvilli and the underlying actin network in the apical cell cortex that was part of the zonula adherens, or adhesion belt, later described by Marilyn G. Farquhar and Palade.[19]

After Porter emigrated to Harvard in 1961, the Rockefeller group under Palade's leadership focused mainly on topics other than cilia and the cytoskeleton; but later, when David Luck established his own laboratory, research on cilia resumed. Luck wished to use the *Chlamydomonas* system, where ciliary mutants had been generated by Ralph Lewin and by Randall's group, to study the proteomics (as we would now call it) of ciliary motility. For about a decade, beginning in the mid-1970s, Luck and his colleagues, especially Gianni Piperno, worked on the protein composition of the axoneme, establishing in a series of significant studies the complexity of functional dynein isoforms, the composition of the radial spokes, and the heterogeneity of ciliary proteins.[20]

The importance of the original studies of Keith Porter and his colleagues and students at Rockefeller in establishing the cilium as a model organelle for cytoskeletal investigations cannot be overlooked. As I have tried to indicate, Porter realized that the cilium originated early in eukaryotic cell evolution and embodied important structural and functional principles. These principles were critical for our current understanding of the basic mechanisms of ciliary motility and of the role that cilia play in human health and in determination of left-right asymmetry in the human body. The early studies of cilia introduced concepts of microtubule structure, polarity, and assembly into cell biology. In addition, the modern concept of microtubule motor molecules, with all its implications for membrane trafficking and for organization of the cell cytoplasm in interphase and mitosis, grew out of such concerns.

9G. The Porter-Bonneville Atlas: *An Introduction to Fine Structure of Cells and Tissues*

Mary Bonneville

After more than four decades since Keith Porter and I published the first edition of *An Introduction to Fine Structure of Cells and Tissues* (1963), which eventually went through four editions and a number of printings, it is a source of both pleasure and some sadness to remember the events that led to this collaboration. Pleasure, because it brings to mind the time when the field of cell biology was new and full of incredible promise and when I had a chance to work with one of the leaders in that field; but also sadness that Keith, my teacher, mentor, colleague, and friend for more than forty years passed away in 1997, so that a memorable, golden era seems, at least to me, to have closed. Nevertheless, it is interesting to recall how persons and happenings seemed to come together by chance and lead to events that changed my life.

My interest in microscopy began when I was at Smith College, from which I graduated in 1953. My favorite courses were histology and embryology, which were then, to a large extent, taught as descriptive subjects; understanding of microscopic anatomy and revelation of the events in development were dependent primarily on the light microscope and on the limits of its resolution. In many ways, the information available in textbooks was tantalizing, but on the other hand, it was often unsatisfactory; for example, the change in the appearance of banding pattern in striated muscle during contraction was not related in any meaningful way to the molecular mechanism of contraction described by physiologists as the interaction of the proteins actin and myosin. The chapter on the thymus took only a few pages and revealed essentially nothing about that organ's function. The nature of the structures at the free surfaces of intestinal epithelial cells and of cells of the proximal convoluted cells of the kidney, referred to as brush borders, was unresolved. The variety of attachments between cells and their functional significance were not well defined. Certain slides were stained in a fashion to visualize mitochondria, which were seen only as small colored dots in light microscope preparations. For the student, the features of mitochondrial structure were not yet correlated with an understanding of mitochondrial function at the molecular level.

This lack of correlation was generally true among the courses I took. For example, when, as a student at Amherst College from 1953 to 1955, I took biochemistry and genetics, these subjects seemed largely unrelated to one another and to the histology course. Genetics was of the classical Mendelian variety, and the biochemical mechanisms we memorized were not very tightly tied to cell structure. The mechanism of protein synthesis was a great mystery. We were very excited about the 1953 publication by James Watson and Francis Crick describing the molecular structure of DNA, a work that seemed to promise that the structure of genes and their action could be understood in biochemical terms. A new era seemed to be dawning.

Amherst, where I was one of only a few women studying for a master's degree, was one of the colleges invited by Detlev Bronk to nominate students to the graduate program he initiated at Rockefeller in 1955. I was fortunate to be named, along with an Amherst senior, Lewis Greene, for admission. Bronk was surprised to receive the name of a woman from Amherst, where all the undergraduates were men. When I lunched with him in New York, he allowed that he would have to sooner or later consider accepting women among the students, and he ultimately admitted me as the only woman in the first group of ten students who came to Rockefeller in the autumn of 1955.

During my interview with Bronk, I expressed an interest in electron microscopy. Keith Porter was not available, and so I was introduced to George Palade, who showed me micrographs of the "small particulate component of the cytoplasm," my first look at structures later known as ribosomes. Palade pointed out that the particles were "free" in the cytoplasm and also attached to membranous sacks. I'm sure Palade had realized that the cell fractionation method of Albert Claude, combined with biochemical analysis of the fractions and with examination of both the fractions and the whole cells from which they were derived, would before long reveal the function of these newly visualized structures. But he said nothing to me about these exciting prospects, and I was too naive to imagine how the work would progress.

I first met Keith Porter that fall when he lectured to us about his discoveries, and later in the academic year we all attended his Harvey Lecture. I had heard about his work on the structure of cilia at a Growth Society meeting held at Amherst while I was a student there, but these presentations in New York were our introduction to the endoplasmic reticulum. This and other discoveries with the electron microscope excited me because it seemed that one could begin, at last, to see clearly the cell structures that were so unsatisfactorily resolved by the light microscope, and perhaps structures not yet dreamed of.

The students were expected to try working in laboratories where the research interested them and eventually to settle in one of them. I was drawn to the Porter/Palade laboratory, but I didn't really know how to become one of its members. Frank Brink, who was dean, realized I was having problems and put me in contact with Maria Rudzinska, who became my counselor and was a steadying influence in a sometimes rocky sea. With her help, and after a rather convoluted series of events, Porter agreed to take me into the laboratory. Although we never did any laboratory work together, Maria remained a valued mentor as long as I was at Rockefeller, and at the graduation ceremony in 1961, she was the one who described my achievements and presented me to Bronk to receive my degree. She gave me invaluable help, and I had made a lifelong friend.

A major question of the day was whether cells could "dedifferentiate," that is, return to an undifferentiated stem cell type and then change into another cell type or at least take on new and different functions as they were subjected to new physiological conditions. Since amphibian metamorphosis, the progression from the larval to the adult form, was under the control of the thyroid hormone, frogs seemed to be the ideal choice to study this question, and I became an expert at raising frogs. I learned that the functional cells of the intestinal epithelium of the larval frog were destroyed, seemingly autodigested by lysosomes, then only recently described, and that the tissue was repopulated by a group of undifferentiated cells that lay at the base of the larval epithelium, waiting for a signal from the thyroid hormone to differentiate and populate the adult intestinal epithelium. Dedifferentiation did not apparently take place since the intestinal lining of the carnivorous frog replaced that of the vegetarian larva.

The Porter/Palade laboratory was in the basement of Smith Hall, where the electron microscopes could benefit from the stability of the bedrock underlying Manhattan. They had mounted a course for students in the second year of the program. It was then that I began to learn the skill necessary to carry out the experiments I would later undertake.

We students came to recognize that Keith was the consummate technician. He could, with seeming ease, cut thin sections on the Porter/Blum microtome, built by Joseph Blum in the Rockefeller Instrument Shop, the prototype of which he helped to design. He seemed to know what pictures to take as he worked on his own research, often spending evenings at the Siemens electron microscope that was off limits to the students. Porter had no patience with a poorly cut section or an out-of-focus micrograph, and he seemed puzzled when we students had trouble matching his laboratory skill. Probably his talent came so naturally to him that he did not realize his skills were a special gift. His micrographs, and those of Palade, set the standard for all students, postdoctoral fellows, guests, and visitors who were drawn to that mecca for electron microscopists, and many outside Rockefeller had difficulty in reaching their technical standard.

Keith's great joy came from interpreting pictures, and he was a master of the art. He was always quickly aware of the information that the picture revealed, astonishingly so. But he was also conscious of the composition of each picture as a work of art. To the uninitiated, micrographs can appear as abstract art and have an aesthetic appeal apart from the information they contain. But beyond their artistic merit, pictures were to him like maps of a world to be discovered, and, indeed, he always enjoyed looking at maps of all kinds. Morphology intrigued him, and I think it was his real love. This may have been because, like me, his academic training had emphasized structural studies in a period when biochemistry was even more separated and remote from the study of cells and tissues. He never revealed his steps in analyzing a micrograph and never gave directions for ferreting out information from it, so I cannot describe the process he followed. I regarded his skill as an inborn talent, a given, and something I had to learn from studying many papers and pouring for long hours over my own micrographs.

When I was conducting the experiments for my thesis, I was required each Friday afternoon to give Keith an oral report of my progress. This was stressful, and he often seemed to regard the meeting as an awkward time as well. When, in 1961, I had finally written a draft of my thesis, we met once a week in a room in Abby Aldrich Rockefeller Hall, where he had me read my work aloud. He was extremely busy, and he seemed not to have the patience to sit down and read the draft himself.

Two moments stand out during these sessions, and both illustrate important talents possessed by Keith. At one point, while reading the draft of the "Discussion," I ventured to say that on morphological observation alone, certain vesicles, apparently derived from the Golgi, were being integrated into the free surface of differentiating intestinal epithelial cells. I wondered whether I could say this with any conviction, or indeed at all, when Keith urged me to include the bolder statement that the Golgi was involved in determining specialized properties of the cell surface. This demonstrated his self-confidence in his interpretations of micrographs and is a good example of how quickly he saw what proved to be a general principle long before it was rigorously proved by subsequent critical experiments. It was clear, then and later, that he took a great deal of pleasure in mining information in micrographs.

As I continued reading the draft, Porter said, at several points, with great surprise in his voice, "This is good!" When Porter told you something was good, you could be sure it really was; he never flattered anyone, and he disliked being flattered. He evidently had more confidence in my ability to write and knew my capabilities better than I did myself. Indeed, I have never met anyone more skilled at evaluating people. After fifteen or twenty minutes he could tell all the strengths and weaknesses of the person he interviewed, and time would invariably prove him correct. The fact that he also asked each person to do what that person did best was a trait that benefitted those who worked for him and made the atmosphere in his lab largely free of stress.

By 1961, studies with the electron microscope had transformed histology courses. Authors were busy integrating the new ideas into texts, and those texts were outdated as soon as they were published. That fall, I was a postdoctoral fellow in the department of anatomy of the College of Physicians and Surgeons at Columbia University, where I assisted in the histology laboratory course for first-year medical students. Keith had just left Rockefeller for a professorship in the biology laboratories of Harvard. There, he had a new Phillips electron microscope (EM), which he loved to use, and a skilled technician, Susan Badenhausen. By this time, he realized that a daunting amount of new information now existed that should be integrated into histology texts and that a guide for students was needed to help them interpret EM images.

Unlike a road map or geological map, the structures seen in electron micrographs have no counterpart in the macroscopic world. At the same time, with all the improvements in techniques, he was producing images of a quality not surpassed even today, and he apparently wanted to publish these pictures, including cells and tissues that he had not himself used in experiments. It was then that he asked me to collaborate in producing what he always referred to as "The Atlas." This would provide a chance to publish pictures of a larger size and of more diverse subject matter than those published in research papers.

One afternoon, while I was visiting him at Harvard, a book agent from Lea and Febiger also visited his office, apparently by chance. Keith described his ideas about design and layout, work we had already started, and the agent immediately agreed to these ideas and produced a contract for us to sign.

As we worked, Keith provided the pictures, and I did almost all the writing to describe them. I also suggested subjects to be included in a book that would supplement the current histology texts, and he and Susan would produce elegant micrographs. After I had written a half dozen or so descriptions, I sent them to Keith with great uneasiness, although he changed very little of what I wrote. Not surprisingly, the only exception was the liver cell, the hepatocyte, which he rewrote, since by that time he had become pretty much the world expert on the microscopic structure of that organ. I eventually gained confidence and lost the student awkwardness that once existed between us. I now look back on this experience as one of the happiest of my life.

We put the finishing touches on the manuscript and checked references in the library of the Marine Biological Laboratory at Woods Hole, Massachusetts, at a time of year when we often had the whole library to ourselves and could talk to each other freely, calling from one floor to another. Finally, after a number of weekends in Woods Hole, where Keith and Elizabeth Porter had a cottage, the book was ready to present to Lea and Febiger. The first edition of the book was published in 1963 as a loose-leaf edition, and it was followed by three hardcover editions, the second edition in 1966, the third in 1968, and the fourth in 1973. The atlas was also published in Great Britain as well as in German, Spanish, Italian, and

Japanese translations. During the preparation for publication, Keith was especially vigilant in assuring himself that the quality of the printed pictures and the overall appearance of the book would be up to his standards. It was due to his care that the atlas remained in print for over two decades.

In later editions, the book was expanded but remained closely tied to the standard histology texts both in its subjects and in the order in which they were presented. Keith went on to produce other atlases: he and Myron Ledbetter produced one of plant cells, and he and Pietro Motta and Peter Andrews produced one of images from the scanning electron microscope. These works, though they were dear to his heart and developed under his demanding gaze, were not associated with courses that were so extensively required of students of biology and medicine. Atlases of mammalian cells appeared by others that were much more exhaustive in pictorial coverage but for the most part did not have the kinds of descriptions that became so useful to students who were being introduced to histology. For example, Don Fawcett, who had known Keith as an undergraduate at Harvard, produced a book with a plethora of excellent electron micrographs. Fawcett, like Keith, was a master of the science and art of electron microscopy, although his atlas was not as specifically targeted to students.

In 1970, I joined the faculty of the University of Colorado in Boulder in the department of molecular, cellular, and developmental biology, which Keith had founded and for which he served as the first chair. There, he added the scanning electron microscope and the high voltage electron microscope (HVEM) to the department's armory of instruments. His work with these microscopes demonstrated their usefulness in biological research and, in the case of the HVEM, is serving biology so well today. The electron micrograph negatives continued to accumulate as long as Keith was active. The Keith R. Porter Endowment for Cell Biology, established in 1981, inherited all these images of his work, including the first interpretable electron micrograph of an intact cell, which Keith took with Albert Claude and Ernest Fullam in 1945. The negatives (over 1,700 pounds of them) now reside in the Porter Archives at the University of Maryland, Baltimore County.

Keith Porter played a major role in revealing cell structure at the macromolecular level, where correlations with the biochemistry of cells could be made. His love of structural biology kept him busy exploring with the electron microscope for the rest of his working life. When I did a series of television interviews with Keith in the 1980s, he clearly showed that cells were close to his heart. Speaking of his longtime acquaintances, he said, "You develop a certain sensitivity after a few years of working with a cell. It's like working many years with an individual. You know what he or she is going to do in response to something that happens, or to something that you say. But you have to be careful not to mention this or that. On the other hand, you know if you want to delight them, you say the right things. When you're studying cells they become persons. You can predict what's going to happen, what the story should be."

It was a happy coincidence that I, too, loved the structural images and was excited by the information they provided to the rapidly developing and expanding field of cell biology. Knowing and working with Keith Roberts Porter was a very great privilege.

9H. Nuclear Pore Complex

Philip Siekevitz

In the late 1940s and early 1950s, a question arose about the nature of the membrane surrounding the nucleus in eukaryotes. The problem became confounded by a series of findings: 1) that protein synthesis occurs on cytoplasmic ribosomes; 2) that messenger and transfer RNA molecules are synthesized in the nucleus as auxiliary "initiators" of protein synthesis; 3) that ribosomal RNA is synthesized in the nucleus, while the ribosomal proteins are synthesized by cytoplasmic ribosomes; and 4) that the precursors to the cytoplasmic ribosomes, combining RNA and proteins, are formed in the nucleolus. The main question then turned to the nature of the trafficking of these molecules between nucleus and cytoplasm. The eventual answer came following the identification of the nuclear pore complex, a huge structure that is the sole mediator of the two-way traffic of proteins and RNAs. Some early discoveries on this structure and its function by Rockefeller cell biologists also contributed to the origins of modern cell biology.

The nucleus, as revealed in Keith Porter's first electron micrograph of an intact cell in 1945, was too thick for detailed electron microscopic study. In the early 1950s, however, when other scientists were studying nuclear membranes that had been teased out of the cell, they found this envelope consisted of two layers, a continuous internal one and a porous external layer; the apparent pores that broke the continuity of the external layer were described as an "annulus," or ring-like structure, in frog[1,2] and sea urchin oocytes.[3]

Shortly thereafter, Michael Watson, a trained microscopist from the biophysics and radiation biology laboratory at the University of Rochester, joined the Porter/Palade laboratory. During 1954, in addition to several cytochemical studies with me on mitochondrial membranes, he also participated in Porter's ongoing studies to look at fine structure. While most studies of fine structure were then focused on submicroscopic organelles in the cytoplasm, Watson used the new techniques of fixation and embedding to look at thin sections of the nucleus. As a result, he provided the first detailed descriptions of the nuclear envelope.[4] He observed that the outer layer (membrane) of the envelope is continuous with the endoplasmic reticulum, the organelle that was just beginning to take on great importance in the laboratory. Following Porter's interest in documenting the universality of cellular organization, Watson also found the pores are prevalent in a wide variety of mammalian tissues and were consistent with those that were being reported by other researchers in amphibia, insects, protozoa, and algae. He, as well as George Pappas, who had just left the laboratory (see Chapter 9C), and Bjorn Afzelius in Sweden, were the first to suggest that the pore was contained within a circular and perhaps octagonal annulus.[3,5] Afzelius conceived the annulus as a cylinder over the pore margins on both the nuclear and cytoplasmic sides that contains material specific to the pore proper and is distinct from the inner nuclear membrane.

Most scientists looked upon these pores—600–1,000 ångströms in diameter and numbering from hundreds to thousands per nucleus—as simple breaks in the nuclear envelope where the outer and inner membranes became continuous at either side of the pore,

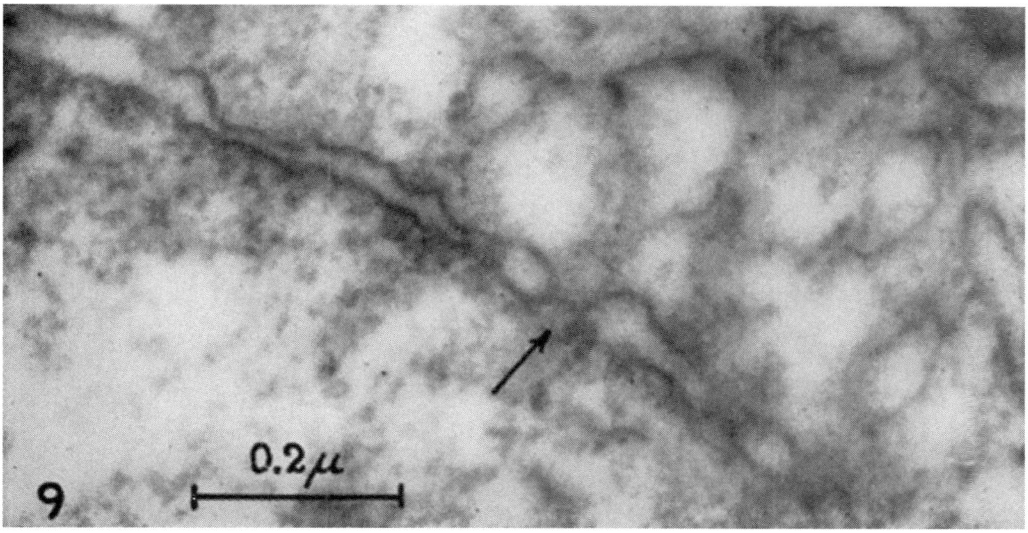

1. A section normal to the nuclear surface of a rat liver parenchymatous cell. The pore, indicated by an arrow, contains a finely granular material concentrated near the waist of the pore and diffused gradually into the matrices of the nucleoplasm and cytoplasm. The inner and outer nuclear membranes are smoothly joined on either side of the pore, which is bridged by a diffuse dark band that may represent a diaphragm. ×120,000. From Watson, 1955. *J. Biophys. Biochem. Cytol.* 1:257–270.

thus creating an opening. However, Watson emphasized strongly that there was material in the channel of the pore and extrusions of channel material into the cytoplasm. He was the one who coined the name "nuclear pore complex" (NPC) for this apparently complicated structure (figure 1). He recognized the difficulty of defining a three-dimensional structure from electron microscopic images because these images illustrated different levels or zones in a complex configuration that pierced the double-membraned envelope: "The lack of satisfactory evidence covering the surface of pores does not necessarily lead to the inference that free diffusion of material can occur through the pores. The high degree of structural organization imperfectly suggested by our observations leads one to emphasize our earlier conclusion that any passage of materials which may take place through the pore complex is probably controlled and not a random process."[6]

During the next few years, many electron microscopists became intrigued by this structure. Images soon appeared that were taken of sectioned material, whole mounts of cells, isolated nuclei, and isolated nuclear membranes. For example, in 1963, Rachele Maggio, George Palade, and I found the complex intact between the two membranes in isolated nuclei, thus indicating the strength of the attachment.[7] This idea of a tightly knit structure was emphasized later (described in the following paragraphs) when several investigators showed that when the nuclear envelope was separated from the nucleus, the NPC was part of this isolated structure. In 1961, Robert W. Merriam at Columbia isolated nuclei from frog eggs and concluded that the annular material in an octagonal array is part of the NPC and that it contains no RNA, thus echoing Watson's finding that the ribosome-like material occasionally seen by others is not a part of the NPC.[8]

In 1970, Herbert Abelson and Gilbert Smith, at the National Cancer Institute, cultured kidney cells in monolayers, fixed them in various ways, treated them with various enzymes, and did serial sectioning and rotational analysis imagery; they observed an annulus extending into the nucleoplasm and cytoplasm, seemingly octagonal and composed of tubular structures with filaments around a dense central element. They were persuaded to include the NPC "among the classical cellular organelles."[9]

Along with this structural work, other experiments attempted to define the function of the complex, namely, the trafficking between nucleoplasm and cytoplasm. Although some workers found small ions could easily penetrate the nuclear envelope, the more important question was the movement of macromolecules. One of the first attempts was by Carl Feldherr in 1962 at the University of Pennsylvania, who injected colloidal gold particles (visible by electron microscopy) into amoeba cytoplasm and observed that they ended up in the nucleus, with some visible in the NPC.[10] In 1975, William Bonner, from the Medical Research Council in Cambridge, showed that when radioactive nuclear proteins were injected into the cytoplasm of frog oocytes, they eventually appeared in the nucleus, whereas injected cytoplasmic proteins were of two types, one remaining in the cytoplasm and the other distributed equally between cytoplasm and nucleus.[11] By that time, it was becoming known that histones and probably other nuclear proteins were synthesized in the cytoplasm, so that the migration from cytoplasm to nucleus was not that surprising.

What was increasingly evident, however, was the complexity of the two-way migrations. For example, it was known that cytoplasmic ribosomal RNA is synthesized in the nucleus but that ribosomal proteins are synthesized in the cytoplasm. Thus, these proteins traverse the nuclear envelope and engage ribosomal RNA precursors in the nucleolus to form precursors to ribosomes, which then have to traverse the nuclear envelope to become the cytoplasmic ribosomes.[12]

The next step involved discerning the biochemical make-up of the NPC, linking it to the structure as visualized with electron microscopy, and trying to solve the puzzle of the two-way communication channels of the NPC. Among the early experimenters were Robert Aaronson and Günter Blobel at Rockefeller. Although the intention of this book has been to describe scientific achievements in cell biology that originated at Rockefeller mainly before 1974, extensive investigations on NPCs continue, many perpetuated by Blobel and his associates. So the rest of this chapter brings the still unraveling picture of the structure and function of these intriguing organelles up to the present day.

Aaronson and Blobel isolated liver nuclei and treated them with the detergent Triton X-100 to remove the nuclear membranes. By removing much of the adjacent material, their electron micrographs showed a more detailed view of the NPCs even though they were still attached to nuclear structures.[13] After the NPC and associated material were further isolated, the resultant electron micrographs showed the NPCs located along a string of fibrous material, called a lamina.[14] This lamina appeared very similar to one first seen ten years earlier in the intact nucleus by Don Fawcett, an early member of the Porter/Palade laboratory, in which fibrous material was located on the nuclear side of the nuclear membrane.[15]

The isolated NPC preparation by Aaronson and Blobel was composed of very little RNA and a distinctive set of many proteins, with three major ones having molecular weights of about 67,000. A year later, in 1976, Werner Franke's laboratory in Heidelberg also showed the NPCs were connected by strands and constituted an "interpore skeleton meshwork," suggesting the lamina provides a rather rigid structure that links the many NPCs to the envelope.[16] Nancy Dwyer and Blobel went on to show the NPC-lamina fraction was not an artifact due to the Triton detergent treatment, for salt treatment produced a similar electron micrograph of pores connected by a fibrous lamina.[17] Scientists at the University of Chicago produced scanning electron micrographs of the surfaces of both untreated and detergent-treated nuclei; their images clearly showed the NPCs attached to the underlying laminar structure, thus also ruling out the possible artificiality of the fixation procedure.[18]

A further advance was made in 1978 by Larry Gerace, a Rockefeller graduate student, with Andrea Blum and Blobel.[19] They purified the NPC-lamina preparation, separated the three major proteins by gel electrophoresis, and made antibodies to them. Immunochemical localization in liver nuclei showed these proteins are components of the peripheral nuclear lamina and are not in the NPC as structural elements. This led them to propose that the lamina provides the structural stability of the NPC as well as interacts with the inner nuclear membrane and possibly with the chromatin found at membrane sites in interphase nuclei; a similar observation was made by the Werner Franke group.[20] The Blobel laboratory further reported that during mitotic cell division, when the nuclear envelope disassembles during prophase, these three lamina proteins, or lamins, assume a different localization throughout the cell. This distribution persists until the telophase part of mitosis, when the proteins become localized at the surface of the daughter chromatin masses.[19]

Gerace and Blobel then used synchronized cell cultures to follow the nuclear envelope disassembly and reassembly more closely and observed a reconstitution of the lamina architecture after the completion of mitosis. They examined particularly the fate of the three lamins, which they found to be phosphoproteins that become reversibly depolymerized during mitosis. They concluded that "reversible enzymatic phosphorylations of the lamins may be involved in modulating the state of phosphorylation of the lamina and its reversible mitotic disassembly."[21]

By the early 1980s, there was enough structural information of the NPC for various workers to construct diagrams of its possible structure. In 1982, Nigel Unwin and Ron Milligan from the Medical Research Council in Cambridge combined electron microscopy of isolated nuclear envelopes and processing of the images by densitometry, Fourier analyses, and rotational analyses to produce diagrams of the NPC having discrete subunits.[22] These subunits were arranged with an octagonal symmetry around a central axis; the whole structure was framed by coaxial rings attached to the two nuclear membranes, and spokes attached to these rings extended inward toward a large particulate central plug. In 1986, Laura Davis and Blobel used immuno-ferritin–coupled electron microscopy and found a 62,000 molecular weight protein in the NPC, most likely at the site of the ring or spokes visualized by Unwin and Milligan.[23] That same year, Christopher Akey of Boston University and Michael Radermacher of the New York State Department of Health presented a three-dimensional structural diagram of membrane-associated and detergent-extracted NPCs, which were imaged in frozen buffers by cryo-electron microscopy. The specimens were tilted at various angles, and the images were scanned densitometrically and computationally averaged. Their diagram presented the NPC as a spoke-ring complex, having an inner spoke-ring formed by domains that bridge the inner domains with adjacent spokes to form an eightfold symmetric annulus (similar to that seen by the earlier microscopists). The outer and inner rings attach to the outer and inner nuclear membranes, forming a large complex with a transporter channel specific for various macromolecules that is linked to adjacent NPCs by the nuclear lamina.[24]

By the late 1990s, several laboratories, including the Blobel group, had purified many NPC proteins, made antibodies to them, and localized them to various sites in the NPC via electron microscopy.[25,26] The most extensive collection of these proteins, from yeast NPC preparations, were characterized in extensive studies by Michael Rout, who joined Blobel's laboratory in 1990.[27] He now heads his own laboratory at Rockefeller, where, with Brian Chait's laboratory and John Aitchison (also an alumnus of the Blobel laboratory), he purified all the NPC component proteins by chromatographic and gel electrophoresis methods and identified them by mass spectrometry. They then used molecular genetic techniques to "tag" all candidate proteins in order to determine which were nucleoporins, the name given to NPC component proteins. Each nucleoporin was then localized within the NPC by immunoelectron microscopy, and its relative abundance was measured to generate, with Andrej Sali, a map of the molecular architecture of the yeast NPC.[28] This map and prior reconstructions based on cryo-electron microscopy of the NPC,[29] along with recent crystallographic structure determinations of various nucleoporins by the Blobel laboratory and others, finally led to the construction of a precise molecular map of the architecture of this elegant and efficient transporter almost fifty years after it was initially observed by Michael Watson.[30]

Chapter 10

Enlarging the Horizons of Modern Cell Biology

Introduction

Carol L. Moberg

Before Scientific Director Herbert Gasser retired in 1953, the future direction of the Institute had been uppermost on the trustees' agenda. After World War II ended, Gasser was able to maintain the research standards and environment for the staff, increase attention to fundamental science, and open new fields for investigation such as the chemistry of proteins, other basic molecules of life, mechanisms of immunology and heredity, the nature of viruses, and, as this book has been demonstrating, the components and energy systems of the cell. While the Institute had once been almost unique in American medical science, this changed after the war when universities began to rival its activities and compete for the most promising investigators. The Institute had contributed to this situation by training hundreds of scientists who went forth to key positions in other institutions. Some scientists came to believe the original mission of the Institute was fulfilled and thus should be discontinued.[1–3]

David Rockefeller, who succeeded his father as chairman of the Board of Trustees in 1950, appointed a committee to consider future directions. The committee was headed by Detlev Wulf Bronk, a member of the Board of Scientific Directors, who was concurrently president of the Johns Hopkins University as well as president of the National Academy of Sciences. Bronk was a distinguished biophysicist and, like his close friend Gasser, was known for his research on the electrical activity of nerve fibers.

During its reappraisal, the committee consulted more than a hundred top-ranking scientists and educators during 1952, including the full members of the Institute, about how to "continue to accomplish its objectives in accordance with the distinguished traditions of the past, and still be in a position to respond to the new opportunities opening up in medical research today."[4] Keith Porter's response to a letter from David Rockefeller clearly reflected Gasser's efforts to focus on the study of fundamental biological problems. To that end, Porter noted, "medicine has made greater strides in the cure of ills which arise from the dysfunction of the component cells of the body. To deal with these latter a better understanding is required of tissue cells, how they are constructed, why they grow, differentiate, and age. If emphasis were to be placed in any line of study, it might be on cell biology." After praising the physical facilities for scientific investigation at the Institute as "near to perfect," he added there was room for improvement in the selection of scientific staff members: "the Institute would naturally benefit from a greater turnover of young people which in turn might be obtained by a special two or three year Institute fellowship available to postdoctoral students."[5]

In ways Porter had not anticipated, the trustees in 1953 acted on their findings that there "was widespread lack of knowledge among leading scientists of the significance of the Institute."[6] They approved the vision of Bronk to formally train young scientists by turning the Institute into a graduate university of science empowered to grant PhD degrees; they also appointed Bronk as president to replace the retiring Gasser.* Initially, only ten students were admitted each year, thus creating a small university that would keep its emphasis on individuals rather than academic departments. There was no tuition, no core curriculum, and no grading system; each student engaged in an independent program of study, working closely with distinguished senior scientists in established laboratories while learning from journals, tutorials, and seminars.

Going much farther in giving direction to the Institute, the trustees urged the new administration to find ways to reduce its isolation and broaden its scope. Their recommendations, in fact, overturned many traditional policies of the Institute. In particular, they advised fostering closer and more extensive relations with other universities so there could be a greater flow of older and younger scientists into and out of the Institute; increasing ways to inform the larger scientific community of the Institute's activities in scientific and professional journals; and, perhaps most surprising, accepting grants from individuals and agencies "for the support of self-initiated activity of the staff."

Perhaps no laboratory profited more quickly from all these recommendations than cell biology. By the mid-1950s, it was not only the largest of the laboratories but also the most visited by outsiders and the fastest growing.

Within the first year, the laboratories of Porter and biophysicist Keffer Hartline received a joint grant of $16,500 from the American Cancer Society to develop and extend the scope of the electron microscope "on the morphology, biochemistry, and biophysics of normal and abnormal cells and on their processes of growth and metabolism."[7] And, within the first decade of the graduate program, twelve students earned their PhDs in the cell biology laboratory of Porter and Palade: Mary Bonneville, Lewis Greene, James Jamieson, Arthur Karlin, Jack Kirsch, David Luck, Lee Peachey, David Sabatini, Peter Satir, Leonard Sauer, Philip Seeman, and Roger Thies. Many other cell biology students followed these individuals. In previous chapters, students Bonneville, Jamieson, Sabatini, and Satir told their scientific stories in the laboratory. Now, in the first section of this chapter, Peter Satir relates his experiences as one of the first students.

To become a viable field of science, however, cell biology needed more than a stable work environment with a continuous flow of new and young talent to enhance its internal strength. One way for the new science to reach the outside world was to create a journal for all cell biologists everywhere to report on their scientific investigations. Moreover, a professional society was needed to further enhance their communication and promote their research among peers. In all of these activities, Porter and Palade found Bronk a constant source of support for enlarging these horizons. The second and third sections of this chapter describe their efforts in founding *The Journal of Biophysical and Biochemical Cytology* and The American Society for Cell Biology.

*The Institute did not officially change its name to The Rockefeller University until 1965, although full members of the scientific staff immediately received the additional title of professor.

10A. Training the First PhDs in Cell Biology

Peter Satir

Detlev Bronk recruited graduate students to The Rockefeller Institute for Medical Research to begin studies in the autumn of 1955. It was a daring move. He faced widespread, although by no means unanimous, opposition by the members who saw their palace of quiet scholarship and serious scientific investigation being invaded by a group of untamed, wet behind the ears, and bumptious, albeit intelligent and perhaps interested, youths. The heads of laboratories, called members, were the equivalent of chairmen of small departments—very prominent, established scientists and senior faculty. They were content with training postdoctoral fellows, as they had done for many years. Postdoctoral fellows were much more knowledgeable about the techniques of scientific research; they were committed to a specific problem and often had accomplished something with that problem in their earlier studies. In short, they were possibly more predictable and certainly more useful.

Bronk and Frank Brink, who was to become the first graduate dean, were not worried. Despite his national responsibilities as president of the National Academy of Sciences and his other institutional responsibilities, Bronk personally invested a massive amount of time and energy recruiting and interviewing potential students and coddling them once they were on campus. The students were well paid: the stipend was essentially three times the going National Science Foundation fellowship rate, and the fringe benefits were extraordinary. Not only did they include subsidized housing and a garage, but once a week the students were invited to have cocktails and dinner with the president and his guests, which normally included the speakers of the week, but also might include movers and shakers from the New York world of music, art, publishing, and so on.

The new students came in two categories: newly minted and more mature. The latter included a group of people who had already earned their MD degrees, for example Howard Rasmussen and Gerald Edelman, and a group who had already spent several years at graduate study, such as Lee Peachey. The former included me, and others like me, who were recruited directly from college. Bronk had asked a number of college presidents to nominate candidates for the new program. At Columbia, that request was passed on to the zoology department of Columbia College, whose faculty nominated a candidate for admittance (me). The procedure was more or less the same for most of my colleagues; usually, the nominee was known to some influential members of the life sciences faculty at the nominating school. To get into the program, we not only had to pass scrutiny by Bronk and Brink, but also to remain on the good side of Mabel Bright, administrator par excellence. We brash youths were little more than twenty years old when accepted into the Rockefeller program; our more mature MD colleagues were nearer thirty. Women comprised a small part of each of the initial classes; unfortunately, there were no minority students.

It was not at all clear what would comprise the new graduate curriculum. Bronk thought that students learned best if left much to their own devices. He believed that they should be exposed to great minds and the important modern scientific concepts that were characteristic of the Institute; in due course, they would find interesting problems and good mentors and produce a thesis. So initially, a required series of lectures for the graduate students was developed. It dealt with currents in contemporary science, ranging from quantum mechanics discussed by Isador I. Rabi to kidney physiology by Homer Smith. There was an impressive series of guest speakers from around the country and the world. Most of the faculty, however, were from the Rockefeller laboratories. One laboratory per week was the rule: Lewis Longsworth, Duncan MacInnes, and Theodore Shedlovsky on physical chemistry in biology; Lyman Craig, Stanford Moore, and William Stein on chemistry of natural products and proteins; Keith Porter and George Palade on the structure and function of cells; and Alfred Mirsky and Vincent Allfrey on biochemistry of the cell nucleus, followed by Fritz Lipmann, Rollin Hotchkiss, and so on through the year to René Dubos and Peyton Rous—every morning from October 1 to May 31. To supplement the formal interactions, there was, of course, that weekly dinner, first in one of the small rooms adjacent to the old main dining hall, and later in Abby Aldrich Rockefeller Hall. If a lot was required of the speakers, there was little required from the students, beyond attendance at the lectures.

The close daily associations between a large faculty and a few students made rigidly defined requirements for study unnecessary. There was no homework and no examinations. A PhD was awarded on the basis of a significant experimental or theoretical research problem, a written dissertation required to have "literary merit," and an oral defense. Some students, particularly the MD group, had come knowing which laboratories interested them; the rest were encouraged to explore various possibilities, either by doing a laboratory project, as I did with Moses Kunitz, or by simply engaging the various heads of the laboratories in conversation about what they were doing. There were about fifty laboratories, but the students were few in number (my class had nine), and their arrival in a laboratory was an event that was likely to bring presidential favor, so the students had easy access to the laboratory heads. They also had their own account at the stockroom, where they could get laboratory supplies or special small items of equipment merely by asking for them. The academic hierarchy trembled and the postdocs were none too happy about this state of affairs, but the students, whether they realized it or not, were in scientific and academic heaven.

Most of my colleagues and I were neither so young nor so naive as to think that this situation would last forever. Most of us soon sorted ourselves out according to our interests. Pretty soon there were several laboratory courses for students to take, notably biochemistry and cytology (as cell biology was originally called). The four-month course on cytology under the direction of Palade occupied several days a week. There were two hours of lectures, five hours of laboratory exercises, and a seminar. The course integrated cytological information obtained by light microscopy, electron microscopy, cytochemistry, and cell fractionation procedures. Given the newness of this field, the principles, advantages, and limitations of these methods in cell research were also discussed. We were introduced to another parade of eminent or soon to be eminent cytologists, including Don Fawcett, Elizabeth Hay, and Jean Paul Revel, from across the street, Helen Deane and Alex Novikoff, from across the city, and Stanley Holt and Daniel Mazia, from far away. It also gave us a chance to interact in seminars and question periods more intensively with the Porter/Palade group. One of my first required presentations in the course was a review of a paper by Thomas Anderson on critical point drying; this made such an impression that I later used the technique to my advantage several times in my career. It was also in this course that Palade first found his affinity for exposition to students, which attracted many people to his mentorship.

At first, the laboratory of cytology, under the direction of newly promoted members Porter and Palade, was somewhat reluctant to take on the younger group of students. It was a different story with Lee Peachey, who had worked at the University of Rochester with Michael L. Watson, one of the associates of the Porter/Palade laboratory in 1955–56. Lee was already introduced into the art of electron microscopy. I wanted to learn the technique to tackle the problem of ciliary motility, on which Porter had written extensively. I was attracted by the beautiful 9+2 pattern of these organelles and especially by the questions Porter posed in his Harvey Lecture of 1956. Despite my preliminary explorations of other laboratories, I had decided where I wanted to work, and all that remained was convincing Keith Porter to take me on.

Because there were so few students, who were thrown together by dint of their common experiences, we grew to know each other reasonably well. When we arrived on campus in October 1956, the graduate student residence hall, together with Abby Aldrich Rockefeller Hall, Caspary Hall, and Caspary Auditorium, were still a gleam in Bronk's eye, but housing was provided for the students in a series of rent-subsidized walk-up double apartments in small buildings near the campus. Fortuitously, Lee and I were roommates, occupying apartment 6H at 1482 York Avenue at a cost of $57.50 per month each. We helped each other along with our respective tasks, his to study muscle structure, mine to learn to take in-focus electron micrographs of cilia and structures related to cilia.

The third student to join the Porter laboratory was Mary Bonneville. Before she came as a member of the first graduate student class, Mary had done some work with embryologists Oscar Schotte, and she was intensely interested in developmental biology. The developmental biology laboratory at Rockefeller was headed by Paul Weiss, who was a superb intellect but a very difficult man for students to work with. In any event, by 1959, Mary was a member of the Porter laboratory and, although Maria Rudzinska became one of her important mentors, she remained in Porter's orbit throughout her graduate studies and for many years afterward.

Since in the late 1950s virtually every organism and cell biological problem was unexplored at the ultrastructural level and Porter and Palade's standards for electron microscopy were the highest in the country (initially rivaled only by Fritiof S. Sjöstrand in the world), the laboratory attracted a cadre of associates eager to explore various aspects of the cell biological world: protozoa, insects, perfused mammalian organs, mitochondria, nuclei, membranes, capillary transport, and on and on. These people, together with the junior faculty in the laboratory, Montrose (Monty) Moses and Philip Siekevitz, had a profound influence on the students, and the mutual interactions enhanced the laboratory experience.

Important technical improvements in electron microscopy were also occurring, and we students wanted to be in on the latest techniques, from glass to diamond knives, new buffers for fixation, new fixatives, and new embedments. Lee and I were caught by the problem of section flattening; he was studying this systematically. Because my sections were so crumpled as to be almost unusable, I discovered that when I wafted xylene vapor over them, the sections magically flattened to their correct length. With Lee's help, a lot of support from Steven Wissig, one of the postdoctoral associates working on the kidney, and with the blessing of

Porter, my finding reached publication together with Lee's extensive study in *The Journal of Biophysical and Biochemical Cytology*, the first publications of the cell biology students.

There were also new microscopes arriving in the basement of Theobald Smith Hall—a Siemens 1A was the pride of the department. The microscope was reserved for Porter and Palade (although Palade often used the older RCA). Advanced graduate students used the RCA; new people, like me, began on the Phillips 100 in the province of Dan Moore, with the technical assistance of Don Jaspar. This was also the general electron microscope for the whole Institute, and there was continuous jockeying for time. There was considerable ferment and discussion, particularly at teatime when new micrographs, hot off the print drier, were passed around for Porter and Palade's comments and suggestions, or in the lunchroom, where members of laboratory usually ate together (some of the students, including me, enjoyed lab hopping at lunchtime and missed many of these sessions).

About this time, Porter and Palade headed in slightly different scientific directions within the field, Porter concentrating on cell motility and Palade concentrating on protein synthesis. Palade attracted two additional students to the laboratory to work on aspects of the synthesis problem: Lewis Greene, who was to demonstrate that purified cellular fractions of the isolated zymogen granules from the exocrine pancreas actually contained the proteolytic enzymes, such as chymotrypsin; and Jack Kirsch, who studied the small particulate component of the cytoplasm that Palade had defined in 1955, which shortly became known as the ribosome.

By 1959, the student cohort in the laboratory was completed with the addition of David Luck, an MD who had served in the Korean War. He entered the graduate program to study basic problems and was captured by the question of mitochondrial replication. The electron microscopy of mitochondrial structure was one of Palade's interests and the subject of one of his most serious disputes with Fritiof Sjöstrand (Palade was right). Of the initial cell biology student group, Luck was the only one who stayed at Rockefeller, rising through the ranks to professor and vice president.

The year 1959 was a momentous one for the student program, since the first students to complete their work were to receive their degrees that June. This would be done with tremendous pomp and circumstance, as befitting the institution and consistent with the vision and style of Bronk. The university presidents of Harvard and Bologna, Lord Adrian of Cambridge, and many other dignitaries would attend and receive honorary degrees. Five students, including Lee, representing the only cell biology student and the only non-MD, would receive their degrees (although Howard Rasmussen had ties to the laboratory). The ceremony would set a pattern for future years.

I was not there for this great event. Because Bronk had a world view of science, he believed that the graduate students should also develop this view by studying abroad for part of their graduate careers. Of course, we would continue to receive the Rockefeller stipend during our leave, making us reasonably affluent in Europe at that time. Many students took advantage of this program to study in some of the best European laboratories. Jack Kirsch studied in Cambridge, England, and Copenhagen, Denmark, and reported that Copenhagen was an exciting place, both scientifically and socially. Kai Linderstrom-Lang, head of the Carlsberg Laboratory in Copenhagen, was one of the lecturers in the orientation series in the 1957–1958 academic year. He was making strong contributions to protein chemistry (he coined the terms "primary," "secondary," and "tertiary" structure) and had a reputation of making his laboratory a special place to work. I thought that it would be necessary to know as much as possible about proteins to understand cell biology, and more particularly ciliary movement, and that I could profit from an experience that broadened my cultural background and from living in a country whose language was not English (although you could get along well in everyday exchanges using English). I arranged to work for a year in Denmark with Linderstrom-Lang, but as it happened, he became ill and I ended up working with Erik Zeuthen at the Biological Institute of the Carlsberg Foundation.

When I returned from Europe in October 1959, Rockefeller had changed. The name of the Institute had been shortened by dropping the phase "for Medical Research." Bronk had introduced philosophers and humanities professors to tread among the scientists. The graduate student residence hall and Caspary and Abby Aldrich Rockefeller halls had been built; the student-faculty club opened its doors. Students were no longer such a novelty, and the faculty began to exert its prerogatives: there were more formal courses and more formal reports. Bronk Laboratory (then called South Laboratory), the first new research building of the university era, was being readied for occupancy, and the laboratory of Porter and Palade was scheduled to move from the old Theobald Smith Hall basement into the fifth floor of the new building in the very near future. The pioneer days of the graduate program as well as cell biology were almost over, with the realization that the constant cell organelle morphology seen by electron microscopy reflected molecular organization.

By 1960, fourteen students had finally completed their theses and graduated; the faculty could now assess the strength of their training more completely. There was considerable dissatisfaction; students would come up to their final examinations—their only examinations—with excellent work, but some were unable to place the work in the widest context that the Institute had hoped to engender. The courses had been taken, the lectures attended, but there had been no time for review, reflection, or consolidation, which usually occurs at a qualifying examination. Bronk had been firm about what examinations were required. Porter came up with a solution to this dilemma, for which he enlisted (or perhaps extracted) my aid. I was asked to "volunteer" for a preliminary examination. Once I passed, the concern regarding preliminary examinations was eliminated.

The great move of the laboratory to South Laboratory took place, but, although a small conference/tea room appeared, the new quarters were less congenial than the old for intermixing. Porter was away a lot. I completed my work on instantaneous fixation of the ciliary beat and began writing my thesis for graduation in June 1961. Mary wrote her thesis on intestinal fine structure during amphibian metamorphosis for the same date. We learned that Porter was moving to Harvard immediately after our graduation, so that we three (Lee, Mary, and I) were to be his only students at Rockefeller. New students began to be recruited to the Palade laboratory: Jim Jamieson and David Sabatini soon appeared on the scene.

The graduating class and their parents were invited to the president's house the evening before the formal convocation to dine with Dr. and Mrs. Bronk and selected trustees. We were the third graduating class, consisting of ten students. Those of us who were touched by the cell biology bug had lived through a special era. To commemorate the occasion, together with our fellow graduates, we presented President Bronk with an antique book on microscopy, Martin Ledermüller's eighteenth-century *Amusement Microscopique*. His most cordial thank you letter of 8 July 1961 observed that our "choice of rare books of historic significance makes it the perfect gift from young scholars to an institution that has sought to nurture scholarly ideals and interests. . . . The Institute will always have a deep interest in your career and will take especial pride in your achievements."

We were a feisty bunch who made an impression on the Institute, as Porter said the next day, when we received our diplomas after short orations by our mentors on our accomplishments, just like the distinguished guests who received honorary degrees. In many ways, we set the pattern for what was to come in the graduate program. I like to think that from the Rockefeller experience, we learned both the spirit of scientific generosity and the rigor of scientific investigation that have informed our lives and careers since those special days.

10B. Starting a New Journal of Cell Biology

Carol L. Moberg

The process of starting a scientific journal required ardor, perseverance, and, not so obviously, political skills.

With the rapid growth during the 1940s of studies in electron microscopy and the isolation of subcellular organelles, a new era of cell research made its appearance. This development raised the issue of suitable journals as outlets for the new research. The American and European organismic or cellular journals being published at that time focused on gross anatomy, histology, general cytology, or light microscopy, but none really conveyed the tenor of current research. Probably the first to devote its contents to the newer aspects of the cell, at least in part, were the Swedish journal *Experimental Cell Research*, founded by Torbjörn Caspersson and John Runnström in 1950, and the *International Review of Cytology*, founded by Geoffrey Bourne and James Danielli in 1952. Next was *The Journal of Biophysical and Biochemical Cytology* (*JBBC*), the precursor name to *The Journal of Cell Biology* (*JCB*) founded in 1954 at Rockefeller.

An impetus for Keith Porter and George Palade to consider creating a journal was their insistence on high quality reproductions of electron micrographs. *The Journal of Experimental Medicine* could have been an ideal outlet. It had published the 1945 paper by Porter, Albert Claude, and Ernest Fullam containing the first electron micrograph of a complete cell. By 1953, this journal also published six other papers by Porter and postdoctoral fellows with micrographs of various cellular structures. In 1947, Porter had gone to lengths to find a special paper for the journal's photographic reproductions of the electron micrographs, telling editor Peyton Rous that it "would not crack or curl and should last a lifetime" and could be "bound in much the same as any extra page."[1] Moreover, on the death of Simon Flexner in 1946, the journal appointed two new, and sympathetic, editors from the Institute: scientific director and physiologist Herbert Gasser and microbiologist René Dubos, whose research was evolving from studies of pathogenic bacteria to those of host-parasite relationships. While balancing the journal's traditional focus on pathology with the newer interests in physiology, the editors wanted to expand its focus beyond diseases. As mentioned in Chapters 3 and 5, Gasser was urging the Institute's scientific staff to study "fundamental life processes at the level of the cell and its constituents," and he strongly supported the work of Porter and Palade by setting up an independent laboratory for them in 1950 upon James Murphy's retirement.[2]

Despite improved techniques in embedding, fixation, and sectioning that led to vastly improved electron micrographs in the early 1950s, Porter and Palade were not satisfied with the quality of the reproductions. Porter and H. Stanley Bennett of the University of Washington published an electron microscopic study of striated muscle in the *American Journal of Anatomy*, and they were greatly disappointed in the quality of the halftones of their electron micrographs.[3] Then in 1953, a paper by Don Fawcett and Porter on the 9+2 structure of cilia was rejected by *The Journal of Experimental Medicine*'s editors Vincent Dole and René Dubos, but not by editors Gasser and Rous.[4] Porter's later explanation of the rejected paper was that their "study involved little that was experimental, that it was not medical, and, moreover, that the editors did not wish to publish morphological papers."[5] Bennett and Porter published the paper elsewhere, but they were again dissatisfied with the poor quality micrographs.[6]

A short time later, Gasser came to the aid of the cell biologists once again and, according to Porter, suggested that he start a new journal. In October 1953, Porter surveyed the situation for Gasser, stating that the major American journals—*Anatomical Record*, *Biological Bulletin*, *Journal of Morphology*, and *American Journal of Anatomy*—were focused on morphology, the European journals were in states of revival, and only the Institute's own journals and the Swedish *Experimental Cell Research* had good reproduction of figures. Importantly, he conveyed Palade's suggestion that the proposed journal "should have the additional mission of bringing together for correlative purposes reports on cytochemistry as well as fine structure." Thus, his survey anticipated two important aspects of shaping a new publication to convey the tenor of the new research: a fresh set of editorial policies in addition to high quality reproduction of electron micrographs.[7]

He tentatively suggested two names for the proposed journal, *Journal of Micromorphology* or *Journal of Cytology*. While the first name narrowly conveys the elements of cell structure, the second choice seemed more justified. Cytology was the name of the Porter and Palade laboratory that had been established by Gasser in 1950, so a journal by this name would readily fit the descriptions of the research they were doing. (Indeed, their laboratory's name was not changed from Cytology to Cell Biology until 1967. This was due partly to Alfred Mirsky and Vincent Allfrey, who in 1963 changed the name of their Rockefeller laboratory from General Physiology to Cell Biology.) Moreover, how the curious use of "cell" as an adjective came to be commonly accepted is unknown. Edmund B. Wilson in the first edition of his famous book *The Cell in Development and Heredity* (1896) used the adjectival form "cellular," which is the more proper usage and corresponds to "molecular" biology. Until the late 1950s, Porter continued to refer to cytology or a new branch of cytology but not the more general area that came to be called "cell biology."

A few weeks after Porter's letter to Gasser, Bennett wrote a similar letter to President Bronk and reiterated the need for a new journal.[8] Perhaps quite by accident, his proposal fell into line with what Bronk had just proposed to the Board of Trustees: the creation of a Rockefeller Institute Press that would consolidate and manage the continued publication of its two long-standing journals, *The Journal of Experimental Medicine* and *The Journal of General Physiology*, as well as the *Studies* (a series of collected reprints by Institute members), the occasional scientific monographs, and annual institutional pamphlets.[9]

A new journal, Bennett wrote, "which might hypothetically be named '*Journal of Cytology*'—derives from the fact that the world is now on the threshold of a classical period of advance in cytological knowledge comparable in significance to the Vesalian advance in anatomy or the classical development of microscopic anatomy in the nineteenth century. The principle instrument responsible for this advance is the electron microscope." After citing some classic studies already made at Rockefeller and elsewhere in this field and the unsuitability of extant journals, Bennett urged that "the resources in substance and scholarly background available to the Institute" were important reasons to sponsor such a journal.[8]

In January 1954, after consulting with Francis O. Schmitt at the Massachusetts Institute of Technology (MIT), Bronk proposed, and the Board authorized, the establishment of a third journal tentatively called *Journal of Cytology*. His proposal stated the journal would deal with this field "from the point of view of the anatomist and histologist; the physiologist; the biochemist, and especially the histochemist and those concerned with the development of the exciting new physical and chemical methods for investigation of the submicroscopic molecular structure of cells." Importantly, according to the proposal, the journal would report research done using a variety of direct and indirect methods, instruments, and biochemical analyses, but it would "not concentrate on such methods inasmuch as the course of cytology will ultimately move on with the aid of other unanticipated, undeveloped methods." In other words, the publication would go beyond being a journal of morphology and electron microscopy.[10,11]

Shortly afterward, Porter sent Bronk a potential list of editors for the new journal.[12] This list was then modified by Schmitt, who was "enthusiastic" about the project but believed the scope of the coverage should be "devoted to 'biology at the molecular (or near-molecular) level.'" Since he found the list too "loaded with morphologists on the more descriptive side," he suggested scientists in other fields who would broaden its interdisciplinary nature.[13] After Bronk met with these nominated individuals on 12 April 1954 at the Federation of Societies for Experimental Biology and Medicine in Atlantic City, he issued a report of the eight scientists who agreed to serve on the journal's editorial board. They were biophysicist Francis Schmitt and crystallographer Richard S. Bear of MIT, biochemist Albert L. Lehninger of Johns Hopkins University, cytogeneticist Franz Schrader of Columbia University, and histochemist Arnold M. Seligman of Beth Israel Hospital in Boston, in addition to Bennett, Porter, and Palade.

Also, during Bronk's lengthy discussion with these editors regarding the scope and desirable content of the journal, the working name *Journal of Cytology* was abandoned for conveying too strong a focus on morphology and not enough on cell functions. Curiously, according to Porter, "No one proposed the name *The Journal of Cell Biology*. At times, disagreements about the title seemed sharp. Realizing that we had not agreed on a title, Bronk threateningly said that if that issue stood in the way, it might be that a new journal was not needed, and left the table. Thereafter the tone of the meeting became more harmonious, and after further conversation, the qualifying adjectives 'biophysical' and 'biochemical' were suggested, to specify and to dignify the general topic 'cytology,' and to convey to all that the context of cytology, as used by the new journal, was not restricted to study of chromosomes. By the end of the lunch, all agreed that *The Journal of Biophysical and Biochemical Cytology* (*JBBC*) was the most appropriate that the assembly could think of." It was also agreed that the journal would be a bimonthly and its first issue would appear in January 1955.[5]

In July 1954, a formal advertisement was sent to major journals and members of appropriate scientific societies announcing the publication and function of a new bimonthly journal and requesting manuscripts. Porter was clearly unhappy with the journal's extended name and, despite Schmitt's protest to Bronk about the cover design, Porter had it (and the announcement) designed in a subtle way so that the words "Journal of Cytology" stood out in large capital letters while the two adjectives appeared in smaller capital letters on a separate line in between.[14]

THE ROCKEFELLER INSTITUTE FOR MEDICAL RESEARCH

announces the publication of a new journal to be known as

THE JOURNAL OF BIOPHYSICAL AND BIOCHEMICAL CYTOLOGY

IT WILL BE THE PURPOSE of this new publication to provide a common medium for the presentation of morphological, biochemical and biophysical studies of the structure of cells and their components and of the functions of these components. The Journal will give special attention to investigations dealing with cellular organization at colloidal and molecular levels. Papers will be favored which integrate information derived from newer approaches to cytology, such as histochemistry, cytogenetics, cytochemistry, electron microscopy and X-ray diffraction.

Because of the significance of photographic evidence, the publishers of the Journal will endeavor to insure excellence of photoengravings and printing.

THE LENGTH OF ARTICLES will be limited to approximately 25 printed pages exclusive of plates. Short notes of 1500 words or less will be accepted; review articles will be published only on invitation.

THE FORMAT AND STYLING used by the *Journal of Experimental Medicine* and the *Journal of General Physiology*, presently published by The Rockefeller Institute, will be followed by this new Journal.

IT IS PROPOSED to publish one volume per year, the issues appearing bimonthly: the price will be $7.50 per volume. In order that manuscripts may be considered for inclusion in the first number which will be published in January 1955, they must be received not later than 1 November 1954.

THE JOURNAL WILL BE EDITED BY

RICHARD S. BEAR	*Massachusetts Institute of Technology*	KEITH R. PORTER	*The Rockefeller Institute*
H. STANLEY BENNETT	*University of Washington*	FRANCIS O. SCHMITT	*Massachusetts Institute of Technology*
ALBERT L. LEHNINGER	*The Johns Hopkins University*	FRANZ SCHRADER	*Columbia University*
GEORGE E. PALADE	*The Rockefeller Institute*	ARNOLD M. SELIGMAN	*Sinai Hospital of Baltimore*

Manuscripts should be addressed to The Editors of *The Journal of Biophysical and Biochemical Cytology*
The Rockefeller Institute for Medical Research, New York 21, New York

Advertisement sent to major journals and members of scientific societies announcing the publication and function of a new bimonthly journal and requesting manuscripts. Courtesy of Keith Porter.

A month before the journal was launched, Porter, as the first managing editor, reported to the editorial board that they had received twenty-five manuscripts. Editorial policy called for Palade to decide which editors would read a manuscript, that all manuscripts be

THE JOURNAL OF

BIOPHYSICAL AND BIOCHEMICAL

CYTOLOGY

EDITED BY

RICHARD S. BEAR
H. STANLEY BENNETT
ALBERT L. LEHNINGER
GEORGE E. PALADE
KEITH R. PORTER
FRANCIS O. SCHMITT
FRANZ SCHRADER
ARNOLD M. SELIGMAN

VOLUME 1
1955

THE ROCKEFELLER INSTITUTE FOR MEDICAL RESEARCH
NEW YORK

The masthead of the initial issue of the new journal.

approved by at least two editors, and that a manuscript should be published four to five months after it was received. Of the first submissions, three were rejected and eight of the first eleven appeared in the first issue. The first print run of 2,500 copies was mailed on schedule in January 1955. Although subscriptions were "nowhere near this number," Porter was confident "that the later subscription orders and orders for back numbers will use up the surplus."[15] Indeed, there were only 192 paid subscriptions, but this number tripled during its first year of publication.

The papers in the first issue were noteworthy contributions to the new science in that they included the first electron microscope observations of synaptic vesicles in neurons by Eduardo De Robertis and H. Stanley Bennett and the first description in liver tissue of what are now called ribosomes by Palade. There were articles on cytochemistry by William Montagna and by David Bloch and Gabriel Godman; biochemistry of isolated mitochondria by Charles Lowe and Albert Lehninger; and insect flight muscle by Bertram Sacktor. In addition, Cecil E. Hall reported on the staining of virus particles, and Sanford L. Palay and Palade described the equation of Nissl substance in neurons with granular endoplasmic reticulum.

Bronk was immediately pleased with the appearance and quality of the *JBBC* and reported to the board that it "has had an enthusiastic reception among scientists; it is recognized as a reliable medium for the furtherance of science."[16] Until the journal became self-supporting, Bronk maintained a close watch over the costs, subscriptions, and scope of this enterprise while reassuring Porter that he would "back him to the limit."

From the beginning, the editors' annual review of the journal's progress revealed two persisting problems, a domination of papers on tissue and fine cell structure and the journal's name. During its first year, Porter reported that "morphological studies had been well represented among the published papers but that papers more properly defined as biophysical and biochemical had been in short supply." The editors representing these domains agreed to pursue more submissions in their fields. They also agreed to consider other efforts, including short reviews, commentary articles, reprints of classical cytology papers, and even translations of current foreign language papers in cytology.[17]

Perhaps complicating these issues, but also greatly enhancing the journal's stature, was the printing of a supplement in 1956 of the proceedings of the Conference on Tissue Fine Structure, which had been organized by Porter and held in January. The U.S. Public Health Service contributed $8,500 to cover printing costs, and Bronk agreed to cover the deficit for what he believed was a "desirable publication."[18] The supplement was a tremendous addition (450 pages and 147 plates), selling more than 200 additional copies to non-subscribers and increasing the journal's visibility. The editors noted, however, that most of the papers in the supplement were on electron microscopy, and this motivated them to attract a greater diversity of papers.*

Within five years, several minor modifications were made. In order to reflect the changing interests in cell biology, Bronk acted on a suggestion from Porter and established a policy to limit tenure for the editors. Many journals, for example *The Journal of Experimental Medicine*, retained the same editors for years, but Bronk suggested this was not advantageous "especially for a young journal that is publishing reports from a new and rapidly expanding field of investigation." In this way, adding or replacing editors meant securing scientists in various specialties who were knowledgeable about the most currently productive lines of cell research. The number of editors was increased to ten, and, with the exception of Porter, who was pressed to stay on, two were replaced each year.[19]

Also during this phase, the number of volumes increased to two per year, the price per volume increased from $7.50 to $15, and reports relying solely on electron microscopy dropped from double those on other topics to less than half. As Porter and Palade noted in

*Three other notable supplements to the journal focused on a translation of Emilio Veratti's 1901 paper on striated muscle fiber (1961), a translation of Walther Flemming's 1880 paper on mitosis (1965), and a special issue devoted to "Discovery in Cell Biology" (1981).

1959, "our contributors and readers now recognize that the Journal is interested in cell biology regardless of the experiment, device or technic [*sic*] used to acquire the information."[20] The format of the journal changed as well, from a one- to two-tone color cover, two columns per page, and the addition of an abstract below the title and authors' names. Porter believed a lag of four to six months between receipt of a manuscript and its publication was justified by the "careful editing"—in some cases, rewriting—of manuscripts and by obtaining high quality, foreign-made engravings of the micrographs. By 1959, the number of subscribers had increased to 1,500, divided almost equally between domestic and foreign subscribers, of which seventy percent were institutional.

A year later, submissions to the *JBBC* had increased, leading to three volumes a year and a backlog of a hundred accepted papers awaiting publication. The question arose whether to reject more papers, publish more frequently, or restrict the scope of the journal. The first two tasks were taking so much time away from Porter and Palade's experimental work that they asked Bronk to appoint a full-time executive editor. Raymond Griffiths, an MD with a PhD in biology who had been the publications editor at the American Cancer Society for many years, assumed his duties in December 1960 and immediately improved the journal's editorial efficiency. The *JBBC* became a monthly publication in January 1961.[21,22]

The third matter revived the problem of the journal's content and name. By this time, other specialty journals were starting up to report on the escalating scientific research after the war and thus were competing for papers on similar topics: *Biochemica et Biophysica Acta* (1947), Fritiof Sjöstrand's *Journal of Ultrastructural Research* (1957), *Journal of Molecular Biology* (1959), *Biochemical and Biophysical Research Communications* (1959), and even Rockefeller's newest publication *Biophysical Journal* (1960). Porter and Palade suggested that while the *JBBC* would "continue to get a few manuscripts that might just as readily be published in one of these others," they noted it should remain broadly focused and "thus favor papers which attempt to draw together and correlate information derived from the several narrower and more highly specialized approaches to cell study."[21]

To better refine this expansive focus, a proposal was made at the May 1961 meeting of the editorial board to change the journal's name to *The Journal of Cell Biology*. According to Griffiths's minutes from that meeting, "it was pointed out that the traditional meaning of 'cytology' largely implies structure only, and most especially chromosome morphology, and that this term is currently being used by clinicians with reference to smears in the diagnosis of cancer." Despite unanimous support for the name change, it was resolved not to make this change hastily.[23]

So Porter formally canvassed the editors for their opinions in a vigorous letter of 23 June 1961.[24] He argued that the predominant contents of the journal currently ranged from "cytochemistry and histochemistry of cells and tissues, to descriptions of cell components and products, studies of cell fine structure, comparative cytology, cell physiology, and experimental cell biology . . . [that] could be included under the blanket term—Cell Biology." Moreover, the purely biophysical and biochemical papers had steadily declined. Noting the increased interest of the cell as a unit and the correlation of cell fine structure and function, Porter urged this journal should balance other journals more focused on greater fragmentation and narrow specialization of scientific disciplines. In other words, he felt a journal of cell biology should not focus on the cell as an object of biological interest but rather on larger aspects of methods and approaches to the study of cells in biological systems.

Lastly, Porter urged the editors to act quickly before "someone else will start a journal with these specific aims and this suggested name." In particular, a new organization—The American Society for Cell Biology—had just been formed (its creation in 1960 and Porter's major role in shaping the ASCB are described in Chapter 10C). Porter was optimistic that by changing the name to cell biology, the new society would be "encouraged more than now to adopt the journal as its official publication," thus securing a solid base of subscribers.

On 16 August 1961, Porter wrote to Bronk and included the replies from the editors, who were nine to one in favor of either *Cell Biology* or *The Journal of Cell Biology*.[25] The only dissent was from Palade, who had published twenty papers in the *JBBC* since 1955 and was interested in its stability. He felt the title cell biology indicated "a withdrawal to a more conservative position centered on cell structure and cell physiology in the traditional sense. Yet cell physiology is already biochemistry and biophysics at the cellular level." While admitting they had not succeeded in integrating various approaches, "still we attracted enough papers to keep in contact with molecular biology at one flank while trying to keep away histology at the other." He worried they would lose contact with the interesting flank and "drift in the opposite direction." To retain the newer trend, he suggested the title 'Cellular and Subcellular Biology,' using the correct adjective form of the word cell while noting it was more relevant than the "catch-word combination like cell biology."[26]

Palade's significant comments, Porter wrote Bronk, had to be weighed against the current competition of journals characterized as biophysical or biochemical while "the majority of effective approaches to the study of cell biology these days can be regarded as biophysical or biochemical." It was more important to emphasize "how cell biology is a recognized field" that uses all the modern methods. A further impetus to changing the name, he concluded, was that cell biology had just become a recognized and defined field of interest for 600 scientists who "wanted to join a new society of that name before its first meeting." Bronk approved the name change to *The Journal of Cell Biology*, beginning with volume twelve in January 1962.

During the following year, The American Society for Cell Biology looked into making an official affiliation with an established journal, considering as candidates *The Journal of Comparative and Cellular Physiology* (Wistar Institute), *Experimental Cell Research* (affiliated with the International Society for Cell Biology), *The Journal of General Physiology* (affiliated with the Society of General Physiologists), and *The Journal of Cell Biology*. In November 1962, the Society's publication committee recommended that its president Alex Novikoff propose to Bronk that it "serve as an official sponsor of *The Journal of Cell Biology*. The society recognizes that some of the most fruitful research in cell biology now deals with the biochemical and biophysical aspects of cell function and organization. The Society as now constituted could contribute measurably to the expression of these significant areas in the Journal."[27] The editorial board would be expanded to twelve members, with two representatives from the ASCB Council and one from The Rockefeller University, all of whose appointments were be subject to approval by the University's president. Bronk agreed to this proposal a month later, and a mail vote by the 1,200 ASCB members during the spring of 1963 overwhelmingly approved this affiliation. The masthead of the January 1964 issue (volume twenty) carried the announcement that the journal was now "edited in cooperation with The American Society for Cell Biology."*

With the first issue called *The Journal of Cell Biology*, Porter confidently removed himself from the editorial board. In less than a decade, the journal had met a real need and reached a highly successful plateau. While it is clear Bronk amply supported Porter and Palade's ideas and was active in many decisions, the journal was now deemed capable of prospering on its own. The increasing number of scientists who were becoming cell biologists had an outlet to become the voice of American cell biology.

On the twenty-fifth anniversary of the *JCB*, the editors and senior scientists who were involved in its formative years produced an entire issue called "Discovery in Cell Biology." In his preface to this 1981 historical review of the advances that had taken place, Philip Siekevitz declared a "veritable revolution" had taken place in the biology of the cell: "When one compares the knowledge then with knowledge now, it hardly seems possible that in so short a time we have come to view the cellular world as familiar terrain. We now know its geography and inhabitants relatively well; we know how it evolved, and even know something about its governance."[28]

In 2005, the *JCB* celebrated its fiftieth year of publication by The Rockefeller University Press. Ira Mellman, its editor in chief, noted that the journal continues to be "an undertaking that both builds and is sustained by the community of cell biologists. As a result, the work published in [its] pages has formed the very definition of cell biology."[29,30]

*Of historical interest, although beyond the scope of these pages, the affiliation of ASCB with *The Journal of Cell Biology* ended in 1995, and the Society continued publishing its own journal, *Molecular Biology of the Cell.*

10C. New Supporting Institutions for Cell Biology

Carol L. Moberg

In addition to starting a new journal, other institutional efforts were made to give prominence to the rapidly evolving field of science. In 1950 the term "cell biology" was still in its infancy. Perhaps the initial appearance of this term came in 1947; the International Society for Experimental Cytology (founded in Germany in 1927) was renamed the International Society for Cell Biology when it became a division of the International Union of Biological Sciences (IUBS). This new society began publishing a journal at first titled *Cell Biological Research* but quickly changed its name to *Experimental Cell Research* (mentioned in Chapter 10B). Since this small society restricted its membership to those with certain research experience, it soon caused resentment among younger cell biologists who were not yet qualified; while it included some Americans, including Keith Porter, this society did not achieve much recognition from the international community of scientists.[1,2]

Many practitioners of these disciplines, especially in the United States, had no milieu among existing biological societies where they could congregate to exchange views. There was an urgency to get together, a hesitancy about how to combine so many disciplines, but never a doubt that a separate organization was needed. Contemplating a society in America devoted to cell biology was not straightforward. The process meant assuring each group that their own interests and memberships would not be diverted, diluted, or compromised. It took two years, from 1958 to 1960, before the American Society for Cell Biology (ASCB) was formed.[3]

The impetus for creating a representative American society came in 1958 from two existing professional institutions. One was the Cell Biology Study Section of the National Institutes of Health (NIH) that was formed to evaluate grant applications having interdisciplinary approaches. The other was the Tissue Culture Association (TCA) that was created in 1946 to foster standardized laboratory media as a research tool; its origins were described in Chapter 5. Keith Porter, as co-founder of the TCA, a member of the NIH study section, and co-founder and editor of *The Journal of Biophysical and Biochemical Cytology* (*JBBC*), played a prominent role in getting scientists from these institutions to form a broader-based organization.

In 1956, the American Cancer Society ended its relationship with the Committee on Growth (COG) at the National Research Council for handling its research grant funds. As described in Chapter 3, The Rockefeller Institute's James Murphy, an original member of this Committee, helped to initiate this affiliation in the belief that it would give scientific credibility to cancer research and would avoid conflict of interest problems at the society. A decade later, however, the COG had become persuaded, in part, that "there were no clear boundaries that could define cancer research."[4] D. Eugene Copeland, a key COG administrator, then moved to the NIH, where he managed research grants for the Morphology and Genetics Study Section. By 1958, the Morphology Section became independent, and within a few days Copeland changed its name to Cell Biology.[5]

One of the first activities of the Cell Biology Study Section was to put together a program committee, with Porter as chairman, that proposed establishing an unusual National Institute of Cell Biology. Their idea was to create research centers at several universities—not a single center in Bethesda, Maryland, with the other institutes—where there were already active programs in cell biology, where interdisciplinary approaches of the investigators would facilitate the increasingly complex studies of the cell, and where administrative costs for individual projects could be consolidated, for example, by sharing the use of such major instruments as electron microscopes. Porter and Copeland queried several institutions about this idea and, based on their favorable responses, recommended that a program called "University Laboratories in Cell Biology" be funded by the NIH. Of historical importance, the committee's proposal provided one of the earliest definitions of modern cell biology: "investigations of [a cell's] parts and their significance, its relation to other cells, its reproduction, its differentiation, its functional activities, its products and their effects on the producing cell and its neighbors, the organization of cells into tissue architecture, and the near and remote influences on proliferation."[6]

The Cell Biology Study Section headed by Elmer Butler formally approved this recommendation at their 27 September 1958 meeting. Nonetheless, Copeland later related, "the proposal created quite a stir at NIH but was eventually talked to a quiet death. It did not neatly fit into the existing granting framework, and, equally important, the concept of large project grants was not yet fully recognized. However, the effort paid big dividends because the term 'cell biology' then became well recognized as denoting a discipline."[5] When a separate Institute of Cell Biology was not founded, Porter said, the study section instead began to award block grants to "a few investigators who were established and productive and who wanted to enlarge their operations. . . . This, too came in for enough criticism to discourage its extensive development. Enough was done, however, to encourage cell biologists generally to think well of their field and to ask for more substantial grants."[7]

Also in late 1958, the TCA was reaching a crossroads because its main objective to standardize criteria for cell cultures had been achieved. Also successful were its annual summer laboratory courses and its publications: a two-volume bibliography of tissue culture research from 1884 to 1950 (1953) and a TCA laboratory manual (1955). The rapid growth of this organization raised the question of whether it should remain independent or seek a relationship with other biological societies. Most members of the TCA had a major professional affiliation in another society. Some board members felt tissue culture was a technique, not a discipline, and therefore they could not justify offering services to satisfy a major professional society. Other members thought a more broadly based society was needed. A committee was formed to explore the issue.[8]

TCA president Morgan Harris, at the University of California in Berkeley, wrote a proposal for a possible reorganization of the TCA into a society "concerned with fundamental problems of cell structure and function, rather than preoccupation with a specific technique for dealing with cell systems (tissue culture)." Harris had in mind a society open to all who were interested and thus emphasized that the word "American" would not restrict international members, but simply act as "an American organization in terms of predominant proportion of active members, locale of meetings and scope of present activities."[9]

In September 1958, he took a poll of TCA members on this possible reconstruction, asking about a possible change of name, relations with other societies, and how the TCA should fit into this venture. Replies were mostly favorable. Don Fawcett, who eventually became the first president of the ASCB, argued that the TCA should be "broadened to include those who are primarily interested in cells no matter what methods they may use to study them."[10] Porter strongly supported the idea that "people interested in cell phenomena—whether known as biophysicists, biochemists, physiologists, morphologists, or cytochemists—should be brought together rather than allowed to wander apart into narrow specialties." He also planted the idea that if such a society were formed, the *JBBC* "might feel privileged to be regarded as its official organ."[11] John Hanks welcomed the expansion of the TCA, believing that its larger purpose "converts the proposition from a shovel to an association of miners [that] would seem wholesome." Hanks also suggested the name Experimental Cytology for the new society.[12]

On 6 April 1959, the U.S. National Committee (USNC) of the IUBS resolved that it "considers highly desirable the establishment of a national society of cell biology to act as a national representative of the International Society for Cell Biology." This committee is responsible for membership of the United States in the various international scientific societies.[13] Its resolution was transmitted to Harris, president of the TCA just as the association was holding its annual meeting on 8 April, and where the creation of a society with a broader scope was being debated. Porter later noted that some tissue culturists felt they would lose their identity in any "marriage with the cell biologists. The truth is that a respected and substantial body of the Tissue Culture Association decided—as they had every right to do—that there was a continuing need for a meeting of people interested in the methodology and phenomenology of cell culture, so they withdrew to their dust-proof rooms."[14]*

Instead of reorganizing and reconstituting itself, the TCA members voted by a margin of four to one to set up a council "with representatives within and without the TCA to formulate proposals for the organization of a new Society of Cell Biology."[15] Harris then asked Porter, "as developer of the cell biology program at the NIH, and of course your role as founding father of the TCA," to serve as chairman of this organizing council.[16] After some hesitation, Porter accepted in September 1959 and a short time later drew up a small list of possible participants representing various fields for this group.[17,18]

Using funds from Elmer Butler's chairman's grant of the NIH Cell Biology Study Section, Porter convened a group of interested biologists at Rockefeller. Their two provisional meetings, 9 January and 28 May 1960, are considered the founding dates of the ASCB. The society was not formally incorporated in New York State until 31 July 1961, in the names of Elmer G. Butler, Keith R. Porter, George E. Palade, Arthur K. Solomon, and Paul Weiss.

For the first provisional meeting, a conscious effort was made to have representatives from several disciplines and existing societies: the NIH Cell Biology Study Section represented by Elmer Butler, Morgan Harris, and Hewson Swift; the TCA by Morgan Harris, Don Fawcett, and Margaret Murray; the IUBS by Paul Weiss; the Society of General Physiologists (SGP) by William McElroy and Teru Hayashi; and the Gordon Conferences on Cell Biology by Teru Hayashi, Montrose Moses, and Arthur K. Solomon. Others present included Heinz Herrmann, Jerome Syverton, Hans Ris, Herbert Stern, and Keith Porter. Of the fifteen scientists present, four were electron microscopists, one was a biochemist (although four had been invited), and one was a bacteriologist/immunologist, while others were general biologists interested in intracellular structure. Thirteen of the group voted to form an

---

*The TCA remained an independent society, although it changed its name in 1994 to the Society for In Vitro Biology.

"American Society for Cell Biology." The two abstentions were from McElroy and Hayashi. A provisional council was then established consisting of the thirteen favorable members, with Porter as chairman and Harris as secretary. A subcommittee, with Arthur K. Solomon as chairman, was instructed to write a constitution and bylaws. Anticipating its first annual meeting, Hewson Swift was selected as program chairman.[19]

However, the SGP, with its 400 members, and the TCA, with its 800 members, remained in a state of flux regarding the interests of their members. The SGP was also evolving its interests toward cell biological structure and function, as opposed to cellular physiology, and they needed further consultation with their members about whether to remain independent or form a federation.[20] There were also two original members of the TCA who protested strongly. Philip White felt that the TCA was being railroaded out of existence.[21] Duncan Hetherington felt it was the beginning of just another federation where "electron microscopists will inevitably settle in one spot to count and explain the beads on their ergastoplasm or the holes in their nuclear membranes; the geneticists will cluster about their DNA spirals; the biochemists will go Krebsing with labelled bait . . . and so on for all the 'ists' . . . intent upon their own problems but nevertheless all under the shade of the cell biology parasol."[22] As a result, Harris and Porter corresponded with others to consider possibilities of a merger in some form: joining of the SGP and TCA with the nascent ASCB into a new federation or creating a sub-affiliation with the ASCB. Due to continuing doubts about the actual form of organization as perceived by the members of the SGP and TCA, Porter called a second provisional meeting in May 1960, where he continued to press for the formal establishment of the new society.

This second provisional council meeting added six more scientists: Clifford Grobstein, David E. Green, Richard Klein, George Palade, J. Herbert Taylor, and Edgar Zwilling. At this meeting, it became evident that all the biological societies wanted to preserve their independent status, thus keeping their special interests and avoiding domination by a parent society. But they were not averse to establishing the new society. Based on these feelings, the council pushed ahead on organizational matters. The constitution and bylaws were given provisional acceptance. In formulating the first annual meeting, it agreed to accept (for this meeting only) all contributed papers, listed possible topics for the symposia, refused to publish abstracts for the time being, and postponed the idea of an official journal. They would await views of the new membership. Annual dues were set at $10.00. Hans Ris, of the University of Wisconsin, was appointed provisional treasurer, and Montrose Moses at Rockefeller was provisional secretary. Finally, an executive committee of the council was appointed: Don Fawcett, David E. Green, Clifford Grobstein, Morgan Harris, George Palade, Keith Porter, Hans Ris, Arthur K. Solomon, Hewson Swift, and Herbert Taylor.[23]

Other activities took place before the first scientific meeting of the ASCB in November 1961. On 3 June 1960, the USNC of the IUBS commended "the establishment of an American Society of Cell Biology which may serve as the U.S. national counterpart of the section of cell biology of the IUBS."[13] On 18 July 1960, the ASCB provisional council formally announced its emergence as a society and its first meeting to be held in 1961. To recruit members, they printed announcements in relevant journals, gathered mailing lists from various societies, and sent out notices to 3,000 scientists. By November, nearly 450 respondents asked for more information, and this resulted in mailing more detailed

ANNOUNCEMENT TO CELL BIOLOGISTS

A group of cell biologists met in New York on January 9th and again on May 28th, 1960, and organized a new society, the purpose of which is to further the development of cell biology and to improve communication among scientists representing diverse disciplines. Membership is expected to include biochemists, biophysicists, cytologists, histologists, microbiologists, physiologists and others having in common an interest in the cell. The organization has been named the American Society for Cell Biology.

Persons having educational or research experience in this area and desirous of joining this new group should request application forms from Dr. Montrose J. Moses, Box 2982, Duke University School of Medicine, Durham, N. C. Membership is not restricted to residents of the U.S.A.

The organizing group which met in New York and which constitutes the provisional council includes the following persons:

DR. ELMER G. BUTLER	DR. MARGARET R. MURRAY
DR. DON W. FAWCETT	DR. GEORGE E. PALADE
DR. DAVID E. GREEN	DR. KEITH R. PORTER
DR. CLIFFORD GROBSTEIN	DR. HANS RIS
DR. MORGAN HARRIS	DR. A. K. SOLOMON
DR. TERU HAYASHI	DR. HERBERT STERN
DR. HEINZ HERRMANN	DR. HEWSON H. SWIFT
DR. RICHARD KLEIN	DR. JEROME T. SYVERTON
DR. WILLIAM D. MCELROY	DR. J. HERBERT TAYLOR
DR. MONTROSE J. MOSES	DR. PAUL A. WEISS

DR. EDGAR ZWILLING

The first annual meeting of the Society will be held in October of 1961.

The announcement sent to 3,000 scientists about the organization of a new society. Courtesy of Keith Porter.

notices to describe the society's purposes, its relationships to other groups, its provisional administrative structure, and a brief version of the constitution and bylaws. Finally, on 28 October 1960, the council made an application for $4,600 to the NIH to fund its first scientific meeting.[24]

Further strategies were adopted to attract interdisciplinary scientists to the newly formed society and to prevent it from becoming identified with any one technique or discipline. At the January 1961 council meeting, the constitution was amended to read "the purpose of the Society is to promote and develop the field of cell biology" and that membership "shall be open to scientists who share its stated purpose and who have educational and research experience in cell biology or in an allied field." At the same meeting, almost a year before the first annual meeting, Moses announced that nearly 1,200 replies had been received in response to the mailed notices, including 427 applications for membership. Most applicants were from general biology, with the exception of botany and biophysics. Consequently, the council voted to send notices to thirty-two additional journals, adding the fields of botany, genetics, biophysics, radiation biology, and chemistry.[25]

At the Chicago meeting in November 1961, three symposia were planned. The topics of cell continuity, cell diversification, and characteristics of cell interfaces as well as their speakers reflected the diversity subsumed in the name cell biology. The first annual meeting was considered a great success: 844 scientists attended, of whom 744 were members. In presiding over the business meeting, Porter reminded the council of their resolution not to

THE AMERICAN SOCIETY FOR CELL BIOLOGY

Announces its First Meeting

November 2, 3, 4, 1961 – Chicago, Illinois

at the Edgewater Beach Hotel

Program will include

A. Symposia on

I. CELL CONTINUITY – Molecular and Genetic Aspects of Replicating Systems

(1) Continuity at the molecular level*
ROLLIN HOTCHKISS (Rockefeller Institute)
(2) Continuity of cytoplasmic organelles
HANS RIS (University of Wisconsin)
(3) The genetic control of cytoplasmic organization
TRACY SONNEBORN (University of Indiana)

II. CELL DIVERSIFICATION – Clonal Diversity and Differentiation

(1) Random and non-random diversification of cell systems
BORIS EPHRUSSI (Lab. de Genetique du CNRS, Paris)
(2) Diversification of cells *in vitro**
MORGAN HARRIS (University of California)
(3) Aspects of differentiation in plants*
F. C. STEWARD (Cornell University)

III. CHARACTERSTICS OF CELL INTERFACES – Chemistry and Fine Structure of Membrane Systems; Problems of Transport

(1) The membrane systems of the cytoplasm
GEORGE PALADE (Rockefeller Institute)
(2) Membrane structure and ion permeability in red cells.
H. PASSOW (University of Hamburg)
(3) The chemical asymmetry of membrane transport processes
PETER MITCHELL (University of Edinburgh)
(4) Effect of pH on reactions at biological interfaces
A. D. McLAREN (University of California)

*Tentative title

B. Sessions for fifteen minute contributed papers, demonstrations and motion pictures.

Abstracts (300 words maximum) of papers, demonstrations and films should be sent to **Dr. Hewson Swift, Department of Zoology, University of Chicago, Chicago 37, Illinois,** *not later than September 1st if they are to be included in the printed program.*

Membership in the Society is not required for participating in this first meeting.

Hotel accommodations should be made directly with the Hotel. Please state that you will be attending the ASCB meeting.

Announcement of the ASCB's first annual meeting, 1961. Courtesy of Keith Porter.

make decisions that would bind the society to long-term policy and gave as an example the question of a journal and other society affiliations. He reported results of a mailed ballot for the first officers of the ASCB: Don W. Fawcett of Harvard Medical School as president, Montrose Moses as secretary, Hans Ris as treasurer, and Alex B. Novikoff as president-elect.

The American Society For Cell Biology

Announces its Second Meeting
November 5, 6, 7, 1962—San Francisco, California
at the Jack Tar Hotel

Program will include

I. Symposia on

A. **PATTERNS OF ORGANIZATION OF GENETIC MATERIAL.** (C. F. Robinow and W. Beerman; H. G. Callan, F. Stahl, C. Yanofsky)

B. **MODIFICATIONS OF GENETIC MATERIAL INDUCED BY RADIATION OR CHEMICAL MEANS.** (H. Fraenkel-Conrat and A. Wacker; L. Lerman, C. S. Rupert, R. Perry)

C. **PLASTIDS—THEIR CHEMISTRY, STRUCTURE AND EVOLUTION.** (M. Calvin and K. Mühlethaler; R. Stanier, J. Schiff, R. Sager)

D. **THE SPERM CELL—COMPARATIVE FINE STRUCTURE, COMPOSITION AND PHYSIOLOGY.** (D. Fawcett and D. Bishop; R. Vendrely, S. Inoué, A. Colwin)

E. **DIFFERENTIATION AND CELL INTERACTION IN PROTISTA.** (V. Tartar and K. Raper; E. Cantino, D. Pitelka, R. Moore)

F. **INTRACELLULAR LOCALIZATION OF SPECIFIC CHEMICAL ACTIVITIES.** (S. J. Holt and J. Bonner; L. Goldstein, E. D. Hay, C. J. Avers)

II. Contributed papers (15 minutes, including discussion) will fall either into one of the categories indicated by the Symposia Topics (A-F), or into one of the following FREE CATEGORIES (G-K):

G. **CONDUCTING AND CONTRACTING SYSTEMS** (cell movement, nerve and muscle structure, chemistry or activity).

H. **MACROMOLECULAR COMPLEXES AND THEIR FUNCTIONS** (ribosomes, conjugated protein complexes, glycogen, collagen, etc.).

I. **MEMBRANE STRUCTURES OF THE CYTOPLASM** (structural and functional organization of cytoplasmic organelles such as mitochondria, ER, Golgi complex).

J. **SURFACE ACTIVITY AND SURFACE INTERACTIONS OF CELLS** (permeability, furrowing, cell cohesion, virus-host cell interaction).

K. **MITOSIS, and NUCLEAR CYTOLOGY** (mitotic apparatus, nucleolus; nuclear metabolism during the cell cycle, etc.)

Abstracts of papers, demonstrations and films must be in the hands of **Dr. Max Alfert, Department of Zoology, University of California, Berkeley, California** before September 7. Further information may be obtained from Dr. Alfert.

Inquiries about the Society should be directed to the Secretary, **Dr. M. J. Moses, Box 2982, Duke University Medical Center, Durham, North Carolina.**

Arrangements for accommodations should be made directly with the hotel. Please state that you will be attending the ASCB Meeting.

Announcement of the ASCB's second annual meeting, 1962. Courtesy of Keith Porter.

The twelve council members elected were Keith R. Porter, George E. Palade, Arthur W. Pollister, and Rollin Hotchkiss for three-year terms; Harry Eagle, Philip Siekevitz, Paul Weiss, and Heinz Herrmann for two-year terms; and Clifford Grobstein, Albert J. Dalton, David M. Prescott, and David E. Green for one-year terms. It was clear that a broad

THE AMERICAN SOCIETY FOR CELL BIOLOGY

Announces its Third Meeting

November 6, 7, 8, 1963 – New York, N. Y.

at the Commodore Hotel

Program will include

I. SYMPOSIA ON

A. REGULATION OF BIOSYNTHESIS.

Chairman: Van R. Potter, University of Wisconsin, Madison.

1. CHANGES IN THE MOLECULAR STATE OF DNA WITH CHANGES IN FUNCTION; Edouard Kellenberger, University of Geneva, Switzerland. *Discussant*: Herbert Stern, University of Illinois, Urbana.
2. REGULATION OF MACROMOLECULAR SYNTHESIS; Sol Spiegelman, University of Illinois, Urbana. *Discussant*: Rollin D. Hotchkiss, The Rockefeller Institute, New York.
3. BIOSYNTHETIC REGULATION IN A VIRUS-ANIMAL CELL SYSTEM; Wolfgang Joklik, Albert Einstein College of Medicine, New York. *Discussant*: John Cairns, John Curtin School of Medical Research, Canberra, Australia.
4. *Summary*: Seymour Cohen, University of Pennsylvania, Philadelphia..

B. TRANSPORT ACROSS THE CELL MEMBRANE

Chairman: A. K. Solomon, Harvard Medical School, Boston.

1. EXCHANGE IN BULK (PINOCYTOSIS AND PHAGOCYTOSIS); John Marshall, University of Pennsylvania, Philadelphia. *Discussant*: Manfred L. Karnovsky, Harvard Medical School, Boston.
2. ION TRANSPORT; Robert Davies, University of Pennsylvania, Philadelphia. *Discussant*: Howard Rasmussen, University of Wisconsin, Madison.
3. TRANSPORT OF SUGARS; Ellis Englesberg, University of Pittsburgh. *Discussant*: Robert K. Crane, Chicago Medical School.
4. *Summary*: George Palade, The Rockefeller Institute, New York.

II. CONTRIBUTED PAPERS

Up to 200 papers (15 minutes each, including discussion), will be arranged into sixteen sessions, with as much continuity as possible within each session. Papers not selected for presentation from the platform will be presented by title. All abstracts, whether presented by title or from the platform, will appear in proceedings of the meeting.

Abstracts of papers, demonstrations and films must be in the hands of the program committee **by August 1, 1963.** *They should be mailed to* **Dr. D. M. Prescott, Biology Division, Oak Ridge National Laboratory, P. O. Box Y, Oak Ridge, Tennessee.** *Further information may be obtained from Dr. Prescott.*

Inquiries about the Society should be directed to the Secretary, **Dr. M. J. Moses, Box 2982, Duke University Medical Center, Durham, North Carolina.**

Arrangements for accommodations should be made directly with the Hotel. Please state that you will be attending the ASCB Meeting.

Announcement of the ASCB's third annual meeting, 1963. Courtesy of Keith Porter.

representation of scientists was wanted by the council and the members. The first president was a leading electron microscopist, and the council included biochemists, physiologists, and general cytologists. Porter previously declined to be nominated for president, stating his responsibilities were finished because "the initial aims of the original group—known as the Provisional Council—have been achieved."[14] He did serve later, however, becoming its seventeenth president in 1978.

During its first years, the society, like the *JBBC*, worked to enhance the participation of scientists from various fields in its memberships, on its governing council, and as its presidents. During its first ten years, the presidents were

Don Fawcett (anatomist and electron microscopist)
Alex B. Novikoff (biochemist)
Hewson Swift (geneticist)
Van R. Potter (biochemist)
David M. Prescott (biophysicist)
Philip Siekevitz (biochemist)
Joseph J. Gall (geneticist)
Montrose Moses (biophysicist)
J. Herbert Taylor (molecular geneticist and biophysicist)
Saul Kit (biochemist and virologist)

The society's emphasis on biological concepts, rather than on methods used, was also visibly projected in the agendas for its annual meetings to attract even more members. For its second annual meeting in San Francisco, one section of symposia focused on genetic topics, and the second section featured speakers from the fields of biochemistry, biophysics, morphology, and zoology. For its third annual meeting in New York, two main symposia focused on biosynthesis and transport of macromolecules, while a third session featured 200 short presentations on a multitude of topics. To bring the findings and ideas of cell biology to a broader community, Siekevitz as president introduced a seminar, with guided tours of exhibits, for high school science teachers and science writers, which set a precedent for having briefings and background sessions at the annual meeting. Moreover, during the spring of 1963, a mail vote by the 1,200 ASCB members overwhelmingly approved the sponsorship by the society of the newly renamed *The Journal of Cell Biology* (see Chapter 10B), and the society then began to nominate individuals for the journal's editorial board.

The organization of the American Society for Cell Biology was a response to new, strong trends in cell research, which, after 1945, had become methodologically more diversified and conceptually more convergent than studies of the cell in the late nineteenth and early twentieth centuries. Forty years after its founding, there were 10,000 members. The concerted activities of so many scientists to organize an independent society now stand as a landmark in the creation of cell biology as a discipline in its own right.

Porter Moves On

Porter played such a key role in founding the ASCB that it seemed inevitable he would also become its first president. He declined the invitation and noted his reluctance in remarks at the banquet during the society's first scientific meeting in Chicago. He was, he explained, "in a restless phase in my career—busy moving from place to place."[14]

During the six years surrounding the creation of a journal and a society (1955–1961), Rockefeller's environment changed considerably. Bronk's vision to expand the Institute into a small university with a few graduate students caused many disruptions; for example, many faculty members refused to take students or lead seminars. Nonetheless, reflecting his continued enthusiasm for research in the cytology laboratory, Bronk promoted both Palade and Porter in 1956 from associate members to full members, thus giving the laboratory autonomy for the first time. Also in 1956, based on Bronk's recommendation, the Board of Trustees announced that the Institute would begin accepting grants from external agencies for research, thus reversing a fifty-five–year policy of financing research solely from its endowment (an exception to this rule is described in Chapter 3).[26] Applying for grants placed another new administrative burden on the faculty.

The cytology laboratory during this transition, according to Palade, retained its "atmosphere of intense activity, remarkable achievements, great excitement, and unlimited optimism. . . . Added to all this excitement was a pervading free spirit—often irreverent, but always helpful, because it acted as an antidote against imagined grandeur. Keith Porter was responsible for a good part of that spirit." The laboratory was buoyed by extremely productive research that included the succession of discoveries about the endoplasmic reticulum and secretory pathway (described in Chapter 7).[27]

As a result of Bronk's changes, there was a rather disagreeable increase in the laboratory's administrative activities. As he had done since 1950, Porter served as official head of the department who managed the daily activities of running a laboratory with very limited means. Now, however, he became burdened with new duties: taking on and training students, organizing seminars for teaching cell biology, writing grants, upgrading equipment, and finding additional space for the expanding laboratory. According to Palade, Porter often "played the role of lord executioner, his preferred victims being carefully selected from among high officials. He showed great wit and intense interest in needling authority, and for this reason acquired the nickname 'Herr Teufel,' more in admiration than in dread." Two examples of Porter's considerable extra efforts as administrator were applied to electron microscopes and an entirely new set of laboratories.

Microscopes. A perennial quest for Porter throughout his career was finding higher and higher resolution electron microscopes. In 1955, the laboratory had only two instruments, the outmoded and feeble RCA EMU-2 purchased in 1948 and a sturdy but low resolution Philips EM 100. Porter then began a campaign with two goals in mind. In letters to Thomas Rivers, the interim director between Gasser and Bronk, he argued that "if we are to retain a prominent position in this field, we must have the best of microscopes."[28] He requested a separate laboratory be created for outsiders using a more rugged instrument so that the present confusion and distraction would be kept out of the cytology workspace.[29] He petitioned both Rivers and Bronk for a Siemens Elmiskop I microscope having a proven resolution of 10 ångströms, which he had seen during a visit to Berlin in 1954; this instrument would be for his personal use in resolving still finer structures.

Both requests were granted, although at greater expenditures of effort for Porter. The Board of Trustees approved the request for $40,000 for the new microscope but made the purchase subject to refunding from a prospective four-year grant written by Porter. Permission was also granted to establish a free-standing Electron Microscopy Service Laboratory. To set up this service, Porter then had to negotiate a trade in of the 1952 Philips EM 100 for an EM 100A, find outside funds to recondition the old RCA microscope ($14,000), purchase a centrifuge and microtome, and air condition a separate room. (Funds were also found for a second-hand RCA EMU-2A for Porter and Palade through the North American Philips Company from the Franklin Institute for $10,000). Once all this equipment was in place and more funds were found for a technician, an official memorandum was circulated in January 1956 to explain the arrangements for using the new service unit: users had to prepare their own specimens, and the technician would take electron micrographs, develop films, and print enlargements. This independent resource center, which was under Porter's administration, charged $10.00 an hour for the beam time and taking of micrographs.[30]

New laboratories. The next year, Porter began a more arduous campaign to remove the cytology laboratory from what Palade once described as their "unattractive dungeon sunk in the third basement" of Theobald Smith Hall.[27] From 1950 to 1957, the cytology laboratory had grown from four to sixteen members, and they were now sharing their original spaces with the service laboratory. So Porter petitioned Bronk to occupy an entire floor in a new laboratory building, the first to be built in thirty years. This request was granted after vibration studies were performed on the floors to determine their suitability for electron microscopy. The senior staff members were asked to design their own laboratories. There was now private office space for the scientists, offices for a secretary and for editors of the *JBBC*, and a small conference room and library. There was space for three electron microscopes, a dark room, and a facility for sectioning and fixation. Palade wanted small rooms for animal containment and operating, and these were also provided. So in early 1960, everyone except electron microscopist Dan Moore, who remained in the service laboratory, moved to the fifth floor of the new building, then called South Laboratory but now named the Detlev W. Bronk Laboratory. According to Siekevitz, who was part of this move, everyone was happy: at last, the windows did not look out on sustaining walls, the sunlight came in, and there were views of the campus and the East River.

On 8 July 1960, after the founding meetings of the society and the move to new laboratories, Porter sent a letter of resignation from Rockefeller to President Bronk; a year later, he became a professor of biology at his alma mater, Harvard University.[31] Palade said, "Everything seemed to be calm, stable, and serene" until Porter surprised everyone with this announcement. "At the time of his departure, we unveiled his portrait, for which he provided, in the old style of the third basement, the inscription: 'Our father who art at Harvard.'"[27]

Porter's reasons for leaving Rockefeller after twenty-two years were many and complex. In his letter of resignation, he told Bronk that his "decision to leave [on 30 June 1961] reflects in no sense my dissatisfaction with the Institute or my colleagues. You personally have been both kind and extremely generous. Actually, the situation forcing the change is self-produced for I have created an environment that is simply taking too much out of me for the results being achieved. I feel that one effective way to reduce the drain on my resources is to change locations."[31]

Later versions of his reasons for leaving were less sanguine. In 1980, he recalled, "they involved living in New York which my wife and I came less and less to enjoy. They involved personal relations with the President and concern with what he was doing to the 'Institute.'"[32] In a 1984 videotaped interview, he admitted that he had "lost patience with the President and did not enjoy his tactics. . . . I suppose some people move for one big fight they have with the President, but that was not my case." In the end, the underlying reason was to return to science: "I felt it was important to put a certain amount of distance between myself and George Palade who was doing three quarters of the research and I was doing three quarters of the administration."[33]

Leaving behind a state-of-the-art laboratory, Porter moved on to many years of extremely productive research. The founding father of biological electron microscopy eventually broadened his research from cellular to developmental and molecular biology. On acquiring a high voltage electron microscope in the 1970s, Porter ultimately realized his earliest goal—to examine whole mounts of cultured cells and to examine the three-dimensional aspects of their internal organization.

Part 3

Cytology Becomes Cell Biology 1961–1974

Carol L. Moberg

Cell biology at The Rockefeller Institute continued to flourish under President Detlev Bronk after Keith Porter's departure. The new science had been given a robust start and was on its way, but it was still far from a solid science. During the 1960s, as George Palade later observed, this cradle of cell biology was transformed into a nursery.

The following chapters take the origins of modern cell biology to a more expansive exploration of how cells carry out the fundamental chores of life itself. This second generation of new scientists was influenced directly by the innovations pioneered by Albert Claude, Porter, and Palade, but they were now positioned to ask and to learn what activities (other than energy production and protein synthesis) were going on inside cells. As a result of their progress, they established within a remarkably short time a large body of knowledge and essential tools for future scientists who would navigate even further into the territory inside cells.

The research described in the following chapters, and written for the most part by those who performed it, took place in several Institute laboratories: the expanded laboratory of Palade and Philip Siekevitz focused on cell membranes and permeability; James Hirsch, Zanvil Cohn, and Ralph Steinman explored the cell biology of the immune system; Alfred Mirsky and Vincent Allfrey studied proteins and properties of the nucleus; and Christian de Duve and Miklós Müller utilized biochemical methods to characterize newly found intracellular particles. Each laboratory took a different path, focused on different types or parts of cells, and discovered previously unknown life processes. Recognizing the great progress in this field, in 1967 Palade changed the name of his laboratory from Cytology, which had been used since 1950, to Cell Biology. This new name complemented a laboratory of the same name that had been used by Mirsky and Allfrey since 1963 for their study of the nucleus.

Within a dozen years, the old cytology was completely superseded by cell biology.

Chapter 11

Adventures in Biochemical Cytology

11A. Exploring Cells with a Centrifuge: Discoveries of the Lysosome and Peroxisome

Christian de Duve

Contrary to most investigators mentioned in this book, I did not do my most important research at Rockefeller, nor was I initially motivated by a desire to dissect cells and understand their structural and functional organization. I was a Belgium-based physiologist-turned-biochemist whose main ambition was to elucidate the mechanism of action of insulin on the liver. Yet, Rockefeller turned out, through a series of chance events, to play a decisive role in my research and, eventually, in my career as a scientist. This chapter summarizes the main steps of these adventures. References are limited to key publications, which may be consulted for any desired additional information.

I crossed for the first time the threshold of the venerable Rockefeller Institute for Medical Research in late January 1948. I had just turned thirty and had behind me some fourteen years of training, which had included medical and chemical studies at the Catholic University of Louvain in Belgium, a postdoctoral stay of eighteen months with Hugo Theorell at the Medical Nobel Institute in Stockholm, and another of six months at Washington University in St. Louis, where I worked with Earl Sutherland under the supervision of Carl and Gerty Cori. I stopped in New York on my way back from St. Louis to Belgium, where I was to take up new academic duties.

My reason for visiting Rockefeller was to meet Albert Claude. Not, however, for scientific reasons. My main interest in those days was carbohydrate metabolism and its regulation by insulin and glucagon. Claude I knew only through his reputation as the best-known Belgian scientist of the time. I wanted to meet this famous countryman of mine and to see the unique research center where so many other great scientists had been doing their work.

Claude, who was soon to return to Belgium, received me very kindly, even though he most likely knew nothing of me and was no more interested in insulin or glucagon than I was in the details of cell structure. He told me all about his work and gave me some reprints, which I religiously read on the long plane journey back to Belgium. The idea that one could separate mitochondria and analyze them biochemically fascinated me. But this was not for me. I had other plans.

Little did I suspect that this visit was going to have far-reaching effects on my research, eventually on my life, and even, to some extent, on the history of cell biology. To explain this, I must go back to my early Belgian years.

Louvain: My Most Fruitful Years

I started my academic career in 1948 on my return from New York as lecturer of physiological chemistry on the medical faculty of my alma mater, the Catholic University of Louvain. There, I opened my first research laboratory, to which I had the good fortune of immediately recruiting a young medical doctor, Géry Hers, and two medical students, Jacques Berthet and Lucie Dupret (the future Mrs. Berthet), soon to be joined by another medical student, Henri Beaufay. The equipment of the laboratory was modest but adequate for our purpose.

As mentioned, my main concern was the action of insulin on liver tissue, a hotly contested topic in those days. Contrary to the majority opinion, I had strong reasons for believing that such an effect existed. But I realized that the question would remain debatable as long as the effect had not been unequivocally demonstrated in vitro. Achieving this and eventually elucidating the biochemical mechanism of the observed effect became my long-term goal.

Carbohydrate metabolism and glucose-6-phosphatase. I chose as our first project to clarify some poorly understood aspects of carbohydrate metabolism in the liver. Those were still the heroic days of dynamic biochemistry. Except for the glycolytic chain and the tricarboxylic acid cycle, metabolism was still an almost uncharted territory. A key contribution by Carl and Gerty Cori, in whose laboratory I had just spent six months, had been to show that glycogen feeds into the glycolytic chain by way of glucose-1-phosphate and glucose-6-phosphate and to characterize the two enzymes involved in this transformation, glycogen phosphorylase and phosphoglucomutase. This pathway exists both in liver and muscle. But liver contains an additional enzyme, identified by the Coris as a hexose-phosphatase of undefined specificity, that catalyzes the hydrolysis of some hexose-phosphate ester and allows the liver to produce free glucose. I selected this hexose-phosphatase as the first target of our research.

Our experiments soon disclosed the presence in rat liver of two distinct phosphatases acting in the acid range. One, with a pH optimum of 6, acted readily on glucose-6-phosphate but neither on glucose-1-phosphate nor b-glycerophosphate, a standard phosphatase substrate. This enzyme most likely was the hexose-phosphatase described by the Coris, which was thereby identified as a glucose-6-phosphatase. The other phosphatase, with a pH optimum of 5, hydrolyzed all three substrates indifferently. It corresponded to the unspecific acid phosphatase known to be present in many tissues.

To purify glucose-6-phosphatase, we used one of the few tools available at that time for the separation of proteins, isoelectric precipitation, and found it was quantitatively precipitated at pH 5, leaving the unspecific acid phosphatase in solution.[1] Unexpectedly, the precipitated enzyme did not redissolve when the pH was brought back to 7, contrary to what was expected in isoelectric precipitation, which should be reversible. Here is where my visit to Claude paid off. I remembered that he had described the agglutination of isolated cytoplasmic granules exposed to pH 5.[2] I concluded that our enzyme was probably bound to some cellular structure and decided further that we should characterize this structure, as it might be a crucial factor in the regulation of the enzyme attached to it and, perhaps, therefore in the mechanism of action of insulin on liver. This is how we came to follow Claude and to engage in cell fractionation.

By the time we entered the field, Claude's procedure had been modified by his pupils, George Hogeboom, Walter Schneider, and George Palade,[3] and later by Schneider,[4] whose fractionation scheme we adopted. In it, the liver was homogenized in the presence of 0.25-molar sucrose, and the homogenate was divided into four fractions by a series of centrifugations and washings at increasing speeds: a nuclear fraction; a large granule fraction, consisting mostly of mitochondria; a small granule fraction, called microsomes by Claude; and a final supernatant. While the mitochondrial nature of the large granules was largely accepted by then, the cytological identity of the microsomes was still being debated. Some workers viewed them as distinct cytoplasmic constituents, whereas others believed them to consist of small mitochondria and mitochondrial fragments. Our first results settled this question.

In our experiments, glucose-6-phosphatase was largely recovered with the microsomal fraction, thus confirming my suspicion of the enzyme being attached to a structure and, at the same time, solving the debate concerning the cytological nature of Claude's microsomes, which clearly originated from a distinct cell component and not from disrupted mitochondria. In fact, it was soon established that the enzyme was most likely associated exclusively with a microsomal component. This finding suggested the use of glucose-6-phosphatase as a microsomal marker, a key tool in our subsequent work. Further characterization showed the enzyme to be tightly bound within phospholipid-protein complexes, presaging its identification as an integral constituent of endoplasmic reticulum (ER) membranes. We might have stopped there, had our experiments not presented us with an intriguing observation.

The riddle of the hidden acid phosphatase. Acid phosphatase, which we had also assayed on the fractions for control reasons, had done a vanishing act, manifesting only about one tenth of the activity we expected in the homogenate. The fractions were a little better, but their combined activities still added up to only about twenty percent of the expected value, with a horrendous recovery of 200% of the activity of the homogenate. Suspecting an assay artifact, we repeated the measurements five days later on the preparations, which had been kept in a refrigerator in the meantime. This time, the activity of the homogenate was found to be in the expected range, and the added activities of the fractions amounted to an acceptable value of eighty-five percent of the activity of the homogenate. This seemed to settle the question, except for two troubling facts. Exactly the same results were obtained with the old reagents, with which we had also repeated the assays in order to find out what might eventually have gone wrong. This rendered the hypothesis of an assay artifact unlikely. Furthermore, the differences between the later and earlier results varied greatly from one fraction to another. The activities of the sedimentable fractions showed high increases, whereas that of the supernatant fraction was hardly changed.[5]

The only explanation was to assume that the association of acid phosphatase with its supporting sedimentable structure went together with a lack of enzymatic activity. I called this phenomenon "structure-linked latency." All our results could be explained if the assumed linkage was broken by disruption in a Waring blender in the presence of distilled water—the technique we had used to prepare liver extracts in our early biochemical work—but remained largely intact when, as in the fractionation experiment, the tissue was homogenized more gently in the presence of isotonic sucrose. The linkage also had to resist damage inflicted in the course of centrifugation and resuspension of the sediments, though not entirely so, thus accounting for the 200% recovery. Finally, the postulated linkage had to spontaneously disintegrate upon cold storage for a few days, thereby explaining the large increases in activity observed in the second series of assays.

There was a simple way of testing this hypothesis: recentrifuge the particulate fractions kept in the refrigerator. If the hypothesis was correct, the high activities present should no longer be sedimentable. This was done, and the anticipated result was obtained. The activities were both free and unsedimentable.[5] The finding was gratifying but immediately raised a new question: what was the nature of the linkage responsible for structure-linked latency?

It took only a few weeks to get the answer. The latent, particle-bound enzyme was found to be simultaneously activated and released in soluble form by a number of physical treatments, which included exposure to distilled water, repeated freezing and thawing, treatment by surface-active agents, or replacing the sucrose in the medium by substances of smaller molecular size, such as NaCl or glycerol. The picture unfolded of a miniature, membrane-bounded, osmotically active bag, within which the enzyme was segregated in fully active form but prevented from acting on added substrate by the interposed membrane. This model implied that the membrane prevented the b-glycerophosphate used as substrate in the assay from entering the bag and reaching the enzyme. In confirmation of this prediction, it was found that b-glycerophosphate could serve as an osmotic protector.[6,7]

Having clarified the enzyme's vanishing act, we were left with the problem of the nature of the little bag within which it was hiding. In our fractionation, the sedimentable acid phosphatase showed a heterogeneous distribution, with roughly two thirds of the activity in the mitochondrial fraction and one third in the microsomal fraction. At first, we assumed that we were dealing with mitochondria, some of which had spilled over into the microsomal fraction, either because they were smaller than the others or because we had made some technical mistake. This interpretation had to be reappraised when it was found, as a result of a minor technical accident, that a particulate fraction isolated by centrifugation at a lower than usual speed possessed a high respiratory activity (indicative of mitochondria) but almost no acid phosphatase activity.[8] If mitochondria were involved, their enzyme content must be very heterogeneous, with cytochrome oxidase (the "respiratory enzyme") being more abundant in the larger mitochondria and acid phosphatase in the smaller ones. But there was also the possibility that acid phosphatase did not belong to mitochondria, but to a distinct group of particles with sedimentation properties such that about two thirds of the particles came down with the mitochondria and one third with the microsomes. Should this be the case, both groups of particles could conceivably be enzymatically homogeneous, and their characteristic enzymes could serve as markers for their intracellular sites.

Experiments designed to distinguish between these two possibilities clearly came out in favor of the second one.[8] Eventually, we worked out a five-fraction scheme in which a small intermediate fraction, containing only a few percent of the cell proteins, was separated between a main mitochondrial fraction, isolated at a lower speed than was customary, and the usual microsomal fraction. This intermediate fraction was considerably enriched in acid phosphatase while containing moderate amounts of both cytochrome oxidase and glucose-6-phosphatase, as was to be expected. Taking into account the amount of protein belonging to the mitochondria and microsomes present in this fraction, as calculated from its content in the relevant marker enzymes, the new particles seemed to represent only a minor part, no more than five percent, and possibly much less, of the total cell proteins.[8]

On the trail of lysosomes. Our findings seemed sufficiently interesting to warrant further exploration. But what the next step should be was not clear. A fortuitous encounter showed the way. This happened in Paris in the summer of 1952, at the Second International Congress of Biochemistry, where I met a young British biochemist, Peter G. Walker, who had obtained results strikingly similar to ours, unfortunately unpublished so far, on another liver enzyme, b-glucuronidase, that had to be compared with acid phosphatase. It was mutually agreed that these experiments were best done in our Belgian laboratory, and Walker later kindly sent me all the information needed for the study of his enzyme. The project was taken over by a newly arrived Canadian postdoctoral fellow, Robert Gianetto, from the Université de Montréal, in collaboration with my two students, Françoise Appelmans and Robert Wattiaux. The investigators also included an acid protease, then simply called cathepsin and now identified as cathepsin D, in their assays, an enzyme shown by other investigators to have a heterogeneous distribution similar to that of acid phosphatase. I was intrigued by the possibility that it might be present in the same particles.

The results of these experiments were very illuminating. All three enzymes displayed structure-linked latency and were released in closely parallel fashion upon graded activation by all the means that were tried.[9,10] This strongly suggested that they were present together in the same particles, a conclusion confirmed by their centrifugal distribution patterns,[11] which were very similar, except for an extra microsomal b-glucuronidase component suggesting a possible second site for this activity. This has, indeed, turned out to be the case, but we have had no part in further studies of this peculiarity.

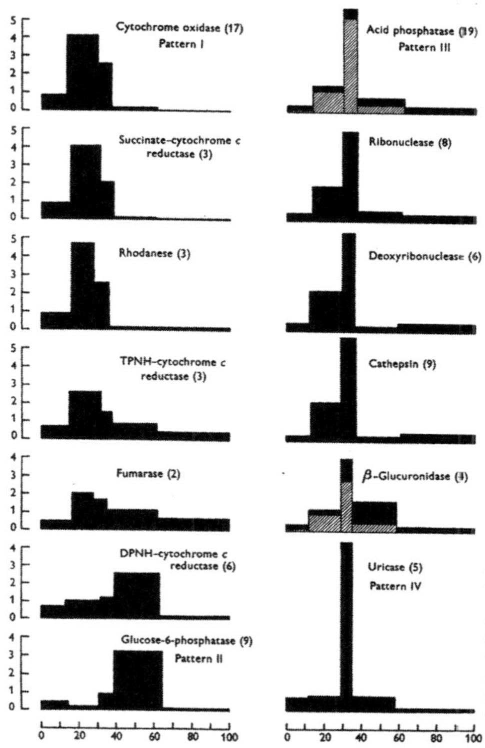

1. Histogram of enzyme distributions obtained with a modified fractionation scheme. This graph replaces indigestible rows of figures with comprehensive patterns that are readily assimilated at a single glance. Specific enzyme activities are plotted against nitrogen or protein content of the fractions, aligned on the abscissa in the order they leave the centrifuge. The ordinate shows the relative specific activity (percent of total activity/ percent of total nitrogen); the abscissa shows the percent of total nitrogen. Once called "duvograms," these patterns were known as "submarines" in our laboratory because of the typical silhouette created by the intermediate peak. From de Duve et al., 1955. *Biochem. J.* 60:604–617.

Stimulated by these findings, I decided to take a closer look at centrifugal fractionation, enlisting for this purpose the collaboration of Berthet, who had first set up the technique in our laboratory.[12] We examined the theoretical foundations of the method, thereby gaining a better appreciation of its limitations and pitfalls, and we surveyed all the enzyme distributions recorded so far, paying special attention to the enzymes that had been found by others to show a heterogeneous distribution similar to that of acid phosphatase and might thus possibly accompany this enzyme in the mysterious new particles. A number of enzymes conforming to this description were singled out for study by means of our newly developed five-fraction scheme.

Experiments by our group yielded varying results.[11] As illustrated in figure 1, some enzymes, in particular the DPNH- and TPNH-specific cytochrome c reductases (DPNH and TPNH were the old abbreviations for NADH and NADPH), showed authentic bimodal distributions, indicative of a dual localization in both the mitochondria and the microsomes. But three additional enzymes revealed a peak in the intermediate fraction. They were ribonuclease, deoxyribonuclease, and urate oxidase, known as uricase in those days. These results were published in *The Biochemical Journal* in 1955[11] as number six in a series entitled "Tissue fractionation studies," which would end with number eighteen in 1964.[13] This paper represents something of a watershed in the history of my laboratory in that, for the first time, I crossed the line between biochemistry and cell biology. Looking at the five enzymes so far identified as components of the new particles, leaving out urate oxidase, which seemed to be a case apart and indeed was, as we learned later—acid phosphatase, b-glucuronidase, cathepsin, ribonuclease, and deoxyribonuclease—I thought I saw a pattern: all five were hydrolytic enzymes with an acid pH optimum. Their substrates, however, were so diverse that the grouping together of such enzymes in a single location could serve only one function: digestion. I suggested as much, proposing for the particles the name lysosomes, Greek for "digestive bodies."

The paper also helped me to clarify a conceptual approach to centrifugal fractionation, inspired essentially by biochemistry and based on two postulates: that of single location, according to which enzymes are situated in single intracellular sites; and biochemical homogeneity, according to which cellular components are biochemically homogeneous. The postulates, as I made clear, were not dogmatic pronouncements, but working hypotheses that need not be true in every case to be useful. Their value was that, when verified, they allowed enzyme activities to be used as quantitative markers; I called them reference enzymes of their intracellular abodes, a major feature in our approach and a much more reliable criterion of identification than morphological examination, which in those days was highly uncertain and subjective.

Finally, 1955 more or less signaled the end of my commitment to insulin and glucagon. Our centrifugation work had opened a host of fascinating new problems to pursue. The most pressing one was morphological. What did the new particles look like? What was their cytological identity?

Our first glimpse of lysosomes. Until then, our approach had been purely biochemical. But some sort of image of the unknown particles had emerged from our studies. They were surrounded by a membrane, as the latency experiments had revealed, and their sedimentation properties indicated a size of the order of 0.5 micrometers. Such entities should be clearly recognizable in the electron microscope and should, in fact, be among the structures already identified by means of existing techniques, which were improving rapidly under the influence of such leaders as Keith Porter and George Palade. The organization of the Third International Congress of Biochemistry in Brussels in 1955 gave me an opportunity to address the morphological problem with the help of an American investigator, Alex Novikoff.

I first came to know Novikoff in 1953 through a paper describing experiments that resembled our own in several respects.[14] Like us, he had fractionated rat liver tissue by unconventional schemes that, in his case, called for the isolation of six or even ten distinct fractions, according to an arbitrarily decided protocol. His choice of enzymes to be assayed had been almost eerily prophetic in that they included what were later to be identified as markers for all major cytoplasmic components, among them 5′-nucleotidase (plasma membranes), succinoxidase and ATPase (mitochondria), esterase (microsomes, i.e., mostly ER fragments), acid phosphatase (lysosomes), and urate oxidase (peroxisomes).

Novikoff's results were similar to ours in that they showed acid phosphatase sedimented more slowly than the mitochondrial enzymes but faster than the microsomal esterase, with, of course, a considerable degree of overlap. He had also found that urate oxidase behaved similarly, which was actually our reason for selecting this enzyme for study in our later work. Novikoff's interpretation of these observations differed from ours, however. He saw his findings as reflecting the enzymatic heterogeneity of mitochondria and microsomes of different sizes. He readily accepted our interpretation when seeing our results and became one of the earliest champions of the lysosome concept, though using a morphological approach, having by then adopted the electron microscope as his main tool.

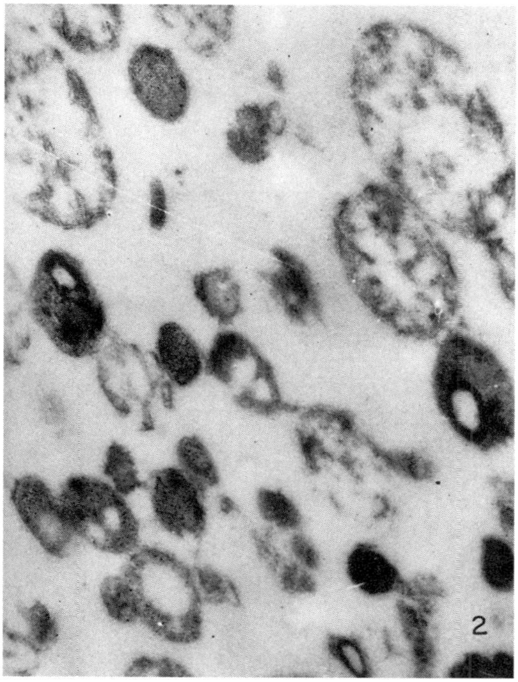

2. First electron micrograph of a lysosome-rich fraction from rat liver. Note the numerous "dense bodies." From Novikoff et al., 1956. *J. Biophys. Biochem. Cytol.* 2:179–184.

I took advantage of the Brussels congress to invite Novikoff to deliver a lecture and to stay a few additional weeks in Belgium so as to look at some of our best lysosome-rich fractions, even though these were still very impure. This work was to be done in collaboration with Claude, who had kindly agreed to put his electron microscope facilities at our disposal. Claude, however, was as exasperatingly slow and painstakingly finicky as Novikoff was ebulliently impatient and apt to cut corners, both technically and intellectually. The mixture was too explosive to last. Novikoff ended up taking the fractions and kept them on ice in a thermos flask on the night train to Paris, where he examined them the next day in the laboratory of Wilhelm Bernhard, one of the pioneers of electron microscopy in France. As chance would have it, this search brought to light a number of "dense bodies" (figure 2), which we now know to be authentic lysosomes, but no "microbodies"—the morphological counterpart of peroxisomes (see next paragraph)—although both entities were probably present in the fractions in comparable amounts. The gods of qualitative morphology chose not to put us on the wrong trail. Novikoff had an opportunity to show these results at an Arden House Conference in January of 1956, thereby presenting lysosomes for the first time to an American audience.[15]

Tracking urate oxidase. I mentioned previously how we had embarked on a systematic study, using our new fractionation scheme, of the distribution of all the enzymes that had been found by other workers to distribute more or less like acid phosphatase in conventional fractionation experiments. In pursuit of this work with several members of the laboratory, we discovered two other acid hydrolases, *N*-acetyl-b-glucosaminidase and b-galactosidase, in the family of lysosomal enzymes, strengthening the notion of an acid digestive function.[16] But two other enzymes, like the previously studied urate oxidase, were not acid hydrolases, namely catalase and D–amino acid oxidase, and showed a revealing distribution pattern. The possibility that the three non-hydrolytic enzymes might belong to a different kind of particle began to deserve serious consideration.

Latency studies confirmed our suspicion. Neither urate oxidase nor D–amino acid oxidase showed any latency. Catalase did, but the barrier responsible for this phenomenon proved different from the lysosomal membrane, being much less sensitive to the disrupting action of digitonin, which is uniquely effective on cholesterol-rich membranes. However, the latency of the enzyme was puzzling in that it seemed rather unlikely that a molecule as small as hydrogen peroxide, the substrate of catalase, could not pass freely through a membrane. Indeed, we found that latency is made possible, in this case, by the fact that catalase shows first-order kinetics and cannot be saturated with substrate. The decreased activity of the imprisoned enzyme is not due to a permeability barrier, but to a diffusion barrier. In any case, these experiments indicated that sedimentable catalase was shielded by a membrane, and they added, as an important piece of information, that this membrane was different from the lysosomal membrane. The possibility that the three enzymes were present together in a special type of particle different from lysosomes could therefore be contemplated.[17] There was a tenuous link between the three enzymes: hydrogen peroxide. This substance arose from oxygen in the oxidation reactions catalyzed by the two oxidases, and it was destroyed by catalase. As will be mentioned later, this association proved significant and eventually formed the basis of the peroxisome concept.

Density gradient centrifugation. Once again, an intriguing hypothesis was in need of experimental testing. The question was, how? Our choice fell on density gradient centrifugation, a technique that was just beginning to be used in cell fractionation.[18] As the name indicates, the technique involves centrifuging material through a medium of increasing density. This gradient was usually constructed with sucrose, but other substances, if sufficiently dense and soluble, could be used.

There were basically two kinds of density gradient centrifugation, according to whether separation depended on differences in sedimentation rate or in density. In the first type, the material to be fractionated was layered over a shallow gradient to stabilize the tube contents against convective disturbances. Centrifugation conditions were chosen so as to cause incomplete sedimentation of the particles under study, and the tube contents were analyzed layer by layer after the run. Particles were distributed along the tube in the order of increasing sedimentation rate. This technique was obviously more refined and less cumbersome than the standard method. Its resolution was also greater, depending only on the thickness of the collected layers. The method was essentially an analytical tool, although it could, under favorable circumstances, also serve for actual separations. It allowed the direct determination of the distribution curves of sedimentation constants of the analyzed constituents.

In the second type of density gradient centrifugation, called isopycnic centrifugation or density equilibration, the gradients were steep and rose to values higher than the densities of the materials under analysis, so that complete sedimentation never occurred. Particles became arrested when they reached a position in the gradient corresponding to their own density. Centrifugation at high speed for a long time was required to reach this situation, since the particles moved with increasing sluggishness as they neared their equilibrium position.[17]

Applying these techniques to cytoplasmic particles raised a number of theoretical and practical problems. Because of the gradient, the two main medium parameters that rule sedimentation rates, density and viscosity, varied along the length of the tubes in a manner that had to be taken into account. The main particle parameters, density and size, also varied as the particles moved and were exposed to increasing concentrations of solute. Their density increased, but in a manner that depended on the permeability of their membrane to the external solute. With a permeable membrane, simple equilibration of the solute between the external and internal water would take place. With an impermeable membrane, osmotic loss of water, depending on the osmolality of the external solute, was to be expected. At the same time, the particle size would shrink. With Berthet and Beaufay, we constructed a theoretical model in which all these parameters were taken into account and used to predict the centrifugal behavior of a particle in a given density gradient. One technical problem was a tendency of the rotor to wobble as it was coming to a stop, causing a certain amount of disturbance of the tube contents, so Beaufay constructed a stabilizer that could be manipulated from the outside. He also built a machine for constructing the gradients and a device for collecting the contents of the tubes layer by layer.[18]

From urate oxidase to microbodies. The first fruit of our technical effort was a clear-cut demonstration that urate oxidase, catalase, and D–amino acid oxidase indeed were present together in a group of particles distinct from lysosomes and, of course, from mitochondria. They equilibrated at a significantly higher density value than the markers for the other two kinds of particles upon isopycnic centrifugation in a sucrose gradient.[13,19] The new particles separated cleanly from the mitochondria but unfortunately not from the lysosomes, which showed a very heterogeneous distribution in equilibrium density. Once again, chance came to the rescue, thanks to experiments that were performed for an entirely different purpose by members of the laboratory who discovered that injection of the mild detergent Triton WR-1339 to rats resulted in a dramatic and selective decrease in the density of hepatic lysosomes.[20] When the animals were treated in this way, the lysosomes remained floating on top of the gradient, the mitochondria equilibrated at their normal density value of 1.18, and the new particles separated as a clean peak around a density of 1.24.

In 1960, we published a note stating that urate oxidase, D–amino acid oxidase, and catalase are present together in a special kind of cytoplasmic particle. The hydrogen peroxide link was hinted at, but I did not go so far as to embody it into a name. One reason for my reluctance was the presence of another enzyme, monoamine oxidase, which I fully expected to accompany the other three. It produced hydrogen peroxide and showed the same heterogeneous distribution in conventional fractionations as the other three. Contrary to my expectation, the distribution of monoamine oxidase turned out to be bimodal and, as later shown by others, an authentic constituent of mitochondria, associated with their outer membrane. Fragments ripped off from this membrane upon homogenization are cosedimented with the microsomal fraction, accounting for the observed presence of part of its monoamine oxidase activity.[21]

The morphological identity of the new particles was soon established as corresponding to entities known in the morphological literature under the uninformative name of "microbodies." The paper in 1965 by Pierre Baudhuin, Beaufay, and me[22] conclusively identified liver lysosomes with "pericanalicular dense bodies" and peroxisomes (not yet named) with "microbodies" (figure 3), a year after publication of the last installment in the "Tissue fractionation

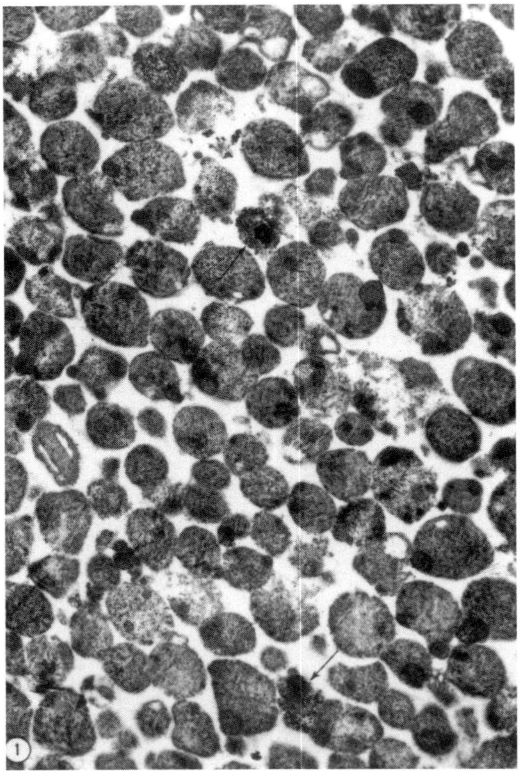

3. Electron micrograph of lysosomes from purified preparations of rat liver. ×20,700. From Baudhuin et al., 1965. *J. Cell Biol.* 26:219–243.

studies" series that had been revealingly titled "Resolution of mitochondrial fractions from rat liver into three distinct populations of cytoplasmic particles by means of density equilibration in various gradients."[13] A fifteen-year adventure had come to a gratifying, as well as completely unanticipated, conclusion.

The centrifuge as a cytological tool. Employing the centrifuge as a cytological tool in our work, we did not simply establish the existence of the three types of particles. We carried out a variety of density gradient experiments, both of the differential sedimentation and isopycnic kinds, using several types of gradients, including sucrose gradients made with either normal or heavy water as solvent and glycogen gradients made with sucrose solutions of different concentrations, that is, osmotic pressures, as medium.[13,19] This provided enough independent data to solve all the variables in our theoretical equation[18] so that we were able to draw a sort of "identikit" picture of the three particles based exclusively on biochemical results. The mitochondria came out as possessing both a sucrose-accessible space (corresponding to the intermembrane space) and a sucrose-inaccessible, osmotically active space (corresponding to the matrix). The peroxisomes had only a sucrose-accessible space, which fitted nicely with the lack of latency of their two oxidases. Lysosomes had an osmotically active space, as predicted from latency experiments. The size distributions of all three particle populations could also be derived from the results. All these findings were summarized in the Harvey Lecture I delivered in New York on 21 November 1963.[17] Eleven years later, I titled my Nobel lecture "Exploring cells with a centrifuge."[23] This, indeed, is what distinguished our approach from that of most other investigators doing similar work. We "saw" the particles with the eyes of biochemists and followed them that way throughout purification before finally seeing them in the electron microscope.

Lysosomes and endocytosis. With lysosomes identified as probable sites of intracellular digestion, we began to speculate on the likely physiological functions of these organelles and on their possible involvement in pathological processes.

The most obvious question concerned the nature and origin of the materials digested in them. Phagocytic uptake was, of course, well known in protists and in white blood cells. But few cell biologists suspected that many other cell types carry out a similar function, especially in higher animals. Illuminating information on this topic came, quite unexpectedly, from studies having an entirely different purpose at the State University of New York at Stony Brook by Werner Straus.

Straus was working in the laboratory of kidney specialist Jean Oliver, studying the renal lesions induced by the injection of egg white. It was known that this treatment causes a pathological condition similar to nephrosis, characterized by the formation of large "hyaline droplets" in the proximal convoluted tubule cells. These structures, which were loaded with ovalbumin, were believed by Oliver to be derived from mitochondria. Straus, using differential centrifugation, could not corroborate this theory. He found the droplets to be different from mitochondria and, by 1954, had already identified two enzymes in them: acid phosphatase and cathepsin D.[24] He was well on his way to discovering lysosomes when our 1955 paper describing the liver particles came out.[11] One year later, Straus had added b-glucuronidase and nucleases to the enzyme content of the kidney droplets, demonstrating their lysosomal nature.[25]

This work was one of the earliest indicating a function of lysosomes in the digestion of materials taken up by endocytosis. It also clarified important aspects of kidney physiology and offered the first example of pathological lysosome overloading. Substances, including small proteins, able to traverse the glomerular barrier are normally reabsorbed by endocytosis by the tubule cells and digested in their lysosomes. If, as was the case, the amount taken up exceeded the digestive capacity of the lysosomes, the substance would accumulate and the lysosomes would enlarge.

Lysosomes and thyroid function. We became interested in this topic through my colleague Michel de Visscher, whose main interest, both scientifically and clinically, was the thyroid gland. Our collaboration with his laboratory allowed the characterization of lysosomes in this gland, a finding that, in turn, led to the discovery of a particularly intriguing adaptation of the digestive function of lysosomes in relation to endocytosis. The thyroid gland manufactures as its principal secretory product the protein thyroglobulin, in which certain iodinated tyrosine residues are rearranged into precursors of the thyroid hormones, triiodothyronine and thyroxin. This protein is discharged into blind enclosures completely fenced off by cells, the lumen of the follicles, within which it accumulates. Upon stimulation by the pituitary thyroid-stimulating hormone (TSH), the cells surrounding the follicles recapture by endocytosis the thyroglobulin they have made and digest it in their lysosomes. Among the digestion products are the thyroid hormones, which are released into the bloodstream.[26]

"Suicide bags" and autolysis. Another notion was that our findings explained what had long been a riddle, namely, how cells can harbor such potentially injurious agents as proteolytic and other lytic enzymes without falling victim to autolysis. Our results suggested this happened because the dangerous enzymes were kept safely segregated within enclosures bounded by a membrane that somehow was resistant to their lytic action. This hypothesis implied, as an interesting corollary, that damage to this membrane leading to the intracellular release of the enzymes could cause cellular injuries, perhaps even cell death. This was the so-called "suicide bag" hypothesis, often misquoted as referring to the sole function of lysosomes.

Directed by this hypothesis, we looked for increases in free lysosomal enzyme activities in liver homogenates from animals subjected to hepatic ischemia and to a variety of hepatotoxic treatments. We also looked for similar evidence in instances of normal autolytic phenomena, as in the involution of the tadpole tail during metamorphosis and of the Müllerian ducts in developing chick embryos. We did indeed find what we sought but ran into a typical chicken-or-egg problem. Were the tissue injuries caused by the lysosomal damage? Or was it the other way around? Up to this day, the possible involvement of lysosomes in physiological and pathological autolytic phenomena remains unclear. The suicide bag hypothesis has slowly receded into the background, not because it is necessarily incorrect, but because it is difficult to test experimentally.[27]

Extracellular release of lysosomal enzymes. The possibility that lysosomal enzymes can cause damage outside the cells was first revealed by experiments carried out at the Strangeways Research Laboratory by British biologist Honor Fell with a young biochemist, John Dingle. A pioneer in organ culture and long interested in vitamin A, Fell had discovered that when an excess of this vitamin was added to embryonic chick cartilage explants kept in vitro, the extracellular matrix of the explants dissolved almost entirely, while the cells remained perfectly viable. Dingle found that this remarkable phenomenon was associated with the discharge of large amounts of proteases and other lytic enzymes, presumably the agents of the observed breakdown, into the extracellular medium.[28] Acquainted with our work, Dingle had identified the lysosomes as the source of the released enzymes. It now seems probable that vitamin A interacts not only with the lysosomal membrane but also with the plasma membrane and causes the two membranes to fuse, leading to the exocytic discharge of the lysosomal enzymes without their coming into contact with the cells' cytoplasm.

For Fell and Dingle, the studies on vitamin A toxicity were the start of a lasting interest in lysosomes, manifested by the creation of a multi-author series, *Lysosomes in Biology and Pathology*, of which seven volumes appeared between 1969 and 1984.[29]

Another participant in this fascinating work was Lewis Thomas, a unique blend of scientist, clinician, administrator, and author of *The Lives of a Cell*.[30] He and his associate Gerald Weissman showed that the effect of vitamin A is inhibited, both in vivo and in vitro, by the anti-inflammatory agent dihydrocortisone. This finding led Thomas and Weissman to develop a theory of inflammation involving lysosomes and inspired by the suicide bag hypothesis. Although in need of reinterpretation, their work, together with that of James Hirsch and Zanvil Cohn on leukocytes and macrophages at Rockefeller, helped to introduce the lysosome concept into the large field of inflammation research (see Chapter 12A).

In Louvain, with Gilbert Vaes, a young physician who had developed an interest in bone physiology, we began an investigation of bone lysosomes and of their possible role in the physiological resorption of the tissue induced by parathyroid hormone (PTH). By means of organ culture, it was shown that the extracellular release of lysosomal enzymes is not just a pathological phenomenon but can also occur physiologically. Briefly summarized, it was demonstrated that osteoclasts stimulated by PTH discharge lysosomal enzymes, together with the acid needed for their activity, into fenced-off zones of bone resorption where the enzymes accomplish their lytic function.[31]

Labilizers and stabilizers. Another set of experiments, inspired by the suicide bag hypothesis, focused on the effects of a variety of fat-soluble substances likely to affect the physical state of membranes on the stability of lysosomes in vitro.[31] We confirmed the results of the Fell group and showed the "labilizers," of which there were many, included vitamin A, whereas dihydrocortisone was one of the rare "stabilizers." Another substance found to stabilize lysosomes was cholesterol, an important constituent of the lysosomal membrane, as indicated by the unique sensitivity of this membrane to the detergent digitonin. This finding, in turn, caused us to become interested in Triton WR-1339, a mild detergent that had been shown by other workers to be a potent stimulator of cholesterol synthesis in vivo. This is how Wattiaux and his colleagues, Baudhuin and Maurice Wibo, came to treat animals with the detergent, in the hope that this treatment might increase the resistance of the animals' lysosomes to injury. What they found instead were much more fragile lysosomes, bloated with detergent and thereby rendered much less dense.[20] I have mentioned how this finding facilitated the purification of peroxisomes.

Inborn lysosomal diseases. One last story related to pathology involves my earliest collaborator, Géry Hers, who remained faithful to our first love, carbohydrate metabolism. Following important contributions to fructose and glycogen metabolism, Hers had become interested in genetic glycogen storage diseases, a set of several inborn, clinically distinct conditions having as a common characteristic the presence of abnormal glycogen deposits in various cells. The enzymatic deficiencies responsible for several of these conditions had been identified in pioneering studies by Gerty Cori and later by Hers. But one form remained unexplained, the rare type II glycogen storage disease, also known as Pompe's disease, from the name of the Dutch pediatrician who first described it. This condition was a biochemical mystery. It was not supposed to exist, as all the enzymes needed for the physiological breakdown of glycogen showed normal levels in the tissues of the afflicted children and these responded normally to glycogenolytic agents such as glucagon. Yet, the victims of the disease rarely lived longer than a few months, usually succumbing to cardiac failure caused by degeneration of the glycogen-overloaded heart cells.

In 1961, Hers ran into one such case, confirming the normal levels of all the enzymes of the glycogenolytic pathway. In trying one last possibility—a reaction involving transglucosylation from maltose—Hers hit upon the missing enzyme. It was not the one he was looking for, but a simple hydrolase acting on maltose and characterized by an acid pH optimum.[32] This enzyme had no place in biochemical schemes and, in another laboratory, might have evoked little more than ineffective puzzlement. But in ours, an acid hydrolase immediately called up lysosomes. And so it was that Hers's acid maltase, or a-glucosidase, came to join the family of lysosomal hydrolases.[33]

If glycogen accumulation resulted from the lack of a lysosomal enzyme, this must mean, so Hers reasoned, that the enzyme plays a role in lysosomal digestion of the polysaccharide and, therefore, that the abnormal deposits must be present inside lysosomes, where they would be both undigested and inaccessible to the cytoplasmic enzymes of glycogen metabolism. When another victim of the disease died, tissue samples taken at autopsy were given to Hers for biochemical analysis and to Baudhuin for morphological examination. As predicted, the stored glycogen was found inside large, membrane-bounded structures.[34] The entrance of the glycogen into these structures, which were rightly assumed to be lysosomes, from the surrounding cytoplasm was attributed to autophagy, a segregation process now beginning to be appreciated. The observation that some of the glycogen-filled vacuoles also contained remnants of cytoplasmic material supported this assumption.

Hers went one step further. It occurred to him that what happens to glycogen could happen to any other segregated cell constituent whose lysosomal digestion is impaired by the genetic deficiency of an enzyme necessary for this process. Thus, there could be as many forms of genetic storage diseases as there are lysosomal enzymes, each characterized by a specific kind of deposit related to the functions of the missing enzyme. In a historic article, published in *Gastroenterology* in 1965, Hers outlined the concept of "inborn lysosomal disease."[35] In one sweeping generalization, he had, as he and many others subsequently abundantly showed, clarified a whole mysterious chapter of genetic pathology, which covered such conditions as Gaucher disease, Tay-Sachs disease, Niemann-Pick disease, Hurler disease, and a host of others. This information was reviewed in a multi-author book, *Lysosomes and Storage Diseases*, edited by Hers and his main collaborator François Van Hoof.[36] Today, progress in this area has proceeded to the point where many of the genetic lesions underlying storage diseases have been pinpointed, prenatal diagnosis and heterozygote detection have become possible, and therapeutic attempts are being carried out with some success.

Lysosomes and therapeutics. As our knowledge of the physiopathological roles of lysosomes developed, we began to envisage possible therapeutic applications of the emerging notions.

In one application, André Trouet and I developed a model of "magic bullet" based on our knowledge of lysosomes.[37] The principle is simple. Take a chemotherapeutic drug whose successful use is thwarted, as in most such cases, by the toxicity of the drug for normal, fast-dividing cells—in bone marrow, skin, or mucosae, for example—other than the target cells. Attach this drug to a substance known to be selectively endocytized by the target cells, using a chemical linkage that 1) renders the drug inactive, 2) is stable in body fluids, and 3) is split by some lysosomal enzyme, and you are in business. The drug-carrier complex will be harmless, except in the cells that take it up by endocytosis, where it will be conveyed to the lysosomes and there hydrolyzed, releasing the drug in active form. When it was first proposed, this model represented one of the very first attempts at drug targeting, now a major research effort in many academic and industrial laboratories.

For reasons that need not be gone into, the first drug-carrier complexes we investigated were much less sophisticated than those postulated in the model. They consisted of the tight complexes that are formed spontaneously between drugs such as daunorubicin and adriamycin (doxorubicin) and high molecular weight DNA. These complexes gave very encouraging results on mice inoculated with L-1210 leukemia cells,[38] so much so that some clinical trials were even initiated. Eventually, however, this project was discontinued in favor of more sophisticated approaches.

End of an era and a fateful decision. In 1963, the Ciba Foundation in London hosted the first major symposium on lysosomes.[39] Chaired by James Danielli, famous for his model of a biological membrane, the meeting had been organized by Novikoff and me and was attended by an international galaxy of experts, including Albert Claude, Jean Brachet, Albert Dalcq, Henri Beaufay, Robert Wattiaux, Werner Straus, George Palade, Zanvil Cohn, Honor Fell, John Dingle, Lewis Thomas, and Heinz Holter. There also was a young Hungarian, Miklós Müller, invited by Novikoff, who had spotted him as the author of some interesting cytochemical work on the food vacuoles of protozoa. This invitation, as will be seen, turned out to be of momentous importance, both for Müller and for me. In a way, this meeting also signaled the culminating point in my Louvain career. Except for two reviews with my main Belgian co-workers, one with Wattiaux on lysosomes[40] and the other with Baudhuin on peroxisomes,[41] my activity shifted to my new position at Rockefeller in New York, where it did, however, remain very much connected with what was happening in Louvain. How this came about deserves to be explained.

By the start of the 1960s, what were to be our main contributions to cell biology had largely been accomplished. Lysosomes had already gained acceptance as intracellular organelles of wide significance, and information on their biological functions was beginning to accumulate. Knowledge of peroxisomes was less advanced, but the main evidence for their existence as distinct organelles in liver cells was already at hand. An emerging "Louvain school" of cell fractionation was in the process of being internationally recognized.

This success was gratifying, but the burden on my shoulders had begun to approach the intolerable. I was still very much involved in all the scientific activities of the laboratory, which included some thirty members by then, and I was almost solely responsible for the administration of the laboratory and for its funding (largely secured in the United States, mostly from The Rockefeller Foundation and the National Institutes of Health [NIH]). The local academic scene was hardly conducive to research on such a scale. My teaching load in Louvain was heavy and included the exhausting chore of interrogating about 1,000 students orally every year. The weight of these responsibilities and distractions was increasingly difficult to bear, especially with so many exciting things happening in the laboratory and crying for my attention.

I had, by that time, become fairly well known in the United States, which I had been able to visit regularly. I had made a number of friends at The Rockefeller Institute, including George Palade, Philip Siekevitz (whom I had met before in Madison, Wisconsin, when he worked at the McArdle Memorial Laboratory with Van Potter), and other members of their group. I had also had many contacts with James Hirsch and Zanvil Cohn, who had done pioneering studies on lysosomes in polymorphonuclear leukocytes and macrophages and shown that these granules discharge their enzymes into phagocytic vacuoles, there to digest the materials taken up by phagocytosis. My fascination with Rockefeller as the mecca of modern cell biological research continued to grow. On every visit, I admired and envied the magnificent surroundings, lavish support, and unique intellectual climate it provided. In contrast with my own situation in Belgium, Rockefeller appeared like heaven, and equally as unattainable.

I had an opportunity to share these feelings with Palade some time in the winter of 1960, adding, "If only I could join The Rockefeller Institute, I would be the happiest man on earth." He looked at me in a peculiar way and asked, "Do you mean this?" No more was said, but a few months later I received a letter from Detlev Bronk, president of the Institute, inviting me to create a new laboratory. As often happens under such circumstances, my Belgian alma mater suddenly discovered some hidden qualities in me. If I agreed to stay on, I was told, I would be relieved of most of my academic chores and allowed to devote myself mainly to research. Negotiations ensued. Bronk came to Louvain and had a long session with the rector of the university. A sumptuous dinner, of the kind higher Catholic clerics are famous for, sealed their accord. For five years, they agreed, I would enjoy a dual appointment, sharing my time more or less equally between the two institutions. After that, I would join the Rockefeller staff full time. This never happened, and what was to be temporary became permanent. I continued commuting across the Atlantic, until I had to give up those long trips.

In retrospect, it was a crazy arrangement, definitely not one that I myself would have approved for a member of my staff when I later became the head of a larger institution. Yet, it turned out to be beneficial for the people involved and, I venture to believe, for science. For me, it was a tremendous bounty.

The Rockefeller Years

By the time I joined Rockefeller, the Institute, soon to be renamed University, was dominated by the personality of its president. Det Bronk, as he was universally known, was a larger-than-life figure who shared a romantically humanistic view of science with a passionate attachment to basic research, an uncompromising regard for excellence, and deep respect for the freedom of investigators, especially young people, of whom he was particularly fond and in whom he saw the true artisans of scientific progress. His ideals were to be my inspiration in the creation of the ICP in Brussels, which I shall mention later.

For its investigators, Rockefeller was a dream place. When I first met members of the administration, I was amazed by the high regard in which they held scientists and by the degree of professional efficiency, dedication, and helpfulness they displayed. Each assured me that it was their job to relieve me of every possible chore they could do in my stead, so that I could spend all my time doing what I had been hired to do: first-class research. The place was still relatively small, and income from the endowment covered a substantial part of laboratory budgets. Applying for outside grants—actually prohibited until Bronk took over the helm—was encouraged but was hardly a problem. Those were the halcyon days when more federal money was allocated for research than was actually requested. For a number of years, I was able to cover all the laboratory expenses not supported by my university budget with a single grant from the National Science Foundation, at the cost of a minimum amount of paperwork.

At the start of my appointment, I received a modest amount of laboratory space in the west end of the seventh floor of the South Laboratory, now the Bronk Laboratory. The east half of the same floor was given to Theodosius Dobzhansky, a world-renowned, Russian-born geneticist whom Bronk had just brought over from Columbia University. The buffer zone between us was progressively filled by my expanding group until, after Dobzhansky's retirement in 1971, my space came to cover the entire floor.

An invitation by Abraham White to spend the autumn of 1961 as visiting professor at the Albert Einstein College of Medicine gave me an opportunity to prepare for my coming duties. I was able to stay in the students' residence and to commute to the Bronx with White, who lived three blocks away from the Institute. I took advantage of this stay to draw the plans of my laboratories. The South building was still a largely empty shell, and I was free to arrange the space as I wished. It included a windowed cold room with a view over the East River, which made it into something of a tourist attraction, as well as two additions of special value in my eyes. One was a large library-cum-meeting room with attending kitchenette, in which workers could retire to read, gather for discussions, share lunch, and attend seminars; in short, forge the many links that make for coherent groups. The other addition was a small but well-equipped shop that would serve for the maintenance and repair of instruments and the construction of special gadgets, an exercise very much in the tradition of my Belgian laboratory. For works requiring greater technical sophistication, we could rely on the celebrated Institute Instrument Shop.

By the spring of 1962, the facilities were ready, and my transatlantic adventure was launched. Running two laboratories separated by the Atlantic Ocean created a number of logistic problems. Fortunately, the main strength of my Louvain laboratory lay in what may be called a specialized fractionation tool box—a set of instruments, techniques, experimental approaches, and computation programs that could readily be exported and applied to

4. The Beaufay rotor,[42] exposed on a stand so as to show its layout inside the centrifuge. On the left is the rotor when allowed to rotate freely, with the filling-emptying device disconnected by raising it. On the right, the filling-emptying device has been lowered to connect with the rotor's axial tube. The device goes through a hermetically sealed opening in the lid of the centrifuge and is operated from the outside without opening the centrifuge. The most delicate part is the connecting piece of the device, which must ensure hermetic passage of fluid under high vacuum while being able to rotate together with the rotor when connected. This piece was perfected in the Rockefeller Instrument Shop with the assistance of Nils Jernberg. Courtesy of Christian de Duve.

problems other than those under study in Belgium. In order to achieve a proper transfer of know-how and to maintain adequate supervision, an airlift with Belgium was organized. Starting with Berthet and Beaufay, who each had several stays at the University, at least one senior member of my Belgian group was always in attendance when I was away. I commuted across the Atlantic five to six times a year. A string of outstanding secretaries and technicians (some from Belgium) provided valuable assistance. On the whole, the experience was a success. Soon, the transfer of technology became reversible. A number of developments achieved in the unique Institute environment were later transferred to Belgium, which eventually became a postdoctoral haven for several of my American co-workers.

The choice of research topic also was something of a problem. Obviously, the Louvain and Rockefeller programs could not be entirely unrelated. Yet, they should not be mutually

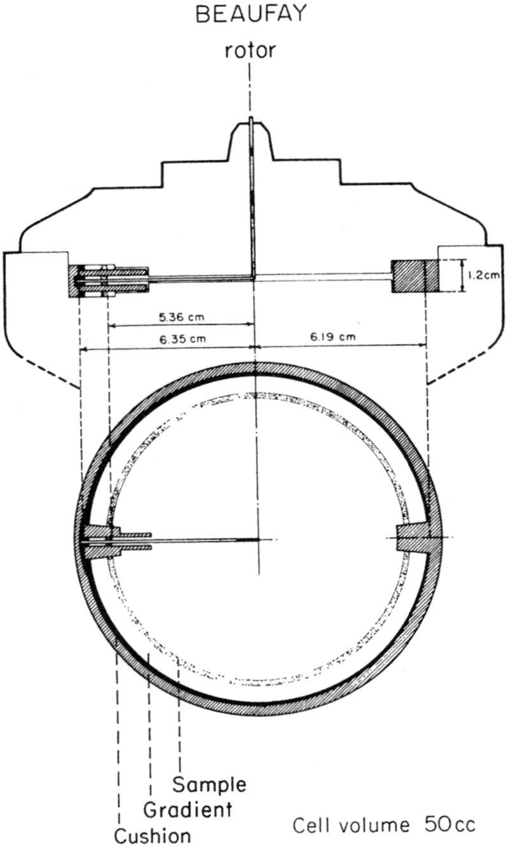

5. Diagram of the Beaufay rotor[42] for density gradient centrifugation, shown in starting condition. The rotor, adapted to density equilibration, has a ring-shaped cell that is filled and emptied through a single axial tube; the direction of flow is determined by the excess of the gas pressure prevailing in the dead space of the rotor over the hydrostatic pressure generated by centrifugation (flow is outward if excess is positive, inward in the opposite case). The rotor is operated entirely from the outside, under uninterrupted high vacuum and temperature control. Due to the high starting radial distance and the short path to be traveled, efficiency is high and hydrostatic pressure remains low, thus causing little damage to fragile cell particles. From de Duve, 1971. *J. Cell Biol.* 50:20D–55D.

competitive. In practice, this never created any real difficulty. There were plenty of questions to be answered, and it was easy to divide them between the two laboratories according to the interests and capabilities of workers on both sides. In a number of cases, the work was intimately collaborative.

New centrifugation equipment. Prominent among our first achievements at Rockefeller was the "Beaufay rotor," a remarkable piece of machinery designed in Louvain by Beaufay for the analysis of microsomal fractions (figure 4). Eventually, he and his group were able to resolve the microsomal fraction into four components: ER fragments, which make up the main part of the material, plasma membranes, Golgi elements, and outer mitochondrial membranes.[21,42] To do this, they had to resort to isopycnic density gradient centrifugation, which, with the conventional swinging-bucket rotor, required overnight spinning at top speed for the small and irregularly shaped microsomal elements to approach their equilibrium position in the highly viscous sucrose gradients used. To obviate this drawback and the considerable amount of heating and consequent damage to the fragile material, Beaufay developed a rotor,[43] which can be spun at up to 39,000 rpm and is particularly adapted to isopycnic density gradient centrifugation (figures 4 and 5). The operation is controlled entirely from the outside under uninterrupted high vacuum and temperature control simply by varying the rotational speed of the rotor. An ingenious device allows the filling unit to be disconnected from the filled rotor before acceleration to a high speed and reconnected after deceleration for emptying. The rotor thus runs freely at high speed between the two operations, with minimal friction problems. More than three times as efficient as a conventional swinging-bucket rotor, the Beaufay rotor avoids all the artifacts associated with swinging buckets and has the additional advantage of developing a much lower hydrostatic pressure and thereby inflicting much less damage on fragile cell organelles. While the delicate filling-emptying device was still posing problems when I joined Rockefeller, these were solved with the skilled collaboration of the head of the Instrument Shop, Nils Jernberg, a Swedish-born engineer.

Our interest in automated rotors also led to a fruitful collaboration with Norman G. Anderson from the Oak Ridge National Laboratory, who had designed a number of highly efficient rotors for zonal centrifugation.[44] His B-XIV rotor with a capacity of 649 cubic centimeters was well suited for the differential sedimentation of layered material through a stabilizing sucrose gradient. Thanks to a contract between the Union Carbide Corporation (which operated the Oak Ridge facilities) and Rockefeller, a B-XIV rotor was put at the disposal of our laboratory. It was widely used in experiments in which fractionation depended on differences in sedimentation velocity, whereas the Beaufay rotor served mostly for fractionations dependent on differences in density. Several important results achieved with the help of the two rotors will be mentioned in the coming pages.

Lysosomes in lymphoid tissues. In Louvain, we had begun to undertake a series of experiments dealing with the physiopathological functions of lysosomes. However, I decided to keep clear of this topic in New York and to concentrate instead on more biochemical aspects of the new particles. For my beginnings, I chose to play it safe, especially since the work was to provide the material for the PhD dissertation of my first graduate student, William Bowers, who looked for lysosomes in spleen and other lymphoid tissues.

This choice was motivated by the work of a Scottish group from the Rowett Institute in Aberdeen, Scotland, who had made detailed studies of a number of spleen enzymes associated with lysosomes in liver. The Scottish workers had expressed doubts about the presence of lysosomes in spleen and questioned the whole lysosome concept. I thought that their results could be explained by the presence of several distinct cell types, each with its specific lysosome population, and had attempted to convince them when I visited Aberdeen, but I found them totally impervious to my arguments. There was thus nothing left but for us to do the necessary experiments. Otherwise, the problem would go on haunting me.

The work turned out not to be, by far, as simple as I anticipated. It was a valuable learning experience, both for Bowers and for me. We encountered a number of obstacles, which had to be overcome by patient, systematic trials, when we tried to adapt our techniques to the spleen. Right from the start, Bill ran into a major problem: clumping. He took advantage of one of my absences to try something that I might not have approved, namely replacing sucrose with potassium chloride in the homogenization medium. In Louvain, salt solutions were rigorously banned because they caused massive agglutination of particles. It turned out to be the opposite with spleen. Sucrose induced clumping, and potassium chloride did not.

The recorded enzyme distributions also raised challenging problems by their unusual complexity, requiring many different experimental approaches for their clarification. Had we not had strong reasons to believe in the reality of lysosomes, we might well have dropped the subject and sided with the Scottish investigators. Fortunately, my young co-worker doggedly pursued his experiments until the results started to make some sense. Spleen, as well as lymph nodes and thymus, did contain bona fide lysosomes but divided into several physically and biochemically distinct populations, presumably originating from different cell types.[45–47] Thanks to his perseverance, Bowers earned a well-deserved PhD degree in 1966.[48]

The digestive potential of lysosomes. Our approach so far had consisted in identifying individual enzymes in lysosomes and deducing the digestive role of the enzymes from their specificities. We had never actually investigated what the lysosomal enzymes were able to accomplish by their collective and concerted action. This opportunity was provided by Trouet's "tritosomes" (lysosomes loaded with Triton),[46] which offered a convenient material for the study of the digestive potential of the whole mixture of lysosomal enzymes. The first studies of this kind were performed by John Coffey, a postdoctoral fellow from Tulane University, who investigated the in vitro digestion of globin and other proteins by lysosomal extracts. He found the process to be optimal between pH 4.4 and 5.6, to require prior denaturation of the proteins, and to degrade these right down to the level of free amino acids and small peptides. The latter were further split at alkaline pH by peptidases present in the high speed supernatant. A coherent picture of in vivo protein digestion, accomplished largely inside lysosomes and completed in the cytosol, emerged from these results.[49]

This work was complemented with the help of John Finkenstaedt, who had worked at Yale University with Joseph Fruton, an expert on cathepsins and who had himself received his training at Rockefeller from Max Bergmann, the initiator of the use of artificial substrates for the study of proteolytic enzymes. With Finkenstaedt, we identified cathepsins B and C as components of lysosomes, in addition to cathepsin D, first recognized in Louvain.

The digestion of hyaluronic acid and of the oligosaccharide side chains of several glycoproteins was similarly investigated by Nathan Aronson. He had observed considerable degradation of the substrates, but with an arrest or marked slowing down at certain branch points in the molecules. As in the case of proteins, digestion required an acid medium, with an optimum around pH 4.0.[50]

It befell Stanley Fowler, who joined the laboratory as a graduate student in 1965, to tackle the particularly arduous task of investigating lipid digestion. In doing this work, Fowler taught himself a great deal of lipid chemistry and enzymology. His 1969 thesis described the extensive digestion of various phospholipids and neutral lipids incubated with rat liver lysosomal preparations at pH 4.3–4.6.[51] He also characterized the main enzymes involved.[52]

Leukocyte granules. With the arrival in 1967 of Marco Baggiolini from Switzerland, a new line of investigation could be initiated in association with James Hirsch, who had, with Zanvil Cohn, done pioneering studies on the granules from rabbit polymorphonuclear leukocytes and shown these granules to be related to lysosomes. However, they had not separated the two types of granules present in the white blood cells, leaving a major uncertainty to be resolved. To this end, Baggiolini applied our technology to the subfractionation of the granules, making exemplary use of the Anderson and Beaufay rotors. In collaboration with Hirsch and his associate Martha Fedorko, who supplied the morphological part of the study, he was able to separate and characterize two main granule populations: large, dense granules identified as the primary, or azurophil, granules and containing the bulk of the myeloperoxidase, one third of the lysozyme, and a number of typical lysosomal hydrolases; and smaller, less dense granules with the morphological properties of secondary, or specific, granules, devoid of lysosomal enzymes but containing alkaline phosphatase and the remainder of the lysozyme.[53,54] A third population of small, dense particles, with the properties of lysosomes, was also characterized biochemically but could not unambiguously be identified morphologically. This problem was later solved in collaboration with Marilyn Farquhar in Palade's laboratory and her associate Dorothy Bainton, who found the unidentified lysosomes originated from mononuclear cells contaminating the peritoneal exudates that provided the leukocytes and not from the leukocytes themselves.[55]

Our work on leukocytes led to a visit by a young Belgian investigator, Pierre Masson, who was associated with my Louvain colleague Joseph Felix Heremans, a distinguished immunologist known for his discovery of immunoglobulin A (IgA) and a Rockefeller alumnus of Henry Kunkel's laboratory. Masson was able to localize lactoferrin, an important leukocyte component, in specific granules, adding one more twist to the "Belgian connection" with Rockefeller.[56]

Autophagic vacuoles. A last contribution to our knowledge of lysosomes was prompted by a finding made by Thomas Ashford in the laboratory of Keith Porter.[57] The investigators had found that glucagon induces the appearance in the liver of numerous sites of "focal autolysis." These sites, formerly described as cytosomes or cytolysomes and later autophagic vacuoles,[39] were membrane-bounded structures apparently arising through some kind of intracellular "self-eating" (cellular autophagy) and containing mitochondria and other cytoplasmic constituents obviously in the process of being degraded. Earlier work by Novikoff and others showing cytochemically demonstrable acid phosphatase in similar structures had suggested that they might be related to lysosomes. Their formation under the influence of glucagon was particularly intriguing to me, in view of my rediscovery of the second pancreatic hormone at the end of World War II. The possible relationship of my former love, glucagon, to my new pet topic, lysosomes, proved irresistible.

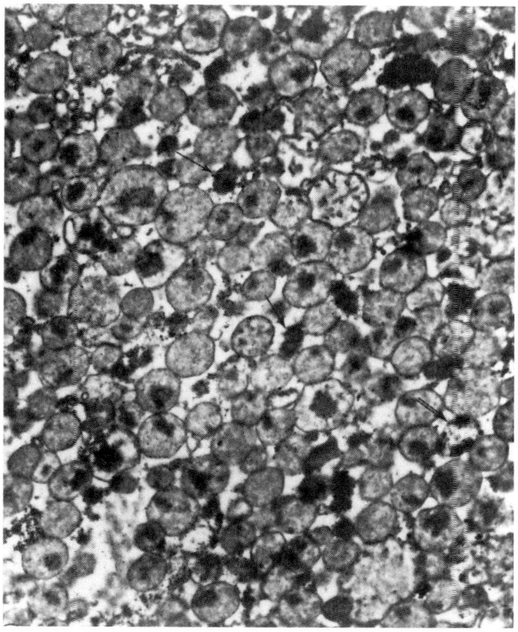

6. Electron micrograph of peroxisomes from purified preparations of rat liver, 1965. ×12,100. From Baudhuin et al., 1965. *J. Cell Biol.* 26:219–243.

When Russell Deter, a postdoctoral fellow from the Baylor University College of Medicine, joined our group in 1964, he used a variety of biochemical methods to show that the hepatic lysosomes of animals treated with glucagon undergo substantial increases in size and fragility.[58] This finding suggested strongly that preexisting lysosomes play an important role in the genesis of autophagic vacuoles under the influence of glucagon. Deter and Baudhuin later confirmed this morphologically in my Belgian laboratory.[59]

Lysosomes in protozoa. In the meantime, an important event, the 1963 Ciba Foundation Symposium on Lysosomes, had taken place that was to have wide-ranging consequences for our research effort.[39] As mentioned, this meeting was attended by Miklós Müller, a young MD attached to the department of histology and embryology at the Medical University of Budapest. Müller had become interested in the application of cytochemical staining methods to protozoa. In the course of extensive investigations on several protozoan species, he had detected acid phosphatase and several other hydrolases in food vacuoles and neighboring organelles, which pointed to a possible relationship of these structures to mammalian lysosomes. I was impressed with the quality of the work he had done under difficult conditions and with his eagerness to acquire additional biochemical training. This led me to offer him a temporary position at Rockefeller. By luck, there was a transient thaw in Hungary at that time, and Müller was allowed to come to New York with his wife and child. After one year at Rockefeller, followed by a year at the Carlsberg Institute in

Copenhagen, he decided not to return to his country and accepted my invitation to come back. Müller assumed increasing authority over the Rockefeller laboratory until he took it over entirely upon my retirement in 1988 and until he retired in 2005. Müller describes his experiences and contributions in Chapter 11B.

Müller did indeed carry out his project. He demonstrated lysosomes in protozoa, especially the ciliate *Tetrahymena pyriformis*, making a detailed biochemical study of these particles and of their functions. He did not pursue this subject further because his research led him to take a major interest in the second group of cytoplasmic particles identified in Louvain.

The peroxisome concept. Joined by Baudhuin from Belgium and Brian Poole, a graduate student who came to us in 1964 after a year spent in the laboratory of Paul Weiss, Müller extended his assays to two enzymes our work in Louvain had shown to be associated, in liver, with the mysterious microbodies described by morphologists: catalase and D–amino acid oxidase, to which he added an L-α-hydroxy acid oxidase. All three enzymes were found to be associated in *T. pyriformis*, as they are in mammalian liver, with a particulate component possessing the unusual distinction of equilibrating at a high density in sucrose gradients.[60] Similar results were obtained on the rat kidney, confirming some observations made in Belgium by Wattiaux. These findings served to overcome my last hesitations with respect to the significance of our observations and convinced me that the association of catalase with hydrogen peroxide–producing oxidases was likely to be of wide biological significance. In 1965, at the Fifth Annual Meeting of The American Society for Cell Biology, I proposed the name "peroxisomes" for these particles.

Large-scale isolation of peroxisomes. Müller's work led to a number of unexpected developments, including the discovery of a new cell organelle and the elucidation of several human genetic diseases. At the time, it rekindled my interest in peroxisomes, which, in mammalian cells, were still organelles in search of a function.

I decided on a detailed study of the composition, turnover, and biogenesis of peroxisomes (figure 6). As these particles were no longer a priority topic in Louvain, the New York laboratory appeared as the natural place to pursue the work. These studies were carried out by an enlarged group that included Beaufay, Baudhuin, Brian Poole, who had decided to stay in my laboratory for his PhD work, and a postdoctoral fellow from Chile, Federico Leighton. Starting in 1968, there was also a new graduate student, Paul Lazarow. During my insulin days, I had known his father, Arnold Lazarow, who had isolated particulate glycogen in Robert R. Bensley's laboratory in Chicago and later distinguished himself by his work on pancreatic islets and insulin secretion.

We set out to put together a set of techniques, as fully automated as possible, for the large-scale isolation, morphological evaluation, and biochemical characterization of rat liver peroxisomes. We took advantage of the effect of Triton WR-1339 injection into animals on the density of lysosomes.[20] When particulate preparations from the livers of animals thus treated, suitably isolated so as to be freed of most of the microsomes, were fractionated by density equilibration in a continuous sucrose gradient, three clean fractions were obtained. They corresponded to the Triton-loaded lysosomes, the mitochondria, and the peroxisomes. Thanks to the Beaufay rotor, preparations from about 100 grams of liver tissue could be processed in a single run. For the first time, we had enough material to "see" the particles on which we had worked for so many years. Isolated mitochondria, instead of being tan colored as described in the literature, turned out to be beautifully pink, the color of reduced cytochromes. Lysosomes were a dirty brown, probably due to their content of lipofuscin, or "age pigment." Peroxisomes were pea green, the color of catalase, which represents fifteen percent of their protein content. I have often described my first sight of these three fractions as the most memorable aesthetic experience of my life.[61]

Several other technical innovations were introduced. Protein and enzyme assays were all performed by automated techniques with equipment manufactured for clinical analytical laboratories by the Technicon company. Morphological examinations were performed with the various techniques worked out by Baudhuin to correlate morphological and biochemical results. Special computer programs were developed for the calculation and plotting of the results. The technology described in a resulting landmark paper was used in a large number of investigations on both sides of the Atlantic.[61]

Peroxisome turnover. In a first application of the new technology, the proteins of purified peroxisomes were fractionated by a variety of chromatographic procedures, and the turnover of the fractions was investigated in animals injected with [^{3}H]leucine. All protein fractions displayed essentially the same half-life, a result that was consistent with en-bloc degradation of the particles, for example, by autophagy.[62,63] Our measured half-life, however, was twice the value other workers had determined for catalase using non-isotopic methods. Tentatively attributed to reincorporation of the labeled leucine released from high turnover proteins, this discrepancy later led Poole to study the kinetics of disappearance and to confirm our reutilization hypothesis.

Peroxisome biogenesis. Our work on turnover led to an interest in how peroxisomes are renewed in the cells. At that time, it was generally accepted that peroxisomes arise as buds from the ER. This theory had been advocated and widely publicized by Novikoff, who claimed to have detected continuities between ER membranes and the surrounding membranes of microbodies. Strong biochemical support for the model had come from Tokuhiko Higashi and Theodore Peters at the Mary Imogene Bassett Hospital in Cooperstown, New York, who studied the biosynthesis of catalase. Using labeled leucine as a tracer, a specific anti-catalase antiserum to purify the labeled protein, and conventional cell fractionation techniques, they followed the pathway of the newly made enzyme from its site of synthesis to its intracellular abode. They found an early association of the labeled protein with the microsomal fraction, followed by its transfer to the mitochondrial fraction, and concluded that catalase is synthesized by membrane-bound ribosomes, to be subsequently transferred to the mitochondria. Except for the fact that the term "mitochondria" had to be replaced by that of "peroxisomes" as a result of our findings, these results fitted perfectly with Novikoff's budding model.[61]

This model had interesting implications for labeling studies. If one assumed progressive growth of the buds up to mature size, followed by their detachment from the ER and subsequent random destruction, one would expect, after a pulse labeling, that the labeled proteins would first be predominantly concentrated in the smallest particles (taken to detach from the ER upon homogenization) and then move progressively to particles of increasing size. Breakdown would affect particles of mature size exclusively. In 1966, Baudhuin and I had analyzed such a model mathematically and found that it could be fitted to the kinetics of catalase turnover, as established at that time by non-isotopic techniques.[41] Our later results, obtained with labeled precursors, were, however, more difficult to reconcile with the budding model. The turnover kinetics were strictly first order for all peroxisomal proteins, indicating that breakdown, while clearly affecting whole particles, was independent of particle size.[62,63]

Poole and I decided to ascertain whether, after a pulse of radioactive leucine, the label at first became more concentrated in the smaller particles originating as budding ER membranes and subsequently developed into particles of larger size as unlabeled proteins continued to be fed into the growing buds. Our results were unequivocal: peroxisomal catalase was equally labeled in particles of all sizes and at all times after injection of the precursor.[64] The budding model was in serious trouble.

This work completed Poole's thesis in 1968.[65] By that time, Lazarow had joined the group, overlapping sufficiently with Poole so that he could take over the project smoothly. His reinvestigation of the biosynthesis of catalase using our improved methodology gave highly revealing results.[66–68] Chemically, catalase, which is a tetrameric hemoprotein, was found to be synthesized as the apomonomer and in a matter of minutes acquired heme and condensed to the tetramer. This was perhaps not entirely unexpected. The truly startling finding was that the newly synthesized apomonomer appeared in the cell cytosol and was recovered in the high speed supernatant. Control experiments excluded leakage from the ER. The apomonomer was rapidly transported into peroxisomes where heme acquisition and tetramerization took place. No evidence of participation of a microsomal component (ER) in the process was obtained. These results formed the substance of Lazarow's PhD thesis in 1972.[69]

As we now know, thanks largely to the later work of Lazarow and his associates, peroxisomal proteins are indeed synthesized by cytosolic ribosomes and transferred into the peroxisomes posttranslationally.[70] The ER, however, may play a role in the biogenesis of peroxisomes, but not in the way originally envisaged by Novikoff. Phospholipids for the peroxisome membrane seem to be made in the ER and transferred to the peroxisomes, possibly by vesicular transport. As to morphological evidence supporting the budding model, the matter was finally settled years later to what I believe was Novikoff's satisfaction when Helen Shio, working with Lazarow, failed to show any sign of confluence between structures stained cytochemically for glucose-6-phosphatase (ER) and peroxisomes stained for catalase.[71]

A parenthesis is in order here. I mentioned my first contact with Alex Novikoff during my early Louvain years. As fate had it, he devoted a good part of his remarkable talent and energies to morphological studies dealing with the two cell organelles that were characterized biochemically in my laboratories, pioneering cytochemical staining procedures for their marker enzymes, the lysosomal acid phosphatase and the peroxisomal catalase. His morphological approach did not always agree with my biochemical findings, but the resulting clashes, although sometimes heated, never interfered with the warm links of friendship that existed between us.[72] On my wall, an electron micrograph of rat kidney, titled "Christian's LP," double stained for lysosomes and peroxisomes, reminds me of our close scientific as well as personal ties.

Peroxisomes in the brain. One last piece of work during the early phase of my Rockefeller years is the thesis work of my last graduate student, George Gaunt, who received his degree in 1973.[73] Intrigued by the reported presence of high amounts of D–amino acid oxidase, a typical peroxisomal enzyme, in cerebellum and other parts of the hindbrain, Gaunt set out on the difficult task of identifying the intracellular site of the enzyme and its possible relation to peroxisomes. Working with commendable care and precision and using a variety of biochemical and morphological approaches, he successfully identified

D–amino acid oxidase as a component of specific particles and characterized a number of properties of these particles. Unfortunately, no other enzyme, with the possible exception of traces of catalase, could be located to the same particles. Our approach was defeated by the cellular heterogeneity of the material investigated. The work ended with the conclusion, "The function of these particles raises an interesting problem, for which there is at present no solution."[74] Such admissions hurt.

Several other research projects launched in New York during that first phase could not be brought to a successful completion or a publication. But it was not enough to make me depart from a firm principle that one should only publish significant contributions to knowledge or, better, to understanding. Science is not just a compilation of data. Time spent at the bench does not automatically entitle one to an equivalent number of printed pages. I am not sure, however, that such a stand would still be defensible under the "publish or perish" pressures of the contemporary scene.

Venturing into cellular pathology. By 1968, when Frederick Seitz succeeded Bronk as president of Rockefeller, I had become smoothly installed in my dual life, with my two laboratories each producing good work independently, while closely collaborating on all issues of common interest. In keeping with the "spirit of 1968" and under the additional pressure of fiscal stringencies, Seitz made many changes to the university, but he never questioned my transatlantic commuting. When offering me the Andrew W. Mellon chair, he refused to hear my protests with the heart-warming comment that "half of me was worth the whole of many others." I myself, however, went through something of a change, partly influenced by the new academic climate and its greater insistence on the social responsibility of scientists.

For the first twenty-five years of my scientific career, I had stuck strictly, almost snobbishly, to "pure" basic research for the furthering of knowledge, the making of discoveries. What beneficial or lucrative applications might be made of the discoveries I did not consider part of my responsibility and left to clinical, pharmaceutical, and industrial investigators to develop.

What made me change my mind was the realization of the immense potential benefits that recent advances of biology held for human health and welfare. I had, after all, first graduated as an MD. My work on lysosomes had taught me how knowledge of these organelles could illuminate the pathogenesis of a wide variety of diseases and serve in the design of new therapeutic or preventive measures. My growing conviction was that it was my responsibility and that of my fellow basic researchers to help convert the knowledge we were producing into beneficial applications if such could be contemplated. Although this was a major departure from my previous attitude, I remained fully committed to untrammeled basic research as the main source of new knowledge and considered it imperative that those who devoted their efforts to this endeavor be free to chart their own course, without having to justify their choice by some prospective beneficial or profitable outcome, impossible to predict in any case since, by definition, the results of authentic basic research are unknown. What was new in my thinking was that this right implied as a counterpart a responsibility toward the society that supported the investigations. Should some basic research discovery lend itself to a useful application, the duty of the discoverers should be to help in the development of this application, all the more so because they were often the only ones with the required competence and technical know-how. Like many others at that time, I began to think in terms of "biomedical" research.

I shall explain later how this concern led to the creation of the ICP in Brussels. In New York, I started my "campaign" with a memorandum recommending the development of a vast program of research in "cellular pathology" at the University. My proposal was endorsed by some, including Palade and Hirsch, but clinical investigators did not unreservedly appreciate my invasion of their territory and my brash affirmations that the future of medical research lay in the development of new approaches based on the advances of cellular and molecular biology. I remember my friend Henry Kunkel gently chiding me, pointing out that the classical clinical approach had generated many discoveries, not only practical but also fundamental. My proposal never had the impact I had hoped, although it was probably used for fund-raising purposes.

I then developed the proposal into a major NIH grant application. Funding my laboratory had not been a problem until then. After being carried by the University for the time needed to become established, I had secured adequate support for my small group from the National Science Foundation and continued enjoying it with a minimum of red tape. But for this new venture, I needed larger funds and decided that the NIH would be the appropriate agency to apply to for a disease-oriented project of this sort. As at Rockefeller, my proposal did not meet with the immediate, enthusiastic approval I had naively expected. Eventually, however, after two supplement applications, two additional site visits, several trips to Bethesda, and large expenditures of time and effort, I finally obtained most of the needed money.

As to the new facilities required for this purpose, it so happened that implementation of the project coincided with Dobzhansky's retirement and move to California in 1971. I was given the space he left vacant on our floor and came to occupy the entire area, doubling our laboratory space. The laboratories were renovated and equipped with the help of grant money. I was also able to add an electron microscope to our equipment. In this venture, I benefited greatly from the valuable and generous assistance of Marilyn Farquhar, in Palade's laboratory two floors below, who took under her wing Helen Shio, the person I hired to run the facility. Helen had little previous experience but turned out to have exceptional qualities. She developed into a full-fledged independent investigator, thanks to the superb training she gained under Marilyn's guidance. After my retirement, she was put in charge of the university's central electron microscope facilities.

To develop the new program, I progressively brought back four of my former graduates whose thorough familiarity with our techniques and approaches was an essential condition and who had acquired valuable new skills during postdoctoral training in other laboratories. The first to rejoin was Poole in 1969, after spending a year in the laboratory of Brachet in Belgium. Bowers returned in 1970, after two years in my Louvain laboratory and one year at Oxford under immunologist James Gowans. He was followed in 1973 by Fowler, who also spent two years in Louvain doing sophisticated immunocytochemical work and then training in experimental pathology under Earl Benditt at the University of Washington in Seattle. The last to return was Lazarow in 1975, who spent a year at the laboratory of Molecular Embryology in Naples under Alberto Monroy and two years at Stanford.

In addition, there was Miklós Müller, who had never left and had long been in charge of an entirely independent group. Although originally associated with my project, Müller soon lost interest in its main thrust, irresistibly drawn back to his pet subject, protistology. As it happened, his work was to take an unexpected turn that led him almost unwittingly into human pathology and therapeutics (see Chapter 11B).

Atherosclerosis. For my part, I chose a topic that had long been on my agenda, namely the investigation of the subcellular events in atherosclerosis. I was, of course, quite ignorant of the field, but thought that our approach might bring new insights into the pathogenic mechanisms responsible for the number one killer in the developed world.

I was helped in this project by a British clinical investigator, Timothy Peters, who joined our group in 1970. He was a gastroenterologist, determined to work on intestinal cells, but I succeeded in diverting his interest to, as we put it, another kind of tube. The project progressed very nicely. With the initial help of Müller, Peters succeeded in miniaturizing our techniques so that they could be applied to the small amount of smooth muscle cells that could be separated from a single rabbit aorta. Using material containing only 0.5–2.0 milligrams of protein, he was able to separate twelve distinct fractions by density gradient centrifugation in the Beaufay rotor and to analyze them for up to seventeen enzymes, in addition to protein and DNA.[75]

Extending these investigations to lipid-laden cells from atherosclerotic aortas of rabbits given a high amount of cholesterol in their diet, Peters obtained evidence of considerable lipid storage in the lysosomes of the diseased cells,[76] a finding confirmed morphologically by Shio in collaboration with Farquhar.[77] A marked increase in the levels of lysosomal enzymes was also observed, as often happens in storage diseases. A side finding, whose possible significance was recognized only after the function of hepatic peroxisomes in fatty acid oxidation was discovered, was the observation of elevated catalase activity and proliferation of "microperoxisomes" in the diseased cells.

An attractive theory accounting for the transformation of arterial smooth muscle cells into the lipid-laden cells described as "foam cells" in the histopathological literature emerged from these findings. Inspired by Hers's discovery of inborn lysosomal storage diseases, the theory pictured the transformation as the consequence of overloading of the lysosomes by undigested cholesteryl esters. So, in collaboration with Tatsuya Takano and William Black, we looked for and successfully identified an acid cholesteryl esterase in the lysosomes of the arterial cells.[78]

Peters returned to Britain in 1972, taking with him the technologies he had developed, together with one of the rare copies of the Beaufay rotor built in the university Instrument Shop. In the meantime, I had been able to recruit Fowler back to our laboratory. With his excellent knowledge of lysosomal lipid degradation, the subject of his PhD dissertation, and with his subsequent training in pathology in one of the best centers of atherosclerosis research under such luminaries as Earl Benditt and Russell Ross, Fowler was ideally placed to take over the project.

In a series of elegant experiments conducted with Nancy Haley and Helen Shio, Fowler prepared suspensions of intact cells from enzymatically disrupted aortas from cholesterol-fed rabbits and applied metrizamide gradient centrifugation to these suspensions to separate foam cells heavily laden with lipid from less injured cells.[79] He distinguished and quantitatively evaluated two kinds of lipid deposits in the foam cells: free droplets and membrane-bounded structures of a lysosomal nature. He also found the lysosomal acid cholesteryl esterase activity to be increased 3.5-fold in the diseased cells.[80] His subsequent work, however, shattered our expectations. The bulk of cholesteryl esters was recovered with the unbounded lipid droplets, whereas the lysosomal deposits consisted mostly of free cholesterol.[81,82] As a result, the pathogenic hypothesis attributing the foamy transformation to an overloading of lysosomes by undigested cholesteryl esters had to be abandoned. Evidently, the lysosomes were overloaded with free cholesterol, possibly as a result of an overstressed transport system. This led to the characterization of a new congenital lysosomal lipid storage disease in NCTR-BALB/c mice, in which free cholesterol is the main form of stored lipid.[83–85]

Protein turnover and lysosomotropism. When Poole returned to our laboratory in 1969, Lazarow had taken over the study on the biogenesis of peroxisomes. Poole focused instead on the general problem of intracellular protein breakdown, which he had first met in his studies on the turnover of peroxisomal proteins and which was of obvious importance, both physiologically and pathologically. There was the fascinating question of the participation of lysosomes in intracellular protein breakdown and of the regulation of this process, a question that was thoroughly examined in a review by Poole on the synthesis and degradation of proteins in relation to cellular structure.[86]

As the object of his investigations, Poole chose cultured fibroblasts. Joined first by Wibo, he set up a technique for the simultaneous measurement of the degradation of proteins of short and of long half-life, making the key discovery that the breakdown of long-lived proteins, but not that of short-lived proteins, is inhibited by fresh serum.[87] Investigating this phenomenon further with Michael Warburton, Poole identified glucose, amino acids, inorganic phosphate, insulin, and fibroblast growth factor among the components participating in the serum effect.[88] In addition, they made the intriguing observation that the inhibition of long-lived protein degradation by serum, which is an immediate effect, is followed fifteen hours later by a stimulation of DNA synthesis. Exposure of the cells to serum for no more than fifteen minutes sufficed to set off the long-term effect on DNA synthesis. This is one of the first observations of the triggering of a complex intracellular cascade of effects, which is so much in the news nowadays.

Another finding Poole made with Wibo was that chloroquine inhibits the breakdown of cellular proteins (both long and short lived) to a moderate extent and, at the same time, causes the cells to be filled with large, microscopically visible vacuoles originating mostly from swollen lysosomes, in which large amounts of chloroquine are stored.[89] The enormous intralysosomal concentration of chloroquine (about 1,000-fold with respect to the extracellular concentration) was attributed to proton trapping, and an adequate model was proposed accounting for the "lysosomotropism" not only of chloroquine, but of weak bases in general, including neutral red, the prototype of "vital stains."[90] According to this model, weak bases diffuse freely across the lysosomal membrane in unprotonated form, become protonated inside the lysosomes, and are retained in that site because the membrane is impermeable to their protonated form.

This model rested on the assumption that the intralysosomal pH is low. Derived from the fact that all lysosomal hydrolases share an acid pH optimum, this assumption had never been verified experimentally. Poole was able to do this, thanks to an ingenious technique based on the pH dependence of the fluorescence excitation spectrum of fluorescein.[91] To get the probe into the lysosomes, Poole together with Shoji Ohkuma coupled the fluorescein to dextran, a carrier that is taken up by cells by endocytosis and accumulates in the lysosomes. They studied the effects of a large number of weak bases on cytoplasmic vacuolation and on the intralysosomal pH in isolated mouse peritoneal macrophages.[92,93] This work provided strong support for the hypothesis that a proton pump exists in the lysosomal membrane. Characterization of this H^+-ATPase was a major achievement of Ohkuma after his return to Japan.

The macrophage system was also used by Poole, Ohkuma, and Jerzy Chudzik to compare the effects of chloroquine and other weak bases, as well as acidic ionophores, on the digestion of exogenous (endocytosed) and endogenous proteins. Their results showed unambiguously that lysosomes, while playing a major role in the breakdown of endocytosed proteins, were only moderately involved in that of endogenous protein.[94]

Poole was fascinated with the nature of the non-lysosomal pathway of intracellular protein breakdown—proteasomes and the role of ubiquitin were not yet known—and had started with Dunne Fong to investigate the preferential breakdown that affects abnormal proteins[95] when he died of a diabetic coma while on a lecture trip in Norway. Poole's wife Elizabeth had started a biochemistry thesis in my Belgian laboratory and a short while later joined my New York laboratory, where she completed her work under Brian's supervision[96] and eventually became his wife. His death was a great loss to his family, friends, and science.[97]

Immunology. During the ten years after his return to New York in 1970, Bowers developed a very active group in immunology. In his first series of experiments, he extended his previous work on rat lymphoid tissues to rat thymus–derived lymphocytes. These cells contain, in addition to typical lysosomes, an unusually located cathepsin D that appears mostly in the soluble fraction, and Bowers showed that this enzyme is present in the main population of small lymphocytes. L-1210 leukemia cells were later shown to possess the same unusual cathepsin D, which seems to be a unique component of lymphoid cells in rodents.[98] This enzyme's activity is five times more sensitive to inhibition by pepstatin than the lysosomal cathepsin D and immunologically distinct from this enzyme. It exists in monomeric form and in a complex, possibly a dimer, that can be dissociated by thiols.

Dendritic cells. Another important finding by the Bowers group was the identification of a special kind of large cell needed for the successful activation of responder cells.[99,100] The workers were able to purify these "accessory" cells 1,000-fold, identified them as dendritic cells (also found in liver and skin), distinctive from macrophages, and showed that, in contrast to the stimulating cells, they could each assist in the activation of a large variety of responding cells. They then produced convincing evidence that the accessory cells act by way of a soluble factor, most likely interleukin 2, that is secreted by the responding lymphocytes when these are stimulated by the accessory cells, perhaps by means of another soluble factor. These investigations complemented the work being done in another Rockefeller laboratory by Ralph Steinman and Zanvil Cohn, who discovered dendritic cells in 1973 (see Chapter 12).

Peroxisomes. The last of my "children" to return to the fold was Lazarow, who set up a very active and productive group that attracted workers from many parts of the world. This offered an opportunity for investigating a possibility that had been at the back of our minds for several years, namely that peroxisomes may be a site—perhaps *the* site—of fatty acid oxidation in the liver. This function had long been attributed to mitochondria, but so were several of the enzymes that were subsequently found to belong to peroxisomes. One reason for doing this was the observation, made in 1965 by workers of the Ciba Company in Basel, Switzerland, that the hypolipidemic drug clofibrate (*p*-chlorophenoxyisobutyrate) causes a remarkable hepatomegaly and proliferation of hepatic peroxisomes in rats.[101]

Lazarow's experiments opened a productive new field of research that even turned out to have important clinical implications. Purified peroxisomes indeed catalyzed the oxygen-dependent reduction of NAD^+ (as well as oxygen consumption and hydrogen peroxide production) in the presence of palmityl-CoA, and this activity was increased about ten-fold in the peroxisomes from rats treated with clofibrate and with two other hypolipidemic drugs.[102,103] Further work conclusively demonstrated the presence of a complete, H_2O_2-linked β-oxidation system in rat liver peroxisomes.[104] This turned out not to be another case of misattribution of a peroxisomal enzyme to mitochondria; rather, both organelles catabolize fatty acids. Later, Jens Kondrup, working with Lazarow, estimated that approximately thirty percent of palmitate β-oxidation in rat liver may occur in peroxisomes.[105]

Peroxisome biogenesis. In a new phase of his research career, Lazarow also went back to the study of peroxisome biogenesis, the subject of his PhD dissertation.[69] With Mariette Robbi, he set up a system for the in vitro synthesis of catalase by a cell-free system and showed that the catalase precursor had the same size[106] and yielded the same peptide map,[107] whether synthesized in vitro or in vivo. Comparison with the mature catalase subunit showed no evidence of modification, indicating that transfer of the precursor to peroxisomes takes place without processing.

In collaboration with Yukio Fujiki, Richard Rachubinski, and Rockefeller graduate student Richard Mortensen, Lazarow investigated the biosynthesis of peroxisomal β-oxidation enzymes, showing, among others, that the acyl-CoA oxidase and the enoyl-CoA hydratase-dehydrogenase, as well as the 3-ketoacyl-CoA thiolase, are synthesized by free polysomes.[108] In contrast with the former two, the latter enzyme is synthesized in the form of a larger precursor, which is processed upon transfer into the peroxisomes.[109] The workers also found that the 22-kilodalton integral membrane protein identified in the peroxisomal membrane is similarly made by free polysomes.[110] These observations led Lazarow and Fujiki to propose that peroxisomes form by growth and division,[70] a hypothesis that has been confirmed subsequently by many investigators.

In addition to the biogenesis of peroxisomes, the structural organization of these organelles also attracted Lazarow's attention. He studied the internal arrangement of the proteins in these particles with the collaboration of Shio and Stefan Alexson[111] and suggested, based on the morphological observations of Shio and others, that some of the particles are associated in a "peroxisome reticulum."[70,112]

This concludes my Rockefeller adventures in "cellular pathology," which, after peaking during the years 1975–80, when my whole flock was around me, were progressively phased out. In retrospect, it appears that the umbrella of cellular pathology covered rather generously a variegated program that, except for the part on atherosclerosis, was focused mainly on basic biological research. Unexpectedly, and gratifyingly, this work led to significant advances in the understanding and treatment of a number of diseases. This is a perfect illustration of the philosophy I have defended over several decades. Carry out free basic research, prompted exclusively by the desire to understand. But, if the results open the possibility of a beneficial application, do not leave it to others to make the necessary translation. Do it yourself, at least to the extent of your abilities. At some stage, if the work is successful, industrial know-how and clinical input may become indispensable. In this way, investigators reconcile their right to freedom with their duty toward society.

Two events, of a more personal nature, occurred during this later phase of my stay at Rockefeller. In 1974, I was awarded the Nobel Prize in Physiology or Medicine jointly with Albert Claude and George Palade. I was in New York when the announcement came, and the event was duly celebrated with the whole group.

The other personal event was my seventieth birthday, which served as the occasion for a one-day symposium, in December 1987, attended by a vast majority of my present and former co-workers coming from all over the world. Thanks to an extraordinary complicity by everybody concerned, including my wife, I was taken completely by surprise and started to become suspicious only when, on an icy-cold Saturday morning, my wife insisted on my going to the University, where Caspary Auditorium, largely filled with the members of my scientific "family," awaited me. It was an unforgettable experience.

I reached emeritus status in 1988. My laboratory of Biochemical Cytology, which I headed for twenty-seven years, gave place to the laboratory of Biochemical Parasitology, directed by my old collaborator and friend Miklós Müller, with whom I am happy and proud to remain associated. We moved from what was then the South Laboratory building and is now the Bronk Laboratory to the fourth floor of Founder's Hall. I find some satisfaction in ending my career in this venerable building, which had seen the birth of this most remarkable institution more than a century ago.

The ICP, a Mini-Rockefeller in Belgium

Before ending, I must briefly mention an enterprise that, although strictly Belgian, has close connections with Rockefeller. In the early 1970s, while I was busy developing my new program in New York, my Belgian alma mater, the Université Catholique de Louvain, was ousted from the old Flemish town of Louvain (Leuven), the university's seat since 1425, to a building in a brand-new city, Louvain-la-Neuve (New Louvain), a mere fifteen miles away from Louvain. Because of the needs of its hospital, the medical school did not accompany the rest of the university but was transferred to a new site on the outskirts of Brussels. The move, which to many was a highly traumatic experience, provided a unique opportunity to create in Belgium a biomedical research center inspired both by the Rockefeller tradition and by my new cellular pathology research philosophy. It came to my mind that several groups that were geographically separated in Louvain could take advantage of the move to become associated in a single building, in which they could share core technical facilities and engage in multidisciplinary collaboration. This project was duly implemented, and the founding groups moved into their new quarters in the spring of 1974, to be followed later by a number of additional groups recruited worldwide, in particular an active branch of the Ludwig Institute for Cancer Research.

The new center was given the rather cumbersome name of "International Institute of Cellular and Molecular Pathology," ICP for short, which combined my original notion of cellular pathology with the new dimension of molecular biology. I must add that, on the occasion of my eightieth birthday, a well-intentioned but, in my opinion, misguided board decided to rename the ICP the "Christian de Duve Institute of Cellular Pathology," now shortened to the "de Duve Institute." Such honors, if at all deserved, are best awarded posthumously.

In its conception and organization, the ICP owed much—in a more modest proportion—to the Rockefeller model of giving priority to quality over all other considerations.[113] Conforming to a founding philosophy similar to the Rockefeller model, the ICP has never had a specific research program, only a general framework within which investigators are free to choose their own objectives. The result is a considerable diversity of research topics, ranging from metabolic regulations and their disorders to human genetic abnormalities, from the biology of endocytosis to physiological and pathological collagen breakdown, from microbial and viral infections to sleeping sickness, and from interleukins and IgA to cancer. This variety is kept together by a common core of conceptual approaches and technical strategies, sometimes conducive to unexpected collaborations.

An important feature, embodied in the term "biomedical," is the combined commitment to untrammeled basic research and to the development of beneficial applications whenever such can be contemplated. At present, almost every group in the ICP may be said to be engaged in both types of research. Today some 275 investigators and supporting personnel represent the major disciplines of modern biomedical research, including biochemistry, cell biology, genetics, molecular biology, immunology, microbiology, and virology.

New Vistas

In recent years, after my retirement on both sides of the Atlantic, I have become increasingly involved with problems of a more general nature. It all started, almost in fun and without any ulterior purpose on my part, when I was invited to deliver the 1976 Alfred E. Mirsky Christmas Lectures at Rockefeller, a series of four lectures to some 550 selected high school students from the larger New York area. I chose to take my young listeners on a tour of the cell. We would shrink to the size of bacteria or, which amounts to the same thing, blow up the cell a million-fold. We could then conveniently walk—or, rather, swim using appropriate scuba-diving gear—through the cell, observe different parts of its structure, and watch them in action to understand their function. Once I got into the spirit of the game, I found it most enjoyable. And my audience of young "cytonauts" was appreciative. Eight years later, the lectures metamorphosed into *A Guided Tour of the Living Cell*, a book of 450 printed pages illustrated by close to 300 pictures, micrographs, drawings, and diagrams of various kinds.[114] I became enamored with the kind of intellectual exercise that had been called for in writing the book, and one thing leading to another, the *Guided Tour* was followed in 1991 by *Blueprint for a Cell*,[115] in 1995 by *Vital Dust*,[116] in 2002 by *Life Evolving*,[117] in 2005 by *Singularities*,[118] and in 2010 by *Genetics of Original Sin*.[119]

Writing those books has been a fascinating voyage of personal research and discovery, which has taken me farther and farther away from cell fractions and biomedical investigation into increasingly distant and unfamiliar territories, including the origin of life, the birth of eukaryotes, the mechanisms of evolution, the emergence of humankind, the functioning of the mind, the future of life, and, finally, the meaning of it all. In each of these areas, I have developed some ideas of my own, not always in conformity with the majority view. And so, quite unexpectedly, I have been fortunate enough to start a new scientific life at an age when it is more customary to look back than forward.

It is unfortunately too late to bring my new interests to the laboratory and put some of my ideas to the crucial test of experiment. At least, I am returning to a dream, conceived some sixty-five years ago when, as a young medical student, I first entered the field of science. What I had not realized then is that there is no room for philosophy in front-line research, which narrows the mind more often than it broadens it. As we dig deeper, our scope shrinks. Only now have I succeeded, by way of writing books, in crawling out of my hole and looking around at the wider horizon.

11B. Discovering the Hydrogenosome

Miklós Müller

My involvement with the single-celled protists began almost sixty years ago. As a young boy in Hungary, I read avidly about astronomy, chemistry, and biology, experimented with a chemistry kit, and loved to use a good microscope belonging to my cousin. We looked at water samples from everywhere, and I am certain that I saw my first *Vorticella* before I was ten. After the Second World War, I read *Arrowsmith* and dreamed of working one day at the real equivalent of the fictional McGurk Institute. Little did I realize that the then impossible dream would come true in less than two decades. When I was a high school student, I regularly spent time in the laboratory of Nándor Gimesi, professor at the Budapest University of Sciences, whose binocular microscope with an oil immersion lens seemed the eighth wonder of the world. I also participated in an essay competition on "Sexuality in Protozoa" and soon became inspired to consider a higher education in biology. During my first year at the Budapest Medical School, Imre Törö, professor of histology and embryology, introduced me to systematic scientific studies on the cytochemistry of protists.

After spending some years in Budapest exploring protist food vacuoles using histochemical methods, I found acid phosphatase in the ciliate *Tetrahymena pyriformis* that I was testing. This led me to assume a correlation between the detected activity and a structural element of the cell. Not versed in biochemical literature, however, I was unaware that this enzyme had become a marker for the biochemically defined digestive structures, the lysosomes, of mammalian cells discovered by Christian de Duve in 1955, but in those days no one associated lysosomes with such "primitive" single-celled organisms. Then in 1960, on my first trip to a western country, I attended the 10th International Congress for Cell Biology in Paris, where Alex Novikoff showed electron micrographs highlighting acid phosphatase in various cells. I suddenly realized I had identified the major digestive structure in protists as lysosomes. When my research was published in 1962, it attracted the attention of Novikoff and de Duve, who were organizing the first international symposium on lysosomes and invited me to present my work.[1] The participants included morphologists, biochemists, and pathologists whose contributions revealed to me the importance of multidisciplinary research. This experience convinced me that to continue my studies with protists, I needed to learn more biochemistry and the new cell fractionation methods. On returning home, I wrote to Heinz Holter at the Carlsberg Laboratory in Denmark and Christian de Duve at The Rockefeller University (then still called Institute)—alias for the McGurk Institute. Both replied favorably. I spent a profitable year in New York and then a year in Denmark. Before leaving for Denmark, de Duve invited me back, so I returned and have spent the rest of my career at Rockefeller, thus seeing my impossible dream come true. He was willing to gamble on possibly transforming a morphologist without biochemical training and experience into someone who would not be lost when confronted with a centrifuge or a spectrophotometer.

On my return to New York, I made a detailed biochemical study of lysosomes in *Tetrahymena* and identified two distinct populations of this organelle, one involved in intracellular digestion and the other similar to secretion granules that discharge their contents extracellularly under starvation conditions.[2,3] A major interest in the laboratory was the peroxisome that de Duve had discovered earlier in mammalian liver tissue; this novel organelle contains enzymes capable of producing and breaking down hydrogen peroxide (see Chapter 11A). With Pierre Baudhuin and Brian Poole in the laboratory, we assayed peroxisomal enzymes in cell fractions obtained with various centrifugation procedures, and we found this organelle in *Tetrahymena*.[4] Then with James Hogg at Queens College, we demonstrated that peroxisomes in this protist unexpectedly contain two enzymes of the glyoxylate bypass of the glyoxylate cycle responsible for gluconeogenesis from acetyl-CoA; the rest of the cycle in this ciliate was localized in the mitochondria, making a most interesting intracellular division of labor.[5]

In 1969, de Duve suggested the possible physiological role and evolution of peroxisomes as the ancestral respiratory organelles. Due to their simple structure, mode of action, and broad distribution across species, he assumed peroxisome respiration predated the establishment of the far more energetically efficient mitochondria.[6] So I set out to search for organisms that might have conserved the ancestral "pre-mitochondrial" situation—in other words, were devoid of mitochondria but possessed peroxisomes. Little did I expect this search would lead to something completely different, another novel organelle about the size of mitochondria, the hydrogenosome.

Through studies of electron microscopic images and our knowledge of comparative biochemistry, we came upon a group of protists called trichomonads, which live in inhospitable environments such as digestive and genitourinary tracts. The human pathogen *Trichomonas vaginalis* and the cattle pathogen *Tritrichomonas foetus* were much-studied species available in bacterium-free cultures and thus amenable to our biochemical and cell fractionation studies. The physiological data indicated the respiration of these species was not of the mitochondrial type because it could not be inhibited with cyanide and other mitochondrial inhibitors, and furthermore no cytochromes could be detected. A biochemical peculiarity of the trichomonads was their essentially anaerobic nature; that is, they were independent of available molecular oxygen as the terminal electron acceptor. At the same time, ultrastructural images were accumulating that revealed the absence of typical christate mitochondria. Instead, the well-known granules of mitochondrial size turned out to resemble microbodies, which were then known to correspond to peroxisomes in many aerobic organisms.

We tested a hypothesis to learn whether the granules seen in trichomonads were related to peroxisomes. Their characteristic morphology and the cyanide-insensitive respiration of the organisms seemed to point in this direction. In 1971, I obtained cultures of *T. foetus* to verify the presence of catalase activity, a marker of peroxisomes, and performed a seemingly obvious experiment. I prepared a homogenate in isosmotic sucrose and separated it by centrifugation into sediment and supernatant, with the expectation that the sediment containing the granules would contain most of the catalase activity. The results were disappointing because most of the activity was recovered in the supernatant. The total recovery of the activity was close to one hundred percent, indicating no problems with the fractionation experiment itself. Two repeat experiments gave the same results. Since I thought I knew that catalase had to be in the organelles, I assumed that they had been broken during homogenization due to their excessive fragility and decided that this approach would not lead me to the desired goal.

Reflecting on this disappointing data, I finally accepted that these organelles were not the subcellular location of catalase. Yet, I believed such a dominant structure occupying a significant portion of the cell volume, as readily seen in the electron microscope, surely must have an important role. Since this was the only large, membrane-bounded entity seen in trichomonads, its relationship to either mitochondria or peroxisomes was still a possibility. An alternative assumption remained open that this structure had an independent origin. The immediate task was to define the *biochemical* nature of the organelle and its possible relationships to either mitochondria or peroxisomes.

Returning to the problem, I looked for typical mitochondrial or peroxisomal enzymes in *T. foetus* cell homogenates and fractions obtained by different centrifugation methods, but all results were negative.[7] It was tempting to think this enigmatic organelle might possibly be connected with its anaerobic nature and somehow involved in the production of hydrogen, a metabolic end product rarely found among eukaryotic organisms.

Donald G. Lindmark, whose thesis work was on hydrogen production in anaerobic prokaryotes, joined our group, and we set out to test this assumption. We assayed homogenates of *T. foetus* for enzymes known to be involved in hydrogen formation in anaerobic bacteria and detected the presence of pyruvate-ferredoxin oxidoreductase and hydrogenase, enzymes fundamentally different from the pyruvate metabolizing enzymes of mitochondria. Assays of subcellular fractions obtained by diverse centrifugation methods localized these enzymatic activities to a large granule fraction that was shown by electron microscopy to be significantly enriched in granules. These results strongly suggested that the granules are indeed the subcellular site of producing hydrogen to fuel the cell (a counterpart to the oxygen-fueled mitochondria in eukaryotic cells). Our first paper in 1973 on the unique nature of these microbody-like particles concluded that in contrast to mitochondria or peroxisomes, in which electron transfer is directed toward molecular oxygen, the particles utilize protons as terminal electron acceptors and thus produce molecular hydrogen. We proposed the term "hydrogenosome" to designate this new biochemically defined subcellular entity (figure 1).[8]

Our definition was based on the biochemical properties of the organelle and not its morphological characteristics, thus leaving open the possibility that hydrogenosomes, if and when detected in other organisms, could display a morphology different from that observed in trichomonads. Consequently, our immediate goals were to see whether hydrogenosomes are a peculiarity of *T. foetus* or are present in other trichomonads and in other protists and to characterize the biochemistry of this novel organelle.

In the first line of inquiry, we found hydrogenosomes in the important human parasite *Trichomonas vaginalis* and *Monocercomonas* sp. from reptiles, the latter regarded as a primitive relative of trichomonads at that time; both species reflected our interest in the possible medical and evolutionary significance of the new organelle. However, in publishing our *T. vaginalis* results, we met unexpected and amusing obstacles. The *Journal of Parasitology*, the most appropriate venue to reach applied and general parasitologists, rejected the paper because its content was deemed self-evident and not new information; this was a time when practically no parasitologist or biologist was even aware of the existence of hydrogenosomes. So, in a brief communication to the same journal, we simply reported hydrogenosomes as a unique type of organelle in trichomonad flagellates without expanding our taxonomic sample.[9]

In biochemical studies, we demonstrated the presence of unusual enzymes that accounted for the production of acetate by the organism, and somewhat later we purified the

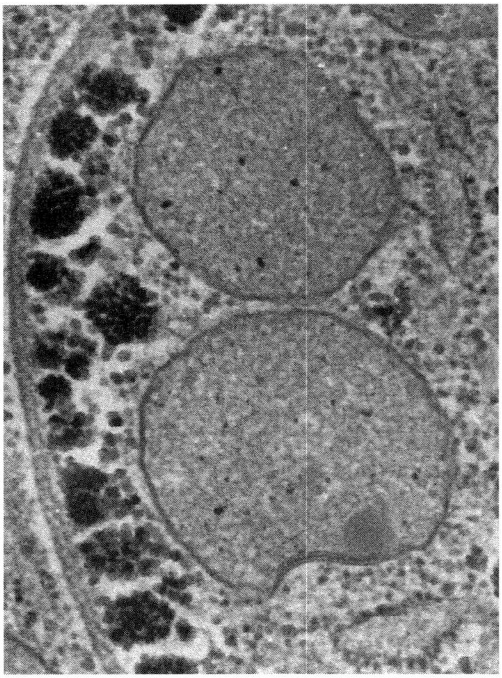

1. Electron micrograph of the *Tritrichomonas foetus* hydrogenosomes. Courtesy of Miklós Müller.

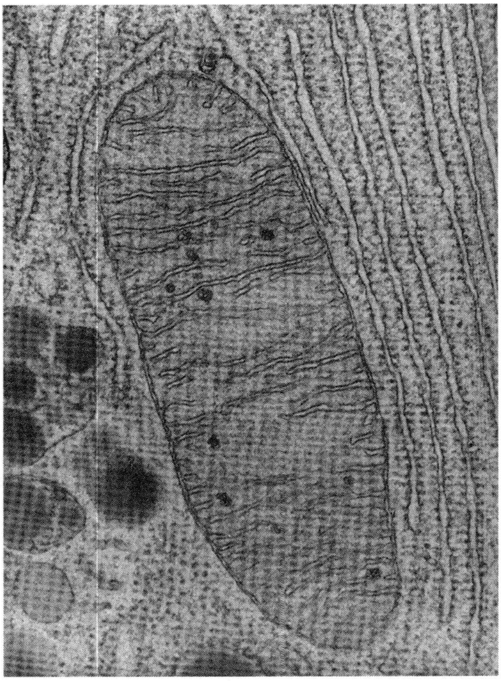

2. Electron micrograph of a mammalian mitochondrion. Courtesy of Keith Porter.

first hydrosomal protein, the small electron transfer protein ferredoxin.[10–12] We also tried to strengthen the evidence for the metabolic differences between hydrogenosomes and mitochondria by lowering the limit of detection of typical mitochondrial components.

Researchers in France and Czechoslovakia also began independent studies of trichomonads. By 1976, all our early results clearly delineated the main features of hydrogenosomes and demonstrated that they differ fundamentally in their core metabolism from mitochondria and peroxisomes.[13] However, their *biological* nature remained an open and vexing question. By looking for properties not related to the core metabolism, we showed that the trichomonad enzyme superoxide dismutase was cyanide insensitive; at the time, this was regarded as a property of the mitochondrial and bacterial enzymes, but not the cytosolic one. Also, only about one sixth of the total enzyme activity was in the organelles, and thus the evidence was not overwhelming and not convincing. The interpretation most consistent with all the data was that hydrogenosomes represented a unique structure unrelated to either mitochondria or peroxisomes. We nevertheless concluded that the presence of superoxide dismutase was an interesting finding that might be a "significant lead in attempts to elucidate the phylogenetic origin of this organelle" (figures 2 and 3).[14]

In 1976, Jean M. and F.R. Whatley at a Royal Society meeting in London gave an overview of the origins of mitochondria and chloroplasts and raised the possibility that hydrogenosomes arose through an independent endosymbiotic event.[15] This idea agreed well with our thinking, and in 1980 we further hypothesized that the hydrogenosome might be the anaerobic equivalent of the mitochondrion. Although the evolutionary origin of mitochondria and chloroplasts was at the time still an open question, I asserted the hydrogenosome-*Clostridium* pair represented a further example in the list of organelles and their free-living prokaryotic ancestor pairs, which includes the mitochondrion–aerobic bacterium and the chloroplast-cyanobacterium.[16] Twelve years later, I went even further to formalize the separate origin in a symposium presentation entitled "Energy metabolism of ancestral eukaryotes: a hypothesis based on the biochemistry of amitochondriate parasitic protists."[17] At that time, this idea seemed to best fit all available information. However, within a few years, convincing data started to appear that showed my hypothesis to be way off the mark, and I did not hesitate to accept the contrary view.

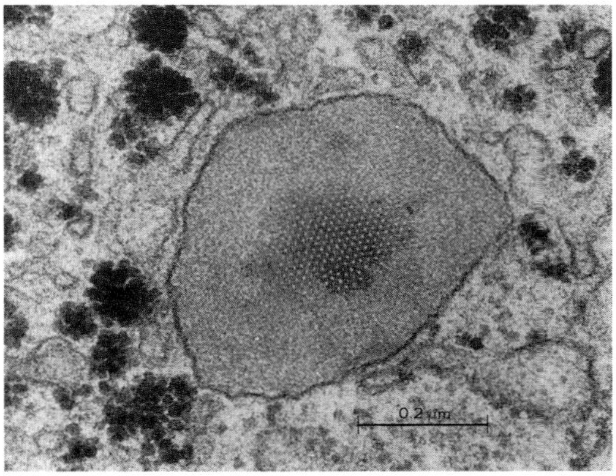

3. Electron micrograph of a peroxisome in guinea pig liver. Courtesy of Christian de Duve.

This story has focused only on the discovery and initial studies of a novel organelle. Although my laboratory continued working on the hydrogenosome until 2005, this organelle and its relatives have subsequently been investigated intensely by many other groups; this is another story to be related another time.

In retrospect, the discovery of hydrogenosomes in 1973 was prompted by a search for an organism containing an organelle, designated the "urperoxisome" by de Duve, that was not to be found. It led instead to the biochemical characterization of a novel organelle in certain protists, the trichomonads, started a hunt for similar organelles in other protists, and finally opened up a Pandora's box full of diverse organelles in eukaryotes.[18,19] A slow and steady increase in taxonomic, morphological, biochemical, and sequence information since the original paper has revealed that some early attempts to understand the biological nature and evolutionary origins of the organelle were misleading, albeit here and there pointing in the right direction. All work since then essentially confirmed the functional interpretations presented in that first paper.[20] Data from the complete sequencing of the genome of *T. vaginalis* not only confirmed these metabolic interpretations but also demonstrated the likely existence of additional functions of the hydrogenosome.[21]

Looking back at this history, it is apparent that in the absence of critical information, the repeated attempts to find the place of this organelle in the overall picture of the eukaryotic cell led to hypotheses now known to be in error. But without exaggeration, it can be stated that they played a significant role in provoking further experimental work on the origins of mitochondria and hydrogenosomes that has exploded during the past two decades. This recent work lead to a consistent picture of the previously unsuspected diversity of mitochondria and firmly placed the trichomonad hydrogenosomes in the big family of organelles derived from the ancestral "protomitochondrion."[19,22–24] We have reached the stage of this story when these organelles can no longer be dismissed, as once was the case. Indeed, they are now taken seriously by cell biologists.

Chapter 12

Defenders: The Endocytic Pathway in the Immune System

Introduction

Carol L. Moberg

Observations that cells are involved in resisting infectious diseases can be traced to mid-1800s microscopic studies of colorless cells that moved in blood and appeared to engulf foreign bodies. In an 1863 study, Friedrich von Recklinghausen, an assistant to pathologist Rudolf Virchow, identified two types of what are now called phagocytic cells after injecting dyes into animals. One wandered freely, and the other was seemingly attached to the inner linings of small vessels of major organs.[1–4]

Various explanations for what these cells did came during the next twenty years from the emerging fields of microbiology, pathology, and embryology. The germ theory, postulated by Louis Pasteur and Robert Koch after they showed that some bacteria were the direct cause of infection, became the basis for microbiology. Even while developing vaccines against several diseases, Pasteur believed that immunity developed only after the pathogens had been exhausted of essential nutrients. In other studies, pathologists of the time generally agreed that inflammation was a deleterious reaction to infections by the pathogens. Some believed inflammation occurred when the abundant phagocytes provided food for infectious agents and served as scavengers to carry away debris from the infected tissues. Others held that phagocytes caused inflammation by transporting and disseminating infectious organisms throughout the body.

Another theory was provided by Russian embryologist Elie Metchnikoff in 1883 at the Pasteur Institute in Paris. He not only found large, highly phagocytic cells in invertebrates and vertebrates, including man, and demonstrated their role in removing cellular debris and foreign bodies, but he concluded phagocytes were the principal and beneficial agent of host resistance against microbial invaders. Inflammation resulted from the antagonism of the irritant and in return facilitated healing. Metchnikoff later classified these free and fixed cells with a single large nucleus as "macrophages" (big eaters, or mononuclear phagocytes), and he distinguished them from circulating cells having a multi-lobed nucleus as "microphages" (small eaters, or polymorphonuclear leukocytes). The observation that phagocytes (or white blood cells) moved into the tissues during inflammation suggested to Metchnikoff that they played a protective role. In response to invading pathogens, he also observed that microphages were the first to pass through the vessel walls, followed later by the macrophages; they seized the intruders and subjected them to intracellular digestion by enzymes. After pursuing these studies for twenty-five years, Metchnikoff gathered abundant experimental evidence to formulate a biological doctrine of cellular defense stressing the central role of phagocytes in the host's responses to microbes and other forms of inflammatory stimuli.[5–8]

A competing theory to Metchnikoff's was introduced in 1890 by Emil von Behring and Shibasaburo Kitasato, who showed that antibodies rather than cells in the blood serum are primary agents of host resistance.[9] This humoral theory of immunity posited that cell-free fluids could kill bacteria without help from phagocytic cells.

In 1897, hematologist Paul Ehrlich advanced the humoral thesis. Using staining methods to differentiate the phagocytes into leukocytes and lymphocytes, he formulated a side chain theory: receptors on cell surfaces, similar to side chains on chemical dyes, handle the chemical interactions between antibodies and antigens in the serum to produce an immune response.[10]

Six years later, Almroth Wright and Stewart Douglas established that phagocytosis occurred with the aid of an opsonin in the serum that coated the pathogenic bacteria and rendered them susceptible to the phagocytes. Consequently, they claimed that cellular and humoral functions are equally important.[11]

Both the cellular and humoral theories of immunity seemed accepted after Nobel Prizes were awarded to von Behring (1901) and to Metchnikoff and Ehrlich (1908). Despite this recognition, however, controversy and debate continued during the first fifty years of the twentieth century. As measured by the volume of extant research literature, most scientists assumed the "antibody held the key to an understanding of immunity, and thus constituted the natural choice for investigative work."[12] The few advances in cellular immunity were essentially directed to pathogens in the overall act of engulfment.

Then, in midcentury, a new approach to phagocytosis was undertaken in the Rockefeller laboratory of René Dubos by scientists who were trained in medical microbiology and medicine. As their research gathered strength and direction, they adapted techniques devised by Albert Claude, Keith Porter, and George Palade to map more cell structures and functions. Eventually, their studies of phagocytic cells led to the identification and understanding of many types of white blood cells working in complex biological and chemical pathways to control host defenses against disease. These pathways differed from the "secretory pathway" that was being unraveled at the same time in Palade's laboratory (see Chapter 7). Within a few years, the modern cell biology of the immune system revealed that what once appeared to be a simple process of phagocytosis, observed by Metchnikoff, was in reality an intricate system now recognized as endocytosis.

12A. From Phagocytosis to Endocytosis in Infectious Diseases: Leukocytes and Macrophages

Carol L. Moberg

Before 1950, four of the original Institute laboratories headed by physicians were instrumental in inspiring the new cell biology of leukocytes that would take years to emerge. After 1950, the young staff scientists who headed the second generation of laboratories followed clues provided by their mentors, made new assumptions, and adopted new techniques to test many unanswered questions about white blood cells.

One of the earliest attempts to look at the cellular side of resistance to disease came in 1912 from experiments by pathologists Peyton Rous and James B. Murphy. As described in Chapter 1, they found abundant lymphocytes in animals that surrounded a foreign graft of a cancerous tumor, a reaction that appeared early and lasted until the transplant was destroyed.[1] Murphy then developed an experimental program to alter the activity of lymphocytes, either creating or abolishing resistance to tumors, cancer, and tuberculosis.[2,3] His findings with tissue fragments, rather than dissociated or free single cells, did not provide critical evidence of an association of lymphocytes with antibodies or other attributes of an immune response. This mystery was solved by others fifty years later. Murphy ended this unproductive research on lymphocytes and pursued biochemical analyses of the origin of cancer in normal cells.

In 1925, Scientific Director Simon Flexner brought to Rockefeller an anatomist, Florence Sabin, who was the first woman to head an Institute laboratory. Although Flexner was more interested in humoral than cellular defense mechanisms, he decided that Sabin was the right scientist to combine investigations of both mechanisms. Sabin had been one of Murphy's professors at Johns Hopkins University and was then well known for her research on the defense role of white blood cells in infection.[4] During her thirteen-year Rockefeller tenure, she and her co-workers extracted chemical fractions from the tubercle bacilli, tested them in laboratory animals, and observed which cells reacted and how they dealt with the fractions. Sabin found the most significant cellular and immunological reactions in tuberculosis involved phagocytic monocytes, and she observed how they evolved into more specialized cells and arranged themselves into tubercles, the characteristic lesion of the disease.[5] Using a dye to trace antigens, she found they were engulfed by many phagocytic cells and appeared in vacuoles within the cells' cytoplasm. Just at the time the cells burst, discharging the dye, antibodies appeared in the bloodstream. She speculated these events might be merely coincidental and that the antigen uptake might be a first step in the production of the antibody by other, yet unknown, cells.[6] Working with circulating cells that could neither be easily controlled nor quantified at that time, Sabin's contributions in the end, much like Murphy's, provided little fundamental knowledge of immunity. Yet her observations prompted the quantitative approaches to tuberculosis and phagocytosis taken by René Dubos a few years later.

During the 1920s and 1930s, two other laboratories helped found a new field of immunochemistry and furnished more concrete influences on cellular immunity.[7,8] One was the laboratory of bacteriology, headed by Oswald Avery in the Rockefeller Hospital, which focused on the chemistry of the bacterium that causes lobar pneumonia, not the phagocytes or defense mechanisms in the host. Over the course of twenty years, three biochemists in this laboratory—Michael Heidelberger, Walther Goebel, and Rollin Hotchkiss—identified specific antigens in the capsules of each type of pneumococcus, mastered the chemistry of their polysaccharide capsules, and began preparing artificial antigens to use as vaccines against pneumonia. Avery's own studies were focused on treating pneumonia patients with serum therapy, using methods devised by von Behring against diphtheria; this vaccine contained antibodies that neutralized the antigens being produced by the pneumonia patients.[9,10] In addition, microbiologist René Dubos unearthed a bacterial enzyme that could decompose the pneumococcal capsule whereby the phagocytes quickly destroyed the bacteria and cured the disease.[11,12]

The other laboratory was that of Karl Landsteiner, who was recognized for his classification of blood groups by a Nobel Prize in 1930. He also conducted elegant chemical studies on blood proteins to understand the specificity of antigen-antibody interactions.[13] In 1942, Merrill Chase, a biologist in Landsteiner's laboratory, set out to study in whole animals how an allergic reaction proceeds when the skin makes contact with irritants termed "contact allergens." He and Landsteiner found to their surprise that this reaction was not transferred by antibodies, which had been the major focus of Landsteiner's research, but instead was transferred from one animal to another by living white blood cells.[14] Chase reported that when "Landsteiner looked through the microscope and saw lymphocytes, he did an abrupt about-face and said with dignity, 'Yes, I thought so!' " He seemed to be recalling Murphy's earlier experiments, Chase continued, in which Murphy associated lymphocytes with graft rejection as well as resistance to experimental tuberculosis.[3,15] Chase's landmark discovery clearly identified cellular immunity and showed that, complementing humoral immunity, it was not innate but adaptive after being activated by an outside stimulus. Although not widely recognized during the war years, his discovery opened a new path for later research by others who eventually pinpointed many types of white blood cells as the body's central safeguards against infection. Chase returned to this work after he joined the laboratory of Dubos, who was quick to sense its significance to tuberculosis.

Revisiting Phagocytosis as an Independent Resistance Mechanism

The microenvironment of tuberculosis. Raising cellular immunity to an experimental subject came from bacteriologists interested in the refractory bacterium *Mycobacterium tuberculosis.* By 1940, evidence was mounting that certain diseases, particularly tuberculosis, were not so readily explained by a phagocytic or opsonic theory of immunity.

Arnold Rich in his classic text *The Pathogenesis of Tuberculosis* (1944) criticized those who concluded tuberculosis was about to be eradicated and stated this "premature optimism . . . takes too little account of the prevalence of the 'sub-clinical' or 'carrier' infections in tuberculosis." In tuberculosis, the bacilli are harbored inside the mononuclear phagocytes within the walled-off tubercles, or lesions, where they are capable of surviving and multiplying and are poised for causing disease.[16]

A very early challenge to Elie Metchnikoff's idea of phagocytosis as a defense mechanism came from Peyton Rous. In 1916, Rous showed how phagocytic cells can also provide a desirable "culture medium" in which they wall off and actively protect microorganisms against circulating antibodies and drugs.[17]

Moreover, in 1942, pathologist Max B. Lurie at the Henry Phipps Institute found little correlation between anti-tubercle antibodies and immunity. He demonstrated that cellular immunity does not depend on increased phagocytosis but instead on an enhanced power of macrophages to inhibit the already engulfed pathogens. Immunity, he noted, is acquired from "forces apart from the inherent properties" of these mononuclear phagocytes.[18]

Rich called for methods to determine the nature of the tissue environment and to identify factors governing the growth of the tubercle bacilli in the body. A major impetus in reviving this challenge came from René Dubos. After spending two years at Harvard Medical School, he returned to Rockefeller in 1944, where he took over the study of tuberculosis in the Smith Hall laboratory disbanded on Sabin's retirement in 1938. After the death of his young wife from tuberculosis in 1942, Dubos developed serious concerns about both external and internal environmental influences on this ancient disease whose virulent microbes remained widespread in the population. In *The Bacterial Cell* (1945), he contended that methods in immunochemistry had "directed the study of infectious diseases along somewhat narrow channels" and urged a return to the main channels of physiology with studies on the host responses in immunity.[19] Instead of studying microorganisms alone, as he had done in Avery's laboratory, Dubos set out to put Rich's suggestion into action and to study host-parasite relationships. He selected a number of tuberculosis researchers, mostly physicians, who would find novel approaches to advance this field.

While planning new techniques for investigation, Dubos reviewed experiments of Louis Pasteur and Elie Metchnikoff that were done at a time before the germ theory of disease and a biochemical view of life had taken precedence in medical science. He learned that both scientists had observed how the outcome of encounters with pathogenic microbes was influenced by the physiological state or health of the host. Metchnikoff, for example, found diabetics were more susceptible to staphylococcal infections and pneumonia while suffering from a cold or influenza. Pasteur found cold body temperatures increased the susceptibility of chickens to anthrax. Both men also observed certain bacteria resisted engulfment by phagocytes. More importantly, they observed that other bacteria ingested by phagocytes could hibernate inside these cells and could be provoked into rapid multiplication and disease by disturbances either in the host's physiology or environment. "The implantation of bacilli," Dubos noted in his biography of Pasteur, "like the planting of seed, does not necessarily insure a growth."[20,21]

Like Pasteur, Dubos viewed infection without disease as the rule rather than the exception. For this reason, he found Metchnikoff's concept of phagocytosis as a defense mechanism an oversimplification. If Metchnikoff had been trained in pathology or medicine, Dubos commented, he would have realized that "most parasites thrive within cells that phagocytize them, and it is not at all certain that those that die are 'devoured' by these cells." He questioned why only some individuals develop tuberculosis when almost everyone is exposed or infected during their lifetime. Dubos advanced instead the concept of a "microenvironment of inflammation." By this he meant the biochemical locales—both the intracellular and inflamed tissue environments sheltering the pathogens—where the individual fates of bacteria are determined. His ecological outlook considered phagocytosis more broadly as capable of either causing death, *or* favoring proliferation, *or* changing the properties of microbial parasites. Similarly, he viewed inflammation as isolating the site of infection, attracting wandering cells, changing the activities of the fixed tissue cells, and concentrating the anti-infectious substances of cellular origin. In intracellular infections, such as tuberculosis, Dubos further hypothesized, the host's immunological defense mechanisms were likely inactivated, weakened, or otherwise unable to breach these walled-off microenvironments.[22,23]

By 1946, one of Dubos's first contributions, with physicians Bernard Davis and Gardner Middlebrook, was devising revolutionary new methods to cultivate the mycobacteria. This culture medium, which remains in use today, produces rapid, dispersed, homogeneous growth and thus provides a standard bacillus for testing. For the first time, scientists could subject mycobacteria to a whole range of experiments, make accurate qualitative and quantitative measurements, and get reproducible results. The Dubos laboratory used the medium to test growth requirements of mycobacteria, environmental effects on growth, and drug sensitivities; they also analyzed the great variability of strains for the BCG (Bacillus of Calmette-Guérin) vaccine against tuberculosis and probed the bacilli's surface for immunogenic components.[24,25]

Alongside numerous other studies in Dubos's laboratory to explore the biochemical nature of the microenvironment of inflammation, described elsewhere,[26] a few preliminary experiments before 1955 pursued the nature of white blood cells. During this time, just as the efforts of Keith Porter and George Palade were expanding to explore the world inside normal cells (see Chapters 5 and 6), so too were efforts in Dubos's laboratory moving toward opening the field of cellular physiology and immunology inside the host's white blood cells.

Cell-mediated immunity in tuberculosis. Merrill Chase realized cell transfer of immunity was not restricted to skin contact allergies. He joined René Dubos's laboratory in 1945 and, in an extension of his study with Landsteiner, applied this concept successfully to tuberculosis and its potential role in immunity to the disease. He transferred a hypersensitivity to tuberculin (a purified protein derived from mycobacteria) from one guinea pig to another, again by injecting sensitized white blood cells.[27] In ensuing studies, described only in the Institute's annual reports and in a few abstracts from 1945 to 1954, Chase pointed to the lymphocyte as the responsible cell in both the delayed type and the anaphylactic reactions. Dubos reported to Director Herbert Gasser and the Board of Scientific Directors that Chase's linking of the lymphocyte to the formation of antibodies "opens an entirely new immunological vista with its evidence of a special intermediate in the process of antibody formation."[28–30]

This work was helped by Cutting Favour, a visiting physician from Harvard Medical School who developed a centrifugation method using silicone tubes to harvest lymphocytes from either the spleen of guinea pigs or the buffy coat of blood cell suspensions; this technique allowed an accurate counting of the white cells. Favour observed a phenomenon he called "lympholysis" in which lymphocytes from tuberculin-sensitive animals release an antibody-like factor that dissolves other lymphocytes, granulocytes, and monocytes, a process that depleted effective cell defenses and propelled the progression of disease.[31,32]

By 1950, aware that Rockefeller colleague Philip McMaster had shown lymph nodes to be prolific sites of antibody formation, Chase asked whether lymphocytes of sensitized animals would also give rise to antibodies. He followed the appearance of antibody in a single guinea pig after injecting a toxin and found that the lymph nodes were the source of cells leading to the first appearance of antibody, followed a few weeks later by the spleen. He further detected two bursts of activity and suggested the first burst was an "active immunization" and the second a "recall phenomenon."[33]

Chase and Dubos then realized that certain white cells were not mere carriers of antibody but were serving in its elaboration and that hypersensitivity to tuberculin was not equivalent to immunity to tuberculosis. McMaster, reviewing this situation in 1951, noted that almost every organ and white cell in the body had been implicated as the site of antibody formation, but the actual sites and cells were yet to be determined.[34] Chase, now recognizing the plurality of immunological functions, stated he no longer wanted to belong to "the school of lympho-maniacs" and at this point established an independent Rockefeller laboratory devoted to allergy and the role of heredity in allergy.[30] What he left to his colleagues in the Dubos laboratory were unanswered questions about the intracellular nature of lymphocytes and whether a previously undetected substance contributed by certain specific cells gives rise to antibody or whether it is the expression of a developmental change leading to the appearance of antibody. Nonetheless, Chase's pioneering transfer experiments opened a new path for exploring two different immune responses, antibody formation and cell-mediated immunity.

Surveying the fate of microorganisms within phagocytes. For Dubos, understanding phagocytosis required a more complete knowledge of the nature and physiological activity of the cells as well as their environmental surroundings. In 1948, Hubert Bloch, a Swiss visiting physician, began quantitative studies of the tubercle bacilli inside phagocytic white blood cells. Bloch followed the fate of virulent tubercle bacilli in the peritoneal cavity of mice and was able to calculate the relative numbers of leukocytes, those that engulfed bacteria and the number of bacteria ingested per phagocyte, as well as the free non-phagocytized bacteria. He observed that both avirulent and virulent forms of the bacilli were taken up first by the polymorphonuclear leukocytes and a few hours later by mononuclear cells. Subsequent events differentiated two forms of bacilli, engulfment of the virulent bacilli often resulted in death or disruption of the phagocytes, a fact not observed with the avirulent forms. This experiment suggested virulence is correlated with the ability of the bacillus to injure the phagocytic cell.[35]

Two other physicians studied the effects of the bacilli on two types of leukocytes. Samuel B. Martin studied the polymorphonuclear phagocytes and found the bacilli slowed their rate of migration and increased their tendency to stick to other leukocytes. His work suggested a possible reason for persistence of bacilli in the host.[36]

Emanuel Suter studied mononuclear phagocytes. He developed an in vitro system to obtain them from the peritoneum of guinea pigs, previously injected with glycogen to attract a large collection of these cells, and made quantitative analyses. Suter found the virulent tubercle bacilli rapidly reached large numbers intracellularly, although macrophages from animals vaccinated with BCG suppressed their growth but did not kill the bacilli. Another arresting finding was that streptomycin, the antibiotic highly effective in killing bacilli outside cells, failed to affect the bacilli inside phagocytes.[37–40]

In vivo studies by Dubos, Cynthia Pierce, and Werner Schaefer, on the other hand, revealed that whatever technique they used to elicit immunity to tuberculosis (primary infection or vaccination), the immune response was insufficient to protect the animals from eventual death. The alarming evidence that the bacilli were not eradicated, only retarded from multiplication, meant latent infections remain capable of causing a relapse of disease when an individual's resistance becomes suppressed.[41]

Secrets in the White Cell

The fact that tubercle bacilli sometimes fail to multiply in closed tuberculous lesions—to establish infection but not disease—is what led Dubos to study the microenvironment of inflammation. His laboratory began to search for broader and more complex forms of host resistance, beyond exposure to the pathogens, as prime determinants of disease processes.

Initiating the cell biology of phagocytes. The arrival of James G. Hirsch in 1950 signaled a new direction in this search. After graduating from Columbia University's College of Physicians and Surgeons, Hirsch spent an internship from 1946 to 1948 under W. Barry Wood Jr. at Barnes Hospital in St. Louis. Wood's research focused on white blood cells, and he was just discovering the phenomenon of "surface phagocytosis" in which most microbes are rapidly eliminated from the circulation before the host produces antibodies to render the bacteria susceptible to cellular phagocytosis. Wood further identified both physical and chemical environmental factors that influenced the phagocytic process.[42]

Hirsch next served on active military duty as chief of medicine in the hospital of Warren Air Force Base in Wyoming, where he was exposed to non-military clinical and scientific research on rheumatic fever. Upon discharge, Hirsch decided it was not enough to be a doctor, so he applied to Wood for research training. There were no vacancies at the time, but Wood advised him to apply to "the best place in the world and work with the best person in the world, namely, The Rockefeller Institute and René Dubos."[43,44]

Within a year of joining the Dubos laboratory on a National Research Council fellowship, Hirsch found natural antibacterial substances released from host tissues that could hinder the growth of tubercle bacilli. One substance he isolated and then crystallized from the kidney was identified as spermine, along with the previously unknown enzyme amine oxidase that was necessary for spermine to exert its activity against the bacilli. He also isolated a group of cationic peptides from thymus that were potent antimycobacterial agents and described their mode of action and limitations.[45–47]

During five years of research under Dubos, Hirsch acquired broad experience in natural host antibiotics, but he became more interested in what the host was doing rather than its effects on parasites. When the opportunity came to undertake independent research in 1955, he turned from tissue-level studies to explore the intracellular world of the freely moving phagocytes. At this time, the study of leukocytes was still in its infancy, and hematology was focused on the erythrocyte.

One outstanding question was whether phagocytes contained preformed bactericidal substances or whether these substances developed after engulfing foreign particles and creating a kind of intracellular inflammation. Dubos pointed to oxygen deficiency, excessive carbon dioxide, accumulated organic acids, and lysozyme as environmental influences that contributed to killing bacilli in lesions. But Hirsch wanted to know how the phagocytes worked and how they killed bacteria inside these cells. "Nobody had any idea what the antibacterial factors were in cells. I simply thought the logical starting place would be to collect the phagocytic cells that are most easily collected—the polymorphs—and try to find

Zanvil A. Cohn, 1957.
Courtesy of The Rockefeller Archive Center.

James G. Hirsch, 1950.
Courtesy of The Rockefeller Archive Center.

out if there is something extractable from them which will kill bacteria."[44] In addition, as Metchnikoff had observed, the polymorphs appear in the first stage of infection and predominate for a short time. Reviewing Metchnikoff's work on phagocytosis, Hirsch realized how little knowledge had been added during the twentieth century, and he set out to find an answer.[48]

Methods to unlock the white cell arsenal. Hirsch's experiment involved several new techniques. The first task was to get enough cells to study. Rabbits were large enough animals to provide the cells, but the laboratory budget could not afford fifty dollars a week to buy and sacrifice ten rabbits to obtain a decent supply of cells. Hirsch then devised a special way to use the same rabbit over and over. He placed the animal on a board with a hole in it and, as Suter was doing to get monocytes, injected glycogen into its peritoneum, which caused white cells to collect at the injection site. A while later he drained its peritoneal fluid into a flask. Although the yield was initially insufficient, by accident he found that repeated injections caused the same rabbit to produce ten times more. Hirsch thought this increase was an allergic response to the glycogen at the site of injection, a hypersensitivity similar to the reactions observed by Chase in his cell transfer studies. Nonetheless, it was a full day's work to collect fluid from four rabbits and harvest large numbers of pure cell populations using a low speed centrifuge.

In the next steps, the polymorphs were disrupted by freezing and thawing, mechanical grinding, or sonic oscillation, and the cells' contents were extracted with salt solutions. The resulting soluble portion manifested strong killing action in vitro against a variety of bacteria, but not against the mycobacteria. When the contents were extracted with acid solutions instead, up to a thousand times more antibacterial agent was recovered, a detail that enhanced Dubos and Metchnikoff's view that phagocytes use acids as their killing chemicals. This technique allowed Hirsch to calculate how many and how rapidly the bacteria were killed. He tried extracting this protein from non-phagocytic cells in skin and liver, but there was none. For this reason, he named the agent "phagocytin." While phagocytin was present in large amounts in the granulocytes, it was absent in macrophages, thus providing an early clue to differences between these two white cell types.[49–51]

The instability of phagocytin made it difficult to further purify and characterize, so Hirsch was unable to test the efficacy of this protein in animals. However, he was able to distinguish phagocytin biochemically from two other known bactericidal substances released from cells, lysozyme and properdin. While he did no further work on phagocytin after 1959, his discovery set off a chain of searches by other laboratories for natural host antibiotics, including what today are called defensins. This work also marked the beginning of the modern cell biology of phagocytes.

A new recruit to phagocytosis research. In 1957, Zanvil A. Cohn joined the Dubos laboratory as a research associate and assistant physician to the Rockefeller Hospital. He had graduated with highest honors from Harvard Medical School, where he also did independent research on the interaction of viruses and chicken eggs. In the introduction to his thesis, he announced what would be a life-long scientific interest, the host-parasite relationships. Also while at Harvard, Cohn heard a provocative series of lectures by Dubos on the microenvironment of inflammation and the fate of microorganisms in vivo, the topics covered in his 1954 monograph *Biochemical Determinants of Microbial Diseases.* After one lecture, Cohn approached Dubos about a position in his laboratory. Dubos was interested and, in one of his occasional overstatements, advised him to forget about studying viruses: "They're *passé.* When you come to Rockefeller we'll work on something important."[52] After an internship and residency at Massachusetts General Hospital, Cohn met his military obligations in the Army Medical Corps at Walter Reed Army Institute. There, he worked in the laboratory of Joseph Smadel exploring respiratory enzymes and the cell wall of the intracellular parasite *Rickettsia tsutsugamushi*, the causative agent of scrub typhus.[53,54]

Cohn's first research project under Hirsch's supervision was a joint effort with graduate student Steven I. Morse to expand studies on phagocyte-microbe interactions. Their particular interest was the ingestion of staphylococci, the bacteria then causing alarm in the medical community due to the emergence of many strains resistant to antibiotics. Using Hirsch's methods to get homogeneous populations of polymorphs, they were able to quantitate two separate phases of phagocytosis: ingestion and killing. And they observed three basic outcomes: phagocytosis with killing (*Staphylococcus albus* and *Escherichia coli*), phagocytosis without killing (*Mycobacterium smegmatis*), and no phagocytosis at all (*S. aureus* and group A streptococci).[55–57] This work complemented experiments by David Rogers, a visiting physician from Cornell Medical College in Dubos's laboratory a few years earlier, who found pathogenic staphylococci can survive in leukocytes.[58] The findings of Rogers and Cohn provided a corollary to Dubos's findings with phagocytes and tubercle bacilli: virulent strains were ingested but not destroyed, whereas nonvirulent forms were destroyed inside these cells.

The arsenal of granules is a cell organelle. During this time, Hirsch and Cohn became involved in a search for the place in the granulocytes (polymorphs) where the chemical weapons are stored and to learn how they are they discharged. They focused on the abundant granules, a major element in the leukocyte cytoplasm; granules had been observed since Paul Ehrlich's staining studies in 1879, but their composition and function were still unknown. In 1960, the year Hirsch became a full professor, he and Cohn published three historic papers on these granules.[59–61]

For this experiment, they used a sucrose solution (not salt or acid as before), the one pioneered by George Palade (see Chapter 4) to get a uniform population of cells. This solution caused the delicate cells to gently liberate their contents. Next, they used a low speed differential centrifugation method, pioneered by Albert Claude (see Chapter 2), to isolate the granules from other cell contents. Each step of this separation was followed by phase microscopy and stained smears. Electron micrographs of the isolated granules, taken by David Luck in the Palade laboratory, revealed a limiting membrane around the granules. In other words, this arsenal was another newly recognized organelle in the cell. Chemical analyses of the granules identified a large number of acid hydrolases as well as phagocytin, lysozyme, and cathepsin, all of which are capable of destroying bacteria.[59]

In an earlier discovery, in 1955, unknown to Cohn and Hirsch for a few years, biochemist Christian de Duve and his associates in Belgium had described the nature of these granules (see Chapter 11A).[62] Using an ultracentrifuge to extract enzymes interacting with insulin in liver cells, this method broke up the cells into five fractions, one more than Claude and Palade obtained to get microsomes. The fifth and smallest fraction contained the granules. Based on biochemical analyses revealing five acidic enzymes and on centrifugal distribution patterns, de Duve identified the granules as the source of intracellular digestion. Calling attention to their richness in hydrolytic enzymes, he named them lysosomes, the Greek term for digestive bodies. He did not "see" the isolated granules until 1956 when Alex Novikoff took electron micrographs of them in Paris.[63] In their first 1960 paper, Cohn and Hirsch noted their granules were "analogous to those described by de Duve of the lysosomes of rat liver." Compared with de Duve's biochemical approach, Cohn and Hirsch adopted the combined cell biological methods of Claude, Porter, and Palade to first isolate a specific organelle and then analyze its biological properties. Their approach was a crucial step to fitting lysosomes into the dynamic context of phagocytosis in intact cells.

Tracing the process of degranulation. Hirsch and Cohn next worked out new methods to see what happened when they fed bacteria to the polymorphs. They applied white cells to glass microscopic slides, followed by a suspension of bacteria that was quickly engulfed by the phagocytes; the preparation was then dried, fixing the cells in their expanded state. Looking through the microscope, they saw that granules disappeared only in those cells engulfing the bacteria. To visualize this degranulation process in a living cell, they used a phase-contrast microscope and selected two larger morsels for digestion, the organism *Bacillus megaterium* and zymosan particles from yeast cell walls. To enhance the difference between granules and morsels, they mixed a drop of white cells with a drop of morsel on a glass microscope slide. They watched as the two drops made contact: the polymorphs' membranes completely surrounded and immobilized the morsel and, within as few as thirty seconds, the granules near this vacuole, or phagosome, vanished. Consequently, clear zones appeared in the cytoplasm next to the ingested material. But, even at a magnification of 1,000, it was impossible to see the granules break down.[60] Cohn and Hirsch then used cell fractionation to prove the granules

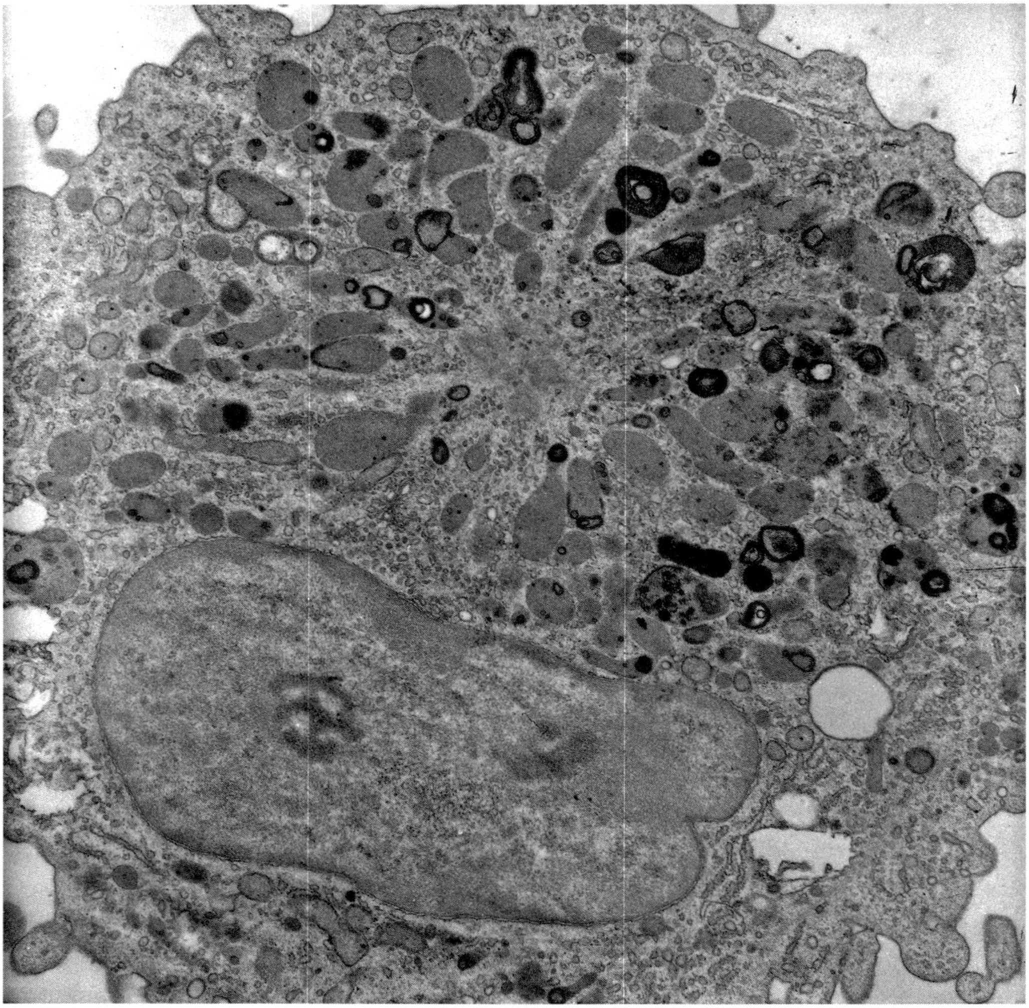

1. Electron micrograph of an alveolar macrophage showing abundant lysosomal granules. Granules of various size, density, and internal structure are arranged in a rosette-like fashion in the centrosphere. ×12,000. From Cohn and Wiener, 1963. *J. Exp. Med.* 118:991–1008.

discharged their contents; the bacteria-phagocytosing cells (but not control cells) released many acid hydrolases and phagocytin from the pellet fraction into a soluble fraction.[61]

Cohn took a biochemical approach and compared how lysosomes in polymorphs and macrophages participate in digesting the engulfed material. He followed radiolabeled nucleic acids and proteins of bacteria through the digestive process until the granules dissolved into soluble products.[64] Then, with postdoctoral fellow Edith Wiener, he isolated the granules from rabbit alveolar and peritoneal macrophages using isopycnic sucrose gradient centrifugation and primed them with BCG and mineral oil. This experiment revealed that macrophage granules, like those of the polymorphonuclear phagocytes, are a particularly rich source of lysosomes. Luck's electron micrographs of the alveolar macrophage showed abundant lysosomal granules arranged in a pleasing rosette pattern and documented their contiguity to the phagosomes (figure 1).[65,66]

Cinemicrography to visualize degranulation. During the same time, Hirsch became engrossed in filming the fast-moving granule release action with a motion picture camera attached to the phase-contrast microscope.

Cinemicrography had been born in the laboratory of tissue culture expert Warren Lewis at the Carnegie Institute. In 1931, Lewis filmed macrophages and other cells taking up extracellular fluid into small vesicles, a process he called pinocytosis or cell drinking (more on this process later).[67,68]

Compared with looking at waves of liquid droplets being ingested, Hirsch found the dense granules more difficult to capture on film. He tried various light sources, films, and film speeds before getting good exposures, but he still could not see the granules breaking. He then filmed much larger polymorphs from chickens and at last saw tiny explosions of the granules as they started to engulf the particles. An ultimate perfectionist, Hirsch labored long and patiently to produce elegant images of live phagocytes on film (figures 2 and 3).[69] These motion picture experiments were also valuable for viewing the actions speeded up, repeated, or reversed at will. They provided Cohn and Hirsch with enough evidence to propose that the granules fused with the membrane around the vacuole containing the engulfed microorganisms, allowing the granule contents to discharge only into the vacuole and not into the cytoplasm of the cell.*

Electron microscopy and degranulation. To record the actual moment of granules fusing with a phagosome, Hirsch needed a more powerful microscope to visualize the membrane structures. The polymorphs, about 15 micrometers in length, and the granules, about three tenths of a micrometer, can be seen in the light microscope. But their membranes are about one hundredth of a micrometer and thus beyond the resolution of this microscope. So Hirsch arranged to buy a Siemens Elmiskop I electron microscope. While waiting more than a year for delivery, he teamed up with hematologist Dorothea Zucker-Franklin at New York University, who was trained in electron microscopy by Johannes Rhodin, a former student of Fritiof Sjöstrand at the Karolinska Institute.

Preparing polymorphs (neutrophils and eosinophils) for electron microscopy was also a more exacting process. This was accomplished by feeding zymosan particles to the polymorphs, applying osmic acid at various intervals to stop the phagocytic process, embedding the cells in epoxy resin blocks, and slicing them with a microtome before viewing in the microscope. Hundreds of slices were examined and photographed before they found a cell in exactly the right position, namely, frozen at the instant when the granule was breaking. The seminal micrographs (at a magnification of 40,000) documented the fused membranes and showed the enzyme content of the granules discharging into the phagocytic vacuole containing the ingested organism (figure 4).[70]

With these experiments, Hirsch and Cohn revealed important secrets about the white cell and thus established a concrete role of phagocytosis in host resistance. This work

*In 1964, Hirsch delivered Rockefeller's sixth annual Christmas Lectures for high school students. He introduced the white cell historically and discussed common features of the various types. In addition to photos of Metchnikoff and Ehrlich, animated drawings, and electron micrographs, he showed his films of white cells in action. A phagocyte as big as the screen engulfing a bacterium caused the students to "literally hold their breath until the contest was over." A year later, these films became the focus of a public television documentary about Hirsch and Cohn. "Secret in the White Cell," in a series called *Experiment*, was hosted by Don Herbert of *Watch Mr. Wizard* fame. The documentary then became a richly illustrated book by the same title, written by Herbert and Fulvio Bardossi, who presented discoveries of the "modern Metchnikoffs" as a scientific detective story.[44] The films are still in circulation and remain an ideal component for many courses in biology.

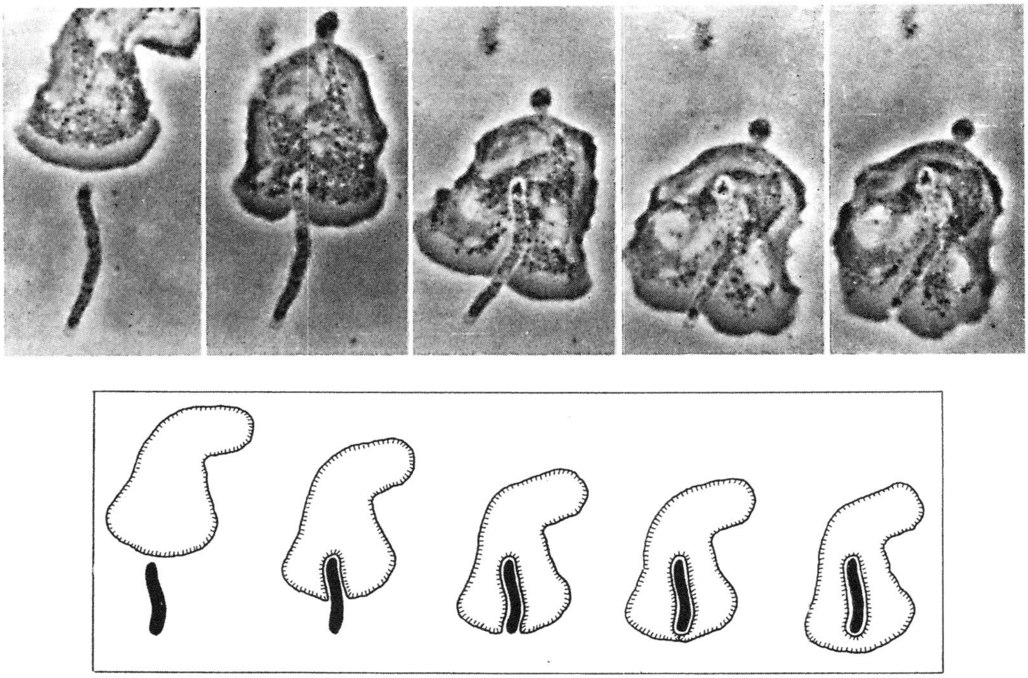

2. Phagocytosis. The top row shows sequences at thirty-second intervals from a motion picture of a human neutrophil engulfing a *Bacillus megaterium*. (Phase-contrast, approximately ×1,000.) The bottom row shows diagrams of the phagocytic process to emphasize the invagination of cell membrane and formation of a digestive pouch or phagosome. From Hirsch, 1965. *Annu. Rev. Microbiol.* 19:339–350. ©Annual Reviews.

stimulated discoveries in many other laboratories because it related to a universal function of cells (not just leukocytes). All cells contain machinery to digest particles and liquids from their environment for their life's functions, whether they are needed for growth or death of tissues, recycling substances, or scavenging and destroying foreign invaders.

Until 1972, when Hirsch became dean of graduate studies, he and his co-workers carried out extensive comparative morphological studies with the electron microscope on the digestive processes of other granulocytes (eosinophils, basophils, and mast cells) as well as monocytes, macrophages, and lymphocytes. Hirsch, Cohn, and Martha Fedorko pursued a 1962 report by Keith Porter and Thomas Ashford, who observed normal cell components were also being degraded inside lysosomes, a process described by de Duve in 1963 as autophagy, or "self-eating."[71] They exposed macrophages to chloroquine, which accumulates in autophagic vacuoles and inhibits acid hydrolase activity, and followed its process in a series of electron micrographs from the development of vacuoles to their disappearance and cell recovery.[72,73] The choice of chloroquine was relevant to ongoing clinical studies in the Rockefeller Hospital of 300 patients with sarcoidosis, an immune disease of unknown origin characterized by multi-organ granulomas. These clinical trials, in collaboration with Merrill Chase, established the Kveim skin reaction test was a valid diagnostic tool, and chloroquine therapy, as a substitute for risky adrenal steroid drugs, produced beneficial anti-inflammatory effects in patients.[74,75] In complementary research, Hirsch and graduate student Sally Zigmond visualized and measured the granulocytes' sensory system that responds, through chemotaxis, and moves toward foreign substances invading their environment.[76,77]

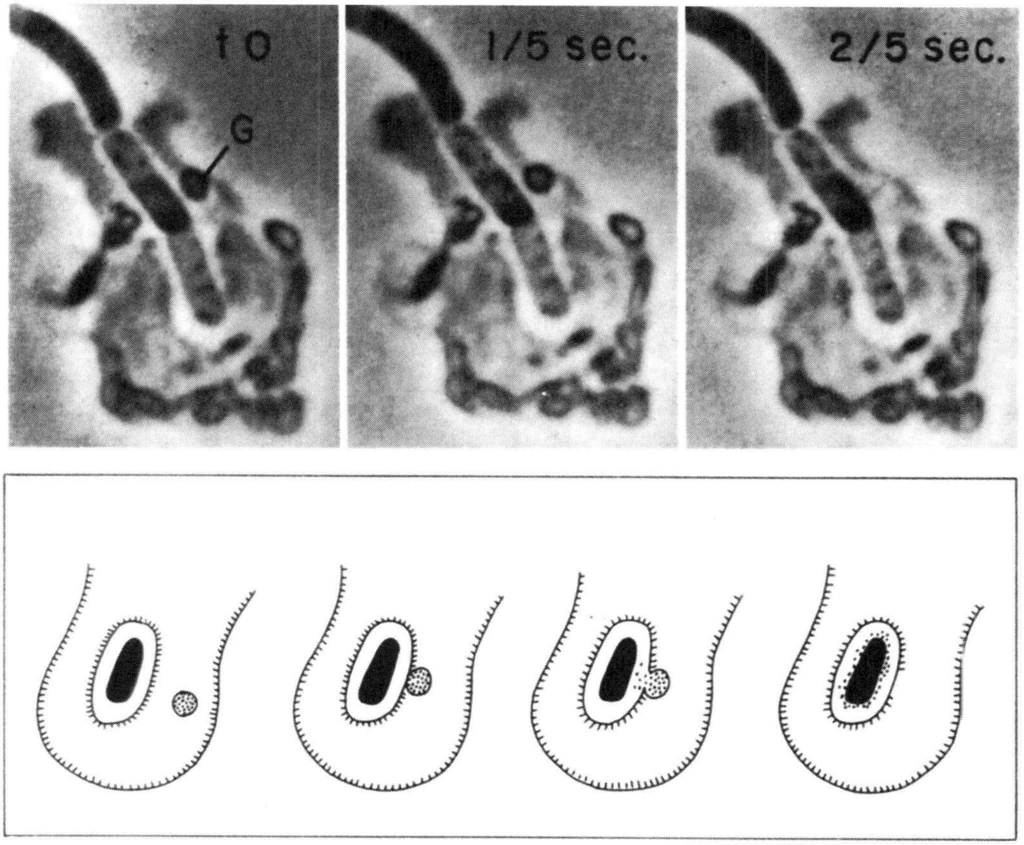

3. Degranulation. The top row shows the lysis of an individual granule (G) in a chicken polymorphonuclear leukocyte engulfing a *Bacillus megaterium*. The granule becomes rounded at one fifth of a second and disappears from view at two fifths of a second. (Phase-contrast, approximately ×1,500.) The bottom row shows diagrams of the membrane fusion mechanism for degranulation. From Hirsch, 1965. *Annu. Rev. Microbiol.* 19:339–350. ©Annual Reviews.

From phagocytosis to endocytosis. Hirsch compared degranulation to the process being studied concurrently by Palade for discharging granules from pancreatic cells (see Chapter 7). Because they both involved membrane fusion, Hirsch introduced the term "exoplasmosis" to describe these two processes of releasing material, whether it moved out of the cell or into the phagosome. However, the potential for confusing terminology was raised at the 1963 "Ciba Foundation Symposium on Lysosomes," where Cohn and Hirsch presented their persuasive films to show granules degrading bacteria. At this meeting, Christian de Duve, a symposium organizer who had just opened his laboratory at Rockefeller, noted a great deal of similarity in words to describe the various engulfing processes (phagocytosis, pinocytosis, reverse pinocytosis, micropinocytosis, and autophagocytosis, among others). Grouping them under the generic name "cytosis" seemed undesirable because this term lacked any connotation or source of the process. de Duve, therefore, coined two words to describe these universal cellular functions: "endocytosis" for engulfment of molecules from the cell's environment and "exocytosis" for secretion of molecules synthesized by the cell.[78] The prevailing understanding of phagocytosis, long dominated by Metchnikoff's view of inflammation and resistance, eventually gave way to endocytosis as scientists continued

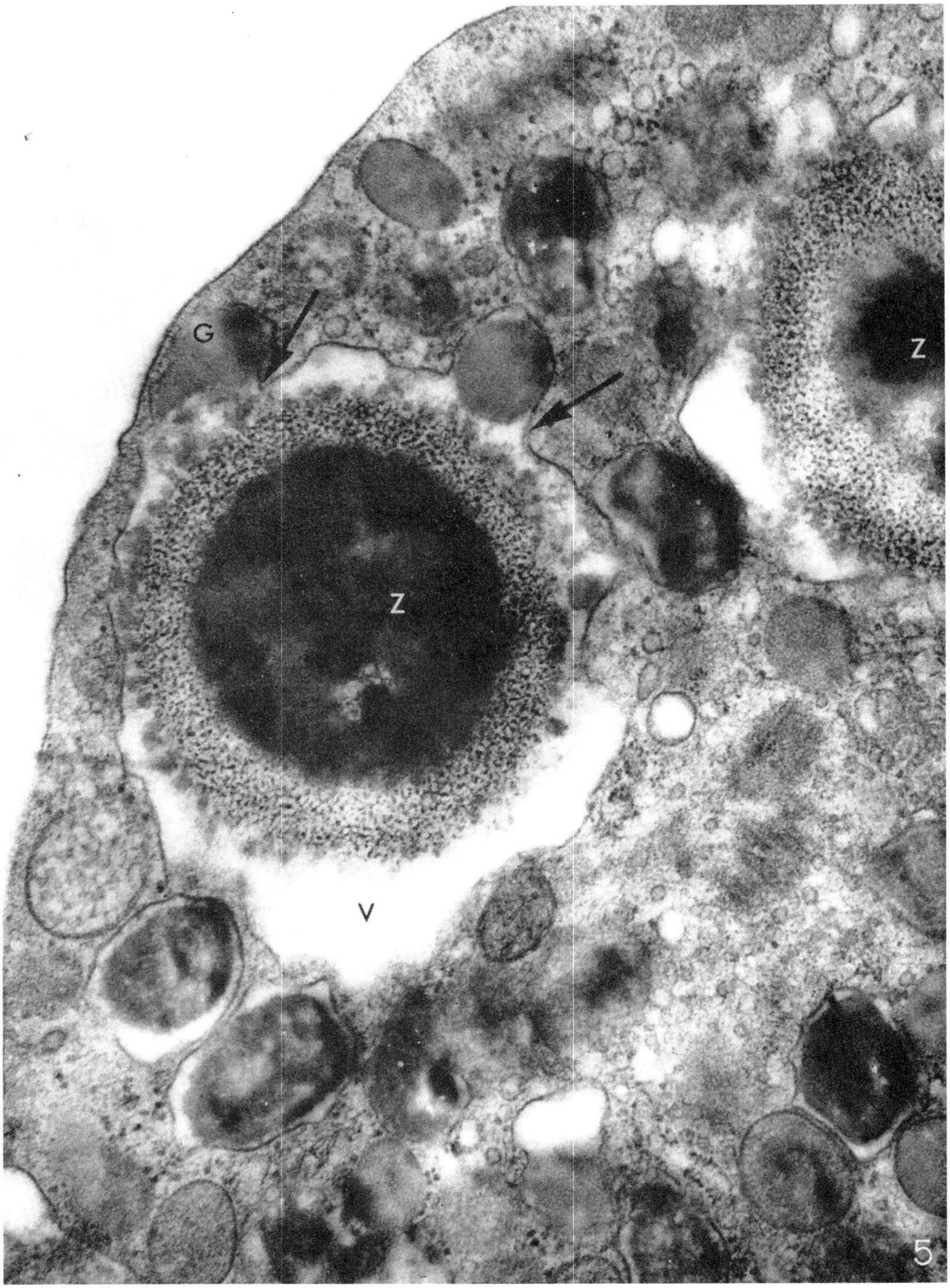

4. Electron micrograph of a rabbit polymorphonuclear leukocyte soon after engulfment of zymosan (Z) particles. At two points, indicated by arrows, the membrane surrounding the engulfed particle is continuous with membranes enclosing cytoplasmic granules, indicating that membrane fusion accounts for discharge of granule contents into the phagocytic vacuole (V). ×32,000. From Zucker-Franklin and Hirsch, 1964. *J. Exp. Med.* 120:569–576.

to unravel more complex functions of cellular intake and output, such as host resistance, immunological reactions, macromolecular transport, and cell nutrition.

The Cell Biology of Macrophages Begins

By 1963, Cohn had turned his attention from polymorphonuclear to mononuclear cells, and for the next thirty years he used this special cell as a tool for important discoveries in cell biology. The monocyte, found in many animals and many tissues, was one of the earliest cell types to be established and maintained under simple conditions in tissue culture. In 1924, Ludwig Aschoff designated the variety of mononuclear phagocytes, which all remove and store certain substances, as belonging to a "reticuloendothelial system."[79,80] A review of the cytology of macrophages by Alexander Maximow in 1932 contains more than 200 references.[81] The three-volume bibliography of tissue culture research from 1884 to 1950, launched by Keith Porter and Margaret Murray as a founding goal of the Tissue Culture Association (see Chapter 5), lists some three dozen pages of references to these cells.[82] Until quantitative techniques became available in the 1950s, however, there was little knowledge of the origin of monocytes, their differentiation into macrophages, their biochemistry, and their phagocytic and microbiocidal powers. This was the uncharted territory that Cohn wanted to explore.

Cohn began research with macrophages by extending his in vitro studies with Morse on phagocytosis in polymorphonuclear leukocytes; he developed in vivo procedures to compare the fate of virulent versus nonvirulent staphylococci in the mouse peritoneal cavity. He found the avirulent organisms were internalized efficiently, whereas the lethal organisms multiplied extracellularly. While the bacteria in the peritoneum were not influenced by outside and humoral factors, they were destroyed by two populations, the mononuclear cells normally present in the cavity and polymorphs entering in response to the infection.[83] Cohn then tested at the cellular level what Dubos was testing at the tissue level: would environmental stresses turn these infections into active disease? He subjected the animals to toxins, x-rays, and cortisone. Importantly, he found these external stresses altered the cellular dynamics and created an insufficient number of phagocytes, thus depressing host resistance and enhancing the course of infection.[84,85]

A new influence on Cohn came from Australian immunologist George Mackaness, who spent 1959 as a postdoctoral fellow in the Dubos laboratory and whose early work at Oxford University with Howard Florey convinced him of the critical role of macrophages in tuberculosis. Mackaness adopted the quantitative methods pioneered by Hirsch, Cohn, and Morse to carry out in vitro experiments comparing the fate of staphylococci in polymorphs with those in macrophages taken from rabbits. His new research overturned previous studies and showed that virulence of the staphylococci did not determine phagocytic activity in either cell type. He also found that the organisms were destroyed more slowly in macrophages and that some macrophages were not destroying the bacteria. Applying dyes to these cells, he detected two types of macrophages: the one he called "unstimulated" had no vacuoles and a few granules, and the other he called "highly stimulated" (later "activated") were filled with vacuoles and granules.[86] Clearly, the macrophage still possessed many unexplained facets.

The pinocytic-lysosome pathway. Macrophages proved to be a useful model. The idea that an outside agent might somehow be activating macrophages and enhancing their microbiocidal activity prompted Cohn to pursue the macrophages to learn, among its many functions, whether there might be a new population of cells or whether the original phagocytes had been greatly transformed. Unlike the short-lived polymorphs with their predominant granular organelle, the lysosome, the longer-living and larger macrophages (12–15 micrometers in diameter) are as complex as any mammalian cell and can be studied as free single cells. Their ability to infiltrate every organ and tissue of the body makes them active participants during any infection or disease. Because mouse peritoneal macrophages readily adhere to glass or plastic surfaces (while other white cells are easily washed away), Cohn had a homogeneous source of competent and abundant phagocytes. With this advance, he could conduct parallel morphological and biochemical studies.

When Cohn and Belinda Benson turned their full attention to culturing macrophages, they found the conversion of a monocyte into a mature macrophage is accompanied by significant changes: cell size increases without cell division, phase-dense granules (lysosomes) are produced and accumulate, mitochondria increase in length, and lipid droplets, or pinocytic vesicles, form.* It soon became clear that lysosomes arose from the endocytosis of these vesicles. Using metabolic precursors and the autoradiography methods just being introduced by Lucien Caro and Robert van Tubergen in the Palade laboratory (see Chapter 7), Cohn showed that these vesicles form at the cell periphery, move into the cell center, and fuse with secondary lysosomes adjacent to the Golgi apparatus to initiate the process of intracellular digestion.[87–90] Membrane fusion and fission maintain both the discontinuity of the acidified vesicle and the flow of substrates and enzymes into them. In concomitant quantitative assays, Cohn described the biochemical factors regulating the process of pinocytosis in macrophages and the formation of lysosomes.[91–94] A distinct capacity of the mature macrophage was now viewed as critical in cellular resistance, inflammatory reactions, and host defense mechanisms.

With the arrival of the electron microscope in the laboratory, Hirsch, Cohn, and Martha Fedorko provided clear views of formation of small pinocytic vesicles, their fusion into larger vacuoles, and their delivery of colloid into macrophage lysosomes.[95,96] Another series of motion pictures was taken to show these activities, but they were less captivating than Hirsch's earlier films showing degranulation during microbial phagocytosis. The new films nevertheless proved more broadly relevant in giving lucid views of the formation and fate of pinocytic vesicles as a widespread cellular activity.

Cohn and Barbara Ehrenreich, his second graduate student, outlined the basic features of the pinocytic-lysosomal pathway. They initially used radiolabeled albumin and hemoglobin as hydrolyzable lysosome markers and followed their intracellular digestion in the granules. Their biochemical analyses showed that macrophages hydrolyzed or digested proteins down to the level of amino acids, which were released into the cytoplasm and utilized for synthesis of new proteins by the cell of origin or adjacent cells. The partially digested fragments were, however, retained within the lysosomes. In subsequent studies, they used hydrolyzable peptides and undigestible saccharides that accumulated in the swollen vacuoles and eventually burst. These in vitro experiments indicated that the extent of digestion depended as much on

*Endocytosis is divided into two categories, phagocytosis (eating) and pinocytosis (drinking). The terms phago- and pinocytosis distinguish between particle- and fluid-containing vacuoles.

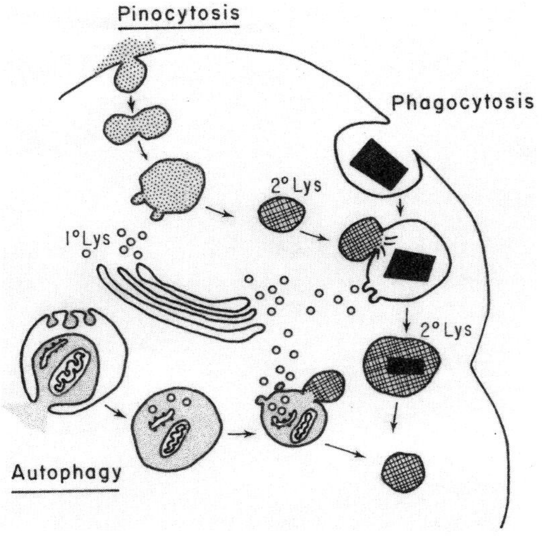

Early drawing by Zanvil Cohn of the endocytic system showing phagocytosis, pinocytosis, and autophagy.
Courtesy of Ralph Steinman.

the permeability of the lysosomal membrane as on the properties of the lysosomal proteases. Warren Lewis, the discoverer of pinocytosis in 1931, speculated that the abundant macrophages in tissues are "always actively engaged in drinking tissue juices, digesting them, and passing the fluid and digestion products back into the tissue fluids." Based on their findings in 1968, Cohn and Ehrenreich now more precisely suggested that pinocytosis contributed to such important activities as cycling of serum proteins, conserving body iron, and providing amino acids to help repair inflammatory sites The universality of the endocytic pathway, as revealed in macrophages, was now open for further experimentation in all cells.[97,98]

Origin of monocytes. Given the different structures of monocytes in various tissues and the uncertainty of their precursors, Cohn then focused on finding the origin of these cells. A starting point was work with William Bennett showing that blood monocytes would develop all of the now typical features of macrophages when placed in culture.[99] Similar events had been noted in 1939 at Oxford University when Robert Ebert and Howard Florey observed the development of extravasated monocytes in situ. When Ralph van Furth, a visiting fellow from the Netherlands, joined Cohn, they used the newly available isotope [^{3}H]thymidine as a DNA marker to trace the pathway from bone marrow to blood to tissues and to make quantitative and direct observations. In this way, they identified the blood monocyte as the precursor for tissue macrophages and the bone marrow as the source of monocytes.[100] They also monitored the movement of monocytes into the peritoneal cavity and elicited a recruitment of new macrophages by means of various stimuli, one being the injection of calf serum. Further studies by van Furth in Leiden monitored the origin of Kupffer cells in the liver, alveolar macrophages in the lung, and spleen macrophages. Later, van Furth, Fedorko, and Hirsch conducted electron microscope studies on colonies of bone marrow containing precursor monoblasts and promonocytes in which they described the formation and differentiation kinetics of these cells.[101]

Observations from these experiments led to a redefinition in 1972 of phagocytic mononuclear cells. Replacing Metchnikoff's "macrophage system" and Aschoff's "reticuloendothelial system," the new classification was called "the mononuclear phagocyte system" and emphasized two new things: the vague term reticular cell was replaced by one special cell type, the mononuclear phagocyte, and this phagocyte was distinguished from endothelial cells.[102] At this point, the notion of a reticuloendothelial system disappeared. The Reticuloendothelial Society became the Society of Leukocyte Biology, and its journal continued as the *Journal of Leukocyte Biology*. The system's well-known capacity to clear colloids, microorganisms, and antigens could now be attributed to the endocytic pathway of mononuclear phagocytes, which included macrophages, monocytes, and their precursors. The continued fruitful interaction with van Furth brought together many investigators into the vastly expanding macrophage universe: he organized and produced five published conference proceedings on *Mononuclear Phagocytes* between 1969 and 1992. The proposal to classify these cells into a unified system was made at its first meeting in 1969 and modified in consultation with numerous specialists worldwide.[103]

Once cell biological criteria were applied to the morphology, function, origin, and kinetics of mononuclear phagocytes, the foundation was in place for the Cohn laboratory to pursue macrophages in the lung, liver, and spleen as well as in other distinct cell types, especially endothelial cells.

The macrophage plasma membrane and membrane recycling. A flurry of new activities emerged once the mouse peritoneal macrophages were readily accessible, and experiments became far less burdensome when tissue culture plastic replaced glassware and carbon dioxide incubators replaced various gassed containers. By 1970, Cohn's interests in the vacuolar system shifted substantially, from contents (acid hydrolases and substrates captured by endocytosis) to containers (the plasma membrane and membranes of the endocytic apparatus).

Cohn wondered whether the membrane is destroyed in the lysosome and new membrane is synthesized or whether it is recycled. He was aware of observations made years earlier in the Porter/Palade laboratory during their electron microscopic survey of the endoplasmic reticulum in various cell types. Palade noted that "the quantities [of fluid] involved in pinocytosis in macrophages, as well as the abundance of 'submicroscopic' pockets and vesicles associated with the cell membrane in many other cell types render unlikely a unidirectional flow of membrane material from the cell surface towards the interior of the cytoplasm. It is more probable that the membrane is repeatedly circulated between the two locations mentioned."[104]

Challenged to develop better methods for analyzing the plasma membrane biochemically and functionally, Cohn tested two peroxidase enzymes to attach radiolabeled iodine covalently to membrane proteins in living cells. Graduate fellow Ann Hubbard and Cohn used lactoperoxidase to document the relatively slow turnover of most plasma membrane proteins in cultured cells;[105] active phagocytosis, however, caused a more rapid turnover of these polypeptides.[106]

Another peroxidase, horseradish peroxidase (HRP), was chosen by postdoctoral fellow Ralph Steinman and Cohn because of the relative ease of quantifying the solute uptake into cultured macrophages.[107] This enzyme did not require the radiolabels and counting of pinocytic vesicles that were used previously. In spite of the vigorous pinocytic activity, they found the amounts of solute internalized in the fluid phase were extremely small, but this could be dramatically enhanced when HRP–anti-HRP immune complexes were administered to be internalized via special Fc receptors on the macrophages.[108]

Adopting an electron microscope technique recently introduced by Richard Graham and Morris Karnovsky at Harvard, they took a stereological approach to measure the size of the intracellular compartments labeled by HRP as well as the kinetics with which these compartments were filled. This technique distinguished between incoming pinocytic vesicles and those that fused with preexisting lysosomes to form secondary lysosomes. They estimated that the area of membrane surrounding incoming pinocytic vesicles was equivalent to an influx of half the macrophage's surface area every thirty-three minutes. Since this represented ten times more fluid and membrane than could be accounted for in the secondary lysosomes, Steinman and Cohn postulated the incoming pinocytic vesicle membranes are recycled back to the cell surface after shrinkage of the newly formed secondary lysosome.[109]

Later, graduate fellow William Muller documented recycling of radioiodinated proteins through the endocytic system and back to the cell surface.[110,111] Then postdoctoral fellow Ira Mellman in the Cohn laboratory documented that unique Fc receptors on macrophages (binding microbes or antigens to these cells) were also recycled rapidly into and out of the endocytic compartments.[112,113] The first mouse and human monoclonal antibodies directed against Fc receptors were developed by Jay Unkeless, also in the Cohn laboratory, and became critical tools for following the intracellular fate of Fc receptors in macrophages.

Recycling of Fc receptors in macrophages was an example of a more general mechanism of receptor-mediated endocytosis that was being observed by other laboratories for many receptors in different systems; for example, the low density lipoprotein receptors in fibroblasts[114] and the galactose-specific receptor in hepatocytes.[115] These observations indicated there are distinct organelles of a lower acidity than the lysosome through which the internalized molecules pass on the endocytic pathway en route to the lysosome. Different terms were coined to describe these intermediate organelles, such as pinosomes, secondary lysosomes, intermediate vacuoles, endosomes, and receptosomes. In the macrophage literature, the term endosome was preferentially used.[116]

After Samuel Silverstein moved from Palade's laboratory to begin work on phagocytosis and viruses in the Cohn laboratory, he addressed the questions of how digestion begins and why macrophages are discriminating eaters. He and Frank Griffin Jr. found the stimulus to phagocytose a particle originates in receptors embedded in the cell surface; the receptors interact with ligands over the entire surface of the particle, much like a key fits into a lock. If precise contact is established, suggesting the particle is a desirable morsel, the receptor sends signals into the cell, and this induces the formation of a gel-like substance in the cytoplasm that moves additional membrane to the surface. The signaling continues until the membrane encloses the entire particle for transport into the cell where the digestion begins. Silverstein described this as the "zipper model of phagocytosis" in which interactions between the receptor and ligands guide the extension of the macrophage pseudopods along the surface of the particle.[117,118] All these discoveries confirmed the process of endocytosis as an enormous flow and recycling of vesicle membranes and their contents into and back out of cells.[109,119,120]

The macrophage as a secretory cell. Another major contribution of the Cohn laboratory was establishing that macrophages are not just "big eaters." In addition to the chemicals they make to kill and degrade the organisms and particles they ingest, they also secrete many molecules into their surrounding environment that in turn affect activities of other cells. Throughout the early 1970s, the laboratory uncovered a diverse array of products secreted by macrophages, not only through the secretory pathway of proteins via the rough endoplasmic reticulum but also at the plasma membrane.[121]

Early studies dealt with three secretory products. With Siamon Gordon, Cohn studied lysozyme (that breaks down bacterial cell walls) and plasminogen activators (that promote and inhibit blood clotting); both were produced continually in the absence of cell activation, although their rate of secretion was enhanced in activated macrophages.[122,123] Carl Nathan and Richard Root, then at Yale, showed that macrophages secrete another class of products, oxygen metabolites, particularly large amounts of hydrogen peroxide, that are potent cellular poisons capable of destroying bacteria and tumor cells.[124] In subsequent investigations with Cohn at Rockefeller, Nathan determined these metabolites require external stimuli for their induction and release; they further noted the specific oxygen metabolite depends on the distinct inflammatory and immune stimuli being received by the macrophages.[125] William Scott, graduate fellow Carol Rouzer, and Cohn pursued a third class of secretions, the arachidonic acid metabolites. They showed the macrophage was a major cellular source of these substances, many known to have major effects in inflammation and on smooth muscle and vascular tone.[126] Over the years, the laboratory catalogued more than fifty substances produced by macrophages.

The importance of a mobile population of macrophages as a traveling factory is its great capacity to generate substances on the spot during early stages of infection, with each substance related to the specific microenvironmental disturbance. However, Cohn recognized that a delicate balance existed between the nature and the amount of these secreted products. As an important part of the inflammatory process, the secreted substances can lead to wound healing and tissue repair. If macrophages are overstimulated and release excessive amounts, they can destroy normal as well as foreign cells and tissues, resulting in such disease states as rheumatoid arthritis, glomerulonephritis, asthma, and even atherosclerosis.

In this context, George Bernard Shaw's 1906 drama *The Doctor's Dilemma*, written two years before Metchnikoff was awarded the Nobel Prize for his theory of phagocytosis, offers a poignant note. The physician, Dr. Ralph Bloomfield Bonnington, insists there is "only one genuinely scientific treatment for all disease, and that is to stimulate the phagocytes. Stimulate the phagocytes. Drugs are a delusion." When these ministrations kill his patient, Bonnginton ponders, "Have we overstimulated the phagocytes?" Phagocytes, as revealed by Cohn's research, are double-edged swords. "The inflammatory process has a beneficial effect," he said. "But if it gets too exuberant, it can produce damage."

Macrophage activation and fate of intracellular bacteria. Before 1950, it was apparent from the work of Arnold Rich, Max Lurie, and Emanuel Suter that some activation of macrophages during infection with intracellular microbial pathogens leads to an enhanced state of microbiocidal activity. In the early 1960s, George Mackaness, after leaving the Dubos laboratory and returning to his native Australia, developed a more specific concept of macrophage activation. He discovered that immune activation by several different facultative bacteria, particularly *Listeria monocytogenes*, caused macrophages to resist or kill intracellular organisms, even those different from the original sensitizing one.[127,128] The mediator of this activation was not known until a few years later, when Barry Bloom (a former graduate fellow of Merrill Chase) at Albert Einstein and John David at New York University demonstrated that antigen-stimulated lymphocytes, later known as the thymus-derived or T cells, could make biologically active factors termed lymphokines and cytokines that attach to macrophage receptors and inhibit their migration; they became known as migration inhibition factors (MIF).[129,130] In subsequent studies, it was demonstrated that antigen-stimulated lymphocytes secrete soluble factors or macrophage-activating factors (MAF) that can change in vitro some macrophage functions, including increased adherence, motility, and phagocytosis.[131] In 1973, Carl Nathan and John David proposed that MAF was identical to MIF.[132]

Work from several laboratories during that time found macrophages stimulated by MAF can become cytotoxic and kill tumors nonspecifically.[133,134] Later, Carl Nathan (who by this time had joined the Cohn laboratory) and Robert Schreiber at the Scripps Research Institute identified interferon gamma as the responsible cytokine that serves as the macrophage-activating factor.[135,136] This cytokine was important in the Cohn laboratory's subsequent clinical efforts to treat diseases in which macrophages play a key role.

The outer membrane of a normal resting macrophage is ruffled, wavy, and in constant flux, turning itself inside out and back again every half hour; in this state, intracellular organisms can thrive and replicate. After activation, however, the macrophage changes shape and becomes smooth and flat; some called it an "angry macrophage" because it looks poised to attack. In a series of quantitative experiments, Cohn and many colleagues measured the various biochemical properties of resting versus activated macrophages. Among the many activities they found associated with activated macrophages, they considered the ultimate property is increased antimicrobial and antitumor activity.[137]

When it became evident that resting macrophages failed to generate the oxygen metabolites, while activated macrophages produced large amounts,[124] Cohn decided to examine this change in more detail. Working with Nadia Nogueira, first a graduate fellow and then faculty member, they studied the in vitro host-parasite interaction of the large parasite *Trypanosoma cruzi*, the causative agent of Chagas disease prevalent in Nogueira's native Brazil. First they learned to isolate the pathogenic trypomastigote form of the parasite and then followed its uptake and intracellular fate at the light and electron microscope levels.[138] Remarkably, they observed the parasites escaped from the phagocytic vacuole and, in the presence of a typical medium, replicated free in the cytoplasm of the macrophage, where they eventually ruptured the cell open and initiated another cycle of infection. However, when they used a corresponding specific antigen to challenge spleen cells from mice (and later humans) already infected with this parasite, the spleen cells produced factors that activated macrophages to destroy a majority of their intracellular parasites.[139] And when they applied a lymphokine-rich medium produced from spleen cells to normal resident macrophages, these phagocytes completely sterilized the infection. A new precedent was set: products of immunologically active spleen or blood cells could change the macrophage from a susceptible to a sterilizing (or killing) cell.[140,141] Subsequently, this model for research was extended to the intracellular parasites *Toxoplasma* and *Leishmania* with similar results and showed a clear association between the formation of hydrogen peroxide and macrophage activation.[125]

From cell biology to cellular immunology. Within two decades, the Cohn laboratory developed major tools of modern cell biology for studying white blood cells. These tools brought new knowledge of the endocytic pathway, origins of macrophages, and their key secretory substances in addition to many previously unknown protective *and* defensive mechanisms against disease. More powerful and multifaceted than Metchnikoff had ever imagined, Cohn redefined the macrophage from a "big eater" to "a versatile element of inflammation." At this point, along with many laboratories around the world, his laboratory moved into elucidating how macrophages interact at the molecular level with the vast array of antigens, antibodies, and other white blood cells in the immune system. As part of this research, more tools and methods developed at the bench also made it possible for Cohn and clinically trained colleagues, including Nogueira, Gilla Kaplan, Marcus Horwitz, and Juliana McElrath, to realize a long-standing goal to study patients suffering from infectious diseases caused by intracellular parasites: tuberculosis, leishmaniasis, trypanosomiasis, leprosy, and AIDS. Their continuing bench research and new clinical studies in cellular immunology, which was cut short by Cohn's untimely death, extend beyond the time frame and focus on the emerging cell biology in this book but have been described elsewhere.[53,54]

Importantly, Cohn remained acutely aware that the cell biology of the macrophage was not yet completely unraveled. "We are still defining the cast of characters," he once said. Perhaps the most important character to emerge was Ralph Steinman's discovery of the dendritic cell, the novel white cell that Cohn declared had been "hiding among the forest of macrophage ruffles" all along. The following chapter describes how a new phase of modern cell biology began in 1973.

12B. Endocytosis and the Discovery of Dendritic Cells

Ralph M. Steinman

My first exposure to New York was during a visit that I made while in college at McGill in Montreal in the 1960s. I was part of a small group that had chosen to major in biochemistry. Some of us decided to come to visit New York to see the places where pioneering events were occurring. Having learned the biosynthetic pathways for most of the amino acids, we wanted to see the buildings where David Rittenberg and Rudolf Schoenheimer[1] were pioneering the use of tracer isotopes at Columbia University to study metabolism and where Stanford Moore and William Stein at The Rockefeller Institute were pioneering the sequencing and structure of proteins.[2] Surprisingly, our undergraduate training had not made us aware of the ferment in cell biology in New York, but that trip probably planted the seeds for my long-standing devotion to the campus, faculty, and educational approaches at Rockefeller.

The freedom to conduct research at McGill made me realize that I wanted to do bench investigation, but I decided to obtain my scientific training at Harvard Medical School, which was (and still is) providing a robust education at the frontiers of physiology and medicine. Some of the eye openers for me came in my first year, 1963–64. There, Don Fawcett, Elizabeth (Betty) Hay, Jean-Paul Revel, Sanford Palay, and Susumu Ito introduced me to cell biology and subcellular organelles. However, little attention was given to lysosomes, the organelles that Christian de Duve and colleagues[3] and my future mentors, Zanvil Cohn and James Hirsch,[4,5] were unraveling at Rockefeller. The neurobiology unit at Harvard was also inspiring with its integrated view of the cell and the nervous system. The unit was headed by Steve Kuffler and included David Hubel, Torsten Wiesel, Ed Potter, and David Furshpan. During the six weeks of continuous neuroscience, the students could attend departmental tea and interact with this remarkable team. I was certain that cell biology was something I wanted to pursue.

Entry of MD students into a PhD program was not encouraged at that time, and MD-PhD programs were still many years away, so I took a year out from the MD program to study cell biology with Hay and Revel. Other classmates who chose to pursue research were Charles Janeway, who began an illustrious career in immunology at that point, and Tom Pollard, who began his insights into the cytoskeleton. I used electron microscopy to study cilia that were developing in explants of amphibian ectoderm and blocked this with colchicine, newly recognized by Keith Porter as a microtubule poison.[6,7] But the year in full-time research was more important for other reasons than the two papers that I published. For one thing, I was shown the advances of Revel and Morris Karnovsky, who were developing tracers that could be used at the electron microscope level, including horseradish peroxidase (HRP), which would prove valuable when I began my postdoctoral work at Rockefeller.[8] Also of importance was the advanced cell biology course. As part of this, George Palade came from Rockefeller to lecture. He was unsurpassed as a lecturer, and this repeatedly was the case when I later attended his classes in New York. Jokes and provocative value judgments were never needed for him to get attention because his subject matter and manner of presentation were simply grand. So by 1966, I was certain that I wanted to use the new cell biological approaches in research, but on what?

It took me many years of reading, listening, and thinking to find my way to what I regarded to be a significant cell biological theme in immunology. Several events helped out during medical training at the Massachusetts General Hospital. One course organized by Kurt Bloch was given late in the day to introduce the community to the new immunology. It is hard to comprehend decades later, but one of the things we were taught in the late 1960s was that macrophages took up antigens and made a "transforming RNA," which then instructed the lymphocyte to produce a corresponding antibody. This did make one's ears perk up concerning the potential of macrophages. A second event was the arrival in my mailbox of a "throw-away" journal called *Hospital Practice*, with a series of articles on the newly recognized field of T cell–mediated immunity, many of them related to an issue of the *British Medical Bulletin* on this subject. A third motivation came from reading MacFarlane Burnet's book, *The Clonal Selection Theory of Acquired Immunity* (1959), that expanded on this beautiful theory and its extensions into medicine.[9] Burnet proposed that the immune system was composed of a repertoire of clones, each with a distinct receptor for antigen. Together, the repertoire of "immunocytes," as he called them (James Gowans later discovered that lymphocytes were the cellular mediators of specific immunity), was capable of recognizing an infinite number of antigens, including medically relevant ones. Clonal selection was a pivotal theory that explained the capacity of the vertebrate immune system to recognize an enormous diversity of antigens each in a specific way.

By 1970, a lot of things were finally gelling for me. I was convinced that cell-mediated immunity would be pivotal in most areas of medicine. However, I was struck by the fact that despite clonal selection theory, it was not possible to initiate clonal expansion and immune responses in a tissue culture system. The experiments in this emerging field were using either polyclonal stimuli, like plant lectins, rather than specific antigens, or already stimulated or primed lymphocytes rather than the naive repertoire. How does the antigen-specific immune response begin? How is clonal selection energized?

The laboratory where I thought cell biological approaches were best being used to study immunological problems was the one at Rockefeller headed jointly by Zan and Jim. It was begun by René Dubos, the premier microbiologist, who had realized the need to study the host during infection. He represented the ideal mentor for Jim and Zan, two physicians devoted to infectious diseases, to pursue their elegant careers with phagocytes and later assume the reins of the laboratory. Luckily, Zan and Jim accepted my application, and I arrived on 1 July 1970.

Cell Biology of Endocytosis in the Cohn/Hirsch Laboratory

The endocytic system was one of the few frontiers in cell biology that was not emphasized during my medical school exposure during the 1960s. The opposite was the case at Rockefeller, where de Duve and his colleagues were defining lysosomes as membrane-bound organelles replete with digestive enzymes that worked best at an acidic pH.[3] Their approaches entailed biochemistry coupled to elegant and quantitative, centrifuge-based cell fractionation. Zan and Jim were using phagocytes (initially polymorphonuclear neutrophils or "polys" and later macrophages)[10] to define the connection of lysosomes to the outside world through two major types of endocytosis: phagocytosis, or particle uptake,[11–13] and pinocytosis, or fluid uptake.[14–18] Endocytosis was the central theme of their research, although not the only one by any means, as summarized in a tribute, "Zanvil Alexander Cohn 1926-1993," in *The Journal of Experimental Medicine.*[19]

One feature of the Cohn/Hirsch legacy is that some of their discoveries became widely available to the community by virtue of brilliant cinematography of living phagocytes and their abundant lysosomes in action. These films energized many future cell biologists and helped portray the captivating powers of science. With these films, as well as additional cell fractionation and electron microscopic approaches, Zan and Jim documented that during phagocytosis by polys, their lysosomal granules fused with an incoming microorganism, which was itself membrane bound in the newly formed phagocytic vacuole. This correlation of structure with function was vital to understanding cellular resistance to infection since many of the powerful digestive and microbicidal activities of polys were exerted within a dynamic membrane-limited endocytic system.

Zan then switched emphasis from polys to macrophages, particularly cells collected by washing out from the abdominal (peritoneal) cavity of mice. Macrophages could be placed on glass coverslips, and later on the new plastics that were amenable to tissue culture, because they selectively adhered firmly to both surfaces. As a result, not only were biochemical studies made on the cultured cells, but also the cells were examined by phase-contrast and electron microscopy to obtain an integrated picture.[14–18] In addition, expressing the results as quantitative data became obligatory. These composite approaches provided a gold mine of information about the basic properties of the endocytic system. The major deficiency was the lack of investigations on comparable in vivo systems because we were all too busy, although these were very much on our minds, as summarized elsewhere.[19]

Zan beautifully analyzed the capacity of cultured macrophages to form pinocytic vesicles continuously. These were membrane-bound vacuoles, 0.1–0.5 micrometers in diameter, without obvious dense content as judged by phase or electron microscopy. Following their formation in the periphery of the cell, the pinocytic vesicles with their imbibed fluids and solutes moved centripetally and fused with lysosomes, which by contrast had quite dense contents. The pinocytic vesicle to lysosome pathway could be demonstrated incisively in the electron microscope by means of 10–50-nanometer colloidal tracers of gold or thorium dioxide. Phagocytosis and autophagy were shown to contribute other substrates for digestion in lysosomes. Zan and Jim were joined in these early studies by several researchers, among them Dorothy Zucker-Franklin, Edith Weiner, and Martha Fedorko. Their investigations revealed the dynamic nature of intracellular digestion and opened the subject for biochemical studies.

Barbara Ehrenreich, during her graduate studies with Zan, did elegant studies on the digestion of proteins (and later saccharides) that had been captured by pinocytosis and delivered to lysosomes.[20–23] In the late 1960s, she and Zan published several papers showing how, over a period of hours, internalized proteins were catabolized right down to the level of amino acids before the latter diffused out of the cell. All this was logical in the sense that if digestion were not completed, osmotically active solutes would accumulate in the endocytic system, forcing the vacuoles to swell and eventually "pop." Indeed, the latter was observed when Barbara and Zan fed macrophages the indigestible disaccharide sucrose that swelled the lysosomes, whereas feeding the digestive enzyme invertase caused them to shrink.

However, the beautifully documented findings of Zan and Ehrenreich seemed counter-intuitive to an idea in immunology at the time that macrophages could retain protein antigens in an *intact, undigested* form for the generation of immunity. This was the subject I chose to investigate in 1970.

Interaction of Horseradish Peroxidase with Mouse Peritoneal Macrophages

During my final year of medical training, I made three visits to Rockefeller to talk with Zan and decide on the subject of investigation upon my arrival in July 1970. The decision was to follow the uptake and fate of HRP in cultured macrophages. Before explaining that this was of greater significance than it sounds, I decided to join Rockefeller despite the fact that it is in New York City, which was not nearly as great then as it is now. Nevertheless, the city could grow on you, and Rockefeller was then in high gear, particularly (from my perspective) the science being pursued in South Laboratory, later renamed Bronk Laboratory.

In this building on the fifth floor was probably the greatest concentration of cell biologists that have ever worked together in a contiguous space. Led by George Palade and Philip Siekevitz, the group in the 1970s included Günter Blobel, David Sabatini, Marilyn Farquhar, Jim Jamieson, Maya and Nicolae Simionescu, Bernie Gilula, and Jean Pierre Kraehenbuhl. The seventh floor was occupied by de Duve and his team that included Miklós Müller, Paul Lazarow, and Brian Poole, who demonstrated how to record the pH of endocytic vesicles in living cells, and Bill Bowers, who was one of the first to study dendritic cells. Sharing the sixth floor with Nobel laureate Fritz Lipmann was David Luck, a close friend of Zan and an incisive intellect; he was studying mitochondrial genetics and, with Gianni Piperno, the biology of microtubules. The Cohn/Hirsch laboratory was on the fourth floor. As mentioned, they were both physicians with a main interest in the host response to infectious diseases. We kept our feet in microbiological waters under the watchful eye of emeritus professor René Dubos, who by this time had switched from investigating microbial ecosystems to environmental science and had just won the 1969 Pulitzer Prize for one of his many valuable books, *So Human an Animal*. This was truly an exceptional and fertile period for cell biology, with phagocyte biology and infectious disease as our laboratory themes.

Before joining Rockefeller, I never had a classmate with my first name, but when I arrived, I was Ralph #4. The first three Ralphs were Ralph van Furth from Holland, who worked with Zan and Jim to define the mononuclear phagocyte system;[24,25] Rolf Seljelid from Norway,[26] who, among other things, accelerated our experiments by bringing us the first disposable tissue culture plastics from Nunc in Scandinavia; and Ralph Nachman from Cornell Medical School/New York Hospital, where he was to become chief of medicine and an inspiration for a vigorous platelet and vascular research program, who was spending a sabbatical year studying the plasma membrane.[27] Sam Silverstein and Siamon Gordon were resident macrophage gurus,[28,29] and they were advisers and sounding boards for my postdoctoral research almost on a daily basis. So too were three graduate students. Zena Werb, a fellow Canadian, was studying cholesterol in macrophages[30] and shared space with me; she was and still is a highly skilled mentor who was essential for instructing novices like me in laboratory techniques. Next door was Sally Zigmond, whose thesis was on chemotaxis of polys[31] and who took movies of living cells during her experiments, an approach that later proved invaluable to me. Ann Hubbard, another neighbor, was pioneering the use of lactoperoxidase to selectively radioiodinate the plasma membrane and to study its composition and sequestration during endocytosis.[32,33] Without her guidance, it would have been difficult for me to apply another peroxidase marker, HRP, in my work. It was obvious upon arrival that graduate students were responsible for projects as sophisticated as those done by the postdocs. The high caliber of the graduate students at Rockefeller has been a major gratifying stimulus throughout my career.

Why was the use of HRP of interest to me? I knew from Revel and Karnovsky, who taught me in medical school, that HRP was a valuable tracer that could be visualized in both the electron microscope and the light microscope, as shown by Werner Straus in Belgium with diaminobenzidine cytochemistry.[34] I also knew from Ann Hubbard and Zan, who in turn were influenced by Zan's friend and colleague Seymour Klebanoff, an expert in the myeloperoxidase and thyroid peroxidase reactions, that one could assess the enzymatic activity of HRP biochemically by using sensitive assays. Both assays have the sensitivity needed to trace the uptake and fate of a protein in macrophages without resorting to the use of radiolabeling. The latter has the limitation of typically marking only a tiny fraction of the molecules in a protein preparation. Furthermore, Stratus Avrameus in France was using HRP as a model antigen to follow the development of specific antibody-producing cells against HRP during an immune response[35] so that it would be possible to determine the frequencies of antibody-producing cells. HRP, therefore, had much potential for clarifying a central immunological problem that challenged me.

I wanted to understand "immunogenicity," or how an immune response begins, namely, what needs to happen for an antigen to induce an immune response. Immunology was blessed with the incisive clonal selection theory of Burnet, mentioned earlier, but a perplexing related feature was the great difficulty, after administering an antigen to cells in tissue culture or to mice, of initiating a specific clone and having it expand and produce antibody. While finishing my medical training, I read new publications concluding that a necessary step might require the retention of antigens in the form of intact proteins on macrophages, including those in vivo termed dendritic reticular cells or "dendritic macrophages" found in the germinal center regions.[36] It was suggested that retention and presentation of the immunogenic protein coax an antigen-specific clone to expand and function in the manner Burnet had proposed. At this time, however, the idea of antigen presentation in terms of processed antigen or peptides affixed to the major histocompatibility complex (MHC) products had not yet been conceived.

We were unable to show that peritoneal macrophages could retain intact HRP, either on or within the cell.[37,38] This was the case whether we studied soluble protein or HRP bound in immune complexes for recognition by Fc receptors on the macrophage surface. Instead, the HRP was taken up and easily visualized in the endocytic system, i.e., in presumptive pinocytic vesicles and lysosomes (figure 1). Following uptake, the enzymatic activity was gradually but completely degraded, and when we used radioiodinated HRP, the only labeled catabolite that we could identify in the culture medium was mono-iodotyrosine, much like Ehrenreich and Zan had described years earlier for radioiodinated bovine serum albumin. In addition to the quantitative assays that the use of HRP permitted, we could also trace the enzyme by electron microscopy, but we could not find a reservoir of intact antigen that persisted on macrophages. This work was facilitated by Jim's acquisition of

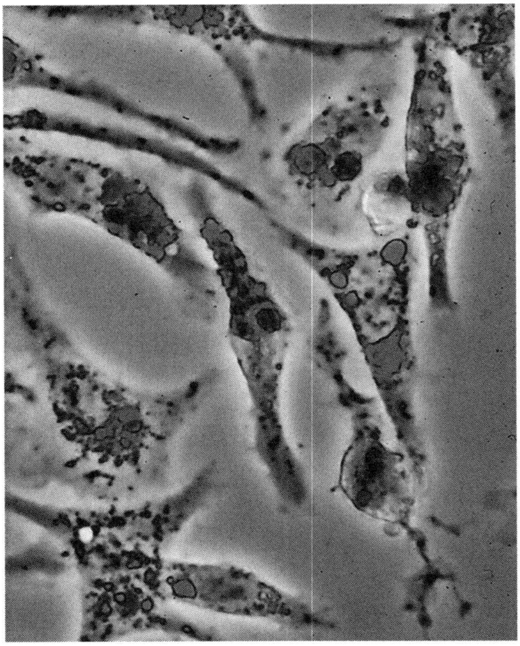

1. Uptake of immune complexes of HRP and anti-HRP by mouse peritoneal macrophages in culture. The HRP is visualized with the diaminobenzidine cytochemical reaction, which could also be extended for use in electron microscopy. Courtesy of Ralph Steinman.

an electron microscope and the presence of Judy Adams, the research assistant who prepared and sectioned the experimental material.

While we did not find what we were looking for, we were able to come up with some interesting results. We could measure the rate of solute uptake into macrophages using HRP as a tracer, a method that proved quite useful in this area of cell biology for many years. We could document large increases in uptake of protein when receptors were engaged, in our case Fc receptor–mediated uptake of HRP–anti-HRP immune complexes. And we performed an analysis of the dimensions of the pinocytic vesicle and lysosome compartments during the sequestration of defined amounts of fluid and vacuolar membrane. This analysis made it evident that, during endocytosis, there had to be a major recycling of internalized plasma membrane and fluid to account for the rate of endocytosis we had measured so as to maintain over a period of hours the size of these compartments.[39,40] We estimated that every hour, the macrophage took up and recycled the equivalent of twice its cell surface area and a quarter of its volume. This membrane recycling plays a role in many cellular processes, particularly as Michael Brown and Joseph Goldstein uncovered for low density lipoproteins, and provides an escalator for cell receptors that continually sequester important ligands and then recycle them to the cell surface to capture more.[41] Later, William Muller, during his graduate studies in the laboratory, documented recycling with direct methods: he followed the rapid movement of radioiodinated proteins in the endocytic system and back to the cell surface.[42–44] Also, Ira Mellman, who arrived for postdoctoral work, documented among other findings the rapid recycling of Fc receptors into and out of the endocytic compartments.[45,46] Nevertheless, from these observations, we were not gaining an understanding of the process whereby a protein was handled by these cells to initiate immunity.

Discovery of Dendritic Cells in Immune Organs

Although I had not yet written up the studies with HRP, I began in 1972 to turn away from investigating the laboratory's "bread and butter," macrophages from the mouse peritoneal cavity, toward studies of mouse spleen cells. I wanted to gain more information about the "accessory cell problem," a subject that seemed potentially important for elucidating immunogenicity. At that time, the evidence indicated that an immune organ like the spleen contained cells that acted as accessories to lymphocytes for the generation of an antibody response. In

1933, Philip McMaster and Stephen Hudak at Rockefeller proved that lymphoid tissues in rodents were the sites where immune responses could begin.[47] Yet it was not until 1966 that Robert Mishell and Richard Dutton, at the Scripps Clinic in California, learned how to generate an immune response from a suspension of mouse spleen cells.[48] Several investigators subsequently found that cells of the spleen, a major organ of the immune system probably accounting for at least half of the peripheral immune system, could be divided into two components: lymphocytes that were unable to adhere to plastic or glass surfaces and were the source of the antibody-producing cells; and adherent cells, presumably macrophages, that acted in some manner as required accessories. These accessory cells could a priori have several functions, perhaps acting to sustain the viability of the lymphocytes. In fact, graduate fellow Chang Chen and Jim were able to show that peritoneal macrophages could sustain lymphocyte viability,[49] but the cells were typically unable to act as accessories for a genuine antibody response. The question remained, what was unique about accessory cells from the spleen?

In contrast to prior studies in the literature, I decided to examine adherent splenic accessory cells by phase-contrast microscopy. Wow! On the very first look, I observed unusual cells that did not resemble any of the macrophages that had been described before (figure 2). The cells I observed had tree-like or dendritic shapes (from the Greek *dendreon*, for tree). They had none of the pinocytic vesicles that from Zan's research and our HRP experiments were so characteristic of macrophages. The phase-dense granules also did not resemble lysosomes and proved to be mostly mitochondria, with only a few typical lysosomes tucked close to the unusually shaped nucleus. Electron microscopy backed up the phase-contrast findings and revealed that these large adherent cells were not laden with the characteristic phagosomes and lysosomes of macrophages but were full of mitochondria and complex vesicular structures (figure 3). The cells were also devoid of some cytochemical features of macrophages such as nonspecific esterase, iron deposits, and a membrane ATPase that Siamon Gordon had been studying. A characteristic feature was the movement of these cells in culture, which we followed under Jim's guidance: they constantly extended and retracted cytoplasmic processes, and their movement was associated with major distortions in the shape of the nucleus. It is exciting that the continuous probing movement of dendritic cells has more recently been observed by my colleague Michel Nussenzweig to also occur in intact immune organs of living mice when examined by intravital two-photon microscopy.[50] We called the new cells "dendritic cells."[51]

The experience with HRP and other tracers also helped to confirm that dendritic cells were a distinct cell type. They were weak at engulfing tracers like HRP or erythrocytes coated with antibody. The presence of active Fc receptors (for antibody-coated substances) was the hallmark of macrophages at the time, as we learned from our long-time colleague, Michel Rabinovitch. The weak endocytic activity was also apparent in vivo when we administered the tracer and then isolated the dendritic cells from the animals. In contrast, when large macrophages were explanted from the spleen, they were actively endocytic, including for immune complexes.[52] These experiments were of course providing a partial picture. Dendritic cells do have endocytic activity, although it is almost always modest when compared with that of macrophages from different sources and best developed in their immature state.

It took six years until we could obtain highly enriched dendritic cells and macrophages from mouse spleen.[53] The reason for the delay was our observation in 1975 that dendritic

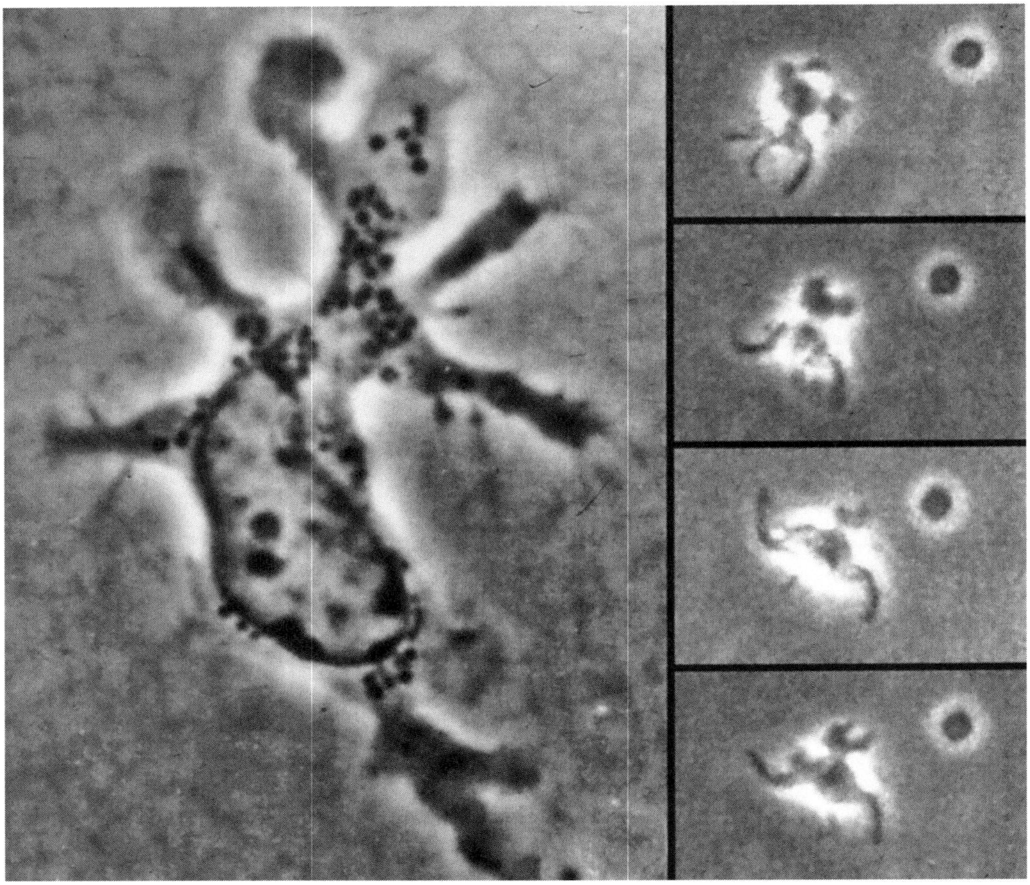

2. Phase-contrast micrograph of dendritic cells, 1973. On the left is a cell from mouse spleen that has been cultured several days in a plasma clot; the cytoplasmic granules are mitochondria. On the right is a cell from human blood showing the rapid changes in cell shape at fifteen-second intervals when the cell is cultured. Courtesy of Ralph Steinman.

cells appeared to stimulate the massive formation of antibody-producing cells, including HRP-specific ones.[54] The purification process started with subjecting spleen cells to de Duve's methods of density gradient centrifugation in a column of bovine serum albumin, on which semi-purified dendritic cells rose to the top and small lymphocytes went to the bottom. After placing this fraction on glass for an hour, all except the adherent dendritic cells and macrophages were washed away. Following an overnight culture, the dendritic cells detached from the glass; the lingering macrophages were removed by adding antibody-coated sheep red blood cells that formed rosettes by virtue of their Fc receptor expression. The rosetted macrophages went to the bottom of a second gradient centrifugation, and the dendritic cells were collected from the light interphase. Unfortunately, the yield of dendritic cells by this method was poor, and for a long time there were never enough cells for all the experiments that we wanted to do.

However, since methods did not exist to distinguish dendritic cells from other cells and to determine their role in cell mixtures, it was difficult to convince anyone that they were really a distinct cell type. The first step in that direction was the development by Michel

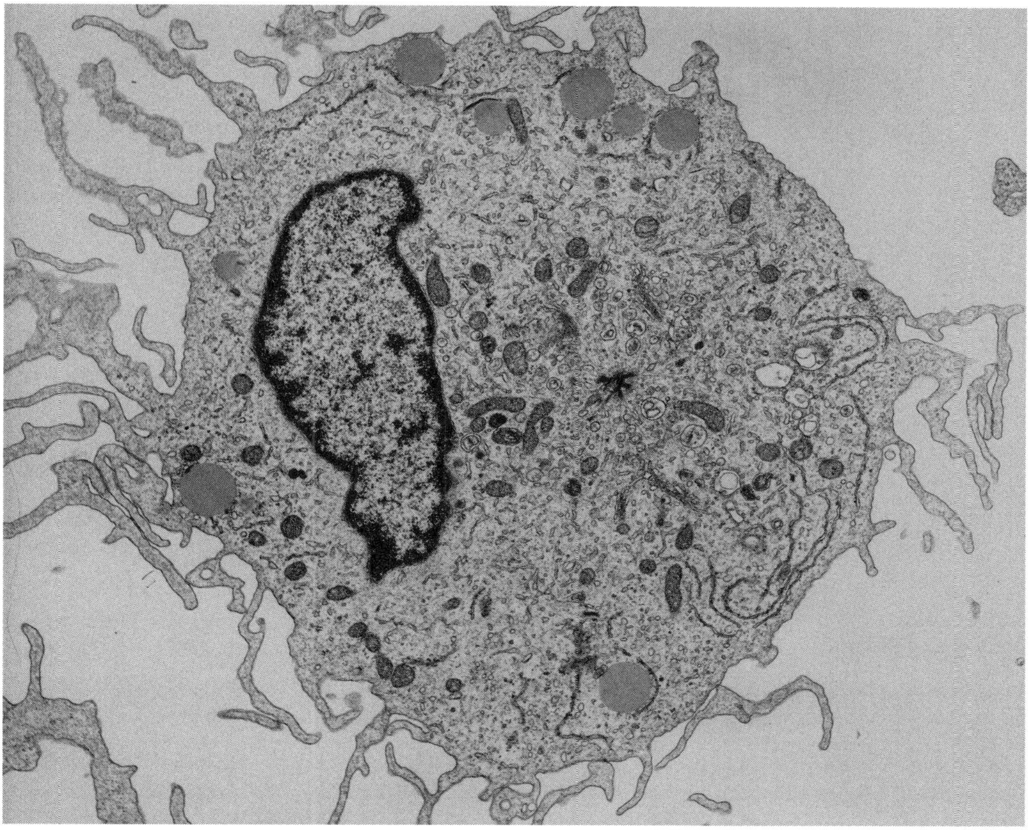

3. Electron micrograph of a dendritic cell showing the numerous thin processes and the many mitochondria as well as endosomal-type vesicles in the centrosphere of the cytoplasm. Courtesy of Ralph Steinman.

Nussenzweig, at the time my graduate student, of the monoclonal antibody 33D1 that was the first molecular marker for dendritic cells.[55] The ability to distinguish and separate macrophages from dendritic cells provided a fruitful path toward identifying their functions, in particular their powerful immunizing properties.

In several studies, we could at last initiate clonal selection to several antigens.[56–59] Not only were dendritic cells capable of initiating T cell–mediated immunity in culture, but they were also unusually potent. On the average, one dendritic cell or less per one hundred lymphocytes was all that was required. In addition, one could simply observe the cultures and see the sizeable aggregates of dendritic cells clustered next to the responding lymphocytes (figure 4). All in all, dendritic cells were the missing link required to convert antigens into immunogens and to initiate a productive immune response.

Dendritic cells were then identified in human blood by Wes Van Voorhis[60,61] as well as in other body sites and animal species by investigators in several countries. Among them was Kayo Inaba from Kyoto University, who recognized the importance of a non-macrophage–type accessory cell during her PhD work.[62] We later teamed up for a life-long collaboration in dendritic cell research. Nevertheless, the Bronk Laboratory, with Zan and his many colleagues, was the ideal place for dendritic cells to be discovered since so many

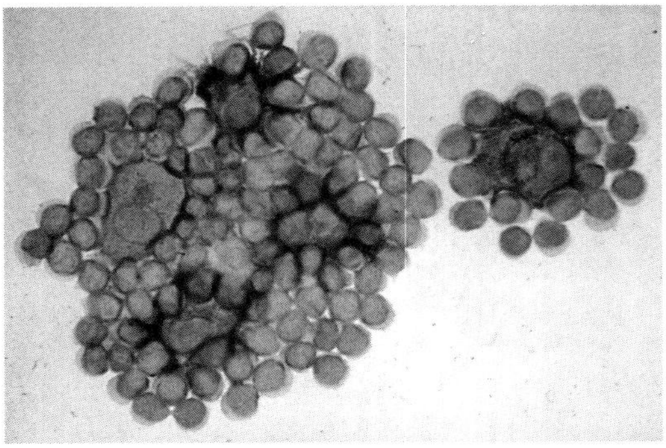

4. Five large dendritic cells (dark stained) surrounded by clusters of small T cells. A cluster is the microenvironment for generating immunity in tissue culture. Courtesy of Nikolaus Romani.

cell biological tools and so much experience were in place to demonstrate the distinct new features of these cells.

An important finding with Maggi Witmer-Pack was the identification of dendritic cells in the lymphoid or immune organs.[63,64] They were most abundant in the T cell areas of organs like the spleen, lymph nodes, and mucosa-associated lymphoid tissue (figure 5). This placed the dendritic cells in the perfect position to select antigen-specific clones of lymphocytes that recirculate continually into and out of the T cell areas. All in all, these early developments provided a new focus for understanding immunogenicity—the dendritic cell—and it has kept us occupied for years.

A Return to Endocytosis, but now in Dendritic Cells

By the mid-1970s, I realized that I had to forego studies on endocytosis and focus on the functional capacities of dendritic cells and their mechanisms. By the 1980s, we also had begun to extend dendritic cell biology into the analysis of human disease, particularly HIV infections and immune therapies. Medicine has always been a motivating force for me and is part and parcel of the general problem of immunogenicity. This is because scientists are often aware of disease-causing antigens, yet one cannot use these antigens to bring about a desired immune response, for example, to increase protection against specific infections and cancer or to decrease unwanted immunity in the case of allergy and autoimmunity. Think of HIV vaccines, where it began to seem possible in 2007 to reliably generate some T cell immunity in healthy volunteers with safe vaccines, even though the genome of HIV and most of its expressed proteins had been known for twenty years. Dendritic cells are repeatedly demonstrated to bring about the specific control of immunity, including to clinically relevant antigens. To a considerable extent, these cells are a missing link to energize clonal selection to bring about immunity, tolerance, and memory.

Interestingly, the biology and medical analysis of dendritic cells have come full circle to our original interest in endocytosis. The endocytic system of dendritic cells seems highly specialized to carry out effective antigen presentation, and it became a focus of our research to develop vaccines that will ensure optimal capture of vaccine antigens by these potent antigen-presenting cells. I often reminisce that we could move more quickly if Zan were emeritus and now sitting where Dubos sat in years gone by.

In part, the return to endocytosis was triggered by the research of Ira Mellman and his group at Yale. They unraveled the endocytic system of dendritic cells, beginning with cells that were expanded from precursors in tissue culture with granulocyte-macrophage colony-stimulating factor (GM-CSF). They repeatedly found that the basic components of the endocytic system in dendritic cells—uptake, acidity, proteolysis, even the turnover of MHC molecules—are highly regulated by what are called maturation stimuli that trigger dendritic cell differentiation to regulate immunity.[65] Overall, the dendritic cell seems to

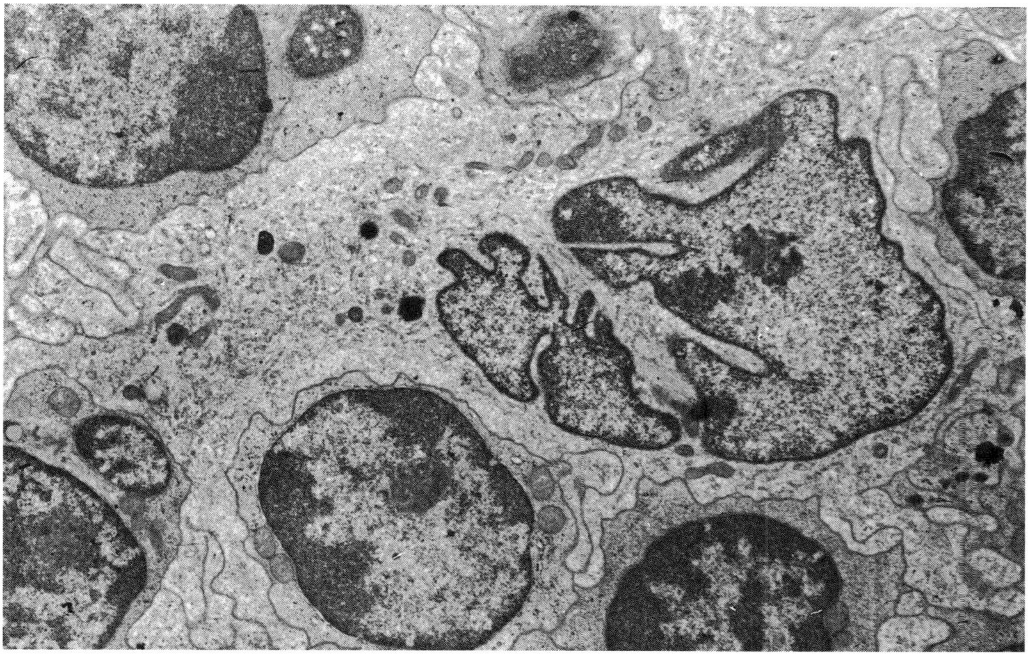

5. Electron micrograph of a dendritic cell in the T cell area of lymphoid tissue showing the irregular-shaped nucleus and the "empty" cytoplasm that ramifies within many cell processes between the lymphocytes. Courtesy of Ralph Steinman.

have an endocytic system that is specialized for antigen processing and generates peptides that bind to different types of antigen-presenting molecules. The dendritic cell endocytic system works very differently from the macrophage system that is specialized for continual antigen scavenging and destruction.

Another trigger for our return to endocytosis involves a robust collaboration with Michel Nussenzweig's laboratory. This began with the cloning of the DEC-205 or CD205 molecule by Michel together with an MD-PhD student, William Swiggard, and postdoctoral fellow Wanping Jiang.[66] This receptor was the first of many that are being identified on dendritic cells in vivo. DEC-205 localizes to coated pits, and its ligation intensifies antigen uptake and presentation more than 100-fold. Karsten Mahnke obtained evidence for a distinct pathway of recycling, which takes DEC-205 into and out of late endosome-type compartments rich in MHC class II products.[67] With the work of Daniel Hawiger, a student in Michel's laboratory, we began emphasizing an approach in which we engineer proteins into an antibody to DEC-205 so that we can target antigens selectively and efficiently to dendritic cells in vivo.[68] This leads not only to proteolysis within the endocytic system, for presentation on MHC II, but also to an unusual consequence of endocytosis called "cross-presentation."[69,70] In this still enigmatic pathway, internalized polypeptides are processed for presentation on MHC I to $CD8^+$ killer T cells. The findings of Laura Bonifaz, Christine Trumpfheller, Leonia Bozzacco, and Angela Granelli-Piperno indicate that delivery of antigens via DEC-205 brings about cross-presentation, e.g., of the HIV gag protein. Such delivery would overcome a major obstacle to the design of protein-based vaccines that have in the past not been able to induce $CD8^+$ T cell immunity.

We are developing fusion (hybrid and chimeric) antibodies to DEC-205 as a means to deliver vaccine proteins selectively and efficiently to numerous dendritic cells in vivo. Along with an appropriate additional stimulus for dendritic cell differentiation or maturation, the targeted antigen induces powerful immune responses.[71,72] We think that the targeted delivery of antigens to dendritic cells will allow researchers to better unravel dendritic cell function in intact animals and people, and it will allow dendritic cells to provide the antigen-specific control of immunity that is so desirable.

Bringing Dendritic Cells into Medicine

A driving force for my entry into immunology was the vital role that immune cells and their products play in disease. The term "translational" research is often used to describe the extension of science to the study of disease in patients. I find the term distracting and even misleading for the type of science that is needed because one is not simply translating what one already knows from more standard research in tractable experimental systems. Rather, the disease in a patient often sets standards for the quality of new information and principles that scientists still need to discover.

For example, I just cited a challenging area of cell biology, which is to understand how dendritic cells are able to cross-present an internalized antigen to elicit killer T cell activity. This form of immunity is thought to be pivotal for resistance to AIDS, cancer, and many other diseases. A lot is being learned in mice about cross-presentation by studying one protein, chicken egg albumin or ovalbumin, but in so doing, scientists are exploiting only one particular peptide and one particular mouse strain. This ovalbumin peptide binds at least a thousand times more strongly to MHC I than many other "good" peptides from microbial proteins. This suggests that mouse models may detect events operating at a tiny fraction of the efficiency required for cross-presentation during human cancer, infection, and other conditions. There is essential knowledge on the pathways to cross-presentation to be gained from studying real vaccines and patients.

For this reason, a significant effort in our laboratory is being directed to research in humans, particularly at The Rockefeller University Hospital, that is designed to provide University scientists, including students, with the opportunity to pursue patient-based studies. We often ask ourselves, why is it that science seems so far away from harnessing the remarkable systems for cell-mediated immunity and host defense that have the potential to bring about resistance to AIDS, malaria, tuberculosis, and cancer? There are many more experiments to be done in mice to gain ideas and identify principles. But there is another side to discovery, to study patients to identify pathways that are pertinent to human microbes and tumors, which are simply different from those found in mouse models. Scientists discover by studying their subjects directly. Why should humans and their diseases not be included? Research on dendritic cells already provides a large number of concepts and strategies to dissect how the immune system defends (or fails to defend) against disease. Now it is necessary to understand the human system well enough so these findings can be directed to the patient's benefit.[73]

Chapter 13

Gatekeepers: Cell Membranes and Permeability

Introduction

Philip Siekevitz

The next phase of research in the laboratory of George Palade and Philip Siekevitz, after concluding work on the secretory pathway, concerned Palade's long fascination with cellular membranes, the gatekeepers of proteins and other molecules transported between cells, tissues, and the bloodstream. In all likelihood, this fascination stemmed from his earliest scientific interest in Romania, where he tried to correlate the structure of capillaries (glomeruli) and nephrons in a dolphin kidney with the mammal's special needs in a saltwater environment.

To introduce this section, a brief summary follows of Palade's initial studies, including those with three postdoctoral fellows. This summary provides the background experiments that set the stage for more detailed work on cellular membranes described in the following five subchapters. More importantly, this research solidified the laboratory's classic approaches to studying structure and function at the cellular level.

Fine structure in endothelial cells. In the early 1950s, while Porter and others in the laboratory began looking at various structures inside cells with the electron microscope (described in Chapter 9), Palade began an independent study of the fine structure of endothelial cells lining blood capillaries. He assumed that was where an exchange of materials between blood and tissues was taking place. In 1953, he very briefly reported his first observations of the capillary wall (accompanied by electron micrographs) in two places with limited visibility: the obligatory, private Annual Report to Scientific Director Herbert Gasser and an abstract for a meeting of the Electron Microscopy Society of America (figure 1). He noted two characteristic structures in these cells: "intracellular fibrils" and a "large number of vesicles concentrated under the cell membrane," including many that appeared open at the cell surface. He speculated that these novel, unexpected structures "may represent a system for actively transporting fluids across the capillary wall."[1,2] Palade named these tiny sacs "plasmalemmal vesicles" because of their relation to the cell membrane. These structures were independently discovered in 1955 by Eichi Yamada at the University of Washington, who called them caveolae (or small caves), the name by which they are known today.[3]

Three years later, Palade and postdoctoral fellow Stephen Wissig continued this work by looking at capillaries in heart muscle and kidney glomeruli. They tested whether large molecules of ferritin or colloidal gold, which can be seen in electron micrographs, would be found inside the vesicles and whether they could serve as markers for tracing the

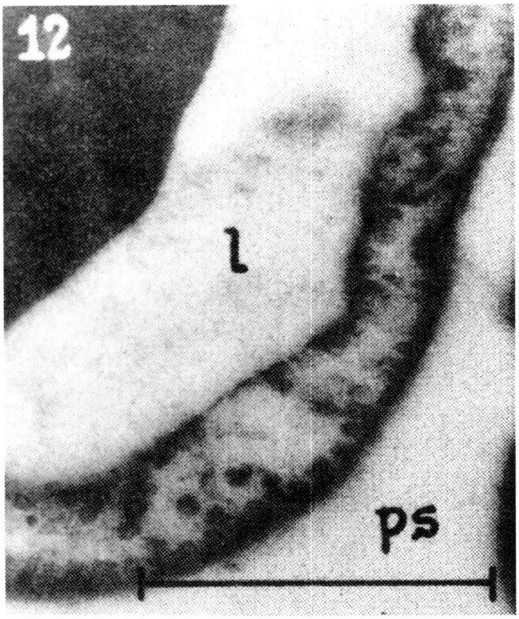

1. Section through wall of rat diaphragm blood capillary. Numerous vesicles below the membrane of the endothelial cell are shown facing both the lumen (l) and the pericapillary space (ps). Courtesy of The Rockefeller Archive Center. From Porter, 1954. RUA, BSD Reports. Volume 42, 195–205.

possible movement of fluids across the endothelium. None of these experiments were published until 1960, and even then only as an abstract, although a few key observations again appeared in the annual reports. The tracer molecules were found in the lumen, the endothelium, basement membrane, and pericapillary spaces. In the endothelium, the particles were enclosed within small pockets or vesicles, while none were found free in the intracellular spaces. This led them to postulate that the vesicles were formed by invagination of the cell membrane and thus acted as carriers of fluids across the endothelium. These findings fit with pre-electron microscope concepts suggesting various size pores exist in the capillary walls to allow various sized molecules to transverse the endothelium in the exchange between blood and tissue. Even though Palade established the existence of a mechanism of transport across the endothelium of these tissues, he was cautious about whether this process existed in blood tissues in general.[4]

Tubules in endothelial cells. Quite by accident, while surveying a large number of electron micrograph sections of rat and human lung tissue, Palade and Ewald Weibel, a postdoctoral fellow from the University of Zurich, came across another new structure in the endothelial cells of these tissues. In 1964, they described a bundle of fine tubules enveloped by a tightly fitting membrane and embedded in a dense matrix. These tubules were disposed parallel to a long axis of rods and were found in capillaries of all arterial, arteriolar, and endocardial endothelia and occasionally in alveolar epithelia. At the time, they had "no clue to their functional significance," and the Palade group did no subsequent work on these structures. Today, these Weibel-Palade bodies are known to store certain components produced and processed by the endoplasmic reticulum in endothelial cells, and these bodies play a dual role in blood coagulation and inflammation.[5]

Vesicular transport in muscle endothelium. In 1961, Palade and Romaine Bruns, a postdoctoral fellow from Massachusetts General Hospital, continued studies on muscle in the diaphragm because of its rich continuous capillary network. Their goal was to learn whether transport of molecules was through membrane pores of different sizes, as some believed,[6,7] or inside vesicles moving across the epithelium.[8] Their experiments followed Palade's earlier approach to the secretory pathway by using a series of electron micrograph images to illuminate a dynamic process. By 1968, they produced three major findings.

Examining serial sections of endothelial tissues on electron micrographs, they discovered the fine structure of the endothelium consisted of three layers: the inner endothelium, a middle basement membrane, and the outer adventitia. Moreover, eighteen percent of these flattened cells were populated with vesicles either singly or fused with others in short chains; some vesicles were open at the surface, and none were connected to other intracellular membranes such as the endoplasmic reticulum or Golgi apparatus.[9]

To obtain evidence for movement of molecules via the vesicles, they injected animals intravenously with ferritin as a marker and followed the course of its movement along and through the vascular bed, using electron micrographs taken at various times over twenty-four hours. They concluded the closed vesicles, which were five to six times more numerous than pores, carry out the function of the theorized large pore system, the basement membrane does not restrain movement of the ferritin, and the transport of ferritin occurs uniformly along the entire capillary.[10]

Following this, they used staining with uranyl acetate to identify the various structural changes of vesicles during stages of transport across the endothelium. Some forms were open at the surface, while others were covered (fenestrated) openings. Some were fused in layers at the outer membrane; some were flask shaped, as if being pinched off to become independent; and still others were fused with the inner membrane. This careful work allowed the Palade group to identify vesicular transport as a dynamic but discontinuous process rather than a system of permanently open pores through the endothelium.[11,12]

Molecular transport in intestinal epithelium. During 1967–1968, Palade and Francesco Clementi, a postdoctoral fellow from the Istituto di Farmacologia in Milan, tested whether the findings in muscle capillaries applied to intestinal mucosa. Unlike muscle, the intestine has a discontinuous epithelium. Others had found differences in capillaries from different tissues, including their construction and ratio of small to large pores.[13–15] This time, Clementi and Palade compared the movement of two molecules of different sizes (the larger ferritin and the smaller horseradish peroxidase) to probe the large and small pore system. They found the molecules leave the capillaries less readily through the vesicles than through the smaller fenestrated openings. In further tests to chemically remove the covers on the openings, they tentatively concluded these capped pores represented a small pore system.[16 17] The full scope and consequences of this work were explored in seminal studies in the laboratory from the late 1950s until 1974, when Palade, Marilyn Farquhar, James Jamieson, and Maya and Nicolae Simionescu moved to Yale University, where they continued this research. The following five subchapters are written by the scientists who did this research at Rockefeller. Two describe structure and function of cell membranes surrounding various organelles as well as those enclosing the cell itself. Three others describe simultaneous studies of the structure and function of endothelial cells that line blood capillaries and how these cells ferry molecules from one side of the membrane to the other.

13A. Vascular Leakage Induced by Inflammatory Mediators

Guido Majno

In 1958, I was assistant professor of pathology at Harvard Medical School, where my laboratory was doing research primarily on cell death and demyelination. Electron microscopy was just appearing on the horizon, and it promised to revolutionize cell biology and medicine, so I decided to join this new movement. Since I did not have access to an electron microscope in the Boston area, my chief, Arthur T. Hertig, made a few phone calls and was advised that the ideal plan would be for me to apply to the laboratory of George E. Palade at The Rockefeller Institute and that the competition would be stiff. So I wrote to Palade, with little hope, because I had a rather short CV and no particular research skills (I had been trained in Geneva as a pathologist). I chose to introduce myself as a friend of the late Dinu Lipatti, a great Romanian pianist who had died in Geneva in 1950. To my delight, Palade offered me a position as guest investigator right away, and I believe that music had something to do with it. It turned out to be a perfect experience and a lifelong blessing.

To keep my lab going while I also worked with Palade in New York, it was decided that I would commute, alternating between Harvard and Rockefeller on a weekly basis (the bus cost $9 and the air shuttle $12). To begin, Palade and I had to choose a topic. In those days it was obvious that all of pathology had to be restudied at the level of electron microscopy. Very little had been done, and, furthermore, there were many skeptics who believed that it was unwise to undertake the study of pathology at the ultrastructural level when normal tissues were still debated or unknown. So choosing a topic was an awesome task—of the kind that makes the heart beat faster. Palade put it this way: he had little experience with nervous tissue (on which I was working), but he had already made some totally new observations on normal capillaries; I was a pathologist, and so why not work on the capillaries in inflammation? There was evidence that histamine and other chemical mediators of inflammation increased capillary permeability, but the mechanism was not known. Permeability, Palade pointed out, is a dynamic notion, so our experiments would be an exciting combination of morphology and physiology. It was a perfect choice.

The next problem was to find an experimental model. The basic plan would be to inject histamine into an appropriate tissue, which would make the vessels leak, and after half an hour we would fix the preparation as quickly as possible. Now, ordinarily Palade could not take the time to do the animal part of the work with all his collaborators, but fortunately he decided to be with me for this initial, practical choice, and his contribution was decisive.

We sacrificed a rat and began looking for a proper anatomical structure, preferably a thin membrane, which would allow quick fixation, a critical requirement for electron microscopy. The eyelids were a possibility, but they were not ideal. I suggested the mesentery, historically the classic model for studies on inflammation in vivo, but Palade pointed out that it contained few blood vessels, a major drawback for studies at the very high powers of electron microscopy. So we decided to look into the internal organs, and I began the dissection as usual: using toothed forceps, I picked up a fold of skin over the upper chest and then cut the base of the fold over the chest, the abdomen, and all the way down over the scrotum. In this way, the testes were exposed, and we noticed that each one was wrapped in a very thin muscular membrane continuous with the abdominal muscles. Palade picked it up with his forceps and said, "The cremaster!" I had never noticed that in rats, each cremaster is a thin sac holding the testis (in humans it is rather like a cord), but Palade's previous experience as a gross anatomist allowed him to identify it instantly. The cremaster proved to be the dream preparation; it is uniformly thin (about 1 millimeter), easy to fix from both sides, and easy to mount between a slide and a coverslip for light microscopic study. Eventually, it was also adapted for studies of the microcirculation in vivo. I would be working on it for the next forty years.

Before I could produce the first electron micrograph, I had to learn all the techniques involved. I was often reminded of one of Dinu Lipatti's favorite sayings, which seemed to apply also to scientists: you must be an accomplished artisan before you can be an artist. The laboratory was full of young and dedicated research fellows with different degrees of seniority, the most experienced helping the recent arrivals. Much of our work was done during the night. The problems at many levels included fixing properly, embedding in methacrylate, preparing glass knives, staining the grids with lead, photographing through the electron microscope, and distinguishing results from artifacts. Nothing was "standard" and little could be bought, so much had to be improvised. In the meantime, Palade would warn us against excessive zeal to make improvements. "In electron microscopy," he would say, "the odds are against you. Do not change anything that seems to work."

While fixation was carried out according to the technique devised by Palade in 1952, embedding was more problematic. Methacrylate tended to produce bubbles as it polymerized, and this made the cutting impossible. We tried polymerization under ultraviolet light, overnight, or under vacuum in big, thick vacuum jars. Eventually somebody suggested that we combine the two methods, which meant placing the specimens in a vacuum jar and then irradiating them through the lid of the jar. This seemed to work, sort of, until a physicist who happened to walk by asked why we were doing this. Had we never learned that ultraviolet light does not go through glass? We were quite embarrassed. And so we struggled, day by day, in a stimulating, electrifying atmosphere.

Eventually I managed to produce some reasonable (or so I thought) ultrathin sections of cremaster in methacrylate, and with some trepidation I showed my first electron micrographs to Palade. He studied them very carefully, as was his style, and after a pause he commented quietly, "This looks like tissue that has been fixed at a pH of 2.5." Note his gentle touch. Instead of telling me that the tissue looked poorly fixed and horrible, he gave me a precise value of the pH to which it corresponded, thus making the mess sound like a real achievement. I went back to work with renewed enthusiasm. That is, until one day Palade knocked at the door of the electron microscopy room where I was working, came in, and gently delivered a shattering message in French (we spoke French in those days). We had been scooped, or so it seemed. A researcher in the department of H. Stanley Bennett in Seattle, John F. Alksne, had applied histamine to the skin of mice and studied the capillaries by electron microscopy.[1] He had concluded that histamine appeared to accelerate the normal process of transport across the capillary wall, namely, transport by the "Palade vesicles" as we called them at the time. It looked like a good paper, and in a roundabout way it actually sounded flattering for Palade because it extended the function of his recently discovered vesicles. "Now we have a choice," Palade said. "We can either change subject, or keep going and hope that we will find something different." We chose the latter path, and once again it was the right decision. So I kept looking. Because physiologists maintained that histamine increased the permeability of the capillaries and because Alksne seemed to confirm it, I concentrated on the smallest vessels that I could identify under the electron microscope.

But no matter how much histamine or serotonin I injected, the capillaries were always structurally intact. There was no trace of increased permeability or malfunction of any kind; it was a strange puzzle. Palade proposed another approach, to inject intravenously a colloidal suspension of tracer particles visible by electron microscopy to see whether we could follow them as they passed through the endothelium. Two possibilities were carbon black (India ink) and colloidal mercuric sulfide. The latter, which had been used by Alksne, was a leftover of the mercury therapy for syphilis; the colloidal particles appeared as black dots on the electron microscope screen, but they had the annoying tendency to disappear while I looked at them because mercuric sulfide sublimates in the vacuum of the column. Carbon black had no such problem.

We began our experiments by injecting first the intravenous colloidal marker and immediately thereafter the subcutaneous dose of histamine. Under the electron microscope, huge deposits of the colloidal marker appeared in some vessels, and at that point we began to see the light: the deposits appeared in the small veins (the venules) and spared quite specifically the capillaries. The mechanistic explanation was that gaps developed between the endothelial cells in the venules, while the basement membrane remained intact; plasma poured out, and the particles of colloidal marker were retained by the basement membrane—like coffee grounds in a filter. Mixed in with the particles of carbon black or mercuric sulfide were also gobs of low density material that turned out to be chylomicrons.

There was an unpredicted bonus. Carbon black is so intensely black that even the smallest deposits in the wall of venules became visible by light microscopy and even to the naked eye. We referred to the blackened venules as "labeled vessels," and thereafter the intravenous injection of carbon black was used for identifying leaky vessels after all manners of insults. Thus, the technique of vascular labeling was born, which became a standard tool for the study of inflammation. The rat cremaster became an ideal tool for studies of vascular leakage because it could be made transparent with glycerin, mounted in glycerin jelly (a technique perfected by Gutta Schoefl), and studied by transillumination. This made it possible to recognize at a glance, and without any need for cutting sections, whether leaky vessels were arterial, capillary, or venular.

While all this was going on, we had lots of visitors (I recall especially Howard Florey, who was also working on the electron microscopy of inflammation). We were quite aware that the new science of cell biology was being born primarily at Rockefeller, and we realized that there were three or four other competing laboratories elsewhere, but our bias was that the underground of the Smith Building was the mecca for anyone working in this field. Visitors came primarily to learn the latest advances and to gather technical hints. The unwritten rule was to be open. Palade received everybody most courteously, shared his experience, and

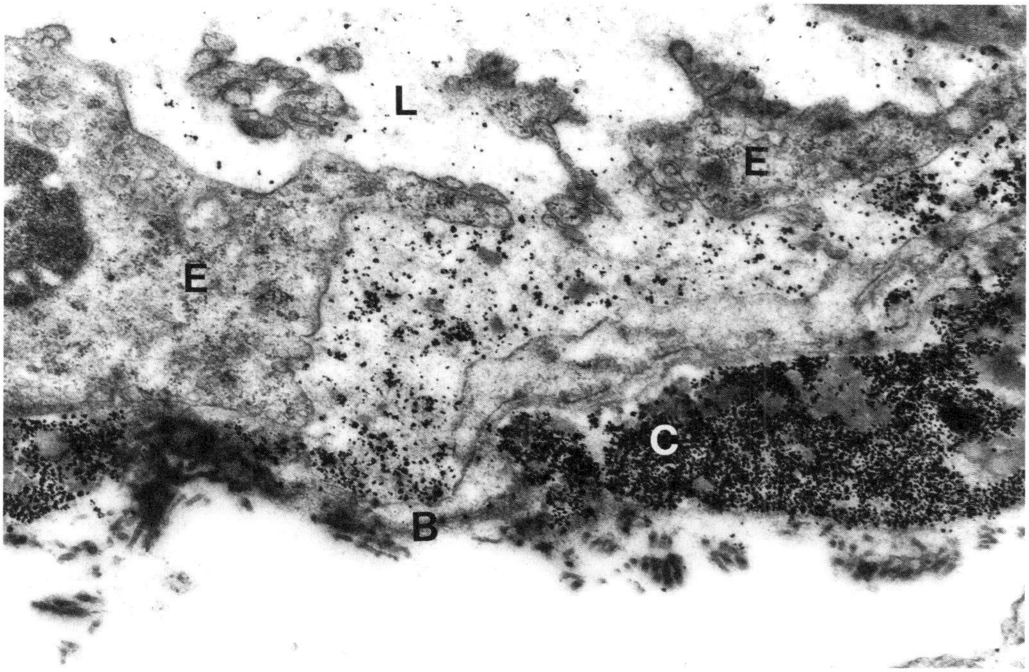

1. Electron micrograph, 1961, showing the wall of a venule three minutes after a local injection of serotonin. L, lumen. A gap has developed between two endothelial cells (E). The black particles (C) represent colloidal mercuric sulfides that are being retained by the basement membrane (B). Courtesy of Guido Majno.

was extremely kind. When visitors came to show their electron micrographs, he always had encouraging comments.

One day the visitor was a foreign scientist, who told Palade that he had been the first to study the lung with the electron microscope, and to prove it he produced a stack of electron micrographs. They were awful, and certainly no improvement over light microscopy; they looked as if somebody had just painted with black ink on photographic paper. But the author was obviously proud of them and finally asked Palade, "What do you think of this work?" As we stood around, we felt that this time Palade had no choice. It was clearly impossible for him to say anything positive, and we were terribly curious to see how he would cope with the situation without compromising his standards. So he looked at the prints very intently, and finally broke the silence. "Well," he said, "now one can say that . . . this has been done." It was a great lesson in kindness.

Eventually he and I produced two manuscripts of our work, one on electron microscopy[2] and the other on the light microscopy of vascular labeling.[3] Figure 1 is a typical example of venular leak induced by serotonin as we saw it in 1960. By today's standards, the electron microscopy is barely acceptable, but the cellular events are obvious.

We sent the manuscripts to Rockefeller's *Journal of Experimental Medicine*. To our surprise, they were both turned down. I received a letter from Peyton Rous (an almost Olympic authority for any experimental pathologist) to the effect that he had a lot to say about those manuscripts and he wanted to talk to me about them. Of course, I ran upstairs to see him. Rous was in a grumpy mood. He recalled that he had worked for years on vascular permeability, using a blue dye (too small to produce vascular labeling), and he mentioned with some irritation that he had never thought of injecting carbon black, thereby missing the discovery of vascular labeling. Turning to our manuscripts, he said that our descriptions of electron microscopy were much too detailed and our quotations from papers published in the 1800s would be of no interest to their readers. If we wanted to publish in *JEM*, we would have to cut out the history. This, he concluded, was the unanimous decision of the four editors.

"But Dr. Rous," I mildly objected, "I distinctly recall a paper of yours from 1930 which begins by quoting Metchnikoff." "Yes, but that was 1930," he replied sternly, and "times have changed." I was quite taken aback and discussed the matter with Palade, who, equanimous as ever, was not too surprised and in fact was somewhat amused. "If you think that the historical references are important, stick to your guns," he said, "and the detailed descriptions are necessary because electron micrographs will be new to most readers." So I submitted the manuscripts to the new journal run by Keith Porter, *The Journal of Biophysical and Biochemical Cytology* (later renamed *The Journal of Cell Biology*; see Chapter 10B). Our papers were accepted essentially by return mail.

Later, I learned that at the prestigious *JEM*, all the manuscripts were read and evaluated by four scientists, including Rous and René Dubos. No matter how knowledgeable, these great patriarchs could not possibly review competently all the "cutting edge" papers submitted, but they thought that they could compensate for their lack of specific competence by the device of "unanimous agreement." Months later, I invited Dubos to Harvard Medical School to give a talk followed by a dinner including abundant French wine. After dinner, Dubos brought up the matter of our manuscripts. He remembered them very well, including the historical references. Rous, he said, was worried and even upset by the extremely fast rate of change in science and felt so overtaken by the wave of new work that any reference to the past made him feel uncomfortable.

In the end, our work was widely quoted (we mailed out about 700 reprints). The paper on labeling was reproduced in a volume on *Benchmark Papers in Vascular Biology*.[4] Vascular labeling became a standard technique for studying vascular permeability, and our experiments were repeated almost identically many times (they are very easy to do, the result is aesthetically pleasing, and it affords instant gratification because the labeling develops a few minutes after the intravenous injection of carbon black).[5] In essence, we never claimed to be the first to describe vascular labeling, for many researchers had noticed that colloidal particles injected intravenously became localized in the wall of small vessels.

I am happy that we were able to acknowledge the careful microscopic studies of our nineteenth-century predecessors. The earliest was probably Julius Arnold in 1875, who published an illustration incredibly similar to ours showing leaks *between* endothelial cells. The difference between earlier studies and ours is that we were able to explain the mechanism of the labeling, thanks to electron microscopy. Previous authors interpreted the localization of the carbon black particles as resulting from endothelial phagocytosis, and some continued to do so even after our papers appeared.

Another result of these two papers is that they ushered in the venules as key contributors to microvascular physiology and pathology. It was most surprising to find that the capillaries were specifically spared by all the inflammatory mediators that we tried. In 2000, there still was not a good explanation.

On returning to Harvard, our laboratory eventually managed to set up a preparation for the study of circulation in the living cremaster muscle[6,7] and made a movie in color of vascular labeling. Further developments from our group included electron microscopic evidence that the endothelial gaps induced in venules by histamine-type mediators are formed by cellular contraction,[8,9] the demonstration of endothelial contraction in vivo[10] and in vitro,[11,12] the demonstration of endothelial cell "tone" in vitro,[13] and the search for new colloidal labels.[14] After so much work on leaky venules, it was especially intriguing to find an experimental model of chronic inflammation that after several hours induces a phase of vascular leakage very specifically limited to capillaries.[15,16] Again, by 2000, the mechanism was still unexplained.

Oddly enough, the slowest to accept our findings on venular leakage were the physiologists, who continued to speak of "capillary leakage induced by histamine" for decades and probably continue to do so. After twenty-five years, one who believed in our findings was George J. Grega, and he graciously invited me to a physiologists' meeting at FASEB to discuss our old findings on the pathogenesis of vascular leakage induced by histamine. They let me talk, and after I finished they went on discussing the mathematics of endothelial permeability as if I had said nothing at all; they obviously had no receptors for morphology. In a last effort, Grega tried to convince his colleagues by publishing a paper in *The American Journal of Pathology* (!) on the correctness of the "Majno-Palade hypothesis," but he too referred to our obvious gaping endothelial holes as a hypothesis.[17] Clearly, for physiologists, visible facts do not count, and the truth must be revealed by numbers. But as Palade's lifetime work has proven, numbers not supported by structure cannot solve the riddles of biology.

Keith Porter's Monument

by Guido Majno

This is the story of a practical joke that I played on Keith Porter in 1959. It could have ruined my career in science, but fortunately everything worked out to near perfection.

In November 1958, a few months after joining George Palade's laboratory, I was busily cutting at the Porter-Blum microtome when an authoritative-looking gentleman walked in. He stretched out his hand and said, "I'm Keith Porter. Who are you?" I introduced myself as an Italian-born, Swiss-trained pathologist from Harvard. Then he told me he would soon be visiting my country to interview a young man in Pavia who wanted to join the laboratory.

"Well," I said, "if you go to Pavia, you must see the monument of Golgi at the university, since you proved the existence of the Golgi apparatus with the electron microscope."

"Golgi has a monument?" Porter replied. "Do you think that the man who discovered the endoplasmic reticulum will ever have a monument erected in his memory?"

That afternoon, I heard Porter telling this story to several colleagues, so I decided to do something about it. As I was going to Italy about the same time Porter was, I commissioned a monument to be made and arranged to have it set up next to Golgi's in time for his visit to Pavia. Giuseppe Enrini, a distinguished sculptor from the Academy of Fine Arts in Brera and a friend of my father, created a beautiful bas-relief in plaster of Paris. The life-sized bust was based on photographs I got from Porter's secretary; it was framed with a pattern of endoplasmic reticulum, covered with a resin to resemble marble, and inscribed in Latin on a plaque below:

Keith Porter monument, now at the University of Pennsylvania. Courtesy of Lee Peachey.

Porterius Canadensis
Lux Rockefellerii
Pater Reticuli Endoplasmici Immortalis

When the rector of the university learned of this monument, he called the anatomy department and learned Porter was quite famous, although he felt the monument was somewhat inappropriate. Nonetheless, on learning of Porter's arrival, the anatomy department turned out enthusiastically, even bringing along octogenarian Antonio Pensa, the only surviving pupil of Golgi. Pensa, who did not speak English, showed Porter the library and the beautiful hand-drawings of the Golgi organelles.

Afterward, Pensa took Porter to pay tribute to the monument of Golgi in the courtyard, where they posed for photographs. In the meantime, I was standing nearby in front of the flag-covered Porter monument. As I pulled the flag off to unveil it, Porter turned so purple that I thought he was going to have a heart attack. Pensa pointed to the words "Pater Reticuli Endoplasmici" and kept saying "molto giusto, molto giusto" (quite right, quite right).

Eventually, Porter went to feel the bas-relief with his fingers. "What are you doing?" I asked. "I am just trying to find out how permanent this is," he replied.

And so it came to pass that Porter had his monument. When all this was over, I shipped the bas-relief to Rockefeller. As far as I know, he kept it in his office and it followed him in subsequent moves, eventually ending up in Philadelphia. One day, he called to ask how much it would cost to have it cast in bronze because he was sure there would never be another monument to his memory.

In the end, I am no longer sure this was a practical joke. Porter took it quite seriously, and the monument itself has nothing funny about it. It was a well-deserved honor, and I still hope that somebody will have it cast in bronze someday before it crumbles into dust.

13B. Membrane Structure

Walther Stoeckenius

After escaping from a Russian prisoner of war camp in Poland early in 1946, I began to study medicine at the University of Hamburg at the age of twenty-five. Instruction in the first two years consisted mainly of lectures or following cookbook recipes in the few laboratory courses. Histology was the only course in which one had the opportunity to do a bit of exploring on one's own by browsing through a large collection of slides available in the Institute of Anatomy. I enjoyed it, and my professor soon asked me whether I might be interested in undertaking a research project on the differentiation of islet cells in the fetal pancreas. I accepted the offer, began to learn classical histology techniques, and produced my first publication, somewhat to the detriment of my further medical education.[1]

I was fascinated by the functionally and largely unexplained structural details visible in cells that could be observed with the light microscope and also frustrated by the resolution limits, the artifact problems, and the obvious limitations of pure morphology to reveal function. My next project tried to establish some connection between the function and morphology of nerve fibers using in vivo fluorescence microscopy. The question arose of whether or not the nerve axon was separated from the sheath by its own membrane, and this problem led me to an intensive study of the literature.

Fortunately, the library of the Institute of Anatomy had survived the war intact and was well equipped; microfilm copies of the missing recent literature were available without charge from the United States Library of Congress. Thus, I became familiar with the relatively recent polarized light and x-ray diffraction work of Wilhelm J. Schmidt and Francis O. Schmitt, the alternating lipid and protein layers on the molecular structure of myelin, and its implications for the structure of cell membranes. Going back to earlier studies, I learned about Ernst Overton's correlation of lipid solubility with permeability and Rudolf Höber's impedance measurements on erythrocytes, both indicating that a thin lipid layer covers the cell surface.[2,3] These papers, together with the measurements on surface films by Evert Gorter and François Grendel[4] and the Hugh Davson and James F. Danielli work with model systems,[5] greatly impressed me because they showed ways to obtain information about the molecular structure of cell components while minimizing the artifact problem. They also convinced me that the bimolecular lipid layer (the "bilayer" only two molecules thick) with adsorbed protein on its hydrophilic surfaces was the best available model for the molecular structure of the cell membrane, especially since the same structure was also obtained for myelin and appeared to be a layer of tightly packed membranes. This conclusion was soon to be confirmed by Betty Geren's electron micrographs demonstrating that the myelin sheath consisted of many layers of cell membrane wrapped around the axon,[6] which David Robertson developed into his general unit membrane model.[7]

This rapid progress was mainly due to the development of osmium tetroxide fixation and thin sectioning for electron microscopy worked out mainly at Rockefeller by Keith Porter and George Palade. The combination of these techniques allowed one to routinely use the high resolving power of the electron microscope for the study of cells, tissues, and organs. The microscope began to reveal a plethora of membrane-bounded organelles inside the cells as well. Even before this happened, what I learned in medical school physiology and pharmacology lectures about selective permeability and the just developing concept of cell surface receptors had already emphasized the importance of membranes as a basic constituent of cells. This further stimulated my interest, and I was hooked. The main problem was how to explain the function of membranes, e.g., excitation and impulse conduction in nerves or selective ion permeability in erythrocytes, on the basis of the molecular structure.

After completing my thesis and passing the final medical examinations, I continued my reading and did some research in pharmacology while doing clinical work as a *Pflichtassistent* (the equivalent of an internship in the United States). However, it was time to find a job, if possible, in research. This was difficult in 1952 because little research money was available and positions were scarce. Therefore, I jumped at the opportunity when my thesis adviser told me about an opening for a postdoctoral fellow to do electron microscopy at the Tropeninstitut in Hamburg, even though the head of the anatomy department, like many other biologists at that time, still dismissed electron microscopy as "nothing but the most expensive way to produce artifacts." So I spent the next two years learning the recently developing preparation techniques for electron microscopy and how to operate the microscope, which at that time usually required more time and effort to maintain the instrument in working order than time devoted to its use. When the electron microscope was operating, we investigated the structure of vaccinia virus, trying to establish the chemical nature of its structural components visible in electron micrographs by selective digestion with nucleases and proteases and contrasting the specimens by shadowing with heavy metals. Meanwhile, my interest in membranes was maintained by following publications on the subject.

I was also married and our first son was born, so it was time to look for a job to sustain the family. I had the choice of an assistant professorship in either the department of gynecology or pathology at Hamburg University. I chose the latter because it offered me complete freedom in the selection of my research topics. However, I was mainly occupied with autopsies, biopsies, and with teaching pathology to not very interested dental students. But I found time for some histopathology research and became interested in the lymphatic system related to the antibody response, which I studied using standard light microscopy techniques. Fortunately, Western Germany was rapidly recovering economically so that more money became available for research at the university, and so in 1956 the department was able to acquire a Siemens electron microscope, which was a vast improvement over the electrostatic AEG/Zeiss instrument I had used before.

At Siemens, Ernst Ruska was still in charge of electron microscope development and production. His latest model, the Elmiskop I, set a new standard for electron microscope performance. He also had established an applications laboratory at the factory in Berlin where new users received intensive training for three days before an instrument was delivered. The laboratory was run by Dr. Cilly Weichan in a very informal and efficient way. One had access to the physicists and engineers, could learn the latest tricks, and later, occasionally, could obtain newly developed parts to try out. It thus became possible to routinely produce electron micrographs, which, due to their high quality, attracted attention. The recently introduced use of the Porter-Blum microtome, glass knives, osmium tetroxide as fixative, and plastic embedding media allowed one with some skill and luck to obtain well-preserved thin tissue sections; any section reasonably free of compression and knife marks revealed a wealth of new and interesting structures.

At the time, the general existence of cell membranes had still not been accepted by all biologists even though the electron micrographs clearly revealed them. Sometimes they could be resolved into triple-layered thin dark lines and, unexpectedly, also around most cell organelles. These observations naturally intrigued me, but the work on membranes was not pursued because I did not immediately see an approach to correlate the triple-layer cell membrane structure in terms of the 1935 Davson-Danielli model that proposed a phospholipid layer between two layers of globular protein.

However, I could continue my work with the electron microscope on the lymphatic system using thin sections of spleens from rabbits that had survived a severe secondary immune reaction. I found in macrophages multilayered structures of regularly spaced concentric dense lines with a 4.0-nanometer repeat period. Based on my earlier reading on the myelin sheath, I guessed these to be myelin figures, i.e., tubular structures made of concentric bimolecular lipid layers, which form spontaneously when membrane lipids are in contact with water. To test this notion, I isolated a phospholipid fraction from brain and prepared thin sections of authentic myelin figures that showed the same structures. Myelin figures were known to consist of concentric bimolecular lipid layers. The contrast in the micrographs was likely due to the deposition of the heavy metal osmium, which I assumed was from the osmium tetroxide used for fixation and had been deposited at the fatty acid double bond where it cross-linked the lipid molecules. An obvious next step was to adsorb protein to the artificially prepared bilayer surfaces and thus create a physical model of the Davson-Danielli membrane. The added protein increased the width and contrast of the dark lines and made the myelin figures now appear like the closely packed cell membranes. The result obtained thus constituted a strong argument in favor of the Davson-Danielli model.

Obviously, this work had taken me rather far afield from classical pathology, but I wanted to continue it. My interest in pathology as practiced in most German university departments at that time had faded because I felt that most of the exciting progress in understanding disease processes had shifted to other disciplines. I also wanted more time for my research than the routine work allowed. Predictably, the department head did not appreciate my point of view, and our formerly good relationship deteriorated to the point where it was agreed that after my "habilitation" I would leave.

Shortly before this happened, I had met at a local histochemistry symposium Professor Frank Hartmann, known for pioneering the use of glass knives,[8] who had come from the University of Minnesota to spend a sabbatical leave at the University of Kiel and had been invited to give a seminar in Hamburg. During his visit, he could not have failed to notice the strain in my relationship with the department head. After I explained the situation to him, he asked whether I would consider coming to the United States, and he generously offered to recommend me to any group there. At the time, the laboratory of Keith Porter and George Palade at The Rockefeller Institute was the mecca of all biological electron microscopists, and unabashedly I named it as my first choice.

A few weeks later, a letter from Porter arrived offering me a one-year postdoctoral position to continue research on the ultrastructure of the liver, which an Italian postdoctoral fellow due to return home was pursuing in his laboratory. Professor Hartmann had also advised me to apply for a Fulbright fellowship, which was granted and essentially consisted of a first-class ticket from Bremerhaven to New York. After acquiring the necessary tuxedo, I embarked on the SS *America* on 28 December 1958.

For someone arriving from the backwater that Hamburg and its university had become after the war, New York and Rockefeller were overwhelming. The laboratory, still located in the third basement of Smith Hall, was crowded, and I could not do much until Porter had found a niche for me. So I had some time to venture forth from the plush surroundings of my temporary quarters and explore Manhattan, mostly on foot. Soon I settled in and work took over.

It was the dawn of a new era in cell biology. The combination of electron microscopy with biochemistry and cell fractionation was allowing one to connect structures and function from the light microscope level to the molecular dimensions. The Porter/Palade laboratory was at the center of these developments, and a steady stream of visitors was passing through, discussing their latest work in seminars and in the famous dining room that was still functioning. Much of the laboratory work was done at night—the microscopes were often signed up until three o'clock in the morning, and one might encounter Palade or Porter in the laboratory well after midnight. The library was open twenty-four hours a day, 365 days a year, and a cafeteria was also open until the early morning hours. It was a unique research environment.

In the summer of 1958, I presented my work on the myelin figures, the membrane model, and the chemistry of osmium tetroxide fixation at the Fourth International Conference on Electron Microscopy in Berlin, where, except for some nasty comments by Frank Schmitt, it was well received. I also briefly met Porter there, but he had found little time to talk to me, and my presentation had been in German. Shortly after my arrival at Rockefeller, he asked me to show him again what I had done, and after listening to my explanations in not exactly fluent English, he concluded our conversation with the remark, "You go on with what you have been doing, I'll find somebody else for the liver." A few weeks later he offered me an assistant professorship, and I stayed at what was just becoming The Rockefeller University for the next eight years, free to pursue my own interests. The situation did not change when Porter left for Harvard and Palade remained head of the laboratory.

I first wrote up the myelin work, but then the exciting environment at Rockefeller proved too distracting.[9] Electron microscopy was much in demand as an auxiliary technique by other groups, and my interest in viruses from the work in Hamburg led to collaborations with Rockefeller scientists Norton Zinder, Purnell Choppin, and several others. We explored the structural details of influenza and tumor viruses, especially their nucleic acid components and the structure of a new filamentous phage.[10–13]

However, my first two years were mainly occupied with expanding, refining, and correcting my bilayer model. While it constituted an independent confirmation for the presence of a lipid bilayer in biological membranes, it conflicted with David Robertson's model in detail.[7,14] I had interpreted the contrast distribution of the myelin figure images and assumed the dark lines represented the site of osmium cross-links between fatty acid chains at the site of double bonds, i.e., the hydrophobic interior of the lipid bilayers. In contrast, he assumed from the electron micrographs of the developing myelin sheath that the dark lines showed the membrane surface. Neither argument was conclusive, but I had to admit that his model was more attractive because mine implied that the cell membrane consisted of two lipid bilayers. The controversy endured for some time.

I found a solution to this nagging problem when I also collected x-ray diffraction data for the myelin figures in order to show that their lamellar structure survived the fixation, dehydration, and embedding procedures. That led to collaborations with Jack Schulman[15] and Vittorio Luzzatti[16] to investigate other liquid-crystalline lipid phases and microemulsions to confirm their deduced structures. During this work, it occurred to me that two hexagonal phases of liquid-crystalline lipid-water mixtures, which could be distinguished by x-ray diffraction, could be used to resolve the problem of contrast distribution in the lipid bilayers. One phase formed a hexagonal array of lipid rods surrounded by water, and the other was the inverted structure, a hexagonal array of water channels surrounded by lipid (figure 1). Therefore, one phase should appear in cross section as a net of hexagons and the other as a hexagonal array of black dots, depending on whether the hydrophobic fatty acid chains or the hydrophilic end-groups at the lipid-water interface produced the contrast. The experiments gave a clear answer: contrast was produced at the hydrophilic lipid surfaces.[17] Apparently, more osmium tetroxide was reduced and/or deposited there than at the double bonds. Porter rarely commented on the quality of my work, but my presentation of these results at a conference elicited the characteristic remark, "I see you also know how to reverse yourself gracefully." Porter would not let an opportunity pass for making a comment, usually friendly but occasionally needling.

1. The arrangement of lipid and water in the hexagonal phase as seen in cross section. The hatched areas represent the spaces occupied by water. The symbol at the bottom is used to represent a phospholipid molecule; the circles mark the sites of the hydrophilic groups. From Stoeckenius, 1962. *J. Cell Biol.* 12:221–229.

The controversy with David Robertson was thus resolved in his favor. However, it became increasingly obvious that to account for the function of biological membranes, a revision of the Davson-Danielli model was required. Trying to connect morphology and function, I selected the erythrocyte ghost as the best-known example of a cell surface membrane and the mitochondrial inner membrane as an example of a highly specialized intracellular membrane with one dominant function, from which functional complexes could be isolated. I expected the basic building principles would be the same. The erythrocyte membrane yielded few results, so I concentrated on the inner mitochondrial membrane. Working with it resulted in contacts and collaboration with Efraim Racker's group at the New York Public Health Research Institute and David Green's group at the University of Wisconsin. I necessarily became involved in the just-beginning battle between the proponents of the chemical and chemiosmotic hypotheses of oxidative phosphorylation, although I did not materially contribute to the controversy until much later. I began to favor the chemiosmotic mechanism rather early, probably in part because I was not biased by a thorough education in biochemistry, but mainly because I found some of the early evidence quite convincing and had an inborn preference for physics over chemistry.

The work led to a modification of the general membrane model. Becca and Sidney Fleischer came from Green's laboratory to work for a year with me. They had shown that totally lipid-depleted mitochondria could be functionally reconstituted by the addition of lipids and wanted to know whether they had reconstituted vesicles as I predicted. We found that they maintain the unit membrane structure of their cristae even in the lipid-depleted state, which led to the conclusion that at least some of the proteins in the inner mitochondrial membrane must span the bilayer to hold the surface protein layers together.[18] This constituted a significant deviation from the classical Davson-Danielli model, which confined protein to the surfaces, but it still retained its central feature of the continuous lipid bilayer. It also made it easier to explain and devise a reasonable test of membrane functions.

In the early sixties, a general dissatisfaction with the Davson-Danielli model spread, prompted by the model's inability to explain specific membrane functions. New models were proposed that discarded the lipid bilayer. They were mainly variations of a basic "subunit model" that depicted membranes as planar arrays of lipoprotein particles. The situation was not helped by the transient appearance of an ill-defined "structural protein" postulated to be present in all membranes. Neither the arguments nor the sparse to nonexistent experimental support for these models looked convincing to me. I did not see how these structures could provide the necessary permeability barrier and continued to favor the bilayer as a central feature of all membranes.

However, in 1963, Austin D. Brown published a paper on halobacteria, the prokaryotes found in saturated brines. He reported that *Halobacterium halobium* does not possess the typical bacterial cell wall; also, its cell membrane disintegrates into a uniform population of macromolecular subunits when the salt concentration in the medium is reduced.[19] The data he presented seemed to support the conclusions of a subunit structure for this membrane. I decided I had to look at it myself. Fortunately, shortly thereafter, Bob Rowen, a microbiologist from the Albert Einstein College of Medicine, came to spend a sabbatical year with me and agreed to start the project. We soon found that the standard preparation techniques for electron microscopy gave poor results with halobacteria but some modifications would preserve the structure well.[20] The micrographs then clearly showed a cell wall and a typical cell membrane, and in the cytoplasm there were surprisingly short membrane pieces with free ends having a typical unit membrane structure. The cell wall, as we showed later, was the source of Brown's subunit fraction.[21] However, the cell membrane did disintegrate in low salt into fragments of different sizes and compositions, and these could be separated in a centrifuge. One of the fractions had a striking deep purple color, another was orange, and a third was nearly colorless.

These results restored my temporarily shaken faith in the universal applicability of the bilayer model. However, I continued work on the halobacteria, partly because the cytoplasmic membranes were intriguing, but mainly because this cell membrane seemed to offer a unique opportunity for resolution and eventual reconstitution of a multifunctional membrane under very mild conditions. This was at a time when the art of solubilizing functional membrane components with detergents was still in its infancy. Influenced by the data on permeability of lipid bilayers obtained with the recently introduced "black lipid films," a method for reconstituting lipid membranes under water, I began to develop a concept of membrane organization that was eventually published in a review article with Don Engelman.[22] We postulated that the lipid bilayer forms the structural backbone for membranes and provides the basic permeability barrier. Bound to and inserted into it are other components, mainly proteins and glycoproteins, that contribute the selective transport functions, energy-transducing systems, and receptors. Generalizing from the mitochondria work, I assumed some of these functional components would form complexes segregated in domains, and I hoped to directly demonstrate such an organization in the halobacterium cell membrane. The concept proved to be quite useful and also led later to isolation of the gap junctions as another such domain.[23] Also intriguing was the unprecedented observation of membrane sheets in the cytoplasm of the halobacteria. They showed the typical triple-layered unit membrane structure and a regularly striated surface when negatively stained.

The halobacteria membrane fractions still contained several morphological components. They were first purified and characterized by Wolf Kunau, a postdoctoral fellow from Ernst Klenk's department at the University of Cologne. He isolated clean orange fractions, prepared a purple fraction free of intracytoplasmic membranes, and removed the contaminating flagella fragments from the colorless intracytoplasmic membrane fraction. None of the fractions contained vesicles. The orange fraction consisted of cell membrane pieces of irregular size and shape, but its lipids could not be hydrolyzed by standard techniques. Fortunately, Morris Kates in Ottawa had just begun to investigate the chemistry of these very unusual lipids and showed that their branched, fully saturated hydrocarbon chains were linked to the glycerol component via ether bonds, thus resolving this puzzle. The purple fraction contained only one main protein, and we named it the "purple membrane." We got some more surprises. Whereas the orange fraction appeared fairly conventional, except for its unhydrolyzable lipids, the purple membrane contained only twenty-five percent lipid and could not be resolved clearly as a triple-layered structure. The intracytoplasmic membranes looked like typical triple-layered unit membranes in sections and contained only protein but no lipid. We would finally identify these as collapsed gas vacuoles with a very intriguing structure and interesting properties that still merit further investigation.[24]

When I first took x-ray diffraction patterns of a concentrated purple fraction to check for bilayer structure, I found to my surprise also a series of sharp reflections indicating the presence of a two-dimensional, well-ordered crystalline lattice in the plane of the membrane. We could not attribute with certainty the lattice to the purple membrane until we obtained a clean fraction. The purple color was shown later with Dieter Oesterhelt to be due to a retinal chromophore in the protein that closely resembled rhodopsin and apparently formed a two-dimensional crystalline lattice in the plane of the membrane. We named it "bacteriorhodopsin." It turned out to be a light-driven proton pump and the main component of a primitive photosynthesis system not based on chlorophyll.[25,26] The lattice was sufficiently ordered to eventually yield the first structure of a membrane-spanning protein at near-atomic resolution.[27–29] Bacteriorhodopsin, together with another light energy transducer, "halorhodopsin," and two light sensors, "sensory rhodopsin I and II," were all found in the same organism and would occupy most of my attention for the next twenty-five years. Halorhodopsin is a light-driven chloride pump, and the sensory rhodopsins govern phototactic reactions of the cells. All four bacterial rhodopsins are closely related proteins.

In 1966, I received a very generous offer from Julius Comroe for a professorship at the University of California Medical School in San Francisco and its Cardiovascular Research Institute, which, in addition to a large, well-equipped laboratory, included $150,000 a year for the next seven years to be used entirely at my discretion. This, together with the attractions of California, proved irresistible, and after eight years I left Rockefeller for California, where I changed fields and started research in photobiology. The unrestricted funds for the first seven years may also have facilitated this work considerably. It probably would have been difficult for an investigator with no proven competence in the field to obtain grants for studying rhodopsin in a bacterium and to look for photosynthesis in the absence of chlorophyll without substantial evidence already in hand.

Looking back, I can see that the eight years at Rockefeller not only brought me up to date in molecular and cellular biology but also taught me some lessons in the sociology of science. The tone in the laboratory was always pleasantly informal and collegial with frequent parties and mostly good-natured practical jokes. For example, the lab had been given a small Czechoslovakian electron microscope for testing, and it was set up by Monty Moses. One day during lunchtime, the power failed, and coming back he found that the oil of the forepump had risen through the small diffusion pump and filled the column to almost the top of the viewing chamber window, whereupon a plaque appeared at the microscope proclaiming it the historic, first oil immersion electron microscope, and we advised him to patent it. Instead, Monty carefully took it completely apart for cleaning and spread out the parts in proper order on a long laboratory bench. During another lunch hour, someone leaving late for lunch added a few screws and other small pieces from the parts bin. Predictably, these were left over after reassembly, and Monty did not know what to do with them. He finally decided to take photographs of the parts and send them to the factory. We never learned whether he got an answer, but after the microscope, which never produced acceptable micrographs, was returned, he received from the manufacturer a pocket camera as a "thank you." He learned the origin of these spare parts at his farewell party.

I fondly remember my time at Rockefeller as a unique and generally very pleasant experience in an exciting environment. However, I also discovered that competition in the general scientific community was fierce. My rather limited experience as a small light among the German community of pathologists and the even smaller group of early electron microscopists had not prepared me for what I saw now. The amount of effort expended and some of the methods used in the struggle to establish priority or to acquire all possible credit at the expense of others shocked me. Some advice from Palade and Porter in this respect was very necessary and—with some delay—appreciated. Fortunately, little of this was noticeable in my immediate surroundings, and it did not compare with what I witnessed in some other laboratories. There was competition, of course, but amazingly few conflicts arose in this group of highly ambitious researchers. The authority of Porter and Palade was never questioned.

The Rockefeller years were a very important period in my career. I developed there the concept for the general structure of membranes with the lipid bilayer functioning as the basic permeability barrier and its organization into functional domains. This result also provided the starting point for the work of the following twenty-five years. The purple membrane was discovered at Rockefeller in attempts to characterize such domains, and my involvement in the investigations of oxidative photophosphorylation prepared me for exploration of its function as a light energy transducer. The collaboration with Efraim Racker resulted later in our demonstration of proton gradient–driven ATP synthesis in lipid vesicles containing bacteriorhodopsin and mitochondrial ATPase, now widely considered the final proof of the chemiosmotic mechanism.[30] All this must to a large part be attributed to the congenial environment I found at Rockefeller.

13C. Glomerular Permeability and the Discovery of Tight Junctions

Marilyn G. Farquhar

Looking back on my career, I often reflect upon how chance directed its course and drew me, step by step, into the field of cell biology.

During the time I was growing up in California, I always dreamed of being a doctor and treating patients. I doggedly pursued this and in 1949 became a medical student at the University of California, San Francisco (UCSF). However, after a year and a half, I left medical school, having decided that clinical work was not for me and that research and teaching might be a more appropriate career. To follow this course, I became a graduate student in experimental pathology at UCSF and was accepted by the professor and chair of pathology, James Rinehart, to become a graduate student under his guidance. Soon after joining his laboratory, he told me that he had an electron microscope (EM) in the basement (an RCA EMU-2)—the only one at UCSF at that time—and that he wanted me to work with it and develop techniques to apply it to the analysis of pathological specimens.

So here was the first instance of how chance played a role in my career by directing me into the new field of biological electron microscopy. I had no idea at the time of the unusual challenges as well as the opportunities ahead. In 1952, the tools available for preparing tissue specimens for examination in the EM were rudimentary. We had available methods for plastic embedding and improved fixation with osmium tetroxide, developed by George Palade. We cut tissues on regular microtomes equipped with crude, special advance mechanisms. Yet it was a time of rapid progress and technical development. During my early graduate career, I mastered the preparation of glass knives, introduced by Harrison Latta and Frank Hartmann.[1] I also ordered from Sorvall and was the first at UCSF to use the Porter-Blum MT-1 microtome developed by Keith Porter at Rockefeller. I remember the excitement when we received the new instrument: I sat down and cut sections of the anterior pituitary gland and looked at them in the EM, all in the same day. These first specimens to roll off the glass knives were thinner and more revealing than any we had prepared before. There was a whole new world to look at! Particularly impressive to me at the time was the detailed organization of the Golgi and secretion granules in the anterior pituitary. I applied these new tools in my PhD work to what turned out to be the first studies on characterization of cell types in the anterior pituitary and organelles, especially the Golgi apparatus, involved in secretion in this gland[2,3] as well as the first studies of the pathology of the kidney glomerulus, research that I still pursue.

I had not quite finished my thesis work when a second chance development intervened in my career. When my husband, John Farquhar, decided to complete his residency in medicine at the University of Minnesota, this provided the opportunity for me to spend a year and a half working in the anatomy department with Frank Hartmann, one of the early pioneers in electron microscopy who, as already mentioned, had developed the glass knife. From him I learned much more about the instrumentation as well as preparatory techniques for the EM and even taught electron microscopy to many faculty members. My most famous student was Joe Gall, then assistant professor in the zoology department at Minnesota. In addition, I had the opportunity with Robert Good and Robert Vernier of the pediatrics department to examine with the increased resolution provided by the EM the first renal biopsies obtained from children with kidney diseases. This was an unprecedented opportunity because the changes in the components of the glomerulus lay just below the resolution of the light microscope and could easily be appreciated at even low magnification in the EM. At that time, the general structure of the glomerular capillaries had been delineated by electron microscopy as a result of the studies by Daniel Pease,[4] B. Vincent Hall,[5] and others, and it was already known that the capillary wall consisted of three distinct layers: the endothelium, the basement membrane, and the epithelial foot processes. However, the changes that occur in glomerular diseases were unknown.

Out of that work came the first descriptions of renal lesions that occur in children with diseases of the glomerulus.[6] Especially intriguing and novel was our description of the changes in the epithelial foot processes, known as foot process "fusion," which is pathognomonic of patients with the nephrotic syndrome.[7] After this brief sojourn in Minnesota, I returned to UCSF, completed my PhD degree in pathology, and spent a few years on the research faculty of that institution carrying out further studies of pituitary secretion and glomerular pathology as seen at the EM level, which included some of the first studies on the Golgi origin of secretory granules[8] and the first studies on diabetic nephropathy.[9]

1959–1962: The Rockefeller Institute for Medical Research

In 1959, chance intervened again when my husband decided to spend a postdoctoral fellowship working at what was then The Rockefeller Institute for Medical Research. This provided an opportunity for me to also extend my research career, and I was lucky enough to be offered a position as a research associate working with George Palade in the Porter/Palade laboratory. Thus, I found myself quite unexpectedly in the American cradle of cell biology.

Space in the Porter/Palade lab was very precious. I remember that on the day I arrived, I was offered a choice of a desk either in the hall or in the dishwashing facility. I chose the desk in the dishwashing facility, where I worked back to back with Eichi Yamada, and Guido Majno, the next to come, was assigned the desk in the hall. To me, progress seemed to be slow because as a working mother with two small children, my time in the lab was limited to eight hours a day, three days a week, as compared with the others who literally worked night and day. Also, each of us did all our own work, making specimens, cutting them on the microtome, and examining them in the EM. Here I had an advantage over the others because of my prior experience, whereas most of the others had to learn all these techniques from scratch.

Palade and I agreed to study glomerular permeability, as it represented a logical mergence of our interests. He had developed an interest in capillary permeability,[10] and my familiarity with the renal glomerulus provided an opportunity to extend this interest to the unique capillaries of the glomerulus. I applied for and received my first grant from the National Science Foundation to work on this topic. Given the developments in biological electron microscopy, there was an unprecedented opportunity to go beyond descriptive studies and define the functions of these different layers in glomerular filtration and eventually to understand the detailed pathogenesis of glomerular diseases.

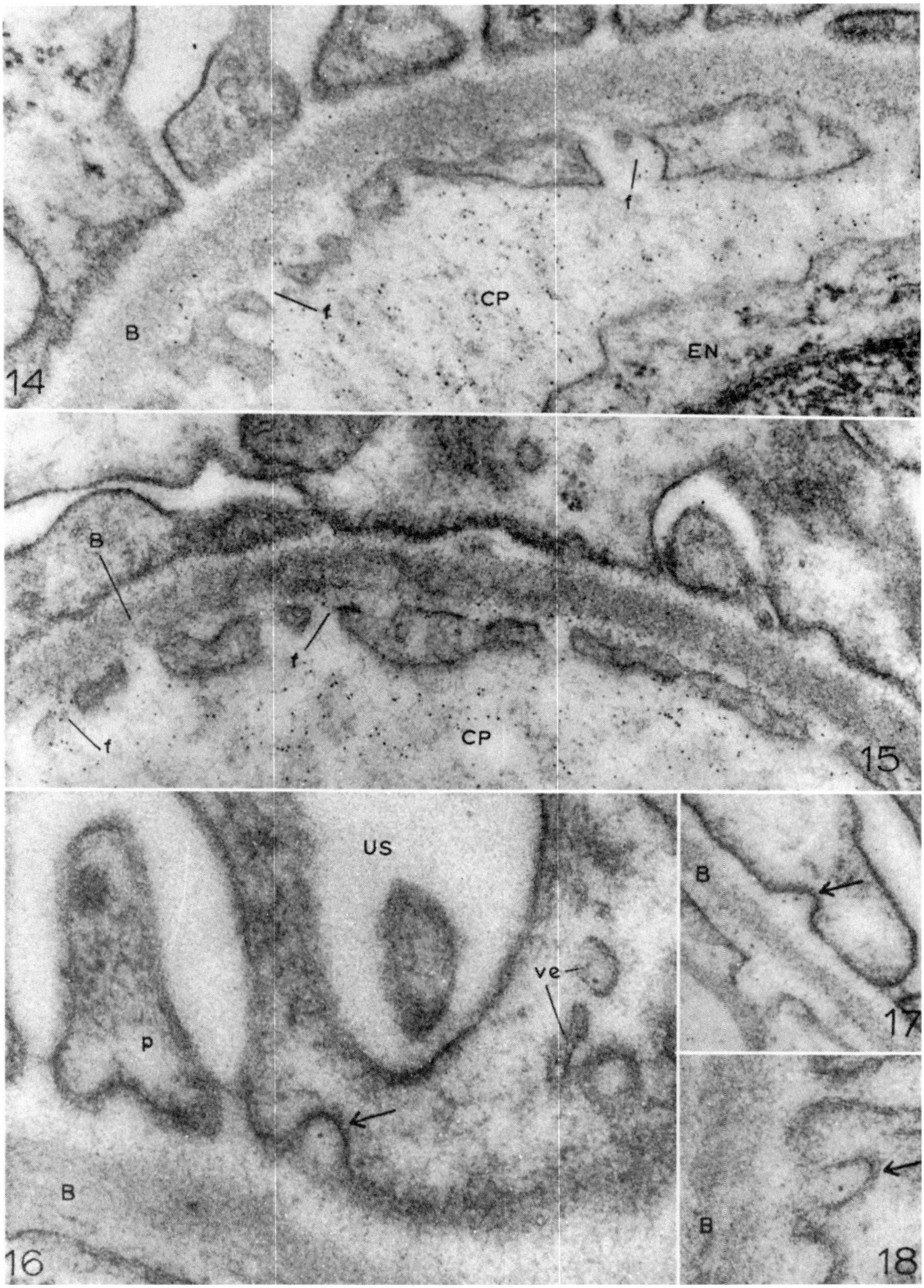

1. Demonstration that the glomerular basement membrane represents the main filtration barrier for large proteins in normal glomerular capillaries. After intravenous injection, most of the ferritin is retained in the lumen of the glomerular capillaries (CP) and does not reach the urinary spaces (US). Ferritin penetrates the fenestrae (f) of the endothelium (EN). Only small amounts cross the basement membrane (B) and are taken up by endocytosis (arrows) in small vesicles (ve) by the glomerular epithelial cells (podocytes). From Farquhar et al., 1961. *J. Exp. Med.* 113:47–66.

Most electron microscopists believed, based on the organization of the glomerulus, that the glomerular basement membrane (GBM) served as the main filter to retain plasma proteins in the circulation. However, the other hypothesis that had surfaced[5] was that the spaces between the foot processes of the podocytes act as "slit pores" that allow free passage of water, small molecules, and electrolytes but restrict the passage of proteins. What was needed was direct experimental evidence on this point as well as to try to bring together physiological studies and the "pore theory" of John Pappenheimer and his colleagues[11] with the morphologic reality that the filtration surface of the glomerular capillary wall did not contain recognizable pores. Instead, it is composed, as already mentioned, of three components—endothelium, basement membrane, and epithelium. What we set out to do was to obtain information on these points by directly testing the permeability of the glomerular layers to proteins by injecting an electron-dense tracer and following its passage across the glomerular capillary wall by electron microscopy. For this purpose, we selected ferritin and pioneered the use of ferritin as an electron-dense tracer. Its globular nature and its structure, with its central dense core of iron rendering it electron dense surrounded by a protein coat, made it ideal for this purpose.

In those days, there was no ferritin in the Sigma catalogue, so we had to prepare our own from horse spleen up until such time as a commercial source became available. Initially, the ferritin we prepared was quite toxic, and intravenous injections killed the rats within a few minutes due to impurities. But eventually, after some months and with the help of Sam Granick at Rockefeller, we succeeded in producing a preparation that could be injected intravenously and was tolerated without problems by the animals. Then we had to work out fixation procedures for retaining the ferritin in situ, which again took some months. But the results were very rewarding: the first successful experiments clearly demonstrated that the injected ferritin penetrated the endothelial fenestrae but did not freely penetrate the GBM. Instead, it accumulated between the endothelium and the GBM (figure 1). Intriguingly, small amounts of ferritin did pass through the GBM and were taken up in small vesicles and segregated into dense protein droplets, now known to correspond to lysosomes. From these observations, we concluded 1) that the endothelial fenestrae are patent, 2) that the GBM is the main filtration barrier to proteins, and 3) that the glomerular epithelium acts as a monitor to partly recover protein that leaks through the GBM. We even postulated that the GBM is not a simple sieve but instead has a gel-like structure, which has proven to be the case.[12] These findings were published in two papers,[13,14] one of which was recently made a landmark in nephrology[15] because it provided the first direct evidence on the functions of the components of the glomerular capillaries in filtration of proteins and because the observations have stood the test of time.

But this was just the beginning. There were a number of subsequent spin-offs of this work. First, we followed up the paper on the normal glomerulus with a second paper in which we used ferritin as a tracer in the nephrotic glomerulus (puromycin aminonucleoside [PAN] model of renal disease).[16] Again, we injected ferritin intravenously and followed its passage across the glomerular capillary wall over time by electron microscopy. We chose to study the nephrotic glomerulus because in this disease there were major changes in the organization of the GBM and especially in the epithelium, where the usual foot process organization was replaced with a continuous layer interrupted only periodically by filtration slits (i.e., foot process fusion). Again, we obtained findings indicating that it was the main filter, but it

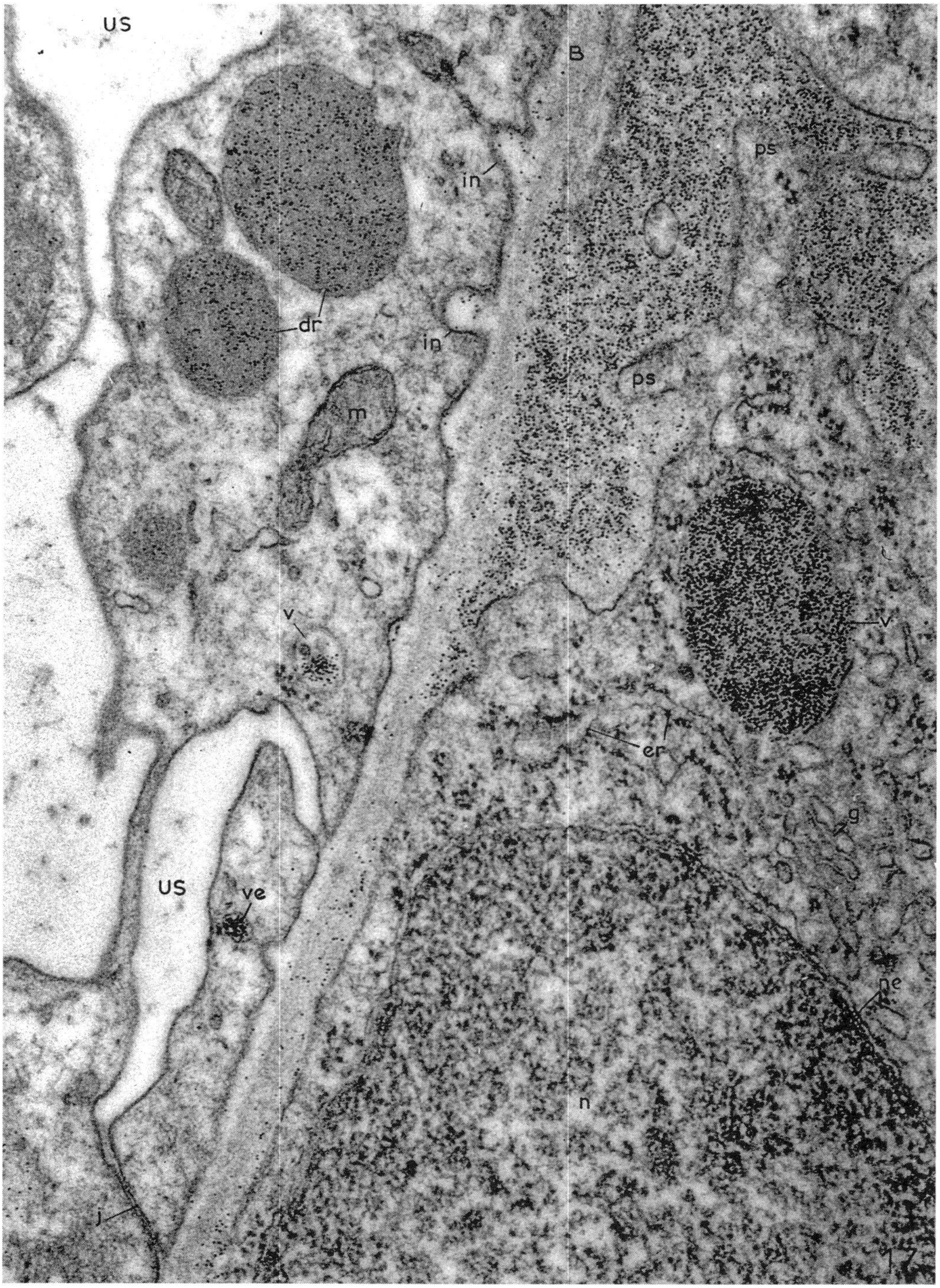

2. The amount of ferritin that crosses the basement membrane and is taken up by the glomerular epithelium is dramatically increased in nephrotic rats. While in normal glomeruli (figure 1), most of the ferritin is retained in the capillary lumen, in nephrotic animals, increased ferritin is found throughout the basement membrane (B) and in endocytic vesicles (ve) and invaginations (in) of the epithelial cell membrane facing the basement membrane. There is a dramatic loss of the epithelial foot process organization and a narrowing of the filtration slits, which are converted into tight junctions (j). From Farquhar and Palade, 1961. *J. Exp. Med.* 114:699–716.

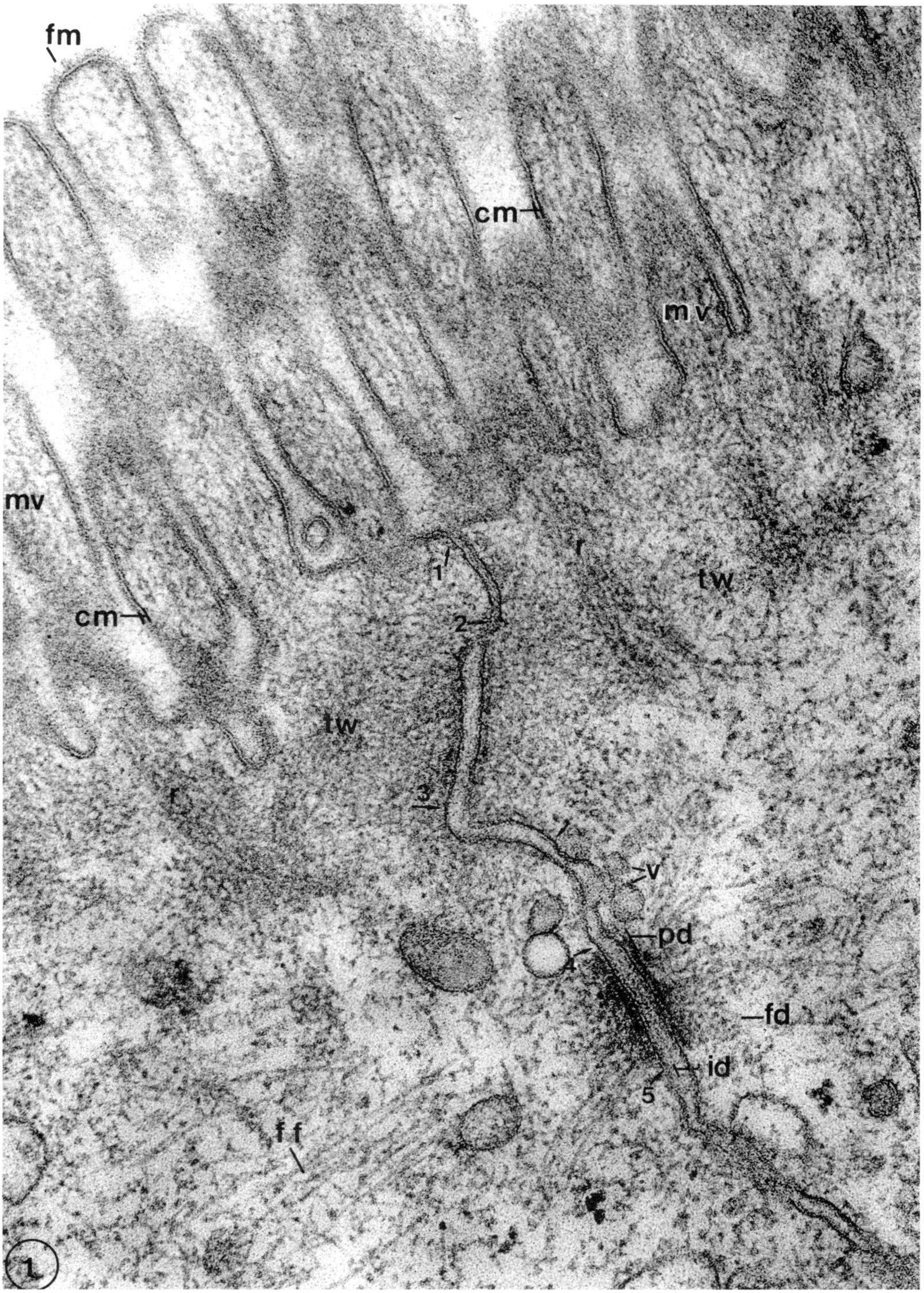

3. Junctional complex between two cells in the epithelium of the intestinal mucosa. The three classical components of the complex are shown: the tight junction or Zonula occludens (between arrows 1 and 2), the adhering junction or Zonula adherens (between arrows 2 and 3), and the desmosome (between arrows 4 and 5). Note that the intercellular space is closed along the tight junction and open at the level of the adhering junction and desmosome. mv, microvilli; cm, cell membrane. From Farquhar and Palade, 1963. *J. Cell Biol.* 17:375–412.

was more leaky, allowing increased tracer to cross (figure 2). As a result, endocytic uptake of tracer by the glomerular epithelium was dramatically increased. This study was particularly satisfying to me, as it allowed me to return to glomerular pathology and to obtain initial insights into the pathogenesis of the nephrotic syndrome at the EM level. Thus, these two papers on glomerular permeability provided the first direct evidence pinpointing the GBM as the site of the main glomerular permeability barrier and the first evidence of protein absorption by the glomerular epithelium.

The second spin-off of the ferritin work was the discovery of the activity of the "deep endothelial cells," now called mesangial cells, in the uptake of filtration residues that accumulate against the GBM. By following the fate of ferritin that accumulated against the GBM by electron microscopy, we noted that it was taken up by endocytosis into these cells and delivered to lysosomes for disposal. This led to a third paper[17] describing the morphology of mesangial cells delineating for the first time their role in clearance of filtration residues, including antibodies.

For cell biologists, probably the most important and most broadly based spin-off of the work on glomerular permeability was the discovery of tight junctions and junctional complexes.[18] This came about because we had noted that, whereas the filtration slits in the normal glomerulus were wide and contained a central dense line,[15] the filtration slits between the foot processes in the nephrotic glomerulus were much tighter.[16] We began to examine by electron microscopy the arrangement of the junctions between other epithelial cells, such as those in the renal tubule and throughout the digestive system, and found in all these epithelia the same apical tripartite junctional complex composed of three types of intracellular junctions: the tight junction (Zonula occludens), adhering junction (Zonula adherens), and desmosomes (Macula adherens). These junctional complexes were located along the lateral intercellular spaces at the point where the apical and lateral cell membranes meet (figure 3). By introducing electron-dense tracer, we also provided functional evidence that the tight junctions are regions where the intercellular spaces are closed and prevent the passage of proteins (such as hemoglobin and zymogens; figure 4). In contrast, we noted that the intercellular spaces are open at the level of adherens junctions and desmosomes. This provided the first description of tight junctions and adherens junctions, of epithelial junctional complexes, and of the role of the tight junction in sealing off the intercellular spaces. As an aside, although this work was published in 1963, it was initially presented to the cell biology audience at the first meeting of the newly organized American Society for Cell Biology (ASCB) by Keith Porter in Chicago in 1961 (see Chapter 10C). The resultant paper[18] still remains my most cited paper and was selected to be included in a special issue of *Molecular Biology of the Cell* devoted to classical papers on the occasion of the fortieth anniversary of the ASCB. In fact, it is still cited, which is remarkable in an era when any paper more than five years old is considered archaic.

1962–1969: University of California, San Francisco

After nearly three years at Rockefeller, I returned to the pathology department at UCSF and initiated my own research program. During the next eight years, I obtained a Research Development Award from the National Institute of General Medical Sciences and my first NIH grant from the National Institute of Diabetes and Digestive and Kidney Diseases. I had to decide which direction to pursue and what new techniques to set up in my own lab. Techniques had just been introduced by Russell Barrnett, Alex Novikoff, and others for carrying out

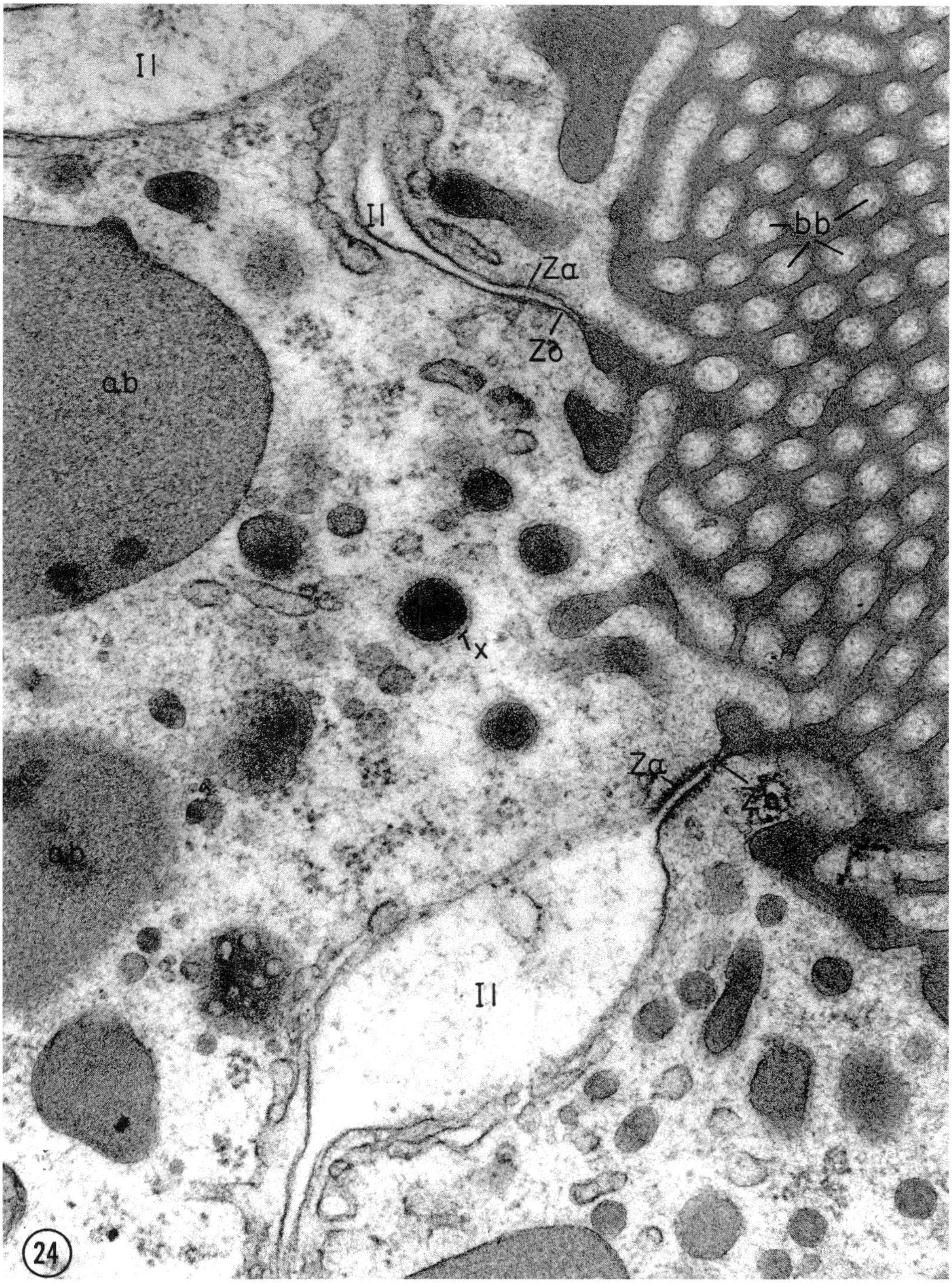

4. Portion of three epithelial cells from the proximal tubule of rat kidney with experimental hemoglobinuria showing tight junctions serve as seals that prevent passage of filtered proteins along the intercellular spaces. Rats were given an intravenous injection of hemoglobin, which is filtered by the glomerulus and fills the lumen of the proximal tubule. It does not appear in the intercellular spaces (Il) below the level of the tight junctions or Zonula occludens (Zo) due to the sealing function of the tight junctions, which are relatively shallow in this epithelium. Some hemoglobin is taken up by endocytosis in clathrin-coated pits (x) and is delivered to endosomes and lysosomes, also called absorption droplets (ab). bb, brush border; Za, Zonula adherens. From Farquhar and Palade, 1963. *J. Cell Biol.* 17:375–412.

enzyme cytochemistry at the EM level. In particular, Novikoff and co-workers[19] modified the Gomori method for acid phosphatase used as a lysosomal marker, which opened new possibilities for carrying out functional studies on lysosomes. I decided to take advantage of the availability of these new methods and to use lead phosphate techniques in my studies. I put together a small group, which included my first graduate student, Robert Smith, and two brilliant postdocs, Daniel Friend and Dorothy Bainton, with whom I moved into the field of enzyme cytochemistry at the EM level. We worked diligently to improve techniques for processing of tissues for enzyme cytochemistry and introduced the Smith-Farquhar Tissue Chopper (ultimately marketed by Sorvall), which had the advantage of allowing us to avoid freezing the specimens, which until then had been cut on a cryostat, requiring freezing and thawing. The outcome of this work was the discovery of crinophagy or the role of lysosomes in degradation of secretory granules,[20] the first description at the EM level of the origin and nature of the characteristic granules of neutrophils,[21–23] basophils,[24] and eosinophils,[25] and the first description of the role of coated vesicles as "primary" lysosomes that carry lysosomal enzymes to lysosomes.[26] This period was exciting and productive, as we contributed to the overall understanding of the biology of lysosomes in several important systems. It was also academically rewarding, as I rose through the ranks to become a full professor.

But things were moving rapidly in the new discipline of cell biology. It soon became evident that morphology alone, including enzyme cytochemistry, was not accepted as conclusive by the increasingly sophisticated cell biological community. It was work at Rockefeller, particularly by Palade and Philip Siekevitz, that set the standard in the field by stressing the need for combining morphology with biochemistry (cell fractionation). This led me to the conclusion that, to remain in the mainstream, I needed to expand my horizons in new directions and in particular to strengthen my familiarity with cell fractionation and biochemical analysis of cell fractions. Prompted by this realization and with the advantage that my Career Development Award allowed support (salary and supplies) to spend time in another lab learning new technologies, I arranged to return to Rockefeller to spend a sabbatical year and was made a visiting professor of cell biology. This was actually the first time in my career after college that I was able to make such plans based on what was best and most timely for my career. I left my laboratory at UCSF in the capable hands of Dan Friend and Dorothy Bainton, who had become assistant professors. With my two sons, Douglas and Bruce, my adventurous technician, Gail Brasil, and a Porter-Blum MT-2 ultramicrotome strapped in the back seat of my car, I set out for Rockefeller in the fall of 1969 for what I thought was going to be a one-year sabbatical.

1970–1973: Return to The Rockefeller University

The environment at Rockefeller was just as exciting when I returned as when I had left eight years before. There were many new faces. Among them were David Sabatini, Jim Jamieson, and Günter Blobel as well as Nicolae and Maya Simionescu, who had just arrived from Romania and had begun their work with Palade on the permeability of muscle capillaries. Jamieson was just beginning his work on secretion in the exocrine pancreas, and Sabatini and Colvin Redman were hard at work on the analysis of microsomes. It was literally a hotbed of ideas and intellectual ferment, but now at a new level where biochemistry and in vitro systems were equal partners with morphology. The period of transition from pure morphology to in vitro systems and biochemistry was difficult and protracted, but I knew it was essential. I had watched many electron microscopists who were highly successful in the heyday of electron microscopy during the 1950s and 1960s gradually lose their grants and drop out of research because they failed to make the transition from electron microscopists to cell biologists. That is, they had not adapted their research activities from pure morphology to a combination that allowed them to assess function and composition of cell structure.

My stay at Rockefeller lasted longer than had been originally anticipated, as George Palade and I were married in 1970. I was appointed professor of cell biology, the only woman professor at Rockefeller at that time, and got my own laboratory, a tiny room of no more than 500 square feet in the foyer of an EM facility in South Laboratory (now Bronk Laboratory).

Upon my arrival, I started working in the same areas that had fascinated me from my days as a graduate student, namely, pituitary secretion, functions of the Golgi apparatus, and glomerular permeability. In my investigations of pituitary secretion, I was fortunate that Colin Hopkins, an associate professor at Liverpool, who had a long-standing interest in pituitary secretion, came to Rockefeller to spend a sabbatical in my laboratory. Together, we decided that to extend our knowledge of secretory processes in pituitary cells, it was necessary to develop and analyze in vitro systems because we had reached the limit of information gleaned from static EMs. Inspired by Jamieson's work on dissociated pancreatic cells, we developed the first preparation of functionally intact dissociated pituitary cells for in vitro analysis. We established that these cells incorporate label, which could be followed by autoradiography, and that they responded to secretagogues by discharging pituitary hormones. We and others used such preparations in our subsequent studies on secretion and recycling in pituitary cells, and this system is still in common use among endocrinologists interested in pituitary secretion.[27,28]

My return to the glomerulus was prompted by a request from a young pathologist, John Caulfield, who came to my lab as a postdoc, and by the fact that our earlier findings delineating the GBM as the main permeability barrier of the glomerulus were being disputed. While the findings with ferritin and the general role of basement membranes as filtration barriers were accepted in cell biological circles, the nephrology audience continued to focus on the filtration slits as the site of the postulated pores. The argument was made, based largely on work carried out by Morris Karnovsky, Manjeri Venkatachalam, and Ramzi Cotran with enzymatic, histochemically detectable tracers such as myeloperoxidase,[29–31] that ferritin is too large to be discriminatory and that while the GBM may serve as the barrier to large molecules such as ferritin, the slit membranes were the main barrier to molecules in the size range of albumin. In other words, there were postulated to be two successive barriers in series, the GBM for large molecules and the epithelial slits for smaller ones.

Therefore, Caulfield and I returned to the problem in the 1970s using dextrans as tracers since they were available in a range of sizes, their behavior was similar to proteins of comparable dimensions, and, as such, they had been used extensively to study capillary permeability by Palade and the Simionescus[32] as well as by physiologists. Repeating the same approach as with ferritin, in work undertaken and nearly completed at Rockefeller, we demonstrated that the GBM serves as the primary barrier to dextrans of three different sizes: molecular weights of 32,000, 62,000, and 125,000, in both normal (figure 5)[33]

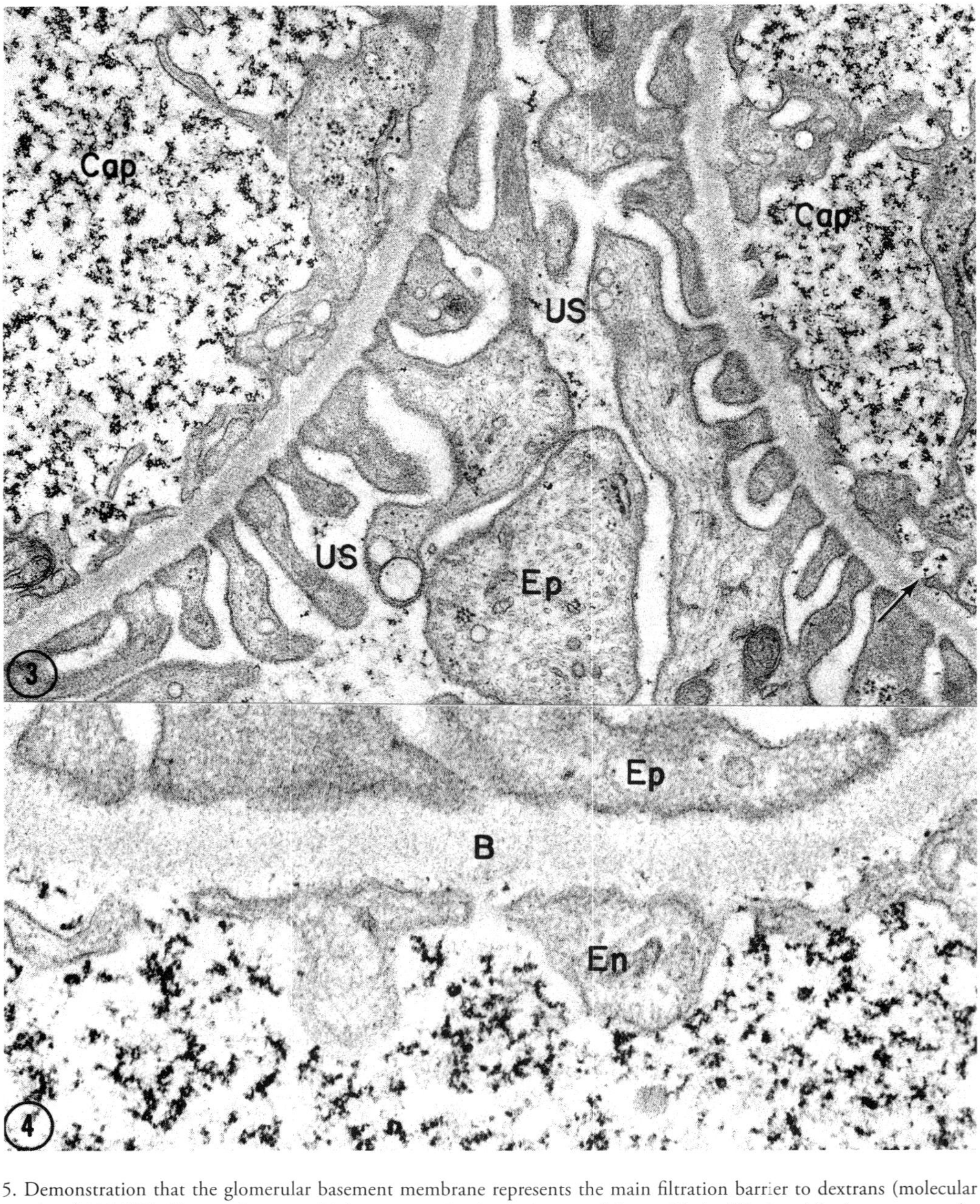

5. Demonstration that the glomerular basement membrane represents the main filtration barrier to dextrans (molecular weight of 125,000). Dextran is seen at high concentration in the lumen of the capillaries (Cap) and freely penetrates the endothelial (En) fenestrae (arrow), but relatively few particles of the dextrans tracer penetrate the basement membrane (B). From Caulfield and Farquhar, 1974. *J. Cell Biol.* 63:883–903.

and PAN[34] nephrotic rats. That is, in all three cases, the dextrans were predominantly retained in the lumen and accumulated against the GBM, but increasing amounts were filtered with decreasing molecular weight. Significantly, no accumulation was seen in the epithelial slits, substantiating with a different tracer the findings obtained in the early 1960s with ferritin.

To our surprise and dismay, these findings did not settle the argument concerning the roles of the GBM and epithelial slits in glomerular filtration. It was never really settled. It continued and only shifted thereafter to a consideration of charge, i.e., where in the glomerulus is the highest negatively charged barrier—epithelium or GBM? To this day, the function of the GBM in filtration continues to be contested, to our chagrin, by many nephrologists who choose to ignore the considerable direct findings obtained initially with ferritin and subsequently with dextrans as well as a wide variety of other tracers.

A third area of investigation initiated at Rockefeller was the analysis of Golgi fractions by combined biochemical and cytochemical methods (see Chapter 7). The Palade group had introduced a method to isolate rat liver Golgi fractions, which relied on the overloading of Golgi elements with very low density lipoproteins (VLDLs) by treatment of the animals with alcohol.[35,36] This facilitates the isolation of Golgi fractions based on their ability to float (due to their VLDL content) in sucrose gradients. The procedure produces three Golgi fractions of differing density. In this work, I had the opportunity to collaborate with John Bergeron and (later on) Kathryn Howell, postdocs in Palade's group, on the cytochemical analysis of such fractions. In one of the first such studies of the kind,[37] we carried out enzyme cytochemistry on Golgi fractions, showing the value of using enzyme markers such as 5′-nucleotidase (assumed to be a plasma membrane marker), acid phosphatase (for lysosomes), thiamine pyrophosphatase (for trans-Golgi cisternae), and glucose-6-phosphatase (for endoplasmic reticulum). Among the contributions of this work was that it established that 5′-nucleotidase is present in Golgi as well as plasma membrane fractions and that acid phosphatase is found in VLDL-filled vacuoles and is indigenous to secretory granules as well as lysosomes. Thus, the work allowed us to distinguish between indigenous and contaminating activities in Golgi fractions.

Another important event during this period at Rockefeller was that I became a consultant for electron microscopy to Christian de Duve's group. Up until that time, the de Duve group, which had previously concentrated almost exclusively on using cell fractionation to investigate cell biological problems, decided that they needed to have an EM facility within their own group and acquired an EM and the necessary supporting equipment. I helped them set up the laboratory and served as consultant, attending all their weekly lab meetings, and helped to train Helen Shio, who served as director of their facility for many years. I also coauthored a paper with this group[38] in which we used cytochemical methods to study the changes that occur in lysosomes in the aorta with atheroma induced by cholesterol feeding.

All in all, my return to Rockefeller was a very stimulating experience that enabled me to accomplish the goal of expanding my horizons to incorporate new emerging technologies into the investigation of important cell biological questions. But new challenges soon arose.

After Leaving Rockefeller

In the summer of 1973, Palade, Jamieson, and I moved to Yale. Others who moved with us included Nicolae and Maya Simionescu and Jean Pierre Kraehenbuhl, Ann Hubbard, David Castle, and Kathryn Howell, who were postdocs at the time but eventually became faculty members. We were attracted to Yale by an offer from Lewis Thomas, then dean of the medical school, with the prospect of setting up a new department for the application of cell biology research to medical problems. We felt that the time was ripe for extending the new discipline of cell biology to the curriculum and graduate programs of medical schools, as it was our vision that cell biology was destined to become the central basic science and research discipline of schools of medicine.

In 1990, another new opportunity and challenge presented itself in the form of an offer to return to the West. Gerald Burrow, a former colleague at Yale, had just been appointed dean of the medical school at the University of California, San Diego (UCSD). Burrow was impressed with what Palade had accomplished at Yale in terms of development of basic sciences and wanted him to participate in the development of basic sciences at UCSD. He also wanted us to set up a new department of cellular and molecular biology in the medical school, which, at that time, had only one existing basic science department. Over the next ten years, we worked to strengthen basic sciences in general and to build a new program that became the Department of Cellular and Molecular Medicine, and I was named its first chair.

By now, my career in science has spanned over fifty years and has included major stays at four prestigious universities. I wish I could say that it followed a preconceived grand master plan. But this was not really the case. Chance was a key factor that put me in the right place at the right time, especially the events that placed me at Rockefeller when it was considered by many to be the American cradle of cell biology. It was there I learned the necessity for a cell biologist to continually change directions, adapt emerging new technologies, and move in new directions like the explorers of old. This was perhaps also in my genes since my great-great-grandfather, Christopher Gist, was a scout who charted new territories for Washington in the French and Indian War and the American Revolution and my own father was an explorer of the wilderness of the Sierra Nevadas. In fact, I often think of the analogy of my career in cell biology with the rapid developments that have taken place with that of my father and mother's generation, that had to adapt to moving from the horse and buggy era to the space age. Similarly, cell biologists of my generation have had to adapt to the greatest and most rapidly moving events in science ever faced by any previous generation. We have witnessed the first beginnings of the application of electron microscopy to the study of biological specimens and adapted to continual changes in technology applied to the study of cell biological problems, including high resolution immunocytochemistry, protein chemistry and structure, molecular biology, and, most recently, genomics, proteomics, and bioinformatics. It makes one wonder what is ahead. Although there are no more new geographical territories to explore, there will always be new territories and new vistas to explore within the cell.

13D. Membrane Biogenesis in Chloroplasts, Endoplasmic Reticulum, and Mitochondria

Philip Siekevitz

Following the laboratory's collaboration on the endoplasmic reticulum (ER), the synthesis of secretory proteins, and their movements through the ER, George Palade and I tackled another problem. During the 1960s, we examined the organization of intracellular membranes, the synthesis of nascent membrane proteins, and their insertions into the membranes. We hypothesized that while the ribosomes attached to these ER membranes were synthesizing proteins destined for secretion, they were also probably synthesizing proteins destined for ER membranes. The question posed was how newly synthesized proteins (and phospholipids) were gathered to form membranes in a system where membrane proliferation was taking place. Various modes of membrane formation were postulated. One was that the future proteins were synthesized at the same time and, together with the phospholipids, were somehow gathered in a single event to form the membrane. Another was that the proteins were synthesized at different times and then inserted into what one might call a "scaffolding" of the membrane system. And yet another was that membranes might arise from preexisting ones.

The laboratory looked at this problem simultaneously in three systems in which membrane proliferation was occurring: chloroplasts, ER in liver, and mitochondria. Evidence from each of these provided important corrections to original hypotheses and essential clues to the working of the intracellular organelles.

Developing membranes of chloroplasts. An impetus for this choice was Palade's experience with developing chloroplast membranes in *Chlamydomonas*, a photosynthetic alga. In 1953, he had examined chloroplast structure in *Euglena* with Jerome Wolken, and they found that when these protozoa are transferred from light into dark surroundings, their chlorophyll disappears and the membrane structure of the chloroplasts is lost; however, both are regained when the cells are transferred back into the light.[1]

A year later, Palade and Ruth Sager found the same thing happens in *Chlamydomonas*, but for this to happen it was necessary to use a mutant of it, one that was isolated by Sager and dubbed "y-1" (y for yellow).[2] Looking further, Sager and Palade found a great similarity in the chloroplast structure of the algae with the one in leaves of higher plants. In both cases, chloroplast membranes were arranged in parallel rows of paired membranes comprising discs, although there was no evidence in the algae of the chloroplast grana found in higher plants. A peculiar feature of *Chlamydomonas* was a single giant cup-shaped chloroplast rather than the many smaller chloroplasts in leaf cells.[3]

The work on chloroplasts in *Chlamydomonas* lay dormant for some ten years. Then Palade had the idea that the y-1 mutant might serve as an excellent material for studying membrane development because both its chlorophyll and membranes could be made to disappear in the dark and reappear in the light. Electron microscope and biochemical evidence could be gleaned by following the two processes. Fortunately, at this time in 1964, postdoctoral fellow Itzhak Ohad joined the laboratory; he was originally from Romania, so he and Palade sometimes conversed with each other in their native language, leaving me momentarily in the dark.

One difficulty, never solved, was isolating an *intact* chloroplast. Since it was very difficult to break the cell wall, many methods were tried. Finally, success came with an instrument known as the French press that first increased pressure on the cells and then, with a quick depressurization, caused an implosion to break the cell open and release its contents. Due to its large size, however, the chloroplast was fragmented in the process. The chloroplast membranes, on the other hand, could be isolated by differential centrifugation in a rather pure form.

As a result, the cytological and biochemical data on the greening and the degreening of the y-1 mutant were forthcoming. Electron micrographs by Palade, superior to those ten years earlier, showed that once placed in the dark, there were several cycles of cell division, and very few of the chloroplast membranes remained in the yellowed cells. Our accompanying biochemical experiments tied into these micrographs and indicated a great reduction in chlorophyll content and photosynthetic activity.[4]

The real interest of this experiment was to determine how the membranes developed, a process that depends wholly on light and takes six to eight hours without cell division. The photosynthetic activities were monitored, and their increases seemed to parallel the increases in chlorophyll content and in new membrane formation that was visible with electron microscopy (EM). The only observed morphological change was the drastic reduction in the light of starch granules that built up during the dark phase. Since all components, visible membranes, and photosynthetic activities seemed to increase in parallel during greening, the evidence seemed to suggest the assembly of membranes was a one-step process.[5]

However, this conclusion did not hold. By this time, results appearing from the work in the laboratory by Gustav Dallner (see following section) on the developing ER system indicated a multistep process. The presumption was that membrane organization in mammalian and plant systems should be similar.

Another postdoctoral fellow, Kenneth Hoober, took up the work in 1966, and he had the clever idea of trying inhibitors of protein synthesis to gain insight into the process. One was chloramphenicol, which was known to inhibit protein synthesis by chloroplast (and mitochondrial) ribosomes. The other was cycloheximide, which was known to inhibit protein synthesis by cytoplasmic ribosomes. This time, the results were startling. Using chloramphenicol increased chlorophyll production and membrane formation, whereas the photosystem activities were greatly decreased, indicating that chloroplast ribosomes synthesized the photosystem proteins. Conversely, using cycloheximide decreased chlorophyll synthesis and membrane disc formation, whereas photosystem activities were the same as the untreated control cells, indicating that cytoplasmic ribosomes synthesized the structural proteins of the membranes. When protein synthesis was measured using incorporation of radioactive leucine into proteins, each inhibitor caused a 50% reduction, while together they reduced it to very low values. This showed that both chloroplastic and cytoplasmic ribosomal systems were involved in generating a fully functional chloroplast membrane system.[6]

It was obvious that the membrane structural proteins, chlorophyll, and cytochrome f of the chloroplast were being formed and assembled in the presence of chloramphenicol without photosynthetic activities. The absence of photosynthetic activities could be remedied by incubating in light using cycloheximide, which inhibited further structural protein formation but permitted photosynthetic protein formation. Thus, the newly synthesized enzymatic proteins were being inserted into an already existing membrane structure.

A multistep membrane assembly process was the new conclusion, in contrast to what was previously reported, although there remained some hesitation concerning possible unforeseen effects by the protein synthesis inhibitors. To test this question, graduate student Seth Schor took over in 1969. He measured the changes at hourly intervals in chlorophyll content (that is, membranes), the contents of both photosystems, and protein cytochromes during light/dark cycles of the y-1 mutant. He again found a stepwise organization of the membrane.[7]

Earlier work with the y-1 mutant had used whole cell homogenates since all the parameters measured were photosynthetic in nature. However, to ensure the interpretations were correct, it was thought necessary to isolate the chloroplast membranes during various parts of the greening cycle and to measure their chemical constituents. This was done in 1970 by Benita de Petrocellis, a postdoctoral fellow from Italy who isolated two fractions by discontinuous density gradient centrifugation, and these were verified by Palade's electron micrographs showing purified membranes with very little contamination. Her results confirmed the earlier work of the laboratory that there was a seemingly clear-cut, stepwise formation of the developing chloroplast membranes.[8]

Developing ER membrane in liver. Since it was possible that the y-1 mutant results were unique to *Chlamydomonas*, another system was chosen in which the laboratory already had extensive experience with its isolation and properties: the ER in rats. An impetus for this choice was a recent report that glucose-6-phosphatase, a microsomal enzyme, appears rapidly just after birth. I had confirmed this finding earlier and also found another microsomal enzyme, NADH–cytochrome c–reductase, that appears very gradually after birth.[9]

In 1964, postdoctoral fellow Gustav Dallner arrived from the Wenner-Grens Institute in Stockholm. He extended these preliminary experiments on the development of the ER membranes in the liver. He isolated smooth and rough ER membranes from perinatal rats and performed biochemical assays on them while Palade took electron micrographs of the entire cell. They found that before birth, the ER increases in volume, remaining predominantly rough, i.e., with ribosomes attached, while after birth, the smooth ER predominated. During the entire period, the RNA/protein ratio decreases while the phospholipid/protein ratio increases, and this agreed with the EM morphological observations. These results suggested that new membranes were synthesized in the rough ER and subsequently appeared as smooth ER.[10]

Dallner next tackled the problem of the differentiation of ER membranes during the ten-day period of rapid growth. He followed biochemically the easily measured enzymes and learned there was a multistep synthesis and insertion of new proteins into already existing membranes. When he fractionated the microsomes into rough and smooth types, it became clear that most of the increases in glucose-6-phosphatase and NADPH reductase were in the rough microsomes. The smooth microsomes lagged behind, indicating again a transition phase from rough to smooth.[11]

Still not completely satisfied with a multistep process in membrane formation, in 1965, Japanese postdoctoral fellow Tsuneo Omura tackled this problem by a different technique that involved measuring the half-lives of radioactive precursors of two ER membrane proteins, the fatty acids, and the glycerol backbone of the membrane phospholipids. The conclusion was that the liver ER membrane, considered as a whole, has a continuous existence throughout the lifetime of a differentiated cell, although its individual molecular components have a much shorter, finite, and discrete existence. In other words, the cytology stays the same while the biochemical components change.[12]

Further experiments by postdoctoral fellows Karl Bock, Yoshiaki Kuryiama, and Tsuneo Omura looked at the stepwise appearance of a complete protein during membrane development. All evidence indicated there were changes in protein composition within discrete time periods.[13–15] Still other experiments by graduate student Andrea Leskes looked at whether enzymes of the ER membrane existed in discrete patches or were scattered randomly over the membranes. She reverted to my earlier work on the rapid appearance of glucose-6-phosphatase in the liver ER of rats right after birth and made a cytochemical examination of the ER using lead phosphate that could be easily seen by the EM. When the micrographs clearly showed a uniform distribution of the lead phosphate all over the ER, we concluded that during the increase in ER membrane mass, there is no new synthesis of membranes different than the preexisting ones and that as new enzymes are synthesized, as measured by the presence of glucose-6-phosphatase, they are inserted all over the preexisting membrane structure. That is, there is no "old" membrane and "new" membrane but a mixture of old and new components.[16,17]

Consequently, I postulated the idea of plasticity of membranes, in which the architecture can change during its lifetime without losing its identity and despite changes in membrane composition. Just as the fertilized egg has membrane structures, which persist during subsequent cell division, its membranes arise from membranes while developmental changes can and do occur. I compared a membrane with a wall made of different colored bricks. The "old" red bricks being shed are replaced by "new" red bricks, old yellow bricks by new yellow ones, old green bricks by new ones, and so forth, and each color changes at different rates of displacement and replacement. During these changes, however, the wall and membrane exist as an identifiable entity.

All the work on chloroplast and ER membranes led to the conclusion that new membranes form through a stepwise process, yet during this time their overall identity persists. While this view of membrane development continues to hold, at that time what still needed to be solved were the intimate biochemical processes by which this occurs.[18–20]

Mitochondrial biogenesis. In this section, what began as part of the laboratory's ongoing experiments on intracellular membranes and their permeability ended with very arresting evidence that pushed back the boundaries of cell biology: the mitochondria contain their own DNA, and it is distinct from the DNA in the nucleus. This research was done by David Luck, who joined our group in 1958 as a graduate student and in 1968 became the first from this laboratory to become a professor at Rockefeller. His thesis project was based on a discovery by Keith Porter of a subclass of microsomal membranes associated with glycogen deposits in the liver. Luck wanted to isolate the membranes to see if he could understand their function in the economy of glycogen in the cells. The rule in the laboratory was that if it was there, it was doing something, so he started out immediately applying the powerful techniques of cell fractionation, monitoring the fractions with the microscope, and doing the biochemistry. It turned out that what he found was not associated with the membranes but with the glycogen. In retrospect, this was a good learning experience, if not slightly sad in terms of a major discovery.[21]

On finishing his thesis in 1962, Luck spent fourteen years studying mitochondria, the organelle that was of great interest to Palade in the early 1950s. By this time, biochemists had resolved important questions to verify mitochondria as the site of cell respiration and where food is converted to chemical energy. A membrane surrounding mitochondria had been suggested as early as 1888 by Albert von Kölliker based on his observations with the light microscope.[22] Claude and Fullam's first electron micrographs of mitochondria in 1945 confirmed this observation.[23] Once thin sectioning techniques were devised by Porter and Palade, it became possible to study the fine structure of mitochondria. Palade then provided electron micrographs showing mitochondria surrounded by a membrane that is infolded to form ridges inside, which he named cristae.[24] At the same time in Sweden, Fritiof Sjöstrand found in his micrographs that the mitochondrion had a double limiting membrane and a number of double membranes inside the organelle, which he named septae.[25] Palade soon confirmed the existence of a double limiting membrane and concluded the cristae are infoldings of the inner membrane (see Chapter 6).

The question of how these membranes are formed was raised in the 1940s, and it was postulated that mitochondria arise de novo by transformation of other cellular structures.[26,27] However, in the early 1950s, it was found that certain properties of two unicellular organisms, *Saccharomyces cerevisiae* and *Neurospora crassa*, were not inherited according to Mendelian laws, and it was suggested that extrachromosomal genetic factors of these organisms might reside in or even be identical with mitochondrial organelles.[28,29] This set the stage for a convergence of genetics and mitochondrial physiology. However, a major hurdle in working on this topic was the reluctance of scientists to accept these single cell organisms as having bona fide mitochondria. Fortunately, our laboratory was one of the few working in this area.[30]

Luck's new project dealt with the question of how mitochondria proliferate during growth and cell division, during which new mitochondria are produced in the daughter cells. And what he introduced to cell biology was the use of genetics to complement the microscopic and biochemical techniques.

During his graduate training, Luck had taken the famous course by Cornelius Van Niel of Stanford University's Hopkins Marine Station on using microorganisms to study basic problems of biology. This was such a powerful experience that Luck decided he was through with rats. Still, he needed to learn how to study genes and how to cultivate microorganisms.

Fortunately, he could walk across the Rockefeller hallway to the laboratory of Nobel laureate Edward Tatum, where he quickly got interested in the power of genetics, especially the single-cell organisms whose rapid growth cycle was critical for analyzing mitochondrial division. It became clear that mutant organisms with deficient mitochondria could serve as incisive analytical tools of biological mechanisms because normal organisms could be compared with these abnormal or deficient organisms. In collaboration with Edward Reich, a member of Tatum's laboratory, Luck was introduced to properties of *Neurospora crassa*, the lowly bread mold that has a venerable history as a subject of scientific research. The Tatum laboratory had a large number of mutant strains of this fungus in its collection, and Luck was able to isolate from them reliable quantities of homogeneous, intact mitochondria that could be verified by electron microscopy.

At the time, there were three possibilities for how new mitochondria were produced: from preexisting ones by growth and division of the existing mitochondria, by de novo synthesis of new mitochondria, or by derivation from non-mitochondrial membrane structures. Luck used two different types of experiments involving cytological, biochemical, and genetic approaches to determine which of these possibilities was correct.

His first approach used a radioactive label for mitochondria or its membranes. Luck selected a *Neurospora* mutant that needed the addition of choline, a constituent of membrane phospholipids, for its growth. He incubated these cells with radioactive choline and later added a non-radioactive choline as a continuing source for membrane phospholipid synthesis during cell growth. After one, two, or three doublings, the cells were harvested, mitochondrial fractions were obtained, the mitochondria were fixed to microscope slides, and quantitative autoradiography was performed according to the improved method of Lucien Caro and Robert van Tubergen, postdoctoral fellows in our laboratory.[31] The electron micrographs showed stable radioactive choline randomly distributed among all the progeny mitochondria during the growth phases; at the end of each doubling phase, the average number of autographic grains was half of what was found at the end of the previous doubling phase. The simplest explanation for these results was there was no de novo formation of new mitochondria. Instead, the mitochondrial mass increased by a continuous process of addition of choline-containing lipids into the existing structural framework, leading to a division of mitochondria for new cell formation.[32,33]

Luck confirmed these results by a second, entirely different type of experiment. He had previously found that the phospholipid to protein ratios of mitochondrial membranes can vary greatly depending on how much choline was added to the medium in which the mutant *Neurospora* was growing. So he isolated mitochondria from cells grown in various concentrations of choline and subjected them to sucrose-density gradient fractionation, which could separate mitochondria from cells grown in low concentration from those grown in a higher one. The centrifuge tubes showed a gradual displacement from the high to low density mitochondria rather than two distinct bands that would be expected if new mitochondria were formed de novo from non-mitochondrial sources. In other words, mitochondrial replication occurs by a division of existing mitochondria. They multiply on their own, and every time a cell divides, its daughter cell acquires a small colony of them, a finding that remains valid until the present day.[34]

Moving further in this new direction, Luck used these experiments on mitochondrial duplication as the cytological basis for biochemical studies to determine whether mitochondria had a distinctive DNA, ribosomes, and a protein-synthesizing system—all the ingredients of a genetic hierarchy.

There were indications in the early 1950s that this was the case.[35,36] In 1963, Sylvan Nass and Margit Nass in Sweden obtained electron microscopic and biochemical evidence of mitochondrial DNA "threads."[37,38] A few months later, in 1964, two papers appeared. One was by Gottfried Schatz and his collaborators in Austria, who separated yeast mitochondria and found a purified peak in a density gradient showing a coincidence of succinate–cytochrome c reductase activity with material containing deoxyribose.[39] The other paper by Luck and Reich on *Neurospora* went further. Using ultracentrifugation with ultraviolet optics, they made a more precise quantification and discerned two peaks of DNA, one from the mitochondria and one from the nucleus. Walter Stoeckenius, who was in the laboratory at the time (see Chapter 13B), provided a vivid electron micrograph of a single molecule of DNA from one of Luck's mitochondrial fractions (figure 1).[40] Shortly thereafter, other scientists found genetic alterations in a yeast mutant that were extrachromosomal and linked to changes in mitochondrial DNA.[41] In addition, Walter Schneider, a former member of our Rockefeller laboratory, found DNA in mammalian liver mitochondria.[42]

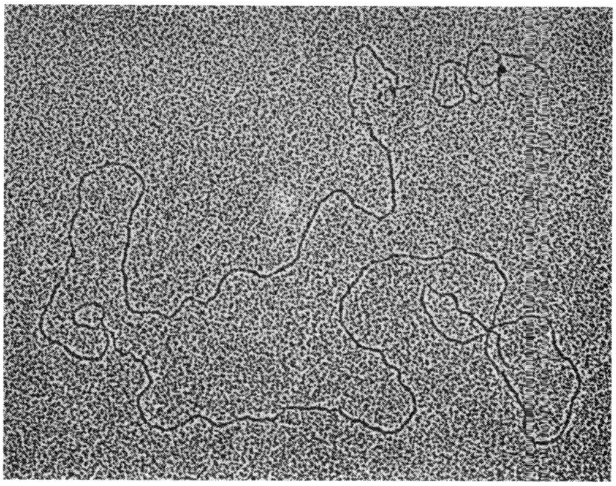

1. Electron micrograph showing a representative single molecule of DNA prepared from a mitochondrial fraction. The length of this molecule is 6.6 micrometers, corresponding to a molecular weight of ~13 million. ×55,000. Micrograph taken by Walther Stoeckenius. From Luck and Reich, 1964. *Proc. Natl. Acad. Sci. USA.* 52:931–938.

Reich and Luck then looked at how mitochondrial DNA was being duplicated. They used the then new, but now classical, technique of Matthew Meselson and Franklin Stahl to distinguish an old strand of DNA from a newly synthesized one.[43] They found in the *Neurospora* mitochondria a semiconservative method of replication since one DNA strand of each daughter is newly synthesized, whereas the other is derived from the parent DNA.[44] In 1967, Luck's laboratory was the first of three to report the presence of ribosomes that synthesize some protein in mitochondria and that differed in size and base composition from ribosomes in the cytoplasm. A significant implication from this finding was the suggestion that there was a marked similarity between mitochondrial ribosomes and those of bacterial origin.[45–47]

That same year, 1967, Lynn Margulis set forth the central elements of the endosymbiotic theory on the origin of mitochondria.[48] As early as 1922, Ivan Wallin had emphasized the similarity in structure and staining reaction between mitochondria and bacteria and even postulated that they were bacterial symbionts.[49,50] Of historic interest for this book, one of the outspoken critics of Wallin was Edmund Cowdry, who at the time was a member of James B. Murphy's laboratory and whose classic monograph on mitochondria influenced Albert Claude's adoption of the centrifuge (see Chapter 2). Cowdry found the generalization so far-reaching that he published an article in 1924 suggesting the kind of knowledge that was needed to prove or disprove the theory.[51] This concept faded with the brilliant demonstration during the next twenty years of the role of nuclear genes in inheritance. The newly developed tools of cell biology and genetics that Luck and many others brought to the discovery of DNA in mitochondria added considerable evidence to the validity of the endosymbiotic theory. Margulis's theory, which is now generally accepted, argues that mitochondria arose long ago from once aerobic free-living bacteria that entered a eukaryotic cell, perhaps resisting digestion or acting as a parasite, and evolved to form an intimate symbiotic relationship with host cells.

In summary, the discovery of mitochondrial DNA brought to cell biology an ever-increasing number of molecular biologists. Luck's laboratory continued to provide substantial molecular evidence on the assembly and function of flagella in *Chlamydomonas* during another two decades of research that goes beyond the time frame of this book. Importantly, his early work led to the insight that basic mechanisms studied first in the simplest of organisms hold true for animal mitochondria, including those in humans.

13E. Vascular Endothelium, More Complex than Previously Thought

Maya Simionescu

I met George Palade in 1968 when he returned to his native country, Romania, after a long lapse of time. He was part of a delegation of the National Academy of Sciences to negotiate an agreement of scientific exchange with the Romanian Academy. At that time, the communist regime of Romania showed apparent signs of independence from the Soviet Union and was opening toward the West. Palade believed that his laboratory at The Rockefeller University "could encourage this tendency."

During this journey, he visited numerous laboratories, including the Institute of Endocrinology, where he met with me and my husband Nicolae Simionescu. We discussed our work on the structural-functional correlates of the endocrine glands in experimental models and human tissues. At that time, we had already started, with some difficulty, to carry out ultrastructural studies using electron microscopy and cytochemistry of thyroid, adrenal, and pineal glands and pancreas in normal and pathological conditions with a special emphasis on their vasculature. Palade did not say too much; when he asked "how many people are involved in your work?" we replied that besides ourselves, there were two technicians. I don't know what impressed him but a year later, totally unexpectedly, he invited us to join his group at The Rockefeller University for one or two years. Without any hesitancy, Nicolae left his position as associate professor and, with some trouble from the government, arrived at Rockefeller on 1 February 1970; five months later, after Palade's pressing requests and letters to the Romanian authorities, I managed to join the Rockefeller group. It was the beginning of a great collaboration that extended over ten years.

From the beginning, we decided to work on the organization and function of the vascular endothelial cells—an exciting extension at a new and different level of our interest in vascular biology. By that time, Palade had discovered the endothelial plasmalemmal vesicles (figure 1a)[1] and their implication in transport of large molecules such as ferritin across the capillary endothelium.[2] But a lot remained to be done to define in more detail the structure and the functional characteristics of the endothelium. The capillary physiologists postulated that all transendothelial exchanges occur through two types of pore system: small pores (about 9 nanometers in diameter) with relatively high frequency and large pores (up to 70 nanometers) at considerably lower frequency associated with intercellular junctions (figure 1a). They considered unlikely or impossible the transport by plasmalemmal vesicles since these structures did not fit the "pore theory." There was a need for new tools and procedures to extend the dimensional spectrum of available tracers, i.e., new tracers of smaller dimensions to detect and localize especially the small pores postulated by capillary physiologists. Thus, in 1970, we began to look for new tools to probe capillary permeability.

New tracers—more data. The search for other tracers led to the use of various dextrans and glycogens, two biological molecules available in various dimensions. These polysaccharides were well tolerated after intravascular injection in mice and rats, induced no vascular leakage,

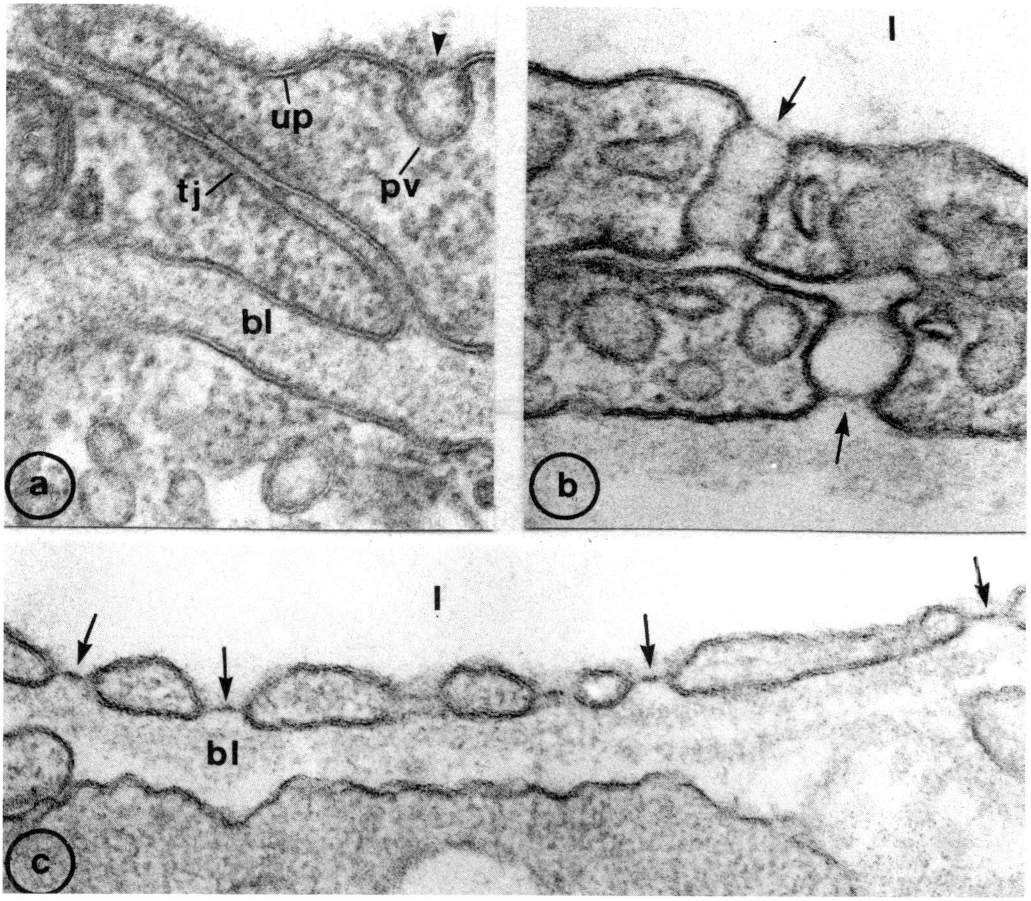

1. Distinctive features of endothelial cells. (a, b) In a continuous capillary, the endothelial cell has characteristic plasmalemmal vesicles (pv) provided with a neck spanned by a thin diaphragm with a central knob (arrowhead), uncoated pits (up), tight junctions (tj), and transendothelial channels made up by one or two vesicles fused simultaneously to both endothelial fronts (arrows). (c) In a fenestrated capillary, the endothelial cell has numerous round fenestrations and also spanned a thin diaphragm (arrows). l, capillary lumen; bl, basal lamina. Figure a: from Simionescu, 1991. Cellular organization of the alveolar capillary unit. *The Pulmonary Circulation and Acute Lung Injury*. S.I. Said, editor. Futura Publishing, Mount Kisco, New York. 13–42. Figure b and c: from Simionescu and Simionescu, 1984. Ultrastructure of the microvascular wall: functional correlation. *Handbook of Physiology*, Volume 4. E.M. Renkin and C.C. Michel, editors. American Physiological Society, Bethesda, Maryland. 41–101.

stained adequately with lead citrate after aldehyde-osmium fixation, and provided a relatively wide range of particulate probes (from 4.5 to 30 nanometers in diameter) for testing the large and small pore system.[3] Various sized dextrans and glycogens were used to study the permeability of the fenestrated capillaries of the intestine that have characteristic openings spanned by a thin diaphragm and channels (figure 1, b and c). To improve the visualization of the tracers, an original "one-step fixation" method for electron microscopy proved to be particularly useful; consisting of a mixture of aldehydes, osmium tetroxide, and lead citrate, it gave faster fixation and better preservation of the tissues and significantly increased contrast of the probes. Using this technique, we found a relatively fast movement of dextrans and glycogens out of the blood through endothelial plasmalemmal vesicles (figure 2), fenestrae (figure 3 inset), and channels. These probes were temporarily retained within

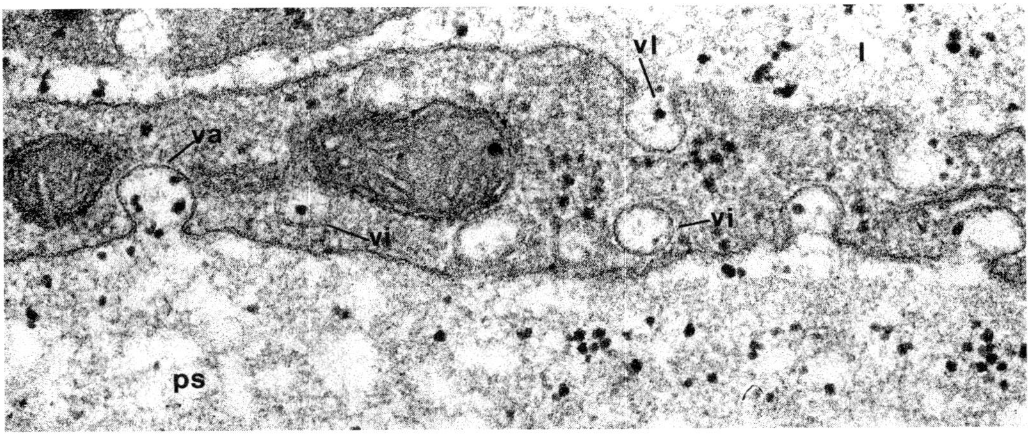

2. Transport of glycogen in a continuous capillary endothelium after intravascular injection in the mouse. After ten minutes, glycogen particles are present in the lumen (l) and in plasmalemmal vesicles open on the luminal front (vl) and are internalized (vi) or open to the abluminal front (va). The probe is present in large numbers in the pericapillary space (ps). From Simionescu et al. Structural-functional correlates in the transendothelial exchange of water-soluble macromolecules. *Thromb. Res.* 8(Supplement):257–269. ©1975, with permission from Elsevier.

the basal lamina, leading to concentration gradients centered on the fenestrated sector of the pericapillary space (figure 3). No probe was found associated with intercellular junctions. Based on time course experiments, we concluded that in visceral capillaries, the fenestrae, channels, and vesicles are parts of a dynamic system that transport macromolecules from the plasma to the interstitia; since the probes were rather large, these endothelial features represented the structural equivalent of the large pore system.[4]

The search continued for smaller probes. In 1973, myoglobin was used as a probe to detect the "pore systems" of muscle capillaries. After appropriate characterization of the tracer behavior in the plasma, a series of time course experiments showed that about thirty seconds after injection, the probe labeled about seventy-five percent of the plasmalemmal vesicles open to the blood front of the endothelium, fifteen percent of those located inside, and none on those on the tissue front. At forty-five seconds, the labeling of vesicles within the cytoplasm of endothelial cells reached sixty percent but remained nil for those on the tissue front. The latter contained the probe after forty-five to sixty seconds, concomitant with the appearance of myoglobin in the pericapillary space. At all times, the intercellular junctions were free of the tracer. By its dimensions, the myoglobin molecule can permeate both pore systems, and because in these continuous capillaries the probe was found only in plasmalemmal vesicles, we concluded that the latter are the structural equivalent of large pores and, at least in part, of a small pore system.[5,6] The data led us to assume that "pores" as hypothesized by capillary physiologists are unlikely to be present in capillary endothelium.

Discovery of transendothelial channels. In 1974, Nicolae and I moved to Yale University with Palade and some members of his Rockefeller laboratory. I do not remember ever wanting to stop the experiments, so we continued our chase for better ways to uncover the mysterious endothelial secrets.

In the search for smaller probes that would be less affected by the molecular sieving when passing through small pores, we selected two heme-peptides, but since these were not commercially available at that time, we had to obtain them by enzymatic hydrolysis

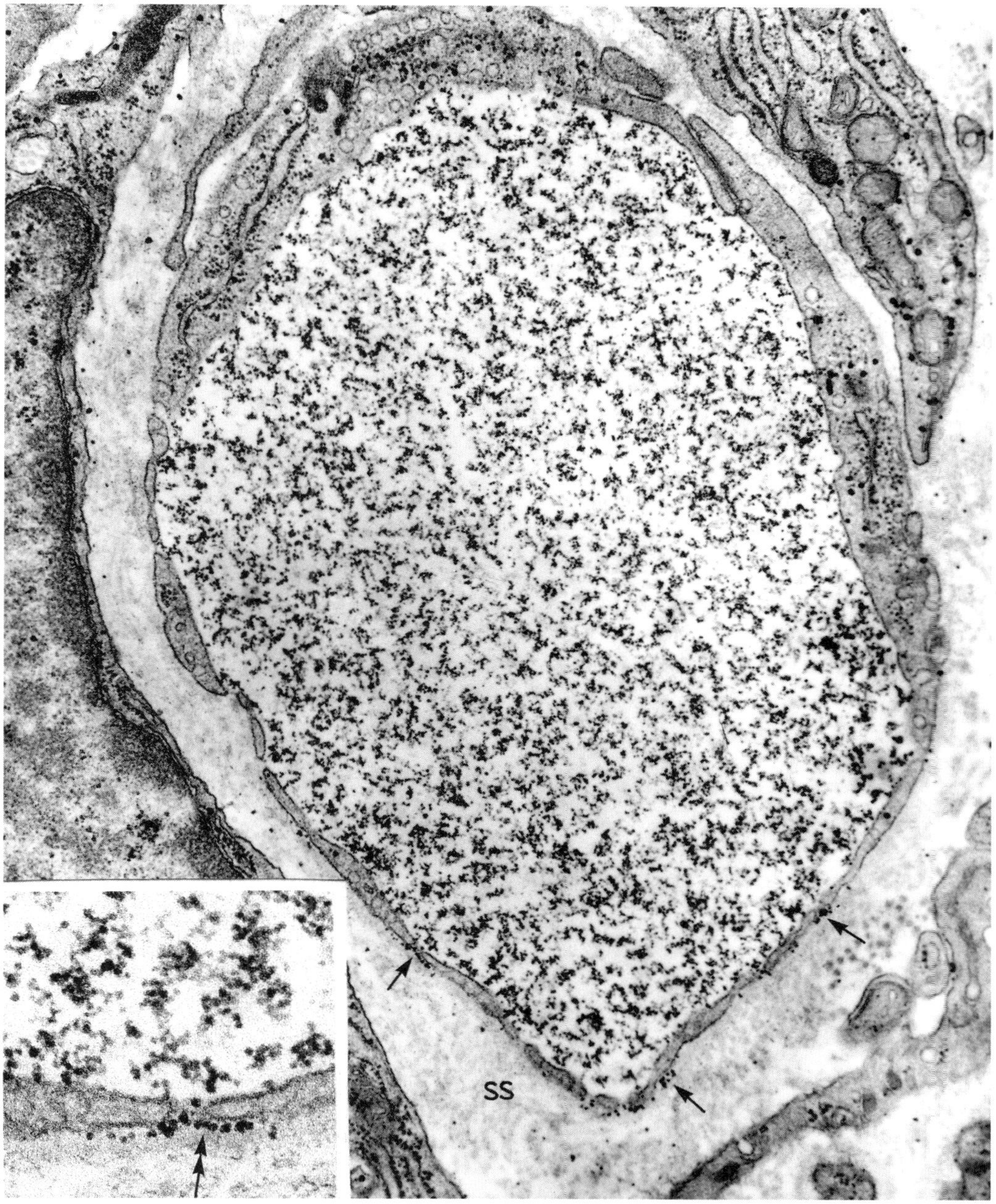

3. Transport of glycogen in fenestrated capillary. Note the accumulation of the probe (arrows) in the subendothelial space (SS), centered on the fenestral diaphragms (inset, double arrow). From Simionescu et al., 1972. *J. Cell. Biol.* 53:365–392.

of cytochrome c. After careful analysis of the behavior of the probes upon injection in the blood, both tracers were localized indirectly (in situ) by the peroxidase reaction enhanced by addition of imidazole in the incubation medium; they crossed the endothelium primarily via plasmalemmal vesicles, which become progressively labeled in three successive, resolvable phases (figure 4). An important bonus of these experiments was the clear evidence for the existence in continuous capillaries of transendothelial channels that were formed

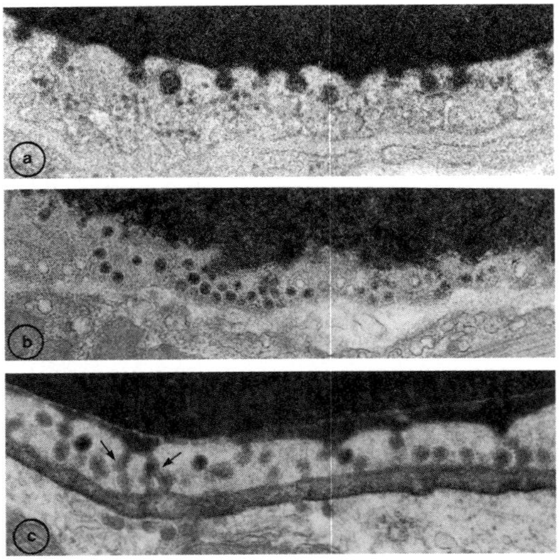

4. Transport of hemeundecapeptide (HUP). Transport, demonstrated via the peroxidase reaction in mouse diaphragm capillaries, takes place in successive resolvable phases: thirty seconds after intravascular injection, the probe is found in vesicles open to the luminal front (a); after forty seconds, HUP labels numerous cytoplasmic vesicles (b); and after forty-five seconds, transendothelial channels (arrows) and vesicles open to the abluminal front (c). From Simionescu et al., 1975. *J. Cell Biol.* 64:586–607.

by one, two, or more vesicles that open simultaneously on both endothelial cell fronts (figure 4c). The openness of these channels was checked by specimen tilting in the electron microscope, and graphic analysis of the images was obtained. No evidence was found for the presence of the probes at the intercellular junctions. We concluded that transendothelial channels are the most likely candidates for the structural equivalents of the small pores (postulated by the physiologists) since they were continuous, water-filled passages and were provided with one or more strictures of 10 nanometers. Altogether, these experiments proved that in continuous endothelium, small molecules cross the vascular wall *only* via plasmalemmal vesicles. The latter can function either as isolated units shuttling in resolvable phases between the two endothelial fronts or can fuse to form transendothelial channels made up of one or more interconnecting vesicles. Similar results were obtained in back diffusion experiments in which the tracers were injected in the interstitial space. We hypothesized that intercellular junctions that were not permeated by these tracers may be permeable for molecules smaller than two nanometers.[7–9]

Segmental differentiation of endothelial intercellular junctions. Since interendothelial junctions were still a subject of high debate and contradictory reports, we postulated that this might be due to variation in the structure of these junctions in different segments of the vessels. Thus, in 1974 and 1975, a systematic study of endothelial junctions in various segments of the vasculature was set up using the alternative method of freeze fracture. The microvessels (arterioles, capillaries, and pericytic venules; figure 5) and the macrovessels (arteries and veins) were dissected out and subjected to freeze-fracture preparation and electron microscopy. The results showed that the intercellular junctions of the endothelium vary characteristically from one vascular segment to another: the arteriole endothelium has an elaborated complex system of continuous tight junctions with interpolated large gap junctions (figure 6a); the capillary endothelium has tight junctions only, which appeared in freeze-fracture preparations as branching or staggered continuous strands (figure 6b); and the postcapillary venules exhibited loosely organized endothelial junctions having discontinuous, low profile ridges and grooves, usually devoid of particles (figure 6c). The endothelium of muscular venules (the segment of microvasculature whose wall is provided with smooth muscle cells) has a similar type of junction to pericytic venules and displays isolated gap junctions of smaller size and lower frequency than in arterioles.[10] In elastic and muscular arteries, the endothelium has a complex

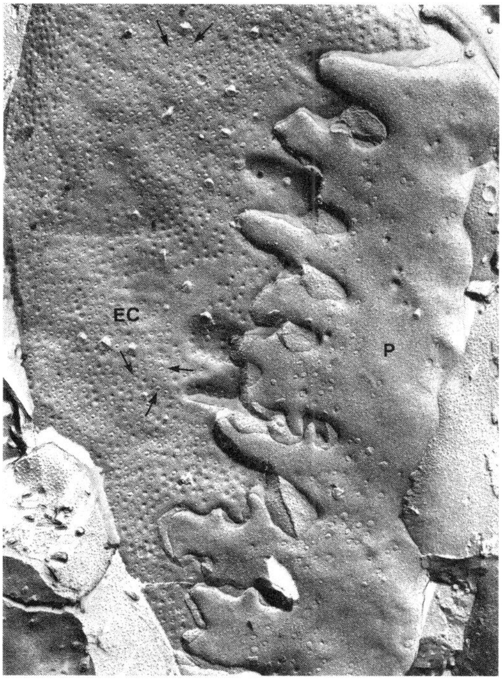

5. In diaphragm capillaries, the endothelial cell (EC) is provided with a large number of plasmalemmal vesicles that can be appreciated in freeze-fracture preparations (arrows). In capillaries as well as in postcapillary (pericytic) venules, the endothelial cells have a close relationship with the underlying pericyte (P), whose processes are tightly bound to the endothelial cell. From Simionescu et al., 1974. *J. Cell Biol.* 60:128–152.

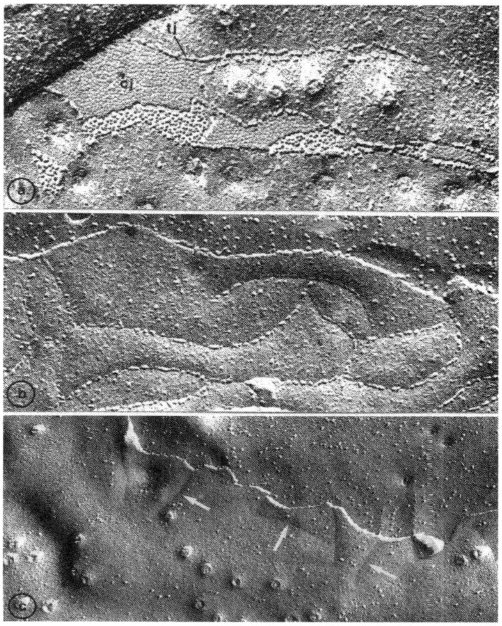

6. Differentiation of endothelial junctions in sequential segments of the microvasculature as seen in freeze-fracture preparations. (a) Arteriolar endothelial junctions are characterized by elaborate tight (occluding) junctions (tj) with intercalated large gap (communicating) junctions (cj). (b) Capillary endothelium is endowed with occluding junctions that only appear as continuous and branching strands. (c) In venular endothelium, tight junctions are discontinuous, appearing as low profile ridges and grooves (arrows). From Simionescu et al., 1975. *J. Cell Biol.* 67:863–885.

of occluding (tight) and gap (communicating) junctions, comparable to yet less elaborate than those found in arterioles. In the venous endothelium (medium and large sized veins), there are long occluding junctions with few associated gap junctions. Two new and different endothelial junctions were detected in arteries and veins.[11] Although these results could be only indirectly correlated with the degree of paracellular permeability, they supported the hypothesis that the features involved in the transendothelial transport might differ along the various segments of the vasculature. We proposed the term communicating junction (*macula communicans*) as a substitute for gap junction since the latter was inappropriate and confusing, especially for the already reported function of direct intercellular communication of gap junctions. However, the term was not appealing to the scientific community because it was too long and too Latin.

Structural basis of permeability in sequential segments of the vasculature. Based on the finding that the organization of endothelial junctions varies significantly along the vasculature, the conflicting data on the permeability of junctions in small vessels identified as capillaries, and the physiological studies suggesting a gradient of permeability with the maximum in venules, we set up experiments to study the permeability along well-identified successive segments of the microvasculature. In the mouse diaphragm, we detected the existence of vascular territories in which the supplying arteriole enters and the draining venule leaves the

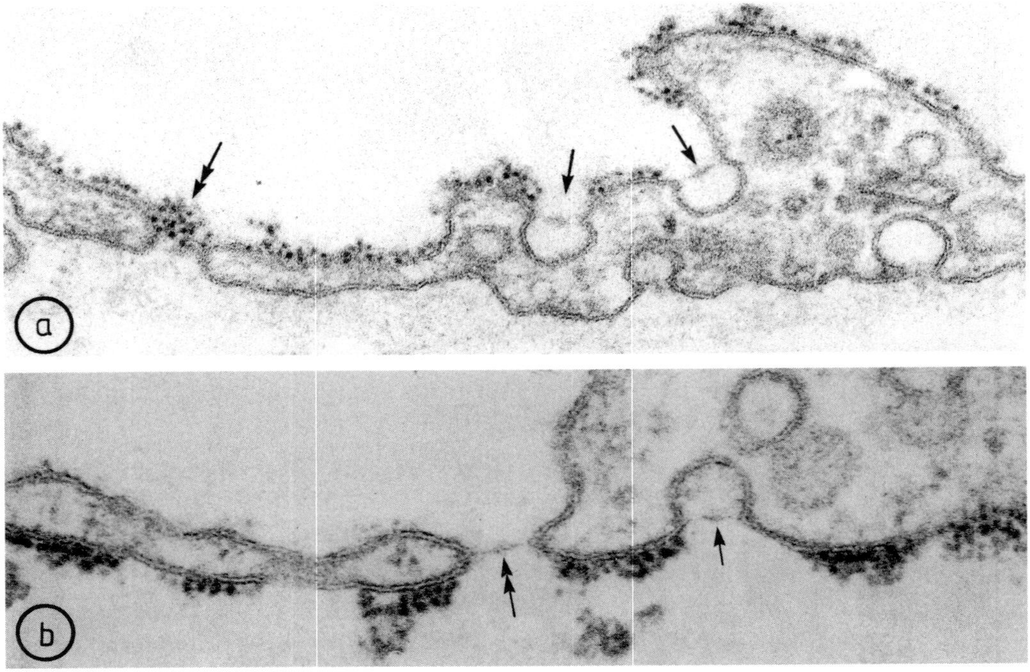

7. Anionic sites detected with cationized ferritin (CF) perfused in situ are unevenly distributed on the surface of endothelial cells. (a) CF accumulates on fenestral diaphragms (double arrow), uniformly decorates the plasmalemma proper, but fails to label the plasmalemmal vesicle membrane and its diaphragm (arrows). (b) The abluminal aspect of endothelial cell exposed to CF reveals the absence of labeling of vesicle and their diaphragms (arrow) as well as the lack of decoration of the abluminal aspect of fenestral diaphragms (double arrow). Figure a: from Simionescu et al., 1981. *J. Cell Biol.* 90:605–613. Figure b: from Simionescu et al., 1982. *J. Cell Biol.* 95:425–435.

capillary bed from opposite ends. These "bipolar microvascular fields" had the advantage that regional differences in the structure and function of arterioles, capillaries, and venules could be investigated.[12] Using hemeundecapeptide as a probe, we found that the vesicular transport was particularly extensive in capillaries and pericytic venules, possibly due to the high number of transendothelial channels found at this level. In arteriolar and middle segments of capillaries, the probe passed only through vesicles and channels.[13] We also tried to estimate the size limit of molecules that can pass through these junctions and found that the size limit for permeant molecules for open venular junctions is about 6 nanometers.[14] These data could explain the conflicting data from other laboratories on capillary transendothelial transport. Based on precise identification, our results indicated that vesicles and channels contribute to the intense transport at the level of capillary. The pericytic venules are the most permeable microvascular segment, having a number of open interendothelial junctions that normally allow rapid passage of small molecules. As Palade and Majno had described earlier,[15] they are also sensitive to vasomediators (i.e., histamine and serotonin), a feature highly significant for the inflammatory process.[16]

Uncovering biochemically differentiated microdomains on endothelial plasmalemma. After establishing the role of molecular size in endothelial permeability, we questioned the effect of molecular net charge in transendothelial transport. We hypothesized that if an electric charge restriction existed for transport, it should operate at the level of the endothelium.

Reported data indicated that the surface of endothelial cells is negatively charged, but there were no reports on the distribution pattern of these charges in relation to the structural features involved in permeability: vesicles, channels, and fenestrae. Upon perfusion in vivo or in situ of cationized ferritin, we detected that the binding of this probe was unevenly distributed on the endothelial surface. Thus, cationized ferritin evenly decorated the plasmalemma proper but labeled the fenestral diaphragms and coated pits at much higher density and affinity. Unexpectedly, cationized ferritin did not bind to the membrane of plasmalemmal vesicles and transendothelial channels or their diaphragm. Alcian blue, a smaller cationic probe, gave the same distribution pattern. This was the first indication that variants in the chemistry of plasmalemma and associated structures may give rise to differentiated microdomains.[17]

To assess the chemical nature of the anionic sites of differentiated microdomains, enzymes (neuraminidase, hyaluronidase, chondroitinase ABC, heparinase, and various proteases) were perfused in situ prior to cationic ferritin. We showed that heparan sulfate and/or heparin contribute primarily to the anionic sites of the fenestral diaphragms, whereas those of the plasmalemma proper were of mixed chemical nature.[18] These data were validated and extended when we investigated the distribution of lectin receptors on the luminal surface of capillary endothelium. Using a battery of lectins that recognize specific monosaccharide residues, we found that plasmalemma vesicles, transendothelial channels, and their associated diaphragms were particularly rich in D-galactosyl residues and *N*-acetylglucosaminyl residues and that receptors for the lectins tested were absent or considerably less concentrated on fenestral diaphragms.[19] Another set of experiments showed that differentiated microdomains were also present on the abluminal surface of the endothelial plasmalemma. At this level, we found that the plasmalemma proper was extensively labeled and coated pits were heavily decorated, whereas the membrane and stomatal diaphragms of vesicles and transendothelial channels were free of cationic ferritin. The interesting finding was that in contradistinction to the heavy labeling of the luminal aspect of fenestral diaphragms (figure 7a), the abluminal aspect of fenestrae was free of cationic ferritin-detectable anionic sites (figure 7b). These findings established the existence of differentiated microdomains on the abluminal endothelial plasmalemma and suggested that the capillary wall selects permeant molecules by their size and charge.[20] The investigation was extended to the continuous endothelium of the microvasculature of the heart, diaphragm, and brain, and the results confirmed the ubiquitous existence of differentiated microdomains.[21]

The important finding was that plasmalemmal vesicles have the surface chemistry suitable for transport of plasma proteins (most of which have a negative charge) and that these chemically differentiated domains corresponded to well-defined structural features involved in endocytosis and transendothelial transport. For a better definition of the process, Nicolae coined the term "transcytosis" (literally, across or through the cell) to describe the mechanisms (fluid phase, adsorptive, and receptor mediated) in which a cell encloses extracellular material in an invagination of the cell membrane to form a vesicle and then moves the vesicle across the cell to eject the material through the opposite cell membrane by the reverse process.[22] This concept of transcytosis was first described for endothelial cells and then became well accepted and further extended for all types of epithelial cells. We concluded that the uptake

Nicolae Simionescu, 1970.
Courtesy of the Rockefeller Archive Center.

and transcytosis of macromolecules depends on the size, electrical charge, and chemical composition of both the permeant molecules and the relevant endothelial features involved in capillary permeability.[23]

An Institute of Cellular Biology and Pathology in Bucharest, Romania. Our interaction with George Palade was both scientifically productive and humanly gratifying. After our first year in the unique climate of his laboratory, we thought that it would be a worthy endeavor to introduce cell biology in Romania by creating an institute where young Romanian scientists would find the intellectual space to foster their talent and desire for research. It was a big challenge at that time. Initially, Palade was skeptical, but he decided to back up our mental dream and accepted our idea as another experiment. During the following eight years, between experiments and writing papers, Nicolae and I designed the layout of the Institute, planned the scientific focus, chose the equipment, and battled with authorities. Palade helped us in many ways, starting with letters to authorities of that time, explaining to many others the importance of financing "the Simionescus's experiment," and ending with a remarkable idea, in 1979, the year we left the United States. He helped make an official arrangement with Romanian authorities by which Nicolae and I would spend three months each year as visiting professors at Yale University to keep our collaboration open and to avoid isolation. Our position as visiting professors in cell biology at Yale University continued for seventeen years, long after Palade moved to the University of California in San Diego.

The Institute of Cellular Biology and Pathology was inaugurated on 3 September 1979 with a scientific workshop at which Palade, Christian de Duve, Günter Blobel, Marilyn Farquhar, David Sabatini, and Werner Franke participated. From its inception, we planned to concentrate on one significant scientific and socioeconomic theme: namely, to uncover at the cellular and molecular level the functioning of the cardiovascular system in health and disease, with particular emphasis on atherosclerosis and diabetes. We envisaged three main steps: cell biology, followed by cell pathology, and then cell pharmacology of the heart and blood vessels. It was a reasonable and necessary extension of our previous studies. The work in the Romanian Institute was difficult, and when all funds for doing science in Romania were cut in 1981, our "survival kit" for twelve years consisted of various NIH grants from the United States. In 1990, the Institute became part of the UNESCO–Molecular and Cell Biology Network. On Nicolae's death in 1995, the Institute was renamed in his honor. In 1998, it was elected as a Centre of Excellence of the Romanian Academy, and in 2000 it was selected as a Centre of Excellence of the European Community. Life at the Institute and the enthusiasm for science continue. Numerous young scientists are enrolled in the PhD program and represent the future of the Institute. Work goes on, and I hope it will never end.

Many things contribute to the life of a scientist, but I think that among the most important are the mentors, the environment, a lot of hard work, and a little bit of luck. Nicolae and I considered ourselves favored to inquire into some of nature's secrets and lucky to replicate the superb American intellectual environment in an institute in our own country. Nicolae often said that our interaction with Palade was a decisive step in our evolution. Personally, I was even more fortunate to have had two mentors, George Palade and Nicolae Simionescu. I am what I am because of them.

Chapter 14

Nuclear Proteins: Control of Gene Structure and Function

Vincent G. Allfrey*

The cell nucleus, central and commanding, is essential for the biosynthetic events that characterize cell type and cell function. It is a vault of genetic information encoding the past history and future prospects of the cell, a firm and purposeful guide to the complex patterns of gene expression during embryonic growth and development. And it is a barometer exquisitely sensitive to the changing demands of the organism and its environment.

Modern cell biology emphasizes the importance of structure in controlling the chemical activities of biological systems. In the case of the nucleus, there is increasing awareness that changes in the structure of chromosomes exert genetic control by altering the accessibility of desoxyribonucleic acid (DNA) sequences to RNA polymerases and different transcription factors. Molecules are sequestered, arranged in complex and interlocking matrices, gathered in clusters or fibrous arrays, aggregated, dispersed, or dissolved. This aspect of structural molecular biology is especially significant in considering the nucleic acids, whose molecular weights range into the millions and whose extended linear dimensions make them primary visible units of nuclear structure. How does chromosome structure permit replication of all the DNA during cell division and yet permit the precise delimitation of function in active and inactive genes? To answer such questions, the biologist interested in nuclear functions is obliged to relate morphology to activity and to search for cytological clues to regulatory mechanisms.

The study of the cell biology of the nucleus at The Rockefeller Institute began after many impressive discoveries of nucleic acids and their chemistry as well as its role in the synthetic activities of the cell and in the transmission and expression of hereditary factors. In fact, the research began just before the 1944 discovery by Oswald Avery, Colin MacLeod, and Maclyn McCarty, also at Rockefeller, that DNA in the chromosomes is the genetic material. The cell biology described in this chapter falls into three phases in the laboratories of Alfred E. Mirsky and Vincent G. Allfrey: early work by Mirsky on nucleoproteins, DNA, and chromosomes; a short period when Mirsky and Allfrey tested the possible existence of a protein-synthesizing system in the nucleus; and research on the role of histones with DNA in the nucleus that was begun by Mirsky but taken over by Allfrey.

Mirsky and Desoxynucleoproteins

When Mirsky came to Rockefeller in 1927, he was primarily interested in protein denaturation. This interest began as a doctoral student at Cambridge University under Joseph Barcroft in studies of the biophysics of oxygen binding to hemoglobin. After spending a postdoctoral

---

*This chapter was begun by Vincent Allfrey and completed after his death in 2002 by Philip Siekevitz, Carol Moberg, Oscar Pogo, Beatriz Pogo, and Bruce McEwen.

year in the biophysics laboratory of the Cancer Commission at Harvard, he joined The Rockefeller Hospital laboratory of cardiologist Alfred E. Cohn. Working with Mortimer Anson, a fellow Cambridge graduate, he studied the denaturation and renaturation of many proteins, concentrating on hemoglobin, and learned that their loss of function was due to a change in the physical state of the protein rather than a change in its amino acid sequence. After an important sabbatical year with Linus Pauling at the California Institute of Technology in 1936, he showed there was a relationship between protein infrastructure and its function and that these structural changes could be altered in a reversible fashion.

In 1937, Mirsky began to study changes in the muscle protein myosin in an unsuccessful attempt to correlate its denaturation with functions such as muscle contraction. In the process, he came across a 1938 paper by Robert R. Bensley, who reported isolating a new protein complex from liver cells, which he thought derived from the protoplasm and so named it "plasmosin."[1] Bensley was the cytologist who in 1934 used maceration and centrifugation to isolate mitochondria from liver tissue and who disagreed with Albert Claude about his identification of the isolated subcellular particles (see Chapter 2).

Mirsky's first attempt to combine chemistry with cell biology began with studies of the chemical nature of plasmosin in 1940. Mirsky enlisted the aid of Arthur W. Pollister of Columbia University, an expert on chromosome morphology and quantitative cytochemical techniques. They isolated and analyzed plasmosins from liver and several other organs and found to their surprise the material contained DNA and the proteins histone and protamine.[2] Their major approach was a gentle procedure involving only successive extractions with neutral sodium chloride solutions, which proved very popular at the time in providing purified deoxynucleoproteins in the form of fibers. From this, DNA was purified by the 1934 method of Maurice Sevag using immiscible organic solvents and centrifugation to produce DNA in solution.[3] Their work made it clear that Bensley's plasmosin was a protein complex located in the nucleus, but not in other parts of the cell, and that the protein was part of the structure of the chromosome. They named it "chromosin" based on its location in the nucleus, but they quickly abandoned the term once they gained a better understanding of its chemical composition.

By this time, Mirsky was considering the possibility that changes in nuclear proteins might alter chromosome structure and thus influence the functions of the genes, but he was not yet committed to the belief that DNA was the sole repository of genetic information. There was evidence that proteins could carry out highly specific activities, but the model of DNA proposed by Rockefeller's biochemist Phoebus A.T. Levene in 1919, a highly repetitive structure containing only four nucleotides in approximately equal proportions, seemed deceptively simple.[4] Mirsky dismissed this model at the 1941 Cold Spring Harbor Symposium, adding, "Other components of the gene are desoxyribonucleic acids and relatively simple basic proteins, the protamines and histones. Our knowledge of the gene will be extended when we know more about the interactions between these components."[5]

Further work of Mirsky and Pollister attempted to isolate chromatin from nuclei of certain animal cells. Of some significance during 1943, Albert Claude was also isolating particles from leukemic cell nuclei. Using the electron microscope at the Interchemical Corporation, he observed that the chromatin threads he had found contained nucleoproteins that were "bead-like structures" spaced along the body of the threads (figures 1 and 2).[6] Claude

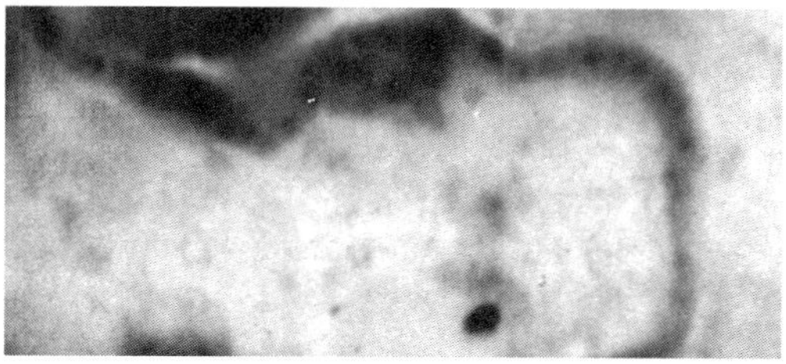

1. Electron micrograph by Albert Claude shows a chromatin thread, 70 micrometers in width, containing the spaced, bead-like structures 20–60 micrometers in diameter. ×18,000. From Murphy, 1943. RUA, BSD Reports. Volume 32, page 70.

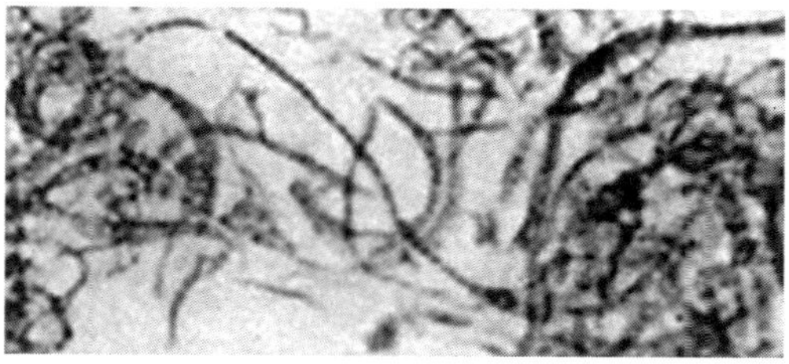

2. Light micrograph of isolated chromatin threads from mouse, revealing a finer structure of granules (histones) in succession along the threads. ×1,600. From Claude and Potter, 1943. *J. Exp. Med.* 77:345.

abandoned this work on nuclei to pursue the identity of the more mysterious subcellular particles he called microsomes. Both Mirsky and Claude spoke at the November 1942 symposium honoring Bensley's seventy-fifth birthday devoted to the organization of the cytoplasm. Claude reported his chemical analysis of cytoplasmic components "separated by the centrifugation technique, indicates that . . . nucleic acids are not only present in the framework of the nucleus but also that they contribute to the constitution of important cytoplasmic structures." This comment reflects Claude's finding of ribonucleoproteins isolated from the cytoplasmic microsomes, the chicken tumor agent, and the nucleus (see Chapter 3).[7] Mirsky concluded that chromatin is found only in the nucleus and exists as "a complex of desoxyribose nucleic acid with a protein" either of the protamine or the histone type. He did not speculate as to the specific chemical nature of the gene but argued that "the genes are physical entities within the threads of chromatin, the chromosomes" and that a "constituent of chromatin contained the substances of the gene." He ended, however, with the leading question, "Can we find in nucleohistone the chemical basis of heredity?"[8]

Mirsky was keenly aware of Oswald Avery's work on identifying the chemical nature of the transforming substance in pneumococci showing that the bearer of genetic information is DNA and not a protein.[9] In fact, there was a brief, early collaboration between the two laboratories when Maclyn McCarty used the Mirsky-Pollister method to purify the deoxynucleoprotein fraction. The fact that the transforming substance could be prepared from pneumococci by using a technique designed to capture DNA from animal tissues was significant, although the yield of pneumococcal DNA was small. According to McCarty, it provided "a considerable boost to the idea that the transforming principle was indeed DNA."[10] While the 1944 paper concluded DNA was the genetic determinant, Mirsky disputed this claim for several years because he believed there was protein contamination in the preparation and suggested instead the protein was the gene. It was not until 1946 when McCarty showed that the addition of purified deoxyribonuclease destroyed the biological activity of the transforming principle that the controversy was seemingly settled for most scientists.[11] Mirsky, however, was still not convinced and for a long time did not accept that a mere four nucleotides could give the great and varied specificity required for genetic material.

Chemical Properties of Chromosomes

On Alfred Cohn's retirement in 1944, Mirsky was absorbed into Rockefeller Director Herbert Gasser's group called Physiology. A year later, Mirsky was joined by Hans Ris, a recent graduate from Pollister's laboratory with expertise in microspectrometry. They developed a method to measure the quantity of DNA in a single nucleus rather than the average in a mass of nuclei. The procedure depends on the Feulgen reaction, the standard staining test for DNA that turns it a brilliant crimson. Cells with nuclei colored by this reaction are placed on a microscope slide, and a photoelectric cell is inserted in the eyepiece of the microscope to measure the amount of light absorbed by the pigment in the nucleus. By determining the amount of DNA in a single nucleus, they calculated and compared the DNA per nucleus in different organs of sixty different species of mammals, fish, birds, and amphibians. They found all somatic nuclei in every organ of a given species contain the same amount of DNA, but this amount differs from one species to another. In contrast, they found the nuclei of the germ cells have only half as much DNA as the somatic cells. These observations directly linked DNA content to chromosome number in the species and supported the concept of DNA as the hereditary material. By 1949, Mirsky had reluctantly begun to concede that "DNA is part of the genic material. This does not mean, however, that the gene consists of nothing but nucleic acid."[12]

Unknown to Mirsky, however, the same conclusion of DNA constancy in different cells had been published the year before in Strasbourg, France, by André Boivin and Collette and Roger Vendrely.[13] I can attest, with regard to the priority of this discovery, that Mirsky had obtained solid evidence for DNA constancy in different somatic cells and for the halving of that amount in sperm cells and oocytes three years before in 1945. I had just returned from military service in Germany and was looking forward to rejoining the Institute and Mirsky's laboratory. When I asked him "What's new?" he immediately gave me an account of how he was able to determine the amount of DNA in a nucleus by measuring the total mass of DNA in a suspension of isolated nuclei and then dividing by the number of nuclei on a hemocytometer slide under the microscope, but he was hesitant about publishing without further confirmatory evidence. Nonetheless, the constancy of DNA in somatic cells and its halving in germ cells became known as the Boivin-Vendrely rule.

Another problem tackled by Mirsky and Ris was the chemical analyses of isolated chromosomes. Early microscopists recognized an equal distribution of chromosomes to both daughter cells during mitosis, but chromosomes were not visible in non-dividing, or interphase, cells unless they were injured or dye stained, thus causing diffuse, filamentous structures to become apparent. Walther Flemming in 1882 was the first to observe these intranuclear structures and named them "chromatin."[14] Today, chromatin encompasses a broader view that includes interactions of DNA with histones, other proteins, enzymes, and regulatory proteins involved in the transcription of DNA into messenger RNA (mRNA).

Mirsky, first alone and then with Ris, isolated threads from cells having prominent nuclei and sparse cytoplasm from trout erythrocytes and calf thymus lymphocytes. Light microscopic examination, chemical tests for DNA and histones, as well as the absence of detectable cytoplasmic contamination seemed to confirm the chromosomal origin of these fragments. The reason why chromosomes are not visible in the interphase cells, Ris commented, is that "in the living nucleus the chromosomes are greatly extended, filling the entire nucleus. Upon injury the chromosomes condense and therefore become visible. At the same time the nuclear volume decreases. This behavior of the chromosomes is connected with their content of DNA."[15,16]

However, W.G.P. Lamb claimed the "interphase chromosome fraction" was an artifact created by the shearing forces in the isolation procedure.[17] Thus, Mirsky and Ris modified their method to use cells with a greater volume of cytoplasm, such as liver, kidney, or pancreas, used a colloidal mill rather than a Waring blender for shearing, and added citric acid in the extraction medium to decrease cytoplasmic debris. As a result, they got a preparation that was identifiable by light microscopy and by cytological staining as chromosomal filamentous material.[18,19] They substantiated their conclusion by referring to studies by Hans H. Pfeiffer, who had compared thin nuclear sections with the electron microscope to adjacent thick sections with the light microscope and observed that the electron-dense regions of the chromatin in electron micrographs corresponded to the chemically stained DNA-rich regions in light micrographs.[20] They also refuted the argument of random fragmentation by referring to their earlier light microscopy studies, which showed that a chromosome carrying the genes for the synthesis of ribosomal RNA (the so-called nucleolus organizing chromosome) was still attached to one end of a particular chromosome among the many visible chromosomal threads.[21] They cautioned that not all the threads were individual chromosomes since "no doubt, sticking together, distorting, and fragmentation do occur. Some of the large structures we observed may be stretched nuclei. The important point is that a high percentage of the threads show the characteristics expected of interphase chromosomes, and above all, individuality of structure that could not be produced by random fragmentation."[18]

Mirsky concluded in his 1950 Harvey Lecture that the constant components of the chromosome ensure its genetic continuity and hypothesized that the variable components may be of greater interest because they "are influenced by the cytoplasm . . . and play a part in the physiological adaptations of the cell."[22] Thus, in other studies of isolated chromosomes, Mirsky and Ris tested the constancy and variability of histones. In contrast to the constancy of DNA, they found great variations in composition, although not in amounts, of these nuclear proteins within a single organism.

Mirsky's unique perspective on the importance of proteins for genomic activity led him for the rest of his career to think about the regulation of variable gene expression by proteins that interact with DNA, and this was long before this notion became fashionable. He recognized, for example, that steroid hormone receptors, discovered in the early 1960s, could, in principle, participate in gene regulation by the mechanisms that he and Allfrey were beginning to uncover via their studies on histone and non-histone chromosomal protein modifications. His advocacy of "variable, environmentally regulated gene expression" played a major role in influencing Bruce McEwen to study steroid hormone effects on the brain.

Early Research on Histones

With the discovery of histones by Nobel laureate Albrecht Kossel in 1884, it was recognized that they are the major protein associated with nucleic acid in many different animal cells.[23] Since the histones are basic in nature and could bind to the acidic nucleic acids, Kossel may have thought of them as providing a structural framework for the nuclein, or nucleic acid, that had been discovered ten years earlier by Friedrich Miescher.[24] Kossel found the histones he isolated from bird erythrocytes differed from protamine, another basic protein that Miescher had isolated from fish sperm. The reason for Kossel's choice of the name histone is not known, although the Greek *histos* appears in words for such interesting structures as a spider web and a mast of a ship, in addition to forming the root of histology, a science devoted to structure. Alternatively, histone's high nitrogen content may have prompted him to recall that the amino acid histidine has the highest percentage of nitrogen of all the amino acids. Kossel later recognized that the high content of the basic amino acids arginine and lysine in histone made it likely that they were interacting with the negatively charged phosphate groups of the nucleic acid, but he had no direct evidence for this idea.[25]

In 1948, when Mirsky became a full member of Rockefeller, he established an independent laboratory called General Physiology. This field reflected the Institute's powerful influence of Jacques Loeb, a German-born physiologist who was recruited to Rockefeller in 1910 by Simon Flexner. Loeb advocated general physiology as the study of biological systems by methods of chemistry and physics, and he founded *The Journal of General Physiology* at Rockefeller in 1918, which was edited by Mirsky from 1950 to 1961.

Mirsky, Ris, and Marie Daly, a new PhD in chemistry from Columbia, undertook a comparative study of basic proteins in isolated chromosomes to learn whether the amino acid composition of histones could differentiate them from other proteins and whether the same amino acid composition occurred in histones from different sources. They isolated histones from calf thymus nuclei, calf liver nuclei, and avian erythrocytes using essentially the method of Kossel, namely breaking the bonds between DNA and histones using weak hydrochloric or sulfuric acid and precipitating them from solution with excess ammonia. What they found were distinctly different basic proteins in different tissues. In agreement with Kossel, they found that arginine and lysine were the predominant amino acids. In disagreement with Kossel, who found a three to one ratio of arginine to lysine, they found a one to one ratio from all three sources. They also determined that the proteolytic enzyme pepsin digests the histones without affecting the chromosome structure, indicating the possible existence of non-histone proteins that hold the chromosome together. The same results were obtained with isolated histones, indicating the accessibility of various peptide bonds when the histones are bound to DNA.[26]

For a brief period, Mirsky and Daly examined the heterogeneity of the histone fractions.[27] This was prompted by two old articles on lysine-rich material. One by Kurt Felix in 1921 reported a higher content of lysine in the basal bodies of intestinal mucosal membranes than had been reported earlier by Kossel.[28] The other paper by Kossel and E.G. Schenck in 1928 found a much higher ratio of lysine to arginine in histones from carp sperm than those reported earlier by Kossel.[29] Daly and Mirsky went further by separating two fractions of histones using a gentler method, first a citric acid/sodium chloride solution to dissociate a lysine-rich fraction from DNA and then an increased salt concentration to extract an arginine-rich fraction. Similar results were obtained whether thymus nuclei, liver nuclei, or avian erythrocytes were used. When pulverized DNA was added to the two fractions, they noticed a difference between them, both in their electrophoretic behavior and in their physical appearances, suggesting the importance of these two basic amino acids in the physical properties of the histones.

During this period, many investigators, including Stan Moore and William Stein at Rockefeller, worked on fractionating histones with about twenty fractions separable by ion-exchange chromatography. Subsequent analyses by Kenneth Murray at the Medical

Research Council showed that the arginine to lysine ratios of these fractions varied greatly.[30] These variations, which gave rise to a grouping of histones into very lysine-rich, moderately lysine-rich, and relatively arginine-rich fractions, partially validated the work of Daly and Mirsky in stressing a high lysine content in histones. However, even though this grouping accentuates the basic amino acids in the binding of histone to DNA, it disregards other amino acids present in histones that could also be important.

Protein Synthesis in the Nucleus

For unknown reasons, Mirsky stopped working on histones from 1955 to 1963 and began to examine whether a system exists within the nucleus capable of synthesizing nuclear proteins. He was joined in this research by Vincent Allfrey, who was just beginning his second position in Mirsky's laboratory. In 1941, while a chemistry student at the City College of New York, Allfrey held a part-time position as a laboratory assistant, where, in addition to washing glassware and dusting books in Alfred Cohn's extensive library, his first jobs for Mirsky were to collect animal organs and blood from New York City's slaughterhouses and poultry and fish markets. It was on a longer such trip to Cold Spring Harbor Laboratory on Long Island, then headed by geneticist Milislav Demerec, that he was treated to a discourse about what they were studying. This made him decide that science could be fun, and he immediately resolved to make it his career. After serving on the battlefront in Europe during World War II, he received a PhD at Columbia University, where his research focused on fluorescence of vitamin B in cells. In 1949, he rejoined Mirsky's laboratory, this time as a member of the scientific staff. He was a versatile chemist, although not a biologist or geneticist, whose abiding opinion of science was that "you always just scratch the surface."

The genetic material of the cell is stored in the nucleus and transcribed into mRNAs, which are then processed and exported to the cytoplasm, where, so the orthodoxy goes, they are "read" by the ribosome and translated into proteins. For Mirsky and Allfrey, the issue of protein synthesis in the nucleus would be the unambiguous demonstration of translation in the nuclear compartment, and they were conceptually strong believers in nuclear translation. Were nuclear proteins synthesized within the nucleus or by the newly recognized cytoplasmic system and then transported into the nucleus? (See Chapter 7 on the discoveries in the cytoplasmic system by the Porter/Palade laboratory). If the former was true, was the nuclear system the same, similar, or dissimilar to the cytoplasmic one?

At first, Mirsky, Allfrey, and Daly examined the incorporation of radioactive amino acids into the proteins of avian erythrocytes as well as liver and pancreas tissues. A short time later, however, Palade and Siekevitz became engrossed in a similar study of cytoplasmic protein synthesis in the pancreas, and a controversy ensued about which group should be studying the pancreas. The matter was resolved in favor of Palade and Siekevitz, who used a combination of electron microscopic and biochemical experiments instead of a solely biochemical approach. Thus, the Mirsky laboratory turned to research on nuclear protein synthesis while adopting the pioneering methods of the Porter/Palade laboratory.

The first step, in 1952, was to isolate the nucleus from the cell using a method first proposed by Martin Behrens in 1932,[31] which involved homogenization in a non-aqueous medium and organic solvents to minimize the loss in water-soluble nuclear components.[32]

They used this technique to measure enzyme activity in the nucleus and to establish criteria for judging the quality of nuclei isolated in aqueous media from various tissues. When they found equal enzymatic activities in thymus nuclei isolated in both media, they decided to use the aqueous method to study protein synthesis in thymus nuclei.[33,34]

In vitro experiments: Biochemical properties. Allfrey published the first paper on nuclear protein synthesis in 1954.[35] He used the method of Siekevitz, who was still at Harvard, to determine the incorporation of radioactive alanine into thymus nuclear protein and found, as Siekevitz did for a liver cytoplasmic system, that a source of energy was necessary for maximal incorporation.[36] Using specific inhibitors, Allfrey found that aerobic oxidation and oxidative phosphorylation were necessary for the uptake of the label into proteins and that the system was unlike that involving protein synthesis in bacteria. He also found that sodium and not potassium ions were needed in the incubation medium, that an oxidative mechanism in the nucleus showed similarities and dissimilarities to the mitochondrial one, and that only the "natural" L form of the radioactive DL-alanine was incorporated.[37] These results strongly indicated that a protein-synthesizing system existed in the nucleus, one that required the presence of DNA and an energy source generated by a peculiarly nuclear oxidative mechanism.

Other scientists questioned the purity of the thymus nuclear preparation since the thymus cell has a predominant nucleus surrounded by a thin rim of cytoplasm, so it seemed likely that the preparation could easily be contaminated with whole thymocytes or tabs of thymus cytoplasm. Trying to resolve this issue by using electron micrographs taken by Mike Watson in the Porter/Palade laboratory, Allfrey and Mirsky counted the number of whole cells and nuclei and determined there was a small percentage of thymocytes in their preparation. They also admitted the "thymus nuclei prepared in sucrose-calcium chloride solution are not entirely free of small amounts of adhering cytoplasm. This type of contamination is more difficult to evaluate."[38] Later experiments to reduce this contamination included varying the homogenization condition, using media of differing ionic strengths, and purifying the nuclei by density gradient centrifugation, all methods relying on the proposition that free nuclei have a higher density than do whole cells or nuclei with tabs of cytoplasm. Finally, they incubated nuclei under conditions in which the incorporation of radioactive amino acids into protein occurs and then purified the preparation using density gradient centrifugation, not in sucrose but in Ficoll, the large high molecular weight polymer of exceeding solubility in water that had been introduced in 1958 by Heinz Holter and Max Møller at the Carlsberg Laboratory in Copenhagen. Radioautography of this preparation showed a predominant number of radioactive grains over inner "condensed" chromatin areas, with just a few at the edges of the nuclei.[39,40] A different set of biochemical experiments also suggested there was a nuclear protein synthesis that was inhibited by DNase treatment in the isolated nuclei, but the energy mechanism necessary for this was still problematic.[41,42]

Isolation and properties of intranuclear ribosomes. The next question about the existence of a protein-synthesizing system concerned whether ribosomes are also found in the nucleus. Nuclei were isolated from calf thymus lymphocytes by the sucrose-calcium chloride method, incubated in solutions containing various radioactive amino acids, and then subjected to chemical extractions and differential centrifugation. Electron micrographs of isolated nuclei, nucleolus, and the isolated subfractions revealed a large number of small, hollow particles similar in size to cytoplasmic ribosomes. The Allfrey group argued these particles were exclusively of nuclear origin because their sizes were close to particles seen by other microscopists in the nucleolus. Radioactive labeling experiments resulted in labeling of the nuclear but not the cytoplasmic "fragments;" nuclei preincubated with either DNase or RNase showed that DNase, but not RNase, lowered the incorporation of radioactive alanine into the proteins of these particles, whereas cytoplasmic ribosome incorporation is sensitive to RNase. Moreover, to incorporate radioactive alanine into protein required the addition of a nuclear protein extract similar to a cytoplasmic extract, an energy source such as glucose, plus ATP and GTP, all of which resembled the mechanisms known of the cytoplasmic ribosome system.[43,44] Allfrey and Mirsky presented these findings at a 1960 conference where Siekevitz presented his work with Palade on in vivo cytoplasmic protein synthesis of secretory proteins in the pancreas.[45]

Skepticism continued on nuclear protein synthesis, so Allfrey and Mirsky refined their methods for isolating ribosomes from thymus nuclei. Magnesium ions were added for stability,[46,47] and ribosomes were purified by sucrose-density gradient centrifugation, a method introduced by Edward Pickels at Rockefeller (see Chapter 2) and modified for nucleoprotein particles by Roy Britten and Richard Roberts at the Carnegie Institution.[48] This preparation gave an ultraviolet peak showing one-third RNA and two-thirds protein and had a sediment constant comparable with that of cytoplasmic ribosomes. A further purification used deoxycholate, as was done for cytoplasmic ribosomes, and this resulted in particles having significantly higher RNA content but similar nucleotide composition compared with cytoplasmic ribosomes. Electron micrographs revealed the nuclear particles were larger than cytoplasmic ones. The preparation obtained from thymus nuclei was quite similar to ribosomal particles obtained by other scientists from various cytoplasmic sources.[49,50]

Energy sources: Comparison of nuclear and cytoplasmic systems. An important adjunct was a determination of the energy source necessary for the synthetic reactions in the nucleus. This work was carried out by Allfrey and Mirsky with graduate student Bruce McEwen (later a professor of neuroendocrinology at Rockefeller).

When McEwen came to Rockefeller, energy metabolism and oxidative phosphorylation were very popular topics through the work of Britton Chance at the University of Pennsylvania and Albert Lehninger at Johns Hopkins, among others. According to McEwen, "I heard the story of cell nuclear energy metabolism and that there was a vacancy in the Allfrey-Mirsky laboratory with the graduation of Johns W. Hopkins III, so I asked whether I could work on the problem. They agreed and accepted me after a trial period. Eric Davidson was in the laboratory along with postdoctoral fellows Oscar and Beatriz Pogo, John Frenster, Atsuhiro Sibatani, and Siwo deKloet, and it was a very lively place. Alfred loved afternoon teas, and his lab helper/dishwasher, Ellie Donahue, would make tea every afternoon at three o'clock. He would hold forth on many topics ranging from embryonic development, puffing of giant salivary gland chromosomes, the cloning work of Robert Briggs and Thomas King, the Mary Lyon X-inactivation hypothesis, and the similarities and differences between identical twins. Needless to say, I learned a great deal that has lasted for many years in shaping my outlook on science. Vince Allfrey was also a great teacher and role model, mentoring me in laboratory techniques and writing, among other contributions."

Their first experiments compared the synthesis of ATP by thymus nuclei with a mitochondrial preparation and revealed that in the nuclear preparation, the synthesis is an aerobic process and that some compounds inhibited both mitochondrial and nuclear systems, while others inhibited only one or the other.[51] Exploring the nature of the substrates in thymus nuclei available for metabolism to produce the synthesis of ATP (either substances of the hexose monophosphate shunt, glycolysis, or the Krebs cycle), they found some enzymes and metabolites of these pathways in both the nuclear and mitochondrial preparations.[52] When they measured ATP content after adding Krebs cycle substrates, such as pyruvic acid that would enter thymus nuclei, they also found that glycolytic and aerobic oxidations of these substrates led to nuclear ATP synthesis. However, since Krebs cycle substrates do not readily penetrate the nucleus, it was surmised that the oxidations of endogenous substrates were responsible for the oxidations and ATP synthesis. But little glycogen and free glucose were found in the isolated nuclei, so the nature of these endogenous substrates remained unknown.[53–55]

Nevertheless, the oxidative cell nuclear metabolism in calf thymus nuclei and intact thymocytes was shown by McEwen to depend on a terminal oxidase different from that in mitochondria because of its lesser sensitivity to carbon monoxide inhibition. As a result, carbon monoxide levels that suppressed mitochondrial oxidative metabolism and suppressed protein synthesis in the intact thymocyte by 60% nevertheless allowed incorporation of radioactive precursors into RNA to proceed without inhibition.[54] This suggests that, at least in immature thymus lymphocytes, RNA synthesis and possibly some protein synthesis depend on energy generated via a non-mitochondrial terminal oxidase that may be associated with the nuclear envelope and appears to be related to cytochrome b5.

Controversy and possible resolution. The postulate that the nucleus has a protein-synthesizing apparatus with the necessary ribosomes, protein, and RNA machinery was not persuasive and even controversial, largely because the thymocyte nuclear preparation could be contaminated by whole cells and/or tabs of cytoplasm containing endoplasmic reticulum–bound ribosomes. Nonetheless, the importance of the idea and the problems associated with it caused many experimenters to work on nuclear protein synthesis.

In 1978, fifteen years after Mirsky and Allfrey began their research, two commentaries appeared on the continuing controversy over nuclear protein synthesis, one by Jo Aline Goidl with supporting evidence for the thesis of nuclear protein synthesis and an adjoining one by W. Ross Allen citing a need for more evidence. Both authors admitted, however, that different interpretations could be made about the results in the papers they cited.[56]

Histones Revisited

Several changes occurred in the Mirsky laboratory in 1963. Vincent Allfrey was promoted to full professor, and he and Mirsky formed a joint laboratory. The laboratory's name was changed from General Physiology to Cell Biology, four years before George Palade and Philip Siekevitz changed their laboratory name from Cytology to Cell Biology. After a lapse of eight years, research was resumed on the role of histones in nuclear physiology and DNA function, and Allfrey became its chief investigator. In addition, the laboratory acquired its own electron microscope and microscopist, Virginia Littau.

Allfrey thought it would be good to continue work on histones because they were the major proteins of chromatin, they could be rather easily purified, and there was much to learn about their chemistry and function. The new studies could take advantage of the cellular biological processes that were being developed in other Rockefeller laboratories.

Also, by 1963, enough work had been done by others on histones that a first World

Conference on Histone Biology and Chemistry was convened in Pasadena, California. Allfrey and Mirsky presented work showing that histones added to thymus nuclei or lampbrush chromosomes caused DNA to condense and inhibited its ability to synthesize ribonucleic acids.[57] This was the first time that a change in function could be linked to a change in structure. Very likely stimulated by other presentations at this meeting, Allfrey returned to work on histones.

Initial studies: Histone modifications and their biochemical effects. The first question they tackled was the process of histone modifications, in which an enzyme attaches an acetyl or methyl group to a specific amino acid on the histone, making some genes more accessible to RNA-synthesizing enzymes that copy the DNA template. Others had reported acetyl groups on the histone N-terminal lysine or arginine, more on lysine than on arginine,[58] as well as the occurrence in thymus histone of a methyl group on the epsilon-N of lysine and the identification of methyl methionine as the methyl donor.[59] The Allfrey group confirmed these findings using radioactive C^{14}-labeled acetate and C^{14}-labeled methionine as acetyl and methyl donors for histone additions and showed these radioactive labels could be incorporated into histone proteins but not into any other thymus nuclear proteins. To verify these findings, the radioactive histones were separated by chromatography and shown to contain the radioactive labels. Going further, they showed that puromycin inhibited incorporation of radioactive lysine into nuclear proteins, but it had no effect either on labeled acetate or methyl into histones, suggesting that the attachment of acetyl and methyl groups into histones occurred after the histones had been synthesized. They also showed that arginine-rich histones greatly inhibited DNA-dependent RNA synthesis while acetylated arginine-rich histones inhibited only slightly. This was the first indication of a functional effect of acetylation on this DNA-mediated reaction; in other words, a connection was made between DNA's physical and structural environment and gene activity.[60] Within two years, other groups reported similar findings on acetylation and methylation of histones.[61–63]

RNA synthesis in isolated nuclei and multiplicity of RNA polymerases. When Allfrey was joined by postdoc Beatriz Pogo and research associate Oscar Pogo, they tested lymphocyte exposure to phytohemagglutinin (PHA) and rat liver regeneration, two physiological processes that produce a dramatic increase in DNA and RNA synthesis.[64,65] One possibility was an increase in the availability of the DNA template to RNA polymerase. The other was histone modifications (acetylation).

The template modifications were tested by the addition of exogenous RNA polymerase (*E. coli* RNA polymerase) and by adding salt and divalent cations such as Mn^2 and/or Mg^2 to activate the endogenous RNA polymerase. The results indicated that the number of DNA sites available for transcription increases during liver regeneration and from PHA stimulation of lymphocytes. Therefore, nuclei from normal liver differ from those of the regenerating organ in their capacity to transcribe the chromatin DNA. Activation of the genome during liver regeneration resulted in other alterations of chromatin. While nuclei of normal liver respond to high salt concentrations by activated template functions as evident in augmented synthesis of RNA high in adenylic and uridylic acid (AU-type mRNA), no such comparable stimulation occurred when salt was applied to nuclei already activated during liver regeneration.[66,67] Combined biochemical and autoradiographic evidence revealed that the types of RNA synthesized by isolated rat liver nuclei and the sites of synthesis can be altered,[68] implying the existence and function of multiple RNA polymerases. For example, a Mn^2-dependent polymerase catalyzed synthesis of DNA-like RNA or mRNA, while a Mg^2-dependent one catalyzed ribosomal RNA synthesis.[67] The existence in eukaryotes of three RNA polymerases is now well established.

Histone modification: Acetylation. Another possibility for increased DNA and RNA synthesis was histone acetylation. By detecting an increase in radioactive acetate incorporation into total arginine-rich or lysine-rich histones, the group had conclusive evidence that histone acetylation modified the DNA-histone interactions and influenced DNA template activity.[69]

The next step was to ascertain where acetylation occurs in the histone molecule. The chemical picture became clearer when Allfrey, with Giorgio Vidali and Edward Gershey, discovered the increase in acetylation involved the formation of *N*-acetyl-lysine, thereby neutralizing the electrical charge, relaxing the constraints on DNA, and leaving DNA accessible and free to perform whatever function was needed.[70] Their methods involved the incorporation of radioactive acetate by calf thymus nuclei, the isolation of highly radioactive arginine-rich histone F2a1, and the determination of which amino acid in the histone was acetylated.[71] They found radioactivity in a single peak and identified for the first time epsilon-*N*-acetyl-lysine as a natural component of histones. By 1969, others had determined the amino acid sequence of the F2a1 histone,[72] showing seven lysines, but the Allfrey group found that only one lysine, at position 16, was radioactively acetylated. This lysine occurs in the most basic part of the histone, indicating a probably profound effect on histone interaction with DNA through a reduction of the basic charge of lysine via acetyl addition.[73,74] This work was followed by studies of turnover in particular histones and included a collaboration with Rockefeller chemist Bruce Merrifield on specific enzymatic sites for the acetylation.[75–77]

Allfrey and others were coming to the conclusion that histone modifications such as methylation and particularly acetylation affect chromatin structure, leading to gene activation. Reviewing differences in histone acetylation and greater RNA synthesis in chromatin fractions, Allfrey used the terms "active" to refer to diffuse chromatin and "inactive" to refer to tightly packed, condensed chromatin.[78] As he commented elsewhere, "DNA's got to have a little elbow room to make DNA; if it's cramped, it can't." When immature avian reticulocytes mature to become erythrocytes, they become practically inactive in RNA synthesis, and this correlated with a condensation of the chromatin. Allfrey found an arginine-rich histone fraction from erythrocytes had only about a fourth the acetyl labeling compared with reticulocytes, and this linked acetylation to a change in chromatin and very likely to the change in synthesis of RNA from the DNA template.[79] After scientists at the Karolinska Institute showed that a chemical acetylation of the basic proteins of the lymphocyte nucleus could change the reactivity of DNA,[80] Allfrey concluded that acetylation of histones "is analogous to opening a file drawer before searching the files for a particular type of information."[79]

At that time, there was no known direct relationship between histone acetylation and RNA synthesis, but, because of its timing, this process was thought to be one of the first molecular events involved in the activation of suppressed regions of the genome. This idea gained favor when Allfrey and Beatriz Pogo found a parallel suppression of RNA synthesis and histone acetylation in granulocytes.[81] However, other results indicated that granulocytes also gradually increased histone deacetylation, and this suggested that acetyl group turnover,

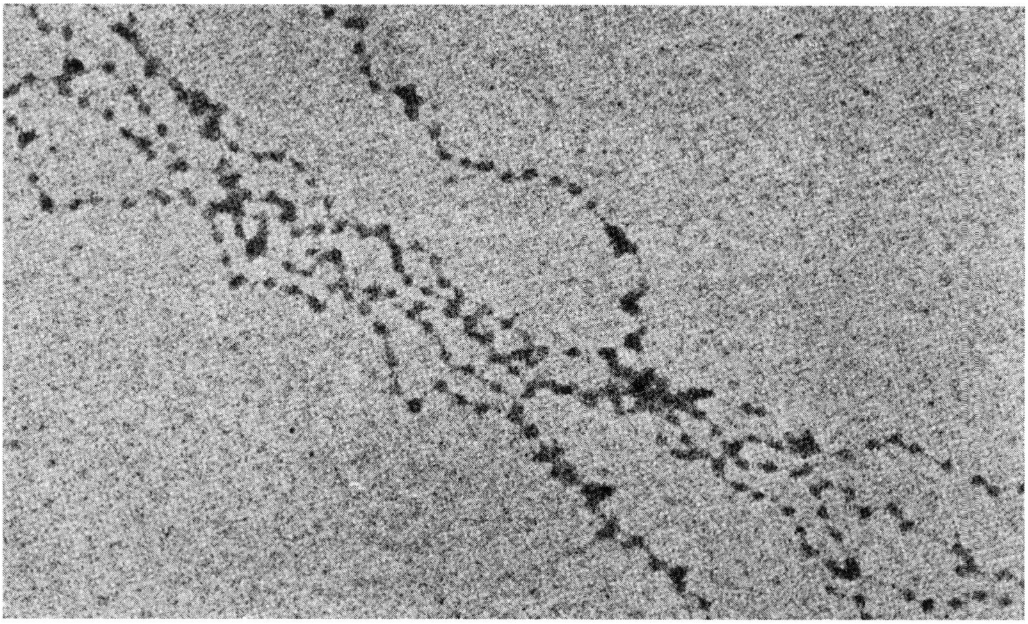

3. Electron micrograph of histone cores along isolated chromatin threads, 1980. Courtesy of Vincent Allfrey.

acetylation and deacetylation, may indicate an impaired nuclear function, such as suppressed RNA synthesis. This conclusion supported earlier Allfrey studies on gene activation during liver regeneration in which the patterns of histone acetylation and RNA synthesis changed in parallel.[66,67] After they observed that histone acetylation takes place in insects, they inferred this might be a fundamental process in any organism, a prescient observation that has now been seen in enough other systems to warrant further investigation.[82] In summary, by 1967, Allfrey and co-workers had a firm foundation for the postulate that histone modification, predominantly acetylation, has a profound effect on the ability of the chromatin to initiate the transcription process of gene activation.

Effects of histones on chromatin. At this point, Mirsky returned to experiments involving the structure of chromatin, particularly looking at how histones interact with DNA in nuclei or in isolated chromatin to open or block accessibility to various enzymes in a particular stretch of DNA. Mirsky had once posed the question, "What controls the controller?" Expanding on his earlier work using unmodified histones, he and a group of postdoctoral fellows showed by electron microscopy that the selective removal of lysine-rich histones loosened the structure of chromatin in metaphase chromosomes, while the re-addition of a lysine-rich fraction resulted in the reappearance of dense chromatin (figure 3).[83] Testing effects of DNase treatment on chromatin itself or on two different preparations of deoxyribonucleoproteins extracted from chromatin, they found that the binding of DNA to nucleoproteins is loose. Mirsky concluded that the DNA available to the enzyme, DNA polymerase, is probably only a small fraction of the total amount of DNA in chromatin, an idea anticipating current concepts that the bulk of DNA is "junk" DNA and not involved in gene activation.[84] According to Mirsky and postdoctoral fellow Burt Silverman in 1973, it was becoming apparent that the lysine-rich histones play "a unique role in the replication of DNA, but that of the effects of both the lysine-rich and arginine-rich histones on the transcription of DNA into RNA are closely tied to interactions with non-histone proteins."[85] These experiments with the higher activity of the lysine-rich histones were among the first to delineate the effects of the two classes of histones on the replicative and transcriptive aspects of the two enzymes on DNA.

Histone modification: Phosphorylation. After other researchers had examined the effect of phosphorylation of proteins on gene activity and observed a rapid phosphorylation of total nuclear proteins in vivo, Allfrey and graduate student Lewis Kleinsmith showed the effects of the phosphorylation of histone.[86–90] Phosphorylated serine in proteins was first discovered in phosvitin, an egg yolk protein, by Fritz Lipmann in 1932 when he was a postdoctoral fellow in the Rockefeller laboratory of Phoebus Levene.[91] When Lipmann returned to Rockefeller in 1960 as a full professor, he and Murray Rabinowitz described a protein kinase from brain or yeast that catalyzes the phosphorylation of serine residues in phosvitin; they also found a protein phosphatase that removes the phosphate, indicating the reversibility of the phosphorylation.[92] What Kleinsmith and Allfrey found in 1969 was that the phosphorylation of serine in proteins was independent of the synthesis of the proteins and that the amino acid serine was the site of most of the phosphorylations. Similar to the acetylation of histones, they discovered spatial and temporal correlations between protein phosphorylation and gene activity regarding RNA synthesis.[93] In another collaboration of John Frenster of the Allfrey laboratory with Thomas Langan of the Lipmann laboratory, the spatial correlations were examined by isolating active and inactive chromatin, and they found that active chromatin had about four times more phosphoprotein per DNA.[94] Since both active and inactive chromatin had similar histone to DNA ratios, it was suggested that the phosphoproteins could complex with histones, perhaps changing the affinity of histones to DNA and the physical state and function of the chromatin. Indeed, experiments by Langan and L.K. Smith showed that nuclear phosphoproteins do form complexes with histones in vitro, and these complexes do not inhibit RNA synthesis to the same extent as observed with the free histones.[95] Based on these findings, Allfrey concluded in 1970 that just as histone acetylation modifies the structure of nuclear acidic proteins and is one of the earlier detectable events of gene activation in the lymphocyte, the increased phosphorylation of nuclear phosphoproteins correlates with a more diffuse state of the chromatin, increased reactivity of the DNA, and subsequent changes in the rate of transcription.[79]

Phosphorylations of the non-histone proteins were also found to have effects on RNA synthesis.[96–102] The Allfrey group added four important findings to this evidence: non-histone phosphorylation is selectively stimulated by the addition of cyclic AMP; the synthesis of liver nuclear acidic proteins can be stimulated by cortisol;[103,104] a fraction of acidic proteins can bind to histones and in doing so selectively precipitates and deacetylates a specific histone fraction, F2a1;[75] and a correlation exists between the phosphoprotein binding to DNA and the enhancement by the phosphorylated proteins of RNA synthesis.[105] All these results strongly indicated the physiological importance of phosphorylations and the connection between histone and non-histone proteins.

After other researchers had found that isolated acidic nuclear proteins could stimulate transcription, synthesizing in vitro RNA from appropriate DNA templates,[106] the Allfrey group extended this line of research to describe the differential binding of nuclear non-histone proteins to various DNA fractions.[107]

In summary, as early as 1964, Allfrey and his group had provided evidence for the importance of histones as transcriptional repressors of DNA, and they emphasized the role of histone acetylation as a modulator of that general repression. Their early focus on the modifications of the histone molecule by acetylation, methylation, and phosphorylation, particularly their relevance on transcription activity of DNA, set the stage for many later studies by others that confirmed Allfrey's results.

By 1977, the Allfrey group determined that the acetylation of a particular histone is much less effective than the non-acetylated form in producing conformational distortions of DNA, though both forms bind to DNA. This was the first direct structural evidence that acetylation of a histone can affect its interaction with the DNA helix.[108–111]

Subsequent work by many laboratories has provided more evidence of how histones are connected in DNA. In 1974, Roger D. Kornberg offered the provocative and largely correct proposal of a structure named the *nucleosome*.[112,113] The five principal classes of histones, which are highly conserved and remarkably invariant across species, combine with DNA into the nucleosome that consists of repeating units of two each of four core histones (H2A, H2B, H3, and H4) with about 200 base pairs of DNA coiled around the histones—a structure more commonly known as "beads on a string." H1 is a linker DNA-associated histone, contacting roughly where DNA enters and exits in the nucleosome and providing additional packaging around the DNA on the outside. These miniature spools both protect and control access to the DNA itself. All the histones are exposed for the modifications described earlier, and they are important for transcriptional regulation. Since the tails of the histones are exposed outside the nucleosome core, it is attractive to consider the possibility that their covalent modification state may contribute to the "plasticity" of the nucleosome or nucleosome arrays and thereby allow the remodeling of the chromatin for transcription of the DNA. Herein lies the structural basis for the acetylation and other modifications of the histones that lead to gene regulation.

More recent work on histone acetylation has shown that some lysines in active chromatin can be reversibly and dynamically acetylated, while there are fewer lysines in inactive chromatin, suggesting that perhaps this pattern is associated with nucleosome assembly. Based on Allfrey's pioneering work, it is now acknowledged that histones are more than passive or static "repressors" of gene activity and are becoming accepted as active downstream participants in gene regulatory pathways. Transcription factors, positive or negative, must engage the chromatin template by binding to nucleosomes that are critically positioned with respect to specific DNA sequences. As articulated by Allfrey and colleagues, the "open" or "closed" state of chromatin may govern transcriptional activity, which is, in turn, brought about by chromatin-modifying enzymes. For example, histone acetyltransferases may add acetyl groups to histones and thereby negate the histone-positive charge, causing a more "open" state; or, in contrast, histone deacetylases may remove acetyl groups, often returning the histone to a more repressive state. Histone methylation, also described by Allfrey and co-workers, has been found to cause both activation and repression of gene expression, depending on the specific residue affected.[114]

Envoi

The Enduring Cell Biology

Carol L. Moberg

> We have entered the cell, the mansion of our birth, and started the inventory of our acquired wealth.
>
> Albert Claude, 1974

In 1974, the Nobel Prize in Physiology or Medicine was awarded to Albert Claude, Christian de Duve, and George Palade for their discoveries concerning the structural and functional organization of the cell. Their award citation noted "there are no earlier Prize Winners in this field, simply because it is one that has been newly created, largely by the Prize Winners themselves. . . . During the last 30 years a new subject has been created, Cell Biology. You have been largely responsible for this development both by creating the basic methodology and by exploiting it to gain insight into the functional machinery of the cell."

Also, by 1974, the original cytology laboratory had disbanded. Three of its early pioneers were making major discoveries elsewhere: Claude left in 1949 to work on cancer biology in Belgium; Porter left in 1961 to penetrate further into the cell's highly organized cytoplasm at Harvard, followed by moves to the University of Colorado and the University of Maryland; and Palade left in 1973 to expand his research on vesicular traffic in cells, first at Yale and then at the University of California in San Diego. Palade observed that his move from Rockefeller was "not a question of having closed possibilities of cell biology, far from it. Notwithstanding the fact that you like to celebrate pioneers, well, pioneers started something but only scratched the surface."

These days, textbook diagrams of cell structures of the nucleus, mitochondria, ribosomes, and endoplasmic reticulum may seem out of date. With new imaging techniques, genome data, and input from other sciences, there is fresh interest in even more intricate cell structures of tubes, vesicles, clumps, filaments, and capsules. Although their functions are not yet clear, the act of observing more delicate landscapes in cells is alive and well. At Rockefeller, the productive science of cell biology endures and prospers as it evolves. Remarkably, two more Rockefeller cell biologists have won Nobel Prizes. In 1999, Günter Blobel was recognized for his discovery that proteins have intrinsic signals that govern their transport and localization in the cell. In 2011, Ralph Steinman was honored for his discovery of the dendritic cell and its role in linking innate and adaptive immunity. Both Blobel and Steinman have described their discoveries in this book.

Entering an Unseen World closes with an adaptation of a speech given by George Palade on the fiftieth anniversary of the 1945 landmark electron micrograph that opened the world of the cell for discovery.

The American Cradle

George E. Palade

In 1995, at the symposium "Journey into the Cell" organized by The Rockefeller University, I portrayed the first electron micrograph of a cultured animal cell as the birth certificate of a vigorous baby, born in 1945, that was baptized a few years later "cell biology." In time, the baby became a remarkably active youngster (a hyperactive kid) and, later on, an enterprising adult, always on the move, still showing no sign of losing momentum even at its now mature age.

The selection of the birth certificate is appropriate because there was nothing better before this electron micrograph and because two of the authors, Albert Claude and Keith Porter, were to become major protagonists in the new field of cell biology, to which they continued to make major contributions for many years.

A newborn baby needs, of course, a friendly, comfortable cradle, which was provided by The Rockefeller Institute for Medical Research and thus became the American cradle of cell biology. This was not only because the early protagonists (Claude, Porter, George Hogeboom, Walter Schneider, and I) were on the research staff of the house but also because it provided a supportive environment that helped the development of new technologies needed to open new fields for research. Lyman Craig's work on separation procedures by countercurrent distribution and Stan Moore and Bill Stein's methods for separating proteins from mixtures (and amino acids from protein hydrolysates) by chromatography date from that period. Cell fractionation and biological electron microscopy, the two technological pillars of the nascent cell biology, were natural additions to this list.

Herbert Gasser, then director of the Institute, had a special flair for recognizing promising developments and the power to support them with carefully controlled generosity. He had changed one of the basic policies of the Institute, according to which the staff of a laboratory was encouraged to leave upon the retirement of its head, and he had decided to retain and support some of the orphans left behind. Porter and I became adopted orphans after our laboratory chief James B. Murphy retired and members of our laboratory had moved to other institutions. This meant in our case that Gasser had decided that cell biology, especially biological electron microscopy, had a future and deserved a friendly cradle. At the beginning, however, the cradle did not have the critical equipment, electron microscopes, and centrifuges that the newly born cell biology needed for its daily activities. Investment in expensive instrumentation had to wait for convincing proofs, and to obtain such proofs, Claude and Porter had to go to an industrial laboratory, and later on we had to use an electron microscope belonging to another institution. The birth certificate micrograph was a convincing proof that, together with a series of other proofs, motivated Gasser to invest in equipment as well as in remodeling a laboratory for us in the third basement of Theobald Smith Hall. The cradle was now in nearly perfect shape. It had three microscopes but no centrifuge.

I should add that, at that time, there were no NIH grants, and all the activities of our group were fully supported by the Institute. So it was not only the cradle that the Institute provided, it was also the daily milk.

So far for the American cradle. What about the newly born cell biology?

As a far-sighted pioneer, Claude had initiated work at Rockefeller not only in electron microscopy, but also in cell fractionation. In 1943, two years before the birth certificate, he had published a review of his findings in *Science*. His cell fractionation protocols used tissue extracts as starting material, differential centrifugation as a basic experimental procedure, and in most cases an analytical approach (as opposed to a preparative approach), which became one of his most important contributions to the field. The fractions prepared from rat liver were generally defined and biochemically characterized along broad lines. Claude realized early on the power inherent in the combined use of these two approaches, but in both cases preparation techniques were still severe limiting factors.

Cell fractionation was improved by our introduction of sucrose solutions to homogenize the isolated organelles. A major immediate need in electron microscopy was to study cells in situ, in their normal habitat in tissues and organs, rather than in culture. Our group contributed a new microtome, designed by Porter and instrument maker Joseph Blum, and a new version of an old fixative; investigators in other laboratories worked out procedures for specimen embedding in plastic and introduced glass knives for cutting thin sections out of plastic-embedded specimens. None of these solutions were final, and all of them were subsequently modified or replaced by better prescriptions. Yet even in their original form, those partial technical improvements were good enough to meet our first objective of making any cell in any tissue or organ of any species amenable to electron microscopy.

Essentially the entire garden of creation was open for study at the new level of resolution. Prospectors for anticipated scientific gold nuggets rushed to the newly open fields in ever-increasing numbers, very much like the forty-niners in the gold fields of California. The spirit was not very different; claims were made and arguments flourished, but the newly open fields were so rich that any diligent microscopist could not avoid making at least a couple of discoveries. The field was in general ferment: old laboratories were revived, and new laboratories sprang into action in this country and abroad. Our French colleagues called themselves *les ultras*, and the Swedish group led by Fritiof Sjöstrand coined the term "ultrastructure." It took years until the spirits calmed down a bit and ultrastructure, meaning beyond structure, was replaced by "fine structure" and finally by just "structure."

Yet the Rockefeller group retained a margin of initiative and on that account attracted professors as trainees in addition to regular postdoctoral fellows. Therefore, we had the unique experience of introducing leading scientists to biological electron microscopy. Quite often they knew more than we did about the topics on which we worked together, and quite a number of important findings were the results of these activities. They became leaders in biological electron microscopy upon returning to their home bases.

Beginning in the early 1950s, there was an ever-increasing flow of new structural information that created a state of euphoria in the field and a self-generated feeling of great historical importance of the discoverers and of their discoveries. From that early period dates the series of papers on the endoplasmic reticulum, the "lace-like reticulum" of the birth certificate, on which Porter and I worked together. Also from that period date discoveries of the fine structure of mitochondria, plasmalemmal vesicles, and the nuclear envelope. In particular, the "small particulate component of the cytoplasm" that Philip Siekevitz and I called ribonucleoprotein particles, later renamed ribosomes, increased the interest of cell biologists when they were shown to be highly complex miniature machines that cells use to synthesize all their proteins.

During this time, we presented our findings to the annual meetings of anatomists and electron microscopists and published them in *The Journal of Experimental Medicine.* But soon it became evident that we were too biological for the electron microscopists of the period and not anatomical enough for the anatomists. It also turned out that by the mid-1950s, cell biology had already acquired a large and active constituency. At this juncture, the organizing genius of Porter came to the fore and led to the founding of The American Society for Cell Biology and a journal to serve that rapidly growing field. We optimistically assumed that we were rapidly heading toward an integration of structural, biochemical, and biophysical approaches in cell research, as expressed in the journal's original title, *The Journal of Biophysical and Biochemical Cytology*, but renamed *The Journal of Cell Biology* a few years later.

Notwithstanding these intense activities in which we were deeply involved, and notwithstanding a few years of often exciting discoveries and apparent successes, there were already some disquieting elements in the general picture. It soon became apparent that the organization of all eukaryotic cells examined involved identical or similar structural elements, presumably subtending common basic functions. This was a major collective achievement and meant that the basic unity of biological organization extended beyond biochemistry all the way to cell organs and cells. The message was reinforced later on by the major contributions made to general cell biology by the work done on yeast cells. But this unity of organization also meant that an extension of the electron microscopy survey to more and more cell types from more and more species was expected to produce a repetitive literature with little novelty and few surprises.

Moreover, cell exploration by electron microscopy had already advanced to the level of macromolecular assemblies; hence, further advances "in depth" had little chance of uncovering new structural elements of general significance. And finally there was the particularly troublesome feeling that electron microscopy was producing clear images of complex structures that were both esthetically pleasing, even enchanting, and intellectually frustrating. We were, in fact, illustrating the works of a great master—evolution. We assumed that the master had perfected these structures for precise, efficient functions, but about those functions we had mostly vague hypotheses and very little real knowledge.

With all this in mind, I felt that it was time to go back to cell fractionation (as done before in the case of mitochondria) to isolate some of the newly discovered subcellular components, study their chemistry, and test, if possible, their functions. Electron microscopy was going to be used in assessing the results of cell fractionation and in monitoring potential structural modulations. While Porter was the head of our laboratory and retained a variety of interests, I chose to concentrate primarily on the cell equipment involved in the synthesis and processing of proteins, based on the premise that protein synthesis is an obligatory function of all cells, which implied that the results of the work would have general applicability. This also meant, however, that some of our promising, initial findings on fine structure were not pursued further in our laboratory. Fortunately, however, these topics became successful research projects in many outstanding laboratories.

We were still in the cradle when Thomas Rivers, the Institute's interim director, gave us funds to recruit the right kind of biochemist, Philip Siekevitz, and to acquire a Spinco preparative centrifuge. Our work was still fully supported by Institute funds. Our new main theme was to define the role played by the ribosomes, endoplasmic reticulum, and other subcellular components in the synthesis and subsequent processing of the proteins produced for export by the pancreatic exocrine cell, a champion cell for pursuing this theme. Detailed accounts of these discoveries have been given in this book both by and about the many scientists who participated in unraveling these two processes.

While we were working on pancreatic cell fractions in our laboratory in the third basement, momentous events were occurring above us. The Institute became a university for graduate studies and led to the arrival of young and eager students on the campus, to an enlargement of the faculty, and to a rapid expansion of its physical plant. As expected, these developments affected in a major way the life of our laboratory as we participated actively in student training and taught a regular course in cell biology. The many students who joined our laboratory for their thesis work and the large number of postdoctoral fellows who came during this period were the most active, imaginative, and enterprising crew in the history of the Rockefeller cradle for cell biology.* I never encountered in my existence (a long existence) a more impressive collection of clever minds and nimble hands.

In the middle of all this excitement, the cradle suffered a major setback. Keith Porter decided to leave for Harvard in 1961 as head of the Biological Laboratories. The call of his alma mater was irresistible. The only thing we could do was to continue the work and try to maintain in his absence the level we had reached under his leadership.

The work on the exocrine pancreas had to go on. During the 1960s, we identified the main compartments involved in intracellular transport, concentration, and storage of secretory products. This work also revealed the involvement of vesicular carriers and the energy requirement at different relays along what became known as the secretory or exocytic pathway. Moreover, this work proved solid enough to become, years later, the foundation for research carried out by one or two successive generations of cell biologists that revealed, on the one hand, the specificity of sorting and delivery of secretory products and membrane components and, on the other hand, the complexity of the molecular interactions involved in the control of transport by vesicular carriers. It appears that we are again facing the works of evolution, but, this time, not in the sturdy construction of subcellular components that we had found so helpful. There is the still bewildering complexity of mechanisms that presumably evolved to prevent errors in delivery in a highly dynamic system of fluid membranes and thereby preserve membrane specificity, which appears to be a critical requirement for cell survival.

By the end of the 1960s, the Rockefeller cradle of cell biology had metamorphosed into a large nursery and was providing professors to Harvard, Yale, Rockefeller, Cornell,

*An appendix contains a roster of the scientific staff, including postdoctoral fellows and students, who worked in the laboratories of cell biology described in this book from 1923 to 1974.

New York University, University of California at San Francisco, Munich, Tokyo, Kansai, Milan, Jerusalem, and Bucharest. But, the romance of the beginning was over. The time of headaches had started.

To close my chronicle, I go back five centuries to Francois Villon, the student poet, and ask with a drop of nostalgia and a touch of elegy, "Où sont de Constantinople les empereurs très honorés? Où est le preux Charlemagne?" [Where are the most honored emperors of Constantinople, and where is the valiant Charlemagne?] Well, many of our princely equivalents have already been carried by Charon over the Styx to the realm of the shadows. But cell biology has gradually metamorphosed into cellular and molecular biology and is more alive than ever. This means that the Rockefeller cradle was truly a great cradle.

Appendix A

Roster of Cell Biology Laboratories 1923–1974

The names of scientific staff in cell biology described in this book are arranged by laboratory affiliations in alphabetical order. Graduate students are distinguished by their class years following their names. Scientists who were affiliated with laboratories that subsequently became jointly headed are listed only under the joint laboratory name.

James B. Murphy
Laboratory of Pathology and Bacteriology, 1923–1950

Auger, Laurier
Barry, Guy
Clark, Harry
Claude, Albert
Duran-Reynals, Francisco
Favilli, Giovanni
Gates, Frederick L.
Gregoire, Charles
Grossfield, Henry
Hance, Robert
Hawkins, James
Helmer, Oscar
Hoffman, Donald
Hogeboom, George
Huseby, Robert
Jacquez, John
Lynch, Clara
MacFadyen, Douglas
MacKenzie, Ronald
Maisin, Joseph
Nakahara, Waro
Palade, George
Palay, Sanford
Porter, Keith
Schneider, Walter
Schober, Bruno
Stewart, Fred
Sturm, Ernest
Tanzer, Radford
Thompson, Helen
Vanamee, Parker

Keith R. Porter and George E. Palade
Laboratory of Cytology, 1950–1961

Ashford, Thomas
Barrnett, Russell
Bonneville, Mary ('61)
Bruni, Carlo
Caro, Lucien
Caulfield, James
Copeland, D. Eugene
Dales, Samuel
Davies, Howard
Davis, Richard
Ernster, Lars
Farquhar, Marilyn
Fawcett, Don
Geer, Jack
Godman, Gabriel
Kallman, Frances
King, Robert
Kirsch, Jack ('61)
Lafontaine, Jean-Gabriel
Ledbetter, Myron
Locke, Michael
Luck, David
Luft, John
Machado, Raul
Maggio, Rachele
Majno, Guido
Meyer, Hertha
Miller, Fritz
Millonig, Guiseppe
Miroff, George
Moore, Dan
Moses, Montrose

Palay, Sanford
Pappas, George
Peachey, Lee ('59)
Randall, John T.
Richter, Goetz
Robinow, Carl
Rosen, Sidney
Rothschild, Joel
Rudzinska, Maria
Sager, Ruth
Satir, Peter ('61)
Scherer, William
Sedar, Albert
Siekevitz, Philip
Smith, David
Sotelo, Jose-Robert
Sreenivasan, Arunachala
Stoeckenius, Walther
Stone, Robert
Takahashi, Noboru
Thies, Roger ('61)
Vanamee, Parker
Vial, Juan
Watson, Michael
Wissig, Steven
Wolken, Jerome
Yamada, Eichi

George E. Palade and Philip Siekevitz
Laboratory of Cell Biology, 1961–1973

Aaronson, Robert P.
Adelman, Mark R.
Amako, Kazunobo
Amsterdam, Abraham
Armstrong, Clara Franzini
Berg, Norton B.
Bergeron, John
Binet, Jacques-Louis
Blobel, Günter
Bock, Karl W.
Borgese, Dominica ('72)
Bruns, Romaine
Caro, Lucien
Cavalier-Smith, Thomas
Chua, Nam-Hai
Clementi, Francesco
Dales, Samuel
Dallner, Gustav
de Harven, Etienne
de Petrocellis, Benita
Ehrenreich, John ('69)
Farquhar, Marilyn
Fleischer, Becca
Fleischer, Sidney
Gilula, Norton
Greene, Lewis ('62)
Hoober, J. Kenneth
Howell, Kathryn E.
Humbel, René
Imai, Yoshio
Jamieson, James ('66)
Karlin, Arthur ('62)
Kern, Horst F.
King, Robert C.
Kraehenbuhl, Jean P.
Kreibich, Gert
Kunau, Wolf-Hubert
Lake, James
Lande, Mauricio A.
Leskes, Andrea ('69)
Lizardi, Paul ('71)
Luck, David ('62)
Maggio, Rachele
Marchesi, Vincent T.
Matile, Philippe
Maylie, Marie-France
Meldolesi, Jacopo
Miller, Fritz
Mok, Winnie
Monneron, Ariane
Moody, Michael F.
Morimoto, Takashi
Nicolaieff, Anatole
Nonamura, Yoshiaki
Ohad, Itzhak
Ohtsuki, Iwao
Omura, Tsuneo

Ploaie, Petru
Popescu, Emil
Redman, Colvin
Rifkin, Mary R. ('69)
Robineaux, Roger
Rothschild, Joel
Rowen, Robert
Sabatini, David ('66)
Sauer, Leonard ('66)
Scheele, George
Schor, Seth L. ('71)
Seeman, Philip ('66)
Seufert, Wolf
Siekevitz, Philip
Silverstein, Samuel
Simionescu, Maya
Simionescu, Nicolae
Simon, Sandra R. ('72)
Srinivasan, Dorothy
Stein, Olga
Stoeckenius, Walther
Sumida, Masahiro
Tartakoff, Alan ('73)
Tashiro, Yutaka
van Tubergen, Robert
Weibel, Ewald
Weisbrod, Roberta
Widnell, Christopher C.
Wood, David D. ('70)
Zaimia, Eleanor R.

James G. Hirsch and Zanvil A. Cohn
Laboratory of Cellular Physiology and Immunology, 1960–1974

Allison, Fred, Jr.
Archer, Gordon T.
Astell, Caroline R.
Axline, Stanton G.
Bennett, William
Blinkoff, Richard C. ('65)
Chen, Chang ('73)
Cross, Nicholas L. ('75)
Edelson, Paul
Ehrenreich, Barbara ('68)
Fedorko, Martha
Gordon, Siamon ('71)
Griffin, Frank
Hoffman, Gary J. ('77)
Holland, Charles P.
Hubbard, Ann L. ('73)
Jones, Thomas C.
Levine, Richard F.
Morse, Stephen I. ('60)
Nachman, Ralph L.
Pachman, Lauren I.
Padgett, George A.
Parr, Earl ('68)
Quie, Paul G.
Rabinovitch, Michel
Roberts, Richard B.
Seljelid, Rolf
Siltzbach, Louis E.
Silverstein, Samuel C.
Steinman, Ralph M.
Stetson, Chandler A.
Strauss, Bella S.
van Furth, Ralph
Werb, Zena ('71)
Wiener, Edith
Zigmond, Sally H. ('72)

Christian de Duve
Laboratory of Biochemical Cytology, 1962–1974

Aronson, Nathan N.
Baggiolini, Marco
Baudhuin, Pierre
Beaufay, Henri
Berthet, Jacques
Black, William J.
Bowers, William ('66)
Chudzik, Jerzy
Coffey, John W.
Deter, Russell L., II
Donaldson, Robert P.
Finkenstaedt, John T.
Fowler, Stanley ('69)

Gaunt, George L. ('73)
Hildebrand, John G. ('69)
Horvat, Agnes
Jacques, Pierre
Kaplan-Bresler, Ann E.
Lazarow, Paul B. ('72)
Leighton, Federico
Lindmark, Donald G.
Müller, Miklós
Panagides, John
Peters, Timothy J.
Poole, Brian ('68)
Schneider, Donald L.
Sun, Alexander
Takano, Tatsuya
Tappel, Aloys L.
Trouet, André
Wibo, Maurice
Wolfson, Elizabeth
Yago, Nagasumi

Alfred Mirsky and Vincent Allfrey
Laboratory of Cell Biology*, 1963–1974

Bachvaroff, Radoslav J.
Bachvarova, Rosemary F. ('66)
Bhandary, Vasanthi
Boffa, Lidia
Burdick, Carolyn J.
Campagne, Roelof N.
Campbell, Ann E.
Crippa, Marco
Davidson, Eric H. ('63)
deKloet, Siwo
Edwards, Gordon S.
Enea, Vincenzo ('76)
Fasy, Thomas M.
Faulkner, Rosemary
Fedoroff, Nina ('72)
Finch, Caleb E. ('69)
Frenster, John H.
Goldberg, Michael E.
Green, Margaret
Hascall, Gretchen K. ('71)
Haslett, Gail W.
Hawks, Andrew
Hough, Barbara R.
Inoue, Akira
Izawa, Mitsuo
Johnson, Edward M.
Karn, Jonathan ('76)
Kleinsmith, Lewis J. ('68)
Kramer, Fred R. ('69)
Lea, Michael A.
Liew, Choong-Chin
Littau, Virginia
Lizardi, Paul M. ('71)
Maclean, Norman
McEwen, Bruce ('64)
Monneron, Ariane
Panda, Nimai Charan
Pogo, A. Oscar
Pogo, Beatriz G.T.
Ruiz-Carillo, Adolfo
Shelton, Keith R.
Sibatani, Atsuhiro
Silverman, Bert
Spencer, E. Martin ('69)
Stacey, Dennis ('76)
Teng, Ching-Sung
Teng, Christina W.T.
Tidwell, J. Thomas ('67)
Vidali, Giorgio
Vidali, Lidea
Wangh, Lawrence J. ('73)

*In 1963, when Allfrey was promoted to professor, the joint laboratory changed its name from General Physiology to Cell Biology.

Appendix B

Illustrations

Notes to Works Cited

References to archival source materials use the following abbreviations.

ASCB, American Society for Cell Biology Papers. Special Collections, University of Maryland, Baltimore County, Catonsville, Maryland.

APS, American Philosophical Society, Philadelphia, Pennsylvania. Collections cited include those of James Bumgardner Murphy, Peyton Rous, and Thomas Anderson.

Bonneville Interviews, Mary Bonneville, "Pioneers in Modern Biology" videotaped and transcribed interviews. Citations refer to the electronic file name of the transcriptions.

Claude Papers, Wisconsin, Albert Claude papers from the private collection of his daughter Philippa Claude, Madison, Wisconsin.

Porter Papers, Keith R. Porter Papers located at three universities: Colorado, Pennsylvania, and UMBC (University of Maryland, Baltimore County). References designate the location and box numbers of unprocessed and unorganized materials that were available for research during 1995. These papers have since been processed by the archives in Colorado and UMBC. The papers seen in Porter's laboratory at the University of Pennsylvania are now located at UMBC.

RF, The Rockefeller Foundation papers of the International Health Division, The Rockefeller Archive Center, Sleepy Hollow, New York.

RUA, The Rockefeller University Archives, The Rockefeller Archive Center, Sleepy Hollow, New York. The following collections are cited in the book; some citations include the Record Group (RG) number of the collection.

> **Administration Files**
>
> **Bronk Papers,** Detlev Wulf Bronk Papers
>
> **BSD Minutes,** Minutes of the Board of Scientific Directors and of the Executive Committee
>
> **BSD Reports,** *Annual Scientific Reports to the Board of Scientific Directors of The Rockefeller Institute for Medical Research*. This unpublished collection comprises forty-two volumes, from 1901 to 1954, and contains detailed annual reports written by each member about research in the laboratory.
>
> **BT Minutes,** Board of Trustee minutes
>
> **Business Manager Files**
>
> **Claude Papers,** Albert Claude Papers
>
> **Gasser Papers,** Herbert S. Gasser Papers
>
> **Granick Papers,** Sam Granick Papers
>
> **History Papers,** Rockefeller History Papers
>
> ***JCB* Papers,** *The Journal of Cell Biology* Papers
>
> **Mirsky Papers,** Alfred Mirsky Papers
>
> **Murphy Papers,** James Bumgardner Murphy Papers
>
> **Prudden Papers,** T. Mitchell Prudden Papers
>
> **Public Affairs Papers**

TCA, Tissue Culture Association Papers. Special Collections, University of Maryland, Baltimore County, Catonsville, Maryland.

Works Cited

Chapter 1

Early Cancer Research

1. Temkin, O. 1950. The European background of the young Dr. Welch. *Bulletin of the History of Medicine.* 24:308–318.
2. Cohen, B. 1950. Comments on the relation of Dr. Welch to the rise of microbiology in America. *Bulletin of the History of Medicine.* 24:319–324.
3. Welch, W.H. 1920. Rudolf Virchow, Pathologist. In *Papers and Addresses, by William Henry Welch.* The Johns Hopkins Press, Baltimore, Maryland. 365–376.
4. Corner, G.W. 1964. *A History of The Rockefeller Institute, 1901–1953: Origins and Growth.* The Rockefeller Institute Press, New York. 575–584.
5. Welch, W.H. 1920. The benefits of the endowment of medical research. In *Papers and Addresses, by William Henry Welch.* The Johns Hopkins Press, Baltimore, Maryland. 74–82.
6. Corner, G.W. 1964. *A History of The Rockefeller Institute, 1901–1953: Origins and Growth.* The Rockefeller Institute Press, New York. 216.
7. Müller, J. 1838. *Ueber den feineren Bau und die Formen der krankhaften Geschwülste.* Reimer, Berlin.
8. Rather, L.J., P. Rather, and J.B. Frerichs. 1986. *Johannes Müller and the Nineteenth-Century Origins of Tumor Cell Theory.* Science History Publications, Canton, Massachusetts.
9. Shimkin, M.B. 1977. *Contrary to Nature: Being an Illustrated Commentary on Some Persons and Events of Historical Importance in the Development of Knowledge Concerning Cancer.* U.S. Dept. of Health, Education, and Welfare, Public Health Service, National Institutes of Health, Washington, D.C.
10. Osler, W. 1892. *The Principles and Practice of Medicine: Designed for the Use of Practitioners and Students of Medicine.* D. Appleton and Company, New York.
11. Osterhout, W.J.V. 1928. Jacques Loeb. *The Journal of General Physiology.* 8:IX–LVIII.
12. Dubos, R.J. 1976. *The Professor, the Institute, and DNA: Oswald T. Avery, His Life and Scientific Achievements.* The Rockefeller University Press, New York.
13. Corner, G.W. 1964. *A History of The Rockefeller Institute, 1901–1953: Origins and Growth.* The Rockefeller Institute Press, New York. 109.
14. Flexner, S., J.W. Jobling, and M.L. Menten. 1910. *Tumors of Animals.* The Rockefeller Institute for Medical Research, New York.
15. Rous, P. 1910. An experimental comparison of transplanted tumor and a transplanted normal tissue capable of growth. *The Journal of Experimental Medicine.* 12:344–366.
16. Rous, P. 1910. A transmissible avian neoplasm (sarcoma of the common fowl). *The Journal of Experimental Medicine.* 12:696–705.
17. Murphy, J.B., to S. Flexner. 1910. Application for a fellowship. RUA, Murphy Papers. Box 1, Folder 1.
18. Murphy, J.B., and P. Rous. 1912. The behavior of chicken sarcoma implanted in the developing embryo. *The Journal of Experimental Medicine.* 15:119–132.
19. Murphy, J.B. 1913. Transplantability of tissues to the embryo of foreign species. Its bearing on questions of tissue specificity and tumor immunity. *The Journal of Experimental Medicine.* 17:482–493.
20. Rous, P. 1911. A sarcoma of the fowl transmissible by an agent separable from the tumor cells. *The Journal of Experimental Medicine.* 13:397–411.
21. Rous, P., and J.B. Murphy. 1914. On the causation by filterable agents of three distinct chicken tumors. *The Journal of Experimental Medicine.* 19:52–68.
22. Murphy, J.B., to J.J. Morton. 21 January 1938. APS, Murphy Papers.
23. Rous, P., and T.M. Prudden. 1924. Interview for an early history of Rockefeller Institute. RUA, History Papers, RG 400. Box 1.
24. Ivanovski, D. 1894. Ueber die Mosaikkrankheit der Tabakspflanze. *Bulletin de l'Académie Impériale des Sciences de St. Pétersbourg.* 3:67-70.
25. Beijerinck, M.W. 1899. Ueber ein Contagium vivum fluidum als Ursache der Fleckenkrankheit der Tabaksblätter. *Centralblatt für Bakteriologie und Parasitenkunde, II.* 5:27–33.
26. Rivers, T.M. 1932. The nature of viruses. *Physiological Reviews.* 12:423–452.
27. van Helvoort, T. 1994. History of virus research in the twentieth century: the problem of conceptual continuity. *History of Science.* 32:185–235.
28. Hughes, S.S. 1977. *The Virus: A History of the Concept.* Science History Publications, New York.
29. Ellerman, V., and O. Bang. 1908. Experimentelle Leukämie bei Hühnern. *Centralblatt für Bakteriologie.* 46:595–609.
30. Rous, P., to T.M. Prudden. 14 October 1922. RUA, Prudden Papers, RG 199.72. Box 1.
31. Rous, P. 1913. Resistance to a tumor-producing agent as distinct from resistance to the implanted tumor cells. Observations with a sarcoma of the fowl. *The Journal of Experimental Medicine.* 18:416–427.

32. Murphy, J.B. 1926. Certain etiological factors in the causation and transmission of malignant tumors. *The American Naturalist.* 60:227–233.
33. Rous, P. 1914. On certain spontaneous chicken tumors as manifestations of a single disease. I. Spindle-celled sarcomata rifted with blood sinuses. *The Journal of Experimental Medicine.* 19:570–576.
34. Murphy, J.B. 30 December 1925. Typescript "Discussion of certain etiological factors in the causation and transmission of malignant tumors." APS, Murphy Papers.
35. Rous, P. June 1916. RUA, BSD Reports. Volume 4.
36. Kevles, D.J. 1995. Pursuing the unpopular: a history of courage, viruses, and cancer. In *Hidden Histories of Science.* R.B. Silvers, editor. The New York Review of Books, New York. 69–112.
37. Weinberg, R.A. 1996. *Racing to the Beginning of the Road: The Search for the Origin of Cancer.* Harmony Books, New York.
38. Andrewes, C. 1970. *Viruses and Cancer.* Weidenfeld and Nicolson, London.
39. Gross, L. 1961. *Oncogenic Viruses.* Pergamon Press, New York.
40. van Helvoort, T. 2004. The start of a cancer research tradition: Peyton Rous, James Ewing, and viruses as a cause of cancer. In *Creating a Tradition of Biomedical Research: Contributions to the History of The Rockefeller University.* D.H. Stapleton, editor. The Rockefeller University Press, New York. 191–209.
41. James, H.J., Jr. 22 July 1915. Endowment, 1913–1949 Henry Rutherford Bequest for Cancer Research. RUA, Administration Files, RG 216.3. Box 5, Folder 1913–1949.
42. Murphy, J.B., to F. Sabin. 7 June 1940. Correspondence. APS, Murphy Papers.
43. Murphy, J.B., to J. Loeb. 1920. Correspondence. APS, Murphy Papers.
44. Rous, P., and J.B. Murphy. 1912. The histological signs of resistance to a transmissible sarcoma of the fowl. *The Journal of Experimental Medicine.* 15:270–286.
45. Murphy, J.B. 1914. Heteroplastic tissue grafting effected through roentgen-ray lymphoid destruction. *Journal of the American Medical Association.* 62:1459.
46. Murphy, J.B. April 1914. RUA, BSD Reports. Volume 3.
47. Loeb, J. 1916. *The Organism as a Whole, from a Physicochemical Viewpoint.* Putnam's Sons, New York. 45.
48. Murphy, J.B., and J.J. Morton. 1915. The lymphocyte as a factor in natural and induced resistance to transplanted cancer. *Proceedings of the National Academy of Sciences of the United States of America.* 1:435–437.
49. Press Clippings Folders. APS, Murphy Papers.
50. Flexner, S. 12 June 1915. RUA, BSD Reports. Volume 4.
51. Corner, G.W. 1964. *A History of The Rockefeller Institute, 1901–1953: Origins and Growth.* The Rockefeller Institute Press, New York. 328.
52. Murphy, J.B., to S. Flexner. 11 December 1917. APS, Murphy Papers.
53. Murphy, J.B. 1926. *The Lymphocyte in Resistance to Tissue Grafting, Malignant Disease, and Tuberculous Infection: An Experimental Study.* Monographs of The Rockefeller Institute for Medical Research. Number 21. The Rockefeller Institute for Medical Research, New York. 1–167.
54. Löwy, I. 1989. Biomedical research and the constraints of medical practice: James Bumgardner Murphy and the early discovery of the role of lymphocytes in immune reactions. *Bulletin of the History of Medicine.* 63:356–391.
55. Silverstein, A.M. 1989. *A History of Immunology.* Academic Press, San Diego.
56. Flexner, S., to J.B. Murphy. 6 May 1924. RUA, Murphy Papers. Box 1, Folder 2.
57. Flexner, S. 1926. The development of the first twenty years of The Rockefeller Institute and an outlook for further growth. RUA, BSD Reports. Volume 14.
58. Anonymous. 1978. *The Legacy of Bessie Dashiell.* Woodstock Foundation, Woodstock, Vermont.
59. Hall, S.S. 1997. *A Commotion in the Blood: Life, Death, and the Immune System.* Henry Holt, New York.
60. Gye, W.E. 1925. The aetiology of malignant new growths. *Lancet.* 2:109–117.
61. Murphy, J.B. 1926. Observations on the etiology of tumors as evidenced by experiments with a chicken sarcoma. *Journal of the American Medical Association.* 86:1270–1271.
62. Rous, P., to J.B. Murphy. 17 January 1927. APS, Murphy Papers.
63. Rous, P., to J.B. Murphy. 28 February 1927. APS, Murphy Papers.
64. Murphy, J.B., to P. Rous. 28 January 1927. APS, Murphy Papers.
65. Murphy, J.B., to H. Zinsser. 1 June 1927. APS, Murphy Papers.

Chapter 2

Taking Cells Apart: Centrifuges and Cell Fractionation

1. Carrel, A., and M.T. Burrows. 1911. Cultivation in vitro of malignant tumors. *The Journal of Experimental Medicine.* 13:571–575.
2. Rous, P., and F.S. Jones. 1916. A method for obtaining suspensions of living cells from the fixed tissues, and for the plating out of individual cells. *The Journal of Experimental Medicine.* 23:549–555.
3. Rous, P., and T.M. Prudden. 1924. Interview for an early history of Rockefeller Institute. RUA, History Papers, RG 400. Box 1.

4. Rous, P., to H. Zinsser. 13 October 1925. APS, Rous Papers.
5. Carrel, A. 1926. Some conditions of the reproduction in vitro of the Rous virus. *The Journal of Experimental Medicine.* 43:647–668.
6. Murphy, J.B. 1927. RUA, BSD Reports. Volume 15.
7. Warburg, O., K. Posener, and E. Negelein. 1924. Über den Stoffwechsel der Carcinomzelle. *Biochemische Zeitschrift.* 152:309–344.
8. Rous, P., to O. Warburg. 5 December 1924. RUA, Administration Files, RG 216.3. Box 15.
9. Okamoto, Y. 1925. Über Anaerobiose von Tumorgewebe. *Biochemische Zeitschrift.* 160:52–65.
10. Murphy, J.B., and J.A. Hawkins. 1925. Comparative studies on the metabolism of normal and malignant cells. *The Journal of General Physiology.* 8:115–130.
11. Murphy, J.B. 1926. Certain etiological factors in the causation and transmission of malignant tumors. *The American Naturalist.* 60:227–233.
12. Murphy, J.B., to J. Ewing. 12 April 1926. APS, Murphy Papers.
13. Duran-Reynals, F., and J.B. Murphy. 1929. Properties of the causative agent of a chicken tumor. I. The specific fixation by tissues of susceptible animals. *The Journal of Experimental Medicine.* 50:315–326.
14. Murphy, J.B. 1931. Discussion of some properties of the causative agent of a chicken tumor. *Transactions of the Association of American Physicians.* 46:182–190.
15. Reynals, F.D. 1928. Studies on the inactivation of vaccine virus and the action of certain substances upon the infecting power of the inactivated virus. *The Journal of Experimental Medicine.* 47:389–402.
16. Pearce, L., and J.B. Murphy. 1927. Further observations on the inability to transmit a rabbit neoplasm by cell-free materials. *The Journal of Experimental Medicine.* 46:205–211.
17. Murphy, J.B., and K.L. Landsteiner. 1925. Experimental production and transmission of tar sarcomas in chickens. *The Journal of Experimental Medicine.* 41:807–816.
18. Sturm, E., F.L. Gates, and J.B. Murphy. 1932. Properties of the causative agent of a chicken tumor. II. The inactivation of the tumor-producing agent by monochromatic ultra-violet light. *The Journal of Experimental Medicine.* 55:441–444.
19. Murphy, J.B. 1928. The nature of the filtrable agent in chicken tumours. In *Report of the International Conference on Cancer, London, 17th-20th July, 1928.* John Wright & Sons for the British Empire Cancer Campaign, Bristol, England. 33–37.
20. Murphy, J.B. 1929. RUA, BSD Reports. Volume 17.
21. Press Clippings Folders. APS, Murphy Papers.
22. Rous, P., to J.B. Murphy. 27 September 1929. APS, Murphy Papers.
23. Rous, P., to S. Flexner. 1929. APS, Rous Papers.
24. Flexner, S., to P. Rous. 1929. APS, Rous Papers.
25. Claude, A. 1974. Autobiography. In *Les Prix Nobel.* Almqvist & Wiksell International, Stockholm. 126–132.
26. Florkin, M. 1972. Pour saluer Albert Claude. *Archives Internationales de Physiologie et de Biochimie.* 80:632–647.
27. de Duve, C. 1987. Albert Claude et les débuts de la biologie cellulaire moderne. *Cellule.* 74:11–19.
28. Claude, A. 1928. Greffes cancéreuses hétérologues. Étude macroscopique de greffes intra-cérébrales. *Comptes Rendus de la Société de Biologie (Belgium).* 99:650–652.
29. Duesberg, J., to S. Flexner. 1 December 1928. RUA, Claude Papers. Box 1.
30. Claude, A. 1928. Grant application to the Fondation Universitaire, Brussels. RUA, Claude Papers. Box 1.
31. Flexner, S., to CRB Educational Foundation. 18 April 1929. RUA, Claude Papers.
32. Claude, A., to J.B. Murphy. 12 August 1929. APS, Murphy Papers.
33. Sturm, E., and F. Duran-Reynals. 1932. Properties of the causative agent of a chicken tumor. VIII. The effect of testicle extract on the rate of growth of chicken tumor I. *The Journal of Experimental Medicine.* 56:711–717.
34. Claude, A., and F. Duran-Reynals. 1934. On the existence of a factor increasing tissue permeability in organs other than testicle. *The Journal of Experimental Medicine.* 60:457–462.
35. Claude, A., and F. Duran-Reynals. 1937. Chemical properties of the purified spreading factor from testicle. *The Journal of Experimental Medicine.* 65:661–670.
36. Murphy, J.B., O.M. Helmer, A. Claude, and E. Sturm. 1931. Observations concerning the causative agent of a chicken tumor. *Science.* 73:266–268.
37. Murphy, J.B., and E. Sturm. 1932. Properties of the causative agent of a chicken tumor. IV. Association of an inhibitor with the active principle. *The Journal of Experimental Medicine.* 56:107–116.
38. Claude, A. 1937. Properties of the causative agent of a chicken tumor. XIII. Sedimentation of the tumor agent, and separation from the associated inhibitor. *The Journal of Experimental Medicine.* 66:59–72.
39. Claude, A. 1939. Properties of the inhibitor associated with the active agent of chicken tumor I. *The American Journal of Cancer.* 37:59–63.
40. Claude, A., and J.B. Murphy. 1933. Transmissible tumors of the fowl. *Physiological Reviews.* 13:246–275.
41. Löwy, I. 1990. Variances in meaning in discovery accounts: the case of contemporary biology. *Historical Studies in the Physical and Biological Sciences.* 21:87–98.
42. Griffith, F. 1928. The significance of pneumococcal types. *The Journal of Hygiene.* 27:113–159.
43. Alloway, J.L. 1932. The transformation in vitro of R pneumococci into S forms of different specific types by the use of filtered pneumococcus extracts. *The Journal of Experimental Medicine.* 55:91–99.
44. Avery, O.T., C.M. MacLeod, and M. McCarty. 1944. Studies on the chemical nature of the substance inducing transformation of pneumococcal types. Induction of transformation by a desoxyribonucleic acid fraction isolated from pneumococcus type III. *The Journal of Experimental Medicine.* 79:137–158.

45. Flexner, S., to J.B. Murphy. 13 June 1933. APS, Murphy Papers.
46. Flexner, S., to J.B. Murphy. 30 March 1933. APS, Murphy Papers.
47. Murphy, J.B., to S. Flexner. 25 April 1933. APS, Murphy Papers.
48. Stanley, W.M. 1935. Isolation of a crystalline protein possessing the properties of tobacco-mosaic virus. *Science.* 81:644–645.
49. Stanley, W.M., and R.W.G. Wyckoff. 1937. The isolation of tobacco ring spot and other virus proteins by ultracentrifugation. *Science.* 85:181–183.
50. Wyckoff, R.W.G., J. Biscoe, and W.M. Stanley. 1937. An ultracentrifugal analysis of the crystalline virus proteins isolated from plants diseased with different strains of tobacco mosaic virus. *The Journal of Biological Chemistry.* 117:57–71.
51. Stanley, W.M. 1942. The concentration and purification of tobacco mosaic virus by means of the Sharples Super-centrifuge. *Journal of the American Chemical Society.* 64:1804–1806.
52. Shope, R.E., and E.W. Hurst. 1933. Infectious papillomatosis of rabbits: With a note on the histopathology. *The Journal of Experimental Medicine.* 58:607–624.
53. Murphy, J.B., to R.E. Shope. 9 January 1933. APS, Murphy Papers.
54. Murphy, J.B. 1936. Relation of filtrable viruses to malignant disease. *Staff Meetings of the Mayo Clinic.* 11:789–790.
55. Rous, P. 1936. The virus tumors and the tumor problem. In *Harvey Lectures.* Series 31. Williams & Wilkins Company, Baltimore, Maryland. 74–115.
56. Rous, P., to H.S. Gasser. 1937. Experiments on the cause of some cancers. APS, Rous Papers.
57. Stehelin, D., H.E. Varmus, J.M. Bishop, and P.K. Vogt. 1976. DNA related to the transforming gene(s) of avian sarcoma viruses is present in normal avian DNA. *Nature.* 260:170–173.
58. Morange, M. 1993. The discovery of cellular oncogenes. *History and Philosophy of the Life Sciences.* 15:45–58.
59. Claude, A. 1935. Properties of the causative agent of a chicken tumor. X. Chemical properties of chicken tumor extracts. *The Journal of Experimental Medicine.* 61:27–40.
60. Claude, A. 1935. Properties of the causative agent of a chicken tumor. XI. Chemical composition of purified chicken tumor extracts containing the active principle. *The Journal of Experimental Medicine.* 61:41–57.
61. Claude, A., and A. Rothen. 1936. Properties of the causative agent of a chicken tumor. XII. Ultraviolet light absorption spectrum of purified chicken tumor extracts containing the active principle. *The American Journal of Cancer.* 26:344–350.
62. Elford, W.J., and C. Andrewes. 1935. Estimation of the size of a fowl tumour virus by filtration through graded membranes. *British Journal of Experimental Pathology.* 16:61–66.
63. Ledingham, J.C.G., and W.E. Gye. 1935. On the nature of the filterable tumour-exciting agent in avian sarcomata. *Lancet.* 1:376–377.
64. McIntosh, J. 1935. The sedimentation of the virus of Rous sarcoma and the bacteriophage by a high-speed centrifuge. *Proceedings of the Pathological Society of Great Britain and Ireland.* 41:215–217.
65. Murphy, J.B., to J. McIntosh. 8 October 1935. APS, Murphy Papers.
66. McIntosh, J., to J.B. Murphy. 24 October 1935. APS, Murphy Papers.
67. Murphy, J.B. 1935. RUA, BSD Reports. Volume 23.
68. Murphy, J.B., to R. Chambers. 21 May 1937. APS, Murphy Papers.
69. Claude, A. 1938. A fraction from normal chick embryo similar to the tumor-producing fraction of chicken tumor I. *Proceedings of the Society for Experimental Biology and Medicine.* 39:398–403.
70. Claude, A., to J.B. Murphy. 1939. Annual report for 1939. Claude Papers, Wisconsin.
71. Stern, K.G. 1943. Studies on macromolecular particles endowed with specific biological activity. In *Frontiers in Cytochemistry. Biological Symposia.* Volume 10. N.L. Hoerr, editor. Jacques Cattell Press, Lancaster, Pennsylvania. 291–321.
72. Claude, A. 1941. Particulate components of cytoplasm. *Cold Spring Harbor Symposia on Quantitative Biology.* 9:263–271.
73. Claude, A. 1939. Chemical composition of the tumor-producing fraction of chicken tumor I. *Science.* 90:213–214.
74. Claude, A., and A. Rothen. 1940. Properties of the causative agent of a chicken tumor. XIV. Relation between a tumor nucleoprotein and the active principle. *The Journal of Experimental Medicine.* 71:619–633.
75. Claude, A. 1940. Particulate components of normal and tumor cells. *Science.* 91:77–78.
76. Claude, A. 1940. R.I. Staff Meeting, 1940. Claude Papers, Wisconsin.
77. Laurence, W.L. 20 April 1941. Hints Cell Itself is a Cancer Cause; Rockefeller Scientist Reports Constituent Plays Big Role as Villain in the Drama. *The New York Times.*
78. Claude, A., to J.B. Murphy. 1942. Annual report for 1942. Claude Papers, Wisconsin.
79. Cowdry, E.V. 1918. The mitochondrial constituents of protoplasm. *Contributions to Embryology.* 25:39–160.
80. Ernster, L., and G. Schatz. 1981. Mitochondria: a historical review. *The Journal of Cell Biology.* 91:227s–255s.
81. von Kölliker, A. 1857. Einige Bemerkungen über die Endigungen der Hautnerven und den Bau der Muskeln. *Zeitschrift für wissenschaftliche Zoologie.* 8:311–325.
82. Flemming, W. 1882. *Zellsubstanz, Kern und Zellteilung.* F.C.W. Vogel, Leipzig.
83. Altmann, R. 1890. *Die Elementarorganismen und ihre Beziehungen zu den Zellen.* Verlag Von Veit, Leipzig.
84. Benda, C. 1898. Ueber die Spermatogenese der Vertebraten und höherer Evertebraten. II. Theil: Die Histogenese der Spermien. In *Archiv für Anatomie und Physiologie: Physiologische Abteilung.* Verlag Von Veit, Leipzig. 393–398.
85. Michaelis, L. 1899. Die vitale Färbung, eine Darstellungsmethode der Zellgranula. *Archiv für mikroskopische Anatomie.* 55:558–575.

86. Palade, G.E. 1964. The organization of living matter. *Proceedings of the National Academy of Sciences of the United States of America*. 52:613–634.
87. Meves, F. 1908. Die Chondriosomen als Träger erblicher Anlagen. Cytologische Studien am Hühnerembryo. *Archiv für mikroskopische Anatomie*. 72:816–867.
88. Regaud, C. 1909. Participation du chondriome à la formation des graines de ségrégation dans les cellules des tubes contournés du rein (chez les ophidiens et les amphibiens). *Comptes Rendus des Séances de la Société de Biologie (Paris)*. 66:1034–1036.
89. Kingsbury, B.F. 1912. Cytoplasmic fixation. *The Anatomical Record*. 6:39–52.
90. Miescher, F. 1871. Über die chemische Zusammensetzung der Eiterzellen. *Hoppe-Seyler's medizinish-chemischen Untersuchungen*. 4:441–460.
91. Warburg, O. 1913. Über sauerstoffatmende Körnchen aus Leberzellen und über Sauerstoffatmung in Berkfeld-Filtraten wässriger Leberextrakte. *Pflügers Archiv European Journal of Physiology*. 154:599–617.
92. Lehninger, A.L. 1964. *The Mitochondrion; Molecular Basis of Structure and Function*. W.A. Benjamin, Inc., New York.
93. Flexner, S. 1922. Report of the Director. RUA, BSD Reports. Volume 10, pages 82–89.
94. Cowdry, E.V., and P.K. Olitsky. 1922. Differences between mitochondria and bacteria. *The Journal of Experimental Medicine*. 36:521–533.
95. Cowdry, E.V. 1926. Surface film theory of the function of mitochondria. *The American Naturalist*. 60:157–165.
96. Bensley, R.R., and N.L. Hoerr. 1934. Studies on cell structure by the freezing-drying method. VI. The preparation and properties of mitochondria. *The Anatomical Record*. 60:449–455.
97. Bensley, R.R. 1910. On the so-called Altmann's granules in normal and pathological tissues. *Transactions of the Chicago Pathological Society*. 8:78–83.
98. Cowdry, E.V. 1924. The value of the study of mitochondria in cellular pathology. *The American Naturalist*. 58:97–114.
99. Cowdry, E.V. 1940. Properties of cancer cells. *Archives of Pathology*. 30:1245–1274.
100. Cowdry, E.V., to J.B. Murphy. 26 March 1940. APS, Murphy Papers.
101. Bensley, R.R., to A. Claude. 24 May 1940. Claude Papers, Wisconsin.
102. Hoerr, N.L., to A. Claude. 21 March 1942. Claude Papers, Wisconsin.
103. Bensley, R.R., to A. Claude. 24 October 1942. Claude Papers, Wisconsin.
104. Bensley, R.R. 1942. Chemical structure of cytoplasm. *Science*. 96:389–393.
105. Hoerr, N.L. 1943. Methods of isolation of morphological constituents of the liver cell. In *Frontiers in Cytochemistry. Biological Symposia*. Volume 10. N.L. Hoerr, editor. Jacques Cattell Press, Lancaster, Pennsylvania. 185–231.
106. Lazarow, A. 1943. The chemical structure of cytoplasm as investigated in Professor Bensley's laboratory during the past ten years. In *Frontiers in Cytochemistry. Biological Symposia*. Volume 10. N.L. Hoerr, editor. Jacques Cattell Press, Lancaster, Pennsylvania. 9–26.
107. Claude, A., to N.L. Hoerr. 20 April 1943. Claude Papers, Wisconsin.
108. Claude, A. 1943. Distribution of nucleic acids in the cell and the morphological constitution of cytoplasm. In *Frontiers in Cytochemistry. Biological Symposia*. Volume 10. N.L. Hoerr, editor. Jacques Cattell Press, Lancaster, Pennsylvania. 111–129.
109. Claude, A., to N.L. Hoerr. 12 May 1943. Claude Papers, Wisconsin.
110. Claude, A. 1943. The constitution of protoplasm. *Science*. 97:451–456.
111. Hoerr, N.L., to A. Claude. 2 June 1943. Claude Papers, Wisconsin.
112. Claude, A. 1944. The constitution of mitochondria and microsomes, and the distribution of nucleic acid in the cytoplasm of a leukemic cell. *The Journal of Experimental Medicine*. 80:19–29.
113. Claude, A. 1946. Fractionation of mammalian liver cells by differential centrifugation. I. Problems, methods, and preparation of extract. *The Journal of Experimental Medicine*. 84:51–59.
114. Claude, A. 1946. Fractionation of mammalian liver cells by differential centrifugation. II. Experimental procedures and results. *The Journal of Experimental Medicine*. 84:61–89.
115. Svedberg, T., and K.O. Pedersen. 1940. *The Ultracentrifuge*. Clarendon Press, Oxford.
116. Pickels, E.G. 1950. Centrifugation. In *Biophysical Research Methods*. F.M. Uber, editor. Interscience Publishers, Inc., New York. 67–105.
117. Elzen, B. 1988. *Scientists and Rotors: The Development of Biochemical Ultracentrifuges*. PhD thesis. University of Twente, Enschede, Netherlands.
118. Gray, G.W. 1951. The Ultracentrifuge. *Scientific American*. 184:42–51.
119. de Duve, C. 1958. Les lysosomes. *Exposés Annuels de Biochimie Médicale*. 20:197–211.
120. Schachman, H.K. 1986. Development of the ultracentrifuge and its application to biological systems. In *The Beckman Symposium on Biomedical Instrumentation*. C.L. Moberg, editor. The Rockefeller University, New York. 37–58.
121. Beams, H.W., and R.G. Kessel. 1987. Development of centrifuges and their use in the study of living cells. *International Review of Cytology*. 100:15–48.
122. de Duve, C. 1971. Tissue fractionation. Past and present. *The Journal of Cell Biology*. 50:20d–55d.
123. de Duve, C., and H. Beaufay. 1981. A short history of tissue fractionation. *The Journal of Cell Biology*. 91:293s–299s.
124. Boestad, G., K.O. Pedersen, and T. Svedberg. 1938. Design and operation of the oil-turbine ultracentrifuge. *The Review of Scientific Instruments*. 9:346–353.
125. Kraemer, E.O., to R.W.G. Wyckoff. 31 August 1937. RF Papers, RG 1.1/200. Box 60, Folder 723.
126. Henriot, E., and E. Hugenard. 1925. Sur la réalisation de très grandes vitesses de rotation. *Comptes Rendus des Seances de l'Academie des Sciences*. 180:1389–1392.

127. Henriot, E., and E. Hugenard. 1927. Les grandes vitesses angulaires obtenues par les rotors sans axe solide. *Journal de Physique et le Radium*. 8:433–443.
128. Pickels, E.G., and J.W. Beams. 1935. High rotational speeds in vacuo. *Science*. 81:342–343.
129. Beams, J.W., and E.G. Pickels. 1935. The production of high rotational speeds. *The Review of Scientific Instruments*. 6:299–308.
130. Pickels, E.G., to T.F. Francis. 24 October 1945. RF Papers, RG 5/4. Box 22, Folder 260.
131. Bauer, J.H., and E.G. Pickels. 1936. A high speed vacuum centrifuge suitable for the study of filterable viruses. *The Journal of Experimental Medicine*. 64:503–528.
132. Svedberg, T., to J.H. Bauer. 25 September 1936. RF Papers, RG 1.1/800. Box 7, Folder 76.
133. Bauer, J.H., and E.G. Pickels. 1940. High-speed ultracentrifuge with air-turbine drive. In *The Ultracentrifuge*. T. Svedberg and K.O. Pedersen, editors. Clarendon Press, Oxford. 191–212.
134. Biscoe, J., E.G. Pickels, and R.W.G. Wyckoff. 1936. An air-driven ultracentrifuge for molecular sedimentation. *The Journal of Experimental Medicine*. 64:39–45.
135. Wyckoff, R.W.G., and J.B. Lagsdin. 1937. Improvements in the air-driven ultracentrifuge for molecular sedimentation. *The Review of Scientific Instruments*. 8:74–77.
136. Bauer, J.H., and E.G. Pickels. 1937. An improved air-driven type of ultracentrifuge for molecular sedimentation. *The Journal of Experimental Medicine*. 65:565–586.
137. Beard, J.W., and R.W.G. Wyckoff. 1937. The isolation of a homogeneous heavy protein from virus-induced rabbit papillomas. *Science*. 85:201–202.
138. Flinn, W.R., to H.S. Gasser. 18 December 1935. RUA, Administration Files, RG 210.3. Box 15.
139. Wyckoff, R.W.G., to E.O. Kraemer. 12 August 1937. RF Papers, RG 1.1/200. Box 60, Folder 723.
140. Bauer, J.H., to G.K. Strode. 14 July 1944. RF Papers, RG 1.1/200. Box 60, Folder 724.
141. Rawson, A.J., H.W. Scherp, and F.E. Lindquist. 1940. A motor-driven, mechanically supported ultracentrifuge for the separation of biological materials. *Journal of Bacteriology*. 40:657–663.
142. Pickels, E.G. 1943. Sedimentation in the angle centrifuge. *The Journal of General Physiology*. 26:341–359.
143. Murphy, J.B. 1940. RUA, BSD Reports. Volume 28.
144. Claude, A., to J.B. Murphy. 1940. Annual report for 1940. Claude Papers, Wisconsin.
145. Pickels, E.G. 1942. An improved type of electrically driven high speed laboratory centrifuge. *The Review of Scientific Instruments*. 13:93–100.
146. 146. Rothen, A. 1940. Molecular weight and electrophoresis of crystalline ribonuclease. *The Journal of General Physiology*. 24:203–211.
147. Anonymous. 1961. The making of instruments for scientific research. *The Rockefeller Institute Quarterly*. 5:1–2.
148. Blum, J. 1945. A new type of self-centering mounting for high speed rotors with flexible motor drive. *The Review of Scientific Instruments*. 16:31–34.
149. Claude, A., to V. Desreux. 29 April 1946. Claude Papers, Wisconsin.
150. Claude, A., to J. Brachet. 25 October 1947. Claude Papers, Wisconsin.
151. Schneider, W.C., A. Claude, and G.H. Hogeboom. 1948. The distribution of cytochrome c and succinoxidase activity in rat liver fractions. *The Journal of Biological Chemistry*. 172:451–458.
152. Hogeboom, G.H., W.C. Schneider, and G.E. Pallade. 1948. Cytochemical studies of mammalian tissues; isolation of intact mitochondria from rat liver; some biochemical properties of mitochondria and submicroscopic particulate material. *The Journal of Biological Chemistry*. 172:619–635.
153. Smith, E.B., to D.M. Nelson. 12 November 1941. RF Papers, RG 5/4. Box 11, Folder 118.
154. Pickels, E.G., to H.E. Pearson. 15 June 1945. RF Papers, RG 1.1/200. Box 60, Folder 724.
155. Baker, W.H., to W.C. Schneider. 3 July 1945. RF Papers, RG 1.1/200. Box 60, Folder 724.
156. Pickels, E.G. 1952. Ultracentrifugation. In *Methods of Medical Research*. A.C. Corcoran, editor. Year Book Publishers, Chicago. 107–133.
157. Porter, K.R. 1985. Interview with Mary A. Bonneville. Bonneville Interviews. UCOLO6C.

Chapter 3

Entering the Realm of the Cell with an Electron Microscope

1. Murphy, J.B. 1935. Experimental approach to the cancer problem. *Bulletin of the Johns Hopkins Hospital*. 56:1–31.
2. Corner, G.W. 1964. *A History of The Rockefeller Institute, 1901–1953: Origins and Growth*. The Rockefeller Institute Press, New York. 328.
3. Little, C.C. 1960. James Bumgardner Murphy, 1884–1950. *Biographical Memoirs of the National Academy of Sciences (U.S.)*. 34:182–203.
4. Weed, L.H., to A. Claude. 8 November 1945. Claude Papers, Wisconsin.
5. Shimkin, M.B. 1979. Committee on Growth, 1945–1956: another noble experiment. *Cancer Research*. 39:262–268.
6. Heller, J.R. 1957. The National Cancer Institute: a twenty-year retrospect. *Journal of the National Cancer Institute*. 19:147–190.

7. Bayne-Jones, S., R.G. Harrison, C.C. Little, J. Northrop, and J.B. Murphy. 1938. Fundamental cancer research. *Public Health Reports*. 53:2121–2130.
8. Murphy, J.B. 1942. An analysis of the trends in cancer research. *Journal of the American Medical Association*. 120:107–111.
9. Rous, P. 1943. The nearer causes of cancer. *Journal of the American Medical Association*. 122:573–581.
10. Claude, A. 1943. The constitution of protoplasm. *Science*. 97:451–456.
11. Murphy, J.B. 1936. Factors involved in the malignant process. *ACTA, International Union Against Cancer*. 1:352–357.
12. Murphy, J.B., to H.G. Coar. 6 November 1940. APS, Murphy Papers.
13. Porter, K.R., to J.B. Murphy. 17 June 1939. Porter Papers, Colorado. Box CM 14.
14. Porter, K.R. 1984. Interview with Mary A. Bonneville. Bonneville Interviews. HARVD1.
15. Porter, K.R. 1984. Interview with Mary A. Bonneville. Bonneville Interviews. YNGYRS1.
16. Duran-Reynals, F. 1939. A note on the action of some carcinogenic hydrocarbons on amphibia. *The Yale Journal of Biology and Medicine*. 11:613–617.
17. Moberg, C.L., and Z.A. Cohn, editors. 1990. *Launching the Antibiotic Era: Personal Accounts of the Discovery and Use of the First Antibiotics*. The Rockefeller University Press, New York.
18. Moberg, C.L. 2005. *René Dubos, Friend of the Good Earth: Microbiologist, Medical Scientist, Environmentalist*. ASM Press, Washington, D.C.
19. Porter, K.R., to J.B. Murphy. 9 May 1941. Porter Papers, Colorado. Box CM 43.
20. Porter, K.R., to J.B. Murphy. 9 October 1941. Porter Papers, Colorado. Box CM 43.
21. Yegian, D., and K.R. Porter. 1944. Some artifacts encountered in stained preparations of tubercle bacilli. I. Non-acid-fast forms arising from mechanical treatment. *Journal of Bacteriology*. 48:83–91.
22. Porter, K.R., and D. Yegian. 1945. Some artifacts encountered in stained preparations of tubercle bacilli. II. Much granules and beads. *Journal of Bacteriology*. 50:563–575.
23. Gasser, H.S., to K.R. Porter. 28 October 1942. Porter Papers, Colorado. Box CM 14.
24. Murphy, J.B., to K.R. Porter. 11 May 1943. Porter Papers, Colorado. Box CM 14.
25. Porter, K.R., and G.E. Palade. 1984. Interview with Mary A. Bonneville. Bonneville Interviews. KC5BC1.
26. Russell, K. 1969. Tissue culture—a brief historical review. *Clio Medica (Amsterdam, Netherlands)*. 4:109–119.
27. Landecker, H. 2007. *Culturing Life: How Cells Became Technologies*. Harvard University Press, Cambridge, Massachusetts.
28. Willmer, E.N. 1965. Introduction. In *Cells and Tissues in Culture: Methods, Biology, and Physiology*. E.N. Willmer, editor. Academic Press, New York. 1–17.
29. Parker, R.C. 1938. *Methods of Tissue Culture*. P.B. Hoeber, Inc., New York.
30. Earle, W.R. 1943. Production of malignancy in vitro. *Journal of the National Cancer Institute*. 4:131–248.
31. Murphy, J.B. 1944. RUA, BSD Reports. Volume 32.
32. Porter, K.R. Transformation of normal cells *in vitro*. Porter Papers, Colorado. Box CM 14, 86.
33. Porter, K.R. 1984. Interview with Mary A. Bonneville. Bonneville Interviews. KC1BC1.
34. Burton, E.F., and W.H. Kohl. 1946. *The Electron Microscope, an Introduction to its Fundamental Principles and Applications*. Second edition. Reinhold Publishing Corporation, New York.
35. Bradbury, S. 1968. *The Microscope: Past and Present*. Pergamon Press, Oxford.
36. Hillier, J. 1950. Electron microscopy. In *Biophysical Research Methods*. F.M. Uber, editor. Interscience Publishers, New York. 381–416.
37. Hawkes, P.W., editor. 1985. *The Beginnings of Electron Microscopy*. Academic Press, Orlando, Florida.
38. Reisner, J.H. 1989. An early history of the electron microscope in the United States. *Advances in Electronics and Electron Physics*. 73:133–231.
39. Newberry, S.P. 1992. *EMSA and Its People: The First Fifty Years*. M. Schumacher, editor. Electron Microscopy Society of America, Milwaukee, Wisconsin.
40. Rasmussen, N. 1997. *Picture Control: The Electron Microscope and the Transformation of Biology in America, 1940–1960*. Stanford University Press, Stanford, California.
41. Anderson, T.F. 1975. The first years of the electron microscope in the United States, 1940–1952 [Rough Draft]. Porter Papers, Pennsylvania. Box UP 14D, pages 1–9.
42. Anderson, T.F. 1975. Some personal memories of research. *Annual Review of Microbiology*. 29:1–18.
43. Pease, D.C., and K.R. Porter. 1981. Electron microscopy and ultramicrotomy. *The Journal of Cell Biology*. 91:287s–292s.
44. Stanley, W.M., and T.F. Anderson. 1941. A study of purified viruses with the electron microscope. *The Journal of Biological Chemistry*. 139:325–338.
45. Marton, L. 1934. Electron microscopy of biological objects. *Nature*. 133:911.
46. Ernst, J. 1976. Dr. Albert Claude: Meeting at Hillcrest. Oral Histories, The Rockefeller Archive Center. Pages 16, 42.
47. Stern, K.G., to A. Claude. 11 April 1940. Claude Papers, Wisconsin.
48. Stern, K.G., to T.F. Anderson. 18 October 1941. APS, Anderson Papers.
49. Anderson, T.F., to K.G. Stern. 3 December 1941. APS, Anderson Papers.
50. Mudd, S., to K.G. Stern. 2 February 1942. APS, Anderson Papers.
51. Stern, K.G., to S. Mudd. 9 February 1942. APS, Anderson Papers.
52. Rivers, T.M., to T.F. Anderson. 18 February 1942. APS, Anderson Papers.
53. Stern, K.G. 1943. Studies on macromolecular particles endowed with specific biological activity. In *Frontiers in Cytochemistry. Biological Symposia*. Volume 10. N.L. Hoerr, editor. Jacques Cattell Press, Lancaster, Pennsylvania. 291–321.

54. Benison, S., editor. 1967. *Tom Rivers: Reflections on a Life in Medicine and Science; An Oral History Memoir.* MIT Press, Cambridge, Massachusetts. 220–221.
55. Green, R.H., T.F. Anderson, and J.E. Smadel. 1942. Morphological structure of the virus of vaccinia. *The Journal of Experimental Medicine.* 75:651–656.
56. Claude, A., and J.S. Potter. 1943. Isolation of chromatin threads from the resting nucleus of leukemic cells. *The Journal of Experimental Medicine.* 77:345–354.
57. Anderson, T.F., to A. Claude. 25 November 1942. Claude Papers, Wisconsin.
58. Green, H., to A. Claude. 28 May 1943. Claude Papers, Wisconsin.
59. Schuster, M. 1946. High-speed microtoming. *Interchemical Review.* 5:31–41.
60. Fullam, D.B., and C.L. Moberg. 1994. Interview with Ernest F. Fullam.
61. Reisner, J.H. 1986. Reflections: Ernest F. Fullam reminisces. *EMSA Bulletin.* 16:35–43.
62. Murphy, J.B. 1944. RUA, BSD Reports. Volume 32, page 70.
63. Fullam, E.F., to A. Claude. 22 July 1943. Claude Papers, Wisconsin.
64. Fullam, E.F., to A. Claude. 22 December 1943. Claude Papers, Wisconsin.
65. Claude, A., to E.F. Fullam. 9 February 1944. Claude Papers, Wisconsin.
66. Claude, A., and E.F. Fullam. 1945. An electron microscope study of isolated mitochondria. Method and preliminary results. *The Journal of Experimental Medicine.* 81:51–62.
67. Claude, A., and E.F. Fullam. 1946. The preparation of sections of guinea pig liver for electron microscopy. *The Journal of Experimental Medicine.* 83:499–503.
68. Porter, K.R., A. Claude, and E.F. Fullam. 1945. A study of tissue culture cells by electron microscopy. Method and preliminary observations. *The Journal of Experimental Medicine.* 81:233–246.
69. Porter, K.R. 1975. Cell Biology at Rockefeller Institute. Porter Papers, Pennsylvania. Pages 3–4.
70. Porter, K.R. 1987. Electron microscopy of cultured cells. In *The American Association of Anatomists, 1888–1987. Essays on the History of Anatomy in America and a Report on the Membership—Past and Present.* J.E. Pauly, editor. William & Wilkins, Baltimore, Maryland. 59–67.
71. Harrison, R.G. 1906. Observations on the living developing nerve fiber. *Experimental Biology and Medicine.* 4:140–143.
72. Carrel, A., and M.T. Burrows. 1911. Cultivation of tissues in vitro and its technique. *The Journal of Experimental Medicine.* 13:387–396.
73. Lewis, W.H., and M.R. Lewis. 1912. The cultivation of chick tissues in media of known composition. *The Anatomical Record.* 6:207–211.
74. Porter, K.R. 1953. Observations on a submicroscopic basophilic component of cytoplasm. *The Journal of Experimental Medicine.* 97:727–750.
75. Gey, G.O., and M.K. Gey. 1936. The maintenance of human normal cells and tumor cells in continuous culture. *The American Journal of Cancer.* 27:45–76.
76. Lewis, W.H. 1935. *Rat malignant cells in roller tube cultures and some results.* Contributions to Embryology. Volume 25. Carnegie Institute, Washington, D.C. 161–172.
77. Paul, J. 1959. *Cell and Tissue Culture.* E & S Livingstone, Edinburgh, Scotland. 7.
78. Strangeways, T.S.P., and R.G. Canti. 1927. Memoirs: The living cell in vitro as shown by darkground illumination and the changes induced in such cells by fixing reagents. *The Quarterly Journal of Microscopical Science.* 71:1–14.
79. Palade, G.E. 1971. Albert Claude and the beginnings of biological electron microscopy. *The Journal of Cell Biology.* 50:5d–19d.
80. Porter, K.R. 1970. Remarks on receiving the Louisa Gross Horwitz Prize with Albert Claude and George Palade. RUA, Public Affairs Papers, RG 365.3. Box 4, Folder 1.
81. Porter, K.R. 1990. The Midwest E.M. Society. Porter Papers, Pennsylvania. Box UP 12.
82. Murphy, J.B., to G.H. Hogeboom. 7 April 1944. APS, Murphy Papers.
83. Murphy, J.B., to G.H. Hogeboom. 2 October 1944. APS, Murphy Papers.
84. Claude, A., to K.R. Porter. 20 July 1944. Porter Papers, Colorado. Box CM 14.
85. Porter, K.R., to A. Claude. 1944. Claude Papers, Wisconsin.
86. Porter, K.R., to A. Claude. 6 September 1944. Porter Papers, Colorado. Box CM 14.
87. Porter, K.R., to D. Yegian. 20 December 1944. Porter Papers, Colorado. Box CM 37.
88. International Health Division to Rockefeller Foundation. 1945. An electron microscope for the International Health Division. RF Papers, RG 1.1/100. Box 9, Folder 78.
89. International Health Division to Rockefeller Foundation. 1945. Electron microscopy. Annual Report. RF Papers, RG 5.3 /100. Box 2, Folder 1945.
90. Carrel, A., and M.T. Burrows. 1911. Cultivation in vitro of malignant tumors. *The Journal of Experimental Medicine.* 13:571–575.
91. Carrel, A. 1926. Some conditions of the reproduction in vitro of the Rous virus. *The Journal of Experimental Medicine.* 43:647–668.
92. Porter, K.R., to J.B. Murphy. 1946. Report of research. Porter Papers, Pennsylvania. Box 14D.
93. Murphy, J.B., to G.H. Hogeboom. 23 March 1945. APS, Murphy Papers.
94. Murphy, J.B., to C.P. Rhoads. 1946. A proposal for study of the structure of normal and malignant human cells as revealed by electron microscopy. Porter Papers, Colorado. Box CM 14.
95. Poteat, D., to R.I.M.R. President. 1950. Budgets for Research Project G-101. RUA, Murphy Papers. Box 1, Folder 3.
96. Murphy, J.B., to A.J. Driessens. 8 April 1947. APS, Murphy Papers.

97. Board of Scientific Directors to Board of Trustees Executive Committee. 1947. Untitled statement requesting a signature to approve grant from American Cancer Society. RUA, Murphy Papers. Box 1, Folder 3.
98. Gasser, H.S., to J.D. Rockefeller Jr. 5 April 1948. RUA, Murphy Papers. Box 1, Folder 3.
99. Murphy, J.B., to L.H. Weed. 18 December 1946. APS, Murphy Papers. Box American Cancer Society.
100. Murphy, J.B., to W. Nakahara. 31 October 946. APS, Murphy Papers.
101. Porter, K.R., to E.G. Pickels. 3 June 1946. Porter Papers, Colorado. Box CM 14.
102. Claude, A., K.R. Porter, and E.G. Pickels. 1947. Electron microscope study of chicken tumor cells. *Cancer Research*. 7:421–430.
103. Williams, R.C., and R.W.G. Wyckoff. 1944. The thickness of electron microscopic objects. *Journal of Applied Physics*. 15:712–716.
104. Passey, R.D., L. Dmochowski, W.T. Astbury, and R. Reed. 1947. Electron microscope studies of normal and malignant tissues of high- and low-breast-cancer strains of mice. *Nature*. 160:565.
105. Graff, S., D.H. Moore, W.M. Stanley, H.T. Randall, and C.D. Haagensen. 1949. Isolation of mouse mammary carcinoma virus. *Cancer*. 2:755–762.
106. Porter, K.R., and H.P. Thompson. 1947. Some morphological features of cultured rat sarcoma cells as revealed by the electron microscope. *Cancer Research*. 7:431–438.
107. Porter, K.R., and H.P. Thompson. 1948. A particulate body associated with epithelial cells cultured from mammary carcinomas of mice of a milk-factor strain. *The Journal of Experimental Medicine*. 88:15–24.
108. Murphy, J.B. 1949. RUA, BSD Reports. Volume 37.
109. Porter, K.R., to J.B. Murphy. 1949. Report of research. Porter Papers, Colorado. Box CM77.
110. Jacquez, J.A., and K.R. Porter. 1949. Electron-microscopy observations on Hodgkin's material. *Cancer*. 2:853–856.
111. Murphy, J.B. 1947. RUA, BSD Reports. Volume 35.
112. Claude, A., K.R. Porter, and E.G. Pickels. 1946. Electron microscope study of chicken tumor cells. *Cancer Research*. 6:502.
113. Porter, K.R., and E.G. Pickels. 1946. Electron microscopy of normal and malignant cells. *Cancer Research*. 6:502.
114. Rous, P. 1946. Comments on "Electron Microscope Study of Chicken Tumor Cells". APS, Murphy Papers.
115. Murphy, J.B., to A. Claude. 28 June 1947. APS, Murphy Papers.
116. Claude, A., to J.B. Murphy. 31 August 1947. APS, Murphy Papers.
117. Claude, A., to J.B. Murphy. 10 September 1947. APS, Murphy Papers.
118. Williams, R.C. 1954. Electron microscopy of viruses. *Advances in Virus Research*. 2:183–239.
119. Gross, L. 1961. *Oncogenic Viruses*. Pergamon Press, New York.
120. Bernhard, W., and C. Oberling. 1953. Échec de la mise en évidence de corpuscules-virus dans les cellules du sarcome de Rous examinées au microscope électronique. *Bulletin du Cancer*. 40:178–185.
121. Williams, R.C. 1961. Virus particles and "virus-like" particles associated with the infected cell. In *Perspectives in Virology*. M. Pollard, editor. Burgess Publishing Co., Minneapolis, Minnesota. 18–37.
122. Haguenau, F., H.L. Febvre, and J. Arnoult. 1961. Ultrastructural morphology of Rous sarcoma grown in vitro. In *Perspectives in Virology*. M. Pollard, editor. Burgess Publishing Company, Minneapolis, Minnesota. 160–170.
123. Dmochowski, L. 1960. Viruses and tumors in the light of electron microscope studies: A review. *Cancer Research* 20:977–1015.
124. Porter, K.R. 2 May 1947. R.I.M.R. Talk. Porter Papers, Colorado. Box CM 8.
125. Palade, G.E. 16 March 1995. The American cradle. Speech delivered at symposium "Journey into the Cell: Celebrating 50 Years of Electron Microscopy and Modern Cell Biology." The Rockefeller University, New York.

Chapter 4

Stalking Chemicals inside Cells

1. Lazarow, A. 1943. The chemical structure of cytoplasm as investigated in Professor Bensley's laboratory during the past ten years. In *Frontiers in Cytochemistry. Biological Symposia*. Volume 10. N.L. Hoerr, editor. Jacques Cattell Press, Lancaster, Pennsylvania. 9–26.
2. Barron, E.S.G. 1943. Some considerations on the application of biological oxidation-reduction reaction systems to the study of cellular respiration. In *Frontiers in Cytochemistry. Biological Symposia*. Volume 10. N.L. Hoerr, editor. Jacques Cattell Press, Lancaster, Pennsylvania. 27–69.
3. Rous, P., to H.S. Gasser. 25 September 1940. APS, Rous Papers.
4. Corner, G.W. 1964. *A History of The Rockefeller Institute, 1901–1953: Origins and Growth*. The Rockefeller Institute Press, New York. 328.
5. Green, D.E. 1940. *Mechanisms of Biological Oxidations*. The University Press, Cambridge, England.
6. Hogeboom, G.H., to J.B. Murphy. 13 September 1940. APS, Murphy Papers.
7. Murphy, J.B. 1942. RUA, BSD Reports. Volume 30.
8. Murphy, J.B. 1943. RUA, BSD Reports. Volume 31.
9. Hogeboom, G.H., and M.H. Adams. 1942. Mammalian tyrosinase and dopa oxidase. *The Journal of Biological Chemistry*. 145:273–279.

10. Dubos, R.J. 1945. *The Bacterial Cell in its Relation to Problems of Virulence, Immunity and Chemotherapy.* Harvard University Press, Cambridge, Massachusetts.
11. Stephenson, M. 1930. *Bacterial Metabolism.* Longmans, Green and Co., London.
12. Holter, H. 1936. Studies on enzymatic histochemistry. XVIII. Localization of peptidase in marine ova. *Journal of Cellular and Comparative Physiology.* 8:179–200.
13. Claude, A. 1944. The constitution of mitochondria and microsomes, and the distribution of nucleic acid in the cytoplasm of a leukemic cell. *The Journal of Experimental Medicine.* 80:19–29.
14. Claude, A. 1943. Distribution of nucleic acids in the cell and the morphological constitution of cytoplasm. In *Frontiers in Cytochemistry. Biological Symposia.* Volume 10. N.L. Hoerr, editor. Jacques Cattell Press, Lancaster, Pennsylvania. 111–129.
15. Claude, A. 1945. Distribution of enzymatic activities in fractions of mammalian liver. *AAAS Research Conference on Cancer.* F.R. Moulton, editor. American Association for the Advancement of Science, Washington, D.C. 223–226.
16. Stern, K.G. 1939. Respiratory catalysts in heart muscle. *Cold Spring Harbor Symposia on Quantitative Biology.* 7:312–322.
17. Hogeboom, G.H., A. Claude, and R.D. Hotchkiss. 1946. The distribution of cytochrome oxidase and succinoxidase in the cytoplasm of the mammalian liver cell. *The Journal of Biological Chemistry.* 165:615–629.
18. Claude, A. 1946. Fractionation of mammalian liver cells by differential centrifugation. I. Problems, methods, and preparation of extract. *The Journal of Experimental Medicine.* 84:51–59.
19. Claude, A. 1946. Fractionation of mammalian liver cells by differential centrifugation. II. Experimental procedures and results. *The Journal of Experimental Medicine.* 84:61–89.
20. Claude, A., to J.B. Murphy. 1944. Annual report for 1944. Claude Papers, Wisconsin.
21. Claude, A., to J.B. Murphy. 1945. Annual report for 1945. Claude Papers, Wisconsin.
22. Potter, V.R., and C.A. Elvehjem. 1936. A modified method for the study of tissue oxidations. *The Journal of Biological Chemistry.* 114:495–504.
23. Potter, V.R. 1941. Studies on the mechanism of hydrogen transport in animal tissues. IV. The succinoxidase system. *The Journal of Biological Chemistry.* 141:775–787.
24. Potter, V.R., to A. Claude. 7 April 1942. Claude Papers, Wisconsin.
25. Claude, A., to V.R. Potter. 22 April 1942. Claude Papers, Wisconsin.
26. Potter, V.R., to A. Claude. 27 April 1942. Claude Papers, Wisconsin.
27. Schneider, W.C., to C.L. Moberg. 1 March 1995.
28. Claude, A., to K.G. Stern. 22 October 1945. Claude Papers, Wisconsin.
29. Potter, V.R., to A. Claude. 2 February 1945. Claude Papers, Wisconsin.
30. Hogeboom, G.H., to J.B. Murphy. 22 April 1945. APS, Murphy Papers.
31. Murphy, J.B., to G.H. Hogeboom. 25 April 1945. APS, Murphy Papers.
32. Murphy, J.B. 1946. RUA, BSD Reports. Volume 34.
33. Schneider, W.C. 1946. Intracellular distribution of enzymes; the distribution of succinic dehydrogenase, cytochrome oxidase, adenosinetriphosphatase, and phosphorus compounds in normal rat tissues. *The Journal of Biological Chemistry.* 165:585–593.
34. Claude, A., to W.C. Schneider. 22 July 1946. Claude Papers, Wisconsin.
35. Schneider, W.C., A. Claude, and G.H. Hogeboom. 1948. The distribution of cytochrome c and succinoxidase activity in rat liver fractions. *The Journal of Biological Chemistry.* 172:451–458.
36. Palade, G.E. 1974. Autobiography. In *Les Prix Nobel.* Almqvist & Wiksell International, Stockholm. 163–166.
37. Palade, G.E. 1987. Interview with Mary A. Bonneville. Bonneville Interviews. GEPAS1.
38. Popa, G.T., to The Rockefeller Foundation. 1 October 1945. Claude Papers, Wisconsin.
39. Palade, G.E., to A. Claude. 21 May 1946. Claude Papers, Wisconsin.
40. Claude, A., to J.B. Murphy. May 1946. Claude Papers, Wisconsin.
41. Murphy, J.B., to H.S. Gasser. 11 June 1946. APS, Murphy Papers.
42. Secretary, Board of Scientific Directors, to G.E. Palade. 28 June 1946. RUA, Administration Files, RG 210.3. Folder Scientific Staff Appointment Letters for 1946.
43. Porter, K.R., and G.E. Palade. 1986. Interview with Mary A. Bonneville. Bonneville Interviews. GEPRU7.
44. Hogeboom, G.H., W.C. Schneider, and G.E. Pallade. 1948. Cytochemical studies of mammalian tissues; isolation of intact mitochondria from rat liver; some biochemical properties of mitochondria and submicroscopic particulate material. *The Journal of Biological Chemistry.* 172:619–635.
45. Porter, K.R., and G.E. Palade. 1986. Interview with Mary A. Bonneville. Bonneville Interviews. GEPRU8.
46. Murphy, J.B. 1947. RUA, BSD Reports. Volume 35.
47. Palade, G.E. 1987. Interview with Mary A. Bonneville. Bonneville Interviews. GEPAS2.
48. Lehninger, A.L. 1964. *The Mitochondrion; Molecular Basis of Structure and Function.* W.A. Benjamin, Inc., New York.
49. Schneider, W.C., and V.R. Potter. 1949. Intracellular distribution of enzymes; the distribution of oxalacetic oxidase activity in rat liver and rat kidney factions. *The Journal of Biological Chemistry.* 177:893–903.
50. Hogeboom, G.H. 1949. Cytochemical studies of mammalian tissues; the distribution of diphosphopyridine nucleotide-cytochrome c reductase in rat liver fractions. *The Journal of Biological Chemistry.* 177:847–858.
51. Schneider, W.C., to C.L. Moberg. 25 April 1995.
52. Silver, R. 1998. Audiotaped Interview with George Palade, Aspen, Colorado. Collected papers of Carol L. Moberg.
53. Hogeboom, G.H., W.C. Schneider, and M.J. Striebich. 1953. Localization and integration of cellular function. *Cancer Research.* 13:617–632.

54. Hogeboom, G.H., and W.C. Schneider. 1950. Cytochemical studies of mammalian tissues. III. Isocitric dehydrogenase and triphosphopyridine nucleotide-cytochrome c reductase of mouse liver. *The Journal of Biological Chemistry*. 186:417–427.
55. Schneider, W.C., and G.H. Hogeboom. 1950. Intracellular distribution of enzymes. V. Further studies on the distribution of cytochrome c in rat liver homogenates. *The Journal of Biological Chemistry*. 183:123–128.
56. Hogeboom, G.H., and G.T. Barry. 1948. Purification of diphosphopyridine nucleotide by counter-current distribution. *The Journal of Biological Chemistry*. 176:935–948.
57. Hogeboom, G.H., and W.C. Schneider. 1952. Cytochemical studies. IV. Physical state of certain respiratory enzymes of mitochondria. *The Journal of Biological Chemistry*. 194:513–519.
58. Hogeboom, G.H., W.C. Schneider, and M.J. Striebich. 1952. Cytochemical studies. V. On the isolation and biochemical properties of liver cell nuclei. *The Journal of Biological Chemistry*. 196:111–120.
59. Hogeboom, G.H., and W.C. Schneider. 1952. Cytochemical studies. VI. The synthesis of diphosphopyridine nucleotide by liver cell nuclei. *The Journal of Biological Chemistry*. 197:611–620.
60. Kuff, E.L., G.H. Hogeboom, and A.J. Dalton. 1956. Centrifugal, biochemical, and electron microscopic analysis of cytoplasmic particulates in liver homogenates. *The Journal of Biophysical and Biochemical Cytology*. 2:33–54.
61. de Duve, C., B.C. Pressman, R. Gianetto, R. Wattiaux, and F. Appelmans. 1955. Tissue fractionation studies. 6. Intracellular distribution patterns of enzymes in rat-liver tissue. *The Biochemical Journal*. 60:604–617.
62. Porter, K.R., to G.H. Hogeboom. 15 September 1952. Porter Papers, Colorado. Box CM 19.
63. Claude, A., to R. Cordier. 22 May 1956. Claude Papers, Wisconsin.
64. Anonymous. 1956. George Hall Hogeboom, 1913–1956. *The Journal of Biophysical and Biochemical Cytology*. 2:ix–xvi.
65. Hogeboom, G.H., and W.C. Schneider. 1956. Biochemistry of cellular particles. *Annual Review of Biochemistry* 25:201–224.
66. Palade, G.E. 1958. A small particulate component of the cytoplasm. In *Frontiers in Cytology*. S.L. Palay, editor. Yale University Press, New Haven, Connecticut. 283–304.

Chapter 5

The New Science Emerges

1. Chase, M.W., and C.C. Hunt. 1995. Herbert Spencer Gasser, July 5, 1888–May 11, 1963. *Biographical Memoirs. National Academy of Sciences (U.S.)*. 67:147–178.
2. Gasser, H.S. 1940. Report on the Activities and Problems of the Rockefeller Institute. RUA, BSD Reports. Volume 28, pages 1–86.
3. Gasser, H.S. 1953. Promotional materials. RUA, Gasser Papers 302.4. Box 1, Folder 16.
4. Schmitt, F.O., to H.S. Gasser. 23 March 1945. RUA, Gasser Papers 302.2.
5. Gasser, H.S. 1946. Future needs. RUA, Gasser Papers 302.4. Box 2, Folder 4. Pages 1–22.
6. Gasser, H.S. 1946. Director's Report to the Corporation. RUA, BSD Reports. Volume 34.
7. Gasser, H.S. 1951. Annual scientific reports: handwritten notes. RUA, Gasser Papers 302.4. Box 1, Folder 1.
8. Pease, D.C., and K.R. Porter. 1981. Electron microscopy and ultramicrotomy. *The Journal of Cell Biology*. 91:287s–292s.
9. Richards, G.A., T.F. Anderson, and R.T. Hance. 1942. A microtome sectioning technique for electron microscopy illustrated with sections of striated muscle. *Proceedings of the Society for Experimental Biology and Medicine*. 51:148–152.
10. von Ardenne, M. 1939. Die Keilschnittmethode, ein Weg zur Herstellung von Mikrotomschnitten mit weniger als 10^{-3} mm Stärke für elektronenmikroskopische Zwecke. *Zeitschrift für Wissenschaftliche Mikroskopie und Mikroskopische Technik*. 56:8–23.
11. Sjöstrand, F.S. 1943. Electron-microscopic examination of tissues. *Nature*. 151:725–726.
12. Bracegirdle, B. 1978. *A History of Microtechnique*. Cornell University Press, Ithaca, New York.
13. O'Brien, H.C., and G.M. McKinley. 1943. New microtome and sectioning method for electron microscopy. *Science*. 98:455–456.
14. Porter, K.R. 1964. Ultramicrotomy. In *Modern Developments in Electron Microscopy*. B.M. Siegel, editor. Academic Press, New York. 119–145.
15. Schuster, M. 1946. High-speed microtoming. *Interchemical Review*. 5:31–41.
16. Gessler, A.E., and E.F. Fullam. 1946. Sectioning for the electron microscope accomplished by the high speed microtome. *The American Journal of Anatomy*. 78:245–283.
17. Fullam, E.F., and A.E. Gessler. 1946. A high speed microtome for the electron microscope. *The Review of Scientific Instruments*. 17:23–35.
18. Reisner, J.H. 1986. Reflections: Ernest F. Fullam reminisces. *EMSA Bulletin*. 16:35–43.
19. Claude, A., and E.F. Fullam. 1946. The preparation of sections of guinea pig liver for electron microscopy. *The Journal of Experimental Medicine*. 83:499–504.
20. Golgi, C. 1898. Sur la structure des cellules nerveuses. *Archives Italiennes de Biologie*. 30:60–71.
21. Golgi, C. 1908. Une méthode pour la prompte et facile demonstration de l'appareil réticulaire interne des cellules nerveuses. *Archives Italiennes de Biologie*. 49:269–274.

22. Worley, L.G. 1946. The Golgi apparatus - an interpretation of its structure and significance. *Annals of the New York Academy of Sciences.* 47:1–52.
23. Whaley, W.G. 1975. *The Golgi Apparatus.* Springer-Verlag, New York.
24. Farquhar, M.G., and G.E. Palade. 1981. The Golgi apparatus (complex)—(1954–1981)—from artifact to center stage. *The Journal of Cell Biology.* 91:77s–103s.
25. Palade, G.E., and A. Claude. 1949. The nature of the Golgi apparatus; parallelism between intercellular myelin figures and Golgi apparatus in somatic cells. *Journal of Morphology.* 85:35–69.
26. Palade, G.E., and A. Claude. 1949. The nature of the Golgi apparatus; identification of the Golgi apparatus with a complex of myelin figures. *Journal of Morphology.* 85:71–111.
27. Porter, K.R., to E.G. Pickels. 3 June 1946. RF Papers, Series 5, Subseries 4. Box 22, Folder 260.
28. Claude, A. 1952. The continued controversy over the Golgi apparatus. Claude Papers, Wisconsin. Pages 1–8.
29. Palade, G.E., to A. Claude. 19 October 1952. Claude Papers, Wisconsin.
30. Dalton, A.J., and M.D. Felix. 1954. Cytologic and cytochemical characteristics of the Golgi substance of epithelial cells of the epididymis in situ, in homogenates and after isolation. *The American Journal of Anatomy.* 94:171–207.
31. Murphy, J.B. 1946. RUA, BSD Reports. Volume 34.
32. Murphy, J.B. 2 June 1945. RUA, BSD Minutes.
33. Palay, S.L., to C.L. Moberg. 9 February 1996.
34. Palay, S.L. 1992. A concatenation of accidents. In *Neurosciences: Paths of Discovery II.* F. Samson and G. Adelman, editors. Birkhäuser, Boston. 191–212.
35. Claude, P., to C.L. Moberg. 22 January 1996.
36. Foster, L.V., to J.A. Scheick. 1945. Microtome knife for thin sections. RUA, Claude Papers. Box 1.
37. Claude, A., to J.A. Scheick. 19 March 1946. RUA, Claude Papers. Box 1.
38. Claude, A., to Schick Service Manager. 24 February 1947. Claude Papers, Wisconsin.
39. Kennedy, T.L., to A. Claude. 4 March 1947. Claude Papers, Wisconsin.
40. Claude, A. 1950. Studies on cells: Morphology, chemical constitution, and distribution of biochemical functions. In *Harvey Lectures.* Series 43. Academic Press, New York. 121–164.
41. Latta, H., and J.F. Hartmann. 1950. Use of a glass edge in thin sectioning for electron microscopy. *Proceedings of the Society for Experimental Biology and Medicine.* 74:436–439.
42. Hillier, J., and R.F. Baker. 1946. The mounting of bacteria for electron microscope examination. *Journal of Bacteriology.* 52:411–416.
43. Claude, A. 1949. Electron microscope studies of ceils by the method of replicas. *The Journal of Experimental Medicine.* 89:425–430.
44. Palay, S.L., and A. Claude. 1949. An electron microscope study of salivary gland chromosomes by the replica method. *The Journal of Experimental Medicine.* 89:431–438.
45. Porter, K.R. 2 May 1947. R.I.M.R. Talk. Porter Papers, Colorado. Box CM 8.
46. Moberg, C.L. 1996. Keith Porter and the founding of the tissue culture association: A fiftieth anniversary tribute, 1946–1996. *In Vitro Cellular & Developmental Biology, Animal.* 32:663–669.
47. Porter, K.R., to R.C. Parker. 28 July 1947. Porter Papers, Colorado. Box CM 61.
48. Paul, J. 1959. *Cell and Tissue Culture.* E. & S. Livingstone Ltd, Edinburgh, Scotland.
49. Porter, K.R., to C.M. Pomerat. 15 January 1946. Porter Papers, Colorado. Box CM 14.
50. Fedoroff, S. 1987. Anatomists' contributions to tissue culture. In *The American Association of Anatomists, 1888–1987: Essays on the History of Anatomy in America and a Report on the Membership—Past and Present.* J.E. Pauly, editor. Williams & Wilkins, Baltimore, Maryland. 208–219.
51. Gey, G.O., to C.M. Pomerat. 17 May 1946. TCA Papers. Box 1, Folder 7.
52. Porter, K.R., to C.M. Pomerat. 17 July 1946. Porter Papers, Colorado. Box CM 14.
53. Conn, H.J., to K.R. Porter. 13 September 1946. TCA Papers. Box 1, Folder 5.
54. Porter, K.R. 1946. The organization of a commission on tissue culture media and procedures. TCA Papers. Box 1, Folder 5.
55. Murray, M.R., and G. Kopech. 1953. *A Bibliography of the Research in Tissue Culture, 1884 to 1950; An Index to the Literature of the Living Cell Cultivated In Vitro.* Academic Press, New York.
56. Porter, K.R. 1946. Minutes of the first meeting of the Tissue Culture Commission. TCA Papers. Box 1, Folder 5.
57. Porter, K.R. 1986. Speech for the opening meeting of IV Decennial Review Conference of the TCA. Porter Papers, Pennsylvania. Box UP, Folder 14C.
58. Gasser, H.S., to K.R. Porter. 2 June 1947. TCA Papers. Box 1, Folder 5.
59. Porter, K.R., to Committee on Growth of the National Research Council. 30 June 1949. Progress report on certification and standardization of centrally-prepared tissue culture media. TCA Papers. Box 1, Folder 10.
60. Murray, M.R., to C.M. Pomerat. 20 December 1948. TCA Papers. Box 1, Folder 32.
61. Porter, K.R. 16 March 1949. Tissue Culture Commission Bulletin on Commercial Media. TCA Papers. Box 32, Folder 1459.
62. Burnett, H.A., to K.R. Porter. 29 April 1949. TCA Papers. Box 1, Folder 21.
63. Porter, K.R., to H.A. Burnett. 21 April 1949. TCA Papers. Box 1, Folder 21.
64. Porter, K.R. 1954. The certification of commercial culture media. *Annals of the New York Academy of Sciences.* 58:1029–1038.

65. Porter, K.R. 1955. Foreword. In *An Introduction to Cell and Tissue Culture.* W.F. Scherer, editor. Burgess Publishing Company, Minneapolis, Minnesota. i–ii.
66. Porter, K.R., and C.V. Hawn. 1947. The culture of tissue cells in clots formed from purified bovine fibrinogen and thrombin. *Proceedings of the Society for Experimental Biology and Medicine.* 65:309–314.
67. Van Zandt Hawn, C., and K.R. Porter. 1947. The fine structure of clots formed from purified bovine fibrinogen and thrombin: A study with the electron microscope. *The Journal of Experimental Medicine.* 86:285–292.
68. Porter, K.R., and C.V.Z. Hawn. 1949. Sequences in the formation of clots from purified bovine fibrinogen and thrombin; a study with the electron microscope. *The Journal of Experimental Medicine.* 90:225–232.
69. Murphy, J.B. 1947. RUA, BSD Reports. Volume 35, pages 69–82.
70. Hillier, J., to K.R. Porter. 17 May 1946. Porter Papers, Colorado. Box CM 14.
71. Hanks, J.H., to M. Murray. 1948. A menagerie: The chiropractor, the pugilist, the pet dog and the tissue cultivator. Porter Papers, Colorado. Box CM 18.
72. Pomerat, C.M., to S. Clark. 12 September 1950. TCA Papers. Box 1, Folder 25.
73. Corner, G.W. 1964. *A History of The Rockefeller Institute, 1901–1953: Origins and Growth.* The Rockefeller Institute Press, New York. 328.
74. Green, R.H., T.F. Anderson, and J.E. Smadel. 1942. Morphological structure of the virus of vaccinia. *The Journal of Experimental Medicine.* 75:651–656.
75. Murphy, J.B., to H.S. Gasser. 21 March 1946. APS, Murphy Papers.
76. Murphy, J.B., to H.S. Gasser. 8 November 1946. APS, Murphy Papers.
77. Gasser, H.S., to J.B. Murphy. 13 November 1946. APS, Murphy Papers.
78. Taylor, R.M., to J.C. Bugher. 22 April 1948. RF Papers, RG 1.1/100. Box 10, Folder 86.
79. Murphy, J.B., to A.M. Chesney. 10 December 1946. APS, Murphy Papers.
80. Murphy, J.B., to H.S. Gasser. 6 October 1947. APS, Murphy Papers.
81. De Robertis, E., and F.O. Schmitt. 1949. An electron microscope study of nerves infected with human poliomyelitis virus. *The Journal of Experimental Medicine.* 90:283–290.
82. Bursar's Statements of Receipts and Disbursements. 30 April 1948. RUA, Business Manager Files, RG 210.4. Box 5, Folder 1947–48.
83. Porter, K.R., and G.E. Palade. 1984. Interview with Mary A. Bonneville. Bonneville Interviews. KC3BC1.
84. Gasser, H.S. 1948. Microphotographs. RUA, Gasser Papers 302.6. Box 20.
85. Gasser, H.S. 1950. Unmedullated fibers originating in dorsal root ganglia. *The Journal of General Physiology.* 33:651–690.
86. Porter, K.R., to H.S. Bennett. 18 October 1949. Porter Papers, Colorado. Box CM 10.
87. Reisner, J.H. 1989. An early history of the electron microscope in the United States. *Advances in Electronics and Electron Physics.* 73:133–231.
88. Porter, K.R., to H. Ericsson. 11 November 1950. Porter Papers, Colorado. Box CM 87.
89. Porter, K.R., to C.J. Woods. 3 December 1949. Porter Papers, Colorado. Box CM 19.
90. Murphy, J.S., to C.L. Moberg. 1996. Personal communication.
91. Palade, G.E., to A. Claude. 4 July 1949. Claude Papers, Wisconsin.
92. Claude, A. 7 July 1949. Institut Jules Bordet, séance spéciale, réponse au bienvenue exprimé par Monsieur R. Ulens, président du comité de gestion de l'Institut. Claude Papers, Wisconsin. Pages 1–3 (typescript, translated from the original French by C.L. Moberg).
93. Gasser, H.S. 1951. Medical research: A look ahead at The Rockefeller Institute for Medical Research. RUA, BSD Reports. Volume 39, pages 1–94.
94. Gasser, H.S. 1949. Teaching at the Rockefeller Institute for Medical Research. RUA, BSD Reports. Volume 37, pages 1–20.

Chapter 6

Cytology: The Tentative Beginnings

1. Palade, G.E., to K.R. Porter. 28 October 1949. Porter Papers, Colorado. Box CM 19.
2. Porter, K.R., to J.E. Ayre. 26 June 1951. Porter Papers, Colorado. Box CM 19.
3. Porter, K.R., to J. Ferrebee. 15 March 1955. Porter Papers, Colorado. Box CM 10.
4. Porter, K.R., to P. Siekevitz. 22 July 1986. Porter Papers, Pennsylvania. Box 14A.
5. Porter, K.R., and G.E. Palade. 1986. Interview with Mary A. Bonneville. Bonneville Interviews. GEPRU7.
6. Rous, P., to K.R. Porter. 9 May 1952. Porter Papers, Colorado. Box CM 42.
7. Porter, K.R., to P. Rous. 13 May 1952. Porter Papers, Colorado. Box CM 42.
8. Palade, G.E. 1951. Intracellular distribution of acid phosphatase in rat liver cells. *Archives of Biochemistry.* 30:144–158.
9. Palade, G.E. 1951. Intracellular localization of acid phosphatase; a comparative study of biochemical and histochemical methods. *The Journal of Experimental Medicine.* 94:535–548.
10. Porter, K.R., to T.M. Rivers. 1 June 1954. Porter Papers, Colorado. Box CM 42.

11. Anderson, T.F. 1975. The first years of the electron microscope in the United States, 1940–1952 [Rough Draft]. Porter Papers, Pennsylvania. Box UP 14D, pages 1–9.
12. Porter, K.R. 1951. RUA, BSD Reports. Volume 39, pages 143–152.
13. Baker, J.R. 1945. *Cytological Technique: The Principles and Practice of Methods Used to Determine the Structure of the Metazoan Cell.* Second edition. Methuen, London.
14. Newman, S.B., E. Borysko, and M. Swerdlow. 1949. Ultramicrotomy by a new method. *Journal of Research of the National Bureau of Standards.* 43:183–199.
15. Newman, S.B., E. Borysko, and M. Swerdlow. 1949. New sectioning techniques for light and electron microscopy. *Science.* 110:66–68.
16. Palade, G.E. 1952. A study of fixation for electron microscopy. *The Journal of Experimental Medicine.* 95:285–298.
17. Palade, G.E., to A. Claude. 10 August 1951. Claude Papers, Wisconsin.
18. Rhodin, J. 1954. *Correlation of Ultrastructural Organization and Function in Normal and Experimentally Changed Proximal Convoluted Tubule Cells of the Mouse Kidney.* PhD thesis. Karolinska Institutet, Stockholm.
19. Zetterqvist, H. 1956. *The Ultrastructural Organization of the Columnar Absorbing Cells of the Mouse Jejunum: An Electron Microscopic Study including Some Experiments regarding the Problem of Fixation and an Investigation of Vitamin A Deficiency.* PhD thesis. Karolinska Institutet, Stockholm.
20. Sjöstrand, F.S. 1956. Electron microscopy of cells and tissues. In *Physical Techniques in Biological Research.* G. Oster and A.W. Pollister, editors. Academic Press, New York. 241–298.
21. Porter, K.R., and F.L. Kallman. 1953. The properties and effects of osmium tetroxide as a tissue fixative with special reference to its use for electron microscopy. *Experimental Cell Research.* 4:127–141.
22. Sabatini, D.D., K. Bensch, and R.J. Barrnett. 1963. Cytochemistry and electron microscopy. The preservation of cellular ultrastructure and enzymatic activity by aldehyde fixation. *The Journal of Cell Biology.* 17:19–58.
23. Hall, C.E., M.A. Jakus, and F.O. Schmitt. 1945. The structure of certain muscle fibrils as revealed by the use of electron stains. *Journal of Applied Physics.* 16:459–465.
24. Hall, C.E. 1955. Electron densitometry of stained virus particles. *The Journal of Biophysical and Biochemical Cytology.* 1:1–12.
25. Caulfield, J.B. 1957. Effects of varying the vehicle for OsO4 in tissue fixation. *The Journal of Biophysical and Biochemical Cytology.* 3:827–830.
26. Watson, M.L. 1958. Staining of tissue sections for electron microscopy with heavy metals. *The Journal of Biophysical and Biochemical Cytology.* 4:475–478.
27. Watson, M.L. 1958. Staining of tissue sections for electron microscopy with heavy metals. II. Application of solutions containing lead and barium. *The Journal of Biophysical and Biochemical Cytology.* 4:727–730.
28. Pease, D.C., and K.R. Porter. 1981. Electron microscopy and ultramicrotomy. *The Journal of Cell Biology.* 91:287s–292s.
29. Gasser, H.S. 1950. Unmedullated fibers originating in dorsal root ganglia. *The Journal of General Physiology.* 33:651–690.
30. Claude, A. 1950. Studies on cells: Morphology, chemical constitution, and distribution of biochemical functions. In *Harvey Lectures.* Series 43. Academic Press, New York. 121–164.
31. Pease, D.C., and R.F. Baker. 1948. Sectioning techniques for electron microscopy using a conventional microtome. *Proceedings of the Society for Experimental Biology and Medicine.* 67:470–474.
32. Bretschneider, L.H. 1950. Die Herstellung von dünnen Gewebeschnitten und ihre Untersuchung im Elektronenmikroskop. *Mikroskopie.* 5:15–31.
33. Hillier, J., and M.E. Gettner. 1950. Sectioning of tissue for electron microscopy. *Science.* 112:520–523.
34. Claude, A., to L.H. Bretschneider. 19 February 1951. Claude Papers, Wisconsin.
35. Claude, A., to G.E. Palade. 6 August 1952. Claude Papers, Wisconsin.
36. Bretschneider, L.H., to A. Claude. 24 February 1951. Claude Papers, Wisconsin.
37. Bretschneider, L.H. 1952. The electron-microscopic investigation of tissue sections. *International Review of Cytology.* 1:305–322.
38. Latta, H., and J.F. Hartmann. 1950. Use of a glass edge in thin sectioning for electron microscopy. *Proceedings of the Society for Experimental Biology and Medicine.* 74:436–439.
39. Palade, G.E., to A. Claude. 19 October 1952. Claude Papers, Wisconsin.
40. Porter, K.R., and J. Blum. 1953. A study in microtomy for electron microscopy. *The Anatomical Record.* 117:685–710.
41. Gettner, M.E., and L. Ornstein. 1956. Microtomy. In *Physical Techniques in Biological Research.* G. Oster and A.W. Pollister, editors. Academic Press, New York. 627–686.
42. Porter, K.R. 1964. Ultramicrotomy. In *Modern Developments in Electron Microscopy.* B.M. Siegel, editor. Academic Press, New York. 119–145.
43. Pease, D.C. 1960. *Histological Techniques for Electron Microscopy.* Academic Press, New York. 95–148.
44. Fernández-Morán, H. 1953. A diamond knife for ultrathin sectioning. *Experimental Cell Research.* 5:255–256.
45. Fernández-Morán, H. 1956. Applications of a diamond knife for ultrathin sectioning to the study of the fine structure of biological tissues and metals. *The Journal of Biophysical and Biochemical Cytology.* 2:29–30.
46. Porter, K.R. 1956. Notes on the use of the diamond knife. Porter Papers, Colorado. Box CM 7.
47. Doherty, J., to C.L. Moberg. 1996. Personal communication.
48. Porter, K.R. 1987. Electron microscopy of cultured cells. In *The American Association of Anatomists, 1888–1987: Essays on the History of Anatomy in America and a Report on the Membership—Past and Present.* J.E. Pauly, editor. Williams & Wilkins, Baltimore, Maryland. 59–67.

49. Watson, M.L. 1953. A new microtome for thin-sectioning for electron microscopy. *Biochimica et Biophysica Acta.* 10:1–4.
50. Gettner, M., and L. Ornstein. 1953. A new microtome for ultrathin sectioning. *Journal of Applied Physics.* 24:113.
51. Sjöstrand, F.S. 1953. A new microtome for ultrathin sectioning for high resolution electron microscopy. *Experientia.* 9:114–115.
52. Manton, I. 1978. Recollections in the history of electron microscopy. *Proceedings of the Royal Microscopical Society.* 13:45–57.
53. Bensley, R.R. 1953. Symposium: the structure and biochemistry of mitochondria. Introduction and greetings. *The Journal of Histochemistry and Cytochemistry.* 1:179–182.
54. Bensley, R.R. 1947. On the nature of the pigment of mitochondria and of submicroscopic particles in the hepatic cell of the guinea pig. *The Anatomical Record.* 98:609–619.
55. Palade, G.E. 1953. Symposium: the structure and biochemistry of mitochondria. Informal discussion. *The Journal of Histochemistry and Cytochemistry.* 1:265–275.
56. Palade, G.E. 1952. The fine structure of mitochondria. *The Anatomical Record.* 114:427–451.
57. Hogeboom, G.H., and W.C. Schneider. 1951. Proteins of liver and hepatoma mitochondria. *Science.* 113:355–358.
58. Sjöstrand, F.S. 1953. Electron microscopy of mitochondria and cytoplasmic double membranes. *Nature.* 171:30–32.
59. Rasmussen, N. 1997. *Picture Control: The Electron Microscope and the Transformation of Biology in America, 1940–1960.* Stanford University Press, Stanford, California.
60. Sjöstrand, F.S., and J. Rhodin. 1953. The ultrastructure of the proximal convoluted tubules of the mouse kidney as revealed by high resolution electron microscopy. *Experimental Cell Research.* 4:426–456.
61. Sjöstrand, F.S., and V. Hanzon. 1954. Membrane structures of cytoplasm and mitochondria in exocrine cells of mouse pancreas as revealed by high resolution electron microscopy. *Experimental Cell Research.* 7:393–414.
62. Porter, K.R. 1953. RUA, BSD Reports. Volume 41, pages 151–190.
63. Palade, G.E. 1953. An electron microscope study of the mitochondrial structure. *The Journal of Histochemistry and Cytochemistry.* 1:188–211.
64. Sjöstrand, F.S. 1956. Recent advances in the biological applications of the electron microscope. In *The Proceedings of the Third International Conference on Electron Microscopy, London 1954.* R. Ross, editor. The Royal Microscopical Society, London. 26–37.
65. Palade, G.E. 1956. The fixation of tissues for electron microscopy. In *The Proceedings of the Third International Conference on Electron Microscopy, London 1954.* R. Ross, editor. The Royal Microscopical Society, London. 129–142.
66. Palade, G.E., to K.R. Porter. 3 August 1954. Porter Papers, Colorado. Box CM 42.
67. Palade, G.E., to K.R. Porter. 30 August 1954. Porter Papers, Colorado. Box CM 42.
68. Palade, G.E. 1956. Electron microscopy of mitochondria and other cytoplasmic structures. In *Enzymes: Units of Biological Structure and Function.* O.H. Gaebler, editor. Academic Press, New York. 185–215.
69. Mitchell, P. 1961. Coupling of phosphorylation to electron and hydrogen transfer by a chemi-osmotic type of mechanism. *Nature.* 191:144–148.
70. Porter, K.R., and G.E. Palade. 1986. Interview with Mary A. Bonneville. Bonneville Interviews. GEPRU8.
71. Sjöstrand, F.S. 1991. Molecular structure of mitochondrial cristae, solid state biochemistry and a simple theory of respiration-phosphorylation coupling. *Journal of Submicroscopic Cytology and Pathology.* 23:465–489.
72. Daems, W.Th., and E. Wisse. 1966. Shape and attachment of the cristae mitochondriales in mouse hepatic cell mitochondria. *Journal of Ultrastructure Research.* 16:123–140.
73. Perkins, G.A., and T.G. Frey. 2000. Recent structural insight into mitochondria gained by microscopy. *Micron (Oxford, England).* 31:97–111.
74. Frey, T.G., and C.A. Mannella. 2000. The internal structure of mitochondria. *Trends in Biochemical Sciences.* 25:319–324.
75. Pullman, M.E., H.S. Penefsky, A. Datta, and E. Racker. 1960. Partial resolution of the enzymes catalyzing oxidative phosphorylation. I. Purification and properties of soluble dinitrophenol-stimulated adenosine triphosphatase. *The Journal of Biological Chemistry.* 235:3322–3329.
76. Penefsky, H.S., M.E. Pullman, A. Datta, and E. Racker. 1960. Partial resolution of the enzymes catalyzing oxidative phosphorylation. II. Participation of a soluble adenosine triphosphatase in oxidative phosphorylation. *The Journal of Biological Chemistry.* 235:3330–3336.
77. Gasser, H.S. 1953. Promotional materials. RUA, Gasser Papers 302.4. Box 1, Folder 16.
78. Palade, G.E., to A. Claude. 17 January 1954. Claude Papers, Wisconsin.

Chapter 7

The Endoplasmic Reticulum Discoveries

7A. Putting the Endoplasmic Reticulum on the Cellular Map

1. Garnier, C. 1900. Contribution à l'étude de la structure et du fonctionnement des cellules glandulaires séreuses. Du role de l'ergastoplasme dans la sécrétion. *Journal de l'Anatomie et de la Physiologie Normales et Pathologiques de l'Homme et Des Animaux.* 36:22–98.

2. Haguenau, F. 1958. The ergastoplasm: Its history, ultrastructure, and biochemistry. *International Review of Cytology*. 7:425–483.
3. Brachet, J. 1933. Recherches sur la synthèse de l'acide thymonucléique pendant le développement de l'oeuf d'oursin. *Archives de Biologie*. 44:519–576.
4. Caspersson, T., and J. Schultz. 1939. Pentose nucleotides in the cytoplasm of growing tissues. *Nature*. 143:602–603.
5. Davidson, J.N., and C. Waymouth. 1943. Ribonucleic acids in animal tissues. *Nature*. 152:47–48.
6. Caspersson, T. 1941. Studien über den Enweissumsatz der Zelle. *Naturwissenschaften*. 29:33–43.
7. Brachet, J. 1942. La localisation des acides pentosenucléiques dans les tissus animaux et les oeufs d'amphibiens en voie de développement. *Archives de Biologie*. 53:207–257.
8. Murphy, J.B. 1946. RUA, BSD Reports. Volume 34.
9. Porter, K.R., and H.P. Thompson. 1947. Some morphological features of cultured rat sarcoma cells as revealed by the electron microscope. *Cancer Research*. 7:431–438.
10. Murphy, J.B. 1947. RUA, BSD Reports. Volume 35.
11. Porter, K.R., and H.P. Thompson. 1948. A particulate body associated with epithelial cells cultured from mammary carcinomas of mice of a milk-factor strain. *The Journal of Experimental Medicine*. 88:15–24.
12. Murphy, J.B. 1948. RUA, BSD Reports. Volume 36.
13. Murphy, J.B. 1951. RUA, BSD Reports. Volume 39.
14. Porter, K.R., and F.L. Kallman. 1952. Significance of cell particulates as seen by electron microscopy. *Annals of the New York Academy of Sciences*. 54:882–891.
15. Ashley, C.A., K.R. Porter, D.E. Philpott, and G.M. Hass. 1951. Observations by electron microscopy on contraction of skeletal myofibrils induced with adenosinetriphosphate. *The Journal of Experimental Medicine*. 94:9–20.
16. Bennett, H.S., and K.R. Porter. 1953. An electron microscope study of sectioned breast muscle of the domestic fowl. *The American Journal of Anatomy*. 93:61–105.
17. Porter, K.R., and G.E. Palade. 1957. Studies on the endoplasmic reticulum. III. Its form and distribution in striated muscle cells. *The Journal of Biophysical and Biochemical Cytology*. 3:269–300.
18. Veratti, E. 1902. Ricerche sulla fine struttura della fibra muscolare striata [Investigations on the fine structure of striated muscle fiber]. *Memorie del Reale Istituto Lombardo*. 19:87–133.
19. Davis, B.D., P.M. Doty, D.W. Fawcett, H.H. Huxley, A.L. Lehninger, D. Mazia, G.E. Palade, K.R. Porter, H. Ris, A.M. Seligman, and R.B. Griffiths, editors. 1961. The Sarcoplasmic Reticulum. *The Journal of Biophysical and Biochemical Cytology*. 10(Supplement):1–226.
20. Porter, K.R. 1952. Untitled lecture at Harvard University. Porter Papers, Colorado. Box CM 73, page 24 (typescript).
21. Porter, K.R. 1953. Observations on a submicroscopic basophilic component of cytoplasm. *The Journal of Experimental Medicine*. 97:727–750.
22. Martin, A., and S.G. Tomlin. 1950. A technique for the cultivation and preparation of tissue cultures for electron microscopy. *Biochimica et Biophysica Acta*. 5:154–165.
23. Oberling, C., W. Bernhard, M. Guérin, and J. Harel. 1950. Images de cellules cancereuses au microscope électronique. *Bulletin de l'Association Française pour l'Étude du Cancer*. 37:97–109.
24. Selby, C.C., and R.E. Berger. 1952. An electron-optical comparison of the cytoplasmic morphology of cultured adult, embryonic, and neoplastic human epithelial cells. *Cancer*. 5:770–786.
25. Hogeboom, G.H., W.C. Schneider, and G.E. Pallade. 1948. Cytochemical studies of mammalian tissues; isolation of intact mitochondria from rat liver; some biochemical properties of mitochondria and submicroscopic particulate material. *The Journal of Biological Chemistry*. 172:619–635.
26. Slautterback, D.B. 1953. Electron microscopic studies of small cytoplasmic particles (microsomes). *Experimental Cell Research*. 5:173–186.
27. Porter, K.R. 1952. RUA, BSD Reports. Volume 41, pages 154–163.
28. Dalton, A.J., H. Kahler, M.J. Striebich, and B. Lloyd. 1950. Finer structure of hepatic, intestinal and renal cells of the mouse as revealed by the electron microscope. *Journal of the National Cancer Institute*. 11:439–461.
29. Dalton, A.J. 1951. Electron micrography of epithelial cells of the gastro-intestinal tract and pancreas. *The American Journal of Anatomy*. 89:109–133.
30. Bernhard, W., F. Haguenau, A. Gautier, and C. Oberling. 1952. La structure submicroscopique des éléments basophiles cytoplasmiques dans le foie, le pancréas, et les glandes salivaires. *Zeitschrift für Zellforschung und mikroskopische Anatomie*. 37:281–300.
31. Porter, K.R. 1954. Electron microscopy of basophilic components of cytoplasm. *The Journal of Histochemistry and Cytochemistry*. 2:346–375.
32. Deane, H.W. 1946. The basophilic bodies in hepatic cells. *The American Journal of Anatomy*. 78:227–243.
33. Palade, G.E., and K.R. Porter. 1954. Studies on the endoplasmic reticulum. I. Its identification in cells in situ. *The Journal of Experimental Medicine*. 100:641–656.
34. Palade, G.E. 1955. Studies on the endoplasmic reticulum. II. Simple dispositions in cells in situ. *The Journal of Biophysical and Biochemical Cytology*. 1:567–582.
35. Palade, G.E. 1955. A small particulate component of the cytoplasm. *The Journal of Biophysical and Biochemical Cytology*. 1:59–68.
36. Watson, M.L. 1955. The nuclear envelope; its structure and relation to cytoplasmic membranes. *The Journal of Biophysical and Biochemical Cytology*. 1:257–270.

37. Afzelius, B.A. 1955. The ultrastructure of the nuclear membrane of the sea urchin oocyte as studied with the electron microscope. *Experimental Cell Research*. 8:147–158.
38. Petermann, M.L., N.A. Mizen, and M.G. Hamilton. 1953. The macromolecular particles of normal and regenerating rat liver. *Cancer Research*. 13:372–375.
39. Petermann, M.L., M.G. Hamilton, and N.A. Mizen. 1954. Electrophoretic analysis of the macromolecular nucleoprotein particles of mammalian cytoplasm. *Cancer Research*. 14:360–366.
40. Chantrenne, H. 1947. Hétérogenéité des granules cytoplasmiques du foie de souris. *Biochimica et Biophysica Acta*. 1:437–448.
41. Palade, G.E., to K.R. Porter. 30 August 1954. Porter Papers, Colorado. Box CM 42.
42. Sjöstrand, F.S. 1953. Electron microscopy of mitochondria and cytoplasmic double membranes. *Nature*. 171:30–32.
43. Sjöstrand, F.S., and V. Hanzon. 1954. Membrane structures of cytoplasm and mitochondria in exocrine cells of mouse pancreas as revealed by high resolution electron microscopy. *Experimental Cell Research*. 7:393–414.
44. Sjöstrand, F.S. 1956. The ultrastructure of cells as revealed by the electron microscope. *International Review of Cytology*. 5:455–533.
45. Sjöstrand, F.S. 1956. Electron microscopy of cells and tissues. In *Physical Techniques in Biological Research*. G. Oster and A.W. Pollister, editors. Academic Press, New York. 241–298.
46. Palade, G.E. 1956. Intracisternal granules in the exocrine cells of the pancreas. *The Journal of Biophysical and Biochemical Cytology*. 2:417–422.
47. Palade, G.E., and P. Siekevitz. 1956. Pancreatic microsomes; an integrated morphological and biochemical study. *The Journal of Biophysical and Biochemical Cytology*. 2:671–690.
48. Mankowski, A. 1902. Ueber die mikroskopischen Veränderungen des Pankreas nach Unterbindung einzelner Theile und über einige mikrochemische Besonderheiten der Langerhans'schen Inseln. *Archiv für mikroskopische Anatomie*. 59:286–294.
49. Rasmussen, N. 1995. Mitochondrial structure and the practice of cell biology in the 1950s. *Journal of the History of Biology*. 28:381–429.
50. Bechtel, W. 1995. Deciding on the data: Epistemological problems surrounding instruments and research techniques in cell biology. In *PSA 1994*. D.L. Hull, M. Forbes, and R.M. Burian, editors. Philosophy of Science Association, East Lansing, Michigan. 167–178.
51. Palade, G.E., to K.R. Porter. 3 August 1954. Porter Papers, Colorado. Box CM 42.
52. Porter, K.R. 1957. The submicroscopic morphology of protoplasm. In *Harvey Lectures*. Series 51. Academic Press, New York. 175–228.
53. Palade, G.E. 1956. The endoplasmic reticulum. *The Journal of Biophysical and Biochemical Cytology*. 2(Supplement):85–98.
54. Palade, G.E. 1975. Intracellular aspects of the process of protein synthesis. *Science*. 189:347–358.

7B. The Microsome Fraction and Ribosomes

1. Siekevitz, P., and D.M. Greenberg. 1949. The biological formation of serine from glycine. *The Journal of Biological Chemistry*. 180:845–856.
2. Siekevitz, P., and D.M. Greenberg. 1950. The biological formation of formate from methyl compounds in liver slices. *The Journal of Biological Chemistry*. 186:275–286.
3. Schneider, W.C., and G.H. Hogeboom. 1950. Intracellular distribution of enzymes. V. Further studies on the distribution of cytochrome *c* in rat liver homogenates. *The Journal of Biological Chemistry*. 183:123–128.
4. Siekevitz, P. 1952. Uptake of radioactive alanine in vitro into the proteins of rat liver fractions. *The Journal of Biological Chemistry*. 195:549–565.
5. Claude, A. 1950. Studies on cells: Morphology, chemical constitution, and distribution of biochemical functions. In *Harvey Lectures*. Series 43. Academic Press, New York. 121–164.
6. Peterson, E.A., and D.M. Greenberg. 1952. Characteristics of the amino acid-incorporating system of liver homogenates. *The Journal of Biological Chemistry*. 194:359–375.
7. Zamecnik, P.C., I.D. Frantz Jr., R.B. Loftfield, and M.I. Stephenson. 1948. Incorporation in vitro of radioactive carbon from carboxyl-labeled dl-alanine and glycine into proteins of normal and malignant rat livers. *The Journal of Biological Chemistry*. 175:299–314.
8. Frantz, I.D., Jr., P.C. Zamecnik, J.W. Reese, and M.I. Stephenson. 1948. The effect of dinitrophenol on the incorporation of alanine labeled with radioactive carbon into the proteins of slices of normal and malignant rat liver. *The Journal of Biological Chemistry*. 174:773–774.
9. Zamecnik, P.C., and E.B. Keller. 1954. Relation between phosphate energy donors and incorporation of labeled amino acids into proteins. *The Journal of Biological Chemistry*. 209:337–354.
10. Mitchell, P. 1961. Coupling of phosphorylation to electron and hydrogen transfer by a chemi-osmotic type of mechanism. *Nature*. 191:144–148.
11. Mitchell, P. 1966. Chemiosmotic coupling in oxidative and photosynthetic phosphorylation. *Biological Reviews of the Cambridge Philosophical Society*. 41:445–502.
12. Siekevitz, P., and V.R. Potter. 1953. Intramitochondrial regulation of oxidative rate. *The Journal of Biological Chemistry*. 201:1–13.

13. Siekevitz, P., and V.R. Potter. 1955. Biochemical structure of mitochondria. I. Intramitochondrial components and oxidative phosphorylation. *The Journal of Biological Chemistry*. 215:221–235.
14. Siekevitz, P., and V.R. Potter. 1955. Biochemical structure of mitochondria. II. Radioactive labeling of intra-mitochondrial nucleotides during oxidative phosphorylation. *The Journal of Biological Chemistry*. 215:237–255.
15. Porter, K.R. 1953. Observations on a submicroscopic basophilic component of cytoplasm. *The Journal of Experimental Medicine*. 97:727–750.
16. Cleland, K.W., and E.C. Slater. 1953. Respiratory granules of heart muscle. *The Biochemical Journal*. 53:547–556.
17. Siekevitz, P., and M.L. Watson. 1956. Cytochemical studies of mitochondria. I. The separation and identification of a membrane fraction from isolated mitochondria. *The Journal of Biophysical and Biochemical Cytology*. 2:639–652.
18. Siekevitz, P., and M.L. Watson. 1956. Cytochemical studies of mitochondria. II. Enzymes associated with a mitochondrial membrane fraction. *The Journal of Biophysical and Biochemical Cytology*. 2:653–669.
19. Keilin, D., and E.F. Hartree. 1949. Activity of the succinic dehydrogenase-cytochrome system in different tissue preparations. *The Biochemical Journal*. 44:205–218.
20. Cooper, C., and A.L. Lehninger. 1956. Oxidative phosphorylation by an enzyme complex from extracts of mitochondria. I. The span b-hydroxybutyrate to oxygen. *The Journal of Biological Chemistry*. 219:489–506.
21. Devlin, T.M., and A.L. Lehninger. 1956. Oxidative phosphorylation by an enzyme complex from extracts of mitochondria. II. The span b-hydroxybutyrate to cytochrome *c*. *The Journal of Biological Chemistry*. 219:507–518.
22. Cooper, C., and A.L. Lehninger. 1956. Oxidative phosphorylation by an enzyme complex from extracts of mitochondria. III. The span cytochrome *c* to oxygen. *The Journal of Biological Chemistry*. 219:519–529.
23. Siekevitz, P., and M.L. Watson. 1957. Some cytochemical characteristics of a phosphorylating digitonin preparation of mitochondria. *Biochimica et Biophysica Acta*. 25:274–279.
24. Siekevitz, P., H. Löw, L. Ernster, and O. Lindberg. 1958. On a possible mechanism of the adenosinetriphosphatase of liver mitochondria. *Biochimica et Biophysica Acta*. 29:378–391.
25. Löw, H., P. Siekevitz, L. Ernster, and O. Lindberg. 1958. Studies on the relation of the adenosinetriphosphate-inorganic phosphate exchange reaction of mitochondria to oxidative phosphorylation. *Biochimica et Biophysica Acta*. 29:392–405.
26. Palade, G.E. 1955. A small particulate component of the cytoplasm. *The Journal of Biophysical and Biochemical Cytology*. 1:59–68.
27. Palade, G.E., and P. Siekevitz. 1956. Pancreatic microsomes; an integrated morphological and biochemical study. *The Journal of Biophysical and Biochemical Cytology*. 2:671–690.
28. Hogeboom, G.H. 1949. Cytochemical studies of mammalian tissues; the distribution of diphosphopyridine nucleotide-cytochrome *c* reductase in rat liver fractions. *The Journal of Biological Chemistry*. 177:847–858.
29. Palade, G.E., and P. Siekevitz. 1956. Liver microsomes; an integrated morphological and biochemical study. *The Journal of Biophysical and Biochemical Cytology*. 2:171–200.
30. Littlefield, J.W., E.B. Keller, J. Gross, and P.C. Zamecnik. 1955. Studies on cytoplasmic ribonucleoprotein particles from the liver of the rat. *The Journal of Biological Chemistry*. 217:111–123.
31. Petermann, M.L., N.A. Mizen, and M.G. Hamilton. 1953. The macromolecular particles of normal and regenerating rat liver. *Cancer Research*. 13:372–375.
32. Taylor, A.R., D.G. Sharp, D. Beard, and J.W. Beard. 1942. Isolation and properties of a macromolecular component of normal chick embryo tissue. *The Journal of Infectious Diseases*. 71:115–126.
33. Taylor, A.R., D.G. Sharp, and B. Woodhall. 1943. Macromolecular components of normal embryonic and adult brain tissue. *Science*. 97:226–227.
34. Petermann, M.L., and M.G. Hamilton. 1952. An ultracentrifugal analysis of the macromolecular particles of normal and leukemic mouse spleen. *Cancer Research*. 12:373–378.
35. Warner, J.R., A. Rich, and C.E. Hall. 1962. Electron microscope studies of ribosomal clusters synthesizing hemoglobin. *Science*. 138:1399–1403.
36. Warner, J.R., P.M. Knopf, and A. Rich. 1963. A multiple ribosomal structure in protein synthesis. *Proceedings of the National Academy of Sciences of the United States of America*. 49:122–129.
37. Chao, F.-C., and H.K. Schachman. 1956. The isolation and characterization of a macro-molecular ribonucleoprotein from yeast. *Archives of Biochemistry and Biophysics*. 61:220–230.
38. Porter, K.R., to D.W. Bronk. 13 August 1956. Porter Papers, Colorado. Box CM 44.
39. Roberts, R.B., editor. 1958. *Microsomal Particles and Protein Synthesis*. Pergamon Press, New York.
40. Palade, G.E. 1958. Microsomes and ribonucleoprotein particles. In *Microsomal Particles and Protein Synthesis*. R.B. Roberts, editor. Pergamon Press, New York. 36–61.
41. Palade, G.E. 1975. Intracellular aspects of the process of protein synthesis. *Science*. 189:347–358.
42. Heidenhain, R. 1875. Beiträge zur Kentniss des Pancreas. *Pflügers Archiv European Journal of Physiology*. 10:557–632.
43. Heidenhain, R. 1883. Die Bauchspeicheldrüse. In *Handbuch der Physiologie*. L. Hermann, editor. F.C.W. Vogel, Leipzig. 173–208.
44. Marshall, J.M., Jr. 1954. Distributions of chymotrypsinogen, procarboxypeptidase, desoxyribonuclease, and ribonuclease in bovine pancreas. *Experimental Cell Research*. 6:240–242.
45. Hokin, L.E. 1955. Isolation of the zymogen granules of dog pancreas and a study of their properties. *Biochimica et Biophysica Acta*. 18:379–388.
46. Palade, G.E. 1956. Intracisternal granules in the exocrine cells of the pancreas. *The Journal of Biophysical and Biochemical Cytology*. 2:417–422.

47. Hogeboom, G.H., W.C. Schneider, and G.E. Pallade. 1948. Cytochemical studies of mammalian tissues; isolation of intact mitochondria from rat liver; some biochemical properties of mitochondria and submicroscopic particulate material. *The Journal of Biological Chemistry*. 172:619–635.
48. Siekevitz, P., and G.E. Palade. 1958. A cytochemical study on the pancreas of the guinea pig. I. Isolation and enzymatic activities of cell fractions. *The Journal of Biophysical and Biochemical Cytology*. 4:203–218.
49. Daly, M.M., and A.E. Mirsky. 1952. Formation of protein in the pancreas. *The Journal of General Physiology*. 36:243–254.
50. Siekevitz, P., and G.E. Palade. 1958. A cytochemical study on the pancreas of the guinea pig. III. In vivo incorporation of leucine-1-C^{14} into the proteins of cell fractions. *The Journal of Biophysical and Biochemical Cytology*. 4:557–566.
51. Daly, M.M., V.G. Allfrey, and A.E. Mirsky. 1955. Synthesis of protein in the pancreas. III. Uptake of glycine-N^{15} by the trypsinogen and chymotrypsinogen of mouse pancreas. *The Journal of General Physiology*. 39:207–210.
52. Allfrey, V., M.M. Daly, and A.E. Mirsky. 1953. Synthesis of protein in the pancreas. II. The role of ribonucleoprotein in protein synthesis. *The Journal of General Physiology*. 37:157–175.
53. Siekevitz, P., and G.E. Palade. 1960. A cytochemical study on the pancreas of the guinea pig. V. In vivo incorporation of leucine-1-C^{14} into the chymotrypsinogen of various cell fractions. *The Journal of Biophysical and Biochemical Cytology*. 7:619–630.
54. Hirs, C.H.W., S. Moore, and W.H. Stein. 1953. A chromatographic investigation of pancreatic ribonuclease. *The Journal of Biological Chemistry*. 200:493–506.
55. Palade, G.E. 16 March 1995. The American cradle. Speech delivered at symposium "Journey into the Cell: Celebrating 50 Years of Electron Microscopy and Modern Cell Biology". The Rockefeller University, New York.
56. Siekevitz, P., and G.E. Palade. 1959. A cytochemical study on the pancreas of the guinea pig. IV. Chemical and metabolic investigation of the ribonucleoprotein particles. *The Journal of Biophysical and Biochemical Cytology*. 5:1–10.
57. Redman, C.M., P. Siekevitz, and G.E. Palade. 1966. Synthesis and transfer of amylase in pigeon pancreatic microsomes. *The Journal of Biological Chemistry*. 241:1150–1158.
58. Redman, C.M., and D.D. Sabatini. 1966. Vectorial discharge of peptides released by puromycin from attached ribosomes. *Proceedings of the National Academy of Sciences of the United States of America*. 56:608–615.
59. Ganoza, M.C., and C.A. Williams. 1969. *In vitro* synthesis of different categories of specific protein by membrane-bound and free ribosomes. *Proceedings of the National Academy of Sciences of the United States of America*. 63:1370–1376.
60. Palade, G.E. 1959. Functional changes in the structure of cell components. In *Subcellular Particles*. T. Hayashi, editor. Ronald Press Company, New York. 64–83.
61. Palade, G.E. 1961. The secretory process of the pancreatic exocrine cell. In *Electron Microscopy in Anatomy*. J.D Boyd, F.R. Johnson, and J.D. Lever, editors. Williams & Wilkins, Baltimore, Maryland. 176–195.
62. Kirsch, J.F., P. Siekevitz, and G.E. Palade. 1960. Amino acid incorporation in vitro by ribonucleoprotein particles detached from guinea pig liver microsomes. *The Journal of Biological Chemistry*. 235:1419–1424.
63. Siekevitz, P., and G.E. Palade. 1960. A cytochemical study on the pancreas of the guinea pig. VI. Release of enzymes and ribonucleic acid from ribonucleoprotein particles. *The Journal of Biophysical and Biochemical Cytology*. 7:631–644.
64. Siekevitz, P., and G.E. Palade. 1962. Cytochemical study on the pancreas of the guinea pig. VII. Effects of spermine on ribosomes. *The Journal of Cell Biology*. 13:217–232.
65. Siekevitz, P., and G.E. Palade. 1966. Distribution of newly synthesized amylase in microsomal subfractions of guinea pig pancreas. *The Journal of Cell Biology*. 30:519–530.
66. Sjöstrand, F.S., and V. Hanzon. 1954. Ultrastructure of Golgi apparatus of exocrine cells of mouse pancreas. *Experimental Cell Research*. 7:415–429.
67. Farquhar, M.G., and S.R. Wellings. 1957. Electron microscopic evidence suggesting secretory granule formation within the Golgi apparatus. *The Journal of Biophysical and Biochemical Cytology*. 3:319–322.
68. Caro, L.G., R.P. van Tubergen, and J.A. Kolb. 1962. High-resolution autoradiography. I. Methods. *The Journal of Cell Biology*. 15:173–188.
69. Warshawsky, H., C.P. Leblond, and B. Droz. 1963. Synthesis and migration of proteins in the cells of the exocrine pancreas as revealed by specific activity determination from radioautographs. *The Journal of Cell Biology*. 16:1–24.
70. Caro, L.G., and G.E. Palade. 1964. Protein synthesis, storage, and discharge in the pancreatic exocrine cell. An autoradiographic study. *The Journal of Cell Biology*. 20:473–495.
71. Palade, G.E., P. Siekevitz, and L.G. Caro. 1962. Structure, chemistry and function of the pancreatic exocrine cell. In *Ciba Foundation Symposium on the Exocrine Pancreas: Normal and Abnormal Functions*. A.V.S. deReuck and M.P. Cameron, editors. J & A Churchill, London. 23–49.
72. Laird, A.K., and A.D. Barton. 1958. Protein synthesis in rat pancreas. II. Changes in the intracellular distribution of pancreatic amylase during the secretory cycle. *Biochimica et Biophysica Acta*. 27:12–15.
73. Morris, A.J., and S.R. Dickman. 1960. Biosynthesis of ribonuclease in mouse pancreas. *The Journal of Biological Chemistry*. 235:1404–1408.
74. Sjöstrand, F.S. 1962. The fine structure of the exocrine pancreas cells. In *Ciba Foundation Symposium on the Exocrine Pancreas: Normal and Abnormal Functions*. A.V.S. deReuck and M.P. Cameron, editors. J & A Churchill, London. 1–19.
75. Ehrenreich, J.H., J.J.M. Bergeron, P. Siekevitz, and G.E. Palade. 1973. Golgi fractions prepared from rat liver homogenates. I. Isolation procedure and morphological characterization. *The Journal of Cell Biology*. 59:45–72.
76. Bergeron, J.J.M., J.H. Ehrenreich, P. Siekevitz, and G.E. Palade. 1973. Golgi fractions prepared from rat liver homogenates. II. Biochemical characterization. *The Journal of Cell Biology*. 59:73–88.
77. Farquhar, M.G., J.J.M. Bergeron, and G.E. Palade. 1974. Cytochemistry of Golgi fractions prepared from rat liver. *The Journal of Cell Biology*. 60:8–25.

7C. Establishing the Secretory Pathway of the Pancreatic Exocrine Cell

1. Palade, G.E. 1961. The secretory process of the pancreatic exocrine cell. In *Electron Microscopy in Anatomy*. J.D. Boyd, F.R. Johnson, and J.D. Lever, editors. Williams & Wilkins, Baltimore, Maryland. 176–195.
2. Hokin, L.E. 1951. The synthesis and secretion of amylase by pigeon pancreas *in vitro*. *The Biochemical Journal*. 48:320–326.
3. de Bold, A.J. 1982. Tissue fractionation studies on the relationship between an atrial natriuretic factor and specific atrial granules. *Canadian Journal of Physiology and Pharmacology*. 60:324–330.
4. de Bold, A.J. 1987. On the shoulders of giants: the discovery of atrial natriuretic factor. *Canadian Journal of Physiology and Pharmacology*. 65:2007–2012.
5. Jamieson, J.D., to P. Siekevitz. 1999. Collected papers of Carol L. Moberg.
6. Jamieson, J.D. 1966. *Intracellular Transport of Secretory Protein: Role of the Golgi Complex*. PhD thesis. The Rockefeller University, New York.
7. Caro, L.G., and G.E. Palade. 1964. Protein synthesis, storage, and discharge in the pancreatic exocrine cell. An autoradiographic study. *The Journal of Cell Biology*. 20:473–495.
8. Jamieson, J.D., and G.E. Palade. 1966. Role of the Golgi complex in the intracellular transport of secretory proteins. *Proceedings of the National Academy of Sciences of the United States of America*. 55:424–431.
9. Jamieson, J.D., and G.E. Palade. 1967. Intracellular transport of secretory proteins in the pancreatic exocrine cell. I. Role of the peripheral elements of the Golgi complex. *The Journal of Cell Biology*. 34:577–596.
10. Jamieson, J.D., and G.E. Palade. 1967. Intracellular transport of secretory proteins in the pancreatic exocrine cell. II. Transport to condensing vacuoles and zymogen granules. *The Journal of Cell Biology*. 34:597–615.
11. Laird, A.K., and A.D. Barton. 1958. Protein synthesis in rat pancreas. II. Changes in the intracellular distribution of pancreatic amylase during the secretory cycle. *Biochimica et Biophysica Acta*. 27:12–15.
12. Morris, A.J., and S.R. Dickman. 1960. Biosynthesis of ribonuclease in mouse pancreas. *The Journal of Biological Chemistry*. 235:1404–1408.
13. Sjöstrand, F.S. 1962. The fine structure of the exocrine pancreas cells. In *Ciba Foundation Symposium on the Exocrine Pancreas: Normal and Abnormal Functions*. A.V.S. deReuck and M.P. Cameron, editors. J & A Churchill, London. 1–19.
14. Castle, J.D., J.D. Jamieson, and G.E. Palade. 1972. Radioautographic analysis of the secretory process in the parotid acinar cell of the rabbit. *The Journal of Cell Biology*. 53:290–311.
15. Castle, J.D., J.D. Jamieson, and G.E. Palade. 1975. Secretion granules of the rabbit parotid gland. Isolation, subfractionation, and characterization of the membrane and content subfractions. *The Journal of Cell Biology*. 64:182–210.
16. Jamieson, J.D., and G.E. Palade. 1968. Intracellular transport of secretory proteins in the pancreatic exocrine cell. III. Dissociation of intracellular transport from protein synthesis. *The Journal of Cell Biology*. 39:580–588.
17. Jamieson, J.D., and G.E. Palade. 1968. Intracellular transport of secretory proteins in the pancreatic exocrine cell. IV. Metabolic requirements. *The Journal of Cell Biology*. 39:589–603.
18. Meldolesi, J., J.D. Jamieson, and G.E. Palade. 1971. Composition of cellular membranes in the pancreas of the guinea pig. I. Isolation of membrane fractions. *The Journal of Cell Biology*. 49:109–129.
19. Meldolesi, J., J.D. Jamieson, and G.E. Palade. 1971. Composition of cellular membranes in the pancreas of the guinea pig. II. Lipids. *The Journal of Cell Biology*. 49:130–149.
20. Meldolesi, J., J.D. Jamieson, and G.E. Palade. 1971. Composition of cellular membranes in the pancreas of the guinea pig. III. Enzymatic activities. *The Journal of Cell Biology*. 49:150–158.
21. Jamieson, J.D., and G.E. Palade. 1971. Condensing vacuole conversion and zymogen granule discharge in pancreatic exocrine cells: Metabolic studies. *The Journal of Cell Biology*. 48:503–522.
22. Palade, G.E. 1959. Functional changes in the structure of cell components. In *Subcellular Particles*. T. Hayashi, editor. Ronald Press Company, New York. 64–83.
23. Jamieson, J.D., and G.E. Palade. 1971. Synthesis, intracellular transport, and discharge of secretory proteins in stimulated pancreatic exocrine cells. *The Journal of Cell Biology*. 50:135–158.
24. Redman, C.M., D. Banerjee, K. Howell, and G.E. Palade. 1975. Colchicine inhibition of plasma protein release from rat hepatocytes. *The Journal of Cell Biology*. 66:42–59.
25. Tartakoff, A.M., L.J. Greene, and G.E. Palade. 1974. Studies on the guinea pig pancreas. Fractionation and partial characterization of exocrine proteins. *The Journal of Biological Chemistry*. 249:7420–7431.
26. Scheele, G.A., and G.E. Palade. 1975. Studies on the guinea pig pancreas. Parallel discharge of exocrine enzyme activities. *The Journal of Biological Chemistry*. 250:2660–2670.
27. Tartakoff, A.M., J.D. Jamieson, G.A. Scheele, and G.E. Palade. 1975. Studies on the pancreas of the guinea pig. Parallel processing and discharge of exocrine proteins. *The Journal of Biological Chemistry*. 250:2671–2677.
28. Kraehenbuhl, J.-P., and J.D. Jamieson. 1972. Solid-phase conjugation of ferritin to Fab-fragments of immunoglobulin G for use in antigen localization on thin sections. *Proceedings of the National Academy of Sciences of the United States of America*. 69:1771–1775.
29. Rothman, S.S. 1975. Protein transport by the pancreas. *Science*. 190:747–753.
30. Scheele, G.A., G.E. Palade, and A.M. Tartakoff. 1978. Cell fractionation studies on the guinea pig pancreas. Redistribution of exocrine proteins during tissue homogenization. *The Journal of Cell Biology*. 78:110–130.

7D. The Palade Model of Combining Structure with Function

1. Palade, G.E. 1975. Intracellular aspects of the process of protein synthesis. *Science*. 189:347–358.
2. Jamieson, J.D., and G.E. Palade. 1977. Production of secretory proteins in animal cells. In *International Cell Biology, 1976–1977*. B.R. Brinkley and K.R. Porter, editors. The Rockefeller University Press, New York. 307–317.
3. Siekevitz, P., and P.C. Zamecnik. 1981. Ribosomes and protein synthesis. *The Journal of Cell Biology*. 91:53s–65s.
4. Farquhar, M.G., and G.E. Palade. 1981. The Golgi apparatus (complex)—(1954–1981)—from artifact to center stage. *The Journal of Cell Biology*. 91:77s–103s.
5. Farquhar, M.G., and G.E. Palade. 1998. The Golgi apparatus: 100 years of progress and controversy. *Trends in Cell Biology*. 8:2–10.
6. Palade, G.E. 1977. Greetings from the President of the American Society for Cell Biology. In *International Cell Biology, 1976–1977*. B.R. Brinkley and K.R. Porter, editors. The Rockefeller University Press, New York. vii–x.

Chapter 8

Ribosome Structure and the Synthesis and Movement of Proteins

8A. Mechanisms that Regulate the Destination of Nascent Proteins

1. De Robertis, E., and D.D. Sabatini. 1958. Mitochondrial changes in the adrenocortex of normal hamsters. *The Journal of Biophysical and Biochemical Cytology*. 4:667–668.
2. Sabatini, D.D., and E.D. De Robertis. 1961. Ultrastructural zonation of adrenocortex in the rat. *The Journal of Biophysical and Biochemical Cytology*. 9:105–119.
3. Sabatini, D.D., E.D. De Robertis, and H.B. Bleichmar. 1962. Submicroscopic study of the pituitary action on the adrenocortex of the rat. *Endocrinology*. 70:390–406.
4. De Robertis, E.D., and D.D. Sabatini. 1960. Submicroscopic analysis of the secretory process in the adrenal medulla. *Federation Proceedings*. 19:70–78.
5. Palade, G.E. 1955. A small particulate component of the cytoplasm. *The Journal of Biophysical and Biochemical Cytology*. 1:59–68.
6. Palade, G.E. 1953. A small particulate component of the cytoplasm. *Journal of Applied Physics*. 24:1419.
7. Siekevitz, P. 1952. Uptake of radioactive alanine in vitro into the proteins of rat liver fractions. *The Journal of Biological Chemistry*. 195:549–565.
8. McQuillen, K., R.B. Roberts, and R.J. Britten. 1959. Synthesis of nascent protein by ribosomes in *Escherichia coli*. *Proceedings of the National Academy of Sciences of the United States of America*. 45:1437–1447.
9. Jacob, F., and J. Monod. 1961. Genetic regulatory mechanisms in the synthesis of proteins. *Journal of Molecular Biology*. 3:318–356.
10. Lasansky, A., and D.D. Sabatini. 1957. Distribution of -SH and -S-S groups in the neurophypophysis and hypothalamus of the toad. *Revista de la Sociedad Argentina de Biología*. 33:177–182.
11. Palade, G.E. 1952. A study of fixation for electron microscopy. *The Journal of Experimental Medicine*. 95:285–298.
12. Fieser, L.F., and M. Fieser. 1956. *Organic Chemistry*. Third edition. Reinhold Publishing Corporation, Boston.
13. Sabatini, D.D., K. Bensch, and R.J. Barrnett. 1963. Cytochemistry and electron microscopy. The preservation of cellular ultrastructure and enzymatic activity by aldehyde fixation. *The Journal of Cell Biology*. 17:19–58.
14. Sabatini, D.D., F. Miller, and R.J. Barrnett. 1964. Aldehyde fixation for morphological and enzyme histochemical studies with the electron microscope. *The Journal of Histochemistry and Cytochemistry*. 12:57–71.
15. Ledbetter, M.C., and K.R. Porter. 1963. A "microtubule" in plant cell fine structure. *The Journal of Cell Biology*. 19:239–250.
16. Porter, K.R. 1966. Cytoplasmic microtubules and their functions. In *Ciba Foundation Symposium: Principles of Biomolecular Organization*. M. O'Connor and G.E.W. Wolstenholme, editors. J & A Churchill, London. 308–345.
17. Tissieres, A., J.D. Watson, D. Schlessinger, and B.R. Hollingworth. 1959. Ribonucleoprotein particles from *Escherichia coli*. *Journal of Molecular Biology*. 1:221–233.
18. Gilbert, W. 1963. Polypeptide synthesis in *Escherichia coli*. II. The polypeptide chain and S-RNA. *Journal of Molecular Biology*. 6:389–403.
19. Tashiro, Y., and P. Siekevitz. 1965. Localization on hepatic ribosomes of protein newly synthesized in vivo. *Journal of Molecular Biology*. 11:166–173.
20. Tashiro, Y., and D.A. Yphantis. 1965. Molecular weights of hepatic ribosomes and their subunits. *Journal of Molecular Biology*. 11:174–186.
21. Tashiro, Y., and P. Siekevitz. 1965. Ultracentrifugal studies on the dissociation of hepatic ribosomes. *Journal of Molecular Biology*. 11:149–165.
22. Sabatini, D.D., Y. Tashiro, and G.E. Palade. 1966. On the attachment of ribosomes to microsomal membranes. *Journal of Molecular Biology*. 19:503–524.

23. Siekevitz, P., and G.E. Palade. 1960. A cytochemical study of the pancreas of the guinea pig. V. *In vivo* incorporation of leucine-1-C^{14} into the chymotrypsinogen of various cell fractions. *The Journal of Biophysical and Biochemical Cytology*. 7:619–630.
24. Palade, G.E., P. Siekevitz, and L.G. Caro. 1961. Structure, chemistry and function of the pancreatic exocrine cell. In *Ciba Foundation Symposium on the Exocrine Pancreas: Normal and Abnormal Functions*. A.V.S. de Reuck and M.P. Cameron, editors. Little Brown, Boston. 23–55.
25. Redman, C.M., P. Siekevitz, and G.E. Palade. 1966. Synthesis and transfer of amylase in pigeon pancreatic microsomes. *The Journal of Biological Chemistry*. 241:1150–1158.
26. Redman, C.M., and D.D. Sabatini. 1966. Vectorial discharge of peptides released by puromycin from attached ribosomes. *Proceedings of the National Academy of Sciences of the United States of America*. 56:608–615.
27. Ernster, L., P. Siekevitz, and G.E. Palade. 1962. Enzyme-structure relationships in the endoplasmic reticulum of rat liver. A morphological and biochemical study. *The Journal of Cell Biology*. 15:541–562.
28. Sabatini, D.D., and G. Blobel. 1970. Controlled proteolysis of nascent polypeptides in rat liver cell fractions. II. Location of the polypeptides in rough microsomes. *The Journal of Cell Biology*. 45:146–157.
29. Sabatini, D.D., G. Blobel, Y. Nonomura, and M.R. Adelman. 1971. Ribosome-membrane interaction: Structural aspects and functional implications. In *Advances in Cytopharmacology*. Volume 1. F. Clementi and B. Ceccarelli, editors. Raven Press, New York. 119–130.
30. Redman, C.M. 1967. Studies on the transfer of incomplete polypeptide chains across rat liver microsomal membranes in vitro. *The Journal of Biological Chemistry*. 242:761–768.
31. Blobel, G., and D.D. Sabatini. 1970. Controlled proteolysis of nascent polypeptides in rat liver cell fractions. I. Location of the polypeptides within ribosomes. *The Journal of Cell Biology*. 45:130–145.
32. Dintzis, H.M. 1961. Assembly of the peptide chains of hemoglobin. *Proceedings of the National Academy of Sciences of the United States of America*. 47:247–261.
33. Malkin, L.I., and A. Rich. 1967. Partial resistance of nascent polypeptide chains to proteolytic digestion due to ribosomal shielding. *Journal of Molecular Biology*. 26:329–346.
34. Petermann, M.L., and A. Pavlovec. 1969. Effects of magnesium and formaldehyde on the sedimentation behavior of rat liver ribosomes. *Biopolymers*. 7:73–81.
35. Ito, A., and R. Sato. 1969. Proteolytic microdissection of smooth-surfaced vesicles of liver microsomes. *The Journal of Cell Biology*. 40:179–189.
36. Omura, T., P. Siekevitz, and G.E. Palade. 1967. Turnover of constituents of the endoplasmic reticulum membranes of rat hepatocytes. *The Journal of Biological Chemistry*. 242:2389–2396.
37. Blobel, G., and D.D. Sabatini. 1971. Dissociation of mammalian polyribosomes into subunits by puromycin. *Proceedings of the National Academy of Sciences of the United States of America*. 68:390–394.
38. Nonomura, Y., G. Blobel, and D. Sabatini. 1971. Structure of liver ribosomes studied by negative staining. *Journal of Molecular Biology*. 60:303–323.
39. Morimoto, T., G. Blobel, and D.D. Sabatini. 1972. Ribosome crystallization in chicken embryos. I. Isolation, characterization, and in vitro activity of ribosome tetramers. *The Journal of Cell Biology*. 52:338–354.
40. Morimoto, T., G. Blobel, and D.D. Sabatini. 1972. Ribosome crystallization in chicken embryos. II. Conditions for the formation of ribosome tetramers in vitro. *The Journal of Cell Biology*. 52:355–366.
41. Byers, B. 1967. Structure and formation of ribosome crystals in hypothermic chick embryo cells. *Journal of Molecular Biology*. 26:155–167.
42. Adelman, M.R., G. Blobel, and D.D. Sabatini. 1973. An improved cell fractionation procedure for the preparation of rat liver membrane-bound ribosomes. *The Journal of Cell Biology*. 56:191–205.
43. Sabatini, D.D., N. Borgese, M.R. Adelman, G. Kreibich, and G. Blobel. 1972. Studies on the membrane associated protein synthesis apparatus of eukaryotic cells. In *FEBS Symposium*. Volume 27. North-Holland Publishing Co., Amsterdam. 147–71.
44. Adelman, M.R., D.D. Sabatini, and G. Blobel. 1973. Ribosome-membrane interaction. Nondestructive disassembly of rat liver rough microsomes into ribosomal and membranous components. *The Journal of Cell Biology*. 56:206–229.
45. Borgese, D., G. Blobel, and D.D. Sabatini. 1973. In vitro exchange of ribosomal subunits between free and membrane-bound ribosomes. *Journal of Molecular Biology*. 74:415–438.
46. Blobel, G., and D.D. Sabatini. 1971. Ribosome-membrane interactions in eukaryotic cells. In *Biomembranes*. Volume 2. L.A. Manson, editor. Plenum Press, New York. 193–195.
47. Walter, P., and G. Blobel. 1981. Translocation of proteins across the endoplasmic reticulum. II. Signal recognition protein (SRP) mediates the selective binding to microsomal membranes of in-vitro-assembled polysomes synthesizing secretory protein. *The Journal of Cell Biology*. 91:551–556.
48. Walter, P., and G. Blobel. 1981. Translocation of proteins across the endoplasmic reticulum. III. Signal recognition protein (SRP) causes signal sequence-dependent and site-specific arrest of chain elongation that is released by microsomal membranes. *The Journal of Cell Biology*. 91:557–561.
49. Gilmore, R., G. Blobel, and P. Walter. 1982. Protein translocation across the endoplasmic reticulum. I. Detection in the microsomal membrane of a receptor for the signal recognition particle. *The Journal of Cell Biology*. 95:463–469.
50. Gilmore, R., P. Walter, and G. Blobel. 1982. Protein translocation across the endoplasmic reticulum. II. Isolation and characterization of the signal recognition particle receptor. *The Journal of Cell Biology*. 95:470–477.
51. Meyer, D.I., E. Krause, and B. Dobberstein. 1982. Secretory protein translocation across membranes—the role of the 'docking protein'. *Nature*. 297:647–650.

52. Borgese, N., W. Mok, G. Kreibich, and D.D. Sabatini. 1974. Ribosomal-membrane interaction: in vitro binding of ribosomes to microsomal membranes. *Journal of Molecular Biology*. 88:559–580.
53. Kemper, B., J.F. Habener, R.C. Mulligan, J.T. Potts Jr., and A. Rich. 1974. Pre-proparathyroid hormone: a direct translation product of parathyroid messenger RNA. *Proceedings of the National Academy of Sciences of the United States of America*. 71:3731–3735.
54. Mach, B., C. Faust, and P. Vassalli. 1973. Purification of 14S messenger RNA of immunoglobulin light chain that codes for a possible light-chain precursor. *Proceedings of the National Academy of Sciences of the United States of America*. 70:451–455.
55. Milstein, C., G.G. Brownlee, T.M. Harrison, and M.B. Mathews. 1972. A possible precursor of immunoglobulin light chains. *Nature: New Biology*. 239:117–120.
56. Tonegawa, S., and I. Baldi. 1973. Electrophoretically homogeneous myeloma light chain mRNA and its translation in vitro. *Biochemical and Biophysical Research Communications*. 51:81–87.
57. Blobel, G., and B. Dobberstein. 1975. Transfer of proteins across membranes. I. Presence of proteolytically processed and unprocessed nascent immunoglobulin light chains on membrane-bound ribosomes of murine myeloma. *The Journal of Cell Biology*. 67:835–851.
58. Blobel, G., and B. Dobberstein. 1975. Transfer of proteins across membranes. II. Reconstitution of functional rough microsomes from heterologous components. *The Journal of Cell Biology*. 67:852–862.
59. Simon, S.M., and G. Blobel. 1991. A protein-conducting channel in the endoplasmic reticulum. *Cell*. 65:371–380.
60. Blobel, G. 1980. Intracellular protein topogenesis. *Proceedings of the National Academy of Sciences of the United States of America*. 77:1496–1500.
61. Sabatini, D.D., G. Kreibich, T. Morimoto, and M. Adesnik. 1982. Mechanisms for the incorporation of proteins in membranes and organelles. *The Journal of Cell Biology*. 92:1–22.
62. Görlich, D., and T.A. Rapoport. 1993. Protein translocation into proteoliposomes reconstituted from purified components of the endoplasmic reticulum membrane. *Cell*. 75:615–630.
63. Kreibich, G., B.L. Ulrich, and D.D. Sabatini. 1978. Proteins of rough microsomal membranes related to ribosome binding. I. Identification of ribophorins I and II, membrane proteins characteristic of rough microsomes. *The Journal of Cell Biology*. 77:464–487.
64. Kreibich, G., C.M. Freienstein, B.N. Pereyra, B.L. Ulrich, and D.D. Sabatini. 1978. Proteins of rough microsomal membranes related to ribosome binding. II. Cross-linking of bound ribosomes to specific membrane proteins exposed at the binding sites. *The Journal of Cell Biology*. 77:488–506.
65. Yu, Y.H., D.D. Sabatini, and G. Kreibich. 1990. Antiribophorin antibodies inhibit the targeting to the ER membrane of ribosomes containing nascent secretory polypeptides. *The Journal of Cell Biology*. 111:1335–1342.
66. Kelleher, D.J., G. Kreibich, and R. Gilmore. 1992. Oligosaccharyltransferase activity is associated with a protein complex composed of ribophorins I and II and a 48 kd protein. *Cell*. 69:55–65.
67. Walter, P., and G. Blobel. 1983. Disassembly and reconstitution of signal recognition particle. *Cell*. 34:525–533.
68. Bernstein, H.D., M.A. Poritz, K. Strub, P.J. Hoben, S. Brenner, and P. Walter. 1989. Model for signal sequence recognition from amino-acid sequence of 54K subunit of signal recognition particle. *Nature*. 340:482–486.
69. Miller, J.D., S. Tajima, L. Lauffer, and P. Walter. 1995. The beta subunit of the signal recognition particle receptor is a transmembrane GTPase that anchors the alpha subunit, a peripheral membrane GTPase, to the endoplasmic reticulum membrane. *The Journal of Cell Biology*. 128:273–282.
70. Lilley, B.N., and H.L. Ploegh. 2004. A membrane protein required for dislocation of misfolded proteins from the ER. *Nature*. 429:834–840.
71. Ye, Y., Y. Shibata, C. Yun, D. Ron, and T.A. Rapoport. 2004. A membrane protein complex mediates retro-translocation from the ER lumen into the cytosol. *Nature*. 429:841–847.

8B. Protein Targeting

1. Palade, G.E. 1975. Intracellular aspects of the process of protein synthesis. *Science*. 189:347–358.
2. Redman, C.M., and D.D. Sabatini. 1966. Vectorial discharge of peptides released by puromycin from attached ribosomes. *Proceedings of the National Academy of Sciences of the United States of America*. 56:608–615.
3. Blobel, G., and D.D. Sabatini. 1971. Ribosome-membrane interaction in eukaryotic cells. In *Biomembranes*. Vol 2. L.A. Manson, editor. Plenum Press, New York. 193–195.
4. Blobel, G., and D.D. Sabatini. 1971. Dissociation of mammalian polyribosomes into subunits by puromycin. *Proceedings of the National Academy of Sciences of the United States of America*. 68:390–394.
5. Blobel, G. 1971. Release, identification, and isolation of messenger RNA from mammalian ribosomes. *Proceedings of the National Academy of Sciences of the United States of America*. 68:832–835.
6. Blobel, G. 1972. Protein tightly bound to globin mRNA. *Biochemical and Biophysical Research Communications*. 47:88–95.
7. Blobel, G. 1973. A protein of molecular weight 78,000 bound to the polyadenylate region of eukaryotic messenger RNAs. *Proceedings of the National Academy of Sciences of the United States of America*. 70:924–928.
8. Adelman, M.R., D.D. Sabatini, and G. Blobel. 1973. Ribosome-membrane interaction. Nondestructive disassembly of rat liver rough microsomes into ribosomal and membranous components. *The Journal of Cell Biology*. 56:206–229.
9. Freienstein, C., and G. Blobel. 1974. Use of eukaryotic native small ribosomal subunits for the translation of globin messenger RNA. *Proceedings of the National Academy of Sciences of the United States of America*. 71:3435–3439.

10. Freienstein, C., and G. Blobel. 1975. Nonribosomal proteins associated with eukaryotic native small ribosomal subunits. *Proceedings of the National Academy of Sciences of the United States of America*. 72:3392–3396.
11. Swan, D., H. Aviv, and P. Leder. 1972. Purification and properties of biologically active messenger RNA for a myeloma light chain. *Proceedings of the National Academy of Sciences of the United States of America*. 69:1967–1971.
12. Milstein, C., G.G. Brownlee, T.M. Harrison, and M.B. Mathews. 1972. A possible precursor of immunoglobulin light chains. *Nature: New Biology*. 239:117–120.
13. Schechter, I., D.J. McKean, R. Guyer, and W. Terry. 1975. Partial amino acid sequence of the precursor of immunoglobulin light chain programmed by messenger RNA in vitro. *Science*. 188:160–162.
14. Blobel, G., and B. Dobberstein. 1975. Transfer of proteins across membranes. I. Presence of proteolytically processed and unprocessed nascent immunoglobulin light chains on membrane-bound ribosomes of murine myeloma. *The Journal of Cell Biology*. 67:835–851.
15. Blobel, G., and B. Dobberstein. 1975. Transfer of proteins across membranes. II. Reconstitution of functional rough microsomes from heterologous components. *The Journal of Cell Biology*. 67:852–862.
16. Sabatini, D.D., and G. Blobel. 1970. Controlled proteolysis of nascent polypeptides in rat liver cell fractions. II. Location of the polypeptides in rough microsomes. *The Journal of Cell Biology*. 45:146–157.
17. Simon, S.M., and G. Blobel. 1991. A protein-conducting channel in the endoplasmic reticulum. *Cell*. 65:371–380.
18. Katz, F.N., J.E. Rothman, V.R. Lingappa, G. Blobel, and H.F. Lodish. 1977. Membrane assembly in vitro: synthesis, glycosylation, and asymmetric insertion of a transmembrane protein. *Proceedings of the National Academy of Sciences of the United States of America*. 74:3278–3282.
19. Lingappa, V.R., F.N. Katz, H.F. Lodish, and G. Blobel. 1978. A signal sequence for the insertion of a transmembrane glycoprotein. Similarities to the signals of secretory proteins in primary structure and function. *The Journal of Biological Chemistry*. 253:8667–8670.
20. Warren, G., and B. Dobberstein. 1978. Protein transfer across microsomal membranes reassembled from separated membrane components. *Nature*. 273:569–571.
21. Walter, P., and G. Blobel. 1980. Purification of a membrane-associated protein complex required for protein translocation across the endoplasmic reticulum. *Proceedings of the National Academy of Sciences of the United States of America*. 77:7112–7116.
22. Walter, P., I. Ibrahimi, and G. Blobel. 1981. Translocation of proteins across the endoplasmic reticulum. I. Signal recognition protein (SRP) binds to in-vitro-assembled polysomes synthesizing secretory protein. *The Journal of Cell Biology*. 91:545–550.
23. Walter, P., and G. Blobel. 1981. Translocation of proteins across the endoplasmic reticulum. II. Signal recognition protein (SRP) mediates the selective binding to microsomal membranes of in-vitro-assembled polysomes synthesizing secretory protein. *The Journal of Cell Biology*. 91:551–556.
24. Walter, P., and G. Blobel. 1981. Translocation of proteins across the endoplasmic reticulum. III. Signal recognition protein (SRP) causes signal sequence-dependent and site-specific arrest of chain elongation that is released by microsomal membranes. *The Journal of Cell Biology*. 91:557–561.
25. Walter, P., and G. Blobel. 1982. Signal recognition particle contains a 7S RNA essential for protein translocation across the endoplasmic reticulum. *Nature*. 299:691–698.
26. Walter, P., and A.E. Johnson. 1994. Signal sequence recognition and protein targeting to the endoplasmic reticulum membrane. *Annual Review of Cell Biology*. 10:87–119.
27. Gilmore, R., P. Walter, and G. Blobel. 1982. Protein translocation across the endoplasmic reticulum. II. Isolation and characterization of the signal recognition particle receptor. *The Journal of Cell Biology*. 95:470–477.
28. Meyer, D.I., E. Krause, and B. Dobberstein. 1982. Secretory protein translocation across membranes—the role of the 'docking protein'. *Nature*. 297:647–650.
29. Tajima, S., L. Lauffer, V.L. Rath, and P. Walter. 1986. The signal recognition particle receptor is a complex that contains two distinct polypeptide chains. *The Journal of Cell Biology*. 103:1167–1178.
30. Connolly, T., and R. Gilmore. 1989. The signal recognition particle receptor mediates the GTP-dependent displacement of SRP from the signal sequence of the nascent polypeptide. *Cell*. 57:599–610.
31. Miller, J.D., S. Tajima, L. Lauffer, and P. Walter. 1995. The beta subunit of the signal recognition particle receptor is a transmembrane GTPase that anchors the alpha subunit, a peripheral membrane GTPase, to the endoplasmic reticulum membrane. *The Journal of Cell Biology*. 128:273–282.
32. Gilmore, R., and G. Blobel. 1983. Transient involvement of signal recognition particle and its receptor in the microsomal membrane prior to protein translocation. *Cell*. 35:677–685.
33. Jackson, R.C., and G. Blobel. 1977. Post-translational cleavage of presecretory proteins with an extract of rough microsomes from dog pancreas containing signal peptidase activity. *Proceedings of the National Academy of Sciences of the United States of America*. 74:5598–5602.
34. Evans, E.A., R. Gilmore, and G. Blobel. 1986. Purification of microsomal signal peptidase as a complex. *Proceedings of the National Academy of Sciences of the United States of America*. 83:581–585.
35. Shelness, G.S., Y.S. Kanwar, and G. Blobel. 1988. cDNA-derived primary structure of the glycoprotein component of canine microsomal signal peptidase complex. *The Journal of Biological Chemistry*. 263:17063–17070.
36. Greenburg, G., G.S. Shelness, and G. Blobel. 1989. A subunit of mammalian signal peptidase is homologous to yeast SEC11 protein. *The Journal of Biological Chemistry*. 264:15762–15765.
37. Shelness, G.S., and G. Blobel. 1990. Two subunits of the canine signal peptidase complex are homologous to yeast SEC11 protein. *The Journal of Biological Chemistry*. 265:9512–9519.

38. Nicchitta, C.V., and G. Blobel. 1990. Assembly of translocation-competent proteoliposomes from detergent-solubilized rough microsomes. *Cell.* 60:259–269.
39. von Heijne, G., and C. Blomberg. 1979. Trans-membrane translocation of proteins. The direct transfer model. *European Journal of Biochemistry.* 97:175–181.
40. Engelman, D.M., and T.A. Steitz. 1981. The spontaneous insertion of proteins into and across membranes: the helical hairpin hypothesis. *Cell.* 23:411–422.
41. Gilmore, R., and G. Blobel. 1985. Translocation of secretory proteins across the microsomal membrane occurs through an environment accessible to aqueous perturbants. *Cell.* 42:497–505.
42. Simon, S.M., G. Blobel, and J. Zimmerberg. 1989. Large aqueous channels in membrane vesicles derived from the rough endoplasmic reticulum of canine pancreas or the plasma membrane of *Escherichia coli. Proceedings of the National Academy of Sciences of the United States of America.* 86:6176–6180.
43. Mueller, P., D.O. Rudin, H.T. Tien, and W.C. Wescott. 1962. Reconstitution of cell membrane structure in vitro and its transformation into an excitable system. *Nature.* 194:979–980.
44. Blobel, G. 1980. Intracellular protein topogenesis. *Proceedings of the National Academy of Sciences of the United States of America.* 77:1496–1500.
45. Simon, S.M., and G. Blobel. 1992. Signal peptides open protein-conducting channels in *E. coli. Cell.* 69:677–684.
46. Stirling, C.J., J. Rothblatt, M. Hosobuchi, R. Deshaies, and R. Schekman. 1992. Protein translocation mutants defective in the insertion of integral membrane proteins into the endoplasmic reticulum. *Molecular Biology of the Cell.* 3:129–142.
47. Görlich, D., S. Prehn, E. Hartmann, K.U. Kalies, and T.A. Rapoport. 1992. A mammalian homolog of SEC61p and SECYp is associated with ribosomes and nascent polypeptides during translocation. *Cell.* 71:489–503.
48. Görlich, D., and T.A. Rapoport. 1993. Protein translocation into proteoliposomes reconstituted from purified components of the endoplasmic reticulum membrane. *Cell.* 75:615–630.
49. Waters, M.G., and G. Blobel. 1986. Secretory protein translocation in a yeast cell-free system can occur posttranslationally and requires ATP hydrolysis. *The Journal of Cell Biology.* 102:1543–1550.
50. Hansen, W., P.D. Garcia, and P. Walter. 1986. In vitro protein translocation across the yeast endoplasmic reticulum: ATP-dependent posttranslational translocation of the prepro-α-factor. *Cell.* 45:397–406.
51. Chirico, W.J., M.G. Waters, and G. Blobel. 1988. 70K heat shock related proteins stimulate protein translocation into microsomes. *Nature.* 332:805–810.
52. Deshaies, R.J., B.D. Koch, M. Werner-Washburne, E.A. Craig, and R. Schekman. 1988. A subfamily of stress proteins facilitates translocation of secretory and mitochondrial precursor polypeptides. *Nature.* 332:800–805.
53. Goldman, B.M., and G. Blobel. 1978. Biogenesis of peroxisomes: intracellular site of synthesis of catalase and uricase. *Proceedings of the National Academy of Sciences of the United States of America.* 75:5066–5070.
54. Erickson, A.H., and G. Blobel. 1979. Early events in the biosynthesis of the lysosomal enzyme cathepsin D. *The Journal of Biological Chemistry.* 254:11771–11774.
55. Erickson, A.H., G.E. Conner, and G. Blobel. 1981. Biosynthesis of a lysosomal enzyme. Partial structure of two transient and functionally distinct NH2-terminal sequences in cathepsin D. *The Journal of Biological Chemistry.* 256:11224–11231.

8C. Structural Studies on Ribosomes

1. Lake, J.A. 1980. Ribosome structure and functional sites. In *Ribosomes, Structure, Function, and Genetics.* G. Chambliss, G.R. Craven, J. Davies, K. Davis, L. Kahan, and M. Nomura, editors. University Park Press, Baltimore, Maryland. 207–236.
2. Lake, J.A., and H.S. Slayter. 1970. Three dimensional structure of the chromatoid body of *Entamoeba invadens. Nature.* 227:1032–1037.
3. DeRosier, D.J., and A. Klug. 1968. Reconstruction of three dimensional structures from electron micrographs. *Nature.* 217:130–134.
4. Cooley, J.W., and J.W. Tukey. 1965. An algorithm for the machine calculation of complex Fourier series. *Mathematics of Computation.* 19:297–301.
5. Crowther, R.A., D.J. DeRosier, and A. Klug. 1970. The reconstruction of a three-dimensional structure from projections and its application to electron microscopy. *Proceedings of the Royal Society of London. Series A: Mathematical and Physical Sciences.* 317:319–340.
6. Lake, J.A., and H.S. Slayter. 1972. Three-dimensional structure of the chromatoid body helix of *Entamoeba invadens. Journal of Molecular Biology.* 66:271–282.
7. Nonomura, Y., G. Blobel, and D. Sabatini. 1971. Structure of liver ribosomes studied by negative staining. *Journal of Molecular Biology.* 60:303–323.
8. Byers, B. 1967. Structure and formation of ribosome crystals in hypothermic chick embryo cells. *Journal of Molecular Biology.* 26:155–167.
9. Lake, J.A., D.D. Sabatini, and Y. Nonomura. 1974. Ribosome structure as studied by electron microscopy. In *Ribosomes.* M. Nomura, A. Tissières, and P. Lengyel, editors. Cold Spring Harbor Laboratory, Cold Spring Harbor, New York. 543–557.
10. Clark, M.W., M. Hammons, J.A. Langer, and J.A. Lake. 1979. Helical arrays of *Escherichia coli* small ribosomal subunits produced *in vitro. Journal of Molecular Biology.* 135:507–512.

11. Clark, M.W., K. Leonard, and J.A. Lake. 1982. Ribosomal crystalline arrays of large subunits from *Escherichia coli. Science*. 216:999–1001.
12. Clemons, W.M. Jr., J.L.C. May, B.T. Wimberly, J.P. McCutcheon, M.S. Capel, and V. Ramakrishnan. 1999. Structure of a bacterial 30S ribosomal subunit at 5.5 Å resolution. *Nature*. 400:833–840.
13. Ban, N., P. Nissen, J. Hansen, M. Capel, P.B. Moore, and T.A. Steitz. 1999. Placement of protein and RNA structures into a 5 Å-resolution map of the 50S ribosomal subunit. *Nature*. 400:841–847.
14. Cate, J.H., M.M. Yusupov, G.Z. Yusupova, T.N. Earnest, and H.F. Noller. 1999. X-ray crystal structures of 70S ribosome functional complexes. *Science*. 285:2095–2104.
15. Leonard, K.R., and J.A. Lake. 1979. Ribosome structure: hand determination by electron microscopy of 30 S subunits. *Journal of Molecular Biology*. 129:155–163.
16. Lake, J.A., M. Pendergast, L. Kahan, and M. Nomura. 1974. Localization of *Escherichia coli* ribosomal proteins S4 and S14 by electron microscopy of antibody-labeled subunits. *Proceedings of the National Academy of Sciences of the United States of America*. 71:4688–4692.
17. Lake, J.A. 1981. The ribosome. *Scientific American*. 245:84–97.
18. Selmer, M., C.M. Dunham, F.V. Murphy IV, A. Weixlbaumer, S. Petry, A.C. Kelley, J.R. Weir, and V. Ramakrishnan. 2006. Structure of the 70S ribosome complexed with mRNA and tRNA. *Science*. 313:1935–1942.

Chapter 9

Early Explorations into the Fine Structure of Cells

Introduction

1. Porter, K.R., to L.D. Peachey. 11 February 1987. Porter Papers, Pennsylvania. Box UP 7.
2. Palade, G.E. 1977. Keith Roberts Porter and the development of contemporary cell biology. *The Journal of Cell Biology*. 75:D3–D19.
3. Porter, K.R. 1955. The fine structure of cells. *Federation Proceedings*. 14:673–682.
4. Porter, K.R. 1957. The submicroscopic morphology of protoplasm. In *Harvey Lectures*. Series 51. Academic Press, New York. 175–228.
5. Porter, K.R. 1956. Preface to the Conference on Tissue Fine Structure. *The Journal of Biophysical and Biochemical Cytology*. 2(Supplement):vi–vii.

9A. Chloroplasts

1. Granick, S. 1938. Quantitative isolation of chloroplasts from higher plants. *American Journal of Botany*. 25:558–561.
2. Granick, S., to L. Michaelis. 26 January 1938. RUA, Granick Papers, RG 450G766. Box 1, Folder 1.
3. Michaelis, L., to S. Granick. 29 January 1938. RUA, Granick Papers, RG 450G766. Box 1, Folder 1.
4. Granick, S., and K.R. Porter. 1947. The structure of the spinach chloroplast as interpreted with the electron microscope. *American Journal of Botany*. 34:545–550.
5. Granick, S. 1949. The chloroplasts: their structure, composition and development. In *Photosynthesis in Plants*. J. Franck and W.E. Loomis, editors. Iowa State College Press, Ames, Iowa. 113–132.
6. Granick, S. 1948. Protoporphyrin 9 as a precursor of chlorophyll. *The Journal of Biological Chemistry*. 172:717–727.
7. Granick, S. 1950. The structural and functional relationships between heme and chlorophyll. In *Harvey Lectures*. Series 44. Charles C. Thomas, Springfield, Illinois. 220–245.
8. Sager, R., and S. Granick. 1953. Nutritional studies with *Chlamydomonas reinhardi*. *Annals of the New York Academy of Sciences*. 56:831–838.
9. Sager, R., and S. Granick. 1954. Nutritional control of sexuality in *Chlamydomonas reinhardi*. *The Journal of General Physiology*. 37:729–742.
10. Sager, R., and G.E. Palade. 1954. Chloroplast structure in green and yellow strains of *Chlamydomonas*. *Experimental Cell Research*. 7:584–588.
11. Granick, S. 1955. Plastid structure, development and inheritance. In *Encyclopedia of Plant Physiology I*. H. Ullrich, editor. Springer-Verlag, Berlin. 507–564.
12. Granick, S. 1957. Chloroplast structure and it relation to photosynthesis. In *Research in Photosynthesis*. H. Gaffron, editor. Interscience Publishers, New York. 459–463.
13. Granick, S. 1961. The chloroplasts: Inheritance, structure and function. In *The Cell*. Volume 2. J. Brachet and A.E. Mirsky, editors. Academic Press, New York. 490–595.
14. Granick, S. 1963. The plastids: Their morphological and chemical differentiation. In *Cytodifferentiation and Macromolecular Synthesis*. M. Locke, editor. Academic Press, New York. 144–174.

15. Gibor, A., and S. Granick. 1964. Plastids and mitochondria: Inheritable systems. *Science*. 145:890–897.
16. Granick, S., and A. Gibor. 1967. The DNA of chloroplasts, mitochondria and centrioles. *Progress in Nucleic Acid Research and Molecular Biology*. 6:143–186.

9B. Protozoa

1. Rudzinska, M.A., to K.R. Porter. 19 November 1952. Porter Papers, Pennsylvania. Box 14A.
2. Rudzinska, M.A., and K.R. Porter. 1954. The fine structure of *Tokophrya infusionum* with emphasis on the feeding mechanism. *Transactions of the New York Academy of Sciences*. 16:408–411.
3. Rudzinska, M.A., and K.R. Porter. 1954. Electron microscope study of intact tentacles and disc in *Tokophrya infusionum*. *Experientia*. 10:460–462.
4. Rudzinska, M.A., and K.R. Porter. 1955. Observations on the fine structure of the macronucleus of *Tokophrya infusionum*. *The Journal of Biophysical and Biochemical Cytology*. 1:421–428.

9C. Collagen

1. Schmitt, F.O., C.E. Hall, and M.E. Jakus. 1942. Electron microscope investigation of the structure of collagen. *Journal of Cellular and Comparative Physiology*. 20:11–33.
2. Porter, K.R., and C.V. Hawn. 1947. The culture of tissue cells in clots formed from purified bovine fibrinogen and thrombin. *Proceedings of the Society for Experimental Biology and Medicine*. 65:309–314.
3. Van Zandt Hawn, C., and K.R. Porter. 1947. The fine structure of clots formed from purified bovine fibrinogen and thrombin: A study with the electron microscope. *The Journal of Experimental Medicine*. 86:285–292.
4. Ruska, H., and C. Wolpers. 1940. The structure of "liquor fibrin". *Klinische Wochenschrift*. 19:695.
5. Porter, K.R., and C.V. Hawn. 1949. Sequences in the formation of clots from purified bovine fibrinogen and thrombin; a study with the electron microscope. *The Journal of Experimental Medicine*. 90:225–232.
6. Porter, K.R., and P. Vanamee. 1949. Observations on the formation of connective tissue fibers. *Proceedings of the Society for Experimental Biology and Medicine*. 71:513–516.
7. Vanamee, P., and K.R. Porter. 1951. Observations with the electron microscope on the solvation and reconstitution of collagen. *The Journal of Experimental Medicine*. 94:255–266.
8. Bahr, G.F. 1950. The reconstruction of collagen fibrils as revealed by electron microscopy. *Experimental Cell Research*. 1:603–606.
9. Porter, K.R., and G.D. Pappas. 1959. Collagen formation by fibroblasts of the chick embryo dermis. *The Journal of Biophysical and Biochemical Cytology*. 5:153–166.
10. Hay, E.D., and J.P. Revel. 1963. Autoradiographic studies of the origin of the basement lamella in *Ambystoma*. *Developmental Biology*. 7:152–168.
11. Robbins, W.C., R.F. Watson, G.D. Pappas, and K.R. Porter. 1955. Some effects of anti-collagen serum on collagen formation in tissue culture: a preliminary report. *The Journal of Biophysical and Biochemical Cytology*. 1:381–384.
12. Porter, K.R. 1953. Observations on a submicroscopic basophilic component of cytoplasm. *The Journal of Experimental Medicine*. 97:727–750.

9D. Neurons

1. Palay, S.L., S.M. McGee-Russell, S. Gordon Jr., and M.A. Grillo. 1962. Fixation of neural tissues for electron microscopy by perfusion with solutions of osmium tetroxide. *The Journal of Cell Biology*. 12:385–410.
2. Palade, G.E., and S.L. Palay. 1954. Electron microscope observations of interneuronal and neuromuscular synapses. *The Anatomical Record*. 118:335–336.
3. Palay, S.L., and G.E. Palade. 1955. The fine structure of neurons. *The Journal of Biophysical and Biochemical Cytology*. 1:69–88.
4. Dalton, A.J., and M.D. Felix. 1953. Studies on the Golgi substance of the epithelial cells of the epididymis and duodenum of the mouse. *The American Journal of Anatomy*. 92:277–305.
5. Palay, S.L., and S.L. Wissig. 1953. Secretory granules and Nissl substance in fresh supraoptic neurones of the rabbit. *The Anatomical Record*. 116:301–313.
6. Palay, S.L. 1956. Synapses in the central nervous system. *The Journal of Biophysical and Biochemical Cytology*. 2:193–202.

9E. Viruses

1. Dales, S., and W.S. Hoar. 1954. Effects of thyroxine and thiourea on the early development of chum salmon (*Oncorhynchus keta*). *Canadian Journal of Zoology*. 32:244–251.

2. Dales, S., and K.C. Fisher. 1958. The effect of luminal on growth, glucose utilization and oxygen consumption in mammalian cells in tissue culture. *Journal of Cellular and Comparative Physiology*. 51:359–384.
3. Dales, S. 1960. Effects of anaerobiosis on the rates of multiplication of mammalian cells cultured in vitro. *Canadian Journal of Biochemistry and Physiology*. 38:871–878.
4. Claude, A., K.R. Porter, and E.G. Pickels. 1947. Electron microscope study of chicken tumor cells. *Cancer Research*. 7:421–430.
5. Joklik, W.K. 1964. The intracellular uncoating of poxvirus DNA. I. The fate of radioactively-labeled rabbit pox virus. *Journal of Molecular Biology*. 8:263–276.
6. Joklik, W.K. 1964. The intracellular uncoating of poxvirus DNA. II. The molecular basis of the uncoating process. *Journal of Molecular Biology*. 8:277–288.
7. Dales, S. 1963. The uptake and development of vaccinia virus in strain L cells followed with labeled viral deoxyribonucleic acid. *The Journal of Cell Biology*. 18:51–72.
8. Caro, L.G., R.P. van Tubergen, and J.A. Kolb. 1962. High-resolution autoradiography. I. Methods. *The Journal of Cell Biology*. 15:173–188.
9. Dales, S., P.J. Gomatos, and K.C. Hsu. 1965. The uptake and development of reovirus in strain L cells followed with labeled ribonucleic acid and ferritin-antibody conjugates. *Virology*. 25:193–211.
10. Silverstein, S.C., and S. Dales. 1968. The penetration of reovirus RNA and initiation of its genetic function in L-strain fibroblasts. *The Journal of Cell Biology*. 36:197–230.
11. Silverstein, S.C., C. Astell, D.H. Levin, M. Schonberg, and G. Acs. 1972. The mechanisms of reovirus uncoating and gene activation in vivo. *Virology*. 47:797–806.
12. Silverstein, S.C., M. Schonberg, D.H. Levin, and G. Acs. 1970. The reovirus replicative cycle: conservation of parental RNA and protein. *Proceedings of the National Academy of Sciences of the United States of America*. 67:275–281.
13. Dales, S. 1963. Association between the spindle apparatus and reovirus. *Proceedings of the National Academy of Sciences of the United States of America*. 50:268–275.
14. Dales, S., and R.M. Franklin. 1962. A comparison of the changes in fine structure of L cells during single cycles of viral multiplication, following their infection with the viruses of Mengo and encephalomyocarditis. *The Journal of Cell Biology*. 14:281–302.
15. Horne, R.W., and J. Nagington. 1954. Electron microscope studies of the development and structure of poliomyelitis virus. *Journal of Molecular Biology*. 1:333–338.
16. Penman, S., Y. Becker, and J.E. Darnell. 1964. A cytoplasmic structure involved in the synthesis and assembly of poliovirus components. *Journal of Molecular Biology*. 8:541–555.
17. Dales, S., H.J. Eggers, I. Tamm, and G.E. Palade. 1965. Electron microscopic study of the formation of poliovirus. *Virology*. 26:379–389.
18. Amako, K., and S. Dales. 1967. Cytopathology of mengovirus infection. II. Proliferation of membranous cisternae. *Virology*. 32:201–215.
19. Dales, S., and P.W. Choppin. 1962. Attachment and penetration of influenza virus. *Virology*. 18:489–493.
20. Morgan, C., and H.M. Rose. 1968. Structure and development of viruses as observed in the electron microscope. VIII. Entry of influenza virus. *Journal of Virology*. 2:925–936.
21. White, J., A. Helenius, and M.J. Gething. 1982. Haemagglutinin of influenza virus expressed from a cloned gene promotes membrane fusion. *Nature*. 300:658–659.
22. Wilson, I.A., J.J. Skehel, and D.C. Wiley. 1981. Structure of the haemagglutinin membrane glycoprotein of influenza virus at 3 Å resolution. *Nature*. 289:366–373.

9F. Cilia and the Cytoskeleton

1. Porter, K.R. 1957. The submicroscopic morphology of protoplasm. In *Harvey Lectures*. Series 51. Academic Press, New York. 175–228.
2. Satir, P. 1997. Keith R. Porter and the first electron micrograph of a cell. *Trends in Cell Biology*. 7:330–332.
3. Fawcett, D.W., and K.R. Porter. 1954. A study of the fine structure of ciliated epithelia. *Journal of Morphology*. 94:221–281.
4. Sedar, A.W., and K.R. Porter. 1955. The fine structure of cortical components of *Paramecium* multimicronucleatum. *The Journal of Biophysical and Biochemical Cytology*. 1:583–604.
5. Manton, I. 1952. The fine structure of plant cilia. *Symposia of the Society for Experimental Biology*. 6:306–319.
6. Sjöstrand, F.S. 1953. The ultrastructure of the outer segments of rods and cones of the eye as revealed by the electron microscope. *Journal of Cellular Physiology*. 42:15–44.
7. De Robertis, E. 1956. Morphogenesis of the retinal rods; an electron microscope study. *The Journal of Biophysical and Biochemical Cytology*. 2(Supplement):209–218.
8. Porter, K.R., and H.S. Bennett. 1981. Introduction: recollections on the beginnings of *The Journal of Cell Biology*. *The Journal of Cell Biology*. 91: ix–xi.
9. Satir, P. 1957. A cytological study of the cerebral vesicle of Amphioxus and its homologies. Unpublished.
10. Miller, W.H. 1960. Visual photoreceptor systems. In *The Cell: Biochemistry, Physiology, Morphology*. Volume 4. J. Brachet and A.E. Mirsky, editors. Academic Press, New York. 325–364.

11. Eakin, R.M. 1968. Evolution of photoreceptors. *Evolutionary Biology.* 2:194–242.
12. Gray, J. 1930. The mechanism of ciliary movement. VI. Photographic and stroboscopic analysis of ciliary movement. *Proceedings of the Royal Society of London (Biology).* 107:313–332.
13. Parducz, B. 1953. Zur Mechanik der Zilienbewegung. *Acta Biologica Academiae Scientiarum Hungaricae.* 4:177–220.
14. von Gelei, J. 1927. Eine neue Osmium-Toluidinmethode für Protistenforschung. *Mikrokosmos.* 20:97–103.
15. Satir, P. 1965. Studies on cilia: II. Examination of the distal region of the ciliary shaft and the role of the filaments in motility. *The Journal of Cell Biology.* 26:805–834.
16. Satir, P. 1961. Cilia. *Scientific American.* 204:108–116.
17. Afzelius, B.A. 1959. Electron microscopy of the sperm tail; results obtained with a new fixative. *The Journal of Biophysical and Biochemical Cytology.* 5:269–278.
18. Gibbons, I.R. 1963. Studies on the protein components of cilia from *Tetrahymena pyriformis. Proceedings of the National Academy of Sciences of the United States of America.* 50:1002–1010.
19. Farquhar, M.G., and G.E. Palade. 1963. Junctional complexes in various epithelia. *The Journal of Cell Biology.* 17:375–412.
20. Luck, J.L., and G. Piperno. 1989. Dynein arm mutants in *Chlamydomonas.* In *Cell Movement: The Dynein ATPases.* F.D. Warner, P. Satir, and I.R. Gibbons, editors. Alan R. Liss, New York. 49–60.

9H. Nuclear Pore Complex

1. Callan, H.G., and S.G. Tomlin. 1950. Experimental studies on amphibian oocyte nuclei. I. Investigation of the structure of the nuclear membrane by means of the electron microscope. *Proceedings of the Royal Society of London. Series B, Containing Papers of a Biological Character.* 137:367–378.
2. Gall, J.G. 1954. Observations on the nuclear membrane with the electron microscope. *Experimental Cell Research.* 7:197–200.
3. Afzelius, B.A. 1955. The ultrastructure of the nuclear membrane of the sea urchin oocyte as studied with the electron microscope. *Experimental Cell Research.* 8:147–158.
4. Watson, M.L. 1955. The nuclear envelope; its structure and relation to cytoplasmic membranes. *The Journal of Biophysical and Biochemical Cytology.* 1:257–270.
5. Pappas, G.D. 1956. The fine structure of the nuclear envelope of *Amoeba proteus. The Journal of Biophysical and Biochemical Cytology.* 2(Supplement):431–434.
6. Watson, M.L. 1959. Further observations on the nuclear envelope of the animal cell. *The Journal of Biophysical and Biochemical Cytology.* 6:147–156.
7. Maggio, R., P. Siekevitz, and G.E. Palade. 1963. Studies on isolated nuclei. I. Isolation and chemical characterization of a nuclear fraction from guinea pig liver. *The Journal of Cell Biology.* 18:267–291.
8. Merriam, R.W. 1961. On the fine structure and composition of the nuclear envelope. *The Journal of Biophysical and Biochemical Cytology.* 11:559–570.
9. Abelson, H.T., and G.H. Smith. 1970. Nuclear pores: the pore-annulus relationship in thin section. *Journal of Ultrastructure Research.* 30:558–588.
10. Feldherr, C.M. 1962. The nuclear annuli as pathways for nucleocytoplasmic exchanges. *The Journal of Cell Biology.* 14:65–72.
11. Bonner, W.M. 1975. Protein migration into nuclei. I. Frog oocyte nuclei in vivo accumulate microinjected histones, allow entry to small proteins, and exclude large proteins. *The Journal of Cell Biology.* 64:421–430.
12. Maul, G.G. 1977. The nuclear and the cytoplasmic pore complex: structure, dynamics, distribution, and evolution. *International Review of Cytology.* (Supplement):75–186.
13. Aaronson, R.P., and G. Blobel. 1974. On the attachment of the nuclear pore complex. *The Journal of Cell Biology.* 62:746–754.
14. Aaronson, R.P., and G. Blobel. 1975. Isolation of nuclear pore complexes in association with a lamina. *Proceedings of the National Academy of Sciences of the United States of America.* 72:1007–1011.
15. Fawcett, D.W. 1966. On the occurrence of a fibrous lamina on the inner aspect of the nuclear envelope in certain cells of vertebrates. *The American Journal of Anatomy.* 119:129–145.
16. Scheer, U., J. Kartenbeck, M.F. Trendelenburg, J. Stadler, and W.W. Franke. 1976. Experimental disintegration of the nuclear envelope. Evidence for pore-connecting fibrils. *The Journal of Cell Biology.* 69:1–18.
17. Dwyer, N., and G. Blobel. 1976. A modified procedure for the isolation of a pore complex-lamina fraction from rat liver nuclei. *The Journal of Cell Biology.* 70:581–591.
18. Kirschner, R.H., M. Rusli, and T.E. Martin. 1977. Characterization of the nuclear envelope, pore complexes, and dense lamina of mouse liver nuclei by high resolution scanning electron microscopy. *The Journal of Cell Biology.* 72:118–132.
19. Gerace, L., A. Blum, and G. Blobel. 1978. Immunocytochemical localization of the major polypeptides of the nuclear pore complex-lamina fraction. Interphase and mitotic distribution. *The Journal of Cell Biology.* 79:546–566.
20. Krone, G., W. Franke, S. Ely, A. D'Arcy, and E. Jost. 1978. Localization of a nuclear envelope-associated protein by indirect immunofluorescence microscopy using antibodies against a major polypeptide from rat liver fractions enriched in nuclear envelope-associated material. *Cytobiologie.* 18:22–38.
21. Gerace, L., and G. Blobel. 1980. The nuclear envelope lamina is reversibly depolymerized during mitosis. *Cell.* 19:277–287.

22. Unwin, P.N.T., and R.A. Milligan. 1982. A large particle associated with the perimeter of the nuclear pore complex. *The Journal of Cell Biology*. 93:63–75.
23. Davis, L.I., and G. Blobel. 1986. Identification and characterization of a nuclear pore complex protein. *Cell*. 45:699–709.
24. Akey, C.W., and M. Radermacher. 1993. Architecture of the *Xenopus* nuclear pore complex revealed by three-dimensional cryo-electron microscopy. *The Journal of Cell Biology*. 122:1–19.
25. Rout, M.P., and S.R. Wente. 1994. Pores for thought: nuclear pore complex proteins. *Trends in Cell Biology*. 4:357–365.
26. Wozniak, R.W., M.P. Rout, and J.D. Aitchison. 1998. Karyopherins and kissing cousins. *Trends in Cell Biology*. 8:184–188.
27. Rout, M.P., J.D. Aitchison, A. Suprapto, K. Hjertaas, Y. Zhao, and B.T. Chait. 2000. The yeast nuclear pore complex: composition, architecture, and transport mechanism. *The Journal of Cell Biology*. 148:635–651.
28. Devos, D., S. Dokudovskaya, R. Williams, F. Alber, N. Eswar, B.T. Chait, M.P. Rout, and A. Sali. 2006. Simple fold composition and modular architecture of the nuclear pore complex. *Proceedings of the National Academy of Sciences of the United States of America*. 103:2172–2177.
29. Beck, M., F. Förster, M. Ecke, J.M. Plitzko, F. Melchior, G. Gerisch, W. Baumeister, and O. Medalia. 2004. Nuclear pore complex structure and dynamics revealed by cryoelectron tomography. *Science*. 306:1387–1390.
30. Alber, F., S. Dokudovskaya, L.M. Veenhoff, W.H. Zhang, J. Kipper, D. Devos, A. Suprapto, O. Karni-Schmidt, R. Williams, B.T. Chait, et al. 2007. The molecular architecture of the nuclear pore complex. *Nature*. 450:695–701.

Chapter 10

Enlarging the Horizons of Modern Cell Biology

Introduction

1. Corner, G.W. 1964. *A History of The Rockefeller Institute, 1901–1953: Origins and Growth*. The Rockefeller Institute Press, New York. 328.
2. Brink, F., Jr. 1979. Detlev Wulf Bronk, 1897–1975. *Biographical Memoirs. National Academy of Sciences (U.S.)*. 50:2–87.
3. Bardossi, F. December 1982. The Most Interesting Thing in the World: A Historical Sketch of The Rockefeller University. Collected papers of Carol L. Moberg.
4. Rockefeller, D., to K.R. Porter. 4 April 1952. Porter Papers, Colorado. Box CM 42.
5. Porter, K.R., to D. Rockefeller. 5 May 1952. Porter Papers, Colorado. Box CM 42.
6. Bronk, D.W. 23 June 1953. Report of the Committee on Nomination and Review. RUA, BT Minutes, RG 130.2.
7. Runyon, M.R., to D.W. Bronk. 29 July 1954. RUA, Administration Files, RG 210.3. Box 1.

10B. Starting a New Journal of Cell Biology

1. Porter, K.R., to P. Rous. 9 July 1947. APS, Rous Papers.
2. Gasser, H.S. 1951. Medical research: a look ahead at The Rockefeller Institute for Medical Research. RUA, BSD Reports. Volume 39, pages 1–94.
3. Bennett, H.S., and K.R. Porter. 1953. An electron microscope study of sectioned breast muscle of the domestic fowl. *The American Journal of Anatomy*. 93:61–105.
4. Porter, K.R. 10 October 1959. The birth of the J.B.B.C. RUA, *JCB* Papers. Box 2, Folder 5.
5. Porter, K.R., and H.S. Bennett. 1981. Introduction: recollections on the beginnings of *The Journal of Cell Biology*. *The Journal of Cell Biology*. 91:ix–xi.
6. Fawcett, D.W., and K.R. Porter. 1954. A study of the fine structure of ciliated epithelia. *Journal of Morphology*. 94:221–281.
7. Porter, K.R., to H.S. Gasser. 26 October 1953. RUA, Gasser Papers. Box 3, Folder 14.
8. Bennett, H.S., to D.W. Bronk. 3 December 1953. RUA, *JCB* Papers. Box 1, Folder 10.
9. Bronk, D.W. 1953. Proposals and Reports by the President Concerning the Development of The Rockefeller Institute: The Press of The Rockefeller Institute. RUA, Bronk Papers. Box 19, Folder 11.
10. Bronk, D.W., to H.S. Bennett. 8 January 1954. RUA, *JCB* Papers.
11. Bronk, D.W. 1954. Proposals and Reports by the President Concerning the Development of The Rockefeller Institute: Establishment of a Journal of Cytology. RUA, Bronk Papers. Box 19, Folder 11.
12. Porter, K.R., to D.W. Bronk. 2 February 1954. Porter Papers, Colorado. Box CM 42.
13. Schmitt, F.O., to D.W. Bronk. 9 February 1954. RUA, *JCB* Papers.
14. Schmitt, F.O., to D.W. Bronk. 2 July 1954. RUA, *JCB* Papers.
15. Porter, K.R., to The Editors. 3 December 1954. RUA, *JCB* Papers.
16. Bronk, D.W. 1955. Proposals and Reports by the President Concerning the Development of The Rockefeller Institute: Publication of *The Journal of Biophysical and Biochemical Cytology*. RUA, *JCB* Papers.

17. Porter, K.R. 15 April 1956. Minutes of Editors' Meeting, Atlantic City, April 15th, 1956. Porter Papers, Colorado. Box CM 80.
18. Bronk, D.W., to K.R. Porter. 1 May 1956. RUA, *JCB* Papers.
19. Bronk, D.W., to The Editors. 1957. RUA, *JCB* Papers.
20. Porter, K.R., and G.E. Palade. 30 March 1959. Annual Report. *The Journal of Biophysical and Biochemical Cytology*. RUA, *JCB* Papers.
21. Porter, K.R., and G.E. Palade. 30 April 1960. Annual Report. *The Journal of Biophysical and Biochemical Cytology*. RUA, *JCB* Papers.
22. Porter, K.R. 26 April 1961. Annual Report. *The Journal of Biophysical and Biochemical Cytology*. RUA, *JCB* Papers.
23. Griffiths, R.B. 5 May 1961. Minutes of the Annual Meeting, J.B.B.C. Board of Editors. RUA, *JCB* Papers.
24. Porter, K.R., to The Editors. 23 June 1961. RUA, *JCB* Papers.
25. Porter, K.R., to D.W. Bronk. 16 August 1961. RUA, *JCB* Papers.
26. Palade, G.E., to K.R. Porter. 27 June 1961. RUA, *JCB* Papers.
27. Novikoff, A.B., to D.W. Bronk. 27 December 1962. RUA, *JCB* Papers.
28. Gall, J.G., K.R. Porter, and P. Siekevitz, editors. 1981. Discovery in Cell Biology. *The Journal of Cell Biology*. 91(Supplement):1–306.
29. Mellman, I. 2005. Fifty years of cell biology. *The Journal of Cell Biology*. 168:15.
30. Gilula, N.B. 1999. Untitled. *The Journal of Cell Biology*. 144:807–808.

10C. New Supporting Institutions for Cell Biology

1. Fell, H.B. 1977. The evolution of the International Federation for Cell Biology. *Cell Biology International Reports*. 1:9–11.
2. Danielli, J.F., to K.R. Porter. 7 September 1949. Porter Papers, Colorado. Box CM 19.
3. Moberg, C.L., and P. Siekevitz. 2000. Founding the American Society for Cell Biology. *The ASCB Newsletter*. 23:2–7.
4. Shimkin, M.B. 1979. Committee on Growth, 1945–1956: another noble experiment. *Cancer Research*. 39:262–268.
5. Copeland, D.E. 1999. Origins of cell biology in the United States. *The FASEB Journal*. 13(Supplement 2):S181–S184.
6. Oliver, C.P., E. Butler, D. Costello, D.W. Fawcett, K.R. Porter, and E. Scharrer to Members of the National Advisory Councils and Study Sections of the National Institutes of Health. 24 October 1958. Memorandum: Recommendation of the Cell Biology Study Section for Special Programming in Cell Biology. ASCB Papers. Box 73-11.
7. Porter, K.R., to R.S. Young. 4 January 1983. ASCB Papers. Box 73-11.
8. Sidman, R.L., to Members of the Tissue Culture Association. February 1968. Memorandum from the Executive Council. Porter Papers, Pennsylvania. Box UP 14C.
9. Harris, M., to Members of the Tissue Culture Association. September 1958. Comments on Plans for Reorganization of the TCA. Porter Papers, UMBC. Box 5-348.
10. Fawcett, D.W., to J. Paul. 29 September 1958. Porter Papers, UMBC. Box 5-348.
11. Porter, K.R., to J. Paul. 3 October 1958. Porter Papers, UMBC. Box 5-348.
12. Hanks, J.H., to J. Paul. 2 October 1958. Porter Papers, UMBC. Box 5-348.
13. U.S. National Committee of the International Union of Biological Sciences. 3 June 1960. Establishment of an American Society for Cell Biology. ASCB Papers. Box 73-1.
14. Porter, K.R. November 1961. ASCB banquet speech. Porter Papers, Colorado. Box CM 87.
15. Hetherington, D.C. March 1959. TCA: Annual report of the secretary-treasurer. ASCB Papers. Box 43.
16. Harris, M., to K.R. Porter. 15 April 1959. ASCB Papers. Box 43.
17. Porter, K.R., to M. Harris. 21 September 1959. Porter Papers, UMBC. Box 5-348.
18. Porter, K.R., to M. Harris. 21 October 1959. Porter Papers, UMBC. Box 5-348.
19. Harris, M. 14 January 1960. Interim report by subcommittee on organization of cell biologists. ASCB Papers. Box 43.
20. Green, J.W., to D.C. Hetherington. 11 February 1960. Porter Papers, UMBC. Box 5-348.
21. White, P., to M. Harris. 27 January 1960. Porter Papers, UMBC. Box 5-348.
22. Hetherington, D.C., to M. Harris. 22 January 1960. Porter Papers, UMBC. Box 5-348.
23. Harris, M. 23 May 1960. Minutes of meeting of provisional council, American Society for Cell Biology. Porter Papers, UMBC. Box 5-348.
24. Ris, H., H. Swift, and K.R. Porter. 28 October 1960. Grant application to the National Institutes of Health. ASCB Papers. Box 70.
25. Moses, M.J. 28 January 1961. Minutes of the meeting of the provisional council of the American Society for Cell Biology. ASCB Papers. Box 73-1.
26. Bronk, D.W. 8 October 1956. Memorandum to members of the faculty: applications for grants. RUA, Mirsky Papers. Box 6, Folder 1.
27. Palade, G.E. 1977. Keith Roberts Porter and the development of contemporary cell biology. *The Journal of Cell Biology*. 75:D3–D19.
28. Porter, K.R., to T.M. Rivers. 25 January 1955. Porter Papers, Colorado. Box CM 42.
29. Porter, K.R., to T.M. Rivers. 8 March 1955. Porter Papers, Colorado. Box CM 42.
30. Flinn, W.R., to K.R. Porter. 16 January 1956. Porter Papers, Colorado. Box CM 42.
31. Porter, K.R., to D.W. Bronk. 8 July 1960. Porter Papers, Colorado. Box CM 44.

32. Porter, K.R., to Chairman of the Search Committee. 31 March 1980. Memorandum and application for fellowship to Fogarty International Center. Porter Papers, Pennsylvania. Box UP 13.
33. Porter, K.R. 1984. Interview with Mary A. Bonneville. Bonneville Interviews. HARVD1.

Chapter 11

Adventures in Biochemical Cytology

11A. Exploring Cells with a Centrifuge: Discoveries of the Lysosome and Peroxisome

1. de Duve, C., J. Berthet, H.G. Hers, and L. Dupret. 1949. Le système hexose-phosphatasique. I. Existence d'une glucose-6-phosphatase spécifique dans le foie. *Bulletin de la Société de Chimie Biologique*. 31:1242–1253.
2. Claude, A. 1946. Fractionation of mammalian liver cells by differential centrifugation. I. Problems, methods, and preparation of extract. *The Journal of Experimental Medicine*. 84:51–59.
3. Hogeboom, G.H., W.C. Schneider, and G.E. Pallade. 1948. Cytochemical studies of mammalian tissues; isolation of intact mitochondria from rat liver; some biochemical properties of mitochondria and submicroscopic particulate material. *The Journal of Biological Chemistry*. 172:619–635.
4. Schneider, W.C. 1948. Intracellular distribution of enzymes; the oxidation of octanoic acid by rat liver fractions. *The Journal of Biological Chemistry*. 176:259–266.
5. Berthet, J., and C. de Duve. 1951. Tissue fractionation studies. I. The existence of a mitochondria-linked, enzymically inactive form of acid phosphatase in rat-liver tissue. *The Biochemical Journal*. 50:174–181.
6. de Duve, C., J. Berthet, L. Berthet, and F. Appelmans. 1951. Permeability of mitochondria. *Nature*. 167:389–390.
7. Berthet, J., L. Berthet, F. Appelmans, and C. de Duve. 1951. Tissue fractionation studies. II. The nature of the linkage between acid phosphatase and mitochondria in rat-liver tissue. *The Biochemical Journal*. 50:182–189.
8. Appelmans, F., R. Wattiaux, and C. de Duve. 1955. Tissue fractionation studies. 5. The association of acid phosphatase with a special class of cytoplasmic granules in rat liver. *The Biochemical Journal*. 59:438–445.
9. de Duve, C., R. Gianetto, F. Appelmans, and R. Wattiaux. 1953. Enzymic content of the mitochondria fraction. *Nature*. 172:1143–1144.
10. Gianetto, R., and C. de Duve. 1955. Tissue fractionation studies. 4. Comparative study of the binding of acid phosphatase, β-glucuronidase and cathepsin by rat-liver particles. *The Biochemical Journal*. 59:433–438.
11. de Duve, C., B.C. Pressman, R. Gianetto, R. Wattiaux, and F. Appelmans. 1955. Tissue fractionation studies. 6. Intracellular distribution patterns of enzymes in rat-liver tissue. *The Biochemical Journal*. 60:604–617.
12. de Duve, C., and J. Berthet. 1954. The use of differential centrifugation in the study of tissue enzymes. *International Review of Cytology*. 3:225–275.
13. Beaufay, H., P. Jacques, P. Baudhuin, O.Z. Sellinger, J. Berthet, and C. de Duve. 1964. Tissue fractionation studies. 18. Resolution of mitochondrial fractions from rat liver into three distinct populations of cytoplasmic particles by means of density equilibration in various gradients. *The Biochemical Journal*. 92:184–205.
14. Novikoff, A.B., E. Podber, J. Ryan, and E. Noe. 1953. Biochemical heterogeneity of the cytoplasmic particles isolated from rat liver homogenate. *The Journal of Histochemistry and Cytochemistry*. 1:27–46.
15. Novikoff, A.B., H. Beaufay, and C. de Duve. 1956. Electron microscopy of lysosome-rich fractions from rat liver. *The Journal of Biophysical and Biochemical Cytology*. 2:179–184.
16. Sellinger, O.Z., H. Beaufay, P. Jacques, A. Doyen, and C. de Duve. 1960. Tissue fractionation studies. 15. Intracellular distribution and properties of β-N-acetylglucosaminidase and β-galactosidase in rat liver. *The Biochemical Journal*. 74:450–456.
17. de Duve, C. 1965. The separation and characterization of subcellular particles. In *Harvey Lectures*. Series 59. Academic Press, New York. 49–87.
18. de Duve, C., J. Berthet, and H. Beaufay. 1959. Gradient centrifugation of cell particles. Theory and applications. *Progress in Biophysics and Biophysical Chemistry*. 9:325–369.
19. Beaufay, H., D.S. Bendall, P. Baudhuin, R. Wattiaux, and C. de Duve. 1959. Tissue fractionation studies. 13. Analysis of mitochondrial fractions from rat liver by density-gradient centrifuging. *The Biochemical Journal*. 73:628–637.
20. Wattiaux, R., M. Wibo, and P. Baudhuin. 1963. Influence of the injection of Triton WR-1339 on the properties of rat-liver lysosomes. In *Ciba Foundation Symposium on Lysosomes*. A.V.S. de Reuck and M.P. Cameron, editors. J & A Churchill, London. 176–196.
21. Beaufay, H., A. Amar-Costesec, D. Thinès-Sempoux, M. Wibo, M. Robbi, and J. Berthet. 1974. Analytical study of microsomes and isolated subcellular membranes from rat liver. III. Subfractionation of the microsomal fraction by isopycnic and differential centrifugation in density gradients. *The Journal of Cell Biology*. 61:213–231.
22. Baudhuin, P., H. Beaufay, and C. de Duve. 1965. Combined biochemical and morphological study of particulate fractions from rat liver. Analysis of preparations enriched in lysosomes or in particles containing urate oxidase, D-amino acid oxidase, and catalase. *The Journal of Cell Biology*. 26:219–243.

23. de Duve, C. 1975. Exploring cells with a centrifuge. In *Les Prix Nobel en 1974*. The Nobel Foundation, Stockholm. 139–160 [reprinted in *Science*, 1975. 189:186–194].
24. Straus, W. 1954. Isolation and biochemical properties of droplets from the cells of rat kidney. *The Journal of Biological Chemistry*. 207:745–755.
25. Straus, W. 1956. Concentration of acid phosphatase, ribonuclease, desoxyribonuclease, β-glucuronidase, and cathepsin in droplets isolated from the kidney cells of normal rats. *The Journal of Biophysical and Biochemical Cytology*. 2:513–521.
26. Herveg, J.P., C. Beckers, and M. De Visscher. 1966. Lysosomal hydrolases in calf thyroid. *The Biochemical Journal*. 100:540–547.
27. de Duve, C. 1964. Lysosomes and cell injury. In *International Symposium on Injury, Inflammation and Immunity*. L. Thomas, J.W. Uhr, and L. Grant, editors. Williams & Wilkins Co., Baltimore, Maryland. 283–311.
28. Dingle, J.T. 1963. Action of vitamin A on the stability of lysosomes *in vivo* and *in vitro*. In *Ciba Foundation Symposium on Lysosomes*. A.V.S. de Reuck and M.P. Cameron, editors. J & A Churchill, London. 385–398.
29. Dingle, J.T., editor [and others]. 1969–1984. *Lysosomes in Biology and Pathology*. Volumes 1–7. North Holland Publishing Company, Amsterdam.
30. Thomas, L. 1974. *The Lives of a Cell: Notes of a Biology Watcher*. Viking Press, New York.
31. Vaes, G. 1966. La résorption osseuse et l'hormone parathyroïdienne. PhD thesis. Imprimerie Warny, Louvain, Belgium.
32. Hers, H.G. 1963. α-Glucosidase deficiency in generalized glycogen storage disease (Pompe's disease). *The Biochemical Journal*. 86:11–16.
33. Lejeune, N., D. Thinès-Sempoux, and H.G. Hers. 1963. Tissue fractionation studies. 16. Intracellular distribution and properties of α-glucosidases in rat liver. *The Biochemical Journal*. 86:16–21.
34. Baudhuin, P., H.G. Hers, and H. Loeb. 1964. An electron microscope and biochemical study of type II glycogenosis. *Laboratory Investigation*. 13:1139–1152.
35. Hers, H.G. 1965. Inborn lysosomal diseases. *Gastroenterology*. 48:625–633.
36. Hers, H.G., and F. Van Hoof, editors. 1973. *Lysosomes and Storage Diseases*. Academic Press, New York.
37. de Duve, C., and A. Trouet. 1973. Lysosomes and lysosomotropic drugs in host-parasite relationship. In *"Non-Specific" Factors Influencing Host Resistance: A Reexamination*. W. Braun and J. Ungar, editors. S. Karger, Basel, Switzerland. 153–170.
38. Trouet, A., D. Deprez-de Campeneere, and C. de Duve. 1972. Chemotherapy through lysosomes with a DNA-daunorubicin complex. *Nature: New Biology*. 239:110–112.
39. de Reuck, A.V.S., and M.P. Cameron, editors. 1963. *Ciba Foundation Symposium on Lysosomes*. J & A Churchill, London.
40. de Duve, C., and R. Wattiaux. 1966. Functions of lysosomes. *Annual Review of Physiology*. 28:435–492.
41. de Duve, C., and P. Baudhuin. 1966. Peroxisomes (microbodies and related particles). *Physiological Reviews*. 46:323–357.
42. Amar-Costesec, A., M. Wibo, D. Thinès-Sempoux, H. Beaufay, and J. Berthet. 1974. Analytical study of microsomes and isolated subcellular membranes from rat liver. IV. Biochemical, physical, and morphological modifications of microsomal components induced by digitonin, EDTA, and pyrophosphate. *The Journal of Cell Biology*. 62:717–745.
43. Beaufay, H. 1966. La centrifugation en gradient de densité. PhD thesis. Ceuterick, Louvain, Belgium.
44. Anderson, N.G., editor. 1966. The development of zonal centrifuges and ancillary systems for tissue fractionation and analysis. *National Cancer Institute Monograph 21*.
45. Bowers, W.E., J.T. Finkenstaedt, and C. de Duve. 1967. Lysosomes in lymphoid tissue. I. The measurement of hydrolytic activities in whole homogenates. *The Journal of Cell Biology*. 32:325–337.
46. Bowers, W.E., and C. de Duve. 1967. Lysosomes in lymphoid tissue. II. Intracellular distribution of acid hydrolases. *The Journal of Cell Biology*. 32:339–348.
47. Bowers, W.E., and C. de Duve. 1967. Lysosomes in lymphoid tissue. III. Influence of various treatments of the animals on the distribution of acid hydrolases. *The Journal of Cell Biology*. 32:349–364.
48. Bowers, W.E. 1966. A biochemical characterization of rat spleen lysosomes and of lysosomes from other lymphoid tissues. PhD thesis. The Rockefeller University, New York.
49. Coffey, J.W., and C. de Duve. 1968. Digestive activity of lysosomes. I. The digestion of proteins by extracts of rat liver lysosomes. *The Journal of Biological Chemistry*. 243:3255–3263.
50. Aronson, N.N., Jr., and C. de Duve. 1968. Digestive activity of lysosomes. II. The digestion of macromolecular carbohydrates by extracts of rat liver lysosomes. *The Journal of Biological Chemistry*. 243:4564–4573.
51. Fowler, S. 1969. Lipolytic activities of rat liver lysosomes. PhD thesis. The Rockefeller University, New York.
52. Fowler, S., and C. de Duve. 1969. Digestive activity of lysosomes. III. The digestion of lipids by extracts of rat liver lysosomes. *The Journal of Biological Chemistry*. 244:471–481.
53. Baggiolini, M., J.G. Hirsch, and C. de Duve. 1969. Resolution of granules from rabbit heterophil leukocytes into distinct populations by zonal sedimentation. *The Journal of Cell Biology*. 40:529–541.
54. Baggiolini, M., J.G. Hirsch, and C. de Duve. 1970. Further biochemical and morphological studies of granule fractions from rabbit heterophil leukocytes. *The Journal of Cell Biology*. 45:586–597.
55. Farquhar, M.G., D.F. Bainton, M. Baggiolini, and C. de Duve. 1972. Cytochemical localization of acid phosphatase activity in granule fractions from rabbit polymorphonuclear leukocytes. *The Journal of Cell Biology*. 54:141–156.
56. Baggiolini, M., C. de Duve, P.L. Masson, and J.F. Heremans. 1970. Association of lactoferrin with specific granules in rabbit heterophil leukocytes. *The Journal of Experimental Medicine*. 131:559–570.
57. Ashford, T.P., and K.R. Porter. 1962. Cytoplasmic components in hepatic cell lysosomes. *The Journal of Cell Biology*. 12:198–202.

58. Deter, R.L., and C. de Duve. 1967. Influence of glucagon, an inducer of cellular autophagy, on some physical properties of rat liver lysosomes. *The Journal of Cell Biology*. 33:437–449.
59. Deter, R.L., P. Baudhuin, and C. de Duve. 1967. Participation of lysosomes in cellular autophagy induced in rat liver by glucagon. *The Journal of Cell Biology*. 35:C11–C16.
60. Baudhuin, P., M. Müller, B. Poole, and C. de Duve. 1965. Non-mitochondrial oxidizing particles (microbodies) in rat liver and kidney and in *Tetrahymena pyriformis*. *Biochemical and Biophysical Research Communications*. 20:53–59.
61. Leighton, F., B. Poole, H. Beaufay, P. Baudhuin, J.W. Coffey, S. Fowler, and C. de Duve. 1968. The large-scale separation of peroxisomes, mitochondria, and lysosomes from the livers of rats injected with triton WR-1339. Improved isolation procedures, automated analysis, biochemical and morphological properties of fractions. *The Journal of Cell Biology*. 37:482–513.
62. Leighton, F., B. Poole, P.B. Lazarow, and C. de Duve. 1969. The synthesis and turnover of rat liver peroxisomes. I. Fractionation of peroxisome proteins. *The Journal of Cell Biology*. 41:521–535.
63. Poole, B., F. Leighton, and C. de Duve. 1969. The synthesis and turnover of rat liver peroxisomes. II. Turnover of peroxisome proteins. *The Journal of Cell Biology*. 41:536–546.
64. Poole, B., T. Higashi, and C. de Duve. 1970. The synthesis and turnover of rat liver peroxisomes. III. The size distribution of peroxisomes and the incorporation of new catalase. *The Journal of Cell Biology*. 45:408–415.
65. Poole, B. 1968. Studies on rat liver peroxisomes. PhD thesis. The Rockefeller University, New York.
66. Lazarow, P.B., and C. de Duve. 1971. Intermediates in the biosynthesis of peroxisomal catalase in rat liver. *Biochemical and Biophysical Research Communications*. 45:1198–1204.
67. Lazarow, P.B., and C. de Duve. 1973. The synthesis and turnover of rat liver peroxisomes. IV. Biochemical pathway of catalase synthesis. *The Journal of Cell Biology*. 59:491–506.
68. Lazarow, P.B., and C. de Duve. 1973. The synthesis and turnover of rat liver peroxisomes. V. Intracellular pathway of catalase synthesis. *The Journal of Cell Biology*. 59:507–524.
69. Lazarow, P.B. 1972. The biogenesis of peroxisomal catalase in rat liver. PhD thesis. The Rockefeller University, New York.
70. Lazarow, P.B., and Y. Fujiki. 1985. Biogenesis of peroxisomes. *Annual Review of Cell Biology*. 1:489–530.
71. Shio, H., and P.B. Lazarow. 1981. Relationship between peroxisomes and endoplasmic reticulum investigated by combined catalase and glucose-6-phosphatase cytochemistry. *The Journal of Histochemistry and Cytochemistry*. 29:1263–1272.
72. de Duve, C. 1987. Alex Novikoff: Reminiscences. *The Journal of Histochemistry and Cytochemistry*. 35:931, 933–934.
73. Gaunt, G.L., Jr. 1973. Subcellular localization of D-amino acid oxidase in rat brain. PhD thesis. The Rockefeller University, New York.
74. Gaunt, G.L., and C. de Duve. 1976. Subcellular distribution of D-amino acid oxidase and catalase in rat brain. *Journal of Neurochemistry*. 26:749–759.
75. Peters, T.J., M. Müller, and C. de Duve. 1972. Lysosomes of the arterial wall. I. Isolation and subcellular fractionation of cells from normal rabbit aorta. *The Journal of Experimental Medicine*. 136:1117–1139.
76. Peters, T.J., and C. de Duve. 1974. Lysosomes of the arterial wall. II. Subcellular fractionation of aortic cells from rabbits with experimental atheroma. *Experimental and Molecular Pathology*. 20:228–256.
77. Shio, H., M.G. Farquhar, and C. de Duve. 1974. Lysosomes of the arterial wall. IV. Cytochemical localization of acid phosphatase and catalase in smooth muscle cells and foam cells from rabbit atheromatous aorta. *American Journal of Pathology*. 76:1–16.
78. Takano, T., W.J. Black, T.J. Peters, and C. de Duve. 1974. Assay, kinetics, and lysosomal localization of an acid cholesteryl esterase in rabbit aortic smooth muscle cells. *The Journal of Biological Chemistry*. 249:6732–6737.
79. Haley, N.J., H. Shio, and S. Fowler. 1977. Characterization of lipid-laden aortic cells from cholesterol-fed rabbits. I. Resolution of aortic cell populations by metrizamide density gradient centrifugation. *Laboratory Investigation*. 37:287–296.
80. Haley, N.J., S. Fowler, and C. de Duve. 1980. Lysosomal acid cholesteryl esterase activity in normal and lipid-laden aortic cells. *Journal of Lipid Research*. 21:961–969.
81. Shio, H., N.J. Haley, and S. Fowler. 1978. Characterization of lipid-laden aortic cells from cholesterol-fed rabbits. II. Morphometric analysis of lipid-filled lysosomes and lipid droplets in aortic cell populations. *Laboratory Investigation*. 39:390–397.
82. Shio, H., N.J. Haley, and S. Fowler. 1979. Characterization of lipid-laden aortic cells from cholesterol-fed rabbits. III. Intracellular localization of cholesterol and cholesteryl ester. *Laboratory Investigation*. 41:160–167.
83. Morris, M.D., C. Bhuvaneswaran, H. Shio, and S. Fowler. 1982. Lysosome lipid storage disorder in NCTR-BALB/c mice. I. Description of the disease and genetics. *American Journal of Pathology*. 108:140–149.
84. Shio, H., S. Fowler, C. Bhuvaneswaran, and M.D. Morris. 1982. Lysosome lipid storage disorder in NCTR-BALB/c mice. II. Morphologic and cytochemical studies. *American Journal of Pathology*. 108:150–159.
85. Bhuvaneswaran, C., M.D. Morris, H. Shio, and S. Fowler. 1982. Lysosome lipid storage disorder in NCTR-BALB/c mice. III. Isolation and analysis of storage inclusions from liver. *American Journal of Pathology*. 108:160–170.
86. Poole, B. 1971. Synthesis and degradation of proteins in relation to cellular structure. In *Enzyme Synthesis and Degradation in Mammalian Systems*. M. Rechcigl Jr., editor. University Park Press, Baltimore, Maryland. 375–402.
87. Poole, B., and M. Wibo. 1973. Protein degradation in cultured cells. The effect of fresh medium, fluoride, and iodoacetate on the digestion of cellular protein of rat fibroblasts. *The Journal of Biological Chemistry*. 248:6221–6226.
88. Warburton, M.J., and B. Poole. 1977. Effect of medium composition on protein degradation and DNA synthesis in rat embryo fibroblasts. *Proceedings of the National Academy of Sciences of the United States of America*. 74:2427–2431.

89. Wibo, M., and B. Poole. 1974. Protein degradation in cultured cells. II. The uptake of chloroquine by rat fibroblasts and the inhibition of cellular protein degradation and cathepsin B_1. *The Journal of Cell Biology*. 63:430–440.
90. Tulkens, P., and A. Trouet. 1978. The uptake and intracellular accumulation of aminoglycoside antibiotics in lysosomes of cultured rat fibroblasts. *Biochemical Pharmacology*. 27:415–424.
91. Ohkuma, S., and B. Poole. 1978. Fluorescence probe measurement of the intralysosomal pH in living cells and the perturbation of pH by various agents. *Proceedings of the National Academy of Sciences of the United States of America*. 75:3327–3331.
92. Ohkuma, S., and B. Poole. 1981. Cytoplasmic vacuolation of mouse peritoneal macrophages and the uptake into lysosomes of weakly basic substances. *The Journal of Cell Biology*. 90:656–664.
93. Poole, B., and S. Ohkuma. 1981. Effect of weak bases on the intralysosomal pH in mouse peritoneal macrophages. *The Journal of Cell Biology*. 90:665–669.
94. Ohkuma, S., J. Chudzik, and B. Poole. 1986. The effects of basic substances and acidic ionophores on the digestion of exogenous and endogenous proteins in mouse peritoneal macrophages. *The Journal of Cell Biology*. 102:959–966.
95. Fong, D., and B. Poole. 1982. The effect of canavanine on protein synthesis and protein degradation in IMR-90 fibroblasts. *Biochimica et Biophysica Acta*. 696:193–200.
96. Davidson, E., and B. Poole. 1975. Fractionation of the rat liver enzymes that hydrolyze benzoyl-arginine-2-naphthylamide. *Biochimica et Biophysica Acta*. 397:437–442.
97. de Duve, C. 1985. Brian Poole, in memoriam. In *Intracellular Protein Catabolism*. E.A. Khairallah, J.S. Bond, and J.W.C. Bird, editors. A.R. Liss, New York. XLI–XLVIII.
98. Bowers, W.E., C.F. Beyer, and N. Yago. 1977. Cathepsin D of mouse leukemia L1210 cells. Unusual intracellular localization and biochemical properties. *Biochimica et Biophysica Acta*. 497:272–279.
99. Klinkert, W.E.F., J.H. LaBadie, J.P. O'Brien, C.F. Beyer, and W.E. Bowers. 1980. Rat dendritic cells function as accessory cells and control the production of a soluble factor required for mitogenic responses of T lymphocytes. *Proceedings of the National Academy of Sciences of the United States of America*. 77:5414–5418.
100. Klinkert, W.E.F., J.H. LaBadie, and W.E. Bowers. 1982. Accessory and stimulating properties of dendritic cells and macrophages isolated from various rat tissues. *The Journal of Experimental Medicine*. 156:1–19.
101. Hess, R., W. Stäubli, and W. Riess. 1965. Nature of the hepatomegalic effect produced by ethyl-chlorophenoxy-isobutyrate in the rat. *Nature*. 208:856–858.
102. Lazarow, P.B., and C. de Duve. 1976. A fatty acyl-CoA oxidizing system in rat liver peroxisomes; enhancement by clofibrate, a hypolipidemic drug. *Proceedings of the National Academy of Sciences of the United States of America*. 73:2043–2046.
103. Lazarow, P.B. 1977. Three hypolipidemic drugs increase hepatic palmitoyl-coenzyme A oxidation in the rat. *Science*. 197:580–581.
104. Lazarow, P.B. 1978. Rat liver peroxisomes catalyze the β oxidation of fatty acids. *The Journal of Biological Chemistry*. 253:1522–1528.
105. Kondrup, J., and P.B. Lazarow. 1985. Flux of palmitate through the peroxisomal and mitochondrial β-oxidation systems in isolated rat hepatocytes. *Biochimica et Biophysica Acta*. 835:147–153.
106. Robbi, M., and P.B. Lazarow. 1978. Synthesis of catalase in two cell-free protein-synthesizing systems and in rat liver. *Proceedings of the National Academy of Sciences of the United States of America*. 75:4344–4348.
107. Robbi, M., and P.B. Lazarow. 1982. Peptide mapping of peroxisomal catalase and its precursor. Comparison to the primary wheat germ translation product. *The Journal of Biological Chemistry*. 257:964–970.
108. Rachubinski, R.A., Y. Fujiki, R.M. Mortensen, and P.B. Lazarow. 1984. Acyl-CoA oxidase and hydratase-dehydrogenase, two enzymes of the peroxisomal β-oxidation system, are synthesized on free polysomes of clofibrate-treated rat liver. *The Journal of Cell Biology*. 99:2241–2246.
109. Fujiki, Y., R.A. Rachubinski, R.M. Mortensen, and P.B. Lazarow. 1985. Synthesis of 3-ketoacyl-CoA thiolase of rat liver peroxisomes on free polyribosomes as a larger precursor. Induction of thiolase mRNA activity by clofibrate. *The Biochemical Journal*. 226:697–704.
110. Fujiki, Y., R.A. Rachubinski, and P.B. Lazarow. 1984. Synthesis of a major integral membrane polypeptide of rat liver peroxisomes on free polysomes. *Proceedings of the National Academy of Sciences of the United States of America*. 81:7127–7131.
111. Alexson, S., H. Shio, and P.B. Lazarow. 1982. Intraperoxisomal localization of β-oxidation system in rat liver. *Hoppe-Seyler's Zeitschrift für Physiologische Chemie*. 363:968–969.
112. Lazarow, P.B., H. Shio, and M. Robbi. 1980. Biogenesis of peroxisomes and the peroxisomal reticulum hypothesis. In *Biological Chemistry of Organelle Formation*. T. Bücher, W. Sebald, and H. Weiss, editors. Springer-Verlag, New York. 187–206.
113. de Duve, C. 1978. The RU-ICP collaborative program in the biomedical sciences - a unique Belgian-American venture. *Belgian-American Trade Review*. 33:15–17.
114. de Duve, C. 1984. *A Guided Tour of the Living Cell*. Scientific American Books, New York.
115. de Duve, C. 1991. *Blueprint for a Cell: The Nature and Origin of Life*. Neil Patterson Publishers, Burlington, North Carolina.
116. de Duve, C. 1995. *Vital Dust: Life as a Cosmic Imperative*. Basic Books, New York.
117. de Duve, C. 2002. *Life Evolving: Molecules, Mind, and Meaning*. Oxford University Press, New York.
118. de Duve, C. 2005. *Singularities: Landmarks on the Pathways of Life*. Cambridge University Press, New York.
119. de Duve, C., with N. Patterson. 2010. *Genetics of Original Sin: The Impact of Natural Selection on the Future of Humanity*. Yale University Press, New Haven, Connecticut.

11B. Discovering the Hydrogenosome

1. Müller, M., P. Röhlich, H. Tóth, and I. Törö. 1963. Fine structure and enzymic activity of protozoan food vacuoles. In *Ciba Foundation Symposium on Lysosomes*. A.V.S. de Reuck and M.P. Cameron, editors. J & A Churchill, London. 201–216.
2. Müller, M., P. Baudhuin, and C. de Duve. 1966. Lysosomes in *Tetrahymena pyriformis*. I. Some properties and lysosomal localization of acid hydrolases. *Journal of Cellular Physiology*. 68:165–175.
3. Müller, M. 1972. Secretion of acid hydrolases and its intracellular source in *Tetrahymena pyriformis*. *The Journal of Cell Biology*. 52:478–487.
4. Baudhuin, P., M. Mueller, B. Poole, and C. de Duve. 1965. Non-mitochondrial oxidizing particles (microbodies) in rat liver and kidney and in *Tetrahymena pyriformis*. *Biochemical and Biophysical Research Communications*. 20:53–59.
5. Müller, M., J.F. Hogg, and C. de Duve. 1968. Distribution of tricarboxylic acid cycle enzymes and glyoxylate cycle enzymes between mitochondria and peroxisomes in *Tetrahymena pyriformis*. *The Journal of Biological Chemistry*. 243:5385–5395.
6. de Duve, C. 1969. Evolution of the peroxisome. *Annals of the New York Academy of Sciences*. 168:369–381.
7. Müller, M. 1973. Biochemical cytology of trichomonad flagellates. I. Subcellular localization of hydrolases, dehydrogenases, and catalase in *Tritrichomonas foetus*. *The Journal of Cell Biology*. 57:453–474.
8. Lindmark, D.G., and M. Müller. 1973. Hydrogenosome, a cytoplasmic organelle of the anaerobic flagellate *Tritrichomonas foetus*, and its role in pyruvate metabolism. *The Journal of Biological Chemistry*. 248:7724–7728.
9. Lindmark, D.G., M. Müller, and H. Shio. 1975. Hydrogenosomes in *Trichomonas vaginalis*. *The Journal of Parasitology*. 61:552–554.
10. Lindmark, D.G. 1976. Acetate production by *Tritrichomonas foetus*. In *Biochemistry of Parasites and Host-Parasite Relationships*. H. Van den Bossche, editor. Elsevier/North Holland Publishing Co., Amsterdam. 15–21.
11. Marczak, R., T.E. Gorrell, and M. Müller. 1983. Hydrogenosomal ferredoxin of the anaerobic protozoon, *Tritrichomonas foetus*. *The Journal of Biological Chemistry*. 258:12427–12433.
12. Gorrell, T.E., N. Yarlett, and M. Müller. 1984. Isolation and characterization of *Trichomonas vaginalis* ferredoxin. *Carlsberg Research Communications*. 49:259–268.
13. Müller, M. 1976. Carbohydrate and energy metabolism of *Tritrichomonas foetus*. In *Biochemistry of Parasites and Host-Parasite Relationships*. H. Van den Bossche, editor. Elsevier/North Holland Publishing Co., Amsterdam. 3–14.
14. Lindmark, D.G., and M. Müller. 1974. Superoxide dismutase in the anaerobic flagellates, *Tritrichomonas foetus* and *Monocercomonas* sp. *The Journal of Biological Chemistry*. 249:4634–4637.
15. Whatley, J.M., P. John, and F.R. Whatley. 1979. From extracellular to intracellular: the establishment of mitochondria and chloroplasts. *Proceedings of the Royal Society of London. Series B. Biological Sciences*. 204:165–187.
16. Müller, M. 1980. The hydrogenosome. *Symposium of the Society for General Microbiology*. 30:127–142.
17. Müller, M. 1992. Energy metabolism of ancestral eukaryotes: a hypothesis based on the biochemistry of amitochondriate parasitic protists. *Bio Systems*. 28:33–40.
18. Müller, M. 1993. The hydrogenosome. *Journal of General Microbiology*. 139:2879–2889.
19. Embley, T.M. 2006. Multiple secondary origins of the anaerobic lifestyle in eukaryotes. *Philosophical Transactions of the Royal Society of London. Series B. Biological Sciences*. 361:1055–1067.
20. Müller, M. 2003. Energy metabolism. Part I: Anaerobic protozoa. In *Molecular Medical Parasitology*. J.J. Marr, T.W. Nilsen, and R.W. Komuniecki, editors. Academic Press, London. 125–139.
21. Carlton, J.M., R.P. Hirt, J.C. Silva, A.L. Delcher, M. Schatz, Q. Zhao, J.R. Wortman, S.L. Bidwell, U.C. Alsmark, S. Besteiro, et al. 2007. Draft genome sequence of the sexually transmitted pathogen *Trichomonas vaginalis*. *Science*. 315:207–212.
22. Martin, W.F., and M. Müller, editors. 2007. *Origin of Mitochondria and Hydrogenosomes*. Springer, New York.
23. Tachezy, J., editor. 2008. *Hydrogenosomes and Mitosomes: Mitochondria of Anaerobic Eukaryotes*. Springer, New York.
24. Ginger, M.L., L.K. Fritz-Laylin, C. Fulton, W.Z. Cande, and S.C. Dawson. 2010. Intermediary metabolism in protists: a sequence-based view of facultative anaerobic metabolism in evolutionarily diverse eukaryotes. *Protist*. 161:642–671.

Chapter 12

Defenders: The Endocytic Pathway in the Immune System

Introduction

1. von Recklinghausen, F. 1863. Ueber Eiter und Bindegewebeskoerperchen. *Virchows Archiv für Pathologische Anatomie und Physiologie und für Klinische Medizin*. 28:157–197.
2. Cameron, G.R. 1952. *Pathology of the Cell*. Oliver and Boyd, Edinburgh, Scotland.
3. Craddock, C.G. 1980. Defenses of the body: the initiators of defense, the ready reserves, and the scavengers. In *Blood, Pure and Eloquent: A Story of Discovery, of People, and of Ideas*. M.M. Wintrobe, editor. McGraw Hill, New York. 417–454.
4. Stossel, T.P. 1999. The early history of phagocytosis. In *Phagocytosis: The Host*. S. Gordon, editor. JAI Press, Stamford, Connecticut. 3–18.

5. Metchnikoff, E. 1892. *Leçons sur la pathologie comparée de l'inflammation*. Masson, Paris [translated as *Lectures on the Comparative Pathology of Inflammation*, by F.A. Starling and E.H. Starling. 1893. Kegan Paul, Trench, Truber and Co., London].
6. Metchnikoff, E. 1901. *L'immunité dans les maladies infectieuses*. Masson, Paris [translated as *Immunity in Infective Diseases*, by F.G. Binnie. 1905. University Press, Cambridge, England].
7. Metchnikoff, O. 1921. *Life of Elie Metchnikoff, 1845–1916*. Houghton Mifflin, Boston.
8. Tauber, A.I., and L. Chernyak. 1991. *Metchnikoff and the Origins of Immunology: From Metaphor to Theory*. Oxford University Press, New York.
9. von Behring, E., and S. Kitasato. 1890. Uber das Zustandekommen der Diphtherie-Immunität und der Tetanus-Immunität bei Thieren. *Deutsche Medizinische Wochenschrift*. 16:1113–1114.
10. Ehrlich, P. 1897. Die Wertbemessung des Diphtherieheilserums und deren theoretische Grundlagen. *Klinisches Jahrbuch*. 6:299–326.
11. Wright, A., and S. Douglas. 1903. An experimental investigation of the role of the blood fluids in connection with phagocytosis. *Proceedings of the Royal Society of London*. 72:357–370.
12. Silverstein, A.M. 1989. *A History of Immunology*. Academic Press, San Diego.

12A. From Phagocytosis to Endocytosis in Infectious Diseases

1. Rous, P., and J.B. Murphy. 1912. The histological signs of resistance to a transmissible sarcoma of the fowl. *The Journal of Experimental Medicine*. 15:270–286.
2. Murphy, J.B. 1914. Heteroplastic tissue grafting effected through roentgen-ray lymphoid destruction. *Journal of the American Medical Association*. 62:1459.
3. Murphy, J.B. 1926. *The Lymphocyte in Resistance to Tissue Grafting, Malignant Disease, and Tuberculous Infection: An Experimental Study*. Monographs of The Rockefeller Institute for Medical Research. Number 21. The Rockefeller Institute for Medical Research, New York. 1–167.
4. McMaster, P.D., and M. Heidelberger. 1960. Florence Rena Sabin, 1871–1953. *Biographical Memoirs. National Academy of Sciences (U.S.)*. 34:271–319.
5. Sabin, F.R. 1932. Cellular studies in tuberculosis. *American Review of Tuberculosis*. 25:153–171.
6. Sabin, F.R. 1941. Cellular reactions to fractions from tubercle bacilli. *American Review of Tuberculosis*. 44:415–423.
7. Heidelberger, M. 1939. Quantitative absolute methods in the study of antigen-antibody reactions. *Bacteriological Reviews*. 3:49–95.
8. Treffers, H.P. 1948. Immunology and immunochemistry. In *Bacterial and Mycotic Infections of Man*. R.J. Dubos, editor. J.B. Lippincott, Philadelphia. 154–188.
9. Avery, O.T., H.T. Chickering, R.I. Cole, and A.R. Dochez. 1917. *Acute Lobar Pneumonia. Prevention and Serum Treatment*. Monographs of The Rockefeller Institute for Medical Research. Number 7. The Rockefeller Institute for Medical Research, New York. 1–110.
10. Dubos, R.J. 1976. *The Professor, the Institute, and DNA: Oswald T. Avery, His Life and Scientific Achievements*. The Rockefeller University Press, New York.
11. Dubos, R.J., and O.T. Avery. 1931. Decomposition of the capsular polysaccharide of pneumococcus type III by a bacterial enzyme. *The Journal of Experimental Medicine*. 54:51–71.
12. Avery, O.T., and R.J. Dubos. 1931. The protective action of a specific enzyme against type III pneumococcus infection in mice. *The Journal of Experimental Medicine*. 54:73–89.
13. Landsteiner, K. 1936. *The Specificity of Serological Reactions*. C.C. Thomas, Springfield, Illinois.
14. Landsteiner, K., and M.W. Chase. 1942. Experiments on transfer of cutaneous sensitivity to simple compounds. *Proceedings of the Society for Experimental Biology and Medicine*. 49:688–690.
15. Chase, M.W. 1985. Immunology and experimental dermatology. *Annual Review of Immunology*. 3:1–29.
16. Rich, A.R. 1944. *The Pathogenesis of Tuberculosis*. C.C. Thomas, Springfield, Illinois.
17. Rous, P., and F.S. Jones. 1916. The protection of pathogenic microorganisms by living tissue cells. *The Journal of Experimental Medicine*. 23:601–612.
18. Lurie, M.B., and with the collaboration of P. Zappasodi. 1942. Studies on the mechanism of immunity in tuberculosis. The fate of tubercle bacilli ingested by mononuclear phagocytes derived from normal and immunized animals. *The Journal of Experimental Medicine*. 75:247–268.
19. Dubos, R.J. 1945. *The Bacterial Cell in its Relation to Problems of Virulence, Immunity and Chemotherapy*. Harvard University Press, Cambridge, Massachusetts.
20. Metchnikoff, E. 1901. *L'immunité dans les maladies infectieuses*. Masson, Paris [translated as *Immunity in Infective Diseases*, by F.G. Binnie. 1905. University Press, Cambridge, England].
21. Dubos, R.J. 1950. *Louis Pasteur: Free Lance of Science*. Little Brown, Boston.
22. Dubos, R.J. 1954. *Biochemical Determinants of Microbial Diseases*. Harvard University Press, Cambridge, Massachusetts.
23. Dubos, R.J. 1955. The micro-environment of inflammation or Metchnikoff revisited. *Lancet*. 269:1–5.
24. Dubos, R.J., and B.D. Davis. 1946. Factors affecting the growth of tubercle bacilli in liquid media. *The Journal of Experimental Medicine*. 83:409–423.

25. Dubos, R.J., and G. Middlebrook. 1947. Media for tubercle bacilli. *American Review of Tuberculosis.* 56:334–345.
26. Moberg, C.L. 2005. *René Dubos, Friend of the Good Earth: Microbiologist, Medical Scientist, Environmentalist.* ASM Press, Washington, D.C.
27. Chase, M.W. 1945. The cellular transfer of cutaneous hypersensitivity to tuberculin. *Proceedings of the Society for Experimental Biology and Medicine.* 59:134–135.
28. Dubos, R.J. 1949. RUA, BSD Reports. Volume 37, pages 46–48.
29. Chase, M.W. 1948. The allergic state. In *Bacterial and Mycotic Infections of Man.* R.J. Dubos, editor. J.B. Lippincott, Philadelphia. 110–153.
30. Chase, M.W. 1953. Immunological reactions mediated through cells. In *The Nature and Significance of the Antibody Response.* A.M. Pappenheimer, editor. Columbia University Press, New York. 156–169.
31. Favour, C.B. 1947. Lytic effect of bacterial products on lymphocytes of tuberculous animals. *Proceedings of the Society for Experimental Biology and Medicine.* 65:269–272.
32. Miller, J.M., C.B. Favour, and with the technical assistance of M.A. Umbarger and B.A. Harrison. 1951. The lymphocytic origin of a plasma factor responsible for hypersensitivity in vitro of tuberculin type. *The Journal of Experimental Medicine.* 93:1–12.
33. Dubos, R.J. 1951. RUA, BSD Reports. Volume 39, pages 65–67.
34. McMaster, P.D. 1951. Sites of antibody formation. In *The Nature and Significance of the Antibody Response.* A.M. Pappenheimer, editor. Columbia University Press, New York. 13–45.
35. Bloch, H. 1948. The relationship between phagocytic cells and human tubercle bacilli. *American Review of Tuberculosis.* 58:662–670.
36. Martin, S.P., C.H. Pierce, G. Middlebrook, and R.J. Dubos. 1950. The effect of tubercle bacilli on the polymorphonuclear leucocytes of normal animals. *The Journal of Experimental Medicine.* 91:381–392.
37. Suter, E. 1952. The multiplication of tubercle bacilli within normal phagocytes in tissue culture. *The Journal of Experimental Medicine.* 96:137–150.
38. Suter, E. 1952. Multiplication of tubercle bacilli within phagocytes cultivated in vitro, and effect of streptomycin and isonicotinic acid hydrazide. *American Review of Tuberculosis.* 65:775–776.
39. Suter, E. 1953. Multiplication of tubercle bacilli within mononuclear phagocytes in tissue cultures derived from normal animals and animals vaccinated with BCG. *The Journal of Experimental Medicine.* 97:235–245
40. Suter, E. 1956. Interaction between phagocytes and pathogenic microorganisms. *Bacteriological Reviews.* 20:94–132.
41. Dubos, R.J., C.H. Pierce, and W.B. Schaefer. 1953. Antituberculous immunity induced in mice by vaccination with living cultures of attenuated tubercle bacilli. *The Journal of Experimental Medicine.* 97:207–220.
42. Wood, W.B., Jr. 1953. Studies on the cellular immunology of acute bacterial infections. In *Harvey Lectures.* Series 47. Academic Press, New York. 72–98.
43. Moberg, C.L., and R.M. Steinman. 2003. James Gerald Hirsch, October 31, 1922–May 25, 1987. *Biographical Memoirs. National Academy of Sciences (U.S.).* 84:182–203.
44. Herbert, D., and F. Bardossi. 1969. *Secret in the White Cell: Case History of a Biological Search.* Harper & Row, New York.
45. Hirsch, J.G., and R.J. Dubos. 1952. The effect of spermine on tubercle bacilli. *The Journal of Experimental Medicine.* 95:191–208.
46. Hirsch, J.G. 1953. Spermine oxidase: an amine oxidase with specificity for spermine and spermidine. *The Journal of Experimental Medicine.* 97:345–355.
47. Dubos, R.J., and J.G. Hirsch. 1954. The antimycobacterial activity of a peptide preparation derived from calf thymus. *The Journal of Experimental Medicine.* 99:55–63.
48. Hirsch, J.G. 1959. Immunity to infectious diseases: review of some concepts of Metchnikoff. *Bacteriological Reviews.* 23:48–60.
49. Hirsch, J.G. 1956. Phagocytin: a bactericidal substance from polymorphonuclear leucocytes. *The Journal of Experimental Medicine.* 103:589–611.
50. Hirsch, J.G. 1956. Studies of the bactericidal action of phagocytin. *The Journal of Experimental Medicine.* 103:613–621.
51. Hirsch, J.G. 1956. Antibacterial protein derived from leucocytes. In *10th Conference on the Plasma Proteins and Cellular Elements of the Blood.* Protein Foundation, Cambridge, Massachusetts. 16–23.
52. Schwartz, J., and F. Bardossi. 1983. *Interview with Dr. Zanvil Cohn.* The Rockefeller University Office of Public Affairs.
53. Moberg, C.L., and R.M. Steinman. 2009. Zanvil Alexander Cohn, 1926–1993. *Biographical Memoirs. National Academy of Sciences (U.S.).* http://www.nasonline.org/publications/biographical-memoirs/memoir-pdfs/cohn-zanvil-a.pdf
54. Steinman, R.M., and C.L. Moberg. 1994. Zanvil Alexander Cohn 1926–1993. *The Journal of Experimental Medicine.* 179:1–30.
55. Cohn, Z.A., and S.I. Morse. 1959. Interactions between rabbit polymorphonuclear leucocytes and staphylococci. *The Journal of Experimental Medicine.* 110:419–443.
56. Cohn, Z.A., and S.I. Morse. 1960. Functional and metabolic properties of polymorphonuclear leucocytes. I. Observations on the requirements and consequences of particle ingestion. *The Journal of Experimental Medicine.* 111:667–687.
57. Cohn, Z.A., and S.I. Morse. 1960. Functional and metabolic properties of polymorphonuclear leucocytes. II. The influence of a lipopolysaccharide endotoxin. *The Journal of Experimental Medicine.* 111:689–704.
58. Rogers, D.E., and R. Tompsett. 1952. The survival of staphylococci within human leukocytes. *The Journal of Experimental Medicine.* 95:209–230.
59. Cohn, Z.A., and J.G. Hirsch. 1960. The isolation and properties of the specific cytoplasmic granules of rabbit polymorphonuclear leucocytes. *The Journal of Experimental Medicine.* 112:983–1004.

60. Hirsch, J.G., and Z.A. Cohn. 1960. Degranulation of polymorphonuclear leucocytes following phagocytosis of microorganisms. *The Journal of Experimental Medicine.* 112:1005–1014.
61. Cohn, Z.A., and J.G. Hirsch. 1960. The influence of phagocytosis on the intracellular distribution of granule-associated components of polymorphonuclear leucocytes. *The Journal of Experimental Medicine.* 112:1015–1022.
62. Appelmans, F., R. Wattiaux, and C. de Duve. 1955. Tissue fractionation studies. 5. The association of acid phosphatase with a special class of cytoplasmic granules in rat liver. *Biochemical Journal.* 59:438–445.
63. Novikoff, A.B., H. Beaufay, and C. de Duve. 1956. Electron microscopy of lysosome-rich fractions from rat liver. *The Journal of Biophysical and Biochemical Cytology.* 2(Supplement):179–184.
64. Cohn, Z.A. 1963. The fate of bacteria within phagocytic cells. I. The degradation of isotopically labeled bacteria by polymorphonuclear leucocytes and macrophages. *The Journal of Experimental Medicine.* 117:27–42.
65. Cohn, Z.A., and E. Wiener. 1963. The particulate hydrolases of macrophages. I. Comparative enzymology, isolation, and properties. *The Journal of Experimental Medicine.* 118:991–1008.
66. Cohn, Z.A., and E. Wiener. 1963. The particulate hydrolases of macrophages. II. Biochemical and morphological response to particle ingestion. *The Journal of Experimental Medicine.* 118:1009–1020.
67. Lewis, W.H. 1931. Pinocytosis. *Bulletin of the Johns Hopkins Hospital.* 49:17–27.
68. Gropp, A. 1963. Phagocytosis and pinocytosis. In *Cinemicrography in Cell Biology.* G.G. Rose, editor. Academic Press, New York. 279–312.
69. Hirsch, J.G. 1965. Phagocytosis. *Annual Review of Microbiology.* 19:339–350.
70. Zucker-Franklin, D., and J.G. Hirsch. 1964. Electron microscope studies on the degranulation of rabbit peritoneal leukocytes during phagocytosis. *The Journal of Experimental Medicine.* 120:569–576.
71. de Duve, C. 1963. The lysosome concept. In *Ciba Foundation Symposium on Lysosomes.* A.V.S. de Reuck and M.P. Cameron, editors. Little Brown, Boston. 1–31.
72. Fedorko, M.E., J.G. Hirsch, and Z.A. Cohn. 1968. Autophagic vacuoles produced in vitro. I. Studies on cultured macrophages exposed to chloroquine. *The Journal of Cell Biology.* 38:377–391.
73. Fedorko, M.E., J.G. Hirsch, and Z.A. Cohn. 1968. Autophagic vacuoles produced in vitro. II. Studies on the mechanism of formation of autophagic vacuoles produced by chloroquine. *The Journal of Cell Biology.* 38:392–402.
74. Morse, S.I., Z.A. Cohn, J.G. Hirsch, and R.W. Schaeder. 1961. The treatment of sarcoidosis with chloroquine. *The American Journal of Medicine.* 30:779–784.
75. Hirsch, J.G., Z.A. Cohn, S.I. Morse, R.W. Schaedler, L.E. Siltzbach, J.T. Ellis, and M.W. Chase. 1961. Evaluation of the Kveim reaction as a diagnostic test for sarcoidosis. *The New England Journal of Medicine.* 265:827–830.
76. Zigmond, S.H., and J.G. Hirsch. 1973. Leukocyte locomotion and chemotaxis. New methods for evaluation, and demonstration of a cell-derived chemotactic factor. *The Journal of Experimental Medicine.* 137:387–410.
77. Hirsch, J.G. 1974. Neutrophil leukocytes. In *The Inflammatory Process.* Second edition. B.W. Zweifach, L. Grant, and R.T. McCluskey, editors. Academic Press, New York. 411–447.
78. Cohn, Z.A., J.G. Hirsch, and E. Wiener. 1963. Lysosomes and endocytosis: The cytoplasmic granules of phagocytic cells and the degradation of bacteria. In *Ciba Foundation Symposium on Lysosomes.* A.V.S. de Reuck and M.P. Cameron, editors. Little Brown, Boston. 126–144.
79. Aschoff, L. 1924. Das reticulo-endotheliale System. *Ergebnisse der Inneren Medizin und Kinderheilkunde.* 26:1–118.
80. Aschoff, L. 1924. Reticulo-endothelial system. In *Lectures on Pathology.* Paul B. Hoeber, Inc., New York. 1–33.
81. Maximow, A. 1932. The lymphocytes and plasma cells. In *Special Cytology: The Form and Function of the Cell in Health and Disease.* Second edition. E.V. Cowdry, editor. Paul B. Hoeber, Inc., New York. 603–650.
82. Murray, M.R., and G. Kopech. 1953. *A Bibliography of the Research in Tissue Culture, 1884 to 1950. An Index to the Literature of the Living Cell Cultivated In Vitro.* Academic Press, New York.
83. Cohn, Z.A. 1962. Determinants of infection in the peritoneal cavity. I. Response to and fate of *Staphylococcus aureus* and *Staphylococcus albus* in the mouse. *The Yale Journal of Biology and Medicine.* 35:12–28.
84. Cohn, Z.A. 1962. Determinants of infection in the peritoneal cavity. II. Factors influencing the fate of *Staphylococcus aureus* in the mouse. *The Yale Journal of Biology and Medicine.* 35:29–47.
85. Cohn, Z.A. 1962. Determinants of infection in the peritoneal cavity. III. The action of selected inhibitors on the fate of *Staphylococcus aureus* in the mouse. *The Yale Journal of Biology and Medicine.* 35:48–61.
86. Mackaness, G.B. 1960. The phagocytosis and inactivation of staphylococci by macrophages of normal rabbits. *The Journal of Experimental Medicine.* 112:35–53.
87. Cohn, Z.A., and B. Benson. 1965. The differentiation of mononuclear phagocytes. Morphology, cytochemistry, and biochemistry. *The Journal of Experimental Medicine.* 121:153–170.
88. Cohn, Z.A., and B. Benson. 1965. The in vitro differentiation of mononuclear phagocytes. I. The influence of inhibitors and the results of autoradiography. *The Journal of Experimental Medicine.* 121:279–288.
89. Cohn, Z.A., and B. Benson. 1965. The in vitro differentiation of mononuclear phagocytes. II. The influence of serum on granule formation, hydrolase production, and pinocytosis. *The Journal of Experimental Medicine.* 121:835–848.
90. Cohn, Z.A., and B. Benson. 1965. The in vitro differentiation of mononuclear phagocytes. III. The reversibility of granule and hydrolytic enzyme formation and the turnover of granule constituents. *The Journal of Experimental Medicine.* 122:455–466.
91. Cohn, Z.A. 1966. The regulation of pinocytosis in mouse macrophages. I. Metabolic requirements as defined by the use of inhibitors. *The Journal of Experimental Medicine.* 124:557–571.
92. Cohn, Z.A., and E. Parks. 1967. The regulation of pinocytosis in mouse macrophages. II. Factors inducing vesicle formation. *The Journal of Experimental Medicine.* 125:213–232.

93. Cohn, Z.A., and E. Parks. 1967. The regulation of pinocytosis in mouse macrophages. III. The induction of vesicle formation by nucleosides and nucleotides. *The Journal of Experimental Medicine.* 125:457–466.
94. Cohn, Z.A., and E. Parks. 1967. The regulation of pinocytosis in mouse macrophages. IV. The immunological induction of pinocytic vesicles, secondary lysosomes, and hydrolytic enzymes. *The Journal of Experimental Medicine.* 125:1091–1104.
95. Cohn, Z.A., J.G. Hirsch, and M.E. Fedorko. 1966. The in vitro differentiation of mononuclear phagocytes. IV. The ultrastructure of macrophage differentiation in the peritoneal cavity and in culture. *The Journal of Experimental Medicine.* 123:747–756.
96. Cohn, Z.A., M.E. Fedorko, and J.G. Hirsch. 1966. The in vitro differentiation of mononuclear phagocytes. V. The formation of macrophage lysosomes. *The Journal of Experimental Medicine.* 123:757–766.
97. Ehrenreich, B.A., and Z.A. Cohn. 1968. Pinocytosis by macrophages. *Journal of the Reticuloendothelial Society.* 5:230–242.
98. Cohn, Z.A., and B.A. Ehrenreich. 1969. The uptake, storage, and intracellular hydrolysis of carbohydrates by macrophages. *The Journal of Experimental Medicine.* 129:201–225.
99. Bennett, W.E., and Z.A. Cohn. 1966. The isolation and selected properties of blood monocytes. *The Journal of Experimental Medicine.* 123:145–160.
100. van Furth, R., and Z.A. Cohn. 1968. The origin and kinetics of mononuclear phagocytes. *The Journal of Experimental Medicine.* 128:415–435.
101. van Furth, R., J.G. Hirsch, and M.E. Fedorko. 1970. Morphology and peroxidase cytochemistry of mouse promonocytes, monocytes, and macrophages. *The Journal of Experimental Medicine.* 132:794–812.
102. van Furth, R., Z.A. Cohn, J.G. Hirsch, J.H. Humphrey, W.G. Spector, and H.L. Langevoort. 1972. The mononuclear phagocyte system: a new classification of macrophages, monocytes, and their precursor cells. *Bulletin of the World Health Organization.* 46:845–852.
103. van Furth, R., editor. 1970. *Mononuclear Phagocytes.* Blackwell Scientific Publications, Oxford, England.
104. Palade, G.E. 1956. The endoplasmic reticulum. *The Journal of Biophysical and Biochemical Cytology.* 2(Supplement):85–98.
105. Hubbard, A.L., and Z.A. Cohn. 1975. Externally disposed plasma membrane proteins. I. Enzymatic iodination of mouse L cells. *The Journal of Cell Biology.* 64:438–460.
106. Hubbard, A.L., and Z.A. Cohn. 1975. Externally disposed plasma membrane proteins. II. Metabolic fate of iodinated polypeptides of mouse L cells. *The Journal of Cell Biology.* 64:461–479.
107. Steinman, R.M., and Z.A. Cohn. 1972. The interaction of soluble horseradish peroxidase with mouse peritoneal macrophages in vitro. *The Journal of Cell Biology.* 55:186–204.
108. Steinman, R.M., and Z.A. Cohn. 1972. The interaction of particulate horseradish peroxidase (HRP)-anti HRP immune complexes with mouse peritoneal macrophages in vitro. *The Journal of Cell Biology.* 55:616–634.
109. Steinman, R.M., S.E. Brodie, and Z.A. Cohn. 1976. Membrane flow during pinocytosis. A stereologic analysis. *The Journal of Cell Biology.* 68:665–687.
110. Muller, W.A., R.M. Steinman, and Z.A. Cohn. 1980. The membrane proteins of the vacuolar system. I. Analysis of a novel method of intralysosomal iodination. *The Journal of Cell Biology.* 86:292–303.
111. Muller, W.A., R.M. Steinman, and Z.A. Cohn. 1980. The membrane proteins of the vacuolar system. II. Bidirectional flow between secondary lysosomes and plasma membrane. *The Journal of Cell Biology.* 86:304–314.
112. Mellman, I.S., R.M. Steinman, J.C. Unkeless, and Z.A. Cohn. 1980. Selective iodination and polypeptide composition of pinocytic vesicles. *The Journal of Cell Biology.* 86:712–722.
113. Mellman, I.S., H. Plutner, R.M. Steinman, J.C. Unkeless, and Z.A. Cohn. 1983. Internalization and degradation of macrophage Fc receptors during receptor-mediated phagocytosis. *The Journal of Cell Biology.* 96:887–895.
114. Goldstein, J.L., R.G.W. Anderson, and M.S. Brown. 1979. Coated pits, coated vesicles, and receptor-mediated endocytosis. *Nature.* 279:679–685.
115. Wall, D.A., and A.L. Hubbard. 1981. Galactose-specific recognition system of mammalian liver: receptor distribution on the hepatocyte cell surface. *The Journal of Cell Biology.* 90:687–696.
116. Helenius, A., I. Mellman, D. Wall, and A. Hubbard. 1983. Endosomes. *Trends in Biochemical Sciences.* 8:245–250.
117. Griffin, F.M., Jr., J.A. Griffin, J.E. Leider, and S.C. Silverstein. 1975. Studies on the mechanism of phagocytosis. I. Requirements for circumferential attachment of particle-bound ligands to specific receptors on the macrophage plasma membrane. *The Journal of Experimental Medicine.* 142:1263–1282.
118. Griffin, F.M., Jr., J.A. Griffin, and S.C. Silverstein. 1976. Studies on the mechanism of phagocytosis. II. The interaction of macrophages with anti-immunoglobulin IgG-coated bone marrow-derived lymphocytes. *The Journal of Experimental Medicine.* 144:788–809.
119. Steinman, R.M., I.S. Mellman, W.A. Muller, and Z.A. Cohn. 1983. Endocytosis and the recycling of plasma membrane. *The Journal of Cell Biology.* 96:1–27.
120. Silverstein, S.C., R.M. Steinman, and Z.A. Cohn. 1977. Endocytosis. *Annual Review of Biochemistry.* 46:669–722.
121. Cohn, Z.A. 1983. The macrophage-versatile element of inflammation. In *Harvey Lectures.* Series 77. Academic Press, New York. 63–80.
122. Gordon, S., J. Todd, and Z.A. Cohn. 1974. In vitro synthesis and secretion of lysozyme by mononuclear phagocytes. *The Journal of Experimental Medicine.* 139:1228–1248.
123. Gordon, S., J.C. Unkeless, and Z.A. Cohn. 1974. Induction of macrophage plasminogen activator by endotoxin stimulation and phagocytosis: evidence for a two-stage process. *The Journal of Experimental Medicine.* 140:995–1010.
124. Nathan, C.F., and R.K. Root. 1977. Hydrogen peroxide release from mouse peritoneal macrophages: dependence on sequential activation and triggering. *The Journal of Experimental Medicine.* 146:1648–1662.

125. Murray, H.W., C.W. Juangbhanich, C.F. Nathan, and Z.A. Cohn. 1979. Macrophage oxygen-dependent antimicrobial activity. II. The role of oxygen intermediates. *The Journal of Experimental Medicine*. 150:950–964.
126. Rouzer, C.A., W.A. Scott, A.L. Hamill, and Z.A. Cohn. 1980. Dynamics of leukotriene C production by macrophages. *The Journal of Experimental Medicine*. 152:1236–1247.
127. Mackaness, G.B. 1962. Cellular resistance to infection. *The Journal of Experimental Medicine*. 116:381–406.
128. Mackaness, G.B. 1964. The immunological basis of acquired cellular resistance. *The Journal of Experimental Medicine*. 120:105–120.
129. Bloom, B.R., and B. Bennett. 1966. Mechanism of a reaction in vitro associated with delayed-type hypersensitivity. *Science*. 153:80–82.
130. David, J.R., H.S. Lawrence, and L. Thomas. 1964. Delayed hypersensitivity in vitro. III. The specificity of hapten-protein conjugates in the inhibition of cell migration. *Journal of Immunology*. 93:279–282.
131. Nathan, C.F., M.L. Karnovsky, and J.R. David. 1971. Alterations of macrophage functions by mediators from lymphocytes. *The Journal of Experimental Medicine*. 133:1356–1376.
132. Nathan, C.F., H.G. Remold, and J.R. David. 1973. Characterization of a lymphocyte factor which alters macrophage functions. *The Journal of Experimental Medicine*. 137:275–290.
133. Evans, R., and P. Alexander. 1970. Cooperation of immune lymphoid cells with macrophages in tumour immunity. *Nature*. 228:620–622.
134. Weinberg, J.B., H.A. Chapman Jr., and J.B. Hibbs Jr. 1978. Characterization of the effects of endotoxin on macrophage tumor cell killing. *Journal of Immunology*. 121:72–80.
135. Nathan, C.F., H.W. Murray, M.E. Wiebe, and B.Y. Rubin. 1983. Identification of interferon-gamma as the lymphokine that activates human macrophage oxidative metabolism and antimicrobial activity. *The Journal of Experimental Medicine*. 158:670–689.
136. Schreiber, R.D., J.L. Pace, S.W. Russell, A. Altman, and D.H. Katz. 1983. Macrophage-activating factor produced by a T cell hybridoma: physiochemical and biosynthetic resemblance to gamma-interferon. *Journal of Immunology*. 131:826–832.
137. Cohn, Z.A. 1978. Activation of mononuclear phagocytes: fact, fancy, and future. *Journal of Immunology*. 121:813–816.
138. Nogueira, N., and Z.A. Cohn. 1976. *Trypanosoma cruzi*: mechanism of entry and intracellular fate in mammalian cells. *The Journal of Experimental Medicine*. 143:1402–1420.
139. Nogueira, N., S. Gordon, and Z.A. Cohn. 1977. *Trypanosoma cruzi*: modification of macrophage function during infection. *The Journal of Experimental Medicine*. 146:157–171.
140. Nogueira, N., S. Gordon, and Z.A. Cohn. 1977. *Trypanosoma cruzi*: the immunological induction of macrophage plasminogen activator requires thymus-derived lymphocytes. *The Journal of Experimental Medicine*. 146:172–183.
141. Nogueira, N., and Z.A. Cohn. 1978. *Trypanosoma cruzi*: in vitro induction of macrophage microbicidal activity. *The Journal of Experimental Medicine*. 148:288–300.

12B. Endocytosis and the Discovery of Dendritic Cells

1. Rittenberg, D. 1950. Dynamic aspects of the metabolism of amino acids. In *Harvey Lectures*. Series 44. Charles C. Thomas Publisher, Springfield, Illinois. 200–219.
2. Moore, S., and W.H. Stein. 1958. Determination of the structure of proteins: studies on ribonuclease. In *Harvey Lectures*. Series 52. Academic Press, New York. 119–143.
3. de Duve, C. 1965. The separation and characterization of subcellular particles. In *Harvey Lectures*. Series 59. Academic Press, New York. 49–87.
4. Cohn, Z.A., J.G. Hirsch, and E. Wiener. 1963. The cytoplasmic granules of phagocytic cells and the degradation of bacteria. In *Ciba Foundation Symposium on Lysosomes*. A.V.S. de Reuck and M.P. Cameron, editors. J & A Churchill, London. 126–144.
5. Hirsch, J.G., and Z.A. Cohn. 1963. Leucocyte lysosomes. In *Cell-Bound Antibodies*. B. Amos and H. Koprowski, editors. Wistar Institute Press, Philadelphia. 45–49.
6. Steinman, R.M. 1968. An electron microscopic study of ciliogenesis in developing epidermis and trachea in the embryo of *Xenopus laevis*. *The American Journal of Anatomy*. 122:19–55.
7. Steinman, R.M. 1970. Inhibitory effects of colchicine on ciliogenesis in ectoderm of *Xenopus laevis*. *Journal of Ultrastructure Research*. 30:423–440.
8. Graham, R.C., Jr., and M.J. Karnovsky. 1966. The early stages of absorption of injected horseradish peroxidase in the proximal tubules of mouse kidney: ultrastructural cytochemistry by a new technique. *The Journal of Histochemistry and Cytochemistry*. 14:291–302.
9. Burnet, F.M. 1957. A modification of Jerne's theory of antibody production using the concept of clonal selection. *Australian Journal of Science*. 20:67–69.
10. Cohn, Z.A., and J.G. Hirsch. 1965. Phagocytic cells. In *Bacterial and Mycotic Infections of Man*. Fourth edition. R.J. Dubos and J.G. Hirsch, editors. J.B. Lippincott Co., Philadelphia. 215–236.
11. Cohn, Z.A., and J.G. Hirsch. 1960. The isolation and properties of the specific cytoplasmic granules of rabbit polymorphonuclear leucocytes. *The Journal of Experimental Medicine*. 112:983–1004.
12. Cohn, Z.A., and J.G. Hirsch. 1960. The influence of phagocytosis on the intracellular distribution of granule-associated components of polymorphonuclear leucocytes. *The Journal of Experimental Medicine*. 112:1015–1022.

13. Hirsch, J.G., and Z.A. Cohn. 1960. Degranulation of polymorphonuclear leucocytes following phagocytosis of microorganisms. *The Journal of Experimental Medicine.* 112:1005–1014.
14. Cohn, Z.A., and B. Benson. 1965. The differentiation of mononuclear phagocytes. Morphology, cytochemistry, and biochemistry. *The Journal of Experimental Medicine.* 121:153–170.
15. Cohn, Z.A., and B. Benson. 1965. The in vitro differentiation of mononuclear phagocytes. I. The influence of inhibitors and the results of autoradiography. *The Journal of Experimental Medicine.* 121:279–288.
16. Cohn, Z.A., and B. Benson. 1965. The in vitro differentiation of mononuclear phagocytes. II. The influence of serum on granule formation, hydrolase production, and pinocytosis. *The Journal of Experimental Medicine.* 121:835–848.
17. Cohn, Z.A., and B. Benson. 1965. The in vitro differentiation of mononuclear phagocytes. III. The reversibility of granule and hydrolytic enzyme formation and the turnover of granule constituents. *The Journal of Experimental Medicine.* 122:455–466.
18. Cohn, Z.A., M.E. Fedorko, and J.G. Hirsch. 1966. The in vitro differentiation of mononuclear phagocytes. V. The formation of macrophage lysosomes. *The Journal of Experimental Medicine.* 123:757–766.
19. Steinman, R.M., and C.L. Moberg. 1994. Zanvil Alexander Cohn 1926–1993. *The Journal of Experimental Medicine.* 179:1–30.
20. Ehrenreich, B.A., and Z.A. Cohn. 1967. The uptake and digestion of iodinated human serum albumin by macrophages in vitro. *The Journal of Experimental Medicine.* 126:941–958.
21. Ehrenreich, B.A., and Z.A. Cohn. 1968. Fate of hemoglobin pinocytosed by macrophages in vitro. *The Journal of Cell Biology.* 38:244–248.
22. Cohn, Z.A., and B.A. Ehrenreich. 1969. The uptake, storage, and intracellular hydrolysis of carbohydrates by macrophages. *The Journal of Experimental Medicine.* 129:201–225.
23. Ehrenreich, B.A., and Z.A. Cohn. 1969. The fate of peptides pinocytosed by macrophages in vitro. *The Journal of Experimental Medicine.* 129:227–245.
24. van Furth, R., and Z.A. Cohn. 1968. The origin and kinetics of mononuclear phagocytes. *The Journal of Experimental Medicine.* 128:415–435.
25. van Furth, R., Z.A. Cohn, J.G. Hirsch, J.H. Humphrey, W.G. Spector, and H.L. Langevoort. 1972. The mononuclear phagocyte system: a new classification of macrophages, monocytes, and their precursor cells. *Bulletin of the World Health Organization.* 46:845–852.
26. Seljelid, R., S.C. Silverstein, and Z.A. Cohn. 1973. The effect of poly-L-lysine on the uptake of reovirus double-stranded RNA in macrophages in vitro. *The Journal of Cell Biology.* 57:484–498.
27. Nachman, R.L., B. Ferris, and J.G. Hirsch. 1971. Macrophage plasma membrane. II. Studies on synthesis and turnover of protein constituents. *The Journal of Experimental Medicine.* 133:807–820.
28. Gordon, S., and Z.A. Cohn. 1973. The macrophage. *International Review of Cytology.* 36:171–214.
29. Silverstein, S.C., R.M. Steinman, and Z.A. Cohn. 1977. Endocytosis. *Annual Review of Biochemistry.* 46:669–722.
30. Werb, Z., and Z.A. Cohn. 1972. Cholesterol metabolism in the macrophage. III. Ingestion and intracellular fate of cholesterol and cholesterol esters. *The Journal of Experimental Medicine.* 135:21–44.
31. Zigmond, S.H. 1974. Mechanisms of sensing chemical gradients by polymorphonuclear leukocytes. *Nature.* 249:450–452.
32. Hubbard, A.L., and Z.A. Cohn. 1972. The enzymatic iodination of the red cell membrane. *The Journal of Cell Biology.* 55:390–405.
33. Hubbard, A.L., and Z.A. Cohn. 1975. Externally disposed plasma membrane proteins. II. Metabolic fate of iodinated polypeptides of mouse L cells. *The Journal of Cell Biology.* 64:461–479.
34. Straus, W. 1963. Comparative observations on lysosomes and phagosomes in kidney and liver of rats after administration of horseradish peroxidase. In *Ciba Foundation Symposium on Lysosomes.* A.V.S. de Reuck and M.P. Cameron, editors. Little Brown, Boston. 151–170.
35. Avrameas, S., and E.H. Leduc. 1970. Detection of simultaneous antibody synthesis in plasma cells and specialized lymphocytes in rabbit lymph nodes. *The Journal of Experimental Medicine.* 131:1137–1168.
36. Nossal, G.J.V., A. Abbot, J. Mitchell, and Z. Lummus. 1968. Antigens in immunity. XV. Ultrastructural features of antigen capture in primary and secondary lymphoid follicles. *The Journal of Experimental Medicine.* 127:277–290.
37. Steinman, R.M., and Z.A. Cohn. 1972. The interaction of soluble horseradish peroxidase with mouse peritoneal macrophages in vitro. *The Journal of Cell Biology.* 55:186–204.
38. Steinman, R.M., and Z.A. Cohn. 1972. The interaction of particulate horseradish peroxidase (HRP)-anti HRP immune complexes with mouse peritoneal macrophages in vitro. *The Journal of Cell Biology.* 55:616–634.
39. Steinman, R.M., S.E. Brodie, and Z.A. Cohn. 1976. Membrane flow during pinocytosis. A stereologic analysis. *The Journal of Cell Biology.* 68:665–687.
40. Steinman, R.M., I.S. Mellman, W.A. Muller, and Z.A. Cohn. 1983. Endocytosis and the recycling of plasma membrane. *The Journal of Cell Biology.* 96:1–27.
41. Goldstein, J.L., R.G. Anderson, and M.S. Brown. 1979. Coated pits, coated vesicles, and receptor-mediated endocytosis. *Nature.* 279:679–685.
42. Muller, W.A., R.M. Steinman, and Z.A. Cohn. 1980. The membrane proteins of the vacuolar system I. Analysis of a novel method of intralysosomal iodination. *The Journal of Cell Biology.* 86:292–303.
43. Muller, W.A., R.M. Steinman, and Z.A. Cohn. 1980. The membrane proteins of the vacuolar system. II. Bidirectional flow between secondary lysosomes and plasma membrane. *The Journal of Cell Biology.* 86:304–314.

44. Muller, W.A., R.M. Steinman, and Z.A. Cohn. 1983. Intracellular iodination of lysosome membrane for studies of membrane composition and recycling. In *Methods in Enzymology. Biomembranes, Part L. Membrane Biogenesis: Processing and Recycling*. Volume 98. S. Fleischer and B. Fleischer, editors. Academic Press, New York. 404–415.
45. Mellman, I.S., R.M. Steinman, J.C. Unkeless, and Z.A. Cohn. 1980. Selective iodination and polypeptide composition of pinocytic vesicles. *The Journal of Cell Biology*. 86:712–722.
46. Mellman, I.S., H. Plutner, R.M. Steinman, J.C. Unkeless, and Z.A. Cohn. 1983. Internalization and degradation of macrophage Fc receptors during receptor-mediated phagocytosis. *The Journal of Cell Biology*. 96:887–895.
47. McMaster, P.D., and S.S. Hudack. 1935. The formation of agglutinins within lymph nodes. *The Journal of Experimental Medicine*. 61:783–805.
48. Mishell, R.I., and R.W. Dutton. 1967. Immunization of dissociated spleen cell cultures from normal mice. *The Journal of Experimental Medicine*. 126:423–442.
49. Chen, C., and J.G. Hirsch. 1972. The effects of mercaptoethanol and of peritoneal macrophages on the antibody-forming capacity of nonadherent mouse spleen cells in vitro. *The Journal of Experimental Medicine*. 136:604–617.
50. Lindquist, R.L., G. Shakhar, D. Dudziak, H. Wardemann, T. Eisenreich, M.L. Dustin, and M.C. Nussenzweig. 2004. Visualizing dendritic cell networks in vivo. *Nature Immunology*. 5:1243–1250.
51. Steinman, R.M., and Z.A. Cohn. 1973. Identification of a novel cell type in peripheral lymphoid organs of mice. I. Morphology, quantitation, tissue distribution. *The Journal of Experimental Medicine*. 137:1142–1162.
52. Steinman, R.M., and Z.A. Cohn. 1974. Identification of a novel cell type in peripheral lymphoid organs of mice. II. Functional properties in vitro. *The Journal of Experimental Medicine*. 139:380–397.
53. Steinman, R.M., G. Kaplan, M.D. Witmer, and Z.A. Cohn. 1979. Identification of a novel cell type in peripheral lymphoid organs of mice. V. Purification of spleen dendritic cells, new surface markers, and maintenance in vitro. *The Journal of Experimental Medicine*. 149:1–16.
54. Steinman, R.M., and Z.A. Cohn. 1975. A novel adherent cell in mouse lymphoid organs. In *Immune Recognition*. A.S. Rosenthal, editor. Academic Press, San Francisco. 571–587.
55. Nussenzweig, M.C., R.M. Steinman, M.D. Witmer, and B. Gutchinov. 1982. A monoclonal antibody specific for mouse dendritic cells. *Proceedings of the National Academy of Sciences of the United States of America*. 79:161–165.
56. Steinman, R.M., and M.D. Witmer. 1978. Lymphoid dendritic cells are potent stimulators of the primary mixed leukocyte reaction in mice. *Proceedings of the National Academy of Sciences of the United States of America*. 75:5132–5136.
57. Nussenzweig, M.C., R.M. Steinman, B. Gutchinov, and Z.A. Cohn. 1980. Dendritic cells are accessory cells for the development of anti-trinitrophenyl cytotoxic T lymphocytes. *The Journal of Experimental Medicine*. 152:1070–1084.
58. Steinman, R.M., B. Gutchinov, M.D. Witmer, and M.C. Nussenzweig. 1983. Dendritic cells are the principal stimulators of the primary mixed leukocyte reaction in mice. *The Journal of Experimental Medicine*. 157:613–627.
59. Inaba, K., R.M. Steinman, W.C. Van Voorhis, and S. Muramatsu. 1983. Dendritic cells are critical accessory cells for thymus-dependent antibody responses in mouse and in man. *Proceedings of the National Academy of Sciences of the United States of America*. 80:6041–6045.
60. Van Voorhis, W.C., L.S. Hair, R.M. Steinman, and G. Kaplan. 1982. Human dendritic cells. Enrichment and characterization from peripheral blood. *The Journal of Experimental Medicine*. 155:1172–1187.
61. Van Voorhis, W.C., J. Valinsky, E. Hoffman, J. Luban, L.S. Hair, and R.M. Steinman. 1983. Relative efficacy of human monocytes and dendritic cells as accessory cells for T cell replication. *The Journal of Experimental Medicine*. 158:174–191.
62. Inaba, K., K. Nakano, and S. Muramatsu. 1981. Cellular synergy in the manifestation of accessory cell activity for in vitro antibody response. *Journal of Immunology*. 127:452–461.
63. Witmer, M.D., and R.M. Steinman. 1984. The anatomy of peripheral lymphoid organs with emphasis on accessory cells: light-microscopic immunocytochemical studies of mouse spleen, lymph node, and Peyer's patch. *The American Journal of Anatomy*. 170:465–481.
64. Metlay, J.P., M.D. Witmer-Pack, R. Agger, M.T. Crowley, D. Lawless, and R.M. Steinman. 1990. The distinct leukocyte integrins of mouse spleen dendritic cells as identified with new hamster monoclonal antibodies. *The Journal of Experimental Medicine*. 171:1753–1771.
65. Trombetta, E.S., and I. Mellman. 2005. Cell biology of antigen processing in vitro and in vivo. *Annual Review of Immunology*. 23:975–1028.
66. Jiang, W., W.J. Swiggard, C. Heufler, M. Peng, A. Mirza, R.M. Steinman, and M.C. Nussenzweig. 1995. The receptor DEC-205 expressed by dendritic cells and thymic epithelial cells is involved in antigen processing. *Nature*. 375:151–155.
67. Mahnke, K., M. Guo, S. Lee, H. Sepulveda, S.L. Swain, M. Nussenzweig, and R.M. Steinman. 2000. The dendritic cell receptor for endocytosis, DEC-205, can recycle and enhance antigen presentation via major histocompatibility complex class II-positive lysosomal compartments. *The Journal of Cell Biology*. 151:673–684.
68. Hawiger, D., K. Inaba, Y. Dorsett, M. Guo, K. Mahnke, M. Rivera, J.V. Ravetch, R.M. Steinman, and M.C. Nussenzweig. 2001. Dendritic cells induce peripheral T cell unresponsiveness under steady state conditions in vivo. *The Journal of Experimental Medicine*. 194:769–779.
69. Bonifaz, L., D. Bonnyay, K. Mahnke, M. Rivera, M.C. Nussenzweig, and R.M. Steinman. 2002. Efficient targeting of protein antigen to the dendritic cell receptor DEC-205 in the steady state leads to antigen presentation on major histocompatibility complex class I products and peripheral $CD8^+$ T cell tolerance. *The Journal of Experimental Medicine*. 196:1627–1638.

70. Bozzacco, L., C. Trumpfheller, F.P. Siegal, S. Mehandru, M. Markowitz, M. Carrington, M.C. Nussenzweig, A.G. Piperno, and R.M. Steinman. 2007. DEC-205 receptor on dendritic cells mediates presentation of HIV gag protein to CD8+ T cells in a spectrum of human MHC I haplotypes. *Proceedings of the National Academy of Sciences of the United States of America*. 104:1289–1294.
71. Boscardin, S.B., J.C. Hafalla, R.F. Masilamani, A.O. Kamphorst, H.A. Zebroski, U. Rai, A. Morrot, F. Zavala, R.M. Steinman, R.S. Nussenzweig, and M.C. Nussenzweig. 2006. Antigen targeting to dendritic cells elicits long-lived T cell help for antibody responses. *The Journal of Experimental Medicine*. 203:599–606.
72. Trumpfheller, C., J.S. Finke, C.B. López, T.M. Moran, B. Moltedo, H. Soares, Y. Huang, S.J. Schlesinger, C.G. Park, M.C. Nussenzweig, A. Granelli-Piperno, and R.M. Steinman. 2006. Intensified and protective CD4+ T cell immunity in mice with anti-dendritic cell HIV gag fusion antibody vaccine. *The Journal of Experimental Medicine*. 203:607–617.
73. Steinman, R.M., and J. Banchereau. 2007. Taking dendritic cells into medicine. *Nature*. 449:419–426.

Chapter 13

Gatekeepers: Cell Membranes and Permeability

Introduction

1. Porter, K.R. 1954. RUA, BSD Reports. Volume 42, pages 195–205.
2. Palade, G.E. 1953. Fine structure of blood capillaries. *Journal of Applied Physics*. 24:1424 (Abstract).
3. Yamada, E. 1955. The fine structure of the gall bladder epithelium of the mouse. *The Journal of Biophysical and Biochemical Cytology*. 1(Supplement):445–458.
4. Palade, G.E. 1960. Transport in quanta across the endothelium of blood capillaries. *The Anatomical Record*. 136:254 (Abstract).
5. Weibel, E.R., and G.E. Palade. 1964. New cytoplasmic components in arterial endothelia. *The Journal of Cell Biology*. 23:101–112.
6. Landis, E.M., and J.R. Pappenheimer. 1963. Exchange of substances through the capillary walls. In *Handbook of Physiology; A Critical, Comprehensive Presentation of Physiological Knowledge and Concepts*. W.F. Hamilton and P. Dow, editors. American Physiological Society, Washington, D.C. 961–968.
7. Wasserman, K., L. Loeb, and H.S. Mayerson. 1955. Capillary permeability to macromolecules. *Circulation Research*. 3:594–603.
8. Palade, G.E., and R.R. Bruns. 1964. Structure and function in normal capillaries. In *Small Blood Vessel Involvement in Diabetes Mellitus*. M.D. Siperstein, editor. American Institute of Biological Sciences, Washington, D.C. 39–49.
9. Bruns, R.R., and G.E. Palade. 1968. Studies on blood capillaries. I. General organization of blood capillaries in muscle. *The Journal of Cell Biology*. 37:244–276.
10. Bruns, R.R., and G.E. Palade. 1968. Studies on blood capillaries. II. Transport of ferritin molecules across the wall of muscle capillaries. *The Journal of Cell Biology*. 37:277–299.
11. Palade, G.E., and R.R. Bruns. 1968. Structural modulations of plasmalemmal vesicles. *The Journal of Cell Biology*. 37:633–649.
12. Palade, G.E. 1968. Small pore and large pore systems in capillary permeability. In *Hemorheology*. A.L. Copley, editor. Pergamon Press, Oxford and New York. 703–718.
13. Bennett, H.S., J.H. Luft, and J.C. Hampton. 1959. Morphological classifications of vertebrate blood capillaries. *The American Journal of Physiology*. 196:381–390.
14. Mayerson, H.S., C.G. Wolfram, H.H. Shirley Jr., and K. Wasserman. 1960. Regional differences in capillary permeability. *The American Journal of Physiology*. 198:155–160.
15. Renkin, E.M. 1964. Transport of large molecules across capillary walls. *The Physiologist*. 60:13–28.
16. Clementi, F., and G.E. Palade. 1969. Intestinal capillaries. I. Permeability to peroxidase and ferritin. *The Journal of Cell Biology*. 41:33–58.
17. Clementi, F., and G.E. Palade. 1969. Intestinal capillaries. II. Structural effects of EDTA and histamine. *The Journal of Cell Biology*. 42:706–714.

13A. Vascular Leakage Induced by Inflammatory Mediators

1. Alksne, J.F. 1959. The passage of colloidal particles across the dermal capillary wall under the influence of histamine. *Quarterly Journal of Experimental Physiology and Cognate Medical Sciences*. 44:51–66.
2. Majno, G., and G.E. Palade. 1961. Studies on inflammation. I. The effect of histamine and serotonin on vascular permeability: an electron microscopic study. *The Journal of Biophysical and Biochemical Cytology*. 11:571–605.
3. Majno, G., G.E. Palade, and G.I. Schoefl. 1961. Studies on inflammation. II. The site of action of histamine and serotonin along the vascular tree: a topographic study. *The Journal of Biophysical and Biochemical Cytology*. 11:607–626.

4. Majno, G., and G.E. Palade. 1974. Studies on inflammation. In *Microcirculation.* M.P. Wiedeman, editor. Hutchinson & Ross, Stroudsburg, Pennsylvania. 302–321.
5. Hurley, J.V. 1972. *Acute Inflammation.* Williams & Wilkins Co., Baltimore, Maryland.
6. Majno, G., V. Gilmore, and M. Leventhal. 1967. A technique for the microscopic study of blood vessels in living striated muscle (cremaster). *Circulation Research.* 21:823–832.
7. Majno, G., V. Gilmore, and M. Leventhal. 1967. On the mechanism of vascular leakage caused by histamine-type mediators. A microscopic study in vivo. *Circulation Research.* 21:833–847.
8. Majno, G., S.M. Shea, and M. Leventhal. 1969. Endothelial contraction induced by histamine-type mediators: an electron microscopic study. *The Journal of Cell Biology.* 42:647–672.
9. Joris, I., G. Majno, E.J. Corey, and R.A. Lewis. 1985. Leukotriene E4 (LTE4)-induced vascular leakage. Evidence of endothelial contraction. *Federation Proceedings.* 44:736 (Abstract).
10. Joris, I., G. Majno, and G.B. Ryan. 1972. Endothelial contraction in vivo: a study of the rat mesentery. *Virchows Archiv B: Cell Pathology.* 12:73–83.
11. Boswell, C.A., K.A. Ostrom, Y. Kitagawa, G. Majno, and I. Joris. 1990. Endothelium, fibroblasts and smooth muscle cells. Do they utilize the same contractile mechanism? *The FASEB Journal.* 4:A904 (Abstract).
12. Boswell, C.A., G. Majno, I. Joris, and K.A. Ostrom. 1992. Acute endothelial cell contraction in vitro: a comparison with vascular smooth muscle cells and fibroblasts. *Microvascular Research.* 43:178–191.
13. Boswell, C.A., I. Joris, and G. Majno. 1992. The concept of cellular tone: reflections on the endothelium, fibroblasts, and smooth muscle cells. *Perspectives in Biology and Medicine.* 36:79–86.
14. Joris, I., U. DeGirolami, K.A. Wortham, and G. Majno. 1982. Vascular labelling with monastral blue B. *Stain Technology.* 57:177–183.
15. Joris, I., H.F. Cuénoud, G.V. Doern, J.M. Underwood, and G. Majno. 1990. Capillary leakage in inflammation. A study by vascular labeling. *The American Journal of Pathology.* 137:1353–1363.
16. Joris, I., H.F. Cuénoud, G.V. Doern, J.M. Underwood, and G. Majno. 1992. Endothelial permeability in inflammation. The role of capillaries versus venules. In *Endothelial Cell Dysfunctions.* N. Simionescu and M. Simionescu, editors. Plenum Press, New York. 233–242.
17. Horan, K.L., S.W. Adamski, W. Ayele, J.J. Langone, and G.J. Grega. 1986. Evidence that prolonged histamine suffusions produce transient increases in vascular permeability subsequent to the formation of venular macromolecular leakage sites. Proof of the Majno-Palade hypothesis. *The American Journal of Pathology.* 123:570–576.

13B. Membrane Structure

1. Ferner, H., and W. Stoeckenius. 1950. Die Cytogenese des Inselsystems beim Menschen. *Cell and Tissue Research.* 35:147–175.
2. Höber, R. 1910. Eine Methode die elektrische Leitfähigkeit im Inneren von Zellen zu messen. *Pflügers Archiv für die gesamte Physiologie.* 133:237–253.
3. Höber, R. 1912. Ein zweites Verfahren, die Leitfähigkeit im Inneren von Zellen zu messen. *Pflügers Archiv für die gesamte Physiologie.* 148:189–221.
4. Gorter, E., and F. Grendel. 1925. On bimolecular layers of lipoids on the chromocytes of the blood. *The Journal of Experimental Medicine.* 41:439–443.
5. Danielli, J.F., and H. Davson. 1935. A contribution to the theory of permeability of thin films. *Journal of Cellular and Comparative Physiology.* 5:495–508.
6. Geren, B.B. 1954. The formation from the Schwann cell surface of myelin in the peripheral nerves of chick embryos. *Experimental Cell Research.* 7:558–562.
7. Robertson, J.D. 1987. The early days of electron microscopy of nerve tissue and membranes. *International Review of Cytology.* 100:129–201.
8. Latta, H., and J.F. Hartmann. 1950. Use of a glass edge in thin sectioning for electron microscopy. *Proceedings of the Society for Experimental Biology and Medicine.* 74:436–439.
9. Stoeckenius, W. 1959. An electron microscope study of myelin figures. *The Journal of Biophysical and Biochemical Cytology.* 5:491–500.
10. Zinder, N.D., R.C. Valentine, M. Roger, and W. Stoeckenius. 1963. F1, a rod-shaped male-specific bacteriophage that contains DNA. *Virology.* 20:638–640.
11. Gomatos, P.J., and W. Stoeckenius. 1964. Electron microscope studies on reovirus RNA. *Proceedings of the National Academy of Sciences of the United States of America.* 52:1449–1455.
12. Choppin, P.W., and W. Stoeckenius. 1964. Interactions of ether-disrupted influenza A2 virus with erythrocytes, inhibitors, and antibodies. *Virology.* 22:482–492.
13. Choppin, P.W., and W. Stoeckenius. 1964. The morphology of SV5 virus. *Virology.* 23:195–202.
14. Robertson, J.D. 1987. The early days of electron microscopy of nerve tissue and membranes. *International Review of Cytology.* 100:129–201.
15. Schulman, J.H., W. Stoeckenius, and L.M. Prince. 1959. Mechanism of formation and structure of micro emulsions by electron microscopy. *Journal of Physical Chemistry.* 63:1677–1680.

16. Luzzati, V., and F. Husson. 1962. The structure of the liquid-crystalline phases of lipid-water systems. *The Journal of Cell Biology.* 12:207–219.
17. Stoeckenius, W. 1962. Some electron microscopical observations on liquid-crystalline phases in lipid-water systems. *The Journal of Cell Biology.* 12:221–229.
18. Fleischer, S., B. Fleischer, and W. Stoeckenius. 1967. Fine structure of lipid-depleted mitochondria. *The Journal of Cell Biology.* 32:193–208.
19. Brown, A.D. 1963. The peripheral structures of Gram-negative bacteria. IV. The cation-sensitive dissolution of the cell membrane of the halophilic bacterium, *Halobacterium halobium. Biochimica et Biophysica Acta.* 75:425–435.
20. Stoeckenius, W., and R. Rowen. 1967. A morphological study of *Halobacterium halobium* and its lysis in media of low salt concentration. *The Journal of Cell Biology.* 34:365–393.
21. Blaurock, A.E., W. Stoeckenius, D. Oesterhelt, and G.L. Scherfhof. 1976. Structure of the cell envelope of *Halobacterium halobium. The Journal of Cell Biology.* 71:1–22.
22. Stoeckenius, W., and D.M. Engelman. 1969. Current models for the structure of biological membranes. *The Journal of Cell Biology.* 42:613–646.
23. Goodenough, D.A., and W. Stoeckenius. 1972. The isolation of mouse hepatocyte gap junctions. Preliminary chemical characterization and x-ray diffraction. *The Journal of Cell Biology.* 54:646–656.
24. Stoeckenius, W., and W.H. Kunau. 1968. Further characterization of particulate fractions from lysed cell envelopes of *Halobacterium halobium* and isolation of gas vacuole membranes. *The Journal of Cell Biology.* 38:337–357.
25. Oesterhelt, D., and W. Stoeckenius. 1971. Rhodopsin-like protein from the purple membrane of *Halobacterium halobium. Nature: New Biology.* 233:149–152.
26. Oesterhelt, D., and W. Stoeckenius. 1973. Functions of a new photoreceptor membrane. *Proceedings of the National Academy of Sciences of the United States of America.* 70:2853–2857.
27. Blaurock, A.E., and W. Stoeckenius. 1971. Structure of the purple membrane. *Nature: New Biology.* 233:152–155.
28. Unwin, P.N.T., and R. Henderson. 1975. Molecular structure determination by electron microscopy of unstained crystalline specimens. *Journal of Molecular Biology.* 94:425–440.
29. Henderson, R., J.M. Baldwin, T.A. Ceska, F. Zemlin, E. Beckmann, and K.H. Downing. 1990. Model for the structure of bacteriorhodopsin based on high-resolution electron cryo-microscopy. *Journal of Molecular Biology.* 213:899–929.
30. Racker, E., and W. Stoeckenius. 1974. Reconstitution of purple membrane vesicles catalyzing light-driven proton uptake and adenosine triphosphate formation. *The Journal of Biological Chemistry.* 249:662–663.

13C. Glomerular Permeability and the Discovery of Tight Junctions

1. Latta, H., and J.F. Hartmann. 1950. Use of a glass edge in thin sectioning for electron microscopy. *Proceedings of the Society for Experimental Biology and Medicine.* 74:436–439.
2. Farquhar, M.G., and J.F. Rinehart. 1954. Electron microscopic studies of the anterior pituitary gland of castrate rats. *Endocrinology.* 54:516–541.
3. Farquhar, M.G., and J.F. Rinehart. 1954. Cytologic alterations in the anterior pituitary gland following thyroidectomy: an electron microscope study. *Endocrinology.* 55:857–876.
4. Pease, D.C. 1955. Electron microscopy of the vascular bed of the kidney cortex. *The Anatomical Record.* 121:701–721.
5. Hall, B.V. 1954. Studies of the normal glomerular structure by electron microscopy. In *Proceedings of the 5th Annual Conference on the Nephrotic Syndrome, Philadelphia, 5–7 November 1953.* National Nephrosis Foundation, New York. 1–39.
6. Farquhar, M.G., R.L. Vernier, and R.A. Good. 1957. An electron microscope study of the glomerulus in nephrosis, glomerulonephritis, and lupus erythematosus. *The Journal of Experimental Medicine.* 106:649–660.
7. Farquhar, M.G., R.L. Vernier, and R.A. Good. 1957. Studies on familial nephrosis. II. Glomerular changes observed with the electron microscope. *The American Journal of Pathology.* 33:791–817.
8. Farquhar, M.G., and S.R. Wellings. 1957. Electron microscopic evidence suggesting secretory granule formation within the Golgi apparatus. *The Journal of Biophysical and Biochemical Cytology.* 3:319–322.
9. Farquhar, M.G., J. Hopper Jr., and H.D. Moon. 1959. Diabetic glomerulosclerosis: electron and light microscopic studies. *The American Journal of Pathology.* 35:721–753.
10. Palade, G.E. 1960. Transport in quanta across the endothelium of blood capillaries. *The Anatomical Record.* 136:254 (Abstract).
11. Pappenheimer, J.R. 1953. Passage of molecules through capillary walls. *Physiological Reviews.* 33:387–423.
12. Farquhar, M.G. 1991. The glomerular basement membrane—a selective macromolecular filter. In *Cell Biology of Extracellular Matrix.* Second edition. E.D. Hay, editor. Plenum Press, New York. 365–418.
13. Farquhar, M.G., and G.E. Palade. 1960. Segregation of ferritin in glomerular protein absorption droplets. *The Journal of Biophysical and Biochemical Cytology.* 7:297–304.
14. Farquhar, M.G., S.L. Wissig, and G.E. Palade. 1961. Glomerular permeability. I. Ferritin transfer across the normal glomerular capillary wall. *The Journal of Experimental Medicine.* 113:47–66.
15. Farquhar, M.G., S.L. Wissig, and G.E. Palade. 1999. Glomerular permeability. I. Ferritin transfer across the normal glomerular capillary wall. *Journal of the American Society of Nephrology.* 10:2645–2662 [reprinted from *J. Exp. Med.* 1961. 113:47–66].

16. Farquhar, M.G., and G.E. Palade. 1961. Glomerular permeability. II. Ferritin transfer across the glomerular capillary wall in nephrotic rats. *The Journal of Experimental Medicine*. 114:699–716.
17. Farquhar, M.G., and G.E. Palade. 1962. Functional evidence for the existence of a third cell type in the renal glomerulus: Phagocytosis of filtration residues by a distinctive "third" cell. *The Journal of Cell Biology*. 13:55–87.
18. Farquhar, M.G., and G.E. Palade. 1963. Junctional complexes in various epithelia. *The Journal of Cell Biology*. 17:375–412.
19. Novikoff, A.B., R. Hirsh, and N. Quintana. 1965. Golgi apparatus of rat epididymis and acid phosphatase activity. *The Journal of Cell Biology*. 27:73A.
20. Smith, R.E., and M.G. Farquhar. 1966. Lysosome function in the regulation of the secretory process in cells of the anterior pituitary gland. *The Journal of Cell Biology*. 31:319–347.
21. Bainton, D.F., and M.G. Farquhar. 1966. Origin of granules in polymorphonuclear leukocytes. Two types derived from opposite faces of the Golgi complex in developing granulocytes. *The Journal of Cell Biology*. 28:277–301.
22. Bainton, D.F., and M.G. Farquhar. 1968. Differences in enzyme content of azurophil and specific granules of polymorphonuclear leukocytes. I. Histochemical staining of bone marrow smears. *The Journal of Cell Biology*. 39:286–298.
23. Bainton, D.F., and M.G. Farquhar. 1968. Differences in enzyme content of azurophil and specific granules of polymorphonuclear leukocytes. II. Cytochemistry and electron microscopy of bone marrow cells. *The Journal of Cell Biology*. 39:299–317.
24. Terry, R.W., D.F. Bainton, and M.G. Farquhar. 1969. Formation and structure of specific granules in basophilic leukocytes of the guinea pig. *Laboratory Investigation*. 21:65–76.
25. Bainton, D.F., and M.G. Farquhar. 1970. Segregation and packaging of granule enzymes in eosinophilic leukocytes. *The Journal of Cell Biology*. 45:54–73.
26. Friend, D.S., and M.G. Farquhar. 1967. Functions of coated vesicles during protein absorption in the rat vas deferens. *The Journal of Cell Biology*. 35:357–376.
27. Hopkins, C.R., and M.G. Farquhar. 1973. Hormone secretion by cells dissociated from rat anterior pituitaries. *The Journal of Cell Biology*. 59:276–303.
28. Farquhar, M.G. 1978. Recovery of surface membrane in anterior pituitary cells. Variations in traffic detected with anionic and cationic ferritin. *The Journal of Cell Biology*. 77:R35–R42.
29. Graham, R.C., Jr., and M.J. Karnovsky. 1966. Glomerular permeability. Ultrastructural cytochemical studies using peroxidases as protein tracers. *The Journal of Experimental Medicine*. 124:1123–1134.
30. Venkatachalam, M.A., M.J. Karnovsky, H.D. Fahimi, and R.S. Cotran. 1970. An ultrastructural study of glomerular permeability using catalase and peroxidase as tracer proteins. *The Journal of Experimental Medicine*. 132:1153–1167.
31. Venkatachalam, M.A., R.S. Cotran, and M.J. Karnovsky. 1970. An ultrastructural study of glomerular permeability in aminonucleoside nephrosis using catalase as a tracer protein. *The Journal of Experimental Medicine*. 132:1168–1180.
32. Simionescu, N., and G.E. Palade. 1971. Dextrans and glycogens as particulate tracers for studying capillary permeability. *The Journal of Cell Biology*. 50:616–624.
33. Caulfield, J.P., and M.G. Farquhar. 1974. The permeability of glomerular capillaries to graded dextrans. Identification of the basement membrane as the primary filtration barrier. *The Journal of Cell Biology*. 63:883–903.
34. Caulfield, J.P., and M.G. Farquhar. 1975. The permeability of glomerular capillaries of aminonucleoside nephrotic rats to graded dextrans. *The Journal of Experimental Medicine*. 142:61–83.
35. Ehrenreich, J.H., J.J.M. Bergeron, P. Siekevitz, and G.E. Palade. 1973. Golgi fractions prepared from rat liver homogenates. I. Isolation procedure and morphological characterization. *The Journal of Cell Biology*. 59:45–72.
36. Bergeron, J.J.M., J.H. Ehrenreich, P. Siekevitz, and G.E. Palade. 1973. Golgi fractions prepared from rat liver homogenates. II. Biochemical characterization. *The Journal of Cell Biology*. 59:73–88.
37. Farquhar, M.G., J.J.M. Bergeron, and G.E. Palade. 1974. Cytochemistry of Golgi fractions prepared from rat liver. *The Journal of Cell Biology*. 60:8–25.
38. Shio, H., M.G. Farquhar, and C. de Duve. 1974. Lysosomes of the arterial wall. IV. Cytochemical localization of acid phosphatase and catalase in smooth muscle cells and foam cells from rabbit atheromatous aorta. *The American Journal of Pathology*. 76:1–16.

13D. Membrane Biogenesis in Chloroplasts, Endoplasmic Reticulum, and Mitochondria

1. Wolken, J.J., and G.E. Palade. 1953. An electron microscope study of two flagellates, chloroplast structure and variation. *Annals of the New York Academy of Sciences*. 56:873–889.
2. Sager, R., and G.E. Palade. 1954. Chloroplast structure in green and yellow strains of *Chlamydomonas*. *Experimental Cell Research*. 7:584–588.
3. Sager, R., and G.E. Palade. 1957. Structure and development of the chloroplast in *Chlamydomonas*. I. The normal green cell. *The Journal of Biophysical and Biochemical Cytology*. 3:463–488.
4. Ohad, I., P. Siekevitz, and G.E. Palade. 1967. Biogenesis of chloroplast membranes. I. Plastid dedifferentiation in a dark-grown algal mutant (*Chlamydomonas reinhardi*). *The Journal of Cell Biology*. 35:521–552.
5. Ohad, I., P. Siekevitz, and G.E. Palade. 1967. Biogenesis of chloroplast membranes. II. Plastid differentiation during greening of a dark-grown algal mutant (*Chlamydomonas reinhardi*). *The Journal of Cell Biology*. 35:553–584.
6. Hoober, J.K., P. Siekevitz, and G.E. Palade. 1969. Formation of chloroplast membranes in *Chlamydomonas reinhardi* y-1. Effects of inhibitors of protein synthesis. *The Journal of Biological Chemistry*. 244:2621–2631.

7. Schor, S., P. Siekevitz, and G.E. Palade. 1970. Cyclic changes in thylakoid membranes of synchronized *Chlamydomonas reinhardi. Proceedings of the National Academy of Sciences of the United States of America.* 66:174–180.
8. De Petrocellis, B., P. Siekevitz, and G.E. Palade. 1970. Changes in chemical composition of thylakoid membranes during greening of the y-1 mutant of *Chlamydomonas reinhardi. The Journal of Cell Biology.* 44:618–634.
9. Ernster, L., P. Siekevitz, and G.E. Palade. 1962. Enzyme-structure relationships in the endoplasmic reticulum of rat liver: A morphological and biochemical study. *The Journal of Cell Biology.* 15:541–562.
10. Dallner, G., P. Siekevitz, and G.E. Palade. 1966. Biogenesis of endoplasmic reticulum membranes. I. Structural and chemical differentiation in developing rat hepatocyte. *The Journal of Cell Biology.* 30:73–96.
11. Dallner, G., P. Siekevitz, and G.E. Palade. 1966. Biogenesis of endoplasmic reticulum membranes. II. Synthesis of constitutive microsomal enzymes in developing rat hepatocyte. *The Journal of Cell Biology.* 30:97–117.
12. Omura, T., P. Siekevitz, and G.E. Palade. 1967. Turnover of constituents of the endoplasmic reticulum membranes of rat hepatocytes. *The Journal of Biological Chemistry.* 242:2389–2396.
13. Bock, K.W., and P. Siekevitz. 1970. Turnover of heme and protein moieties of rat liver microsomal cytochrome b5. *Biochemical and Biophysical Research Communications.* 41:374–380.
14. Bock, K.W., P. Siekevitz, and G.E. Palade. 1971. Localization and turnover studies of membrane nicotinamide adenine dinucleotide glycohydrolase in rat liver. *The Journal of Biological Chemistry.* 246:188–195.
15. Kuriyama, Y., T. Omura, P. Siekevitz, and G.E. Palade. 1969. Effects of phenobarbital on the synthesis and degradation of the protein components of rat liver microsomal membranes. *The Journal of Biological Chemistry.* 244:2017–2026.
16. Leskes, A., P. Siekevitz, and G.E. Palade. 1971. Differentiation of endoplasmic reticulum in hepatocytes. I. Glucose-6-phosphatase distribution in situ. *The Journal of Cell Biology.* 49:264–287.
17. Leskes, A., P. Siekevitz, and G.E. Palade. 1971. Differentiation of endoplasmic reticulum in hepatocytes. II. Glucose-6-phosphatase in rough microsomes. *The Journal of Cell Biology.* 49:288–302.
18. Siekevitz, P., G.E. Palade, G. Dallner, I. Ohad, and T. Omura. 1967. The biogenesis of intracellular membranes. In *Organizational Biosynthesis; A Symposium.* H.J. Vogel, J.O. Lampen, and V. Bryson, editors. Academic Press, New York. 331–362.
19. Siekevitz, P. 1972. On membranes. In *Horizons in Bioenergetics.* A. San Pietro and A. Gest, editors. Academic Press, New York. 181–211.
20. Siekevitz, P. 1972. Biological membranes: the dynamics of their organization. *Annual Review of Physiology.* 34:117–140.
21. Luck, D.J.L. 1962. Glycogen synthesis from uridine diphosphate glucose: the relation of the enzyme to glycogen deposits and membranes in the liver cell. PhD thesis. The Rockefeller University, New York.
22. von Kölliker, A. 1888. Zur Kenntnis der quergestreiften Muskelfasern. *Zeitschrift für wissenschaftliche Zoologie.* 47:689–710.
23. Claude, A., and E.F. Fullam. 1945. An electron microscope study of isolated mitochondria. Method and preliminary results. *The Journal of Experimental Medicine.* 81:51–62.
24. Palade, G.E. 1952. The fine structure of mitochondria. *The Anatomical Record.* 114:427–451.
25. Sjöstrand, F.S. 1953. Electron microscopy of mitochondria and cytoplasmic double membranes. *Nature.* 171:30–32.
26. Harvey, E.B. 1946. Structure and development of the clear quarter of the *Arbacia punctulata* egg. *The Journal of Experimental Zoology.* 102:253–275.
27. Zollinger, H.U. 1950. Mitochondria; study with phase contrast microscope. [In French.] *Revue d'hématologie (Paris).* 5:696–745.
28. Ephrussi, B. 1952. The interplay of heredity and environment in the synthesis of respiratory enzymes in yeast. In *Harvey Lectures.* Series 46. Charles C. Thomas Publisher, Springfield, Illinois. 45–67.
29. Mitchell, M.B., and H.K. Mitchell. 1952. A case of "maternal" inheritance in *Neurospora crassa. Proceedings of the National Academy of Sciences of the United States of America.* 38:442–449.
30. Ernster, L., and G. Schatz. 1981. Mitochondria: a historical review. *The Journal of Cell Biology.* 91:227s–255s.
31. Caro, L.G., R.P. Van Tubergen, and J.A. Kolb. 1962. High-resolution autoradiography. I. Methods. *The Journal of Cell Biology.* 15:173–188.
32. Luck, D.J.L. 1963. Genesis of mitochondria in *Neurospora crassa. Proceedings of the National Academy of Sciences of the United States of America.* 49:233–240.
33. Luck, D.J.L. 1963. Formation of mitochondria in *Neurospora crassa.* A quantitative radioautographic study. *The Journal of Cell Biology.* 16:483–499.
34. Luck, D.J.L. 1965. Formation of mitochondria in *Neurospora crassa.* A study based on mitochondrial density changes. *The Journal of Cell Biology.* 24:461–470.
35. Ephrussi, B., and H. Hottinguer. 1951. On an unstable cell state in yeast. *Cold Spring Harbor Symposia on Quantitative Biology.* 16:75–85.
36. Mitchell, M.B., H.K. Mitchell, and A. Tissieres. 1953. Mendelian and non-Mendelian factors affecting the cytochrome system in *Neurospora crassa. Proceedings of the National Academy of Sciences of the United States of America.* 39:606–613.
37. Nass, M.M.K., and S. Nass. 1963. Intramitochondrial fibers with DNA characteristics. I. Fixation and electron staining reactions. *The Journal of Cell Biology.* 19:593–611.
38. Nass, S., and M.M.K. Nass. 1963. Intramitochondrial fibers with DNA characteristics. II. Enzymatic and other hydrolytic treatments. *The Journal of Cell Biology.* 19:613–629.

39. Schatz, G., E. Haslbrunner, and H. Tuppy. 1964. Deoxyribonucleic acid associated with yeast mitochondria. *Biochemical and Biophysical Research Communications*. 15:127–132.
40. Luck, D.J.L., and E. Reich. 1964. DNA in mitochondria of *Neurospora crassa*. *Proceedings of the National Academy of Sciences of the United States of America*. 52:931–938.
41. Mounolou, J.C., H. Jakob, and P.P. Slonimski. 1966. Mitochondrial DNA from yeast "petite" mutants: specific changes in buoyant density corresponding to different cytoplasmic mutations. *Biochemical and Biophysical Research Communications*. 24:218–224.
42. Schneider, W.C., and E.L. Kuff. 1965. The isolation and some properties of rat liver mitochondrial deoxyribonucleic acid. *Proceedings of the National Academy of Sciences of the United States of America*. 54:1650–1658.
43. Meselson, M., and F.W. Stahl. 1958. The replication of DNA in *Escherichia coli*. *Proceedings of the National Academy of Sciences of the United States of America*. 44:671–682.
44. Reich, E., and D.J.L. Luck. 1966. Replication and inheritance of mitochondrial DNA. *Proceedings of the National Academy of Sciences of the United States of America*. 55:1600–1608.
45. Rifkin, M.R., D.D. Wood, and D.J.L. Luck. 1967. Ribosomal RNA and ribosomes from mitochondria of *Neurospora crassa*. *Proceedings of the National Academy of Sciences of the United States of America*. 58:1025–1032.
46. Küntzel, H., and H. Noll. 1967. Mitochondrial and cytoplasmic polysomes from *Neurospora crassa*. *Nature*. 215:1340–1345.
47. Dure, L.S., J.L. Epler, and W.E. Barnett. 1967. Sedimentation properties of mitochondrial and cytoplasmic ribosomal RNA's from *Neurospora*. *Proceedings of the National Academy of Sciences of the United States of America*. 58:1883–1887.
48. Sagan, L.M. 1967. On the origin of mitosing cells. *Journal of Theoretical Biology*. 14:225–274.
49. Wallin, I.E. 1922. On the nature of mitochondria. III. The demonstration of mitochondria by bacteriological methods. IV. A comparative study of the morphogenesis of root-nodule bacteria and chloroplasts. *The American Journal of Anatomy*. 30:451–471.
50. Wallin, I.E. 1923. The mitochondria problem. *American Naturalist*. 57:255–261.
51. Cowdry, E.V. 1924. The value of the study of mitochondria in cellular pathology. *American Naturalist*. 58:97–114.

13E. Vascular Endothelium, More Complex than Previously Thought

1. Marchesi, V.T., and G.E. Palade. 1967. The localization of Mg-Na-K-activated adenosine triphosphatase on red cell ghost membranes. *The Journal of Cell Biology*. 35:385–404.
2. Bruns, R.R., and G.E. Palade. 1968. Studies on blood capillaries. II. Transport of ferritin molecules across the wall of muscle capillaries. *The Journal of Cell Biology*. 37:277–299.
3. Simionescu, N., and G.E. Palade. 1971. Dextrans and glycogens as particulate tracers for studying capillary permeability. *The Journal of Cell Biology*. 50:616–624.
4. Simionescu, N., M. Simionescu, and G.E. Palade. 1972. Permeability of intestinal capillaries. Pathway followed by dextrans and glycogens. *The Journal of Cell Biology*. 53:365–392.
5. Simionescu, N., M. Simionescu, and G.E. Palade. 1973. Permeability of muscle capillaries to exogenous myoglobin. *The Journal of Cell Biology*. 57:424–452.
6. Simionescu, M., N. Simionescu, and G.E. Palade. 1974. Morphometric data on the endothelium of blood capillaries. *The Journal of Cell Biology*. 60:128–152.
7. Simionescu, N., M. Siminoescu, and G.E. Palade. 1975. Permeability of muscle capillaries to small heme-peptides. Evidence for the existence of patent transendothelial channels. *The Journal of Cell Biology*. 64:586–607.
8. Simionescu, M., N. Simionescu, and G.E. Palade. 1976. Characteristic endothelial junctions in different segments of the vascular system. *Thrombosis Research*. 8(Supplement):247–256.
9. Simionescu, N., M. Simionescu, and G.E. Palade. 1976. Structural-functional correlates in the transendothelial exchange of water-soluble macromolecules. *Thrombosis Research*. 8(Supplement):257–269.
10. Simionescu, M., N. Simionescu, and G.E. Palade. 1975. Segmental differentiations of cell junctions in the vascular endothelium. The microvasculature. *The Journal of Cell Biology*. 67:863–885.
11. Simionescu, M., N. Simionescu, and G.E. Palade. 1976. Segmental differentiations of cell junctions in the vascular endothelium. Arteries and veins. *The Journal of Cell Biology*. 68:705–723.
12. Simionescu, N., M. Simionescu, and G.E. Palade. 1978. Structural basis of permeability in sequential segments of the microvasculature of the diaphragm. I. Bipolar microvascular fields. *Microvascular Research*. 15:1–16.
13. Simionescu, N., M. Simionescu, and G.E. Palade. 1978. Structural basis of permeability in sequential segments of the microvasculature of the diaphragm. II. Pathways followed by microperoxidase across the endothelium. *Microvascular Research*. 15:17–36.
14. Simionescu, N., M. Simionescu, and G.E. Palade. 1978. Open junctions in the endothelium of the postcapillary venules of the diaphragm. *The Journal of Cell Biology*. 79:27–44.
15. Majno, G., and G.E. Palade. 1961. Studies on inflammation. 1. The effect of histamine and serotonin on vascular permeability: an electron microscopic study. *The Journal of Biophysical and Biochemical Cytology*. 11:571–605.
16. Simionescu, M., N. Simionescu, and G.E. Palade. 1982. Biochemically differentiated microdomains of the cell surface of capillary endothelium. *Annals of the New York Academy of Sciences*. 401:9–24.

17. Simionescu, N., M. Simionescu, and G.E. Palade. 1981. Differentiated microdomains on the luminal surface of the capillary endothelium. I. Preferential distribution of anionic sites. *The Journal of Cell Biology.* 90:605–613.
18. Simionescu, M., N. Simionescu, J.E. Silbert, and G.E. Palade. 1981. Differentiated microdomains on the luminal surface of the capillary endothelium. II. Partial characterization of their anionic sites. *The Journal of Cell Biology.* 90:614–621.
19. Simionescu, M., N. Simionescu, and G.E. Palade. 1982. Differentiated microdomains on the luminal surface of capillary endothelium: distribution of lectin receptors. *The Journal of Cell Biology.* 94:406–413.
20. Simionescu, M., N. Simionescu, and G.E. Palade. 1982. Preferential distribution of anionic sites on the basement membrane and the abluminal aspect of the endothelium in fenestrated capillaries. *The Journal of Cell Biology.* 95:425–434.
21. Simionescu, M., N. Simionescu, F. Santoro, and G.E. Palade. 1985. Differentiated microdomains of the luminal plasmalemma of murine muscle capillaries: segmental variations in young and old animals. *The Journal of Cell Biology.* 100:1396–1407.
22. Simionescu, N. 1979. The microvascular endothelium: segmental differentiations, transcytosis, selective distribution of anionic sites. *Advances in Inflammation Research.* 1:61–70.
23. Simionescu, M., and N. Simionescu. 1991. Endothelial transport of macromolecules: transcytosis and endocytosis. A look from cell biology. *Cell Biology Reviews.* 25:5–78.

Chapter 14

Nuclear Proteins: Control of Gene Structure and Function

1. Bensley, R.R. 1938. Plasmosin. The gel- and fiber-forming constituent of the protoplasm of the hepatic cell. *The Anatomical Record.* 72:351–369.
2. Mirsky, A.E., and A.W. Pollister. 1942. Nucleoproteins of cell nuclei. *Proceedings of the National Academy of Sciences of the United States of America.* 28:344–352.
3. Sevag, M.G. 1934. Eine neue physikalische Enteiweissungsmethode zur Darstellung biologisch wirksamer Substanzen. Isolierung von Kohlenhydraten aus Huehnereweiss und Pneumococcen. *Biochemische Zeitschrift.* 273:419–429.
4. Levene, P.A. 1919. The structure of yeast nucleic acid. *The Journal of Biological Chemistry.* 40:415–424.
5. Mirsky, A.E. 1941. Some observations on protein folding and unfolding. *Cold Spring Harbor Symposia on Quantitative Biology.* 9:278–282.
6. Murphy, J.B. 1943–1944. RUA, BSD Reports. Volume 32, pages 69–70.
7. Claude, A. 1943. Distribution of nucleic acids in the cell and the morphological constitution of the cytoplasm. *Biological Symposia.* 10:111–129.
8. Mirsky, A.E., and A.W. Pollister. 1943. Fibrous nucleoproteins of chromatin. *Biological Symposia.* 10:247–260.
9. Avery, O.T., C.M. MacLeod, and M. McCarty. 1944. Studies on the chemical nature of the substance inducing transformation of pneumococcal types. Induction of transformation by a desoxyribonucleic acid fraction isolated from pneumococcus type III. *The Journal of Experimental Medicine.* 79:137–158.
10. McCarty, M. 1985. *The Transforming Principle: Discovering that Genes are Made of DNA.* Norton, New York. 147.
11. McCarty, M., and O.T. Avery. 1946. Studies on the chemical nature of the substance inducing transformation of pneumococcal types. II. Effect of desoxyribonuclease on the biological activity of the transforming substance. *The Journal of Experimental Medicine.* 83:89–96.
12. Mirsky, A.E., and H. Ris. 1949. Variable and constant components of chromosomes. *Nature.* 163:666–667.
13. Boivin, A., R. Vendrely, and C. Vendrely. 1948. L'acide desoxyribonucléique du noyau cellulaire, dépositaire des caractères héréditaires; arguments d'ordre analytique. *Comptes Rendus Hebdomadaires des Séances de l'Académie des Sciences, Paris.* 226:1061–1063.
14. Flemming, W. 1882. *Zellsubstanz, Kern, und Zelltheilung.* F.C.W. Vogel, Leipzig.
15. Mirsky, A.E. 1947. Chemical properties of isolated chromosomes. *Cold Spring Harbor Symposia on Quantitative Biology.* 12:143–146.
16. Mirsky, A.E., and H. Ris. 1947. The chemical composition of isolated chromosomes. *The Journal of General Physiology.* 31:7–18.
17. Lamb, W.G.P. 1949. Chromatin threads from cell nuclei. *Nature.* 164:109.
18. Mirsky, A.E., and H. Ris. 1951. The composition and structure of isolated chromosomes. *The Journal of General Physiology.* 34:475–492.
19. Ris, H., and A.E. Mirsky. 1951. Isolated chromosomes. *Experimental Cell Research.* 2:263–267.
20. Pfeiffer, H.H. 1950. Experiments with chromosomes isolated from intermitotic nuclei. *Experientia.* 6:334–335.
21. Ris, H., and A.E. Mirsky. 1949. The state of the chromosomes in the interphase nucleus. *The Journal of General Physiology.* 32:489–502.
22. Mirsky, A.E. 1952. The chemical composition of chromosomes. In *Harvey Lectures.* Series 46. Charles C. Thomas Publisher, Springfield, Illinois. 98–115.
23. Kossel, A. 1884. Über einen peptonartigen Bestandteil des Zellkerns. *Zeitschrift für Physiologische Chemie.* 8:511–515.

24. Miescher, F. 1871. Über die chemische Zusammensetzung der Eiterzellen. *Hoppe-Seyler's medizinisch-chemischen Untersuchungen.* 4:441–460.
25. Kossel, A. 1928. *The Protamines and Histones.* Longmans, Green, and Co., New York.
26. Daly, M.M., A.E. Mirsky, and H. Ris. 1951. The amino acid composition and some properties of histones. *The Journal of General Physiology.* 34:439–450.
27. Daly, M.M., and A.E. Mirsky. 1955. Histones with high lysine content. *The Journal of General Physiology.* 38:405–413.
28. Felix, K. 1921. Über Eiweissderivate basischer Natur. *Zeitschrift für Physiologische Chemie.* 116:150–165.
29. Kossel, A., and E.G. Schenck. 1928. Untersuchungen über die basischen Eiweissstoffe, ein Beitrag zu ihrer Entwicklungsgeschichte. *Zeitschrift für Physiologische Chemie.* 173:278–308.
30. Murray, K. 1964. Heterogeneity of the histones. In *The Nucleohistones.* J. Bonner and P. Ts'o, editors. Holden-Day, San Francisco. 21–35.
31. Behrens, M. 1932. Untersuchungen an isolierten Zell- und Gewebebestandteilen. I. Mitteilung: Isolierung von Zellkernen des Kalbsherzmuskels. *Zeitschrift für Physiologische Chemie.* 209:59–74.
32. Allfrey, V.G., H. Stern, A.E. Mirsky, and H. Saetren. 1952. The isolation of cell nuclei in non-aqueous media. *The Journal of General Physiology.* 35:529–554.
33. Stern, H., and A.E. Mirsky. 1953. Soluble enzymes of nuclei isolated in sucrose and nonaqueous media; a comparative study. *The Journal of General Physiology.* 37:177–187.
34. Schneider, R.M., and M.L. Petermann. 1950. Nuclei from normal and leukemic mouse spleen. I. The isolation of nuclei in neutral medium. *Cancer Research.* 10:751–754.
35. Allfrey, V.G. 1954. Amino acid incorporation by isolated thymus nuclei. I. The role of desoxyribonucleic acid in protein synthesis. *Proceedings of the National Academy of Sciences of the United States of America.* 40:881–885.
36. Siekevitz, P. 1952. Uptake of radioactive alanine in vitro into the proteins of rat liver fractions. *The Journal of Biological Chemistry.* 195:549–565.
37. Allfrey, V.G., A.E. Mirsky, and S. Osawa. 1955. Protein synthesis in isolated cell nuclei. *Nature.* 176:1042–1049.
38. Allfrey, V.G., and A.E. Mirsky. 1955. On the supposed contamination of thymus nuclear fractions by whole cells. *Science.* 121:879–880.
39. Allfrey, V.G., V.C. Littau, and A.E. Mirsky. 1964. Methods for the purification of thymus nuclei and their application to studies of nuclear protein synthesis. *The Journal of Cell Biology.* 21:213–231.
40. Holter, H., and K.M. Møller. 1958. A substance for aqueous density gradients. *Experimental Cell Research.* 15:631–632.
41. Allfrey, V.G., A.E. Mirsky, and S. Osawa. 1957. The nucleus and protein synthesis. In *A Symposium on the Chemical Basis of Heredity.* W.D. McElroy and B. Glass, editors. Johns Hopkins Press, Baltimore, Maryland. 200–231.
42. Mirsky, A.E., S. Osawa, and V.G. Allfrey. 1956. The nucleus as a site of biochemical activity. *Cold Spring Harbor Symposia on Quantitative Biology.* 21:49–73.
43. Frenster, J.H., V.G. Allfrey, and A.E. Mirsky. 1960. Metabolism and morphology of ribonucleoprotein particles from the cell nucleus of lymphocytes. *Proceedings of the National Academy of Sciences of the United States of America.* 46:432–444.
44. Frenster, J.H., V.G. Allfrey, and A.E. Mirsky. 1961. *In vitro* incorporation of amino acids into the proteins of isolated nuclear ribosomes. *Biochimica et Biophysica Acta.* 47:130–137.
45. Allfrey, V.G., and A.E. Mirsky. 1961. Amino acid transport into the cell nucleus and reactions governing nuclear protein synthesis. In *Protein Biosynthesis.* R.J.C. Harris, editor. Academic Press, New York. 49–81.
46. Pogo, A.O., B.G.T. Pogo, V.C. Littau, V.G. Allfrey, A.E. Mirsky, and M.G. Hamilton. 1962. The purification and properties of ribosomes from the thymus nucleus. *Biochimica et Biophysica Acta.* 55:849–864.
47. Petermann, M.L., and M.G. Hamilton. 1955. A stabilizing factor for cytoplasmic nucleoproteins. *The Journal of Biophysical and Biochemical Cytology.* 1:469–472.
48. Britten, R.J., and R.B. Roberts. 1960. High-resolution density gradient sedimentation analysis. *Science.* 131:32–33.
49. Littlefield, J.W., E.B. Keller, J. Gross, and P.C. Zamecnik. 1955. Studies on cytoplasmic ribonucleoprotein particles from the liver of the rat. *The Journal of Biological Chemistry.* 217:111–123.
50. Petermann, M.L., and M.G. Hamilton. 1957. The purification and properties of cytoplasmic ribonucleoprotein from rat liver. *The Journal of Biological Chemistry.* 224:725–736.
51. McEwen, B.S., V.G. Allfrey, and A.E. Mirsky. 1963. Studies on energy-yielding reactions in thymus nuclei. I. Comparison of nuclear and mitochondrial phosphorylation. *The Journal of Biological Chemistry.* 238:758–766.
52. McEwen, B.S., V.G. Allfrey, and A.E. Mirsky. 1963. Studies on energy-yielding reactions in thymus nuclei. II. Pathways of aerobic carbohydrate catabolism. *The Journal of Biological Chemistry.* 238:2571–2578.
53. McEwen, B.S., V.G. Allfrey, and A.E. Mirsky. 1963. Studies on energy-yielding reactions in thymus nuclei. III. Participation of glycolysis and the citric acid cycle in nuclear adenosine triphosphate synthesis. *The Journal of Biological Chemistry.* 238:2579–2586.
54. McEwen, B.S., V.G. Allfrey, and A.E. Mirsky. 1964. Dependence of RNA synthesis in isolated thymus nuclei on glycolysis, oxidative carbohydrate catabolism and a type of "oxidative phosphorylation." *Biochimica et Biophysica Acta.* 91:23–28.
55. Allfrey, V.G. 1970. Biosynthetic reactions in the nucleus. In *Aspects of Protein Biosynthesis.* C.B. Anfinsen Jr., editor. Academic Press, New York. 247–365.
56. Goidl, J.A., and W.R. Allen. 1978. Does protein synthesis occur within the nucleus? *Trends in Biochemical Sciences.* 3:N225–N228.

57. Allfrey, V.G., and A.E. Mirsky. 1964. Role of histone in nuclear function. In *The Nucleohistones*. J. Bonner and P. Ts'o, editors. Holden-Day, San Francisco. 267–288.
58. Phillips, D.M.P. 1963. The presence of acetyl groups of histones. *The Biochemical Journal*. 87:258–263.
59. Murray, K. 1964. The occurrence of epsilon-N-methyl lysine in histones. *Biochemistry*. 3:10–15.
60. Allfrey, V.G., R. Faulkner, and A.E. Mirsky. 1964. Acetylation and methylation of histones and their possible role in the regulation of RNA synthesis. *Proceedings of the National Academy of Sciences of the United States of America*. 51:786–794.
61. Nohara, H., T. Takahashi, and K. Ogata. 1966. Acetylation of histones by pigeon liver enzymes. *Biochimica et Biophysica Acta*. 127:282–284.
62. Kim, S., and W.K. Paik. 1965. Studies on the origin of epsilon-N-methyl-L-lysine in protein. *The Journal of Biological Chemistry*. 240:4629–4634.
63. Comb, D.G., N. Sarkar, and C.J. Pinzino. 1966. The methylation of lysine residues in protein. *The Journal of Biological Chemistry*. 241:1857–1862.
64. Pogo, B.G.T., V.G. Allfrey, and A.E. Mirsky. 1966. RNA synthesis and histone acetylation during the course of gene activation in lymphocytes. *Proceedings of the National Academy of Sciences of the United States of America*. 55:805–812.
65. Bucher, N.L.R. 1963. Regeneration of mammalian liver. *International Review of Cytology*. 15:245–300.
66. Pogo, A.O., V.G. Allfrey, and A.E. Mirsky. 1966. Evidence for increased DNA template activity in regenerating liver nuclei. *Proceedings of the National Academy of Sciences of the United States of America*. 56:550–557.
67. Pogo, A.O., V.C. Littau, V.G. Allfrey, and A.E. Mirsky. 1967. Modification of ribonucleic acid synthesis in nuclei isolated from normal and regenerating liver: some effects of salt and specific divalent cations. *Proceedings of the National Academy of Sciences of the United States of America*. 57:743–750.
68. Perry, R.P., P.R. Srinivasan, and D.E. Kelley. 1964. Hybridization of rapidly labeled nuclear ribonucleic acids. *Science*. 145:504–507.
69. Pogo, B.G.T., A.O. Pogo, V.G. Allfrey, and A.E. Mirsky. 1968. Changing patterns of histone acetylation and RNA synthesis in regeneration of the liver. *Proceedings of the National Academy of Sciences of the United States of America*. 59:1337–1344.
70. Vidali, G., E.L. Gershey, and V.G. Allfrey. 1968. Chemical studies of histone acetylation. The distribution of epsilon-N-acetyllysine in calf thymus histones. *The Journal of Biological Chemistry*. 243:6361–6366.
71. Phillips, D.M.P., and E.W. Johns. 1965. A fractionation of the histones of group F2a from calf thymus. *The Biochemical Journal*. 94:127–130.
72. DeLange, R.J., D.M. Fambrough, E.L. Smith, and J. Bonner. 1969. Calf and pea histone IV. II. The complete amino acid sequence of calf thymus histone IV; presence of epsilon-N-acetyllysine. *The Journal of Biological Chemistry*. 244:319–334.
73. Gershey, E.L., G. Vidali, and V.G. Allfrey. 1968. Chemical studies of histone acetylation. The occurrence of epsilon-N-acetyllysine in the f2a1 histone. *The Journal of Biological Chemistry*. 243:5018–5022.
74. Vidali, G., E.L. Gershey, and V.G. Allfrey. 1968. Chemical studies of histone acetylation. The distribution of epsilon-N-acetyllysine in calf thymus histones. *The Journal of Biological Chemistry*. 243:6361–6366.
75. Vidali, G., L.C. Boffa, and V.G. Allfrey. 1972. Properties of an acidic histone-binding protein fraction from cell nuclei. Selective precipitation and deacetylation of histones F2A1 and F3. *The Journal of Biological Chemistry*. 247:7365–7373.
76. Inoue, A., and D. Fujimoto. 1970. Histone deacetylase from calf thymus. *Biochimica et Biophysica Acta*. 220:307–316.
77. Krieger, D.E., R. Levine, R.B. Merrifield, G. Vidali, and V.G. Allfrey. 1974. Chemical studies of histone acetylation. Substrate specificity of a histone deacetylase from calf thymus nuclei. *The Journal of Biological Chemistry*. 249:332–334.
78. Frenster, J.H., V.G. Allfrey, and A.E. Mirsky. 1963. Repressed and active chromatin isolated from interphase lymphocytes. *Proceedings of the National Academy of Sciences of the United States of America*. 50:1026–1032.
79. Allfrey, V.G. 1970. Changes in chromosomal proteins at times of gene activation. *Federation Proceedings*. 29:1447–1460.
80. Killander, D., and R. Rigler. 1969. Activation of deoxyribonucleoprotein in human leucocytes stimulated by phytohemagglutinin. I. Kinetics of the binding of acridine orange to deoxyribonucleoprotein. *Experimental Cell Research*. 54:163–170.
81. Pogo, B.G.T., V.G. Allfrey, and A.E. Mirsky. 1967. The effect of phytohemagglutinin on ribonucleic acid synthesis and histone acetylation in equine leukocytes. *The Journal of Cell Biology*. 35:477–482.
82. Allfrey, V.G., B.G.T. Pogo, V.C. Littau, E.L. Gershey, and A.E. Mirsky. 1968. Histone acetylation in insect chromosomes. *Science*. 159:314–316.
83. Mirsky, A.E., C.J. Burdick, E.H. Davidson, and V.C. Littau. 1968. The role of lysine-rich histone in the maintenance of chromatin structure in metaphase chromosomes. *Proceedings of the National Academy of Sciences of the United States of America*. 61:592–597.
84. Mirsky, A.E. 1971. The structure of chromatin. *Proceedings of the National Academy of Sciences of the United States of America*. 68:2945–2948.
85. Silverman, B., and A.E. Mirsky. 1973. Addition of histones to histone-depleted nuclei: effect on template activity toward DNA and RNA polymerases. *Proceedings of the National Academy of Sciences of the United States of America*. 70:2637–2641.
86. Kleinsmith, L.J., V.G. Allfrey, and A.E. Mirsky. 1966. Phosphoprotein metabolism in isolated lymphocyte nuclei. *Proceedings of the National Academy of Sciences of the United States of America*. 55:1182–1189.

87. Ord, M.G., and L.A. Stocken. 1966. Metabolic properties of histones from rat liver and thymus gland. *The Biochemical Journal.* 98:888–897.
88. Davidson, J.N., S.C. Frazer, and W.C. Hutchison. 1951. Phosphorus compounds in the cell. I. Protein-bound phosphorus fractions studied with the aid of radioactive phosphorus. *The Biochemical Journal.* 49:311–321.
89. Johnson, R.M., and S. Albert. 1953. Incorporation of P32 into the phosphoprotein fraction of mammalian tissue. *The Journal of Biological Chemistry.* 200:335–344.
90. Kleinsmith, L.J., and V.G. Allfrey. 1969. Nuclear phosphoproteins. I. Isolation and characterization of a phosphoprotein fraction from calf thymus nuclei. *Biochimica et Biophysica Acta.* 175:123–135.
91. Lipmann, F., and P.A. Levene. 1932. Serine phosphoric acid obtained on hydrolysis of vitellinic acid. *The Journal of Biological Chemistry.* 98:109–114.
92. Rabinowitz, M., and F. Lipmann. 1960. Reversible phosphate transfer between yolk phosphoprotein and adenosine triphosphate. *The Journal of Biological Chemistry.* 235:1043–1050.
93. Kleinsmith, L.J., V.G. Allfrey, and A.E. Mirsky. 1966. Phosphorylation of nuclear protein early in the course of gene activation in lymphocytes. *Science.* 154:780–781.
94. Langan, T.A. 1968. Phosphorylation of protein in cell nucleus. In *Regulatory Mechanisms for Protein Synthesis in Mammalian Cells.* A. San Pietro, M.R. Lamborg, and F.T. Kenney, editors. Academic Press, New York. 101–118.
95. Langan, T.A., and L.K. Smith. 1967. Phosphorylation of histones and protamines by a specific protein kinase from liver. *Federation Proceedings.* 26:603–605.
96. Allfrey, V.G. 1971. Functions and metabolic aspects of DNA-associated proteins. In *Histones and Nucleohistones.* D.M.P. Phillips, editor. Plenum Press, London. 241–294.
97. Kamiyama, M., B. Dastugue, N. Defer, and J. Kruh. 1972. Liver chromatin non-histone proteins. Partial fractionation and mechanism of action on RNA synthesis. *Biochimica et Biophysica Acta.* 277:576–583.
98. Dokas, L.A., and L.J. Kleinsmith. 1971. Adenosine 3′,5′-monophosphate increases capacity for RNA synthesis in rat liver nuclei. *Science.* 172:1237–1238.
99. Langan, T.A. 1969. Action of adenosine 3′,5′-monophosphate-dependent histone kinase in vivo. *The Journal of Biological Chemistry.* 244:5763–5765.
100. Frenster, J.H. 1965. Nuclear polyanions as de-repressors of synthesis of ribonucleic acid. *Nature.* 206:680–683.
101. Benjamin, W., and A. Gellhorn. 1968. Acidic proteins of mammalian nuclei: isolation and characterization. *Proceedings of the National Academy of Sciences of the United States of America.* 59:262–268.
102. Murthy, L.D., D.S. Pradhan, and A. Sreenivasan. 1970. Effects of hydrocortisone upon metabolism of histones in rat liver. *Biochimica et Biophysica Acta.* 199:500–510.
103. Johnson, E.M., and V.G. Allfrey. 1972. Differential effects of cyclic adenosine-3′,5′-monophosphate on phosphorylation of rat liver nuclear acidic proteins. *Archives of Biochemistry and Biophysics.* 152:786–794.
104. Shelton, K.R., and V.G. Allfrey. 1970. Selective synthesis of a nuclear acidic protein in liver cells stimulated by cortisol. *Nature.* 228:132–134.
105. Teng, C.S., C.T. Teng, and V.G. Allfrey. 1971. Studies of nuclear acidic proteins. Evidence for their phosphorylation, tissue specificity, selective binding to deoxyribonucleic acid, and stimulation effects on transcription. *The Journal of Biological Chemistry.* 246:3597–3609.
106. Shea, M., and L.J. Kleinsmith. 1973. Template-specific stimulation of RNA synthesis by phosphorylated non-histone chromatin proteins. *Biochemical and Biophysical Research Communications.* 50:473–477.
107. Allfrey, V.G., A. Inoue, J. Karn, E.M. Johnson, and G. Vidali. 1974. Phosphorylation of DNA-binding nuclear acidic proteins and gene activation in the HeLa cell cycle. *Cold Spring Harbor Symposia on Quantitative Biology.* 38:785–801.
108. Allfrey, V.G. 1977. Post-synthetic modifications of histone structure. A mechanism for the control of chromosome structure by the modulation of DNA-histone interactions. In *Chromatin and Chromosome Structure.* H.J. Li and R.A. Eckhardt, editors. Academic Press, New York. 167–199.
109. Johnson, E.M., and V.G. Allfrey. 1978. Postsynthetic modifications of histone primary structure. Phosphorylation and acetylation as related to chromatin conformation and function. In *Biochemical Actions of Hormones.* Volume 5. G. Liwack, editor. Academic Press, New York. 1–51.
110. Allfrey, V.G. 1980. Molecular aspects of the regulation of eukaryotic transcription. Nucleosomal proteins and their postsynthetic modifications in the control of DNA conformation and template function. In *Cell Biology: A Comprehensive Treatise.* Volume 3. L. Goldstein and D.M. Prescott, editors. Academic Press, New York. 347–437.
111. Adler, A.J., G.D. Fasman, L.J. Wangh, and V.G. Allfrey. 1974. Altered conformational effects of naturally acetylated histone f2al (IV) in f2al-deoxyribonucleic acid complexes. Circular dichroism studies. *The Journal of Biological Chemistry.* 249:2911–2914.
112. Kornberg, R.D. 1974. Chromatin structure: a repeating unit of histones and DNA. *Science.* 184:868–871.
113. Kornberg, R.D., and Y. Lorch. 1999. Twenty-five years of the nucleosome, fundamental particle of the eukaryote chromosome. *Cell.* 98:285–294.
114. Grunstein, M. 1997. Histone acetylation in chromatin structure and transcription. *Nature.* 389:349–352.

Index

Page numbers in *italics* represent illustrations.
Numbers in brackets indicate pages of that individual's contributions in the book.

Carol L. Moberg is on the faculty of The Rockefeller University. She has a PhD from Columbia University in comparative literature. Her articles on scientists, science history, and biomedical science have been published in *Scientific American, Science, The Journal of Experimental Medicine,* and *Perspectives in Biology and Medicine.* She was co-editor and contributor to *Launching the Antibiotic Era* and is the author of *René Dubos, Friend of the Good Earth.*

This book was offset printed by Sheridan Books, Inc., Ann Arbor, MI on 60 pound white opaque paper using Adobe Garamond Pro typefaces. Cover inset photograph of the first micrograph of an intact cell by Keith Porter, Albert Claude, and Ernest Fullam. Cover and interior design by Erinn A. Grady.